NEW TRIBOLOGICAL WAYS

Edited by **Taher Ghrib**

New Tribological Ways
Edited by Taher Ghrib

Published by InTech
Janeza Trdine 9, 51000 Rijeka, Croatia

Publishing Process Manager Ivana Lorkovic

Technical Editor Teodora Smiljanic

Cover Designer InTech Design Team

Additional hard copies can be obtained from orders@intechopen.com

New Tribological Ways, Edited by Taher Ghrib
p. cm.
ISBN 978-953-307-206-7

For EU product safety concerns:
IN TECH d.o.o., Prolaz Marije Krucifikse Kozulić 3, 51000 Rijeka, Croatia,
info@intechopen.com or visit our website at intechopen.com.

Meet the editor

Taher Ghrib is currently a doctor and aggregated teacher in physic, at the the Institute of Computer Science and Telecommunications at Hammam Sousse-Tunisia. He obtained his habilitation in Physics and Chemistry at the Preparatory Institute for Scientific and Technical studies of El Marsa on 2001. He prepared his thesis in Tunisia that covers "investigation of the thermally treated steel by the Photothermal deflection technique" and presented on 2008. He worked as a Course teacher of all branches of first and second year in preparatory school and the Institute of Computer Science. He worked on steel and metal alloys in his thesis; he demonstrated a very practical method for measuring the mechanical properties of steels such as hardness by measuring the thermal conductivity by means of mathematical models and experimental nondestructive measurements. Almost all of his work is published in numerous publications in various journals and books. His current research is based on thin layers of metal type and is involved in the project ETRERA (Empowering Tunisian Renewable Energy Research Activities) which object is to realize a fuel cell of high electrical performance.

Contents

Preface

The tribology term comes from the Greeks of the words " tribos" meaning " friction ", and "logos" meaning " law ". Therefore Tribology is defined as " a science which studies surfaces moving one compared to the other " and also a field of science studying lubrication, friction, and wear.

This book aims to recapitulate old informations available and brings new informations that are with the fashion research on an atomic and nanometric scale in various fields by introducing several mathematical models to measure some parameters characterizing metals like the hydrodynamic elasticity coefficient, hardness, lubricant viscosity, viscosity coefficient, tensile strength .... It uses measurement techniques very developed and nondestructive such as the optical and thermal methods like those used with chapters I, II.

The principal distinction of this book is that it brings practical manners to model and to optimize the cutting process using various parameters and different techniques, namely, using water of high-velocity stream, tool with different form and radius, the cutting temperature effect, that can be measured with sufficient accuracy not only at a research lab and also with a theoretical forecast.

Therefore this book aims to minimize and eliminate the losses resulting from surfaces friction and wear which leads to a greater machining efficiency and to a better execution, fewer breakdowns and a significant saving.

A great part is devoted to lubrication, of which the goal is to find the famous techniques using solid and liquid lubricant films applied for giving super low friction coefficients and improving the lubricant properties on surfaces.

To recapitulate the book contains twenty three chapters which can be arranged in four sections entitled as follows:

- **Tribology of Thin Layers**
- **Tribology in cutting**
- **Tribology and low friction**
- **Tribology and Lubrication**

The sessions which are quoted before and their contents are enumerated and developed below:

Part 1 Tribology of Thin Layers

CHAPTER 1 STUDY OF CRALN MULTILAYRED THIN FILMS.

This chapter stipulates that the characterizations microstructural, morphological and physicochemical of a various multi-layer coatings deposited on surface are essential, of which the objective is it to connect the mechanical properties of these deposits to their macroscopic study like the forces of adhesion and microscopic like the pores surface and the cells sizes by using methods of optic and thermal characterizations like the AFM and the PTD technique.

CHAPTER 2 OPTICAL CHARACTERIZATION OF ELASTOHYDRODYNAMIC LUBRICATION PRESSURE WITH SURFACE PLASMON RESONANCE

This chapter completes recent advances of optical techniques for Elastohydrodynamic lubrication (EHL) studies and reports a recent application of surface plasmon resonance (SPR) sensing for the imaging of EHL point contact. A nondestructive technique based on the interference of two luminous rays, one collimated beam incidents and the other reflects on the sensing surface at the resonance angle (~65°). After enlargement the SPR image is further processed by an image analysis program for two dimensional Hue profile extraction and processing and using an experimental calibration curve, the Hue profile is further converted into corresponding refractive index distribution. Finally, the map of SPR resonance minimum is converted to corresponding RI distribution with the help of a calibration curve it could noticed that only a small increase in refractive index is found in the outermost region of the dimple, which corresponds to the contact zone between the steel ball and the glass surface.

CHAPTER 3 TRIBOLOGICAL PROPERTIES OF FLUORINATED AMORPHOUS CARBON THIN FILMS

In this chapter the atomic flat layers on surface samples are manufactured with very significant tribological properties and which are used for significant applications such as hard drive disk coating. Among the deposition it shows the importance of the technique known in the literature as diamond-like carbon (DLC), which can only be obtained at conditions far away from equilibrium, and so it is not spontaneously found in nature. The control of the growth parameters broadens the field of microstructures and, in parallel, the properties of this material, which can only be synthesized in thin film form.

The fastest growing market for DLC simplify the manufacture of multilayers tribological coatings, this coating it implanted with protective films, which can reduce corrosion and wear, may extend the lifetime of implants by preventing or alleviating the aforementioned issues. The friction coefficient of the sample depends to the etching duration it varies from 13% for 15 min to 26% for 30 min this variation may be ascribed to the trapping effect of the pits defined by the micro-pillars that appears on the surface.

CHAPTER 4 DEPOSITION AND TRIBOLOGY OF CARBON AND BORON NITRIDE NANOPERIOD MULTILAYER HARD AND SOLID LUBRICATING FILMS

The aim of this chapter is focused on developing a new solid lubricant film with an extremely low friction which usual simple film and mixed film material lacked of carbon and boron nitride nanoperiod multilayer (C/BN)n whish are deposited by RF sputtering using two semicircular targets. Nanoindentation and microwear tests were carried out to investigate the nanometer-scale deformation properties, macroscopic friction coefficient and sliding endurance of these films.

CHAPTER 5 FERROFLUID SEALS

This chapter intends to explain how ferrofluid seals are formed in magnetic structures by presenting a simple analytical model to describe their static behavior. The seal shape and performances are described with regard to the magnetic structure. The evaluation of the seal static capacity is given. Moreover, the seal shape changes when the seal is radially crushed by the inner cylinder: these changes are described and calculated and the radial force exerted by the ferrofluid on the moving part is determined as well as the stiffnesses associated. Then, various magnetic structures are presented and studied to illustrate the magnet role and deduct some design rules for ferrofluid seals with given mechanical characteristics.

Part 2 Tribology in Cutting

CHAPTER 6 CHARACTERISTICS OF ABRASIVE PARTICLES AND THEIR IMPLICATIONS ON WEAR

This chapter studies the characteristics affecting the wear rate caused by the abrasive particles such as Geometry, hardness and size and their implications on wear. The effect of abrasive particle size in wear on mass loss of mottled cast iron with different retained austenite contents was studied, a high increase of mass loss with abrasive size can be observed, up to approximately 70 microns. After this, a scheme shows how the increase of particle size can mean a change only in the height roughness parameter with no variation in the slope of surface. Specific energy for cutting as a function of tool radius, it is high for small particles.

An experimental study is undertaken to low-carbon steel and the measurement of its mass losses and friction coefficients, it is possible to calculate the abrasion factor, f_{ab}, by using the relationship: $f_{ab} = K_A / \mu_P$ where K_A is the wear coefficient and μ_P is the ploughing component of the friction coefficient and after it make that possible to conclude that the λ_q roughness which discriminating the effect of particle size under severe wear, which shows that the abrasive characteristics are changed a little during the mechanical contact.

CHAPTER 7 TOPOGRAPHICAL CHANGE OF ENGINEERING SURFACE DUE TO RUNNING-IN OF ROLLING CONTACTS.

This chapter studies analytically, experimentally and numerically the topographical change due to running-in of rolling contact. A running-in model of rolling contact by considering the deterministic contact of the engineering surface has been proposed which performs a good agreement with the experiments. Finite element simulations of two-dimensional rolling contact model contribute in illustrating the truncation of the highest asperity, the contacted stress and the residual stress due to plastic deformation.

The comparison of the repeated static contact and the repeated rolling contact leads un-similarity in material transfer direction and asperity deformation. This investigation can contribute in predicting the initial, the change, and the final of the surface topography for a success running-in stage. Running-in plays an important role in plastic deformation, friction and wear of tribology systems during the steady-state period.

CHAPTER 8 TRIBOLOGY IN WATER JET PROCESSES

This chapter describes the history of water jetting technology, high-pressure pumps and water jet machining systems, various water jets used in water jet processes, and material removal mechanisms of water jets. Several points influencing the efficiency, reliability, and lifetime of the high-pressure pump such as Friction and wear between the cylinder and the piston were studied; and also it shows that the corrosion and erosion in valves and nozzles are serious problems that affect the reliability of water jetting systems.

CHAPTER 9 THE ELLIPTICAL ELASTIC-PLASTIC MICROCONTACT ANALYSIS.

This work accounted the relationship between the geometry and material to study the behavior of an ellipsoid deformable coming into contact with a rigid apartment elastoplastic. The material is modeled with elliptic symmetry and analyzed with the finished analysis of element (FEA). In order to validate the model, the mesh density is iteratively increased until the contact force and contact area differed by less than 1% between iterations. In the finite element analyses, the resulting meshes consist in framing elements corresponding to ellipticity by varying the radii of curvature at the ellipsoid tip used the contact deformation.

CHAPTER 10 METHODS OF CHOOSING HIGH-STRENGTHENED AND WEAR-RESISTANT STEELS ON A COMPLEX OF MECHANICAL CHARACTERISTICS

This chapter presents the methods which make it possible to choose a steel of high strength and resistant to wear while exploiting a complex of the mechanical characteristics. The analysis results of an extensive experimental data show that for providing the best indexes of wear resistance at mechanical outwearing it is necessary to combine three components: high static toughness, hardness and plasticity. Only the combination of such characteristics provides the best results regarding the wear resistance increase. This task in defined aspects is finished up to an engineering decision and can be used in designer's practice for choosing the wear-resistant steels and alloys. The tendered methods allow not only to produce an estimation of suitable steels for different conditions of wear and external forces of uploading, but also to orient metallurgists to melting new steels with quite defined mechanical properties and their combinations.

CHAPTER 11 A COMPARISON OF THE DIRECT COMPRESSION CHARACTERISTICS OF ANDROGRAPHIS PANICULATA, EURYCOMA LONGIFOLIA JACK, AND ORTHOSIPHON STAMINEUS EXTRACTS FOR TABLET DEVELOPMENT.

This chapter presents a comparison of the direct characteristics of compression introduced by Andrographis paniculata, Eurycoma longifolia Jack, and Orthosiphon stamineus and which is checked on selected powders of grass extract of Malaysie and discuss the properties of such powders. The compressions were performed at room temperature between 23 °C and 26 °C with the humidity between 37 and 42% RH. The comparison of the efficacity of the various techniques amounts comparing the properties tribological such as tensile strength, grain size of tablets manufactured. IT proves that *Eurycoma longifolia* Jack extract powder was the easiest of the three herb powders to compress, and it underwent significant particle rearrangement at low compression pressures, resulting in low values of yield pressure.

Part 3 Tribology and Low Friction

CHAPTER 12 FRICTIONAL PROPERTY OF FLEXIBLE ELEMENT.

The chapter presents the theory of belt buckle by considering the self-locking mechanism generated by wrapping the belt on belt friction, presents the frictional property of belt wrapped on an axis two and three times through deriving the formulas corresponding to an each condition and finally discus frictional property of flexible element wrapped on a hard body with any contour. The frictional force can be calculated by the curvilinear integral of the curvature with respect to line element along the contact curve. Utilizing the self-locking property of belt, a novel one-way clutch was developed. The problem of this clutch is how to get the smaller and stable coefficient of belt-belt friction for long time use.

CHAPTER 13 SURFACE FRICTION PROPERTIES OF HUMAN SKIN AND THE FRICTION BETWEEN FABRICS AND SKIN.

This chapter studies the properties external of friction of the skin and materials of clothing, using twelve fabrics consists of various fiber, yarn structure, and the yarn density. The experimental apparatus measuring the surface friction properties and the measuring results of fabrics are explained and the Friction between human skin and fabrics are measured. The effects of the friction of fabrics and the moisture regain of human skin are shown.

CHAPTER 14 INVESTIGATION OF ROAD SURFACE TEXTURE WAVELENGTHS.

In this paper, experimental texture measurements and Data Dependent Systems (DDS) modeling methodology were introduced to analyze the real road pavement surfaces. The elevation profiles collected from real road core surfaces were composed of 1 micron step size scan and 45 micron step size scan by an accurate laser sensor and were modeled by DDS program and it mentioned that after more factors that affect the surface roughness are considered, a better correlation of roughness with friction can be obtained.

CHAPTER 15 ADHESION THEORY FOR LOW FRICTION ON ICE

This chapter proposes a study of the frictions on the surface of the ice and it seeks the various causes which are responsible for its low values, it concludes that the increase in the friction is caused by the ploughing of ice at the contact area and it noticed that the minimum friction is shifted because the higher temperature generated following the reduction of the heat propagation velocity and with the water vapor pressure on the surface.

CHAPTER 16 TRIBOLOGY OF 2-MERCAPTOBENZOTHIAZOLE IN LITHIUM COMPLEX GREASE

This chapter proposes a study of the properties of 2-Mercaptobenzothiazole and the effect of its addition on a complex grease of lithium. The developed complex grease was tested for other regular characteristics in addition to tribological properties. It summarized that 2- Mercaptobenzothiazole has exceptional load carrying capacity compared to conventional EP additives and the grease show very low coefficient of friction and good additive response.

Part 4 Tribology and Lubrication

CHAPTER 17 NO CONVENTIONAL FLUID FILM BEARINGS WITH WAVED SURFACE

In this chapter, the wave bearings lubricated with incompressible fluids, commonly known as fluid film wave bearings, are analysed. The performance of both journal and thrust bearings is examined and it examines that one of most important properties of the wave journal bearings compared to other types of journal bearings is their improved stability. The wave thrust bearings can be used for axially positioning the rotor or to carry a thrust load. For this reason, the first part of the chapter is devoted to the study of the dynamic behaviour of the journal wave bearingsand the second part executes the steady state of the thrust wave bearings. It stipulates that wave journal bearing steady-state and dynamic performance can be precisely predicted with computer codes validated by experiments on dedicated test rigs and the wave amplitude, the minimum film thickness, oil supply pressure and temperature, as well as the number of waves can be used to maximize the bearing performance for a particular application.

CHAPTER 18 IDENTIFICATION OF DISCHARGE COEFFICIENTS OF ORIFICE-TYPE RESTRICTORS FOR AEROSTATIC BEARINGS AND APPLICATION EXAMPLES

The chapter describes an experimental study conducted in order to identify the supply hole discharge coefficients of externally pressurized gas bearings. Tests were carried out over specific hole, feed pocket and air gap size ranges on pneumatic pads with two types of air feeding systems: annular orifices and simple orifices with feed pocket. For simple orifices with feed pocket, the flow characteristics can be described using two experimental discharge coefficients: $C_{d,c}$ for the circular section of the orifice and $C_{d,a}$ for the annular section of the air gap in correspondence of the pocket diameter.

Air consumption and pressure distributions were measured as a function of supply pressure and air gap height. Discharge coefficients were approximated by an experimental formula based on the Reynolds number and the feeding system geometry. The validity of the formulation found in the study was verified by comparing the numerically calculated pressure distribution with the experimental distribution measured on different pad types.

CHAPTER 19 INVERSE APPROACH FOR CALCULATING TEMPERATURE IN THERMAL ELASTO HYDRODYNAMIC LUBRICATION OF LINE CONTACTS

This research proposes a thermal elastohydrodynamic lubrication (TEHL) inverse approach to estimate the pressure, temperature rise, and apparent viscosity distributions in an EHL line contact, it requires a lot of measured points of film thickness to establish the amplitude and location of the pressure and temperature spikes where the pressure and estimated film thickness distributions can be calculated from force balance and elastic deformation theories.

CHAPTER 20 CONSTRUCTION OF VARIOUS SELF ASSEMBLED FILMS AND THEIR APPLICATION AS LUBRICANT COATINGS

In this chapter, the tribological behaviors of self-assembled nanofilms (SANFs), including self-assembled monolayers (SAMs), self-assembled multilayer films (SAMFs), self-assembled inorganic films (SAIFs) and self-assembled organic-inorganic composite films (SAO-ICFs), are reviewed, aiming at discovering the basic "microstructures-properties" correlation. It is expected that the extracted "structures-properties" correlation can serve as the guidance to direct the further designing of lubricant coatings for MEMS/NEMS and other devices in molecule-level. Theses films these films can generated a great importance not only for its simple preparation procedure but also for its wide potential applications in many fields, such as surface modification, boundary lubricant coatings, sensors, photoelectronics, and functional bio-membrane modelling.

CHAPTER 21 A NOVEL TOOL FOR MECHANISTIC INVESTIGATION OF BOUNDARY LUBRICATION: STABLE ISOTOPIC TRACERS

This chapter proposes a new tool for mechanistic investigation in tribo-chemistry. The main subject of this investigation is based on the secondary ion mass spectroscopy study (SIMS) of surfaces rubbed with isotope-labeled lubricants. It begins by explaining the technique of SIMS and its advantage analysis for tribo-chemistry, followed by the tribological properties and lubrication mechanism of monomolecular films with friction modifiers, the tribo-chemical reaction of diamond-like carbon with water and finished by an analysis of the boundary film yielded from additives in lubricants.

CHAPTER 22 FEM APPLIED TO HYDRODYNAMIC BEARING DESIGN.

The present chapter is aimed to provide the theoretical foundation of Finite-Element Method (FEM), mass- and energy-conserving models as well as to report their application to the ThermoHydroDynamic (THD) and ThermoElastoHydroDynamic (TEHD) analysis of different bearing types. The application examples are relevant to both journal and axial bearings with fixed and tilting pads, in order to demonstrate the high flexibility of the method. The most used FEM formulations of the mass-conserving lubrication problem have been proved, while an original approach to the thermal problem has been explained. Whereas, TEHD models are very sensitive to boundary conditions, its choice is particularly difficult in all of the multi-physics simulations. The scope of the chapter is anyway limited to the analysis of steadily-loaded bearings working in laminar lubrication regime.

CHAPTER 23 COMPARISON BETWEEN DIFFERENT SUPPLY PORT CONFIGURATIONS IN GAS JOURNAL BEARINGS.

The aim of this chapter is to compare three externally pressurized gas journal bearings at given air consumption rates. The idea was to investigate which offers the best spatial distribution of supply orifices under the same pneumatic power. The study compared radial stiffness and pressure distribution for the three bearing types, also evaluating the damping factor and the whirl ratio of the shaft. The stability threshold was calculated for different restriction parameters so that the proposed bearing types could be compared.

Taher Ghrib

Part 1

Tribology of Thin Layers

1

Study of CrAlN Multilayred Thin Films

Tlili Ibrahim[1] and Taher Ghrib[2]
[1]*Laboratoire Bourguignon des Matériaux et Procédés (LaboMap), CER ENSAM de Cluny 71250*
[2]*Technopole Borj Cedria*
[1]*France*
[2]*Tunisia*

1. Introduction

The optimization of mechanical and tribological properties of monolayer coatings developed and studied is a way of reaching and important research, because of their potential application in various mechanical fields. Hardness and tenacity are the first two principal characteristics of the deposits which it is necessary to optimize them to improve their wear resistance. From this, the layers Cr, CrN and CrAlN are superimposed with different levels of constraints residual (Pc, Pt or combined) to obtain multilayer coatings that fulfill better the industrial requirements.
The characterizations microstructural, morphological and physicochemical of the various multi-layer coatings are essential within the framework of this study, of which the objective is it to connect the mechanical properties of these deposits to their macroscopic and microscopic study.
A variety of multilayer systems such as TiN/CrN [1], TiAlN/CrN [2], TiN/TiAlN [3], etc., have been studied extensively. However, there are very few reports on the multilayer coatings based on CrN and CrAlN [4]. The addition of Al to CrN system permits to work at higher temperatures where the oxidation occurs [5]. CrAlN coatings have been reported to be stable up to a temperature of 900°C depending upon the Al content in the coatings [6]. CrAlN coatings also exhibit higher hardness and a lower friction coefficient compared to CrN coatings [7]. The new ternary film structure brought about significant advances in coating designs, such as the decrease of the grain size and the formation of grain boundaries between the two phases. As a consequence, it is reported that CrAlN films exhibited excellent mechanical properties and oxidation resistance owing to their solid solution structure [8]. Furthermore, it is well known that most properties of these solid solution composite films are influenced by certain factors such as crystalline structure and micro-structure. These factors led to the study of the multilayers coatings of CrN/CrAlN and Cr/CrN/CrAlN, in which the properties of Cr, CrN and CrAlN can be combined. The Cr underlayer is considered as a bonding layer.
In this work, we have investigated the effect of thickness on residual stress and hardness of the monolayers. Moreover, the relationship between mechanical properties and the stress state of the multilayers coatings was also discussed and established.

2. Structural, microstructural and morphological characterization

By taking account of the optimal conditions obtained on the mono-layers, we developed six multi-layers coatings Cr/CrN/CrAlN and CrN/CrAlN with various residual stresses levels in mono-layers which constituents them.

The films composition, thickness and roughness are given respectively by the EDS or WDS analysis and the 3d optical profilmeter. (fig. 1).

The analysis carried out on the multi-layer coatings Cr/CrN/CrAlN or CrN/CrAlN shows the homogeneity of various chemical compositions at the several surface points, the principal characterizations results carried out are given in table 1.

According to analyses EDS one found that in the studied coatings the N/Cr report/ratio is about 1.17 to 1.2 and that they contain only 2.2% atomic of oxygen to the maximum. The roughness of films is characterized by two parameters (Ra and Rt) whose measured values are high, which leads us to await a dispersion of hardness by tests of nano indentation.

Concerning the analysis by DRX (fig. 2), one will limit the study on the films PVD_1 and PVD_4 since the total thickness of other films is low (< 35nm).

These multi-layers is elaborate with low internal stresses (Pt). The coating PVD_4 is presented by the peaks of principal diffraction [(111) and (200) respectively with 36.8° and 51.27°, to compare with file PDF 76-2494, PCPDFWIN version 2.3, JCPDS-ICDD (2002)] [9]. The addition of an under-layer Cr to the coating PVD_4 one obtains the film PVD_1 and this gives a peak to position $2\theta = 43.6°$ allotted to the presence of a cubic phase CrN (111) identical to the ref [9-11].

In addition, it is established that the peaks Cr_2 NR (111) and CrN (200) are confused they is because of the effect of under-stoechiometric coatings. Another remark, one can conclude according to the spectra obtained on the multi-layers PVD_1 and PVD_4 that the CrN layer crystallizes on the surface according to the preferential orientation (311) whereas it crystallizes in volume according to the preferential orientation (200) [12]. The quantitative

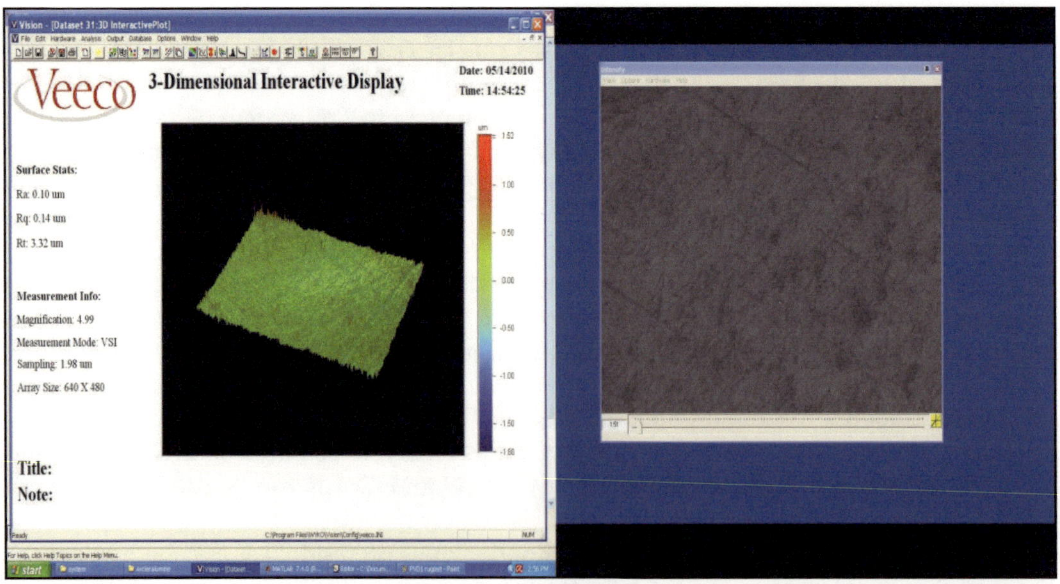

Fig. 1. Topography of coating PVD_1 obtained by the optical profilometer.

Revêtements	N (at. %)	Al (at. %)	Cr (at. %)	N/Cr	Total thickness (nm)	Rugosité (µm)	
						Rugosité arrithmétique (R_a)	Quadratic roughness (R_q)
Cr/CrN/CrAlN ($P_t/P_t/P_t$):PVD_1	51.8	4.2	42.9	1,2	1500	0.1	0.14
CrN/CrAlN (P_t/P_c):PVD_2	50.8	4	43.1	1.17	300	0.12	0.16
Cr/CrN/CrAlN ($P_t/P_c/P_c$):PD_3	50.5	4.7	42.5	1.18	300	0.11	0.15
CrN/CrAlN (P_t/P_t):PVD_4	52	5	41.3	1.2	1500	0.13	0.15
CrN/CrAlN (P_c/P_c):PVD_5	49.7	4.9	43.6	1.18	256	0.12	0.16
Cr/CrN/CrAlN ($P_c/P_c/P_c$):PVD_6	50.1	4.2	44.5	1.17	356	0.09	0.13

Table 1. Chemical composition and topography

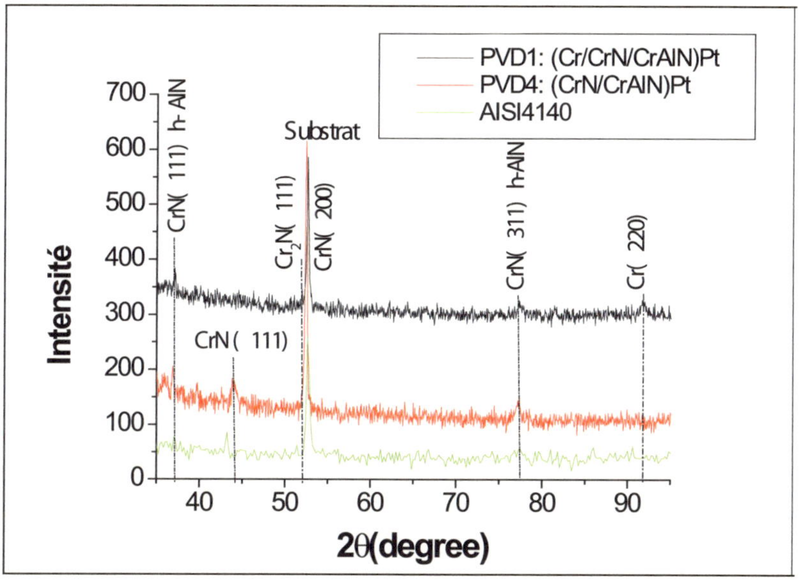

Fig. 2. Diffraction spectra of x-rays obtained with the multi-layer PVD1, PVD4 and steel 42CrMo4.

analysis of the layer revealed the coexistence of the CrN phases and Cr2N, this result is confirmed by C Nouveau [13, 14]. An observation in HRTEM and diffraction stereotype confirmed the presence of a mixture of CrN (200) phases in the Cr2N layer.

We remark that by adding a Cr underlayer to the PVD_4 coating, the peak CrN (111) disappeared completely.

This can be explained by the fact why the plans (311) have a speed of growth more significant than the plans (111) and by preferential pulverization of the atoms of nitrogen of the plans (111) compared with those of the plans (311).

In addition, the presence of a crystallographic structure is related to the column-like morphology of the mono-layers Cr, CrN and CrAlN observed by SEM on a cross section after a polishing mirror (fig. 3).
The growth of the structure in columns took place perpendicular to surface. The found results are comparable with that obtained in work of Harish C. Barshilia and M. Okumiya, M. Griepentrog [15, 16] carried out on the CrN/CrAlN layers.
In order to conclude this study from obtaining the multi-layers, have determined the profiles of the chemical compositions versus the pulverization time (fig. 4). The interface

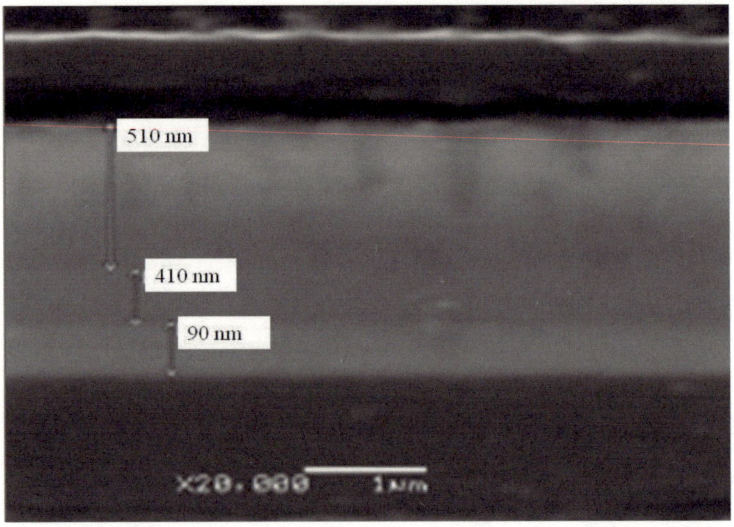

Fig. 3. SEM of a transverse cut followed by a polishing mirror coating Cr/CrN/CrAlN.

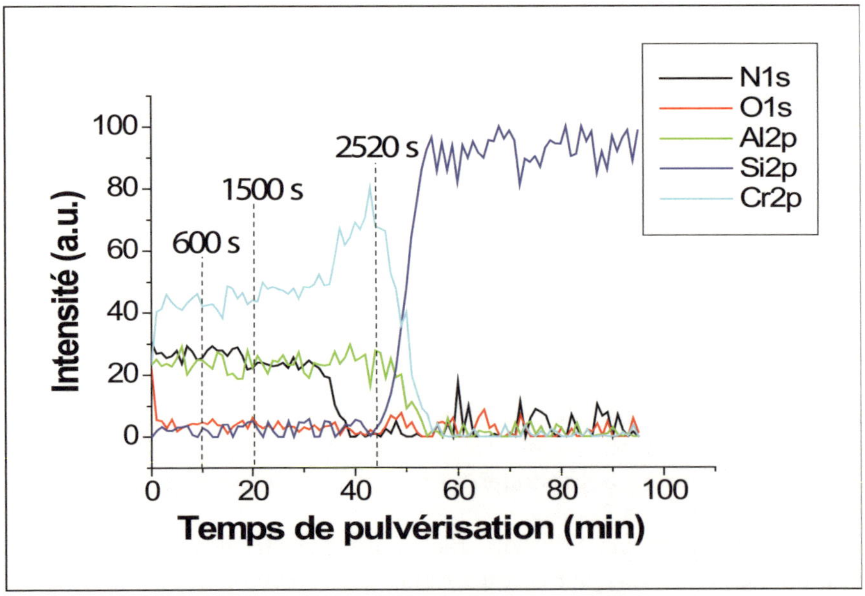

Fig. 4. Compositions Profiles of the Cr/CrN/CrAlN layer obtained by XPS.

between the individual layers is not very clear, which proves that these layers are very miscible. Moreover, there is a contamination of surface of this coating is highlighted by the presence of species oxidized of chromium on a depth of a few tens of nanometers. Aluminium is presented by a small proportion in the surface of CrAlN layer with 5% which imply that it is not possible to distinguish the variation from its profile in the film.
By taking account dimensions of the aluminium atoms and its kinetic energy. It is well doped in the CrN layer, which to allow him to reach a significant depth, from where the interface between the two layers CrN/CrAlN is not remarkable. The profile of the flexible layer (Cr) is more intense close to the interface layer/substrat which is coherent with the superposition of muti-layers (Cr/CrN/CrAlN) on silicon. The interface is relatively narrow and the involved species do not seem to penetrate in the substrate what attests the layer good quality.
One selected two moments over the pulverization time (600s and 1500s), of which the goal to know the evolution of the specific spectrum of the studied multi-layer Cr/CrN/CrAlN. The Fig. 5 (1) schows that spectra N_{1s} of the studied layer revealed the presence of a characteristic peak of chromium nitride centered at 397 eV. This result is found in the references [17 - 18]. The width with middle height (WHM) for the two moments 600s and 1500s is about 1.45 eV and 1.62 eV respectively.

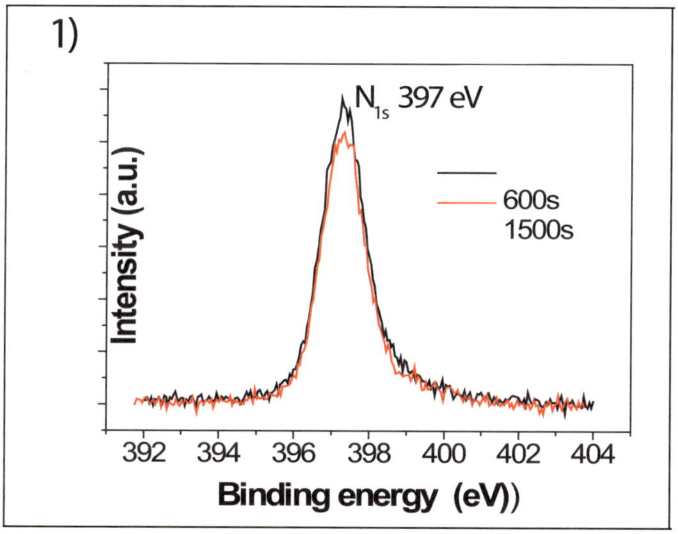

Fig. 5.1.

The binding energy corresponds to the peak O_{1s} is 531.5 eV, and the width with middle height is about 2 eV (fig. 5 (2)) for the two moments.
The peak associated to Cr (fig. 5(3)) is composed of two peaks centered at 574,8 and 584 eV, who correspond to Cr $_{2P3/2}$ and Cr $_{2P1/2}$ respectively.
The widths with middle height (FWHM) for the two peaks are measured in the two selected moments, 600 and 1500 and the values obtained are as follows: the peak $Cr_{2P3/2}$ (1.3 and 2.3) and the peak Cr $_{2P1/2}$ (2.25 and 2.75), respectively.
The spectra Al_{2p} and Cr_{3s} (fig. 5(4)) are confused and the corresponding binding energy is 74.3 eV [15, 18 - 20]. The width with middle height is 3.75eVand 4.2eV for 600s and 1500s respectively.
Investigations XPS on the figure I.1f, shows the presence of the elements O, N, Al and Cr in the Cr/CrN/CrAlN coating. From the intensity of the peaks N_{1s}, O_{1s}, Cr_{2p} and Al_{2p} + Cr_{3s},

we can estimate the film composition. There is oxygen with a feeble proportion (< 3%) in the coating. coherent with the results obtained by WDS, it is attributed to the incorporation of

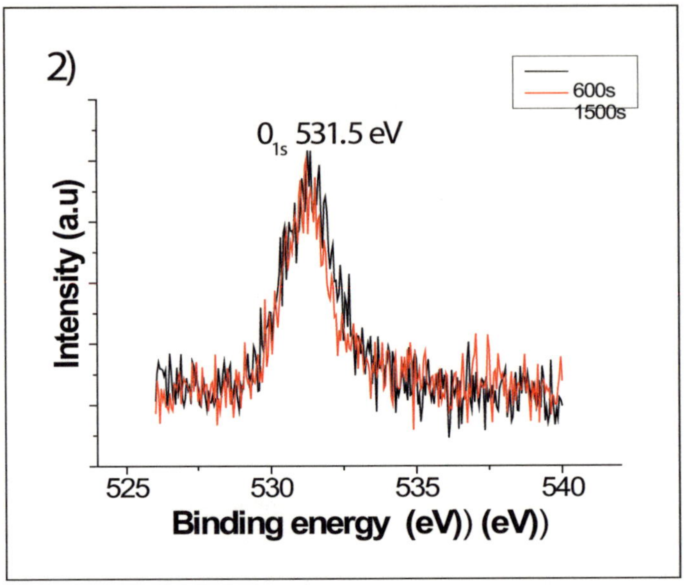

Fig. 5.2.

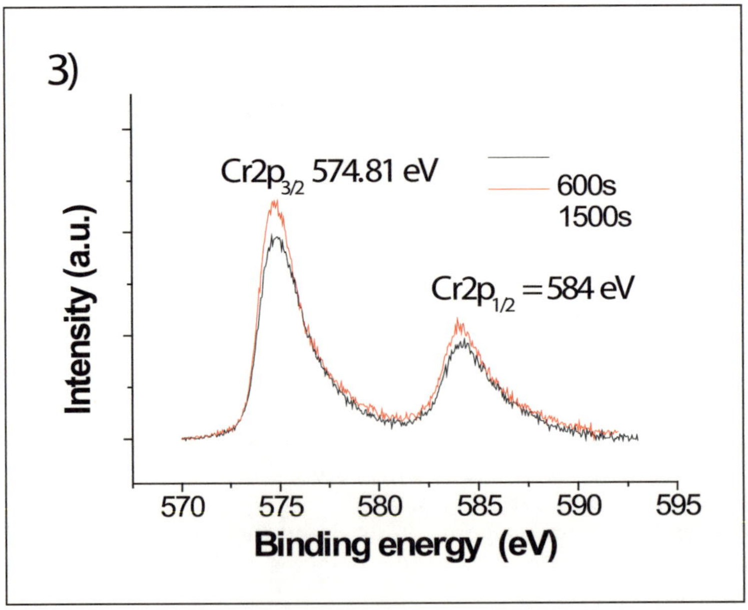

Fig. 5.3.

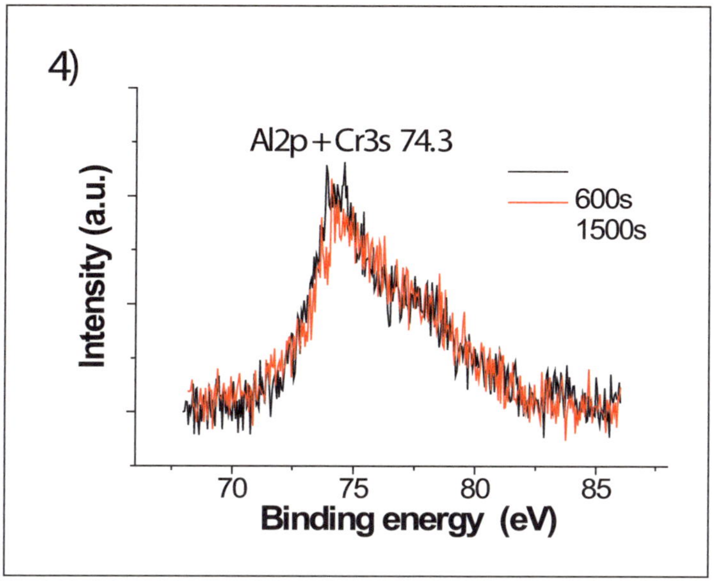

Fig. 5.4.

oxygen in plasma during the pulverization of the target of Cr. These results are coherent with the other work completed by Lappitz and Hubert [20]. The narrow spectrum of metal chromium is for 574.81 eV, but its incorporation with nitrogen changes the spectrum with 575.7±0.5 eV [21]. The origin of this change in fact that the film of nitride is composed of phases mixture like CrN, Cr_2N, CrN_xO_y and $CrN(O_2)_x$.

The peak O_{1s} relates to chromium oxides Cr_2O_3 and CrO_3 [21]. The analysis by atomic force microscopy AFM of a surface section for each multi-layer makes it possible to determine the morphology properties of the Cr/CrN/CrAlN and CrN/CrAlN coating (fig. 6). Surfaces are

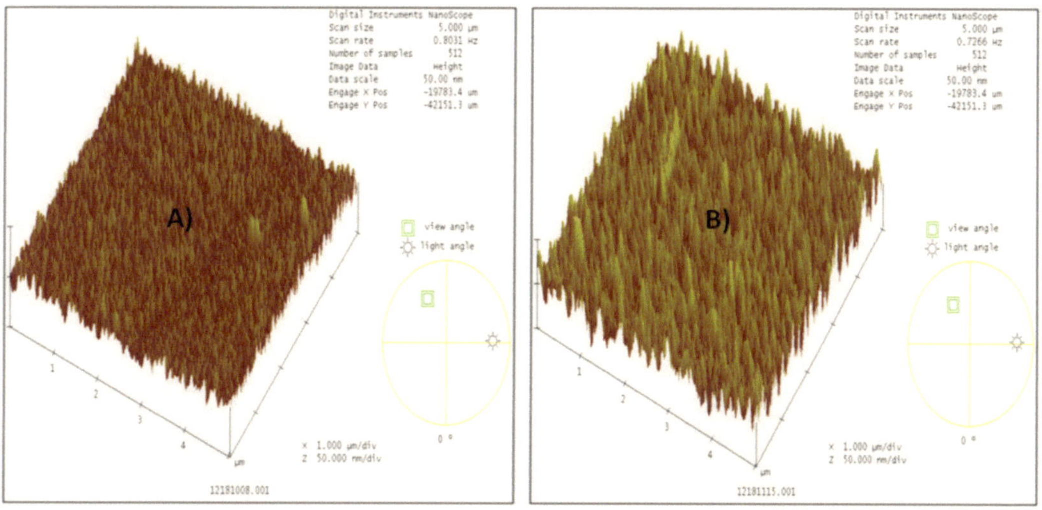

Fig. 6. Morphologies of the multi-layer coatings obtained by AFM: A) CrN/CrAlN and B) Cr/CrN/CrAlN.

generally uniform, they present domes and peaks of which average radii and the maximum height are respectively about 150 and 41 nm for Cr/CrN/CrAlN, and approximately of 178 and 26 nm for CrN/CrAlN. In addition the roughness of surface is characterized by the parameters Ra and Rz who are in order 7.18 and 25.85 nm for CrN/CrAlN (fig 7-A), and 12.31 and 40.83 nm for Cr/CrN/CrAlN (fig 7-B). Following this analysis one can say that the CrN/CrAlN coating has a structure finer than that of the Cr/CrN/CrAlN coating. Thus, the morphology of CrN/CrAlN is characterized by a denser and uniform structure.

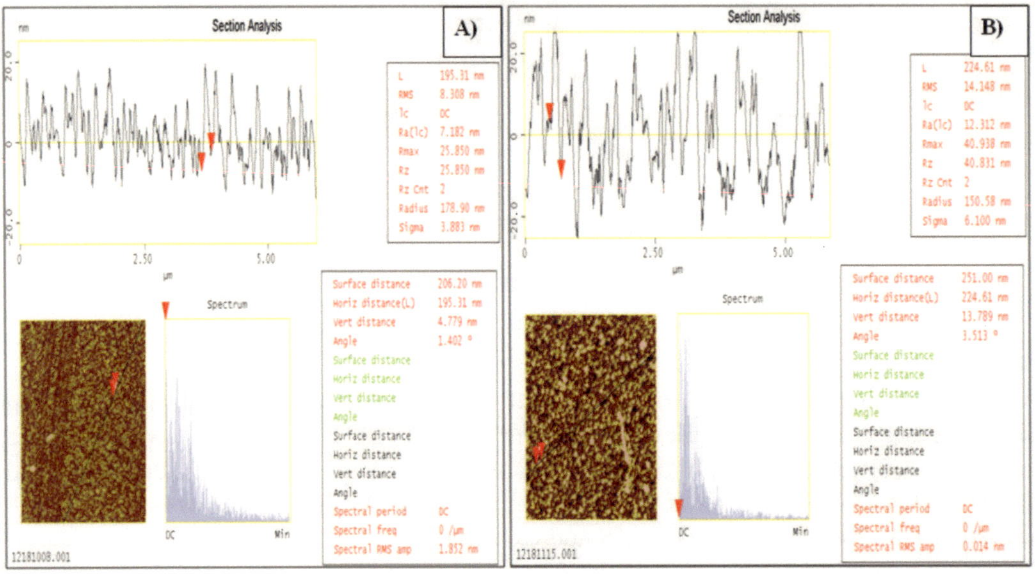

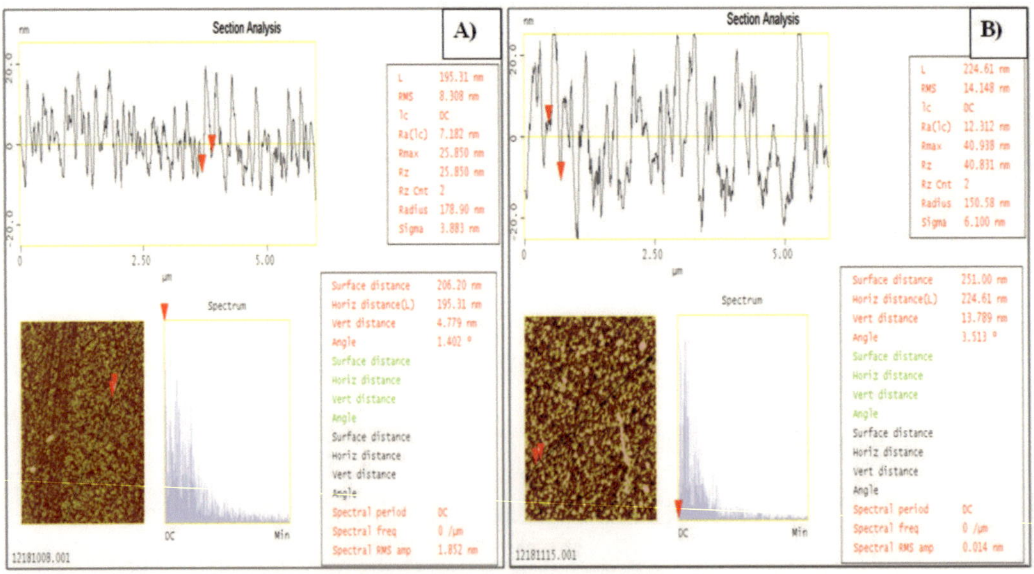

Fig. 7. Analyze by x AFM of a multi-layer section coatings:
A) CrN/CrAlN and B) Cr/CrN/CrAlN.

2. Thermal properties

2.1 Principle of the PTD technique

The thermal properties such as the thermal conductivity and the thermal properties are determined by the PTD technique. This method consists in heating a sample with a modulated light beam of intensity $I = I_0(1+\cos\omega t)$. The thermal wave generated by the optical absorption of the sample will propagate in the sample and in the surrounding fluid (air in our case). The thermal wave in the fluid will induce a temperature gradient then a refractive index gradient in the fluid which will cause the deflection ψ of a probe laser beam skimming the sample surface. This deflection may be related to the thermal properties of the sample. The sample is a stack of n layers, we write the heat equations in these areas and in the two surrounding fluids which are the air in by designating K_i, D_i and l_i, respectively the thermal conductivity, the thermal diffusivity, and the thickness of the layer i (Fig.8).

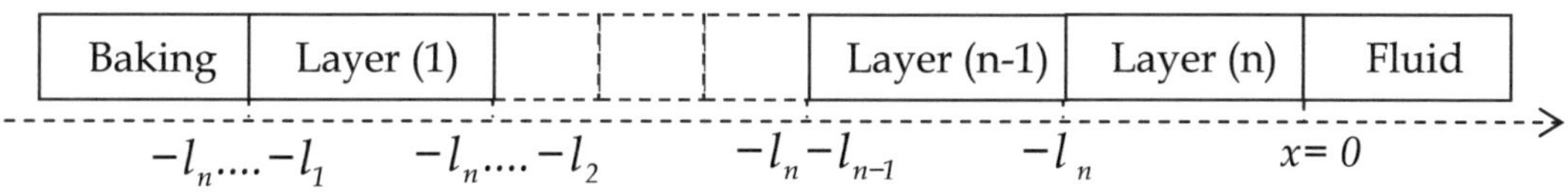

Fig. 8.

2.2 Theoretical model

In the case of an uniform heating we can use a 1-dimensional approximation, and the amplitude $|\psi|$ and phase φ of the probe beam deflection ψ are given by:

$$|\psi| = \frac{\sqrt{2}\,L}{n\,\mu_f}\frac{dn}{dT_f}|T_0|\,e^{-\frac{x}{\mu_f}} \quad \text{and} \quad \varphi = -\frac{x}{\mu_f} + \theta + \frac{5\pi}{4}$$

Where l is the width of the pump beam in the direction of the probe laser beam, n, μ_f and T_f are respectively the refractive index, the thermal diffusion length and the temperature of the fluid.. $|T_0|$ and θ are respectively the amplitude and phase of the temperature T_0 at the sample surface which are function of the thermal properties of the different media, and x is the distance between the probe beam axe and the sample surface.

Before the calculation of the probe beam deflection, one must know the expression of the surface temperature T_0 that obtained by writing the temperature and heat flow continuity at the interfaces $x_i = -l_n - l_{n-1} \ldots -l_i$:

$$T_0 = \left[((1+b)\,\eta_4 e^{\sigma_1 l_1} - (1-b)\eta_2 e^{-\sigma_1 l_1})E_n + (r_1 - b)\,e^{-\alpha_1 l_1} E_1\right] \Big/ \left[(1-b)\eta_1\, e^{-\sigma_1 l_1} - (1+b)\,\eta_3 e^{\sigma_1 l_1}\right]$$

With

$$E_i = \frac{\alpha_i}{2K_i(\alpha_i^2 - \sigma_i^2)},\ \sigma_i = (1+j)\sqrt{\frac{\pi f}{D_i}},\ b = \frac{K_b\sigma_b}{K_1\sigma_1},\ g = \frac{K_f\sigma_f}{K_n\sigma_n}\ et\ r_1 = \frac{\alpha_1}{\sigma_1}$$

and $\eta_1, \eta_2, \eta_3, \eta_4$ are constant whose expressions depend on thermal properties and numbers of layers [17].

2.3 Experimental set-up

The sample is heated by an halogen lamp light of power 100W modulated thanks a mechanical chopper at a variable frequency (Fig. 9). A He-Ne laser probe beam skimming the sample surface at a distance z is deflected. This deflection can be detected by a four quadrant photo-detector and converted to an electrical signal which is measured by a lock-in amplifier. Through the intermediary of interfaces, the mechanical chopper and the lock-in amplifier a microcomputer will set the desired modulation frequency and read the values of the amplitude and phase of the photothermal signal and then draw their variations according to the square root modulation frequency.

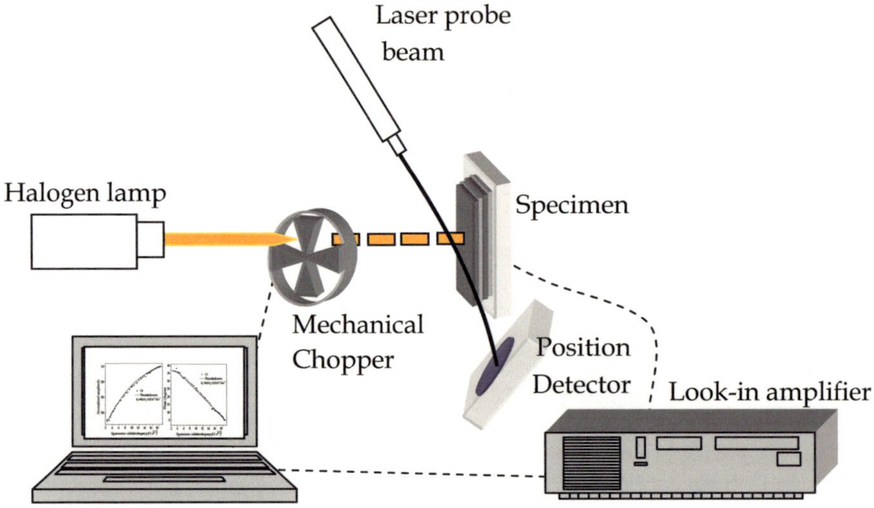

Fig. 9. Experimental set-up

2.4 Determination of the thermal properties

By applying the model developed with N layers to our coatings multi-layer Cr/CrN/CrAlN, Cr/CrN and mono-layer Cr by taking account of the thickness of each layer constituting the coating [22,23]. The obtained experimental results are presented by the figures 10-a, b, c. The thermal properties of each layer are deduced when there is a coincidence between the experimental and theoretical curves.
The results obtained are given in the table 2:

Revetment	Thermal conductivity ($W.m^{-1}.K^{-1}$)	Thermal diffusivity ($10^{-4}m^2.s^{-1}$)	Equivalent thermal Conductivity ($W.m^{-1}.K^{-1}$)
Cr	93,9	0,79	
CrN	11	0,23	
CrAlN	2.8	0,052	
Cr/CrN			13.1
Cr/ CrN/CrAlN			4.6

Table 2. Thermal properties of different multi-layer.

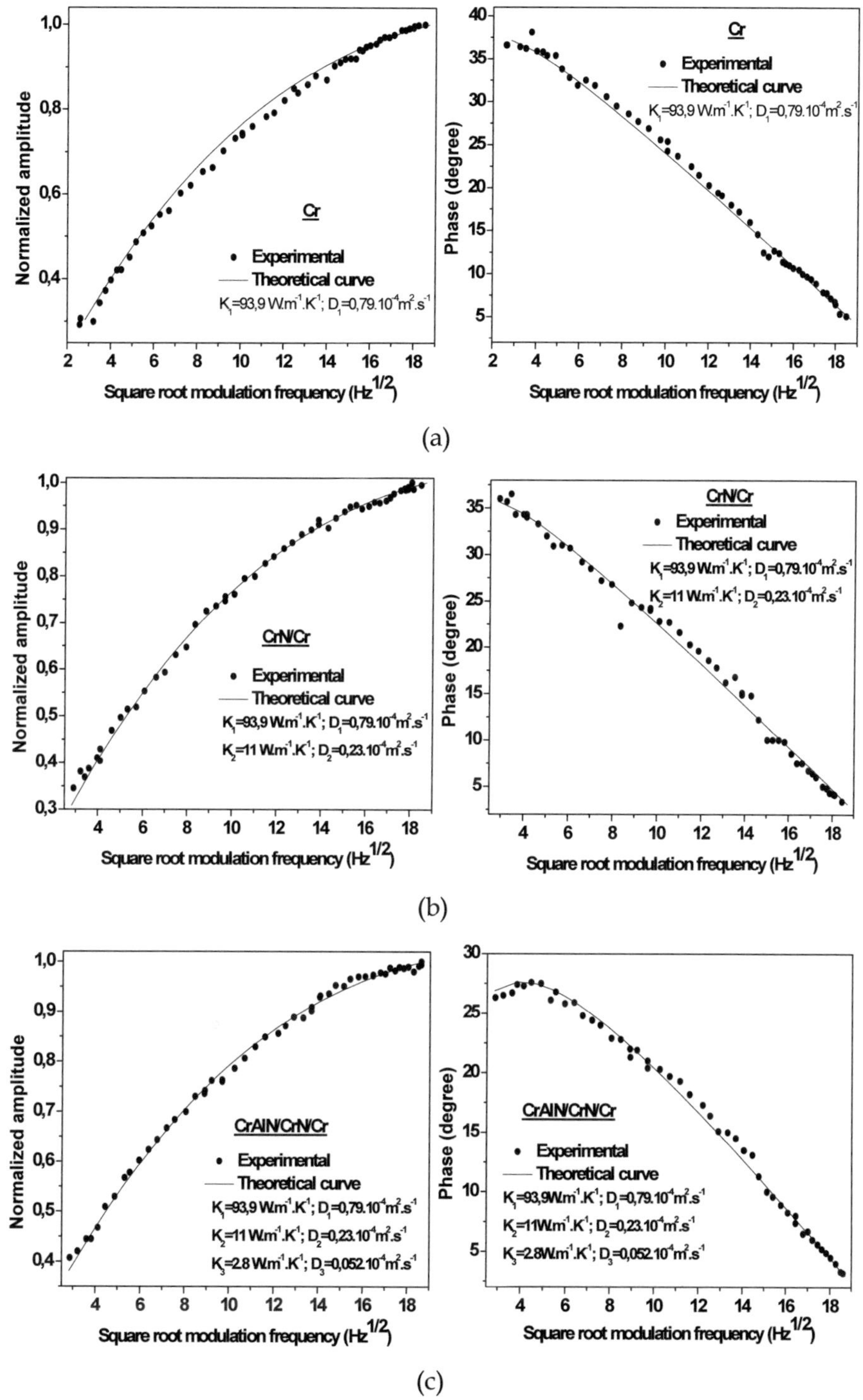

Fig. 10. Amplitude and phase of the signal for a sample of Silicon on which is deposited the coatings: a) Cr, b) Cr/CrN et c)Cr/CrN/CrAlN

These values show that the thermal properties of the CrAlN layer are weak, compared to the CrN layers and Cr taken separately, therefore it can act like a heat insulator. That can be explained by a column microstructure which is characterized by the lengthened shape of the pores inter-column-like which are mainly aligned perpendicular to the plan of the layer [24,25], and since the heat flow is primarily parallel to these gaps inter-columns, the transfer of heat is effectively significant. These fine gaps inter-column-like contribute to a moderate reduction of thermal conductivity and are generally opposed to the heat propagation.

For this reason, the thermal conductivity of the CrN. layer is also clearly lower than the thermal conductivity of the Cr layer which is dense. This result suggests that the structure of coatings Cr and CrN close to the interface to the metal substrate have a very different thermal conductivity. The difference in the structure of multi-layer coatings Cr/CrN / CrAlN and Cr/CrN can coarsely be divided into two zones, the interior zone with fine grains (CrN/Cr) and the zone external with large columns (CrAlN). The thermal conductivity of the interior zone with fine grain is much higher than the thermal conductivity of the external zone.

Thermal conductivity is dominated by the defect of the grain boundary in this part of the coating, it results from it a lower conductivity of about 2,5 $W.m^{-1}.K^{-1}$ at the ambient temperature.

Moreover, the reduction in the thermal conductivity of the multi-layer coating Cr/CrN/CrAlN is allotted to the increase in total porosity, We can conclude for thin films that the thermal conduction is a comprehensive process, which is due to the vibration of the crystal lattice, and the realization of a stacking of several layers provide a thermal conductivity which verifies the general relationship $\frac{1}{K}=\frac{1}{l}\sum_i \frac{l_i}{K_i}$, where l and K are respectively the total thickness and the equivalent conductivity of the stack; l_i and K_i are respectively the thickness and thermal conductivity of different layers constituting the stack.

3. Study of the mechanical characteristics

3.1 Residual stresses

A similar study of the residual stresses is carried out on the multi-layer coatings.

Fig. 11, shows that the residual stresses in the multi-layers are lower than those of constituent mono-layers.

This reduction of the residual stresses is mainly du to the interfaces created between the various superposed layers. Indeed, the total constraint is 3.4 times higher if one compares PVD_1 (all the mono-layers are with P_t) with PVD_6 (all the mono-layers are with P_c). The value of the highest constraints is obtained in multi-layer PVD_6. In addition, the same results are obtained when one compares PVD_4 and PVD_5; the constraints are also increased by 3.5 times. In fact, if one compares multi-layer PVD_1 and PVD_4 or PVD_5 and PVD_6, it is obvious that the underlayer of Cr has a major effect, because the connection Cr of the layer, has exploited a principal role in the constraints intensity, what proves that the level of these constraints on the coating PVD_6 (-1.2 GPa) is higher than in PVD_5 (- 0.7GPa). We can also note that if one compares PVD_2 and PVD_5, the level of the constraints remains almost the same whatever the state of stress in the CrN layer (~ 0.65 - 0.7GPa).

These results mean that the underlayer, Cr or CrN improves the constraints intensity of these multi-layer that if they are at their point P_T as the surface layer. If not, if the underlayers are at their P_T point and the surface layers at their P_c point, then the increase in constraints is due to the fact that these layers have their highest constraints. Lastly, these results would be interesting in the choice of multi-layer according to their later applications (mechanical, thermal, wear, etc...).

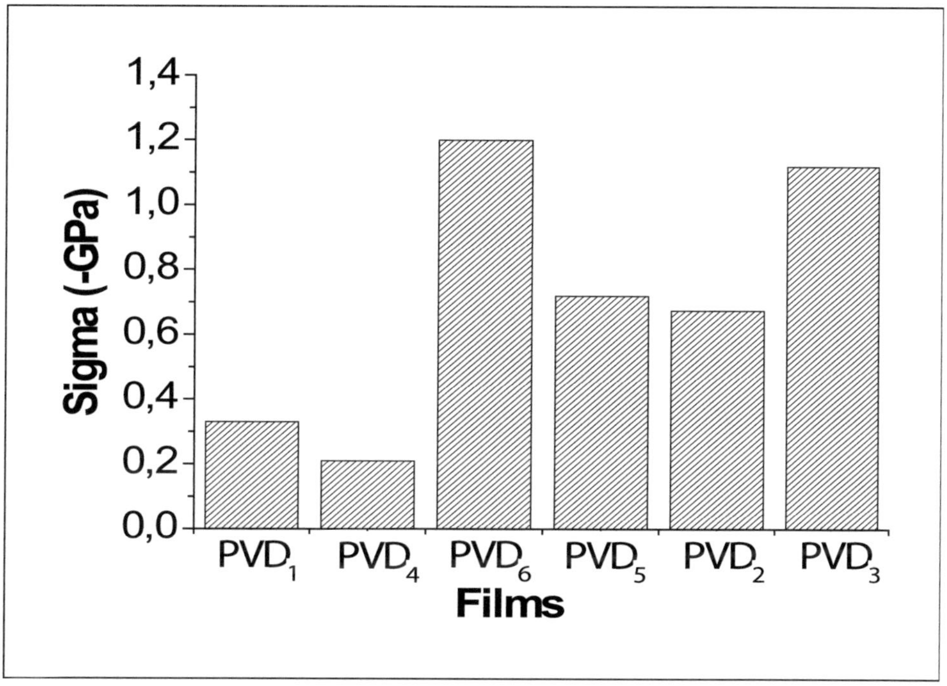

Fig. 11. Multi-layers Residual stresses

3.2 Adhesion

Tests of scratching are carried out by scratch-test on the six multi-layer coatings (fig. 12).

The films are compared two to two according to their total thicknesses and of level of the residual stresses (P_T and P_c) in the mono-layers. The stripes of various films are studied and investigated.

Generally, all the films have an average adherence with the substrate steel of $42CrMo_4$

Exceptionally PVD_1 has a high adherence characterized by a critical effort F_c = 78 N (fig II.2a). This fact, whatever the level of the residual stresses (P_t, P_c or combined) of the mono-layers, good adherence in films is always associated to the presence of a fixing under-layer Cr : Fc_{-PVD1}> Fc_{-PVD4} ; Fc_{-PVD3}> Fc_{-PVD2} et Fc_{-PVD6}> Fc_{-PVD5} (fig 12-a, b et c). These films PVD_1, PVD_3 and PVD_6 are also characterized by high residual stresses (fig. 11).

Analysis of this adherence performance can be connected to the rupture of the columns growth of the mono-layers and a propagation envisaged of the cracks will be limited only on the surface layers. If one takes account only to films thicknesses, the coatings PVD_3 and PVD_6 are less thick than PVD_1, on the other hand this last has a better adherence although these coatings have the same constituent layers. It is known that if the thickness increases the adherence decreases, but the wear resistance increases. The greatest critical load associated with PVD_1 can be related to the reduction at the same time to the average intern energy and to the constraints to the interface of the mono-layer [24].

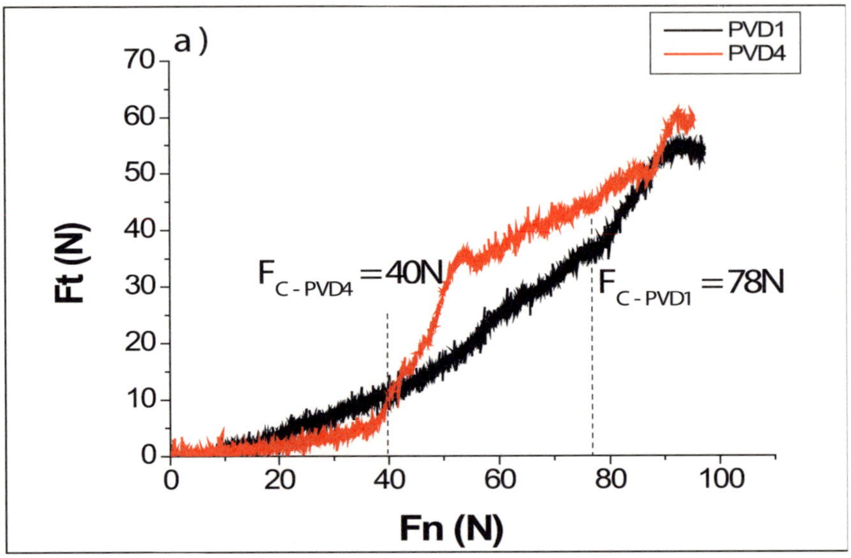

(a)

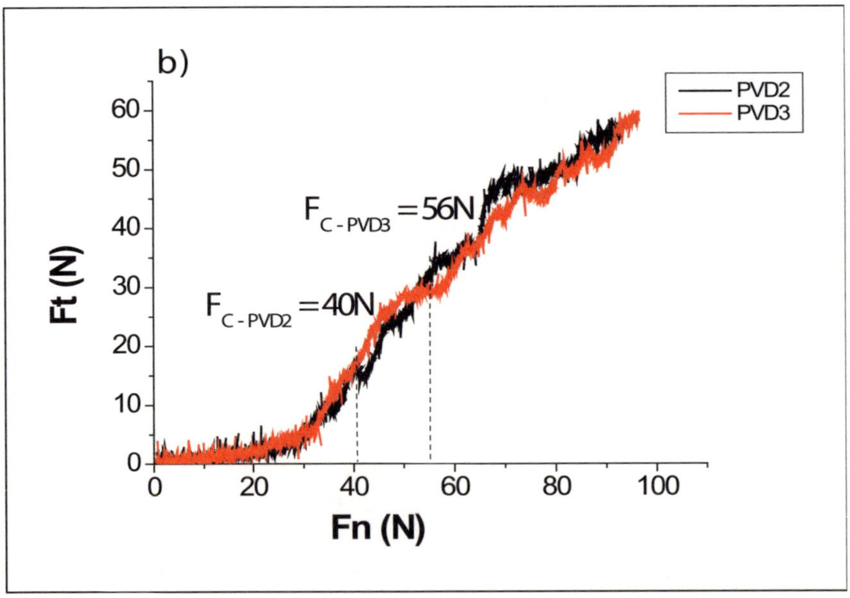

(b)

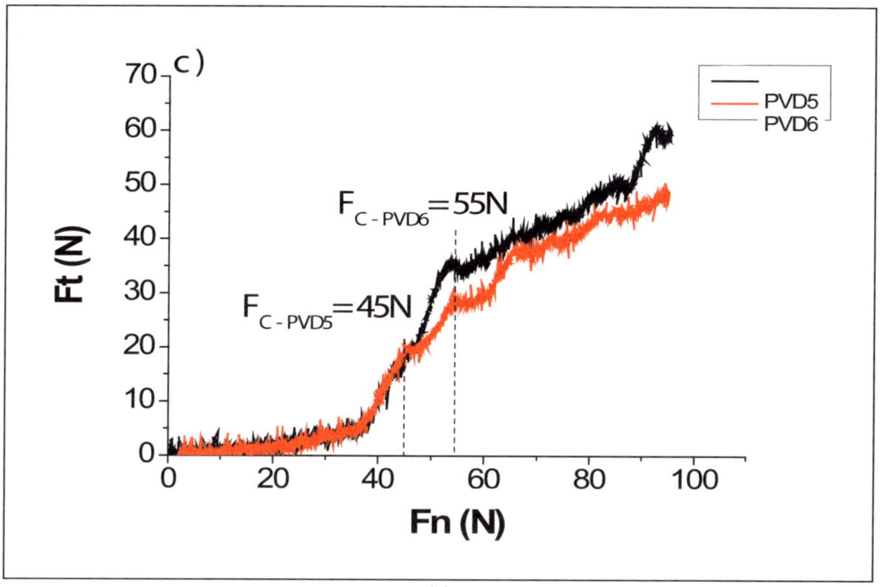

(c)

Fig. 12. Scratch-tests carried out on the multi-layers:
a) PVD_1 et PVD_4, b) PVD_2 et PVD_3 et c) PVD_5 et PVD_6

Moreover, if one takes account only for constraints levels in the mono-layers which constitute film. It is established that, more this level is weak P more the adherence is significant (table 3). On the other hand the absence of a flexible underlayer Cr in films supports a low adherence whatever the thickness and the constraints levels associated to the established coatings.

3.3 Study of behavior to wear

The wear of the various deposits is quantified by using the method of the average values in order to estimate the total profile of the worn surface. After each test of friction the damage creates is studied. By taking account of films thicknesses and roughness associated to their surfaces. Worn volumes of the coating and the basic substrate are given separately after each tribological test (fig 13-a). By comparing the wear of the films two to two and by taking account of the intensity of the residual stresses and film thickness, the multi-layer coatings developed with an underlayer Cr have a better wear resistance than those developed without this underlayer ($V_{PVD1} > V_{PVD4}$, $V_{PVD6} > V_{PVD5}$ and $V_{PVD3} > V_{PVD2}$).

Coatings	PVD1	PVD2	PVD3	PVD4	PVD5	PVD6
Fc (N)	78	40	56	40	45	55

Table 3. Adhesion Forces F_c of multi-layer deposited on steel 42CrMo4.

If one takes account only of the constraints intensity in the partial layers, constituent of film, it appears that when the constraints levels are combined (P_t and P_c), the film has a better wear resistance than those presented by levels P_t or P_c only, it is the case of the PVD_3 coating. So one can say that worn volume is proportional to the depth of wear for each tribological test and the results found on the wear depth are coherent with those found for worn volume. In addition, all the coatings improve the wear resistance compared to the basic substrate (42CrMo4).

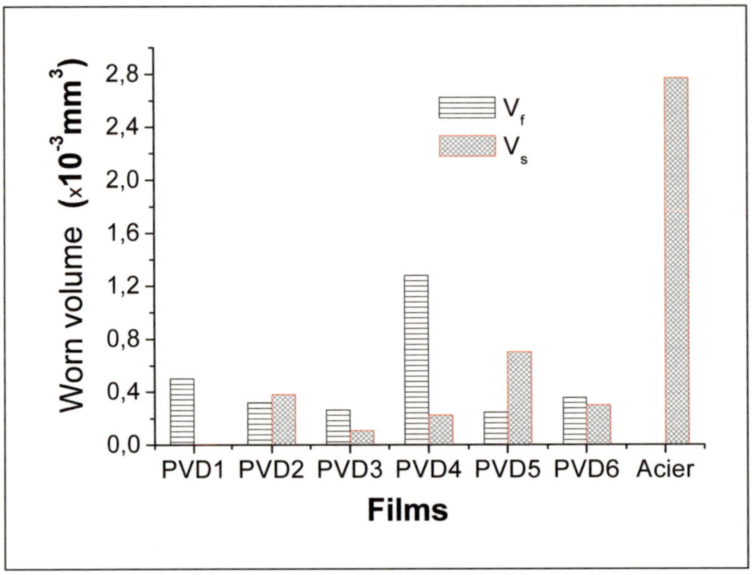

Fig. 13-a. Worn volume of the various multi-layer coatings: t = 15min, Fn = 5N et δ = 5mm

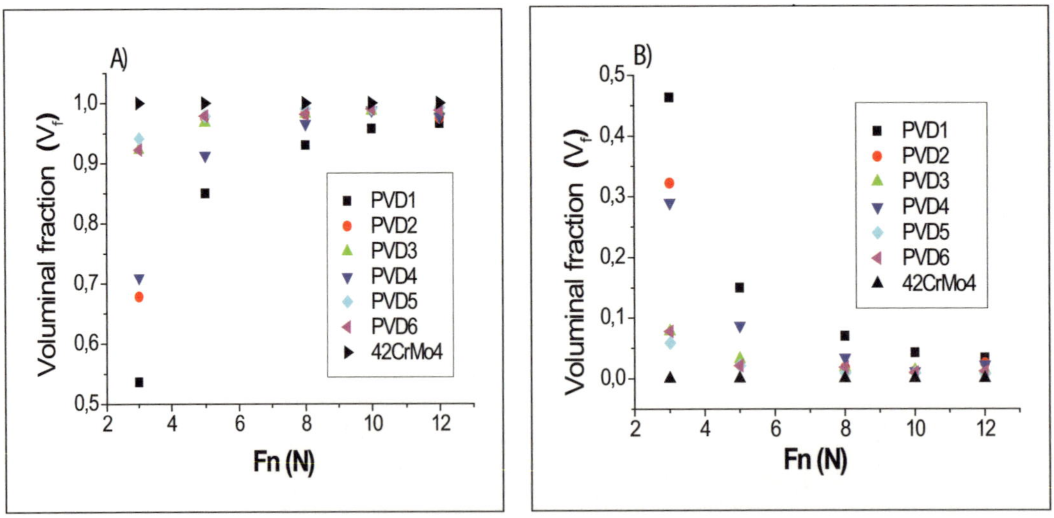

Fig. 13-b. Voluminal fraction of the multi-layer coatings and the substrate:
A) films and B) Substrate.

To study the influence of the operational parameters on the removed matter quantity, after each tribological test the voluminal fraction is represented according to the applied normal load Fn (fig. 13-b). The found results show that the beneficial effect of films to resist wear is detected for applied loads $F_n \leq 5N$, high this value one brings closer more and more the tribological behavior of the two antagonists in contact alumine/substrat, from where the beneficial effect of films is detected for the weak applied loads.

4. Conclusion

We synthesized coatings having a composition report of N/Cr close to 1 with a dual magnetron for 20% of nitrogen in plasma. These coatings are well crystallized. Analyses XPS and SIMS showed that the multi-layer coatings carried out by pulverization magnetron present quite clear interfaces between the mono-layers and who do not present inter-diffusion substrate-coating. A good homogeneity of the layers on all their depth was highlighted. In addition, the oxygen contamination of the multi-layer coatings is minimal, less than 5% at in volume on a few ten nanometer of depth. The analyses by x-rays diffraction highlighted the strong texture of the multi-layer deposits CrN/CrAlN and Cr/CrN/CrAlN, which is the consequence of the high residual stresses in the mono-layers. These constraints are decreased in the multi-layers under the effect of the interfaces presented in film. Moreover, the thermal properties of multi-layer are refined under the presence of the surface layer CrAlN, which plays the role of a thermal barrier. In addition, all the multi-layers developed in this work have a good adherence with steel 42CrMo4, and adhesion is better in the case of presence of an underlayer Cr (P_T) in film.

5. References

[1] H.A. Jehn, M.E. Baumggartner, *Surf. Coat. Technol.* 108, 54-55 (1992)

[2] M. Panjan, S. Šturm, P. Panjan, M. Čekada, *Surf. Coat. Technol.* 202, 815-819 (2007)

[3] L.A. Dobrzański, M. Polok, M. Adamiak, *Journal of Materials Processing Technology* 164-165, 843-849 (2005)

[4] Youn J. Kim, Tae J. Byun, Jeon G.Hon, *Superlattices and Microstructures* 45, 73-79 (2009)

[5] P.H. Mayrhofer, H. Willmann, C. Mitterer, *Surf. Coat. Technol.* 222, 146 -147 (2001)

[6] M. Kawate, A.K. Hashimoto, T. Suzuki, *Surf. Coat. Technol.* 165,163 (2003).

[7] Y. C. Chim and al., *Thin Solid films* 517, 4845- 4849 (2009)

[8] M. Uchida, N. Nihira, A. Mitsuo, K. Toyoda, K. Kubota, T. Aizawa, *Surf. Coat. Technol.* 627, 177-178 (2004).

[9] L. WANG et al. Surf. Coat. Technol. (2010).

[10] MAGNUS Odén, CLAES Ericsson, GREGER Hakansson, HENRIK Ljunerantz, Surf. Coat. Technol. 114(1999)39-51

[11] B. TLILI, Mustapha. NASRI, C. Nouveau, Y. BENLATRECHE, G. GUILLEMO, M. LAMBERTIN, Vacuum 84 (2010) 1067-1074

[12] C. NOUVEAU, Thèse de doctorat n°21-2001, ENSAM de Cluny

[13] C. NOUVEAU, M. A. DJOUDI, C. DECES-PETIT, Surf. Coat. Technol.174-175 (2003)-455.

[14] B. WARCHOLINSKI, A. GILEWICZ, Z. KUKLINSKI, P. MYSLINSKI, Surf. Coat. Technol. 204(2010), 2289-2293.

[15] Harish C. BARSHILIA, B. DEEPTHI, N. SELVAKUMER, Anja JAIN, K.S. RAJAM; Applied Surface Science 253 (2007) 5076-5083.
[16] A. LIPPITZ, T. HUBERT. Surf. Coat. Technol. 200 (2005) 250.
[17] H.C. BARSHILIA et al. Surface & Coatings Technology 201 (2006) 2193-2201
[18] K. J. Lee, J. W. Kang, P. S. Jeon, H. J. Kim, J. Yoo, Thermochimica Acta, 455 (2007) 60–65.
[19] A. LAPPITZ. Th. HUBERT, Surf. Coat. technol. 200 (2005) 250.
[20] S.R PULUGURTHA et al. Surf. Coat. technol. 202 (2007) 755-761.
[21] Taher GHRIB, Sahbi BEN SALEM, Noureddine YACOUBI, Tribology International, Volume 42, Issue 3, March 2009, Pages 391-396.
[22] Taher GHRIB, Noureddine YACOUBI, Faycel Saadallah, Sensors and Actuators A 135 (2007) 346-354.
[23] Li CHEN, Yong DU, S.Q. WANG, Jia LI, International Journal of Refractory Metals & Hard Materials 25 (2007) 400-404.
[24] Yu CHUNYAN, Tian LINHAI, Wei YINGHUI, Wang SHEBIN, Li TIANBAO, Xu BINGSHE, Applied Surface Science 255 (2009) 4033-4038.
[25] J. ALMER, M. ODEN, G. HAKANSSON, Thin Solid Films 385 (2001) 190.

2

Optical Characterization of Elastohydrodynamic Lubrication Pressure with Surface Plasmon Resonance

C.L. Wong[1], X. Yu[2], P. Shum[3] and H.P. Ho[4]
[1]*Division of Microelectronics, School of Electrical & Electronic Engineering Nanyang Technological University, 50 Nanyang Drive,*
[2]*Singapore Institute of Manufacturing Technology (SIMTech), 71 Nanyang Drive,*
[3]*Division of Communication Engineering, School of Electrical & Electronic Engineering Nanyang Technological University, 50 Nanyang Drive,*
[4]*Department of Electronic Engineering, The Chinese University of Hong Kong, Shatin, N.T.,*
[1,2,3]*Singapore*
[4]*Hong Kong*

1. Introduction

Elastohydrodynamic lubrication (EHL) is the dominant mode of lubricant in highly stressed machine elements including gears, rolling contact bearings, cams and tappets. EHL studies describe the mechanism of lubrication in highly pressurized non-conformal contacts and they contribute to the effective operation of gears and other highly stressed machine elements. Grubin and Vinogradov [1] found that the heavily loaded EHL line contacts are governed by hydrodynamics, elastic deformation of the metal surface and the lubricate viscosity variation under extreme pressure (viscosity-pressure characteristics). Theoretical studies and numerical solutions to EHL contacts have been reported by Martin [2], Downson and Higginson [3], Crook [4], Cameron and Gohar [5] and significant contributions have been made towards the complete understanding of the EHL mechanism. The theoretical revelations were later confirmed by experimental EHL techniques, especially in optical measurements. In this chapter, we review recent advances of optical techniques for EHL studies. In addition, we report the recent application of surface plasmon resonance (SPR) sensing for the imaging of EHL point contact [6-8].

2. Optical characterization of elastohydrodynamic lubrication (EHL) contacts

Optical interferometry is by far the most widely used experimental measurement method for EHL contacts. Pioneer works on optical interferometry were carried by Archard and Kirk [9, 10], Gohar, Cameron and Paul [11-13]. Archard and Kirk [9, 10] used white light to investigate the lubrication property change in crossed cylinders. The speed of the conjunction was found to be related to the colour interference fringes shift. Gohar and Cameron [11] and Cameron and Gohar [12] used the impact between a steel ball and a glass disc to generate a high pressure micro-chamber in GPa pressure level and optical interferometry measurement was performed for the EHL lubricant film. Paul and Cameron [13] used two coherent light beams to incident on an EHL contact at two different incident angles. By analyzing the fringe orders, the refractive index and lubricant film thickness profiles were estimated.

Larsson et al. [14] applied the optical interferometry technique to investigate the pressure fluctuation when soap lumps entered a lubricated elastohydrodynamic lubrication contact. Yagi et al. [15] demonstrated the experimental study on the viscosity wedge effect in point contact EHL under high slip ratio condition. Color optical interferometry technique was applied for lubricant film thickness measurement. Nonishi et al. [16, 17] used the optical interferometry technique to study the influence of sliding on the elastohydrodynamic film. The film thickness formed between a crowned roller and a sapphire disk was measured with optical interferometry. Yang et al. [18] studied the dimple phenomena in EHL point contact formed by a glass disk and a steel ball with optical interferometry. Félix-Quinonez et al. [19] combined the optical interferometry technique with a high-speed color video camera for automatic anlaysis of interferograms, which allowed rapid imaging for the variations of interference fringe orders and film thickness. Kaneta et al. [20] used optical interferometry and Newtonian thermal EHL analyses to study the effects of the thermal conductivity of contacting surfaces on point EHL contact. They found that the distributions of pressure and film thickness were greatly affected by the velocity and slide-roll ratio. In addition, the temperature-viscosity wedge action, the heat produced by the compression work and the shearing of the lubricant were found to be the major causes of the phenomena. Jang [21] developed an image processing method for the analysis of EHL images obtained from a monochromatic optical interferometer. In their studies, interference fringe images were formed by the reflection beam from a Cr coating surface and a metal ball surface. During data processing, the interference images were converted into digital data and the interference information of the whole contacting area was collected. In the same year, Jang [22] applied the optical interferometry technique for in-situ measurement of lubrication (EHL) film, while viscosity index improver was added to the base oil. A resolution of 5nm was reported for film thickness measurement. Guo et al. [23] presented an experimental observation of a dimple-wedge elastohydrodynamic lubricating film with optical interferometry. Anomalous EHL films, a wedge shape together with a tiny dimple at the inlet region, were observed with optical interferometry in pure sliding conditions (with ultra slow speeds of 3-800 μm/s). Yang et al. [24] investigated the effect of a transversely/longitudinally oriented surface bump/groove on the lubricating performance and dimple phenomena with optical interferometry. They also obtained numerical solution of thermal EHL in the study. Wang et al. [25] used optical interferometry and theoretical analysis to study the behavior of point contact EHL films in pure rolling short stroke reciprocating motion. Jang et al. [26] modified the conventional optical interferometry technique and developed a coaxial aligning trichromatic incident light filtering system. Compared to the conventional monochromatic/dichromatic optical interferometry, their new system could provide more spectra of color components for film thickness calculation. Furthermore, the film thickness was finely digitized and nanometer scale measurement was possible. Wang and Guo et al. [27] used optical interferometry to observe the formation of abnormal EHL film with inlet-dimple in ball-on-disc and spherical roller-on-disc contacts. It was found that the inlet-dimple only appeared within a limited entrainment speed range at particular load levels, entrainment direction and kinematic condition. Kaneta et al. [28] studied the behaviors of point contact EHL films under different impact loads using optical interferometry. They also investigated

the effects of initial impact gap, impact at oily Hertzian contact and the effect of impact loading under rolling/sliding conditions. Guo et al. [29] used the optical interferometry technique to investigate the change between EHL and hydrodynamic lubrication (HL) under different entrainment speeds and applied loads. A log-log scale linear relationship was found in two lubrication regions between the film thickness and the entrainment speed (or load). In addition, they also reported the presence of a transition region, which was significantly affected by relative sliding, between these two regions. In the same year, the research group of Fu, Guo and Wong [30] further presented a stratified-layer model for the numerical analysis of optical EHL contact. The simulation results were found correlating with their previous experimental findings. Li and Guo [31] then studied the influence of spinning on EHL films under the condition of wall slippage with optical interferometry. Spin-slide ratio was introduced to represent different spinning levels and the spin-slide ratio was found to be related to the EHL film thickness, film shape, depth of inlet dimple and the speed index of the minimum film thickness. Guo and Wong et al. [32] also applied optical interferometry to identify some of the boundary slip phenomena of highly pressurized polybutenes in EHL contacts. The post-impact lateral movement of the entrapment was studied under pure rolling, pure glass block sliding and pure ball sliding conditions. Their experimental findings supported the existence of wall slippage, which was the cause of the abnormal EHL film profile with inlet dimple. Young et al. [33] studied the start-up friction behavior of EHL entrapments with optical interferometry. It was found that entrapment substantially decreased the start-up friction. The short-lived entrapment was also found providing the greatest reduction in start-up friction. Recently, the optical interferometric technique was used by Myant et al. [34] for the investigation of fluid film thickness in sliding, isoviscous elastohydrodynamic contacts (I-EHL). A monochromatic two-beam interferometer was used to measure the lubricant film thickness at different entrainment speeds and applied loads. Experimental results showed film thickness profile with a convergent wedge shape, which was believed to be responsible for the fluid pressure and load carrying capacity of the sliding contacts.
In addition to optical interferometry, other optical imaging techniques have been reported for the studies of EHL contacts in recent years. Bongaerts et al. [35] demonstrated the application of in situ confocal Raman microscopy for friction measurements in EHL contacts. Optical imaging on the EHL contact and the determination of film thickness has been demonstrated. Optical images of the contact also showed that the starvation of the contact occurred above a critical value of the product of the entrainment speed and the lubricant viscosity for single-phase aqueous Newtonian lubricants in the EHL regime. Recently, Reddyhoff et al. [36] applied the fluorescence detection technique for lubricant flow mapping in EHL contact formed between a steel ball and a glass disc. The testing lubricant was dyed with a fluorescent species. During measurement, it was illuminated with laser light and a fluorescence intensity map of the lubricant flow was produced. The fluorescence intensity results were found correlating well with conventional optical interferometric film thickness measurements under the same condition.

3. Surface plasmon resonance (SPR) imaging of EHL point contact

Ho and Wong et al. [6-8] presented the optical characterization of EHL point contact with the surface plasmon resonance (SPR) imaging technique. This technique makes use of the spectral characteristics variation associated with the SPR effect. The hydrostatic pressure inside the EHL contact causes a localized refractive index change of the lubricant. This results in a shift of the SPR absorption dip in the spectrum, thereby producing corresponding color changes in the resultant spectral SPR image. The refractive index profile inside the entire EHL contact therefore can be imaged by the color variation in the spectral SPR image. Comparing to conventional optical interferometry, more than two order of magnitude resolution improvement has been demonstrated in the refractive index measurement inside EHL point contact [8]. In addition, the SPR imaging allows direct measurement of refractive index profile and it is unnecessary to rely on general pressure-density model. In light of these unique advantages, the SPR detection approach should be an attractive alternative for optical EHL contact characterization.

SPR imaging was first introduced by Yeatman and Ash et al. [37] and Rothenhausler and knoll [38] as a microscopy technique in late 80's. This imaging technique has further been developed for various bio-sensing applications, such as the detection of DNA [39-41], protein-protein [42], protein-carbohydrates [43] and antibody-antigen binding interactions [44-46]. Common SPR imaging approach replies on the resonance shifts in the wavelength domain and a board band light source is used for surface plasmons excitation. As described by the Fresnel model [47], a spectral absorption curve is produced by the surface plasmons excitation and the absorption dip position is related to the refractive index change in the dielectric medium. In conventional intensity based SPR imaging technique [39-43], the boardband light source is replaced by a laser source (with narrowband wavelength) and the variation of absorption dip position is transformed into corresponding intensity changes in the reflection SPR image [46]. However, the useful information in other spectral range has been neglected in this approach. Furthermore, the signal to noise ratio is found to be limited [48]. In this chapter, a novel approach of SPR imaging technique, which makes use the full spectral information in the visible range for the image formation, is presented. With a calibration curve obtained from fluids with known refractive index, the spectral variation in the spectral SPR image is further converted into corresponding two dimensional (2D) refractive index profile. This imaging technique is applied to the study of point elastohydrodynamic lubrication (EHL) contact for the 2D imaging of refractive index and pressure profile in highly pressurized EHL dimple.

3.1 Experimental set-up

An experimental set-up based on spectral SPR imaging for studying elastohydrodynamic lubricant (EHL) dimples was formed using the impact EHL microviscometer as previously described elsewhere [49-51]. The SPR imaging results were further compared with those obtained using conventional optical interferometry for refractive index determination inside the point EHL contact [52].

Formation of impact EHL dimple

As shown in **Figure 1**, the impact EHL microviscometer operates by having a steel ball impinging on a glass plate covered by a thin layer of lubricant fluid. This results in a tiny quantity of lubricant being trapped between the contacting surfaces and thus the formation of a dimple. The dimension of a typical dimple is about 300 microns in diameter and a few microns in the central film thickness. Meanwhile, the entrapped fluid dimple acts as a microscopic high pressure chamber with pressure level reaching GPa.

Spectral SPR and interference imaging system

The experimental scheme is given in **Figure 1**. Two halogen lamps with spectral range in between 400 and 850nm (emission peak at 596nm) are used as the broad band light sources in the experiment. The optical outputs are collimated and the beam diameters are approximate 30mm. In the SPR imaging arm, a linear polarizer is used to extract the p-polarized component of the incident beam for surface plasmon excitation. Silver film (50nm) is deposited on the total internal reflection surface of a dove prism with n_d equals to 1.92 RIU. The collimated beam incidents on the sensing surface at the resonance angle (~65°). The reflection SPR image is magnified with a 10x objective lens and a CMOS imaging chip is used for image capturing. The SPR image is further processed by an image analysis program (home built) for two dimensional Hue profile extraction and processing. With the use of an experimental calibration curve, the Hue profile is further converted into corresponding refractive index distribution. During the experiment, an optical fiber spectrometer (Ocean Optics, HR2000) is used to record the SPR absorption curves for different samples. In the optical interference imaging arm, a monochromater is used to select particular wavelength from the boardband light source. At normal incident, the partial reflection beam from the silver thin film surface interferes with the reflection beam from the steel ball surface and the interferogram of the EHL dimple is formed. The interferogram is captured by another CMOS imaging chip.

3.2 Experimental results

3.2.1 EHL dimple measurement by spectral SPR imaging

PB2400 lubricant dimple

In the measurement experiment, spectral SPR imaging was applied to studying EHL dimples produced from polybutenes lubricant (PB2400). PB2400 is a high viscosity lubricant oil with a high refractive index of n_d-1.5053 RIU. The high viscosity characteristic of this lubricant may assist the formation of EHL dimple.

Spectral SPR image

As shown in the experimental configuration, by projecting a 30mm diameter beam on the EHL dimple surface, corresponding spectral SPR images were obtained and shown in **Figure 2**. **Figure 2a** is the SPR image before EHL dimple formation. We can recognize it as a pure contact between the steel ball and the glass surface. Since no SPR absorption is taking place inside the contact, the image shows the original color of the incident beam.

Figure 2b is the SPR image after EHL dimple formation. PB2400 lubricant was trapped inside the EHL dimple and experienced extremely high pressure (GPa). It changed the refractive index of the lubricant. From the SPR image, we could observe that different SPR absorptions were taking place in the dimple. The reflected color changed from the color of the incident beam to red and finally became green in color at the centre region of the dimple. As indicated in the figure, the outermost ring is believed to be the contacting region between the steel ball and the glass surface. The whole middle region is believed to be the EHL dimple trapped inside the contact. It is believed that the color variation in the SPR image is due to the increasing refractive index and corresponding shifts of SPR absorption dip. Therefore, SPR has clearly revealed the RI distribution caused by the entrapment of lubricant under EHL condition.

In **Figure 2c**, an area of the SPR image which is taken under atmosphere pressure serve as a control image. By going through the same analysis process as the dimple SPR image, it provides the uncertainty. **Figure 2d** shows the optical interferogram obtained from the EHL dimple, which was taken simultaneously during the capture of the SPR image (**Figure 2a**).

3.2.2 Image analysis of 2D spectral SPR imaging

The image coding system

In image analysis, the spectral SPR image was treated as the linear combination of three primary colors, namely red (R), green (G) and blue (B). The CIE (Commission International de I'Eclarage) defines the RGB imaging model with three standard primary components, R, G and B, which refer to the monochromatic spectral energies at wavelengths 700nm, 546.1nm and 435.8nm respectively. The color shown in an imaging pixel can therefore be represented by a linear combination of these primary components [53] (**Figure 3a**). Besides, the XYZ, CMYK, YIQ and HSV models are several most commonly used color coding systems [49, 50], which are designed to provide equivalent image representation for different engineering applications. Nevertheless, only the Hue component in the HSV coding model directly refers to the dominant wavelength of a color [54, 55]. In our experiment, the color variations in the spectral SPR image are caused by the shift of resonance absorption dip at particular wavelength, the use of Hue value thus is a more appropriate approach to represent the spectral information in the spectral SPR image.
The HSV color space model was first introduced by Smith [55] in 1978. It was originally used for the digital control of color television monitors. Hue (H), saturation (S) and value (V) are the three dimensions in the HSV color space and the hexcone model of HSV color space is shown in **Figure 3b**. **Hue (H)** refers to the color variation in the color circle. When Hue changes from 0 to 1, the colors change from red through yellow, green, blue, magenta, and back to red. **Saturation (S)** indicates the departure of a Hue from achromatic. When the saturation value varies from 0 to 1, the whiteness component of the Hue decreases. **Value (V)** refers to the departure of a Hue from black. The blackness component in the color increases, when V decreases from 1 to 0. Smith has also introduced a RGB to HSV algorithm, which given in [55], to transform colors from RGB color space model to HSV hexcone model.

Hue distribution extraction

Hue value distribution is extracted from the SPR image and shown in **Figure 4**. A 2D map of Hue is shown in **Figure 4a**. As shown in the figure, the hue values decrease from (~ 220-250 units) in the outer region to (~ 150-180 units) in the centre region. In the control image (**Figure 4b**), the standard deviation over the whole map is 1.26 units and the average value is 13 units.

3.2.3 Calibration curve

A calibration curve between the Hue values obtained from the spectral SPR image and refractive index is established with measurements on a range of lubricants and their mixtures.

SPR images

During the SPR imaging experiment, different fluid samples were coated on the silver sensor surface and the corresponding spectral SPR images were acquired and shown in **Figure 5a-5v**. SPR effect is polarization dependent and only p-polarization light suffers the SPR absorption, s-polarization can serve as a reference beam. **Figure 5a** shows the spectral SPR image, it were obtained when the polarizer was turned into p- polarization pass mode. Beside, **Figure 5b** shows the s-polarization reference image of the same surface. Various colors are shown in **Figure 5a** for different lubricant samples: PB900 (n_d - 1.4950 RIU), 5P4E (n_d - 1.6294 RIU) and PB1300 (n_d - 1.4992 RIU). No significant colour change is observed in the s-polarization image and the spectral content resembles the original profile of the light source. This confirms that the colour variation in SPR image is due to the shift of spectral absorption profile caused by SPR effect. The SPR images of PB2400 (n_d - 1.5053 RIU) and PB 680 (n_d – 1.4903 RIU) are shown in Figure 5c and 5e respectively. In Figure 5g to 5v (except Figure 5k and 5l), the lubricant samples were obtained from the mixture of the standard lubricant samples and their refractive index values (n_d, 1.5437-1.675RIU) were measured with Abbe refractometer. The Hue values were determined by converting the RGB intensities of individual image pixels using the Hue extraction function in the MatLab program. The Hue values are plotted in the calibration curve (**Figure 7**). The Hue values are extracted from the average value within 50 x 50 pixels on the image, the standard deviation within this area provided the error bar of the data.

SPR absorption spectrum

In addition, the corresponding SPR absorption spectra were also recorded by the spectrometer. In order to remove the background spectral variation of the light source and any effect from the system spectral transfer function, each SPR spectrum was divided by the spectral obtained from the s-polarization which did not support SPR. The raw spectral plots were fitted by nine variable regression curves. SPR absorption minimums were then extracted from the fitted curves for different lubricant samples, and the corresponding plots are shown in **Figure 6**. Data shown in **Figure 6a** (n_d - 1.4903-1.5437 RIU) and **Figure 6b** (n_d - 1.5498-1.6294 RIU) reveal that the resonance minimum has moved to a longer wavelength upon increasing the refractive index. In order to determine the uncertainty of SPR resonance minimum, SPR measurements were repeated for five lubricant samples for 5 times: 5P4E (n_d - 1.6294 RIU), PB2400 (n_d - 1.5053 RIU), H 1900 (n_d -1.5052 RIU), PB 900 (n_d – 1.4950 RIU) and PB 680 (n_d – 1.4903 RIU). The uncertainty of SPR resonance minimum, which is equal to 1.2nm, is obtained from the average of five standard deviations.

Refractive index of the lubricant samples

To consider the frequency-dispersion refractive indices of the lubricants, the following equation [7] is used for the refractive indices calculation of the lubricants.

$$n_s = \sqrt{\frac{\varepsilon_m(\lambda) n_p(\lambda)^2 \sin^2\theta}{\varepsilon_m(\lambda) - n_p(\lambda)^2 \sin^2\theta}} \quad (1)$$

where n_s = refractive index of lubricant, $n_p(\lambda)$ = refractive index of glass prism as a function of wavelength, $\varepsilon_m(\lambda)$ = relative permittivity of metal film as a function of wavelength and θ = angle of incidence.

In the experiment, silver thin film is used as the sensing layer and the glass material of the dove prism is PBH72 O'Hara glass. The corresponding permittivity ε_m (λ) and refractive index n_p (λ) value can be found from the literature [57, 58]. At a fixed incident angle of 65°, the refractive index values of lubricants are calculated at particular SPR dip wavelengths.

Calibration curve

While the Hue, SPR resonance minimum values and refractive index values for standard lubricant samples have been obtained, a calibration curve can be constructed to give the relationship between Hue value and corresponding refractive index. The calibration curve is shown in **Figure 7**. This plot enables direct conversion from SPR image colour (Hue) to refractive index value. Therefore, each single pixel in a SPR image can be converted into corresponding refractive index value. In the calibration plot, we have modified the Hue (0-255) into relative hue values (0-285). For hue values smaller or equal to 30, an additional value of 255 is added. (This modification provides us with a continuous curve in the relationship between the relative hue and SPR resonance minimum.

3.2.4 Refractive index distribution calculation

The Hue value in each pixel of **Figure 4a** is converted to spectral SPR resonance minimum value with the calibration curve (**Figure 7**). Consequently, a continuous map of resonance minimum is obtained and shown in **Figure 8a**. A standard deviation of 1.7nm and an average value of 491.2nm are recorded in the control image (**Figure 8b**). Finally, the SPR resonance minimum maps are further converted to corresponding RI distributions with the help of the calibration curve and they are shown in **Figure 8c and 8d**. Only a small increase in refractive index is found in the outermost region of the dimple, which corresponds to the contact zone between the steel ball and the glass surface. However, a much larger refractive index increase is found near the centre region of the dimple. Using a RI value of 1.4664 RIU for PB 2400 lubricant at atmospheric pressure, the Hue value obtained from the centre of the dimple, where the pressure is at its maximum, the RI value has increased by 5.3% to 1.5447 RIU. In the control image, the average value over the map is 1.47 RIU, which corresponds to the refractive index value of PB 2400 lubricant at atmosphere pressure. In addition, the standard deviation value in the map also gives us the error bar of the detection, which is 0.00079 RIU. The centre region of an EHL dimple acts as a microscopic high pressure chamber with pressure level reaching the order of GPa. Therefore, the molecular density and corresponding refractive index of the entrapped fluid are increased.

3.2.5 Pressure distribution

Figure 9 shows the experimental pressure-refractive index relationship of PB 2400 lubricant that we previous reported in [6]. From this plot, the refractive index map can be further converted to pressure map of the EHL dimple. The corresponding 2-D pressure map as converted using this plot is shown in **Figure 10a**. As shown in the figure, a rapid increase of pressure (maximum of 0.76 GPa pressure) is observed in the center part of the EHL dimple. In the control image (Figure 5.10b), the measured pressure variation over the surface is equal to 0.0024 GPa, which indicates the measurement uncertainty of the system. In addition, the average value is found to be 0.158 GPa. This represents the baseline drift of the system. We have also done the measurements with other lubricants besides PB2400. Measurements on EHL dimple produced with another kind of high viscosity lubricant H1900 (n_d -1.5052 RIU) have been carried out also. Experimental results from H1900 agree with the findings in PB2400 dimple.

3.2.6 Sensitivity estimation

As reported in [59, 46], the sensitivity and resolution of a SPR sensor can be calculated from the following equations,

$$S = \Delta x / \Delta n_s \tag{2}$$

$$R = S \cdot U \tag{3}$$

where S and R are the sensitivity and resolution respectively, Δx is the sensor response and U is the measurement uncertainty.

The calibration plot (**Figure 7**) shows the Hue value response (ΔHue) in spectral SPR image for particular refractive index variations (Δn_s). As shown in **Figure 7**, $\Delta Hue = 120.6$ units is measured in between 1.5656 and 1.4609 RIU. From equation (2), the system sensitivity is found to be 8.7 $\times 10^{-4}$ (RIU/Hue unit). For the standard deviation in PB 2400 lubricant measurement (1.08 Hue units) is used as the measurement uncertainly, the detection resolution of the spectral SPR imaging technique is found to be 9.4×10^{-4} RIU.

4. Comparison with optical interferometry

As described in section 2, optical interferometry is by far the most widely used measurement technique for EHL contacts [9-34]. Wong and Guo et al. [52] presented the measurement of the refractive index of a liquid lubricant at high pressure within an EHL impact dimple using optical interferometry technique. An accuracy of $\pm 10\%$ in refractive index determination (resolution of 0.17 RIU) was achieved. **Table 1** compares the refractive index measurement resolution between spectral SPR imaging and optical interferometry technique. Experimental results show that the SPR techniques can provide more than 180 times resolution improvement over the conventional technique in refractive determination inside the EHL dimple. The spectral SPR imaging technique is based on the shift of the resonance minimum in the spectral curve. As indicated by **Figure 6b**, a change as low as 0.06 RIU in refractive index value causes a shift of over 100nm in the SPR dip. Meanwhile a relatively low measurement uncertainty of 1.2nm has been demonstrated.

In conventional optical interferometry, the interference fringe pattern image cannot provide pixel-wise two-dimensional RI mapping for the EHL contact, which means that some information has been excluded. On the other hand, spectral SPR imaging offers complete information related to the two dimensional RI distribution of an EHL dimple. With known pressure-refractive index relations, one can readily obtain *in situ* pressure mapping for EHL dimple.

In conclusion, the spectral SPR imaging technique has overcome several major detection shortcomings of current measurement techniques for studying EHL. In addition, ,significant improvement in measurement resolution has been demonstrated.

	Optical Interferometry	Spectral SPR Imaging sensor
Resolution	0.17 RIU	9.4×10^{-4} RIU
Resolution Improvement	-----	181 times
		2D RI mapping

Table 1. Measurement resolution comparison between spectral SPR image and optical interferometry.

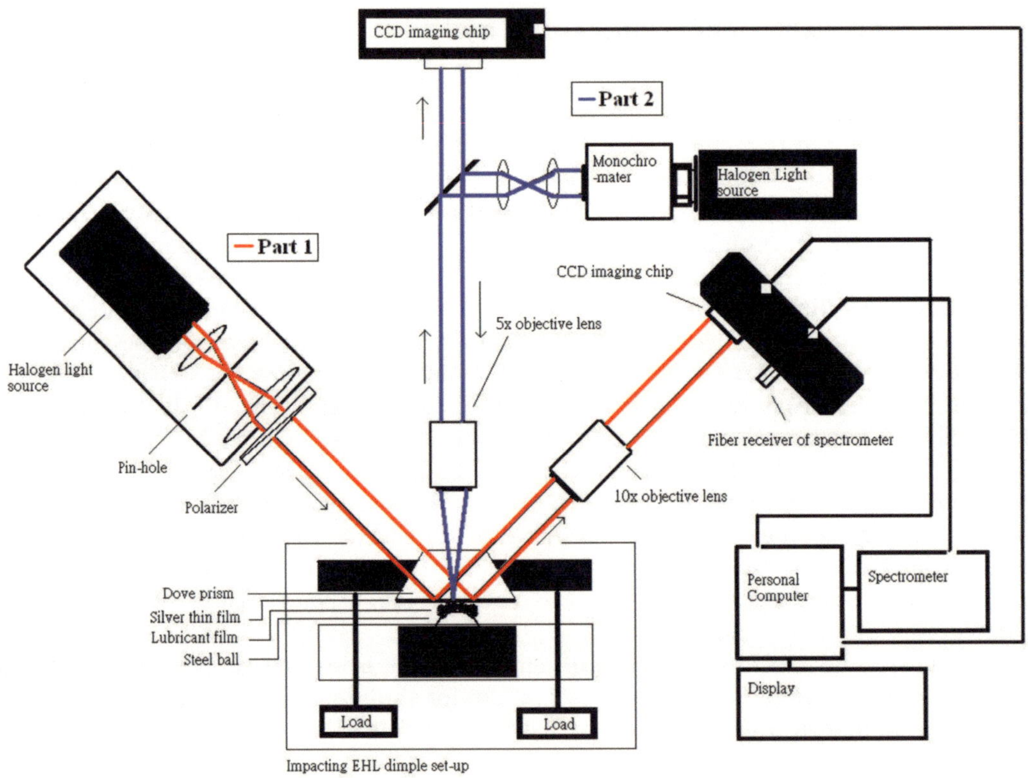

Fig. 1. Experimental scheme of the EHL dimple detection with spectral SPR sensing technique. Part 1 is the spectral SPR imaging set up (the red optical path) and part 2 is the conventional optical interferometry set up (the blue optical path).

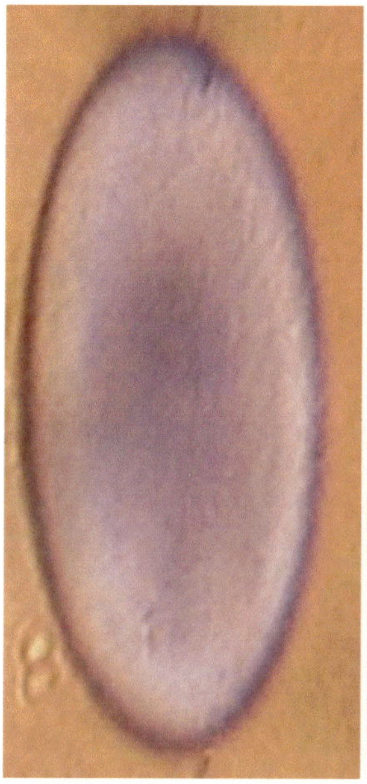

(a)

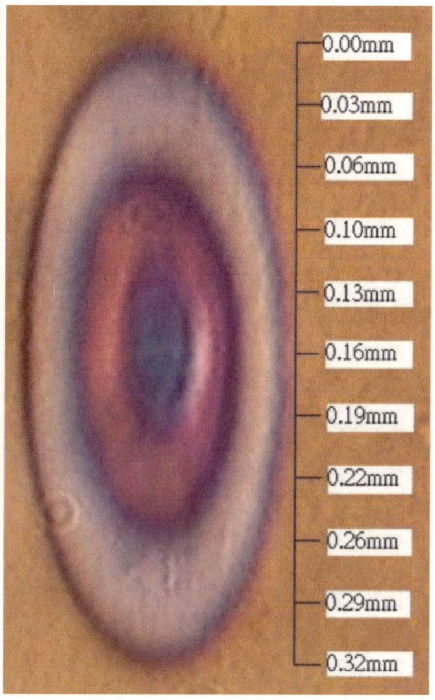
0.00mm
0.03mm
0.06mm
0.10mm
0.13mm
0.16mm
0.19mm
0.22mm
0.26mm
0.29mm
0.32mm

(b)

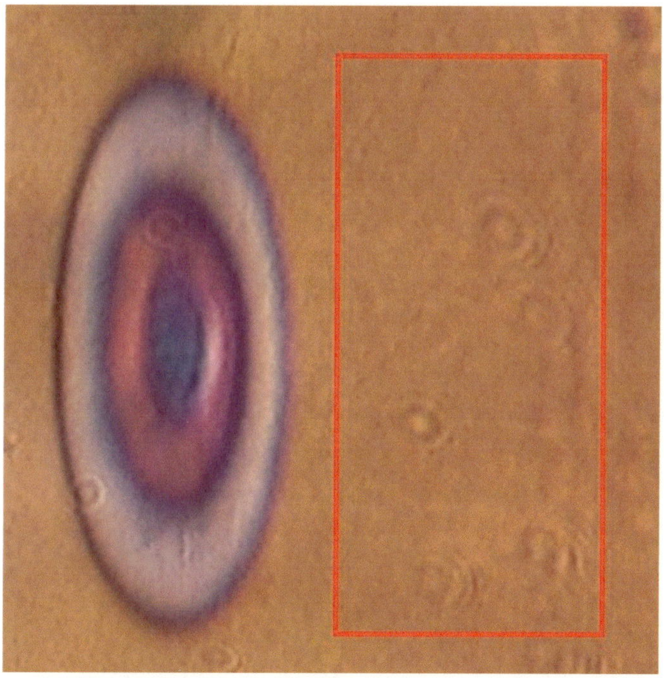

(c)

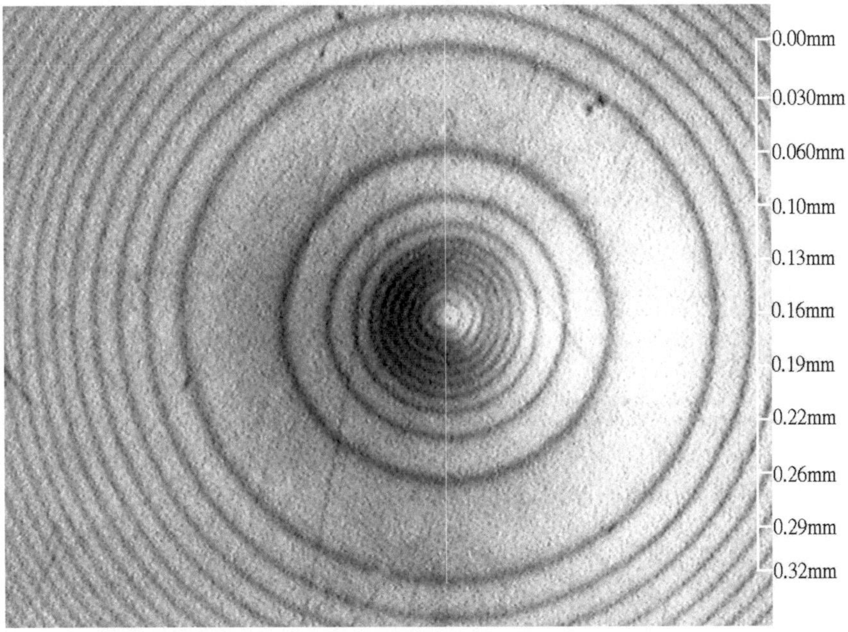

(d)

Fig. 2. Spectral SPR imaging of lubricant PB 2400 dimple: a) Image before dimple production, b) Image after dimple formation c) Image obtained under atmosphere pressure (control image). d) Optical interference image of the EHL dimple.

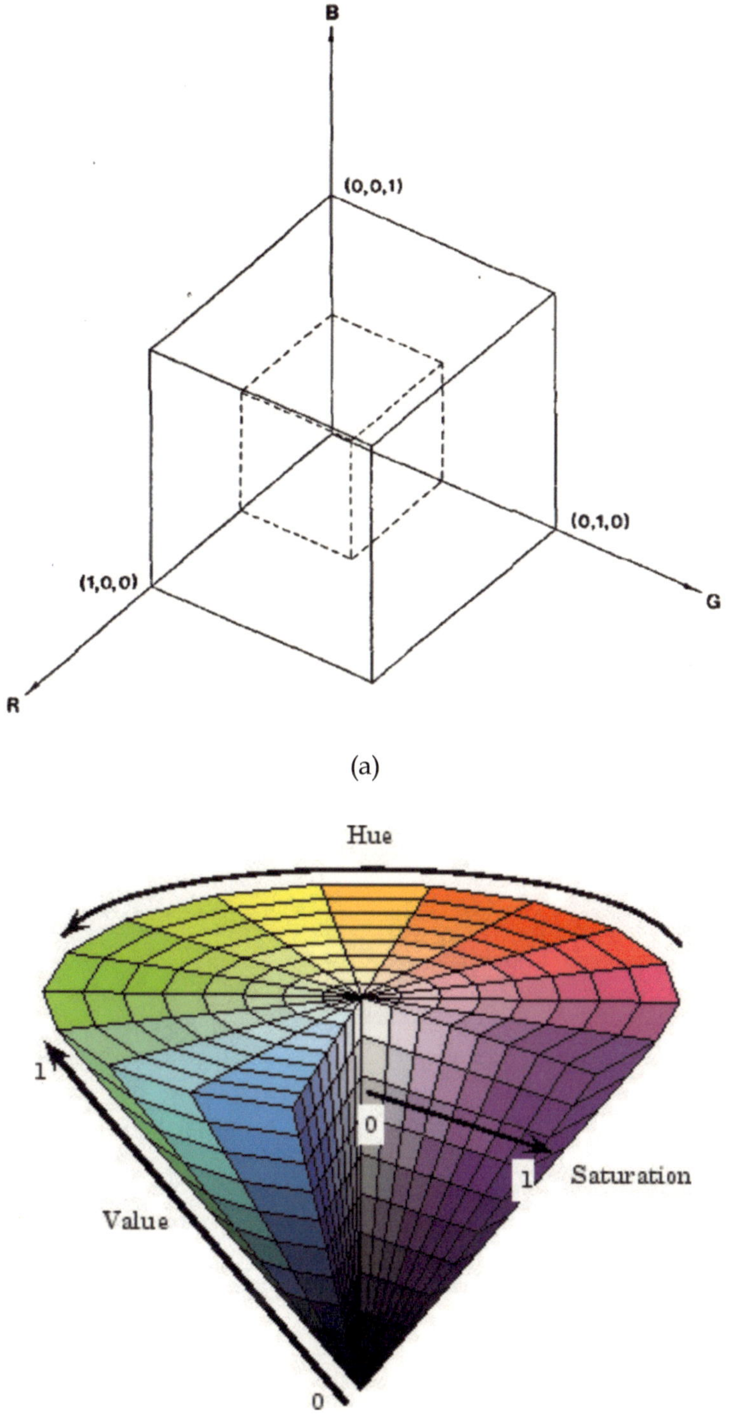

(b)

Fig. 3. a) RGB colorcube model [53], b) HSV hexcone model [56].

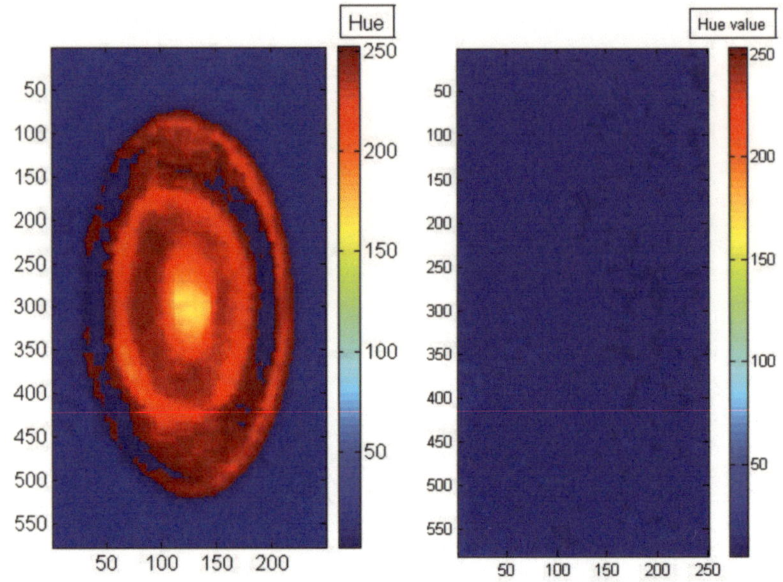

(a) (b)

Fig. 4. a) Hue profile of lubricant PB 2400 dimple, b) Hue profile of control image

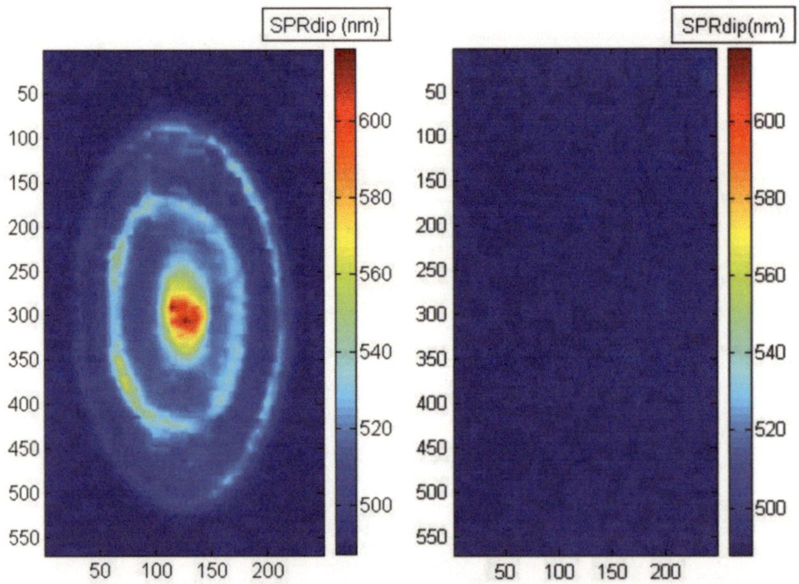

(c) (d)

Fig. 4. c) SPR wavelength minimum profile of lubricant PB 2400 dimple d) SPR wavelength minimum profile of control image.

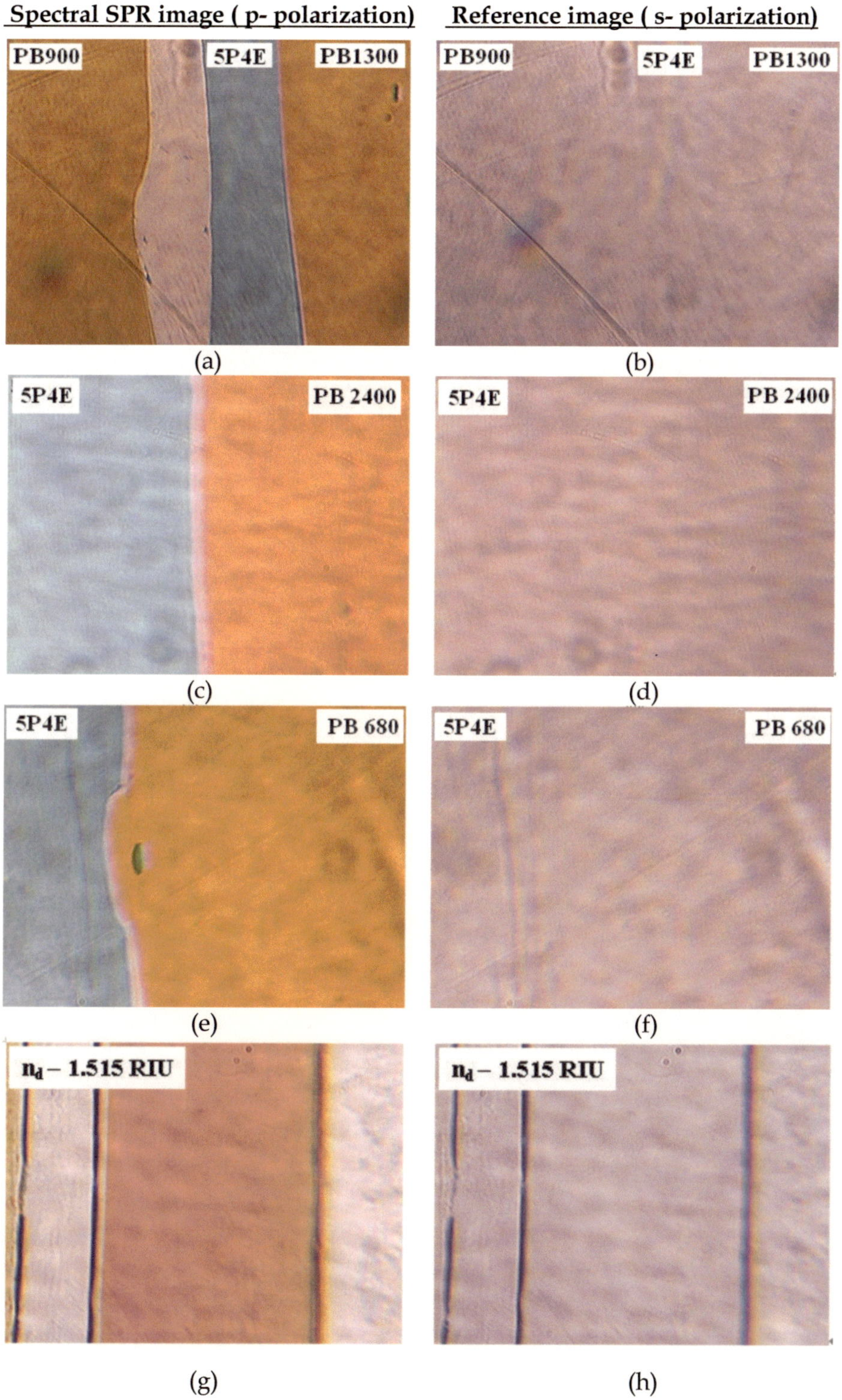
Spectral SPR image (p- polarization)
Reference image (s- polarization)
PB900
5P4E
PB1300
(a)
(b)
PB 2400
(c)
(d)
PB 680
(e)
(f)
n_d – 1.515 RIU
(g)
(h)

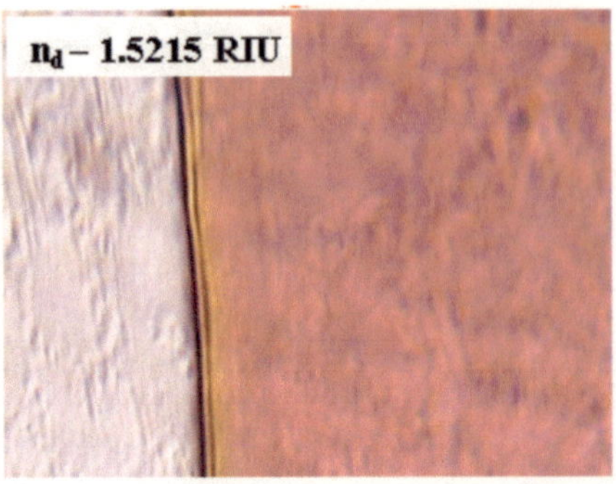
n_d – 1.5215 RIU

(i)

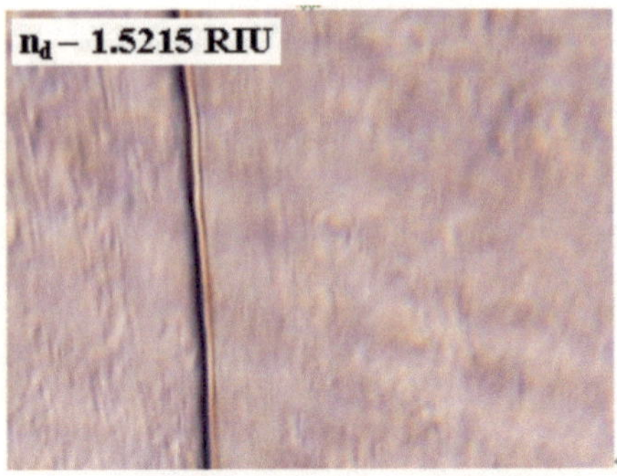
n_d – 1.5215 RIU

(j)

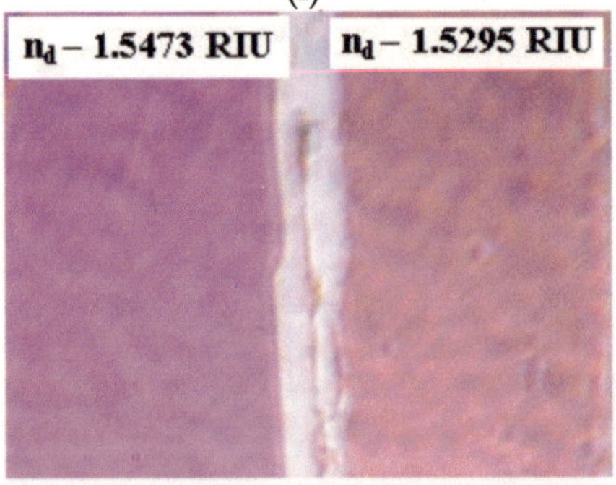
n_d – 1.5473 RIU
n_d – 1.5295 RIU

(k)

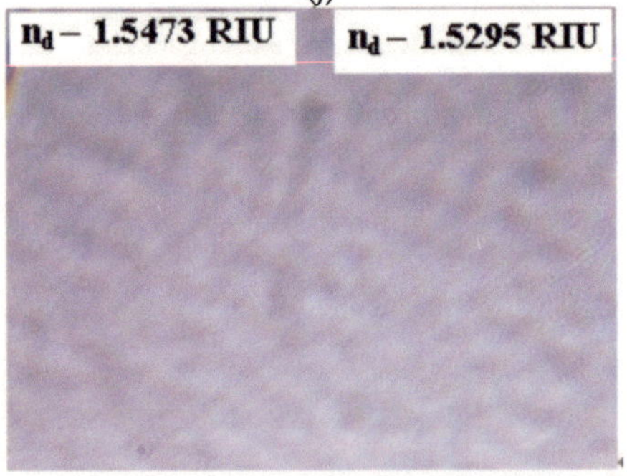
n_d – 1.5473 RIU
n_d – 1.5295 RIU

(l)

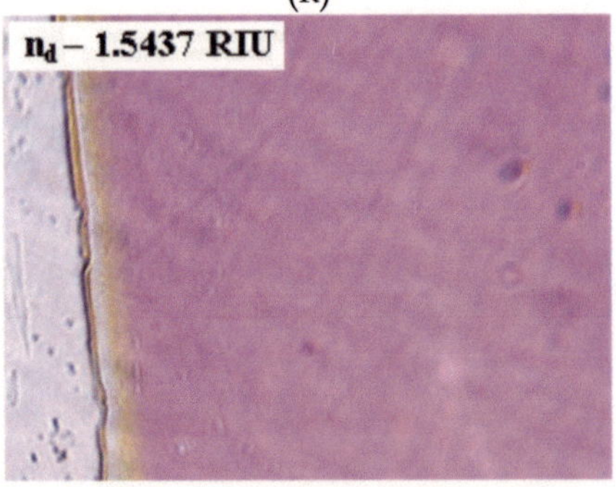
n_d – 1.5437 RIU

(m)

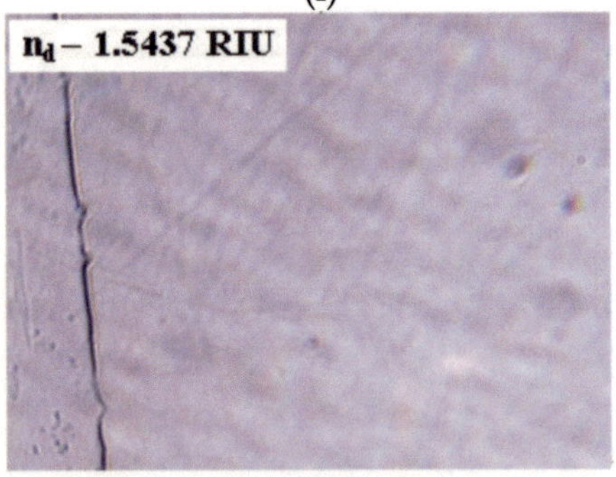
n_d – 1.5437 RIU

(n)

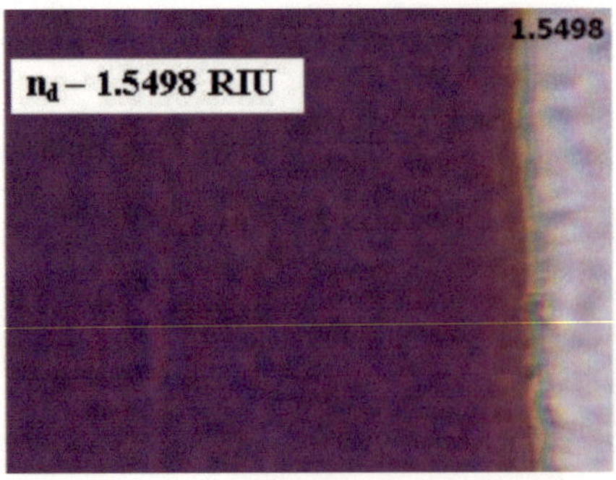
1.5498
n_d – 1.5498 RIU

(o)

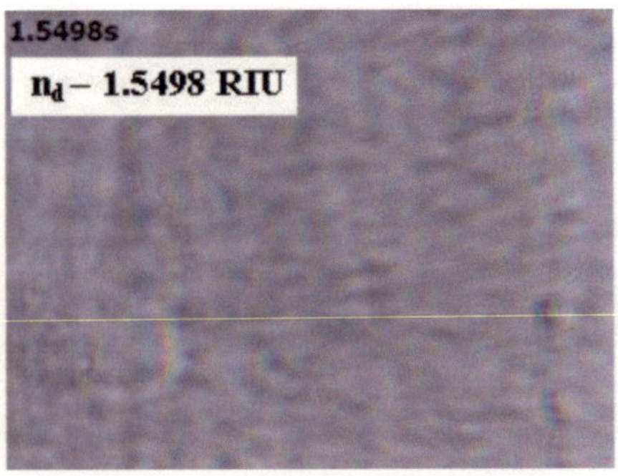
1.5498s
n_d – 1.5498 RIU

(p)

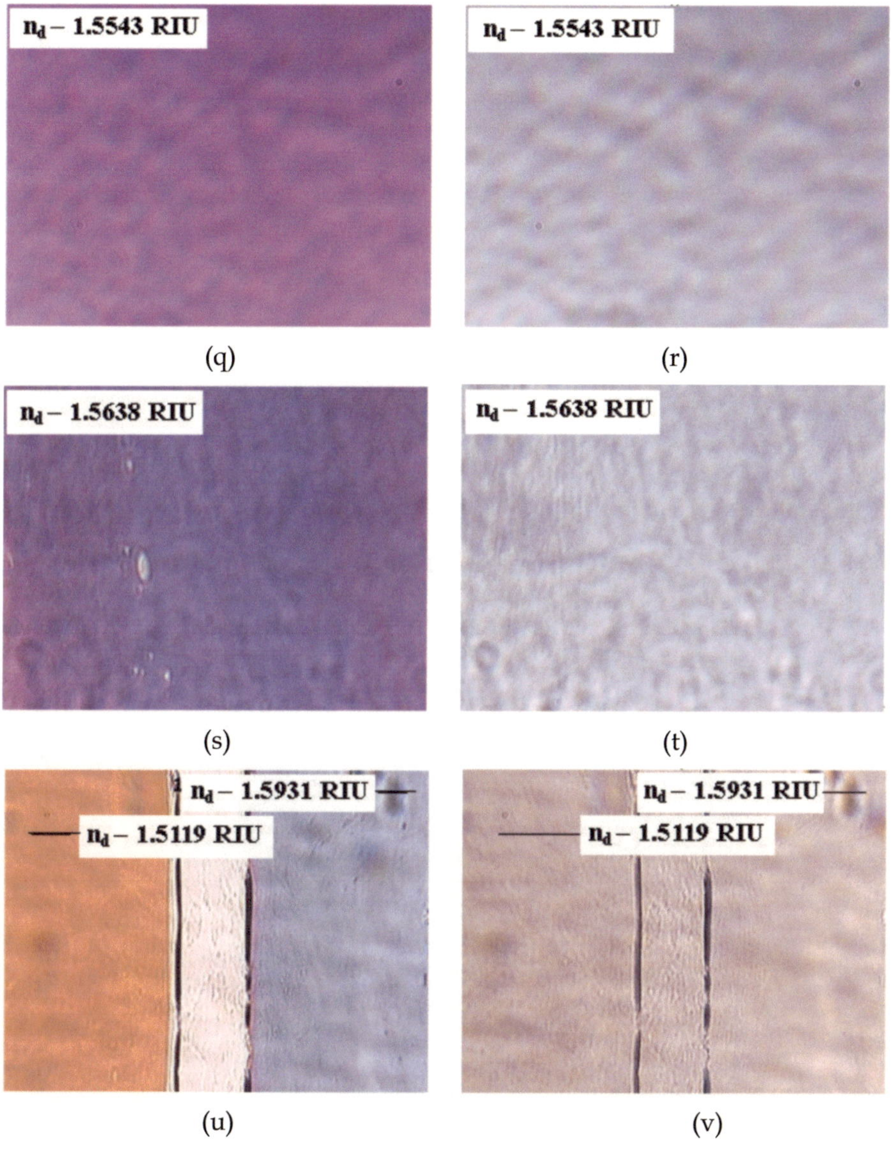

Fig. 5. Spectral SPR images obtained from lubricant samples with known refractive index values for calibration. PB 900 (n_d - 1.4950 RIU), 5P4E (n_d - 1.6294 RIU), PB1300 (n_d - 1.4992 RIU), PB2400 (n_d - 1.5053 RIU) and PB 680 (n_d - 1.4903 RIU). In g)-v) (expect k) and l)), the lubricant samples are obtained from the mixture of the standard lubricant samples and their refractive index values (n_d, 1.5437-1.675RIU) were measured with Abbe refractometer.

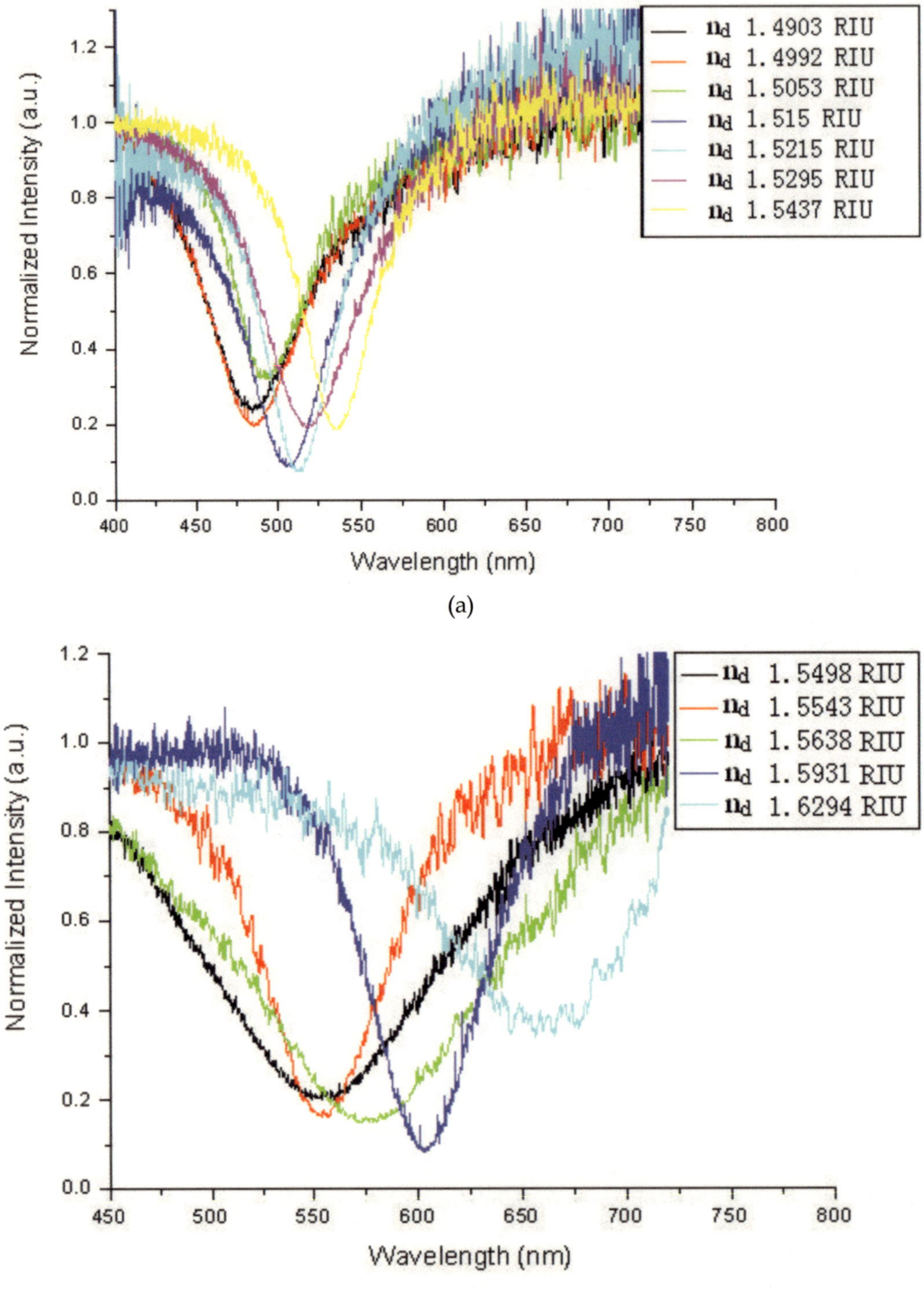

Fig. 6. SPR absorption spectra of lubricant samples for calibration.

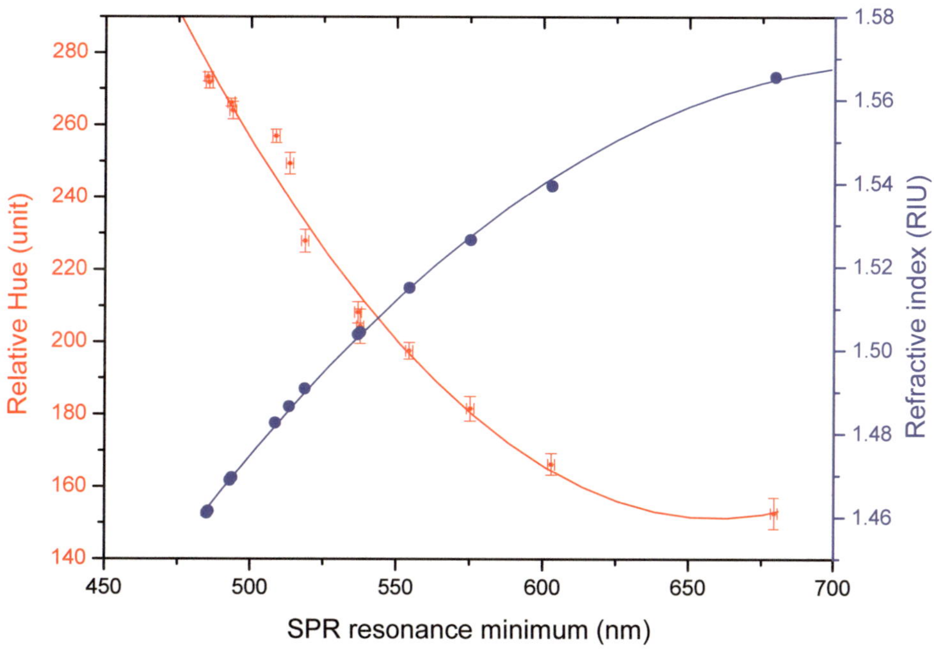

Fig. 7. Calibration curve for image conversion from refractive index value.

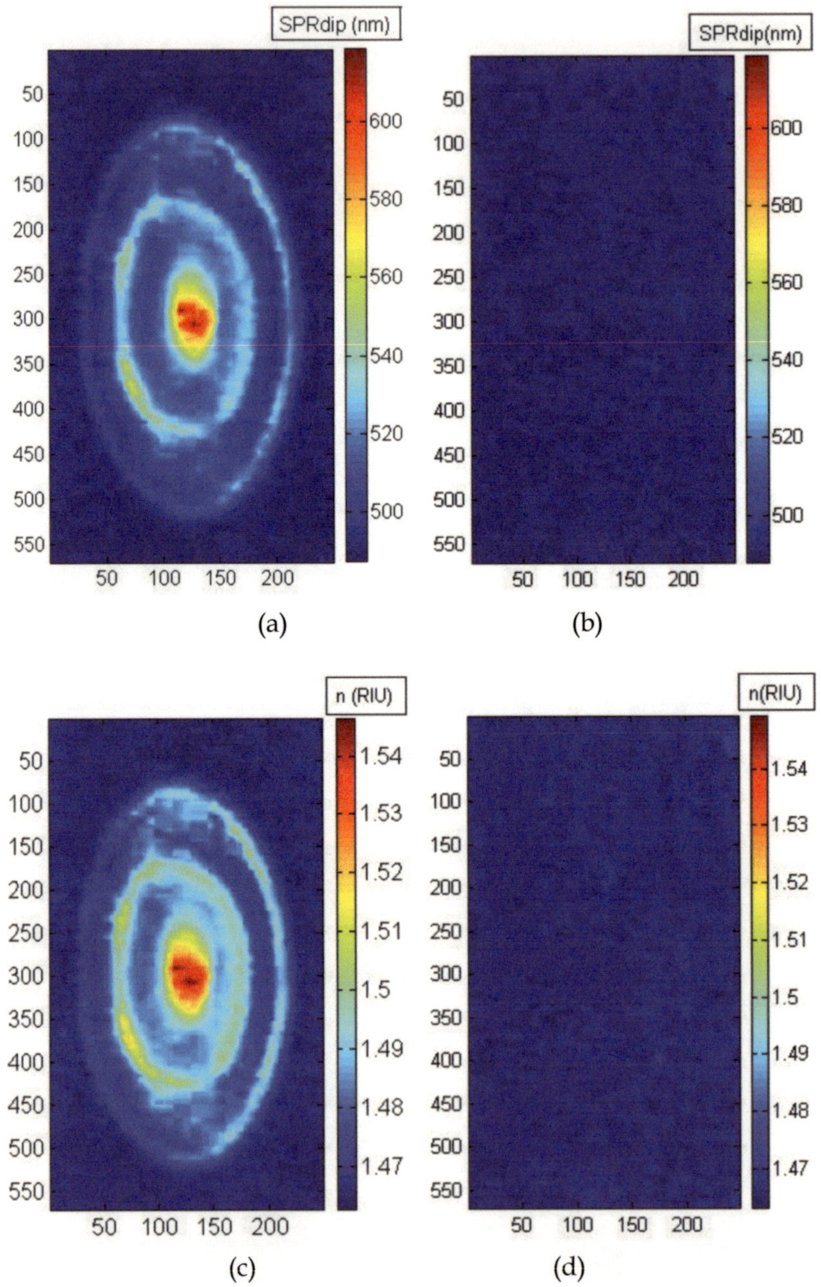

Fig. 8 a) SPR wavelength minimum profile of lubricant PB 2400 dimple b) SPR wavelength minimum profile of control image c) Refractive index profile of lubricant PB 2400 dimple d) Refractive index profile of control image.

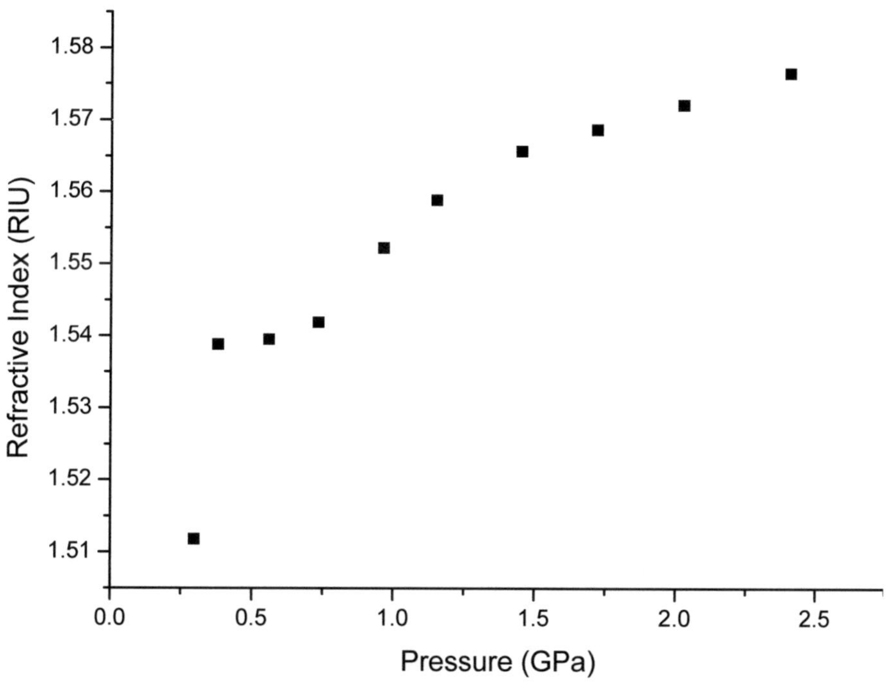

Fig. 9. Pressure – refractive index relation of lubricant PB 2400 [6].

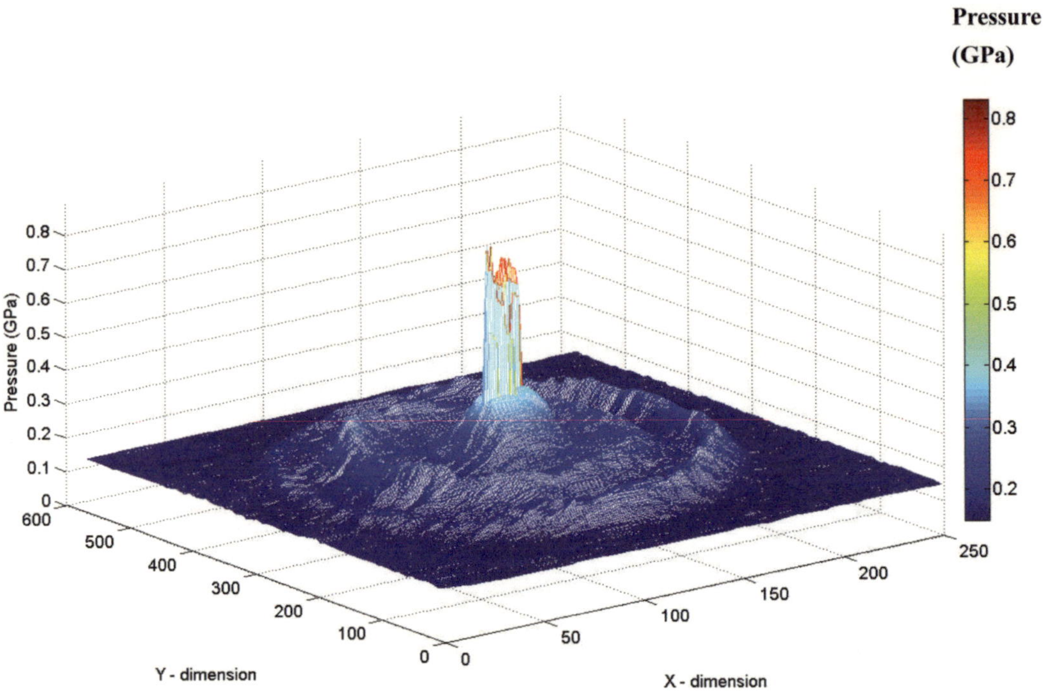

(a)

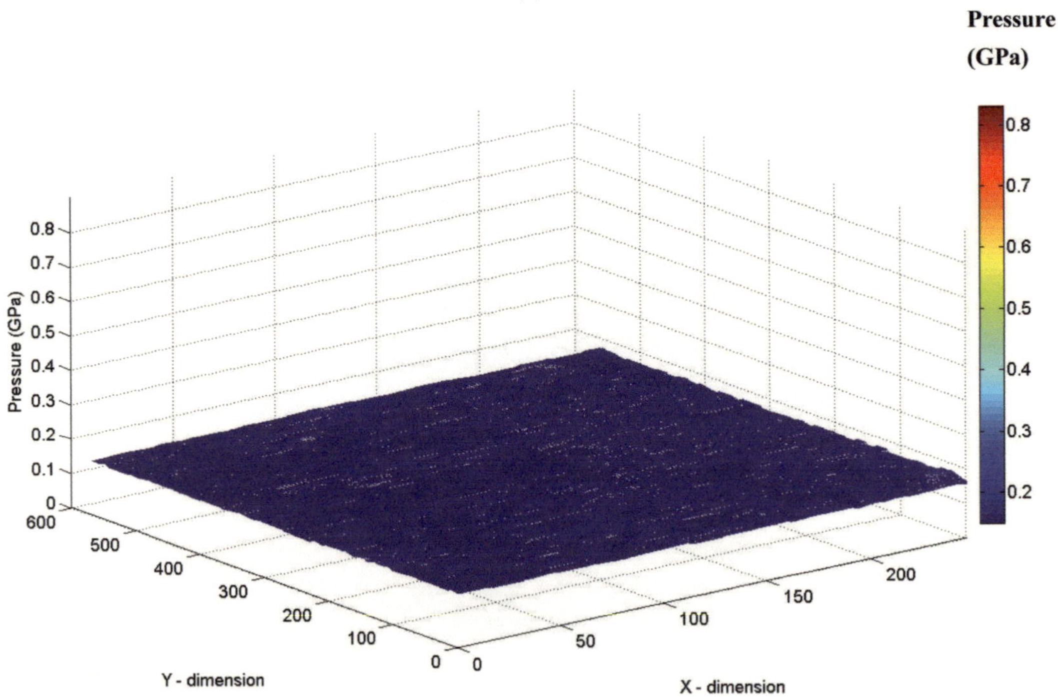

b)

Fig. 10. a) Pressure profile of lubricant PB 2400 dimple, b) Pressure profile of control image.

5. Reference

[1] A.N. Grubin and I.E. Vinogradova, Central Scientific Research Insfitute for Technology and Mechanicul Engineering 30, Moscow, (1949). (DSIR Translation No. 337).

[2] H.M. Martin, Lubrication of gear teeth, London 102, pp.199, (1916).

[3] D. Dowson and G.R. Higginson, Elastohydrodynamic Lubrication, Pergamon Press, Oxford, (1977).

[4] A.W. Crook, Electohydrodynamic Lubrication of rollers, Nature 190, pp. 1182, (1961).

[5] A. Cameron and R. Gohar, Optical measurement of oil film thickness under electo-hydrodynamic Lubrication, Nature 200, pp. 458-459, (1963).

[6] C.L. Wong, H.P. Ho, K.S. Chan, and S.Y. Wu, Application of surface plasmon resonance sensing to studying elastohydrodynamic lubricant films, Applied Optics 44, pp. 4830-4837, (2005).

[7] H.P. Ho, C.L. Wong, K.S. Chan, S.Y. Wu and C. Lin, Application of 2-D spectral surface plasmon resonance to the imaging of pressure distribution in elastohydrodynamic (EHD) lubricant films, Applied Optics 45, pp. 5819-5826, (2006).

[8] C.L. Wong, H.P. Ho, K.S. Chan, P.L. Wong, S.Y. Wu, C. Lin, Optical characterization of elastohydrodynamic lubricated (EHL) contacts using surface plasmon resonance (SPR) effect, Tribology International 41(5), pp. 356-366, (2008).

[9] J.F. Archard and M.T. Kirk, Lubrication at point contacts, Proc. R. Sot. London, Ser. A 261, pp. 532-550, (1961).

[10] J.F. Archard and M.T. Kirk, Influence of elastic modulus on the lubrication of point contacts, Proc. Lubrication and Wear Convention, Inst. Mech. Eng., London 15, pp. 18 l-189, (1963).

[11] R. Gohar and A. Cameron, The mapping of elastohydrodynamic contacts, ASLE Trans., pp. 215-225, (1966).

[12] A. Cameron and R. Gohar, Theoretical and experimental studies of the oil film in lubricated point contact, Proc. R. Sot. London, Ser. A 291, pp. 520-536, (1966).

[13] G.R. Paul and A. Cameron, An absolute high-pressure microvisometer based on refractive index. Proc. R. Soc. A 331, pp. 171-84, (1972).

[14] P.O. Larsson, R. Larsson, A. Jolkin, O. Marklund, Pressure fluctuations as grease soaps pass through an EHL contact, Tribology International 33 (3-4), pp. 211-216, (2000).

[15] K. Yagi, , K. Kyogoku, T. Nakahara, Temperature measurements of oil film and surface in point contact EHL under high slip ratio conditions, Japanese Journal of Tribology 46 (5), pp. 290-291, (2001).

[16] T. Nonishi, S. Oda, Decrease of elastohydrodynamic minimum film thickness with increase of a slide, Japanese Journal of Tribology 46 (1), pp. 53-55, (2001).

[17] T. Nonishi, S. Oda, Decrease of Elastohydrodynamic Minimum Film Thickness with Increase of a Slide, Toraibarojisuto/Journal of Japanese Society of Tribologists 46 (2), pp. 171-177, (2003).

[18] P. Yang, M. Kaneta, S. Masuda, Quantitative comparisons between measured and solved EHL dimples in point contacts, Journal of Tribology 125 (1), pp. 210-214, (2003).

[19] A. Félix-Quinonez, P. Ehret, J.L. Summers, New experimental results of a single ridge passing through an EHL Conjunction, Journal of Tribology 125 (2), pp. 252-259, (2003).
[20] M. Kaneta, P. Yang, Effects of thermal conductivity of contacting surfaces on point EHL contacts, Journal of Tribology 125 (4), pp. 731-738, (2003).
[21] S. Jang, Monochromatic image analysis of elastohydrodynamic lubrication film thickness by fringe intensity computation, KSME International Journal 17 (11), pp. 1704-1713, (2003).
[22] S. Jang, Experimental study on the shear thinning effects of viscosity index improver added lubricant by in-situ optical viscometer, Korea Australia Rheology Journal 15 (3), pp. 117-124, (2003).
[23] F. Guo, P.L. Wong, Experimental observation of a dimple-wedge elastohydrodynamic lubricating film, Tribology International 37 (2), pp. 119-127, (2004).
[24] P. Yang, J. Cui, M. Kaneta, H. Nishikawa, Influence of a surface bump or groove on the lubricating performance and dimple phenomena in simple sliding point EHL contacts, Journal of Tribology 126 (3), pp. 466-472, (2004).
[25] J. Wang, T. Hashimoto, H. Nishikawa, M. Kaneta, Pure rolling elastohydrodynamic lubrication of short stroke reciprocating motion, Tribology International 38 (11-12 SPEC. ISS.), pp. 1013-1021, (2005).
[26] S.Y. Jang, H.S. Kong, A study on the monitoring of nano scale EHL film formation with ultra low aspect ratio, Key Engineering Materials 297-300 I, pp. 257-262, (2005).
[27] X.F. Wang, F. Guo, Z.X. Fu, P.R.Yang, Experimental study of an abnormal EHL film with inlet-dimples, Mocaxue Xuebao/Tribology 27 (3), pp. 269-273, (2007).
[28] M. Kaneta, S. Ozaki, H. Nishikawa, F. Guo, Effects of impact loads on point contact elastohydrodynamic lubrication films, Proceedings of the Institution of Mechanical Engineers, Part J: Journal of Engineering Tribology 221 (3), pp. 271-278, (2007).
[29] F. Guo, Z. Fu, P.L. Wong, P. Yang, Experimental study on lubrication regime variation in point contacts, Tribology International 41 (6), pp. 451-460, (2008).
[30] Z. Fu, F. Guo, P.L. Wong, Theoretical study on the interferometry of thin EHL film measurement, Tribology Letters 31 (1), pp. 57-65, (2008).
[31] X.M. Li, F. Guo, Influence of spinning on elastohydrodynamic lubrication film with wall slippage, Mocaxue Xuebao/Tribology 29 (2), pp. 123-127, (2009).
[32] F. Guo, P.L. Wong, M. Geng, M. Kaneta, Occurrence of wall slip in elastohydrodynamic lubrication contacts, Tribology Letters 34 (2), pp. 103-111, (2009).
[33] A. Young, S. Bair, Experimental investigation of friction in entrapped elastohydrodynamic contacts, Tribology International 43 (9), pp. 1615-1619, (2010).
[34] C. Myant, M. Fowell, H.A. Spikes, J.R. Stokes, An investigation of lubricant film thickness in sliding compliant contacts, Tribology Transactions 53 (5), pp. 684-694, (2010).
[35] J.H.H. Bongaerts, J.P.R. Day, C. Marriott, P.D.A. Pudney, A.M. Williamson, In situ confocal Raman spectroscopy of lubricants in a soft elastohydrodynamic tribological contact, Journal of Applied Physics 104 (1), art. no. 014913, (2008).

[36] T. Reddyhoff, J.H. Choo, H.A. Spikes, R.P. Glovnea, Lubricant Flow in an Elastohydrodynamic Contact Using Fluorescence, Tribology Letters 38, pp. 207–215 (2010).
[37] E. Yeatman and E. Ash, Surface plasmon microscopy, Electronics Letters 23, pp. 1091, (1987).
[38] B. Rothenhäusler and W. Knoll, Surface-plasmon microscopy, Nature 332, pp. 615, (1988).
[39] E.A. Smith, M.G. Erickson, A.T. Ulijasz, B. Weisblum and R.M. Corn, Surface plasmon resonance imaging of transcription factor proteins: interactions of bacterial response regulators with DNA arrays on gold films, Langmuir 19, pp. 1486–1492, (2003).
[40] E.A. Smith, M. Kyo, H. Kumasawa, K. Nakatani, I. Saito and R.M. Corn, Chemically induced hairpin formulation in DNA monolayers, Journal of the American Chemical Society 124, pp. 6810–6811, (2002).
[41] H.J. Lee, T.T. Goodrich and R.M. Corn, SPR imaging measurement of 1-D and 2-D DNA microarrays created from microfluidic channels on gold thin films, Analytical Chemistry 73, pp. 5525–5531, (2001).
[42] G.J. Wegner, H.J. Lee, G. Marriott and R.M. Corn, Fabrication of histidine-tagged fusion protein arrays for surface plasmon resonance imaging studies of protein-protein and protein-DNA interactions, Analytical Chemistry 75, pp. 4740–4746, (2003).
[43] E.A. Smith, W.D. Thomas, L.L. Kiessling and R.M. Corn, Surface plasmon resonance imaging studies of protein-carbohydrate interactions, Journal of the American Chemical Society 125, pp. 6140–6148, (2003).
[44] C.L. Wong, H.P. Ho, Y.K. Suen, S.K. Kong, Q.L. Chen, W. Yuan, S.Y. Wu, Real-time protein biosensor arrays based on surface plasmon resonance differential phase imaging, Biosensors and Bioelecrtonics 24 (4), 606-612 (2008).
[45] C.L. Wong, H.P. Ho, T.T. Yu, Y.K. Suen, W.W.Y. Chow, S.Y. Wu, W.C. Law, W. Yuan, W.J. Li, S.K. Kong, and C. Lin, Two-dimensional biosensor arrays based on surface plasmon resonance phase imaging, Applied Optics 46, 2325-2332 (2007).
[46] C.L. Wong, Imaging surface plasmon resonance (SPR) photonic sensors, Thesis (Ph.D.), The City University of Hong Kong, 2007.
[47] P. Yeh, Optical waves in layered media, Wiley, New York, (1988).
[48] J. Homola, H. Vaisocherová, J. Dostálek, M. Piliarik, Multi-analyte surface plasmon resonance biosensing, Methods 37 (1), 26–36 (2005).
[49] P.L. Wong, S. Lingard and A. Cameron, A simplified impact microviscometer, Tribology International 25 (6), pp 363-366, (1992).
[50] R. Gohar and A. Cameron, The mapping of elastohydrodynamic contacts, ASLE Transaction 10, pp. 215–225, (1966).
[51] A. Cameron and R. Gohar, Theoretical and experimental studies of the oil film in lubricated point contact, In: Proc. R. Soc. London, Ser. A 291, pp. 520–536, (1966).
[52] P.L. Wong, F. Guo and C. Feng, The measurement of the refractive index of a liquid lubricant at high pressure within an EHL impact dimple using a single optical interferogram, Tribology International 36 (7), pp. 497-504, (2003).

[53] F.V.D. Heijden, Image Based Measurement system, John Wiley & Sons, pp. 76-77, (1994).
[54] T. Bose, Digital signal and image processing, Wiley, pp. 606-609, (2004).
[55] A.R. Smith, Color gamut transform pairs, Computer Graphics 12(3), pp. 12-19, (1978).
[56] Matlab image processing toolbox, http://www.mathworks.com/access/helpdesk/help/toolbox/images/f8-22706.html.
[57] E.D. Palik, Handbook of Optical Constants of Solids, Academic, pp. 275–358, (1985).
[58] Department of Applied and Computer Optics of the St. Petersburg State Institute of Fine Mechanics and Glass, Glass bank, http://glassbank.ifmo.ru.
[59] C.L. Wong, H.P. Ho, K.S. Chan, S.Y. Wu, and C. Lin, Application of spectral surface plasmon resonance to gas pressure sensing, Optical Engineering, vol. 44, pp. 124403, (2005).

3

Tribological Properties of Fluorinated Amorphous Carbon Thin Films

Miguel Rubio-Roy, Carles Corbella, José-Luís Andújar, Enric Bertran
Universitat de Barcelona
Spain

1. Introduction

1.1 Amorphous carbon characteristics

The peculiar electronic configuration of carbon atoms, $1s^2\,2s^2\,2p^2$, and the small energy difference between their 2p and 2s orbitals, compared to the binding energy of the carbon bonds, allow the electrons to rearrange in s and p mixed orbitals that enhance the binding energy with other atoms. This process is called hybridization and produces three different types of orbitals: sp = s + p, sp^2 = s + p + p and sp^3 = s + p + p + p.

Each different bonding state corresponds to a certain structural arrangement: sp bonding gives rise to chain structures (with two σ bonds and two π bonds), sp^2 bonding conforms onto planar structures (three σ bonds and one π bond) and finally sp^3 bonding produces tetrahedrical structures (four σ bonds). The p orbitals that form π bonds overlap less than the orbitals forming σ bonds. The reduced overlapping makes π bonds weaker than σ bonds. However, a number of scenarios are possible. Sometimes, as in ethene (C_2H_4), a σ and π bond combine producing a stronger bond between carbon atoms. This is called a double bond: C=C. Triple bonds consist of a σ bond and two π bonds, as in ethyne (C_2H_2). Although chemically stronger thanks to double bonds, the mechanical stability obtained with sp^2 hybridization in solids is limited, due to the planar geometry. Instead, sp^3 hybridization allows the creation of a three dimensional network of σ bonds.

Due to this variety of possible bonding configurations, carbon has a number of allotropes: graphene (sheet of sp^2 bonded carbons: σ bonds plus delocalized π bonds), carbon nanotubes and fullerenes (graphene sheets rolled over themselves forming cylinders or spheres, respectively), graphite (Bernal stack of graphene sheets), diamond (network of sp^3 bonded carbons) and amorphous carbon (cross-linked and non-organized carbon matrix with a mixture of sp^2 and sp^3 bonds). It is to the modification of the latter with fluorine that this chapter is devoted to.

The International Union of Pure an Applied Chemistry (IUPAC) defines amorphous carbon as "A carbon material without long-range crystalline order". It also states that "Short range order exists, but with deviations of the interatomic distances and/or interbonding angles with respect to the graphite lattice as well as to the diamond lattice."

Depending on the ratio of sp^2 and sp^3 bonds in the matrix, amorphous carbon (a-C) presents a variety of well-reviewed mechanical properties [Silva, 2003]. Tetrahedral amorphous carbon films (ta-C or TAC) present the highest hardness, with a high degree of sp^3 bonding and without hydrogen. It is almost exclusively deposited by filtered cathodic vacuum arc

(FCVA), [Martin et al., 1988; McKenzie et al., 1991; Fallon et al., 1993; Polo et al., 2000] although other techniques such as mass-selected ion beam (MSIB) [Kaukonen & Nieminen, 1992; Miyazawa et al., 1984] or laser deposition [Scheibe & Schultrich, 1994; Voevodin & Donley, 1996] can also be used. For the hydrogenated version of ta-C, plasma beam source (PBS) [Weiler et al., 1996] and electron cyclotron wave resonance (ECWR) [Morrison et al., 1999] have been used. Despite the high resemblance of ta-C to diamond on its short range microstructure and mechanical properties, its optoelectronics properties are controlled by the small percentage of sp^2 electrons with energies close to the Fermi level. [Silva, 2003]

On the other hand, films with comparable degrees of sp^3 bonding but soft are named polymer like carbon (PLC). These films have high H content (above 40%), which terminates single bonds and therefore reduces carbon-carbon cross-linking. The quantity of sp^3 bonds is high due to hydrogen's preferential bonding to carbon in sp^3 configuration, but this is no longer related to enhanced hardness. This kind of films present a small amount of defects compared to others and have a wide bandgap [Silva et al., 2002].

Films with intermediate hydrogen content can be either hydrogenated ta-C or hydrogenated a-C (a-C:H). The latter is the most studied type of a-C films and has been known in the literature as diamond-like carbon (DLC). It is softer than ta-C(:H) but still harder than most materials: values ranging from 20 to 40 GPa. Density is also slightly smaller than that of ta-C(:H). Its main advantage is the simple setup for its deposition. A wide range of deposition techniques is available compared to the complex setup for ta-C deposition [Aisenberg & Chabot, 1971; Zou et al., 1989; Koidl et al., 1990; Kessels et al., 1998]. Also, DLC has a wider range of optical gaps than ta-C depending on the deposition conditions, which enables it for applications such as UV nanoimprint molds, where middle or near UV light transmission is necessary. For shorter wavelengths, thin films allow to further extend the working range.

Figure 1 shows the space of film types depending on the sp^2:sp^3 ratio and H content. Figure 1a is an updated version of the popular diagram by [Ferrari & Robertson, 2000] which includes hydrogenated films. Figure 1b by [Silva, 2003] presents a revised diagram based on a wider review.

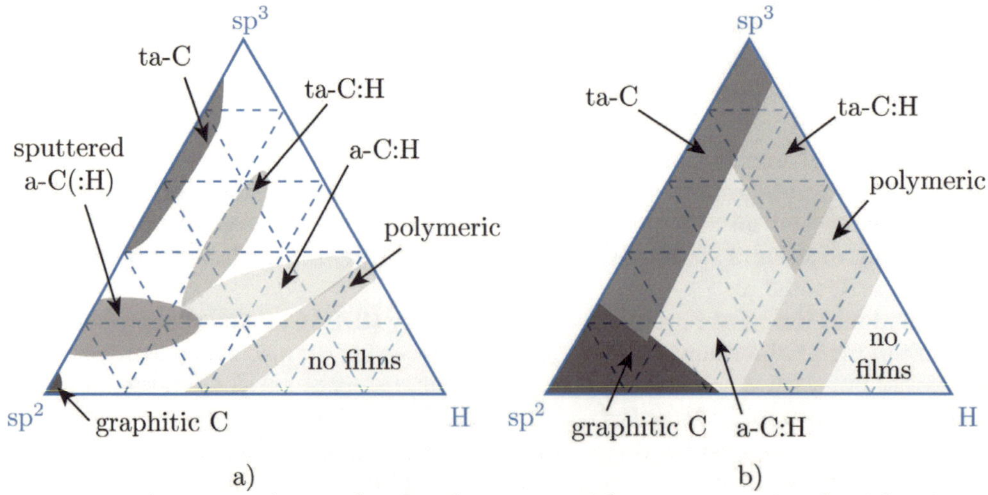

Fig. 1. Ternary diagrams of amorphous carbon types with respect to sp^2 and sp^3 fraction, and H content. a) by [Robertson, 2002]; b) by [Silva, 2003].

High hardness and elastic modulus are the main features of a-C:H and ta-C(:H) films. However, the rigid network of the films produces a high intrinsic compressive stress of up to 12 GPa [Ferrari et al., 2002]. This is an important drawback in the production of DLC coatings, since it limits the thickness depending on the adhesion to the substrate. For badly adhered or too thick films, cracks can appear and finally peel off the film if they progress, or otherwise dramatically reduce wear resistance. Thin buffer layers between substrate and film can be used to avoid this problem. Several transition metals and compounds can be used to this respect (W, Ta, TiC, Cr, SiC, etc.) depending on the substrate nature [Lee et al., 2000; Bahl et al., 2000; Kiuru et al., 2003].

With regard to surface properties, water contact angle has usually been examined providing relatively high values from 55° to 70°, depending on the presence of low energy C–H groups on the surface and the oxygen contamination of the film. The introduction of elements like F, Fe, Al and others has proven to be a useful method to increasing contact angle values up to more than 100° [Grischke et al., 1995; Chen et al., 2001].

1.2 Deposition methods

DLC can only be obtained at conditions far away from equilibrium, and so it is not spontaneously found in nature. The control of the growth parameters broadens the field of carbon microstructures and, in parallel, the properties of this material, which can only be synthesized in thin film form.

The production of a-C thin films is carried out by a large variety of deposition techniques. Some of them are adequate for laboratory studies while some others are more addressed to industrial production. Almost all processes work with low pressure of precursors or inert gases. Under these conditions, the increased mean free path of the species allows adequate particle energies, in order to obtain good mechanical and tribological properties. DLC films, and hard coatings in general, are very sensitive to particle bombardment during their growth. Therefore, this is a key parameter that must be controlled in any deposition method. Contamination by oxygen can affect importantly the properties of DLC. For this reason, clean processes in chambers capables of reaching low base pressures are used. The different methods to deposit DLC exhibit three characteristics [Silva, 2003]:

i. A plasma is used as the source of reactive species.
ii. The substrate is found, in general, at room temperature. At least there is no need to maintain a high substrate temperature during the process unlike in classical CVD processes.
iii. Energetic particles (ions and/or neutrals) participate in the deposition process.

Ion beam deposition

Ion beam deposition was the first method used for the synthesis of DLC [Aisenberg & Chabot, 1971]. Carbon and hydrocarbon ions coming from a graphite cathode are condensed in an energetic beam and produce sp^3 states as they impinge on the growing film. Ion beam sources operate efficiently within an ion energy range of 100–1000 eV. Mass selected ion beam (MSIB) techniques are addressed to obtain ion beams with a certain energy. Ions pass through a magnetic filter that selects charged species with a fixed charge to mass ratio. Unfortunately, the ion beam is usually weakly ionized and, thus the mechanical properties of DLC are not optimal.

Sputtering

Magnetron sputtering is a well-known physical vapor deposition technique (PVD). It is one of the most used techniques, in industrial environments, to grow DLC. The impinging

species are commonly sputtered from a graphite target, which is RF- or DC-powered, in an Ar glow discharge [Rubio-Roy et al., 2007a]. The energy of the ion species being deposited can be increased or adjusted by applying a DC bias to the substrate or by using an unbalanced magnetron whose magnetic field substantially reaches the substrate.
Reactive magnetron sputtering processes are also possible if hydrogen or hydrocarbon source gases are added to Ar. An alternative sputtering process can be performed without plasma: an Ar ion beam from an ion gun sputters a carbon target, say graphite, while another beam bombards the growing film to promote sp^3 formation. This method is called ion beam assisted deposition (IBAD).

Cathodic arc

Cathodic arc constitutes one of the oldest vacuum deposition techniques [Randhawa, 1988; Sanders & Anders, 2000]. The arc discharge is concentrated at the cathode surface, where high ion densities ($\approx 10^{12}$ A/m^2) form non-stationary spots. The ionization of species reaches 100% of the total flux, and the maximized parameter is the electron current to the anode, which is called arc current. The ion current is about 10−100 times lower than the arc current. This method is really appropriate to obtain hard DLC films, with a significant sp^3 fraction, but certain issues have limited its use. First of all, films may present serious inhomogeneities due to the instabilities caused by the location of the arcs in the cathode. Secondly, the installation of a filtering device is necessary to prevent the deposition of macroparticles. Filtered cathodic vacuum arc (FCVA) solves the last problem by means of a magnetic filter, either in single bend or S-bend configuration [Robertson, 2002].

Pulsed laser deposition

Pulsed laser deposition (PLD) provides energetic ions, as in cathodic arc processes. Therefore, it is possible to deposit DLC films with high sp^3 content with this technique, without the need of substrate biasing or great current discharges [Voevodin & Donley, 1996]. The intense energy pulses (15-30 ns) of a laser beam are able to vaporize a carbon target, which shows major advantages when it is graphite due to minimization of droplets emission. Fluxes of neutrals, ions and particulates are ejected from the target through an intense plasma, named plasma plume. The first PLD setups used Nd:YAG lasers, operating at a wavelength of 1064 nm. Nowadays, excimer lasers have progressively substituted them because higher-quality DLC is obtained using shorter wavelengths, for instance 193 nm and 248 nm (ArF and KrF lasers, respectively). The structures of DLC samples grown by PLD can be nanocrystalline diamond and ta-C.

Plasma-enhanced chemical vapor deposition (PECVD)

PECVD is a very popular technique to produce DLC films. A plasma is ignited in a reactor, between two electrodes of different areas, giving place to an asymmetrical discharge. The supplied power cannot be DC, since DLC may be electrically insulating. Thus, PECVD processes employ capacitively coupled RF power or pulsed-DC power. As depicted in figure 2, power is driven to the small electrode (cathode), where the substrate is placed, and the reactor walls are usually grounded (anode) [Rubio-Roy et al., 2007b].
The negative bias of the substrate cathode accelerates the bombarding ions to create sp^3 sites. The enhancement of ion-to-neutrals ratio in the discharge needs low-pressure plasmas, but too low pressures will avoid a self-maintained discharge regime. This appropriate equilibrium is accomplished by raising the supplied power or by coupling an external magnetic field to the sheaths. The latter is performed in inductively coupled plasmas (ICP).

Electron cyclotron resonance systems (ECR) use both microwaves and magnetic field for a resonant transfer of energy to the electrons in the plasma. The simultaneous use of two different plasma sources, besides increasing the ionization degree, permits to decouple the particle flux and the bombarding energy of the ions [Corbella et al., 2008].

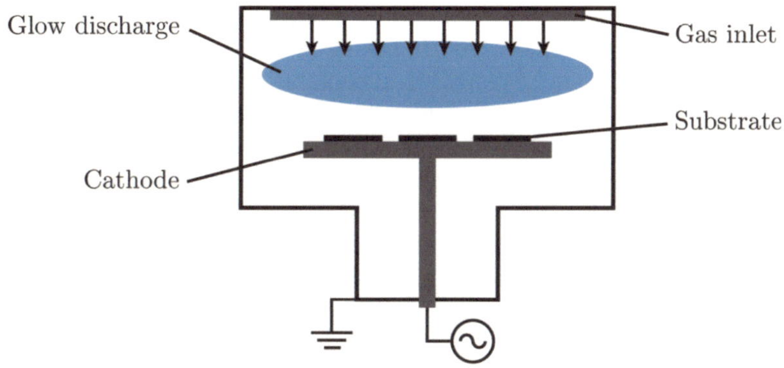

Fig. 2. Layout of a PECVD reactor.

Electron cyclotron wave resonance (ECWR)

When trying to couple low frequency electromagnetic radiation to a plasma, the latter behaves as a conductive medium and totally absorbs the incident waves within a characteristic absorption depth ("skin" depth). The plasma becomes transparent to the electromagnetic waves only when the excitation frequency exceeds the plasma frequency or when its conductivity is reduced (e.g.: with higher plasma density). However, RF excitation frequencies require very low plasma densities.

ECWR consists on applying a weak magnetic field (≈20 G) that inhibits plasma conductivity by confining electrons in cyclotron orbits around the field lines. Hence, the electromagnetic radiation can be distributed across the plasma, which contributes to increase the plasma density and generates propagating electromagnetic modes [Morrison et al., 1999; Rodil, 2000]. The dimensions of the chamber are set so that one of these modes matches the resonant condition.

1.3 Tailoring of amorphous carbon properties

DLC shows special surface properties, and in some cases, they must be tuned depending on the selected application. The physical and chemical properties can be tailored by adding new elements into the a-C matrix as nanocomposites or by structuring the material in multilayers or gradient coatings. This provides the so-called modified a-C. Another strategy consists on patterning its surface, as explained in section 3.1.

Regarding the mechanical properties of DLC, both hardness and elastic modulus reach high values, and can be tailored by controlling the sp^3 fraction and hydrogen content, if any. Although hardness of a-C:H ranges from 20 to 40 GPa, the rigid network of the films provides an intrinsic compressive stress ranging from 0.5 to 12 GPa. This high stress is an important drawback in the production of DLC coatings, since it limits the film thickness to less than 1 μm. In fact, the accumulation of micro-strains within the amorphous matrix leads to a global compressive stress that is large enough to damage the film by development of cracks and strains that prevent the adhesion to substrates and reduce film stability. DLC films, deposited by RF-PECVD, thicker than 50 nm are generally so stressed that they lose their consistency and delaminate.

Besides improving adhesion to substrate, another strategy to avoid film peeling is stress relaxation. This is possible through doping with N, Si or metals, as well as by post-annealing or incorporation of nanometric particles (DLC nanocomposites), but these operations may degrade the mechanical properties [Zhang et al., 2002; Chen & Hong, 2003; Corbella et al., 2009a]. Metal incorporation in a DLC matrix has been proved to reduce stress below 1 GPa and maintain hardness and wear rate within acceptable values [Dimigen et al., 1987]. Low stress values are also achieved by depositing alternating hard and soft a-C layers [Ager et al., 1997; Logothetidis et al., 2000]. The resulting multilayered structures show a lower overall stress, and hardness is almost unaffected. Metal and a-C multilayers (Me/a-C) [Bertran et al., 2003], as well as a-CN/a-C structures [Pino et al., 2001], have been also deposited with similar results. Bias-graded deposition of DLC provides the best tribological performance through surfaces showing high sp^3/sp^2 ratios, which gradually decrease from the surface to the interface in order to enhance adhesion [Zhang et al., 2004]. Another way to improve coating adhesion consists in growing a buffer layer (Ti, Cr) prior to DLC deposition. Figure 3 summarizes the above mentioned strategies.

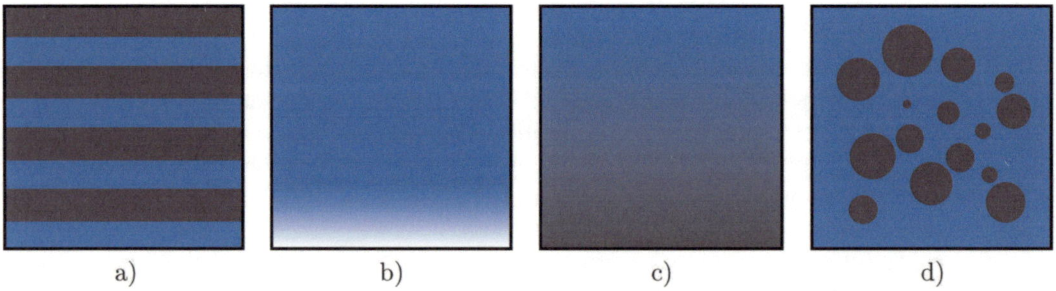

Fig. 3. DLC modified structures: a) multilayers; b) buffer layers; c) composition-graded coatings; and d) nanocomposites.

DLC films with hydrogen are mostly exploited in tribological applications due to their low friction coefficient and low wear rate. In all environments, the tribological behavior of DLC is controlled by an interfacial transfer layer formed during friction [Voevodin et al., 1996]. Thus, DLC is a valuable alternative as solid lubricant for applications where liquids are not convenient.

Besides wear and corrosion resistant, DLC is also known to be a hydrophobic material. The C–H groups located on the surface diminish the wettability, which is parametrized by the contact angle held between a liquid drop and the film surface. DLC typically exhibits a contact angle of 55–75° for water, although it can be increased by adding metals such as Fe or Al. In the case of Al-DLC, the contact angle exceeds 100° [Chen et al., 2001]. The high hydrophobicity shown by a-C:H:Si:O and a-C:H:F films is also advantageous for anti-sticking and self-cleaning applications [Grischke et al., 1998; Rubio-Roy et al., 2007b]. Highest values of contact angle are reached when combining hydrophobic films together with special surface architectures [Kim et al., 2007]. On the other hand, a-C:H:Si:O films find also applications as ophthalmic and decorative coatings due to their comparatively high optical transmittance. The resistance to higher temperatures makes a-C:H:Si ideal as protective coating in components for thermal processes.

Low roughness is another intrinsic surface property that makes DLC films suitable for certain tribological applications. DLC layers show a flat and smooth surface, with their roughness being minimized when deposited by filtered arc evaporation. RMS surface roughness below 0.05 nm has been reported for DLC deposited either by sputtering or PECVD [Peng et al., 2001]. These atomically flat surfaces show improved tribological properties of importance for applications such as hard drive disk coating. Moreover, surface roughness of ultra-thin DLC films is revealed to follow fractal scaling laws, i.e. it is related to film thickness through an exponential function [Casiraghi et al., 2004].

1.4 Applications

The wide variety of applications of DLC is mainly addressed to hard coatings with special tribological properties. These properties can be tuned by modifying DLC structure and by incorporating other elements [Hauert & Patscheider, 2000], whereas other properties ensure its performance as optical filter or for components with electronic functionality.

The fastest growing market for DLC application during the previous years was the automotive industry [Lampe et al., 2003], in which DLC, ta-C, W-DLC and WC/C multilayers were being preferentially developed as hard and tribological coatings [Hauert, 2004]. Automotive industry demands solutions for preparing low friction and wear resistant coatings for lubricant performance on components. DLC fulfils these requirements and has been tested on a large number of automotive components: gears, wrist pins, valve filters and fuel injector parts. Specifically, serial production of metal-containing DLC (Me-DLC) coatings by physical vapor deposition has been performed on diesel injection systems [Gahlin et al., 2001]. The modern high-pressure injection systems were developed thanks to DLC. Also, racing cars were the first targets in automotive applications of DLC, where it was used to maximize engine power by friction reduction [Johnston & Hainsworth, 2005].

The combination of elevated hardness and wear resistance of DLC is appropriate for cutting and abrasive wear instruments, like drills and other machines used in material manufacturing. Optimal frictional properties of DLC coatings increase efficiency and lifetime of ultrasonic motors (USM), which have applications in robotics and automation equipments [Ko et al., 2005] where liquid lubricants are not an option. Also, dry lubricants are indispensable for pharmaceutical and food processing systems because of requirements of chemical inertness and low friction. In this way, any machine component is protected by DLC: worm gears, lead screws, roller bearings, compressors and air bearings. Concerning textile industry, the existence of residual substances and adding compounds in yarn motivate the coating with DLC of components of textile machines in contact with yarn, e.g. spinning rings, rapiers and needles. Anti-sticking properties of DLC are used to obtain better product quality in injection molding. Gillette Mach3 blade edges constitute a successful application of ta-C. These edges, coated by 150 nm thick DLC, appeared in 1998 and consolidated razor blades as one of the basic applications of this material.

DLC and modified DLC present elevated hydrophobicity, which is a basic requirement for surfaces of fluid-contacting components and devices. This material is useful to improve surface properties of inner walls of pipes and tubes. Presently, extensive research is being carried out to use DLC as gas barrier on the inside surfaces of bottles and packaging due to its high atomic density [Chhowalla, 2003; Abbas et al., 2005; Ozeki et al., 2009].

DLC coatings are applied in optical storage devices in order to protect the disk. The main conditions are low wear rate and transparency in the recording wavelength and low deposition temperature to allow the use of polymeric substrates.
The high transparency in the infrared (IR) range makes DLC suitable for surface-protection applications in IR optics [Rusliet al., 2003]. In the visible range, sunglasses from Ray-Ban and glass plates for laser barcode scanner windows demonstrate the performance of DLC as scratch resistant coating for light transmitting surfaces [Hauert, 2004]. The control of interferential colors suggested DLC films for their use as decorative coatings. Electroluminescent and phase-shifting are other applications of DLC.
Protection to corrosion and wear of magnetic hard disks is one of the most widespread uses of ultra-thin DLC coatings [Morrison et al., 1999; Bhushan, 1999]. They must be ultra-thin to minimize the gap between the disk and the recording head, to achieve the highest areal storage densities. Magnetic media are currently protected by an approximately 2 nm thick DLC overcoat. Atomic smoothness (<0.15 nm) and absence of pinholes are of importance in order to protect the disk from corrosion [Jiang et al., 2010; Casiraghi et al., 2004].
Current research on DLC or modified DLC is mainly motivated by its biocompatibility [Grill, 2003; Hauert, 2004; Roy & Lee, 2007; Ma et al., 2007]. This property, together with antithrombogenicity, makes DLC-based films special candidates for medical applications where wear-resistant coatings, such as in prosthesis, are required. The surfaces of the implants are exposed to body cells and fluids, and are also exposed to potentially corrosive medium. These effects can provide cellular damage, infections, blood coagulation and failure of the implants. Coating the implants with protective films, which can reduce corrosion and wear, may extend the lifetime of implants by preventing or alleviating the aforementioned issues. Further biomedical applications include surgical tools and other devices in contact with blood or organic tissues [Dearnaley & Arps, 2005].

2. Modification with fluorine

2.1 Fluorine incorporation

In addition to metals, the introduction of Si, O, N or F in a-C films has also been investigated. In this case, O and N increase the surface free energy (SFE), while Si and F reduce it. Interestingly, addition of both Si and O can reduce the SFE more than only with Si [Grischke et al., 1995]. Among these elements, fluorine incorporation exhibits the highest water contact angles (higher than 100°) and the lowest surface energies (about 20 mJ/m^2).
A number of recent studies regarding the introduction of F in DLC films (FDLC) can be found in the literature. However, one must distinguish between growing fluorinated films and fluorinating films or other surfaces. The former corresponds to the growth of films containing fluorine in bulk, while the latter only adds fluorine functionalization to the surface. This chapter focuses on the growth of fluorinated films.
Fluorination of surfaces is easily achieved by etching with common fluorocarbon plasmas. This consequence has sometimes been seen as an adverse effect and cleaning procedures have been designed to remove fluorinated groups from the surface [Simko et al., 1991]. Others have intentionally used fluorocarbon plasmas to functionalize surfaces [Miyamoto et al., 1991; Korotkov et al., 2007; Roy et al., 2007].
The advantage of growing fluorinated films is that even after being worn out to some extent, the new surface preserves bulk's chemical composition. In addition, fluorine not only changes surface properties but also provides interesting bulk properties. Fluorinated

amorphous carbon has been extensively studied as a low dielectric constant (low-κ) material for electronic devices where parasitic capacitance must be avoided, because it can be deposited with microelectronic compatible methods. Static dielectric constant values as low as 2.1 have been obtained by RF-PECVD with CH_4 and CF_4, with increasing ε_r with RF power [Endo & Tatsumi, 1995]. As a comparison, the traditional dielectric layer in microelectronics, SiO_2, has ε_r = 4.0 and current dielectrics used in 90 nm, 65 nm and 40 nm lines of Intel® processors have ε_r = 3.0 [Grill, 1999]. The other major property of fluorinated amorphous carbon is its reduced surface energy. Compared to Si, fluorine not only reduces its polar component but also the dispersive contribution. The introduction of fluorine in the film displaces hydrogen, and bigger $-CF_2$ and $-CF_3$ groups are formed. This reduces the network density therefore reducing the dispersive component [Grischke et al., 1995]. When the appropriate growing conditions are used, low surface energy coatings (20 mJ/m^2) with improved hardness (5 to 20 GPa) over PTFE (18 mJ/m^2 and 0.3 GPa) can be obtained. These conditions are reported to fall into the moderate fluorination conditions ($R_F^{film} \equiv [F]/([F]+[C]) < 0.2$) [Donnet, 1998].

The deposition process of fluorinated a-C films is usually performed by PECVD although sputtering has also been used in some studies. The deposition by sputtering can be performed in reactive mode by using a graphite target and some fluorocarbon gas (e.g. CF_4, CHF_3, etc.) [Trippe et al., 2004; Jiang & Ning, 2006; Guerino et al., 2007] or by directly sputtering a PTFE target [Gonon & Sylvestre, 2002; Tang et al., 2005].

Instead, more work can be found in the literature about PECVD. There is a number of possible precursors with different C:F ratio which can be used to deposit fluorinated amorphous carbon. As reported, film's fluorine content is primarily determined by the gas precursor chemistry. Generally, more fluorine in the process gas results in higher fluorine concentration in the film; adding hydrogen to the feed gas reduces the film F:C ratio [Mountsier & Samuels, 1998]. CF_4 is one of the highest F:C ratio precursors available. Combined with CH_4, it is reported to produce lower dielectric constant than other precursors, and Vicker's hardness values from 5 to 10 GPa [Jacobsohn et al., 2000]. Other studies, show results for the same gases with hardness values from 16 to 19 GPa and surface free energy values descending from 40 mJ/m^2 for films deposited from pure CH_4 to about 28 mJ/m^2 for 4:1 CF_4:CH_4 gas flow ratio [Yu et al., 2003].

The results presented in this chapter were obtained using CHF_3, a barely studied gas for PECVD processes. Only a few papers, concerning deposition of low-κ dielectrics, can be found in the literature and diamond-like properties have virtually not been sought with this precursor. Among them, an article from Winder and Gleason shows lower roughness values and higher fluorine incorporation for films deposited from CHF_3 than $C_2H_2F_4$ or CH_2F_2 [Winder & Gleason, 2000]. Xin et al. have also published work about growth of a-C:F:H films deposited with CHF_3 and using ECWR-PECVD. Neither the mechanical properties nor the surface energy of the films were examined [Xin et al., 2003].

However, more work can be found dealing with the use of CHF_3 as precursor for plasma treatments, including treatments of previously grown DLC films [Kawasaki et al., 2004]. Schvartzman and Wind make use of the modified DLC for NIL molds and show slightly lower surface free energy values for surfaces treated with CHF_3 than with C_4F_8 [Schvartzman & Wind, 2009].

2.2 Surface and chemical properties of FDLC

X-Ray Photoelectron Spectroscopy (XPS) [Goldstein et al., 1992] is an adequate technique to quantify fluorine to carbon ratio in the films, although it is not able to detect hydrogen. The analyzed films always show lower F to C ratio in the film than in the precursor. For example, for films deposited with pure CHF_3, F to C ratio in the gas is 3 while in the film is one. The results of F:C ratio in the film are shown in figure 4 as a function of $R_{CHF_3}^{gas} \equiv \Phi_{CHF_3}/(\Phi_{CHF_3} + \Phi_{CH_4})$, where Φ_x represents the gas flow of x.

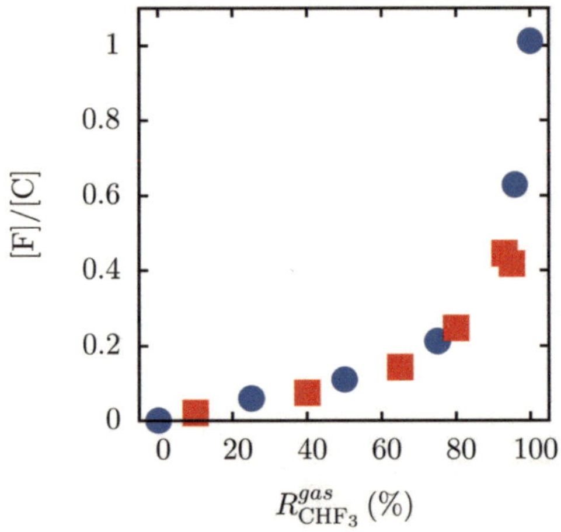

Fig. 4. Atomic F to C ratio in the film as determined from XPS analysis for two series of samples (deposited with 10 sccm of total gas flow, red squares; and 25 sccm blue circles).

The diamond-like characteristics of the films for low $R_{CHF_3}^{gas}$ suggest that the growth mechanism is ruled by the subplantation model [Robertson, 2002; Lifshitz et al., 1989, 1990, 1994]. Under these conditions, there is an intense energetic ion bombardment which subplants energetic species, either locally increasing stress (according to McKenzie et al. [McKenzie et al., 1991]) and stabilizing diamond-sp^3 phase, or by locally increasing density which would tend to increase sp^3 bonding as well (according to Robertson [Robertson, 1993]). In any case, ion bombardment causes the loss of hydrogen and fluorine from the surface and subsurface.

Ion bombardment is carried out by positively charged species that are accelerated in the plasma sheath towards the sample. At low $R_{CHF_3}^{gas}$, fluorine (and probably hydrogen) incorporation into the film is low while for low methane concentrations the F:C ratio rises significantly. The high hydrogen content provided by CH_4 probably etches away fluorine atoms back to the plasma, which fits with optical emission spectroscopic actinometry results, from the plasma, for atomic F. For a-C:H films, it is believed that atomic H, either neutral or ion, reacts with C–H bonds, abstracting H and creating H_2. According to these results, when fluorine is added, atomic H probably reacts with F in the film, as well, creating volatile HF.

As $R_{CHF_3}^{gas}$ is increased, hydrogen introduced as part of the feed gas is reduced and, eventually, fluorine content in the films starts to increase. Therefore, the reduction of atomic hydrogen increases the chances of fluorine to remain in the film. Fluorine, the substitute of hydrogen when $R_{CHF_3}^{gas}$ is increased, does not produce the same effect as H, due to its strong electronegativity, therefore avoiding the bombardment of F atoms. In addition, ion energy measurements have shown that the progressive substitution of CH_4 by CHF_3 reduces, and finally eliminates, the population of high energy ions [Corbella, 2009b]. The combination of hydrogen reduction, exchange by a strong electronegative atom (F) and reduction of the density of energetic ions bombarding the growing surface, would explain the non-linear increase of [F]/[C].

2.3 Friction of FDLC

Several factors influence the frictional behavior of DLC thin films: the degree of sp^2 or sp^3 carbon bonding, the amount of hydrogen in the film, the eventual alloying elements, the chemical and mechanical interactions between film and ball, the ball composition or the film and ball roughness [Erdemir & Donnet, 2006]. Moreover, the sliding conditions of speed and load [Heimberg et al., 2001] or debris production can also determine the possible interactions in the experiment. The interpretation of friction data is therefore not straightforward.

The results presented in this section correspond to measurements carried out with a nanotribometer (50 mN load) in a controlled N_2 + H_2O atmosphere as detailed elsewhere [Rubio-Roy, 2009].

Measurements show that friction coefficient is reduced for increasing relative humidity (RH) between 20% and 60%. Beyond that point, it does not suffer significant variation. Moreover, for high RH, friction does not seem to depend on the fluorination degree of the films. Regarding the effect of fluorine introduction, non-fluorinated films provide, on average, the lowest friction, while films deposited with $R_{CHF_3}^{gas} = 10\%$ show the highest friction (0.4 for 20% RH). For higher F concentrations friction coefficient is reduced again. Low-fluorinated films see their friction more affected with humidity than non-fluorinated or highly-fluorinated DLC films.

Optical microscope photographs (see figure 5b) show fiber-like debris particles at both sides of the sliding track. This phenomenon occurs when discontinuous wear is given contrary to what happens when a transfer layer is present between the sliding counterpart and the film surface [Erdemir & Donnet, 2006].

Hydrogen terminated surfaces provide low friction characteristics due to the passivation of σ and π carbon bonds on the surface [Erdemir & Donnet, 2006]. As the WC ball slides over the surface, H atoms are mechanically removed from the film by strain effects thus leaving superficial dangling bonds [Li et al., 2005]. Depending on the experimental conditions, and specifically on the gas environment, hydrogen and hydroxyl groups originated from water will passivate these bonds again. In our experiment, for RH between 20% and 60% the results would justify this casuistic, so that bonds would react with water to restore themselves to a low energy state. However, for higher presence of water (from 60% to 80%), friction is stabilized for all films and independently of humidity or fluorine content (see figure 6). For these values of RH, a layer of physisorbed water could be formed on the surface, as already suggested by other authors [Erdemir & Donnet, 2006; Erdemir, 2001], which would make friction independent of the films' variations.

Regarding the rather high friction values for $R_{CHF_3}^{gas} = 10\%$ in comparison to the rest, this behavior could be due to at least two competitive mechanisms. One of them would be the

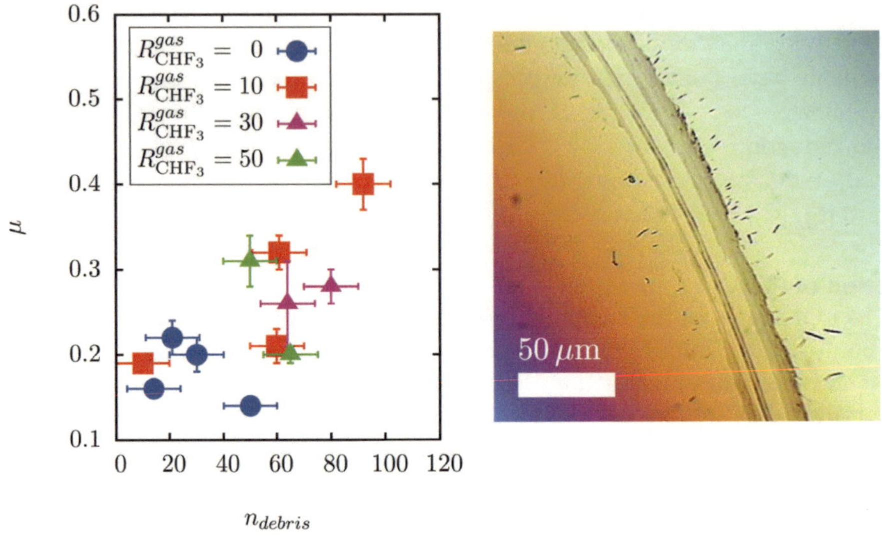

a) b)

Fig. 5. a) Friction vs number of debris particles visible at 50× (200×200 μm).The dependence of friction with number of particles is only clear for high friction values; b) 50× optical microscope picture of a typical sliding track, corresponding to a film with $R^{gas}_{CHF_3} = 10\%$ measured CHF_3 at 60% RH. Fiber-like debris particles are present at both sides of the sliding track.

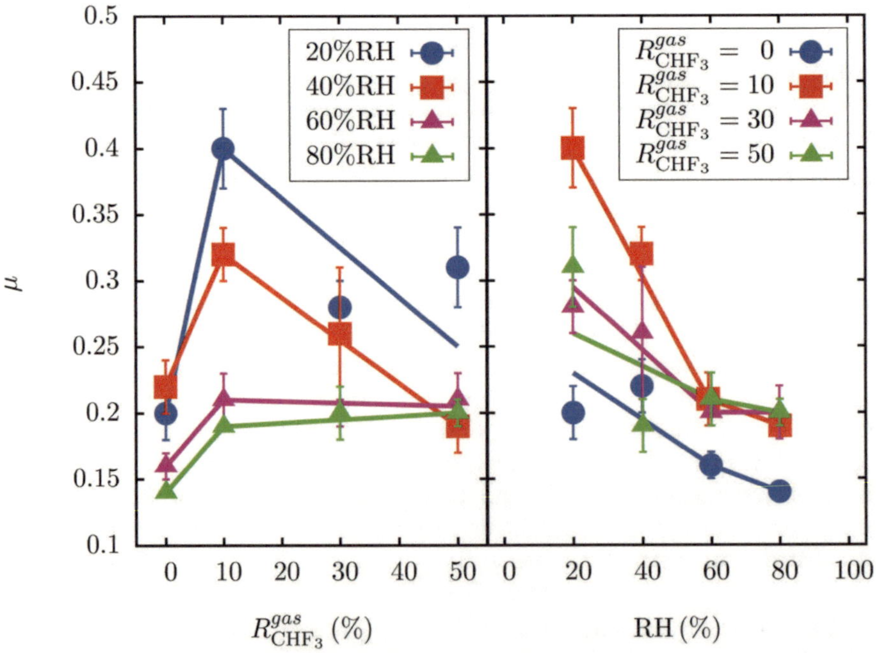

Fig. 6. Friction dependence on humidity and fluorination degree.

somewhat beneficial effect of F presence to reduce friction by reducing adhesion to the sliding counterpart and by passivating carbon with a high-energy bond. The other would be the disadvantageous effect of a lower C atomic density (F is bigger than H), the reduced reservoir of H inside the film that would otherwise relink dangling bonds on the surface, and the probable decrease of H in the film as expected from the reduction in hydrogenated species in the plasma, from optical emission spectroscopic actinometry. The increase of friction with fluorine for humidity values below 40% is in agreement with other analogue works [Gilmore & Hauert, 2001].

The statistical analysis of 50× micrographs, taken after the tests, reveals that the influence of debris production in friction must also be taken into account as friction increases, where a stronger correlation between debris and friction is found. However, for low friction experiments, debris influence is limited and strongly coupled to microscopic effects (tests with similar debris production have different frictions and the other way round).

Films' roughness (as measured from atomic force microscopy) is sufficiently low so that it is a minor factor in friction phenomena for the tested $R_{\mathrm{CHF_3}}^{gas}$ range), not producing interlocking and/or asperity-asperity interactions during sliding.

2.4 Abrasive wear of FDLC

Abrasive wear measurements were carried out with a 1:3 suspension of alumina particles 1 μm in diameter in glycerin. Such setup is highly hygroscopic and minimum changes in ambient humidity can affect the results. All measurements must be done during a short period of time (to avoid water absorption) and at a controlled temperature and relative humidity (here 23°C and 33%).

CSM Calowear equipment was used to measure the abrasive wear resistance of the amorphous carbon films. The measurement setup consists on a plastic or hardened stainless steel ball leaning against the surface of interest (which is tilted a known angle) and spun by a rotating shaft at a known distance from the surface. As the ball center does not move relatively to the surface, the footprint of the test is very small. More details about the setup can be found elsewhere [Gee et al., 2003].

Two different balls with equal diameter were used to test the wear resistance: a steel ball applying a force of 0.54 N and a plastic (polyacetal) ball producing 0.092 N of normal load.

The deposited films showed high wear resistance up to $R_{\mathrm{CHF_3}}^{gas} = 80\%$ (see figure 7), with values in the range of 0.2×10^3 μm^3/Nm to 0.4×10^3 μm^3/Nm, one order of magnitude more resistant than monocrystalline Si (100) in the same conditions (2.1×10^3 μm^3/Nm). Fluorine content slightly increases wear rate, which is consistent with the observed changes in composition. The overall stability of the films is excellent. Data in figure 7 at $R_{\mathrm{CHF_3}}^{gas} = 10\%$, 40%, 60% and 80% correspond to two-year-old films, in contrast to the rest of data, which corresponds to recent (weeks old) films. Contact angle measurements also confirm the stability of the films.

For more fluorinated films, this setup worn the film too much, so that the rate could not be measured. This is probably a consequence of both the reduced cohesive forces of the films and a loss of adhesion between film and buffer layer.

However, using the plastic ball during shorter periods of time, it could be determined that fluorinated films are up to 5 times more resistant than Si up to $R_{\mathrm{CHF_3}}^{gas} = 80\%$. Beyond that point, Si is more resistant, which evidences either a dramatic reduction of adhesion between film and buffer layer, or a strong structural and chemical change of the film.

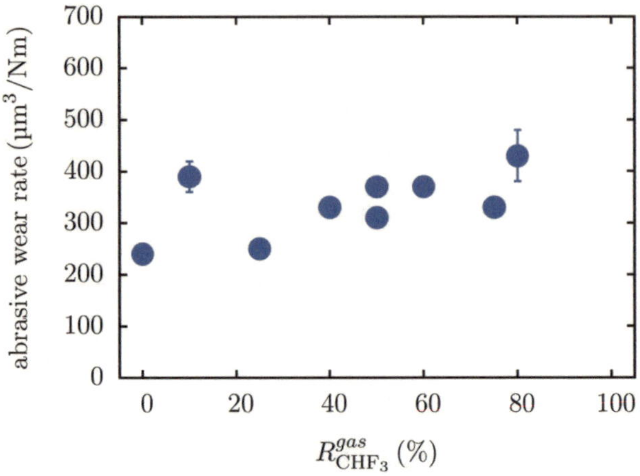

Fig. 7. Abrasive wear rate of the films after 6 min with a 1:3 suspension of alumina particles 1 µm in diameter in glycerin, with a load of 0.54 N

3. Modification by surface structuring

3.1 Structuring strategies

Tribological measurements of patterned surfaces involve the use of macroscopic probes, in order to sufficiently average the effect of the structures. Therefore, large-area capable patterning techniques are required. Here, we report on the tribological properties of samples prepared by colloidal lithography (bottom-up approach), although other structures have also been prepared by electron beam lithography (top-down).

Colloidal lithography is a fast, cost-efficient technique able to produce periodic nanometric features in large-scale (up to some centimeters or more) with high quality. The technique consists on the arrangement of particles on a substrate so that they can be used as a template/mask for subsequent etching (see figure 8). Although CL is less versatile than EBL or other lithographic techniques, regarding non-periodic patterns, the reduction of time and resources for the production of large-area periodic nanopatterns represents important advantages for certain applications or fields such as photonic crystals, [López, 2003] thermophotovoltaics [Mao & Ye, 2010] or optical biosensors [Dorfner et al., 2009]. The differences in the structuring produced by ion beam etching with different angles have been studied as a method for nanoimprint mold fabrication [Portal et al., 2009].

On the other hand, the realization of sub-100 nm features in large areas with e-beam lithography (EBL) requires the stitching of several exposure fields. Proper alignment of the fields is complicated and, depending on how critical good stitching is, specific design of the equipment is required. Laser interferometer driven stages provide state-of-the-art stitching, but even with them a small thermal expansion/contraction of the sample after introduction on the e-beam writer is to be expected even for small temperature differences. This change in size can affect the stitching quality if enough time is not allowed for stabilization.

With these precautions taken into account, EBL is a suitable technique for large-area patterning, although expensive equipment is required, and patterning is slow. Other faster techniques such as those based on light (X-Ray or extreme UV lithography) can be much faster but also even more expensive.

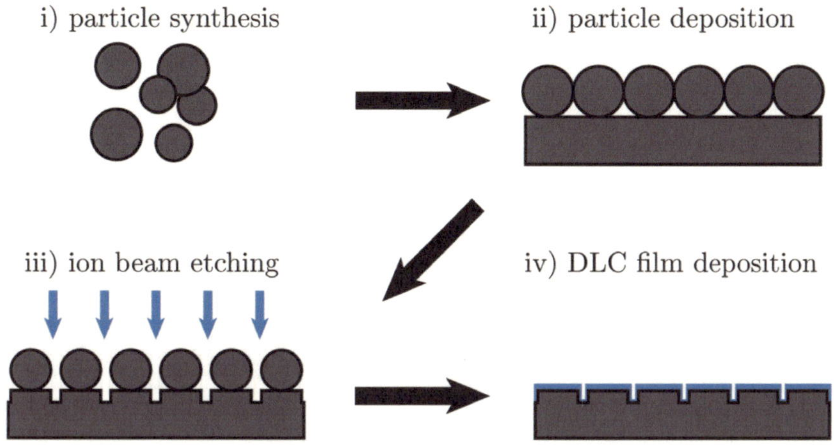

Fig. 8. Fabrication process of DLC films with patterned surface: i) Synthesis of silica particles; ii) Particle deposition by Langmuir-Blodgett (colloidal template); iii) Pattern transfer to the substrate surface (Ar+ beam etching); iv) DLC film deposition on the patterned substrate.

In the following subsection we report on the friction and wear behavior of structured DLC films.

3.2 Friction and wear of patterned films

As described above, surface structuring leads to a variation in surface properties, such as adhesion, wear resistance, friction, wettability and gas adsorption. Textured coatings must show superior tribological properties in applications leading to strong interfacial interactions like nanoimprint molds, self-cleaning surfaces, MEMS and data storage devices.
A series of DLC samples with modified surface topographies were prepared by colloidal lithography following the procedure schematized in figure 8. After DLC deposition on the patterned c-Si samples, the surface topography was studied by scanning electron microscopy (SEM) and atomic force microscopy (AFM), and friction tests were performed on them.
Coefficient of friction was measured by means of a nanotribometer from CSM Instruments in ball-on-disk configuration. The ST-117 cantilever was used with a WC ball 3 mm in diameter and 100 mN of load was applied on the sample, performing circular tracks of 1.6 mm in diameter at 1 mm/s during 6250 cycles. The measurements were carried out in a nitrogen atmosphere at 40% relative humidity, which was controlled by means of a PID system equipped with a water bubbler. More details can be found elsewhere [Rubio-Roy et al., 2009].
SEM micrographs in figure 9 show the surface morphology of DLC films deposited on substrates that underwent different times of ion beam etching. In these samples, film thickness is of the order of the pit depth after etching, i.e. 50 nm. The shadowing effect of the silica particles (SP) led to the formation of a pillar network. The substrate was only etched at the interstices of the 2-D crystal monolayer, which acted as a sacrificial template during the substrate lithography. Thus, the resulting topography reproduced the pattern of the CL template, where the pillar top coincides with the substrate region protected by the particle. In general, the pillar height increases with etching time. However, the surface structure

substantially changes when the etching time surpasses a threshold between 30 and 45 min. According to figure 9, the pillar network vanished with 45 min of substrate etching. At this etching time, the SP could not be removed from the patterned c-Si probably due to an effect of particle implantation. Furthermore, when the IBE is performed during 60 min, the SP are totally etched, i.e. the particle monolayer that constitutes the etching mask is removed by the bombarding ions and only the resulting rough surface remains.

Fig. 9. SEM micrographs of DLC-coated (50 nm) substrates patterned by colloidal lithography for different etching times: (a) 15, (b) 30, (c) 45 and (d) 60 min. [Corbella et al., 2010]

Figure 10a shows the evolution of coefficient of friction of a flat DLC film measured with the nanotribometer. This parameter increased gradually from 0.12 to 0.23 passing through different stages. As reported elsewhere, the curve is divided into three sections [Corbella et al., 2009b]: run-in period, transition regime and steady state. The variations in the first stage are ascribed to surface ploughing and mechanically induced desorption of hydrogen. During the transient regime, surface modification conformal to the ball shape takes place, as well as interactions with wear particles. Finally a steady state is achieved around cycle #4000. This scenario changes in the case of structured DLC samples (figures 10b and 10c). The coefficient of friction, which is already stabilized around or prior to cycle #1000, ranges between 0.17 and 0.20. The measured samples achieved lower and more stable values than in the flat DLC coating. Spikes in the friction coefficient (not shown in figure 10 due to data smoothing) could come from momentary interactions of the wear particles with the ball of the nanotribometer. These particles, which are debris removed from the sample during the tests, contribute to enhance the friction via several mechanisms, as for instance dragging, ploughing or adhesion.

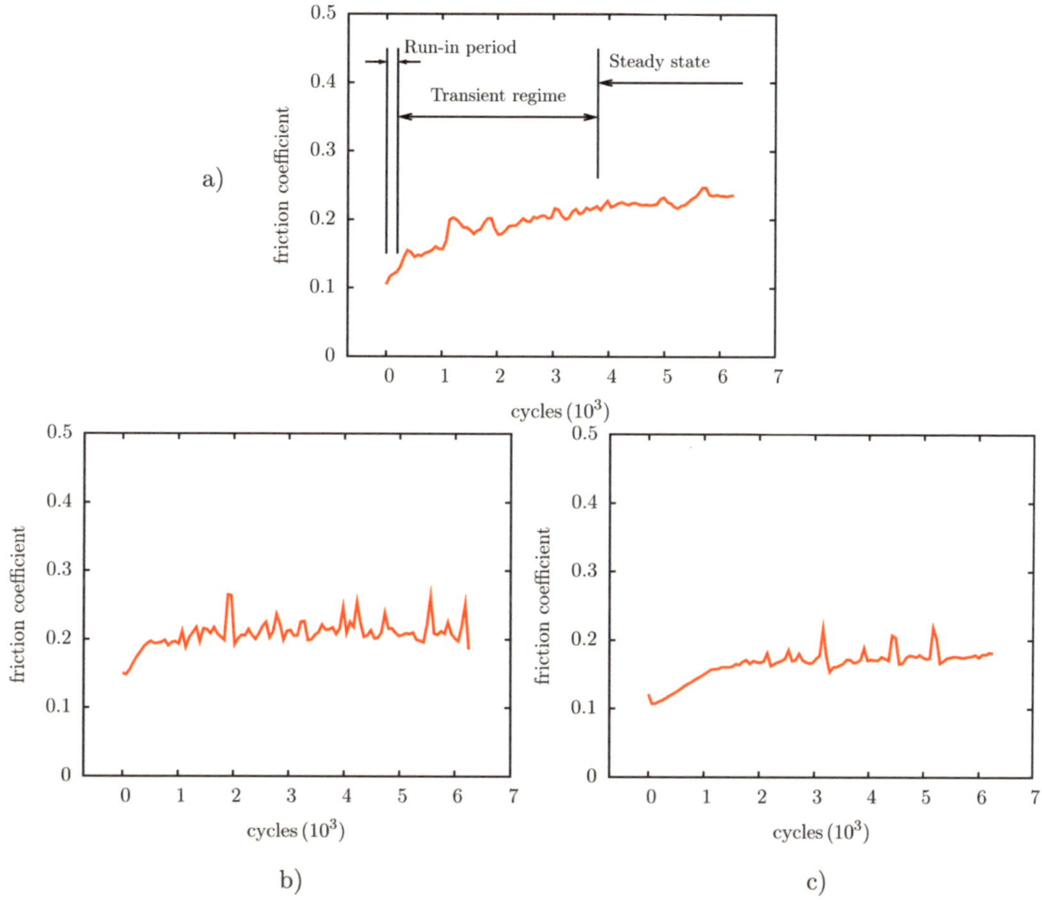

Fig. 10. Evolution of coefficient of friction of DLC films deposited on a) flat, b) 15 min, and c) 30 min etched substrates. In the first plot, we have indicated the run-in period, transition region and steady state.

The wear effects on surface structure due to nanotribometry are evidenced by the morphological changes of the pillars located on the wear track. These pillars lost the concave ending observed in figure 9. Thus, the pillar edges constitute the initial contact region between the ball and sample surface. Since the pillars are progressively worn by the WC ball, the friction increases until the pillar top becomes flat. From this moment onwards, the coefficient of friction of the patterned DLC stabilizes into the values shown in figure 10. The wear rate was quantified from AFM profilometry of the volume removed by the ball. A value of 10^{-7} $mm^3N^{-1}m^{-1}$ was measured for flat DLC from the wear track left by the nanotribometer, whereas textured DLC exhibited a rate 10 times higher. The reduced contact area in the case of a patterned surface with respect to a flat surface may cause this

increase in wear rate. Therefore, more robust structures are required to improve the wear resistance.
The coefficient of friction of the sample etched during 15 min (figure 10b) is reduced 13% with respect to flat DLC, whereas the reduction corresponding to the sample etched for 30 min is 26% (figure 10c). This behavior is in agreement with the results from Jung et al., where the role of asperity size on friction was discussed [Jung et al., 2006]. The reduction in the stable values of coefficient of friction may be ascribed to the trapping effect of the pits defined by the micro-pillars. These interstices could store the wear particles generated during the tribological tests. The effect of these wear particles on friction force is thus alleviated in comparison with systems formed by lower pillars [Dumitru et al., 2003]. Also, the higher pillars lead to a weaker interaction with the sample due to the larger distance between the WC ball and the base surface. This explanation supports the monotonic reduction in friction coefficient as the pillar height increases. In this scenario, the preparation of a non-close-packed 2D crystal could give place to weaker friction forces due to a more important particle trapping. A more complete modeling of the tribological phenomena at the nanoscale should also account for the elastic response of the micro-pillars to the friction force.

4. References

Abbas, G.A.; Roy, S.S.; Papakonstantinou, P. & McLaughlin, J.A. (2005) Structural investigation and gas barrier performance of diamond-like carbon based films on polymer substrates. *Carbon* 43, 303-309

Ager, J.W.; Anders, S.; Brown, I.G.; Nastasi, M. & Walter, K.C. (1997) Multilayer hard carbon films with low wear rates. *Surface and Coatings Technology* 91, 91-94

Aisenberg, S. & Chabot, R. (1971) Ion-beam deposition of thin films of diamondlike carbon. *Journal of Applied Physics* 42, 2953-2958

Bahl, R.; Vedawyas, M.; Patel, D.; Kumar, A. & Shamsuzzoha, M. (2000) Evaluation of mechanical properties of DLC-TiC microlaminate coatings, In: *Thin Films-Stresses and Mechanical Properties VIII*, Vinci, R.; Kraft, O.; Moody, N.; Besser, P. & Shaffer, E. (Ed.), 331–336, Materials Research Society.

Bertran, E.; Corbella, C.; Pintol, A.; Vives, M. & Andújar, J.L. (2003) Comparative study of metal/amorphous-carbon multilayer structures produced by magnetron sputtering. *Diamond and Related Materials* 12, 1008-1012

Bhushan, B. (1999) Chemical, mechanical and tribological characterization of ultra-thin and hard amorphous carbon coatings as thin as 3.5 nm: recent developments. *Diamond and Related Materials* 8, 1985-2015

Casiraghi, C.; Ferrari, A.C.; Ohr, R.; Chu, D. & Robertson, J. (2004) Surface properties of ultra-thin tetrahedral amorphous carbon films for magnetic storage technology. *Diamond and Related Materials 13*, 1416-1421

Chen, J.S.; Lau, S.P.; Sun, Z.; Chen, G.Y.; Li, Y.J.; Tay, B.K. & Chai, J.W. (2001) Metal-containing amorphous carbon films for hydrophobic application. *Thin Solid Films* 398-399, 110-115

Chen, L.Y. & Hong F.C.N. (2003) Diamond-like carbon nanocomposite films. *Applied Physics Letters* 82, 3526-3528

Chhowalla, M. (2003), In: *Properties of Amorphous Carbon*, Silva S.R.P. (Ed.), INSPEC, The Institution of Electrical Engineers

Corbella, C.; Bialuch, I.; Kleinschmidt, M. & Bewilogua, K. (2008) Modified DLC coatings prepared in a large-scale reactor by dual microwave/pulsed-DC plasma-activated chemical vapour deposition. *Thin Solid Films* 517(3), 1125-1130

Corbella, C.; Echebarria, B.; Ramírez-Piscina, L.; Pascual, E.; Andújar, J.L. & Bertran, E. (2009a) Growth kinetics of nanometric dendrites in metal-carbon thin films. *Acta Mater* 57(17), 4948-4956

Corbella, C.; Rubio-Roy, M.; Bertran, E.; Polo, M.C.; Pascual, E. & Andújar, J.L. (2009b) Low friction and protective diamond-like carbon coatings deposited by asymmetric bipolar pulsed plasma. *Diamond and Related Materials* 18(5-8), 1035-1038

Corbella, C.; Portal, S.; Rubio-Roy, M.; Vallvé, M.A.; Ignés-Mullol, J.; Bertran, E. & Andújar, J.L. (2010) Surface structuring of diamond-like carbon films by colloidal lithography with silica sub-micron particles. *Diamond and Related Materials* 19(7-9), 1124-1130

Dearnaley, G. & Arps, J. H. (2005) Biomedical applications of diamond-like carbon (DLC) coatings: A review. *Surface and Coatings Technology* 200, 2518-2524

Dimigen, H.; Hübsch, H. & Memming, R. (1987) Tribological and electrical properties of metal-containing hydrogenated carbon films. *Applied Physics Letters* 50, 1056-1058

Donnet, C. (1998) Recent progress on the tribology of doped diamond-like and carbon alloy coatings: a review. *Surface and Coatings Technology* 100-101, 180–186

Dorfner, D.; Zabel, T.; Hürlimann, T.; Hauke, N.; Frandsen, L.; Rant, U.; Abstreiter, G. & Finley, J. (2009) Photonic crystal nanostructures for optical biosensing applications. *Biosensors and Bioelectronics* 24(12), 3688–3692

Dumitru, G.; Romano, V.; Weber, H.P.; Pimenov, S.; Kononenko, T.; Hermann, J.; Bruneau, S.; Gerbig, Y. & Shupegin, M. (2003) Laser treatment of tribological DLC films. *Diamond and Related Materials* 12(3-7), 1034-1040

Endo, K. & Tatsumi, T. (1995) Fluorinated amorphous carbon thin films grown by plasma enhanced chemical vapor deposition for low dielectric constant inter-layer dielectrics. *Journal of Applied Physics* 78(2), 1370–1372

Erdemir, A. (2001) The role of hydrogen in tribological properties of diamond-like carbon films. *Surface and Coatings Technology* 146-147, 292–297

Erdemir, A. & Donnet, C. (2006) Tribology of diamond-like carbon films: recent progress and future prospects. *Journal of Physics D: Applied Physics* 39(18), R311

Fallon, P. J.; Veerasamy, V. S.; Davis, C. A.; Robertson, J.; Amaratunga, G. A. J.; Milne, W. I. & Koskinen, J. (1993) Properties of filtered-ion-beam-deposited diamondlike carbon as a function of ion energy. *Physical Review B* 48(7), 4777–4782

Ferrari, A. C. & Robertson J. (2000) Interpretation of Raman spectra of disordered and amorphous carbon. *Physical Review B* 61(20), 14095–14107

Ferrari, A. C.; Rodil, S. E.; Robertson, J. & Milne, W. I. (2002) Is stress necessary to stabilise sp^3 bonding in diamond-like carbon? *Diamond and Related Materials*, 11(3-6), 994–999

Gahlin, R.; Larsson, M. & Hedenqvist, P. (2001) Me-C:H coatings in motor vehicles. *Wear* 249, 302-309

Gee, M. G.; Gant, A.; Hutchings, I.; Bethke, R.; Schiffman, K.; Van Acker, K.; Poulat, S.; Gachon, Y. & von Stebut, (2003) J. Progress towards standardisation of ball cratering, *Wear* 255(1-6), 1-13

Gilmore, R. & Hauert, R. (2001) Control of the tribological moisture sensitivity of diamond-like carbon films by alloying with F, Ti or Si. *Thin Solid Films* 398-399, 199–204

Goldstein, J. I.; Newbury, D. E.; Echlin, P.; Joy, D. C.; Fiori, C. & Lifshin, E. (1992) *Scanning Electron Microscopy and X-Ray Microanalysis.* Plenum Press, ISBN 0-306-44175-6

Gonon, P. & Sylvestre, A. (2002) Dielectric properties of fluorocarbon thin films deposited by radio frequency sputtering of polytetrafluoroethylene. *Journal of Applied Physics* 92(8), 4584–4589

Grill, A. (1999) Diamond-like carbon: state of the art. *Diamond and Related Materials* 8(2-5), 428–434

Grill, A. (2003) Diamond-like Carbon coatings as biocompatible materials - an overview. *Diamond and Related Materials.* 12, 166-170

Grischke, M.; Bewilogua, K.; Trojan, K. & Dimigen, H. (1995) Application-oriented modifications of deposition processes for diamond-like-carbon-based coatings. *Surface and Coatings Technology* 74-75, 739–745

Grischke, M.; Hieke, A.; Morgenweck, F. & Dimigen, H. (1998) Variation of the wettability of DLC-coatings by network modification using silicon and oxygen. *Diamond and Related Materials* 7(2-5), 454-458

Guerino, M.; Massi, M. & Mansano, R. (2007) The influence of nitrogen and fluorine on the dielectric constant of hydrogenated amorphous carbon (a-C:H) films. *Microelectronics Journal* 38(8-9), 915–918

Hauert, R. & Patscheider, J. (2000) From alloying to nanocomposites - Improved performance of hard coatings. *Advanced Engineering Materials* 2, 247-259

Hauert, R. (2004) An overview on the tribological behavior of diamond-like carbon in technical and medical applications. *Tribology International* 37, 991-1003

Heimberg, J. A.; Wahl, K. J.; Singer, I. L. & Erdemir, A. (2001) Superlow friction behavior of diamond-like carbon coatings: Time and speed effects. *Applied Physics Letters* 78(17), 2449–2451

Jacobsohn, L. G.; Franceschini, D. F.; da Costa, M. E. H. M. & Freire, J. F. L. (2000) Structural and mechanical characterization of fluorinated amorphous-carbon films deposited by plasma decomposition of CF_4–CH_4 gas mixtures. *Journal of Vacuum Science & Technology A: Vacuum, Surfaces, and Films* 18(5), 2230–2238

Jiang, M. & Ning, Z. (2006) Influence of deposition pressure on the structure and properties of fluorinated diamond-like carbon films prepared by RF reactive magnetron sputtering. *Surface and Coatings Technology* 200(12-13), 3682–3686

Jiang, W. F.; Diao, D. F.;(2010) The Critical Conditions for Tribo-Demagnetization of Perpendicular Magnetic Recording Disk Under Sliding Contact. *Journal of Tribology* 132(2), 021901

Johnston, S.V. & Hainsworth, S.V. (2005) Effect of DLC coatings on wear in automotive applications. *Surface Engineering* 21, 67-71

Jung, Y.C. & Bhushan, B. (2006) Contact angle, adhesion and friction properties of micro- and nanopatterned polymers for superhydrophobicity. *Nanotechnology* 17, 4970-4980

Kaukonen, H. P. & Nieminen, R. M. (1992) Molecular-dynamics simulation of the growth of diamondlike films by energetic carbon-atom beams. *Physical Review Letters* 68(5), 620–623

Kawasaki, D.; Tsuchimura, D.; Choi, W.; Iseri, Y.; Ando, T. & Tomokage, H. (2004). Scanning probe field emission current measurements on diamond-like carbon films treated by reactive ion etching. *Journal of Physics: Condensed Matter* 16(2), S301

Kessels, W. M. M.; Gielen, J. W. A. M.; de Sanden, M. C. M. V.; Ijzendoorn, L. J. V.; Dekempeneer, E. H. A. & Schram, D. C. (1998) A model for the deposition of a-C:H using an expanding thermal arc. *Surface and Coatings Technology* 98(1-3), 1584–1589

Kim, T.Y.; Bialuch, I.; Bewilogua, K.; Oh, K.H. & Lee, K.R. (2007) Wetting behaviours of a-C:H:Si:O film coated nano-scale dual rough surface. *Chemical Physics Letters* 436(1-3), 199-203

Kiuru, M.; Alakoski, E.; Tiainen, V. M.; Lappalainen, R. & Anttila, A. (2003) Tantalum as a buffer layer in diamond-like carbon coated artificial hip joints. *Journal of Biomedical Materials Research Part B: Applied Biomaterials*, 66B(1), 425–428

Ko, H.P.; Kim, S.; Kim, J.S.; Kim, H.J. & Yoon, S.J. (2005) Wear and dynamic properties of piezoelectric ultrasonic motor with frictional materials coated stator. *Materials Chemistry and Physics* 90, 391-395

Koidl, P.; Wild, C.; Dischler, B.; Wagner, J. & Ramsteiner, M. (1990) Plasma deposition, properties and structure of amorphous hydrogenated carbon films. *Materials Science Forum* 52, 41–70

Korotkov, R.; Goff, T. & Ricou, P. (2007) Fluorination of polymethylmethacrylate with SF6 and hexafluoropropylene using dielectric barrier discharge system at atmospheric pressure. *Surface and Coatings Technology* 201(16-17), 7207–7215

Lampe, Th.; Eisenberg, S. & Rodríguez Cabeo, E. (2003) Plasma surface engineering in the automotive industry - trends and future prospective. *Surface and Coatings Technology* 174-175, 1-7

Lee, K. R.; Eun, K. Y.; Kim, I. & Kim, J. (2000) Design of W buffer layer for adhesion improvement of dlc films on tool steels. *Thin Solid Films* 377-378, 261–268

Li, K. Y.; Zhou, Z. F.; Bello, I.; Lee, C. S. & Lee S. T. (2005) Study of tribological performance of ECR-CVD diamond-like carbon coatings on steel substrates: Part 1. The effect of processing parameters and operating conditions. *Wear* 258(10), 1577–1588

Lifshitz, Y.; Kasi, S. R. & Rabalais, J. W. (1989) Subplantation model for film growth from hyperthermal species: Application to diamond. *Physical Review Letters* 62(11), 1290–1293

Lifshitz, Y.; Kasi, S. R.; Rabalais, J. W. & Eckstein, W. (1990) Subplantation model for film growth from hyperthermal species. *Physical Review B* 41(15), 10468–10480

Lifshitz, Y.; Lempert, G. D. & Grossman, E. (1994) Substantiation of subplantation model for diamondlike film growth by atomic force microscopy. *Physical Review Letters* 72(17), 2753–2756

Logothetidis, S.; Charitidis, C.; Gioti, M.; Panayiotatos, Y.; Andrea, M. & Kautek, W. (2000) Comprehensive study on the properties of multilayered amorphous carbon films. *Diamond and Related Materials* 9, 756-760

López, C. (2003) Materials aspects of photonic crystals. *Advanced Materials* 15(20), 1679–1704

Ma, W. J.; Ruys, A. J.; Mason, R. S.; Martin, P. J.; Bendavid, A.; Liu, Z.; Ionescu, M. & Zreiqat, H. (2007) DLC coatings: Effects of physical and chemical properties on biological response. *Biomaterials* 28, 1620-1628

Mao, L. & Ye, H. (2010) New development of one-dimensional Si/SiO_2 photonic crystals filter for thermophotovoltaic applications. *Renewable Energy* 35(1), 249–256

Martin, P. J; Filipczuk, S. W.; Netterfield, R. P.; Field, J. S.; Whitnall, D. F. & McKenzie, D. R. (1988) Structure and hardness of diamond-like carbon films prepared by arc evaporation. *Journal of Materials Science Letters* 7(4), 410–412

McKenzie, D. R.; Muller, D. & Pailthorpe, B. A. (1991) Compressive-stress-induced formation of thin-film tetrahedral amorphous carbon. *Physical Review Letters* 67(6), 773- 776

Miyamoto, T.; Kaneko, R. & Miyake, S. (1991) Tribological characteristics of amorphous carbon films investigated by point contact microscopy. *Journal of Vacuum Science & Technology B* 9(2), 1336–1339

Miyazawa, T.; Misawa, S.; Yoshida, S.; & Gonda, S. (1984) Preparation and structure of carbon film deposited by a mass-separated C+ ion beam. *Journal of Applied Physics* 55(1), 188–193

Morrison, N. A.; Rodil, S. E.; Ferrari, A. C.; Robertson, J. & Milne, W. I. (1999) High rate deposition of ta-C:H using an electron cyclotron wave resonance plasma source. *Thin Solid Films* 337(1-2), 71–73

Mountsier, T. W. & Samuels, J. A. (1998) Precursor selection for plasma deposited fluorinated amorphous carbon films. *Thin Solid Films* 332(1-2), 362–368

Ozeki, K.; Nagashima, I.; Ohgoe, Y.; Hirakuri, K. K.; Mukaibayashi, H. & Masuzawa, T. (2009) Gas barrier properties of diamond-like carbon films coated on PTFE. *Applied Surface Science* 255, 7286-7290

Peng, X.L.; Barber, Z.H. & Clyne, T.W. (2001) Surface roughness of diamond-like carbon films prepared using various techniques. *Surface and Coatings Technology* 138, 23-32

Pino, F.J.; Bertran, E.; Polo, M.C. & Andújar, J.L. (2001) Microstructural and mechanical properties of nanometric-multilayered a-CN/a-C/.../a-CN coatings deposited by rf-magnetron sputtering and nitrogen ion-beam bombardment. *Diamond and Related Materials* 10, 952-955

Polo, M.C.; Andújar, J.L.; Hart, A.; Robertson, J. & Milne, W.I. (2000) Preparation of tetrahedral amorphous carbon films by filtered cathodic vacuum arc deposition. *Diamond and Related Materials* 9(3-6), 663-667

Portal, S.; Rubio-Roy, M.; Corbella, C.; Vallvé, M.; Ignes-Mullol, J. & Bertran, E. (2009) Influence of incident ion beam angle on dry etching of silica sub-micron particles deposited on Si substrates. *Thin Solid Films* 518(5), 1543–1548

Randhawa, H. (1988) Cathodic arc plasma deposition technology. *Thin Solid Films* 167(1-2), 175-185

Robertson, J. (1993) Deposition of diamond-like carbon. *Philosophical Transactions of the Royal Society of London. Series A: Physical and Engineering Sciences* 342(1664), 277–286

Robertson, J. (2002) Diamond-like amorphous carbon. *Materials Science and Engineering R* 37, 129-281

Rodil, S.E. (2000) PhD Thesis: *Preparation and characterization of carbon nitride thin films*, University of Cambridge, Cambridge.

Roy, R. K.; Choi, H. W.; Park, S. J. & Lee, K. R. (2007) Surface energy of the plasma treated Si incorporated diamond-like carbon films. *Diamond and Related Materials*, 16(9), 1732–1738

Roy, R. K. & Lee, K. R. (2007) Biomedical Applications of Diamond-Like Carbon Coatings: A Review. *Journal of Biomedic Materials Research B: Applied Biomaterials* 83, 72-84

Rubio-Roy, M.; Corbella, C.; García-Céspedes, J.; Polo, M.C.; Pascual, E.; Andújar, J.L. & Bertran, E. (2007a) Diamondlike carbon films deposited from graphite target by asymmetric bipolar pulsed-DC magnetron sputtering. *Diamond and Related Materials* 16(4-7), 1286-1290

Rubio-Roy, M.; Bertran, E.; Pascual, E.; Polo, M.C. & Andújar, J.L. (2007b) Fluorinated DLC deposited by pulsed-DC plasma for antisticking surface applications. *Diamond and Related Materials* 17(7-10), 1728-1732

Rubio-Roy, M.; Corbella, C.; Bertran, E.; Portal, S.; Polo, M.; Pascual, E. & Andújar, J. (2009) Effects of environmental conditions on fluorinated diamond-like carbon tribology. *Diamond and Related Materials* 18(5-8), 923–926

Rusli; Yoon, S.F. & Huang, Q.F. (2003), Hydrogenated a-C optical coatings, In: *Properties of Amorphous Carbon*, Silva S.R.P. (Ed.), INSPEC, The Institution of Electrical Engineers

Sanders, D.M. & Anders, A. (2000) Review of cathodic arc deposition technology at the start of the new millennium. *Surface and Coatings Technology* 133-134, 78-90

Scheibe, H.J. & Schultrich, B. (1994) DLC film deposition by laser-arc and study of properties. *Thin Solid Films* 246(1-2), 92–102

Schvartzman, M. & Wind, S. J. (2009) Plasma fluorination of diamond-like carbon surfaces: mechanism and application to nanoimprint lithography. *Nanotechnology* 20(14), 145306

Silva, S.; Carey, J.; Khan, R.; Gerstner, E. & Anguita, J. (2002) Amorphous carbon thin films, In: *Handbook of Thin Films*, Nalwa, H. S. (Ed.), 403–506. Academic Press, ISBN 978-0-12-512908-4, Burlington.

Silva, S.R.P. (2003), Microstructure of a-C, In: *Properties of Amorphous Carbon*, Silva S.R.P. (Ed.), INSPEC, The Institution of Electrical Engineers

Simko, J. P.; Oehrlein, G. S. & Mayer, T. M. (1991) Removal of fluorocarbon residues on CF_4/H_2 reactive-ion-etched silicon surfaces using a hydrogen plasma. *Journal of The Electrochemical Society* 138(1), 277–284

Tang, G.; Ma, X.; Sun, M. & Li, X. (2005) Mechanical characterization of ultra-thin fluorocarbon films deposited by R.F. magnetron sputtering. *Carbon* 43(2), 345–350

Trippe, S. C.; Mansano, R. D.; Costa, F. M. & Silva, R. F. (2004) Mechanical properties evaluation of fluor-doped diamond-like carbon coatings by nanoindentation. *Thin Solid Films* 446(1), 85–90

Voevodin, A.A. & Donley, M.S. (1996) Preparation of amorphous diamond-like carbon by pulsed laser deposition: a critical review. *Surface and Coatings Technology* 82, 199-213

Voevodin, A.A.; Phelps, A.W.; Zabinski, J.S. & Donley, M.S. (1996) Friction induced phase transformation of pulsed laser deposited diamond-like carbon. *Diamond and Related Materials* 5, 1264-1269

Weiler, M.; Sattel, S.; Giessen, T.; Jung, K.; Ehrhardt, H.; Veerasamy, V. S. & Robertson, J. (1996) Preparation and properties of highly tetrahedral hydrogenated amorphous carbon. *Physical Review B* 53(3), 1594–1608

Winder, E. J. & Gleason, K. K. (2000) Growth and characterization of fluorocarbon thin films grown from trifluoromethane (CHF_3) using pulsed-plasma enhanced cvd. *Journal of Applied Polymer Science* 78(4):842–849

Xin, Y.; Ning, Z. Y.; Ye, C.; Lu, X. H.; Xiang, S. L.; Du, W.; Huang, S.; Chen, J. & Cheng, S. H. (2003) Optical emission study of CH_4+CHF_3 ECR plasma and properties of a-C:F:H films. *Surface and Coatings Technology* 173(2-3), 172–177

Yu, G. Q.; Tay, B. K.; Sun, Z. & Pan, L. K. (2003) Properties of fluorinated amorphous diamond like carbon films by pecvd. *Applied Surface Science* 219(3-4), 228–237

Zhang, P.; Tay, B.K.; Sun, C.Q. & Lau, S.P. (2002) Microstructure and mechanical properties of nanocomposite amorphous carbon films. *Journal of Vacuum Science and Technology A* 20, 1390-1394

Zhang, S.; Bui, X.L.; Fu, Y.; Butler, D.L. & Du, H. (2004) Bias-graded deposition of diamond-like carbon for tribological applications. *Diamond and Related Materials* 13, 867-871

Zou, J. W.; Reichelt, K.; Schmidt, K. & Dischler, B. (1989) The deposition and study of hard carbon films. *Journal of Applied Physics* 65(10), 3914–3918

4

Deposition and Tribology of Carbon and Boron Nitride Nanoperiod Multilayer Hard and Solid Lubricating Films

Shojiro Miyake[1] and Mei Wang[2]
[1]Department of Innovative System Engineering, Nippon Institute of Technology, Saitama
[2]Department of Research and Development, OSG Corporation, Aichi
Japan

1. Introduction

In the tribology application of thin films, in order to improve wear resistance properties, super hard films composed of boron, carbon and nitrogen have attracted much attention in recent years as known boron, carbon and nitrogen are used as basic component elements (Enomoto. Miyake, 1994; Miyake, 1999; Miyake et al., 1994). As also, it is expected to realize the application of solid lubricant films which will be provided super low friction coefficients to improve the lubricant properties of surfaces.

More and more so-called solid lubricant films in stead of lubricant oil are expected to work under such extreme conditions as high-temperature and vacuum operations in which lubricant oil is unqualified to run. In these solid lubricant materials, the representatives are soft metals as Au and Ag, crystallizing layered inorganic compounds as graphite, h-BN, MoS_2 (molybdenum disulfide) and WS_2 (tungsten disulfide) and polymers as PTFE (polytetrafluoroethylene) et al. (Enomoto & Miyake, 1994). On the other hand, a wear-resistant, not the conventional solid lubricant film, is desired to reduce atomic-scale wear and friction force fluctuation (Miyake et al., 1994; Miyake, 1994; Miyake et al., 1992). One candidate is carbonaceous films such as diamond and diamond like carbon films, which have wear-resistant and lubricating properties (Miyake et al., 1994; Miyake, 1994; Miyake & Kaneko, 1992).

The volume module of a hypothetical material, ß-C_3N_4, which is structurally analogous to ß-C_3N_4, may be larger than or equal to that of diamond (Liu & Cohen, 1989). The hardness of this stoichiometric crystalline carbon nitride is expected to exceed that of diamond. Partially crystalline carbon nitride films have been deposited by reactive sputtering, and their structure and compositions have been investigated. In our previous studies, the use of an atomic force microscope enabled accurate evaluation of the effect of nitrogen inclusions on nanoindentation hardness and microwear (Miyake et al., 1994).

Research on the development of nanoperiod multilayer films has recently been carried out, to improve the film hardness. For example, hard multilayer films were deposited, and it was found that the hardness of these multilayer films was significantly improved compared with that of one monolayer of them. As a super hard film, the superlattice film laminated with thin layers by several nanometers was investigated. Concretely, superlattice films such as CNx/TiN (Li et al., 1995) and TiN/AlN (Nakayama, 1997) were performed. In our studies, the (CN/BN)n multilayer film was proposed to deposit using boron nitride and carbon nitride layer by layer (Miyake et al., 1999) to form a novel superlattice film and its properties were investigated. The hardness enhancement mechanism of a superlattice film is based on the restriction of dislocation movement within and between layers in the superlattice film (Donohue et al., 1995). In our previous study, a nanoperiod (CN/BN)n multilayer film was deposited in order to improve its hardness (Miyake & Sekine, 2000). The nanoindentation hardness and modulus of dissipation of a 4-nm-period multilayer film are the highest and the lowest among other multilayer films (Miyake, 2003). The dependence of atomic-order wear properties on sliding cycles indicates that the wear resistance increased at the interface between the CN and BN layers of the multilayer films (Miyake, 2003).

Generally, friction force is proportional to contact area A and shearing strength S, i.e. F = AS. Friction coefficient μ is defined as friction force F divided by load W, i.e. $\mu = F/W = AS/W$. Hardness is defined as $H = W/A$. Therefore, $\mu = S/H$. i.e. the friction coefficient μ is approximately proportional to the ratio of shearing strength S to hardness H (Bowden et al., 2004). To develop new low-friction solid lubricant films, the hardness H must be increased and the shearing strength S must be decreased, as shown in Fig. 1 (Miyake et al., 2004). Therefore, nanoperiod multilayer films composed of low-shearing-strength graphite and hexagonal boron nitride (h-BN) that have solid lubricating properties, are expected to have a low friction coefficient because the hardness is increased due to the nanoperiod multilayered structure, while the shearing strength is decreased due to the inclusion of the low-shearing-strength C and BN layers (Miyake et al., 2004).

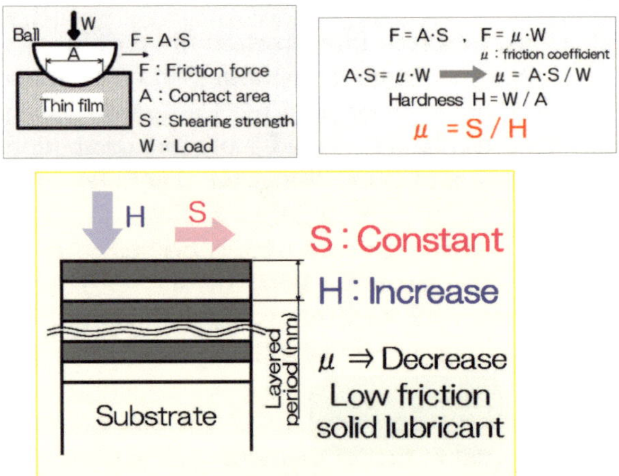

Fig. 1. Low-friction model of nanoperiod multilayer solid film: as the shearing strength S is constant, μ decreases with an increase in hardness.

In this study, by utilizing the superlattice film technique above, the hard films were proposed to deposit and their hardness in perpendicular direction were significantly increased with increasing their elastic moduli. It is considered that the friction coefficient is decreased due to the effect of the hardness increases as keeping the shear resistance of a sliding plane paralleling its substrate. Our aim is focused on developing a new solid lubricant film with an extremely low friction which usual simple film and mixed film material lacked of carbon and boron nitride nanoperiod multilayer films were deposited by RF sputtering using two semicircular targets. Auger electron spectroscopy (AES), X-ray photoemission spectroscopy (XPS) and Fourier transform infrared spectrometer (FTIR) analysis were used to observe the coatings structure. Nanoindentation and microwear tests were carried out to investigate the nanometer-scale deformation properties, macroscopic friction coefficient and sliding endurance of these films.

2. Experimental methods

2.1 Film deposition

As shown in Fig. 2, films were deposited by a RF sputtering system, in which both targets and substrate were supplied with high frequency RF power. Then, two semicircular graphite and boron nitride sinters were used as targets, and they were fixed in the target holder. A film was deposited by rotating a substrate, and two kinds of the semicircular targets were set to alternately face the substrate in an Ar gas atmosphere as shown in Fig. 2. To deposit C and BN layers with the same thickness, the time of setting the substrate opposite to the graphite target was nearly twice that of the h-BN target.

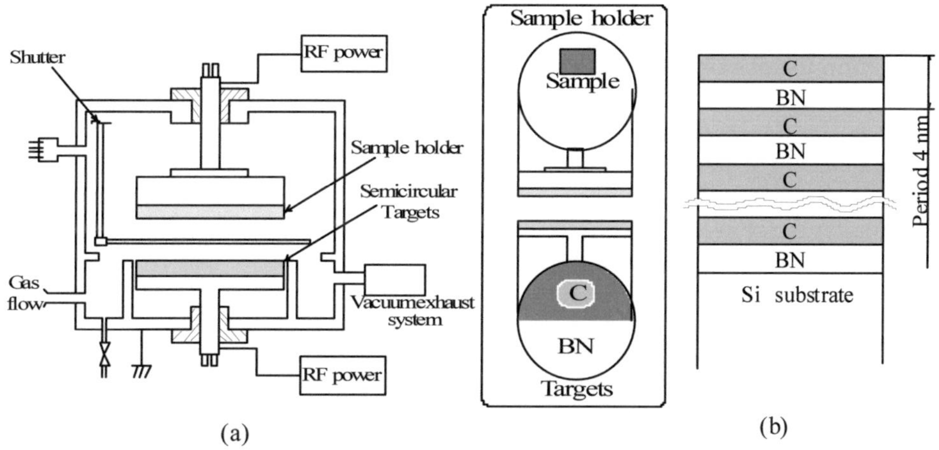

Fig. 2. Multilayer film deposition method of nanoperiod multilayer film using semicircular targets (a) and Model of 4-nm-period multilayer film

A (C/BN)n film was deposited by rotating a substrate . The BN layer was deposited first, and then the carbon layer was deposited. The topmost of the (C/BN)n film was deposited to be a carbon layer. In this study, 2-, 4-, 6-, 8- and 10-nm-period multilayer films were deposited by controlling deposition time. To compare the multilayer (C/BN)n film with other films like as the 4-nm-period (CN/BN)n film, C (carbon) single-layer film, BN single-layer film and 4-nm-period multilayer (CN/BN)n were prepared. The thickness of the C and BN single-layer films, (CN/BN)n and (C/BN)n multilayer films was proposed to be nearly 200 nm. Before the deposition of a film, a substrate was pretreated by Ar sputtering in order to clean its surface.

2.2 Tribology evaluation methods

The micro Vickers hardness tester and nanoindentation experiment system with AFM (atomic force microscope) were employed to determine the hardness of the (C/BN)n multilayer, BN and C films. Micro-Vickers hardness was measured at a load of 0.05 N and the load duration of 10 s. Nanoindentation test were performed at a maximum load of 100 μN. The hardness was geometrically calculated from the Ar section area of plastic deformation using the plastic deformation depth and P_{max} maximum load. Plastic deformation depth was determined from the point of intersection of the straight line fitted to the appropriate unloading curve and the x axis to the point of the maximum deformation depth on x axis. The hardness H is evaluated from Eq. (1)

$$H = Pmax / Ar \quad (1)$$

The nanoindentation hardness and surface deformation properties of the films were investigated using AFM (Digital instruments Nanoscope III) together with a nanoindentation measurement system (Hysitron Inc.). The nanoindentation hardness was measured by AFM with a Bercovich-type diamond indenter (a three-sided pyramid diamond tip) with a radius of about 100 nm, under 20~25 °C temperature and 30~50 % humidity conditions. The diamond tip is regularly checked using a quartz standard item. These films were as thin as 200 nm, therefore, the nanoindentation hardness value evaluated under higher load was influenced by the substrate hardness. As a result, the hardness value of these films evaluated by this method under higher load is lower than that of real film. The test was performed with a loading time of 20 s at 60-100 μN load. Here, the hardness is evaluated from the plastic deformation depth. The plastic deformation depth was evaluated from the intersection point of the y-axis and the straight line fitting from the appropriate unloading curve (Doerner & Nix, 1986). In this study, the Vickers and nanoindentation hardness values of the silicon substrate were 1500 and 13 GPa, respectively.

To evaluate the deformation behavior using the nanoindentation test, energy analysis of the nanoindentaion curve was performed. Total deformation energy was calculated using the integral of the loading curve as shown in Fig. 3. The storage energy was calculated using the integral of the unloading curve. The dissipated energy was evaluated from the remainder of the total energy minus the storage energy. The modulus of dissipation was calculated by dividing the dissipated energy by the total energy (Donohue et al., 1995; Miyake et al., 2004; Miyake, 2005).

The hardness was evaluated from the plastic deformation depth which was determined from the point of intersection of the straight line fitted to the appropriate unloading curve and the x axis (Doerner & Nix, 1986; Syed Asif et al., 1999; Miyake & Kim, 2003). The total deformation energy of nanoindentation was calculated as the integral of the loading curve. Storage energy was calculated as the integral of the unloading curve. As shown in Fig. 3, dissipated energy was evaluated as the total energy minus storage energy.

To clarify the tribological characteristic of an extremely thin film, an AFM with a diamond tip with a radius of about 50 nm was employed to observe the surface profile and microwear properties. A schematic representation of the evaluation is shown in Fig. 4. The oscillation scratch tests were performed with gradually increasing loads using the diamond tip indenter under setting conditions of a 6.7 μN/s loading speed, a 200 μN maximum load. In addition, the oscillation scratch test was carried out to evaluate the fracture characteristics and adhesion strength of the film by measuring an acoustic emission (AE) signal which can indicate a critical load in the test.

Wear-resistance of the films was determined with a ball-on-disk tribometer opening under RT ~400 °C. As shown in Fig. 5, a disk specimen deposited a multilayer film was tested.

The opposing stainless-steel ball (radius of 3.2 mm) was slid against the specimen being set rotating. Friction force was measured by the strain gauge attached to the plane spring mounted the tribometer. Data on the friction coefficient were obtained without lubricant and a normal load of 9.8 N, a sliding speed of 94 mm/s and a room temperature of 25, 200 and 400 °C, and a relative humidity of approximately 40~60 %. After the friction test, the wear tracts of the sample were observed by AFM and using an optical microscope and three-dimensional profile meter. The friction test was performed more than three times, and the mean and typical data were discussed. The variation of saturated friction value is less than 15 %.

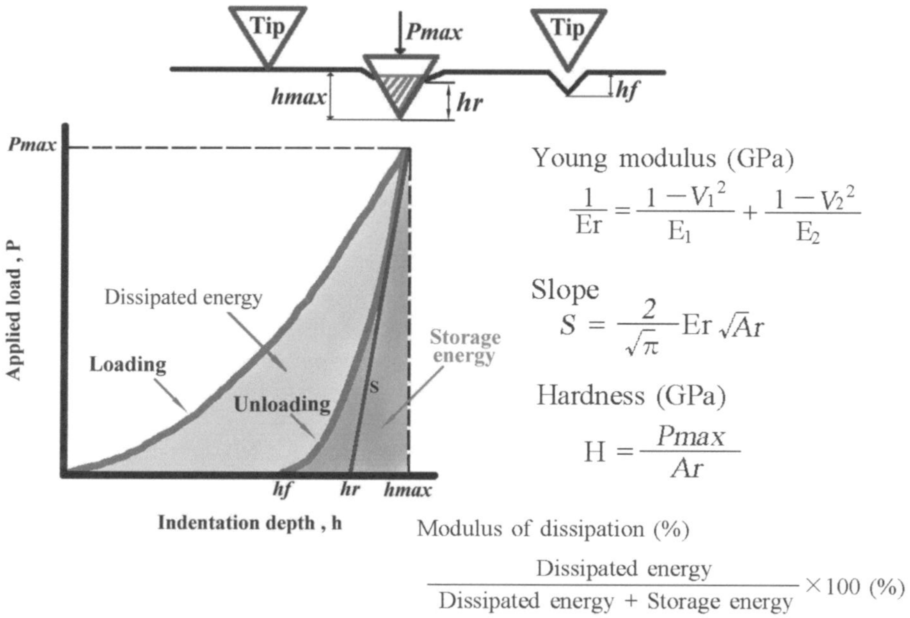

Fig. 3. Evaluation method of nanoindentation test

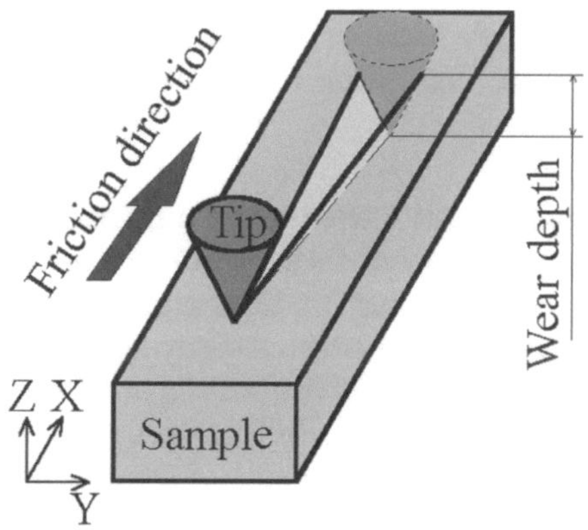

Fig. 4. Schematic representation of a load gradual increase nanoscratching test

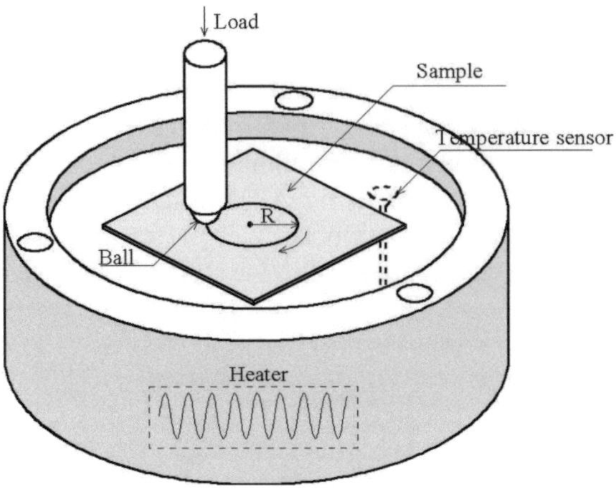

Fig. 5. High-temperature tribometer

3. Results and discussion

3.1 Composition and structure of deposited films

The total thicknesses of all the films tested were set at nearly 200 nm. A crystalline structure cannot be detected by X-ray diffraction analysis, FTIR and TEM. These single-layer and multilayer films are found to be amorphous from these microanalyses (Miyake, 2005). From the Raman spectrum, the differences between these carbon-containing films were not clearly evaluated (Miyake, 2005).

In a previous paper, a transmission-electron microscopy (TEM) image of a nano-meter-period multilayered film composed of alternate C-N and B-N layers was obtained and the structure of the period-layered CN/BN was clarified. The thicknesses of both the C-N and B-C-N layers were controlled so as to be nearly identical. The period of this TEM image of the multilayer film is 8 nm, therefore the thickness of each C-N and B-N layer is about 4 nm. Even though the layers are very thin, each layer can be distinguished (Miyake, 2003). Moreover, atomic concentrations of boron, carbon and nitrogen elements classified as super-hard film materials were detected according to Auger electron spectroscopy (AES) measurement of the depth profile of the multilayered films (Miyake et al., 2000). The C/BN multilayers discussed in this study were deposited by the same deposition method as that mentioned above. By controlling deposition time and using different targets, the (C/BN)n films were deposited as expected.

3.2 Micro Vickers hardness and nanoindentation hardness

The results of micro Vickers hardness of these thin films are shown in Fig. 6. The precise measurement could not be carried out for the BN simple layer film because there was some frictional peeling on the substrate after the test. Exactly as a expected effect of solid lubricant films, the micro Vickers hardness of (C/BN)n multilayer films synthesized in our study is changed with changing the multilayer period of the films. The hardness of the 4 nm (C/BN)n multilayer film shows the highest in all (C/BN)n films.

The nanoindentaion curves show a relationship between indenter penetration depth and load of these films obtained at 100 μm as shown in Fig. 7. Therefore, the nanoindentaion hardness can be evaluated from the nanoindentaion curves and the results are shown in Fig.

8. In the case of the BN film, as the indentation depth cased by an indenter was drastically fluctuated in the nanoindentation test, the result of the BN film in this test can not be reproducible. According to the indentation curves of the films, the maximum indentation depths of C simple film, 2- and 8-nm-period $(C/BN)_n$ films are quite similar and close to 15 nm, and it can be confirmed that the plastic deformation depth is nearly 3 nm. Upon the curves, it is observed that the deformation depth of the 4-nm-period $(C/BN)_n$ film shows a tendency to decrease significantly. The curve of indentation depth of the 4-nm-period (C/BN)n film is near that of the 4-nm-period $(CN/BN)_n$ film and shows the same tendency to decrease on its indentation depth. It is to say that elastic deformation mainly occurred on the surface of these multilayer films in the indentation test and the films exhibit excellent hardness. In particularly, the indentation hardness of the 4-nm-period $(C/BN)_n$ film is highest in all $(C/BN)_n$ films.

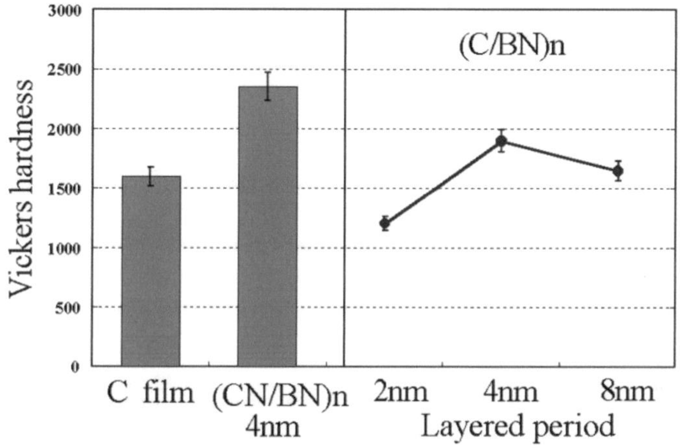

Fig. 6. Vickers hardness of super lattice films

When measuring the films in the micro Vickers test, the hardness could not be really measured by an indenter because the indenter would destroy those films and pierce through them, then reached the substrate of Si wafer. Hence, it was considered that the hardness of the film can not be exactly measured due to the influence of the substrate. In contrast, the hardness measurement will be hardly affected by the substrate employed as the groundwork in the nanoindentation test, because the indentation depth of the indenter is only about 3 to 17 nm. Therefore, the hardness of these films could be really investigated and measured by the nanoindentation test. Upon the indentation test, it is obvious that the $(C/BN)_n$ multilayer film with a layer period of nearly 4 nm show the highest hardness.

Furthermore, the indentation hardness of C and BN monolayer and (C/BN)n multilayer films was evaluated and compared at maximum applied loads of 60, 80, and 100 μN. Compared with the elastic-plastic properties of the (C/BN)n multilayer films, the maximum indentation depth (h_{max}) of the 4-nm-period multilayer film is less than that of the other films. The estimated average indentation hardness values in this study are summarized in Fig. 9. The hardness of all multilayer films is higher than that of the C and BN single layer films. It is observed that the nanoindentation hardness of the multilayer film changes with a change in a layer period. The multilayer film with the 4-nm-period film shows the highest hardness among all multilayer films in the study, indicating that the tribological properties of the multilayers, such as low friction coefficient and high wear-resistance, i.e. long wear life, were improved compared with those of the single layer (Miyake, 2005). Corresponding to this, the 4-nm-period film shows the highest hardness.

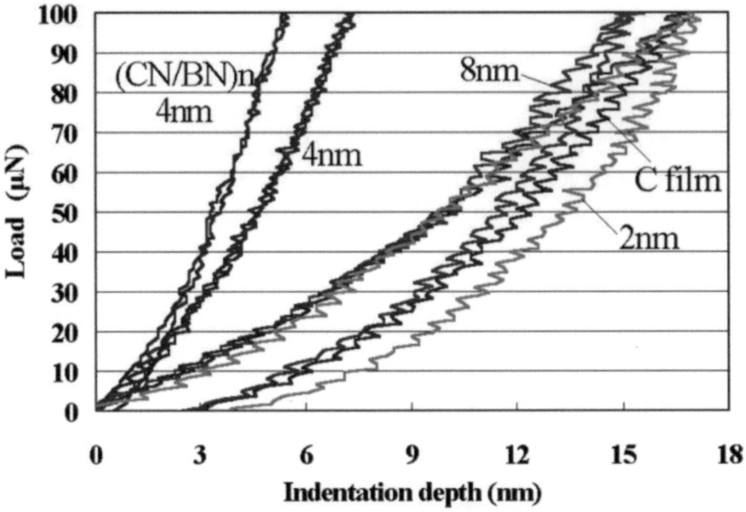

Fig. 7. Nanoindentation curves of 4-nm-period (CN/BN)n, 2-, 4- and 8-nm-period (C/BN)n super lattice films

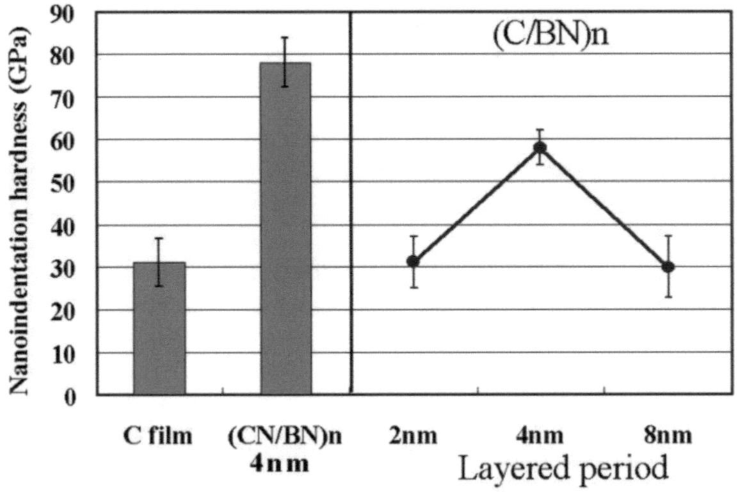

Fig. 8. Nanoindentation hardness of C film and (CN/BN)n and (C/BN)n super lattice films

The relationship between the modulus of dissipation, nanoindentation hardness (H) and the material factor of index of plasticity (E/H) for the monolayer and multilayer films is shown in Fig. 10. E/H is a quantity value which is used to express the plastic deformation of material in the plastic index of material constant (Taniguchi, 1996). The modulus of dissipation and the E/H decrease with an increase in nanoindentation hardness. The dissipated and storage energies of these films were evaluated. Compared with the single layer films, the total energy of the nanoperiod multilayer films decreased, and in particular, the dissipated energy of the multilayer films significantly decreased. Figure 10 shows the modulus of dissipation, which is calculated from the ratio of dissipated energy to total energy of the indentation. It is evident that the modulus of dissipation of the multilayer films was reduced to less than 50 % due to the nanoperiod multilayer structure.

As shown in Fig. 10, the modulus of dissipation decreases with increasing hardness. It is considered that the deformation of multilayer films is mainly elastic deformation, along

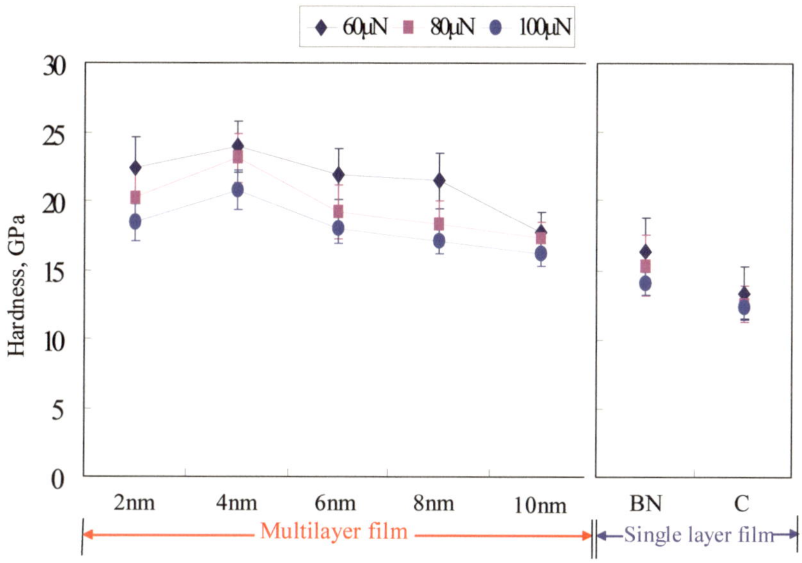

Fig. 9. Nanoindentaion hardness as a function of layer-periods for single layer and multilayer films (error bars: 3σ)

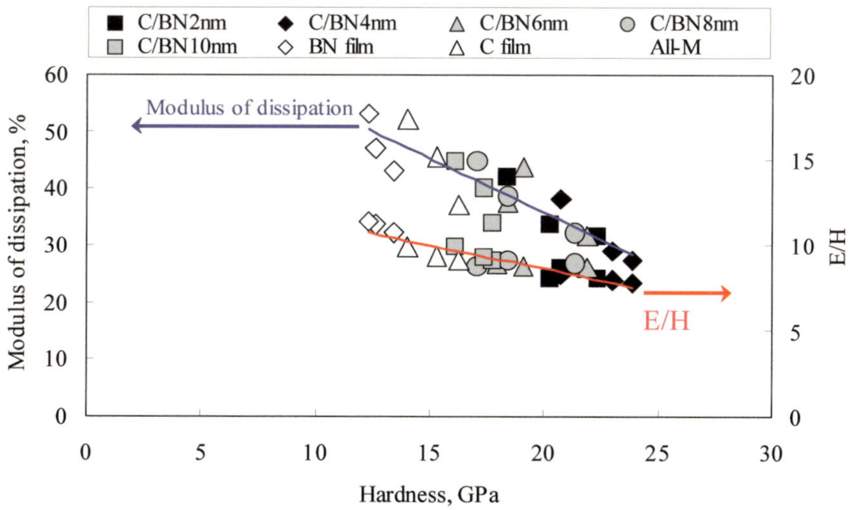

Fig. 10. Modulus of dissipation and the material factor of index of plasticity (E/H) as a function of nanoindentation hardness for single layer and multilayer films

with a low dissipation energy. The material factor of index of plasticity (E/H) was evaluated as Young's modulus E divided by hardness H (Miyake et al., 2004). The nanoindentation hardness is defined as the resistance to plastic deformation per unit area. On the other hand, the modulus of dissipation and E/H corresponds to the ease of plastic deformation. By indentation testing, the hardest 4-nm-period multilayer film shows the lowest modulus of dissipation and E/H, i.e. it has less dissipated permanent deformation due to indentation. This is considered that the nucleation and propagation of the lateral cracks beneath the indentation were limited due to the nanostructure of the multilayer (Bhushan, 1995). Therefore, the relationship among nanoindentation hardness, the modulus of dissipation and E/H was expressed approximately as a simple empirical formula for investigation and analysis of the nanoscale mechanical properties of film.

3.3 Tribological properties

The measured results of the frictional force and friction coefficient in the nano scratch test are shown in Fig.11. Upon the results, the frictional force of the BN film varies drastically and shows the highest value at the loads of 0 to 20 μN. It can be considered that the hardness of the BN film is lower than those of other films, therefore, the destruction on its surface easily occurred even from the beginning of the test and the frictional force of its surface shows to increase sharply. On the other hand, upon investigating the friction dependence of the (C/BN)n multilayer films on the period, it could be confirmed that the effect of the reduction in the frictional force of the (C/BN)n multilayer film with the period of 4 nm is significantly remarkable, and its hardness similarly exhibits a maximum value. With respect to the friction coefficients of these films, the initial values of the friction coefficients display drastically fluctuated, but they become steady with a load being over 40 μN. Upon the results, the friction coefficient of BN simple layer film is as high as $\mu = 0.45$ from 40 μN load and those of 2- and 8-nm-period (C/BN)n films are $\mu = 0.28$ and $\mu = 0.24$, respectively. The friction coefficients of the carbon film and 4-nm period (C/BN)n film are as low as $\mu = 0.22$ and $\mu = 0.17$, respectively. It could be found that the friction coefficients of films would show a tendency to decrease with the hardness increasing. As the hardness of the 4-nm-period (C/BN)n film is increased at the lamination direction and the shear strength of its scratch plane is nearly similar to that of the simple layer film which has the characteristic value, it can be consider that the friction coefficient μ of the (C/BN)n film is able to decrease due to its hardness increase according to the Eq. (2) when its shear strength is almost the same as that of the simple layer film. The equation shows as follows,

$$\mu = S / H \tag{2}$$

Here, S is a shear resistance and constant and H is hardness.

Figure 12 shows the section profiles of the scratch scars of all deposited films after the nanoscratch test. The maximum wear depths of the films are shown in Fig. 13. The wear scar of the 4-nm-period (C/BN)n film is the smallest. In the nanoscratch test, the destruction of the simple layer film continuously becomes degraded with gradually increasing the test load, whereas that of the superlattice film can be significantly deterred due to the restraint effect of boundary plane of the multilayer film.

Moreover, it is obvious that the maximum wear depth of the film also depends on the layer period and the wear depth of the 4-nm-period (C/BN)n film is reduced to a minimum. These results are corresponding to those of the nanoindentation hardness. Therefore, it is confirmed that the wear depth of the 4-nm-period (C/BN)n film decreases because its elastic modulus increases at the lamination direction.

The results of oscillating scratch test are shown in Fig. 14. The AE (acoustic emission) numbers could be obtained in this scratch test and the inclination of the scratch force curve is corresponding to the friction coefficient. As observing the curves, the inclination of the 4-nm-period (C/BN)n film shows comparatively low, that is to say, the 4-nm-period (C/BN)n film has a low friction coefficient and this result is corresponding to that in the nanoscratch test and its AE also shows very small. When the load was rapidly increased, the AE of the 4-nm-period (C/BN)n film would increase drastically and a critical load was obtained and rather high because its surface layer destruction occurred difficultly. It became clear that the critical load of the 4-nm-period (C/BN)n film is significantly improved more than those of the simple films and other multilayer period films, and the 4-nm-period (C/BN)n film shows the high critical load and superior adhesion property.

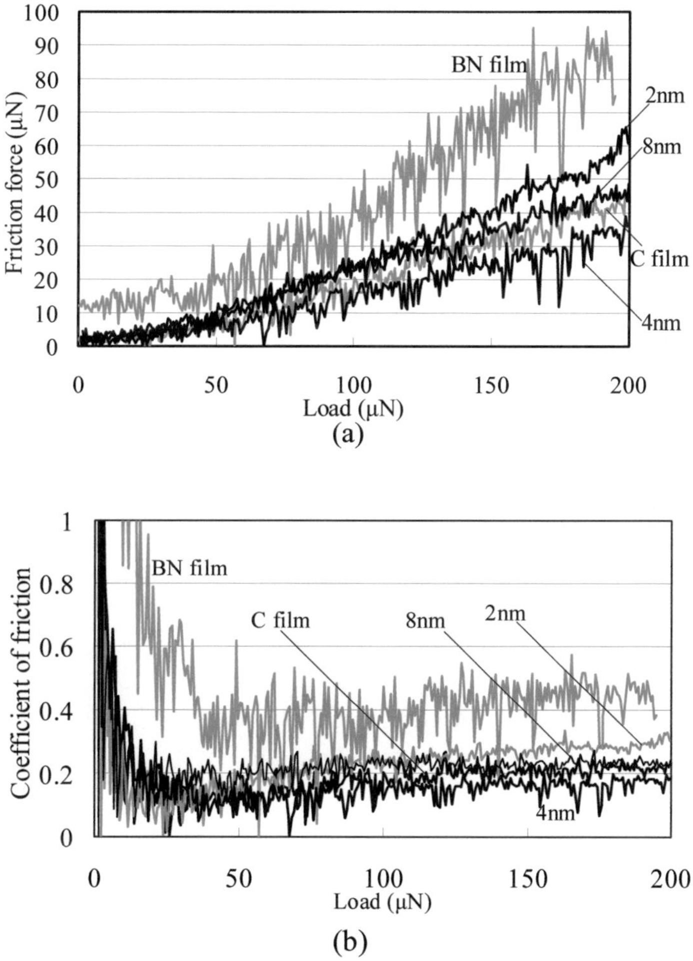

Fig. 11. Friction force (a) and coefficient of friction (b) of C, BN and (C/BN)n films in a load gradual increase nanoscratching test (diamond tip , load 0~200μn , frictional rate 6.67μn/s)

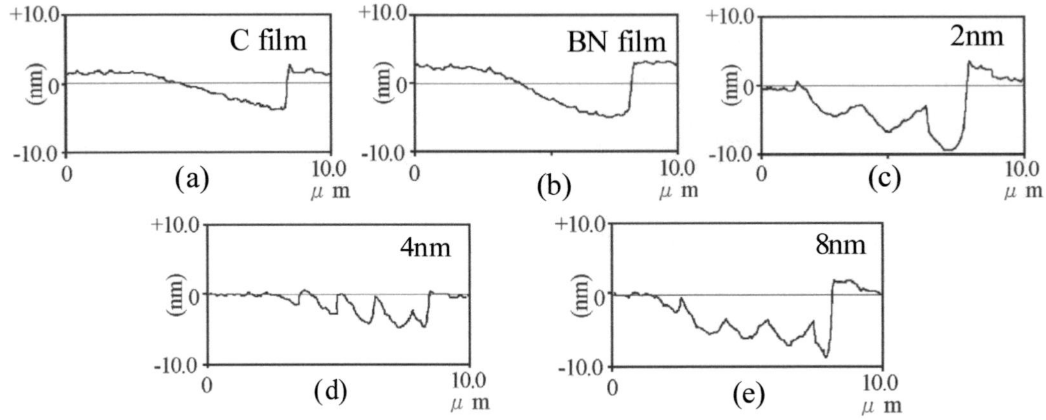

Fig. 12. Wear marks on (a) C film, (b) BN film, (c) 2-nm-period (C/BN)n, (d) 4-nm-period (C/BN)n and (e) 8-nm-period (C/BN)n film, obtained by a load gradual increase nanoscratching test

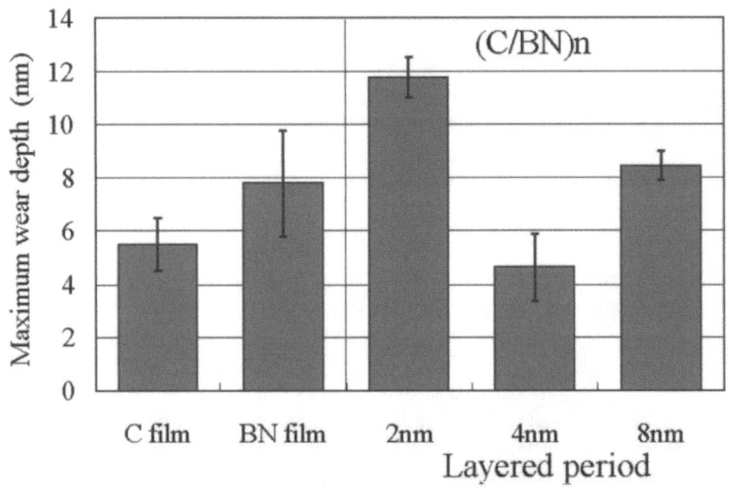

Fig. 13. Maximum wear depth by a load gradual increase nanoscratching test

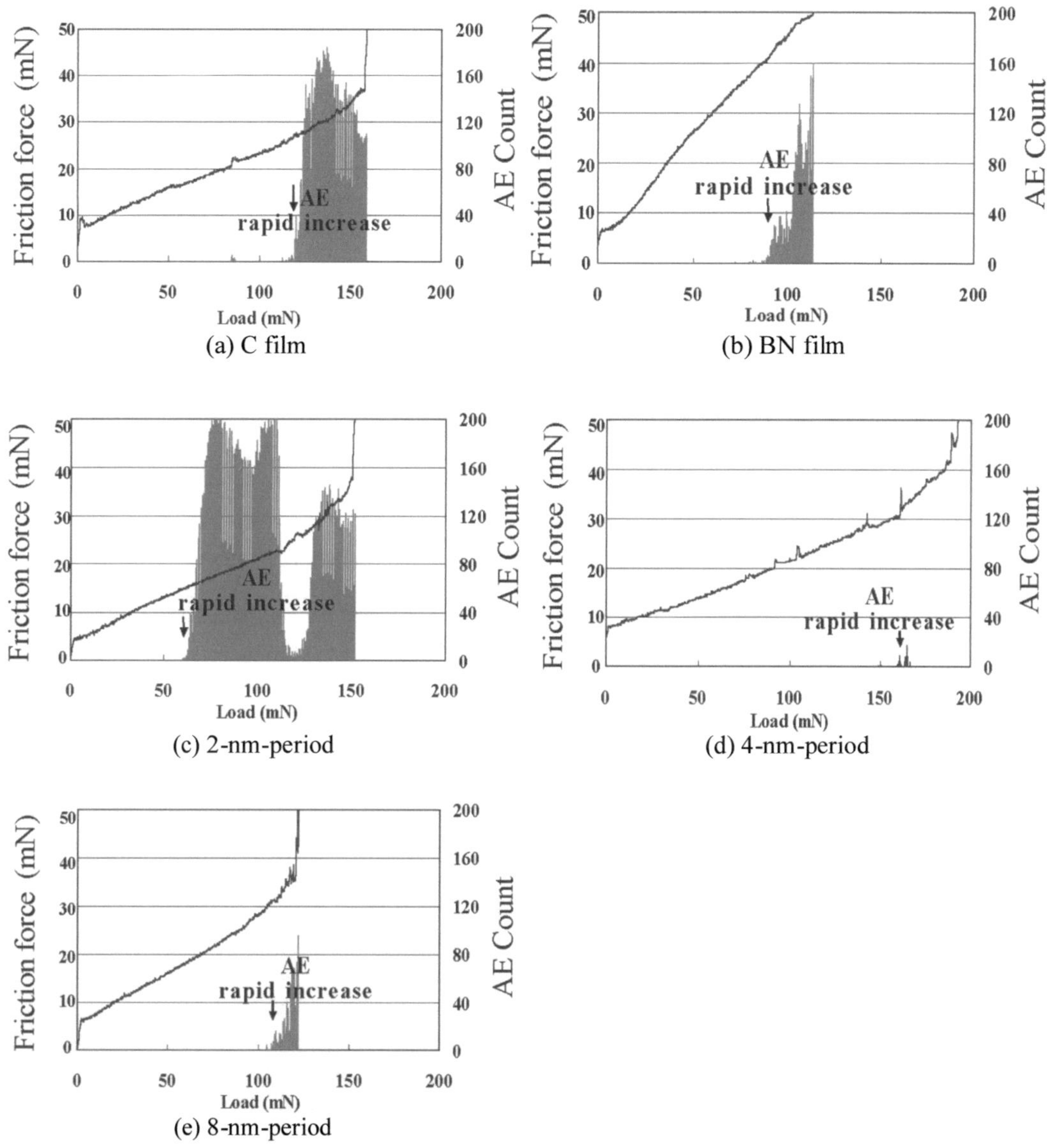

Fig. 14. AE (Acoustic emission) and friction force as functions of indentation load for (a) C film, (b) BN film, (c) 2-nm-period (C/BN)n, (d) 4-nm-period (C/BN)n and (e) 8-nm-period (C/BN)n film, obtained by scratching test

The dependence of the friction coefficient of C and BN monolayer and (C/BN)n multilayer films on sliding cycle was investigated under a load of 9.8 N, a sliding speed of 94 mm/s and a room temperature of 25 °C, as shown in Fig. 15. The friction coefficient of the C film is $\mu = 0.3$ high at sliding cycles from 0 to 1000 and sharply increased to $\mu = 0.6$ at sliding cycles of 1000 and above. Corresponding to this, the friction coefficients of the 4- and 8-nm-period (C/BN)n films remain comparatively low at nearly $\mu = 0.15$. Wear is hardly observed on the surface of 4- and 8-nm-period (C/BN)n films. As the surface of these films was magnified to observe, damage mark could be found on the 8-nm-period (C/BN)n film but not on the 4-nm-period (C/BN)n film, suggesting that the 4-nm-period (C/BN)n film has the lowest friction coefficient and superior wear resistance.

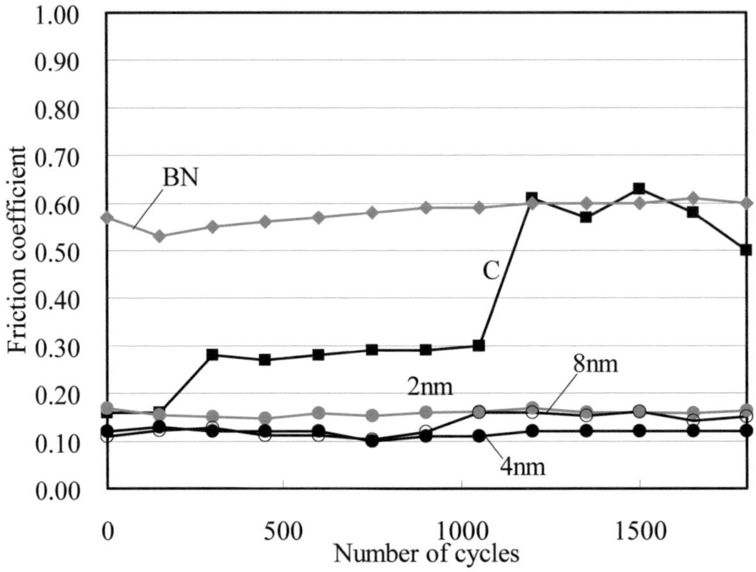

Fig. 15. Dependence of friction coefficient on the number of sliding cycles tested at 25 °C

The dependence of the friction coefficient of the nanoperiod (C/BN)n film on sliding cycles was evaluated at an environment temperature of 200°C, as shown in Fig. 16. The friction coefficients of the C and BN single-layer films are high and markedly increase with sliding cycles. However, the nanoperiod multilayer films show a low friction coefficient. In particular, the friction coefficient of the 4-nm-period multilayer film is low and stable, and it is less than 0.35 after 1800 sliding cycles.

In the tribotest performed at 400 °C temperature, it is found that at sliding cycles of less than 600, the friction coefficient of the 4-nm-period multilayer (C/BN)n is lower than that of the C, BN and other (C/BN)n multilayer films as shown in Fig. 17. Observed the surface of the films, it is found that for the C, BN and 2-nm-period films, the film was immediately removed by friction at the beginning of the test and the substrate was exposed, whereas the 4- and 8-nm-period multilayer films show a low friction coefficient. This result shows that the lubrication property of the multilayer films was improved due to its microstructure formed with an appropriate layer-period. The reason why the 4-nm-period multilayer film shows a longer life is considered to be as follows. For the single-layer films, the surface of the C or BN films was lubricated by the single layer during friction test. Since the C and BN single-layer films have a poor elastic deformation, i.e. a high modulus of dissipation, it was easily removed by friction test. Corresponding to this, since the multilayer film exhibits a low modulus of dissipation and defect elongation is prevented by the interface formed between layers (Miyake, 2003; Doerner & Nix, 1986), deep plastic deformation difficultly initiate, resulting in the wear-resistance of the film being improved. The shearing damage due to friction is suppressed due to the extremely thin layer. As a result, a longer life of the surface was obtained due to the multilayer microstructure.

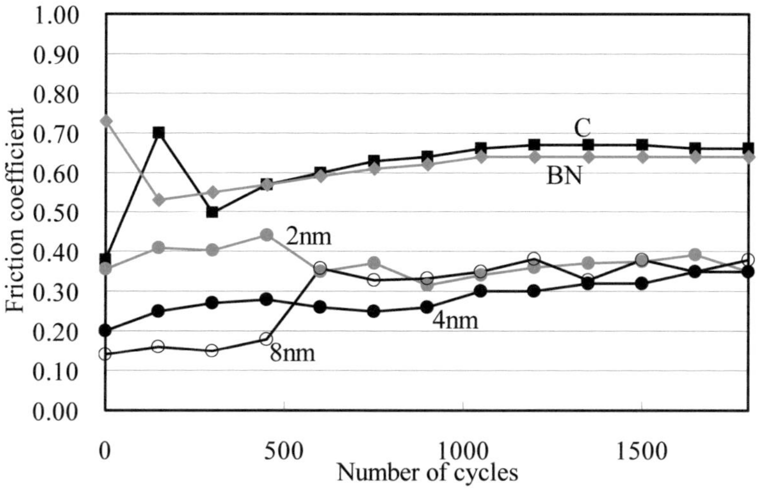

Fig. 16. Dependence of friction coefficient on the number of sliding cycles tested at 200 °C

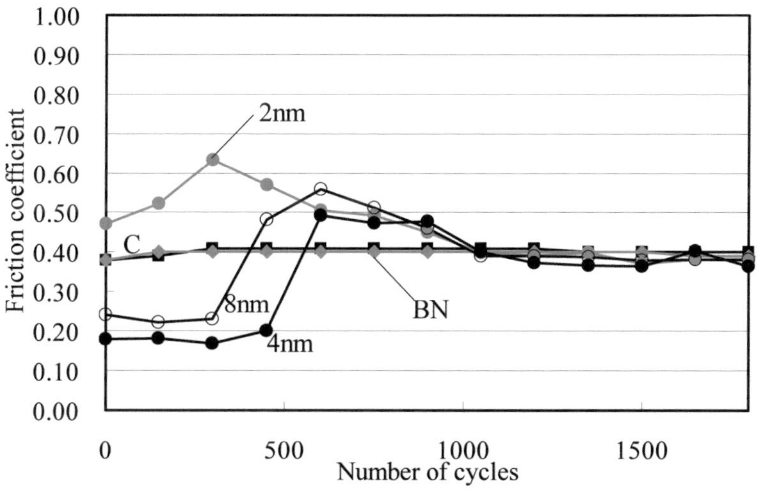

Fig. 17. Dependence of friction coefficient on the number of sliding cycles tested at 400 °C

Figure 18 shows the wear profiles of C and BN single layers and 2-, 4- and 8-nm-period multilayers after the ball-on-disk tribotest, respectively after 1000 cycles of sliding at 25 °C. In the case of the C and BN single layers, the deep wear grooves exposed the substrate surface during sliding, producing a rough wear scar and leading to a rapid increase in the friction. Corresponding to this, the wear of the 2-, 4- and 8-nm-period (C/BN)n multilayers is very small, and therefore, after 1500 cycles of sliding, the multilayered films remained on the sliding surface and still acted as a solid lubricant to prevent wear propagation. It is considered that since the multilayer film has a low modulus of dissipation, and defect elongation was prevented by the interface between layers, such as those formed in carbon nitride and boron nitride nanomulitlayer films (Miyake, 2003), and therefore, deep plastic deformation is difficult to occur. The 4-nm-period film shows a better wear-resistance, this corresponds to that it has higher hardness and lower friction coefficient compared with other films in this study.

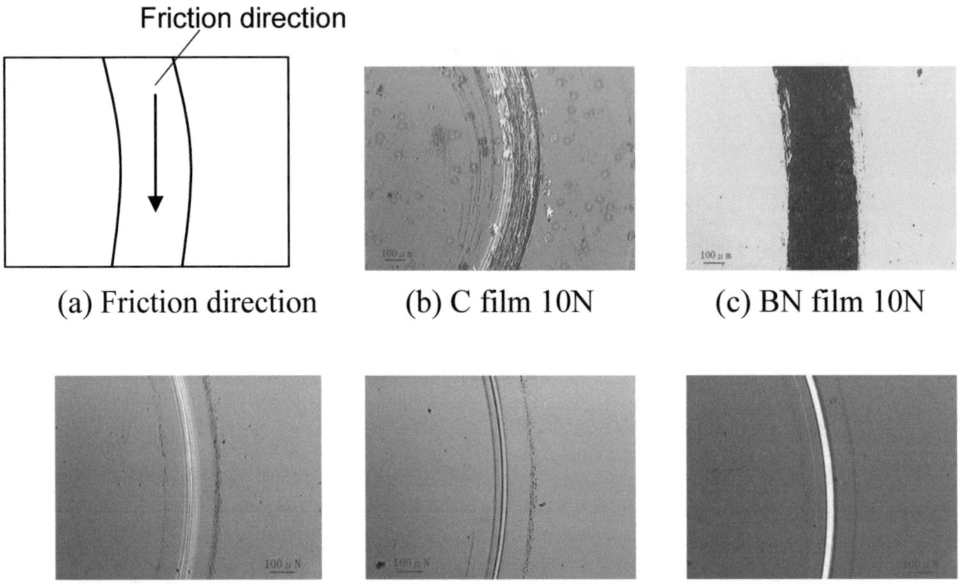

(a) Friction direction (b) C film 10N (c) BN film 10N

(d) (C/BN)n 2nm 10N (e) (C/BN)n 4nm 10N (f) (C/BN)n 8nm 10N

Fig. 18. Microwear morphology of C and B single layers and (C/BN)n multilayers coated on Si substrates after 1000 cycles of sliding.

4. Conclusion

Nanoperiod multilayer films composed of carbon and boron nitride layers, which have a structure expected to provide solid lubrication, such as the h-BN and graphite structures, were respectively deposited, and their nanoindentation and frictional properties were evaluated. The following results were obtained.

1. Micro Vickers and nanoindentaion hardness of the 4-nm-period multilayer film are the highest of all (C/BN)n films. The nanoindentation hardness of the nanoperiod (C/BN)n multilayer films changed with the layer period. A 4-nm-period multilayer film shows the highest hardness among all films in this study. It is considered that since the 4-nm-period multilayer has the lowest modulus of dissipation and E/H, which corresponds to the ease of plastic deformation, the 4-nm-period multilayer film has mainly elastic deformation due to indentation.
2. In nanoscratch test, the friction coefficient of 4-nm-period (C/BN)n film is the lowest. As the elastic modulus increases in the laminated direction and the change of the shear resistance of sliding plane is small in the case of 4nm-period (C/BN)n film, therefore, its friction coefficient seems to decrease.
3. The improvement effects of 4-nm-period (C/BN)n multilayer film on the decrease of the friction coefficient and damage of the wear surface were significantly remarkable and evaluated in both of the reciprocation and ball on disk type tribo tests.

4. The friction coefficient of the 4-nm-period (C/BN)n film was the lowest, as determined by the ball-on-disk tribotest at environment temperatures of 25, 200 and 400 °C. The effect of the 4-nm-period (C/BN)n multilayer film on improvement of friction coefficient and wear-resistance was experimentally confirmed.

5. Acknowledgement

This research was partly supported by a Grant-in-Aid to the Scientific Research Foundation from the Ministry of Education, Culture, Sports, Science and Technology (B21360076).

6. References

Enomoto, Y. & Miyake, S. (1994). *Tribology of thin films,* Univ. of Tokyo (In Japanese), ISBN 4-13-061150-X, Tokyo

Miyake, S. (1999). Zero-Wear Films, *J. Jpn. Soc. Mech. Eng.*, (in Japanese) Vol.102, No.962 (1999, 01), pp.20-24. ISSN 0021-4728

Miyake, S.; Watanabe, S.; Miyazawa, H.; Murakawa, M.; Kaneko R. & Miyamoto T. (1994). Micro Scratch Hardness Increasing of Ion Plated Carbon Film by Nitrogen Inclusion Evaluated by Atomic Force Microscope, *Appl. Phys. Lett.*, Vol.65, No.9 (1994) pp.3206-3208, ISSN 0003-6951

Miyake, S. (1994). Tribological improvements of polished chemically vapor deposited diamond films by fluorination, *Appl. Phys. Lett.*, 65, 9, pp.1109-1112. ISSN 0003-6951

Miyake, S.; Watanabe, S.; Murakawa, M.; Kaneko, R. & Miyato, T. (1992). Tribological study of cubic boron nitride film, *Thin Solid Film.* Vol. 212, 1-2, pp.262-266. ISSN 0040-6090

Miyake, S. & Kaneko, R. (1992). Micro-tribological properties and potential applications of hard, lubricating coatings, *Thin Solid Films,* Vol. 212, pp.256–261. ISSN 0040-6090

Liu, A.Y. & Cohen, M.L., (1989). Prediction of New Low Compressibility Solids, *Science.,* Vol. 245, 4920, pp. 841-842. ISSN 0036-8075

Miyake, S.; Watanabe, S.; Miyazawa, H.; Murakawa, M.; Kaneko, R. & Miyamoto, T. (1994). Improved microscratch hardness of ion - plated carbon film by nitrogen inclusion evaluated by atomic force microscope, *Appl. Phys. Lett.*, Vol. 65, 9 (1994) pp. 3206-3209. ISSN 0003-6951

Li, D.; Chu, X.; Cheng, S.C.; Lin, X.W.; Dravid, V. P.; Chung, Y.W.; Wong M.S. & Sproul W.D. (1995). Synthesis of superhard Carbon Nitride Composite Coatings, *Appl. Phys. Lett. Vol.* 67, pp.203-206. ISSN 0003-6951

Nakayama, A. (1997). A Mystery of Superlattice Thin Film Hardness, *J. of the Jpn. Soc. of Mech. Eng.,* (in Japanese) Vol.100, No.943, pp.593-595. ISSN0021-4728

Miyake, S.; Sekine, Y. & Watanabe, S. Deposition and Microtribology of Carbon Nitride and Boron Nitride Super Lattice Films, (1999). *Trans. Jpn. Soc. Mech. Eng.,* (in Japanese) Vol.65, No.639, C pp. 4496-4501. ISSN 0287-7066

Donohue, L.A.; Cawley, J.; Lewis, D.B.; Brooks, J.S. & Munz, W.D. (1995). Investigation of superlattice coatings deposited by a combined steered are evaporation and unbalanced magnetron sputtering technique, *Surf. Coat. Technol.* Vol 76-77, pp.149-158. ISSN 0257-8972

Miyake, S. & Sekine, Y. (2000). Improvement of Tribological Properties of Carbon Films due to Other Material Addition, Proc. of Int. Tribol. Conf., Nagasaki, 2000, pp.59-64.

Miyake, S. (2003). Improvement of mechanical properties of nanometer period multilayer films at interfaces of each layer, *J. of Vac. Sci. & Technol.* Vol. 21B, 2, pp.785-790. ISSN 0734-211X

Bowden, F.P.; Moore, A.J.W. & Tabor, D. (1943). The Ploughing and Adhesion of Sliding Metals, *J. Appl. Phys.* Vol. 14, 2, pp.80-92. ISSN 0021-8979

Miyake, S; Sekine, Y.; Noshiro, J. & Watanabe, S. (2004). Low-Friction and Long-Life Solid Lubricant Films Structured of Nanoperiod Tungsten Disulfide and Molybdenum Disulfide Multilayer, *Jpn. J. Appl. Phys.*, Vol. 43, 7A, pp.4338-4343. ISSN 0021-4922

Miyake, S.; Sekine, Y. & Watanabe, S. (2004). Low friction model of nanoperiod multilayer solid film: as the shearing strength S is constant, μ decreases with an increase in hardness, *Jpn. Soc. Mech. Eng. Int. J.* Vol. 47C, 1, pp.377. ISSN 0021-4728

Doerner, M.F. & Nix, W.D. (1986). A method for interpreting the data from depth-sensing indentation instruments, J. Mater. Res. Vol. 1, 4, pp.601-609. ISSN 0272-9172

Miyake, S. (2005). Tribology of carbon nitride and boron nitride nanoperiod multilayer films and its application to nanoscale processing, *Thin Solid Films*, Vol. 493, 1-2, pp.160-169. ISSN 0040-6090

Syed Asif, S.A.; Wahl, K.J. & Colton, R.J. (1999). Nanoindentation and contact stiffness measurement using force modulation with a capacitive load-displacement transducer, *Rev. Sci. Instrum.* Vol. 70, 5, pp.2408-2414. ISSN 0034-6748

Miyake, S. & Kim, J. (2003). Nanoprocessing of Carbon and Boron Nitride Nanoperiod Multilayer Films, *Jpn. J. Appl. Phys.* Vol. 42, 3B, pp.L322-325. ISSN 0021-4922

Miyake, S.; Sekine, Y.; Kim, J. & Yamamoto, H. (2000). Development of Nanofabrication Technique Based on Wear Properties of Nanoperiod Multilayerd Films, (in Japanese), *Journal of the Japan Society of Precision Engineering.*, Vol. 66, 12, pp.1958-1962. ISSN 0912-0289

Taniguchi, N. (Ed.), (1996). *Nanotechnology: Integrated Processing Systems for Ultra-Precision and Ultra-Fine Products*, Oxford Univ Pr on Demand, ISBN-10 0198562837, Oxford Science Publications

Bhushan, B. (Ed.), (1995). *Handbook of Micro/Nanotribology*, CRC Press, Inc., ISBN 0-8493-8401-X, USA

5

Ferrofluid Seals

V. Lemarquand[1] and G. Lemarquand[2]
[1]*LAPLACE UMR CNRS 5213, IUT Figeac, Universite de Toulouse, Avenue de Nayrac, 46100 Figeac*
[2]*Laboratoire d'Acoustique de l'Universite du Maine UMR CNRS 6613, Avenue Olivier Messiaen, 72085 Le Mans Cedex 9*
France

1. Introduction

Ferrofluids are very peculiar materials. Indeed, a stable colloidal suspension of magnetic particles in a liquid carrier is something special. These magnetic particles, of about 10 nanometers in diameter, are coated with a stabilizing dispersing agent that prevents their agglomeration. The liquid can be either water or synthetic hydrocarbon or mineral oil. But this material class, discovered in the 1960s, proves specific various chemical and physical properties, whose increasing knowledge leads to ever more numerous technological applications.
Indeed, they are efficiently used in various engineering areas such as heat transfers, motion control systems, damping systems (1), sensors (2)(3). Their use to design fluid linear pumps for medical applications seems also very promising (4)(5). However, they are more commonly used as squeeze films in seals and bearings for rotating devices. Tarapov carried out some pioneering work regarding the ferrofluid lubrication in the case of a plain journal submitted to a non-uniform magnetic field (6) but more recent works show and discuss the recent trends in such a use (7)-(13). Moreover, ferrofluid dynamic bearings have been regularly studied and their static and dynamic characteristics have been described theoretically (14)-(21).
The various properties of ferrofluids enable them to fulfill such functions as heat transfer, ensuring airtightness, working as a radial bearing. Therefore they are used in electrodynamic loudspeakers. Moreover, a ferrofluid seal can replace the loudspeaker suspension and leads to a better linearity of the emissive face movement (22)-(26). This chapter intends to explain how ferrofluid seals are formed in magnetic structures by presenting a simple analytical model to describe their static behavior (27)(28). The originality lies in the fact that the considered structures are made of permanent magnet only, without any iron on the static part. The moving part is a non magnetic cylinder. The seal shape and performances are described with regard to the magnetic structure. The evaluation of the seal static capacity is given. Moreover, the seal shape changes when the seal is radially crushed by the inner cylinder: these changes are described and calculated and the radial force exerted by the ferrofluid on the moving part is determined as well as the stiffnesses associated.
Then, various magnetic structures are presented and studied to illustrate the magnet role and deduct some design rules for ferrofluid seals with given mechanical characteristics.

2. Structure and method description

This section presents the basic ironless structure used to create a magnetic field which has the double function of trapping and fixing the ferrofluid to form a seal.
The device is cylindrical and constituted of a static outer part made of stacked ring permanent magnets separated from the inner non magnetic moving part by an airgap. The number of ring magnets is an issue and will be discussed later on. The simplest structure has a single ring, but the performances are better for two or three rings, and even more, depending on the intended values. The magnet polarization direction is also an issue and can be either axial or radial. The trick may be to associate correctly two kinds of polarization.
The ferrofluid is located in the airgap and forms a seal between the moving and the static parts.

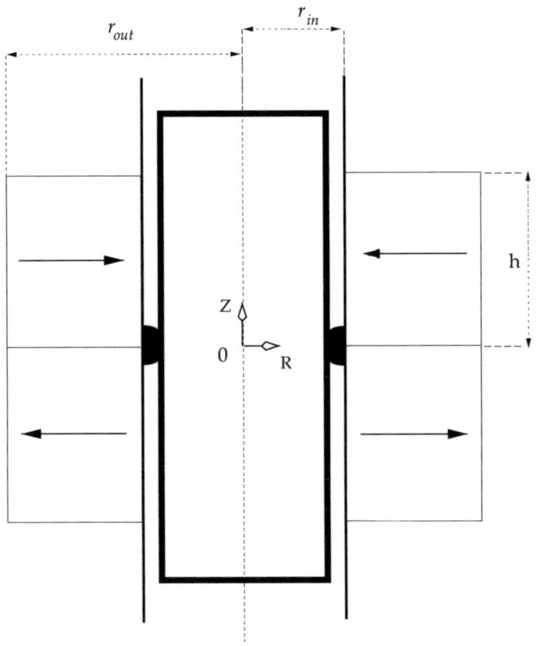

Fig. 1. Geometry : two outer ring permanent magnets and an inner non-magnetic cylinder with a ferrofluid seal between them; the ring inner radius is r_{in}, the ring outer radius is r_{out}, the height of a ring permanent magnet is h.

The ring inner radius is r_{in}, the ring outer radius is r_{out} and the ring permanent magnet height is h. The z axis is a symmetry axis.
The first step of the modelling is to calculate the magnetic field created by the permanent magnets. Exact formulations for the three components of the magnetic field created by axially or radially polarized permanent magnets have been given in the past few years. They are based either on the coulombian model of the magnets or their amperian one. Both models are equivalent for the magnet description but aren't for calculating: one may be more adapted to lead to compact formulations in some configurations where the other will be successful in others. The calculations of this chapter were carried out with formulations obtained with the coulombian model of the permanent magnets.
The location and the shape of the ferrofluid seal will be deducted from the magnetic field value by energetic considerations. Nevertheless, the conditions of use of the ferrofluid have to be

given here, as they differ from the ones encountered in their usual applications. Indeed, the magnetic field created by the magnets, which are considered to be rare earth ones (and rather Neodymium Iron Boron ones), is higher than 400 kA/m and the ferrofluid is consequently saturated, as the highest saturation field of the available ferrofluids are between 30 and 40 kA/m. This means that the field created by the ferrofluid itself won't really modifiy the total field and therefore it can be neglected. This is a great difference with most of the usual applications of the ferrofluids. Moreover, as the ferrofluid is completely saturated, its magnetic permeability is equal to one. Its magnetization is denoted M_s.
Furthermore, all the particles of the saturated ferrofluid are aligned with the permanent magnet field. So, the ferrofluid polarization has the same direction as the magnet orienting field. In addition, the sedimentation in chains of the ferrofluid particles is omitted (29).
Some other assumptions are made: the thermal energy, E_T, ($E_T = kT$ where k is Boltzmann's constant and T is the absolute temperature in degrees Kelvin) and the gravitational energy, E_G, ($E_G = \Delta\rho V g L$ where V is the volume for a spherical particle, L is the elevation in the gravitational field, g is the standard gravity, $\Delta\rho$ is the difference between the ferrofluid density and the outer fluid) are neglected. In addition, the surface tension exists but its effects can be omitted as the considering of both values of the surface tension coefficient, A, (A equals $0.0256 kg/s^2$ for the considered ferrofluids) and the radius of curvature leads to the conclusion that it won't deform the free boundary surface.
One of the aims of this chapter is to describe how the ferrofluid seals are formed and which is their shape. It has to be noted that the ferrofluid location depends on the value of the magnetic field in the airgap. Furthermore, the seal shape is the shape of the free boundary surface of the ferrofluid, which is a result of the competing forces or pressures on it. And the predominant pressure is the magnetic one. Therefore, the calculation of the magnetic field will be explained first and then the concept of magnetic pressure will be detailed.

3. Magnetic field calculation

3.1 Basic equations

The magnetic field created by the ring permanent magnets can be determined with a fully analytical approach. Let us consider the four fundamental Maxwell's equations:

$$\vec{\nabla}.\vec{B} = 0 \tag{1}$$

$$\vec{\nabla} \wedge \vec{H} = \vec{j} \tag{2}$$

$$\vec{\nabla}.\vec{D} = \rho \tag{3}$$

$$\vec{\nabla} \wedge \vec{E} = -\frac{\partial \vec{B}}{\partial t} \tag{4}$$

where $\vec{B}$ is the magnetic induction field, $\vec{H}$ is the magnetic field, $\vec{j}$ is the volume current density, $\vec{D}$ is the electric flux density, $\vec{E}$ is the electrostatic field and ρ is the electrical charge. The currents are nil in the considered structures as the magnetic field is created only by the permanent magnets. The vector fields $\vec{B}$ and $\vec{H}$ are defined for all points in space with the following relation:

$$\vec{B} = \mu_0 \vec{H} + \vec{J} \tag{5}$$

where μ_0 is the vacuum magnetic permeability and $\vec{J}$ is the magnet polarization. When the magnetic field is evaluated outside the magnet, $\vec{J} = \vec{0}$. The analogy with the Mawxell's equations leads to write that:

$$\vec{\nabla}.\vec{H} = -\frac{\vec{\nabla}.\vec{J}}{\mu_0} = \frac{\sigma^*}{\mu_0} \quad (6)$$

where σ^* corresponds to a fictitious magnetic pole density. On the other hand, the magnetic field $\vec{H}$ verifies:

$$\vec{\nabla} \wedge \vec{H} = \vec{0} \quad (7)$$

Thus, $\vec{H}$ can be deducted from a scalar potential $\phi(\vec{r})$ by

$$\vec{H} = -\vec{\nabla}\left(\phi(\vec{r})\right) \quad (8)$$

For a structure with several ring permanent magnets, (6) and (7), lead to:

$$\phi(\vec{r}) = \frac{1}{4\pi\mu_0}\left(\sum_i \int\int_{S_i} \frac{\vec{J_k}.d\vec{S_i}}{\left|\vec{r} - \vec{r'}\right|} + \sum_j \int\int\int_{V_j} \frac{-\vec{\nabla}.\vec{J_k}}{\left|\vec{r} - \vec{r'}\right|} dV_j\right) \quad (9)$$

where $\vec{J_k}$ is the magnetic polarization of the k ring permanent magnet and $\left|\vec{r} - \vec{r'}\right|$ is the distance between the observation point and a magnetic charge contribution. Then the magnetic field created by the ring permanent magnets is determined as follows:

$$\vec{H} = -\vec{\nabla}.\left(\frac{1}{4\pi\mu_0}\left(\sum_i \int\int_{S_i} \frac{\vec{J_k}.d\vec{S_i}}{\left|\vec{r} - \vec{r'}\right|} + \sum_j \int\int\int_{V_j} \frac{-\vec{\nabla}.\vec{J_k}}{\left|\vec{r} - \vec{r'}\right|} dV_j\right)\right) \quad (10)$$

3.2 Magnetic field created by ring permanent magnets

The coulombian model of the magnets is used to determine the magnetic field created by the ring magnets (30)-(33). Moreover, the devices dimensions are supposed to be chosen so that the volume pole density related to the magnetization divergence can be neglected: the rings are assumed radially thin enough. Indeed, its influence has been discussed by the authors in some complementary papers.

Consequently, each permanent magnet is represented by two charged surfaces. In the case of a radially polarized permanent magnet the magnetic poles are located on both curved surfaces of the ring and the magnetic pole surface density is denoted σ^* (Fig. 2). In the case of an axially polarized permanent magnet, the magnetic pole surface density σ^* is located on the upper and lower faces of the ring (Fig. 3).

The three magnetic field components have been completely evaluated in some previous papers. As the structure is axisymmetrical, only two components of the magnetic field created by the magnets have to be evaluated: the axial one and the radial one, and they only depend on both dimensions z and r.

The radial component $H_r(r,z)$ of the magnetic field created by the permanent magnet is given by (11).

$$H_r(r,z) = \frac{\sigma^*}{\pi\mu_0} i(1+u)\left(\varsigma(u_1) - \varsigma(u_2)\right) \quad (11)$$

where the parameter i is the imaginary number ($i^2 = -1$), with

$$
\begin{aligned}
\varsigma(u) \quad = \quad & \frac{\xi_1(-(a_1 d + b_1(c+e_1)))F^*\left[i\sinh^{-1}[\frac{\sqrt{-c+d-e_1}}{\sqrt{c+e_1+du}}], \frac{c+d+e_1}{c-d+e_1}\right]}{d\sqrt{-c+d-e_1}e_1\sqrt{\frac{d(1+u)}{c+e_1+du}}\sqrt{1-u^2}} \\
& + \frac{\xi_1(b_1 c - a_1 d)\Pi^*\left[\frac{e_1}{c-d+e_1}, i\sinh^{-1}[\frac{\sqrt{-c+d+e_1}}{c+e_1+du}], \frac{c+d+e_1}{c-d+e_1}\right]}{d\sqrt{-c+d-e_1}e_1\sqrt{\frac{d(1+u)}{c+e_1+du}}\sqrt{1-u^2}} \\
& + \frac{\xi_2\left(-(a_2 d + b_2(c+e_2))\right)F^*\left[i\sinh^{-1}[\frac{\sqrt{-c+d-e_2}}{\sqrt{c+e_2+du}}], \frac{c+d+e_2}{c-d+e_2}\right]}{d\sqrt{-c+d-e_2}e_2\sqrt{\frac{d(1+u)}{c+e_2+du}}\sqrt{1-u^2}} \\
& + \frac{\xi_2(b_2 c - a_2 d)\Pi^*\left[\frac{e_2}{c-d+e_2}, i\sinh^{-1}[\frac{\sqrt{-c+d+e_2}}{c+e_2+du}], \frac{c+d+e_2}{c-d+e_2}\right]}{d\sqrt{-c+d-e_2}e_2\sqrt{\frac{d(1+u)}{c+e_2+du}}\sqrt{1-u^2}} \\
& - \frac{\eta_3\left((a_3 d - b_3 e_3)\right)F^*\left[i\sinh^{-1}[\frac{\sqrt{-d-e_3}}{\sqrt{e_3+du}}], \frac{-d-e_3}{d+e_3}\right]}{d\sqrt{-d-e_3}(-c+e_3)\sqrt{\frac{d(1+u)}{e_3+du}}\sqrt{1-u^2}} \\
& - \frac{\eta_3(b_3 c - a_3 d)\Pi^*\left[\frac{-c+e_3}{d+e_3}, i\sinh^{-1}[\frac{\sqrt{-d+e_3}}{e_3+du}], \frac{-d+e_3}{d+e_3}\right]}{d\sqrt{-d-e_3}(-c+e_3)\sqrt{\frac{d(1+u)}{e_3+du}}\sqrt{1-u^2}} \\
& - \frac{\eta_4\left(a_4 d - b_4 e_4\right)F^*\left[i\sinh^{-1}[\frac{\sqrt{-d-e_4}}{\sqrt{e_4+du}}], \frac{-d+e_4}{d+e_4}\right]}{d\sqrt{-d-e_4}(c+e_4)\sqrt{\frac{d(1+u)}{e_4+du}}\sqrt{1-u^2}} \\
& - \frac{\eta_4(b_4 c - a_4 d)\Pi^*\left[\frac{-c+e_4}{d+e_4}, i\sinh^{-1}[\frac{\sqrt{-d-e_4}}{e_4+du}], \frac{-d+e_4}{d+e_4}\right]}{d\sqrt{-d-e_4}(-c+e_4)\sqrt{\frac{d(1+u)}{e_4+du}}\sqrt{1-u^2}}
\end{aligned} \tag{12}
$$

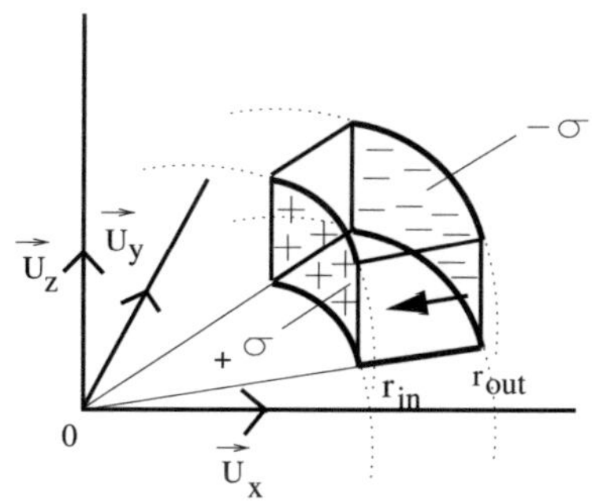

Fig. 2. Radially polarized tile permanent magnet: the inner curved face is charged with the magnetic pole surface density $+\sigma^*$ and the outer curved face is charged with the magnetic pole surface density $-\sigma^*$, the inner radius is r_{in}, the outer one is r_{out}

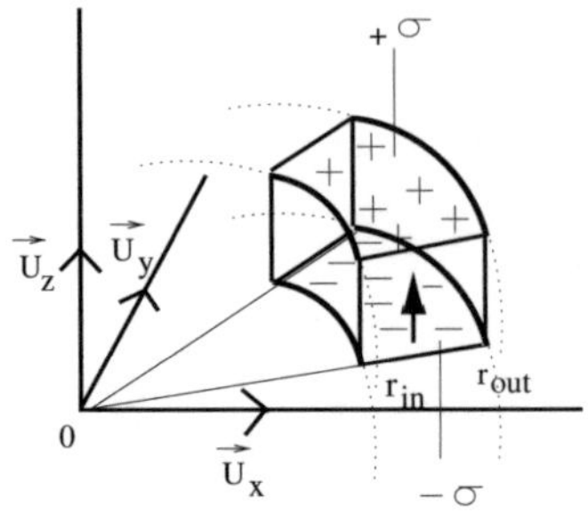

Fig. 3. Axially polarized tile permanent magnet: the upper face is charged with the magnetic pole surface density $+\sigma^*$ and the lower face is charged with the magnetic pole surface density $-\sigma^*$, the inner radius is r_{in}, the outer one is r_{out}

The axial component of the magnetic field created by the ring permanent magnet is given by (13).

$$\begin{aligned} H_z(r,z) &= \frac{\sigma^*}{\pi\mu_0}\left(-r_{in}\frac{K^*\left[\frac{-4rr_{in}}{(r-r_{in})^2+z^2}\right]}{\sqrt{(r-r_{in})^2+z^2}}\right) \\ &+\frac{\sigma^*}{\pi\mu_0}\left(r_{in}\frac{K^*\left[\frac{-4rr_{in}}{(r-r_{in})^2+(z-h)^2}\right]}{\sqrt{(r-r_{in})^2+(z-h)^2}}\right) \\ &-\frac{\sigma^*}{\pi\mu_0}\left(r_{in}\frac{K^*\left[\frac{-4rr_{in}}{(r-r_{in})^2+z^2}\right]}{\sqrt{(r-r_{in})^2+z^2}}\right) \\ &+\frac{\sigma^*}{\pi\mu_0}\left(r_{in}\frac{K^*\left[\frac{-4rr_{in}}{(r-r_{in})^2+(z+h)^2}\right]}{\sqrt{(r-r_{in})^2+(z+h)^2}}\right) \end{aligned} \tag{13}$$

$$\xi_i = \sqrt{\frac{d(-1+u)}{c+e_i+du}} \tag{14}$$

$$\eta_i = \sqrt{\frac{d(-1+u)}{e_i+du}} \tag{15}$$

where $K^*[m]$ is written in terms of the incomplete elliptic integral of the first kind by (16)

$$K^*[m] = F^*[\frac{\pi}{2}, m] \tag{16}$$

$F^*[\phi, m]$ is written in terms of the elliptic integral of the first kind by (17):

$$F^*[\phi, m] = \int_{\theta=0}^{\theta=\phi}\frac{1}{\sqrt{1-m\sin(\theta)^2}}d\theta \tag{17}$$

Parameters	
a_1	$r_{in}rz$
b_1	$-r_{in}^2z$
c	$r^2+r_{in}^2$
d	$-2rr_{in}$
e_1	z^2
a_2	$-r_{in}r(z-h)$
b_2	$r_{in}^2(z-h)$
e_2	$(z-h)^2$
a_3	$r_{in}rz$
b_3	$-r_{in}^2z$
e_3	$r^2+r_{in}^2+z^2$
a_4	$r_{in}r(-z-h)$
b_4	$-r_{in}^2(-z-h)$
e_4	$r^2+r_{in}^2+(z+h)^2$

Table 1. Definition of the parameters used in (12)

$\Pi^*[n,\phi,m]$ is written in terms of the incomplete elliptic integral of the third kind by (18)

$$\Pi^*[n,\phi,m] = \int_0^{\phi} \frac{1}{(1-n\sin(\theta)^2)\sqrt{1-m\sin(\theta)^2}}d\theta \tag{18}$$

The parameters used in (12) are defined in Table 1. As a remark, an imaginary part, which has no physical meaning, may appear because of the calculus noise of the calulation program (Mathematica). Therefore, the real part only of $H_r(r,z)$ must be considered.

4. The magnetic pressure

The magnetic pressure determines the shape of the free boundary surface of the ferrofluid. Moreover, the assumptions for the calculations have been described in the method description section (2).
Then, the magnetic pressure is defined as follows:

$$p_m(r,z) = \mu_0\mathbf{M_s}.\vec{H}(r,z) = \mu_0 M_s\sqrt{H_r(r,z)^2 + H_z(r,z)^2} \tag{19}$$

where the evaluation of both magnetic field components $H_r(r,z)$ and $H_z(r,z)$ have been given in the previous section and where M_s is the magnetization of a magnetic particle of the ferrofluid. Thus, the magnetic pressure is the interaction of the magnetic field created by the permanent magnets and the particle magnetization. Eventually, for hydrodynamic pressures which equal zero or have low values, the seal free boundary surface is a magnetic iso-pressure surface.
Fig. 4 shows a three-dimensional representation of the magnetic pressure created by two in opposed directions radially polarized ring permanent magnets. This magnetic pressure can also be seen as a magnetic energy volume density, and can be given either in N/m^2 or in J/m^3.

The magnetic pressure $p_m(r,z)$ has been evaluated with (19). Figure 4 shows that the magnetic pressure is higher next to the ring magnets, especially where both the magnetic field and its gradient are the strongest.

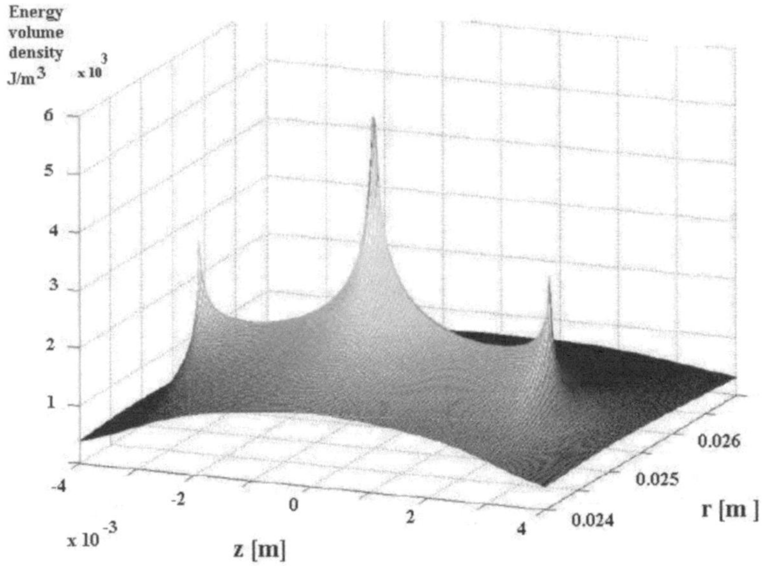

Fig. 4. Three-dimensional representation of the magnetic pressure in front of two in opposed directions radially polarized ring permanent magnets.

This representation also shows that the potential energy is concentrated in a very small ferrofluid volume. As a consequence, it gives information on what quantity of ferrofluid should be used to create a ferrofluid seal. When a large quantity of ferrofluid is used, then the ferrofluid seal is thick and the potential energy increases. But the viscous effects become an actual drawback with regard to the dynamic of the inner moving cylinder. When too small an amount of ferrofluid is used, then the viscous effects disappear but the main properties of the ferrofluid seal (damping, stability, linearity,...) disappear as well. So, an adequate quantity of ferrofluid corresponds to a given geometry (here two ring permanent magnets with an inner non-magnetic cylinder) in order to obtain interesting physical properties with very little viscous effects.
The concept of potential energy thus appears, which is defined by (24):

$$E_m = -\int\int\int_{(\Omega)} p_m(r,z)dV \tag{20}$$

where (Ω) is the ferrofluid seal volume. Indeed, this potential energy, given in J, allows the calculation of the seal mechanical properties and will be used throughout the remainder of this chapter.

5. Shape of the ferrofluid seal

As the shape of the seal depends on the magnetic pressure in the structure it naturally depends on the magnetic structure which creates the magnetic field. This section intends to describe some structures and discuss the corresponding seals.

5.1 Basic structure

Figure 5 shows the structure constituting the base of all the devices presented. It consists of three outer stacked rings, of an inner non-magnetic piston and of ferrofluid seals. The piston is radially centered with the rings. The rings' inner radius, r_{in}, equals 25 mm and their outer radius, r_{out}, equals 28 mm. The rings can be either made with permanent magnet -as here the middle ring- or with non-magnetic material -like the upper and lower rings-. The ferrofluid seals are located in the air gap between the piston and the rings. The whole section will discuss the seal number, their position and the polarization direction of the ring magnets. Furthermore, the radial component of the magnetic field created by the ring permanent magnets is also presented for each studied configuration in order to illustrate the link with the seal shape.

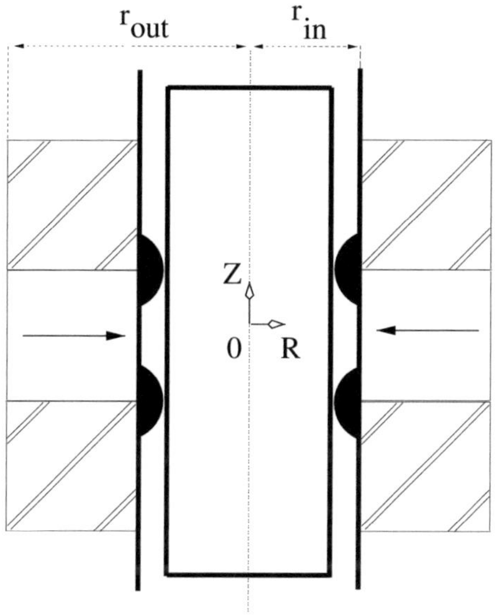

Fig. 5. Basic structure: three outer rings (permanent magnet or non-magnetic) radially centered forming an air gap with an inner non-magnetic piston. Ferrofluid seals located in the air gap. $r_{in} = 25mm, r_{out} = 28mm$.

5.2 Single magnet structures

The first structure considered corresponds exactly to the configuration shown in Fig.5, which is the simplest one which can be used. All the rings have the same square cross-section with a 3 mm side. The middle ring is a radially polarized permanent magnet and the upper and lower rings are non-magnetic. The magnetic field created by the magnet in the air gap is calculated along the Z axis at a 0.1 mm distance from the rings and its radial component H_r is plotted versus Z (Fig.7). As a remark, H_r is rather uniform in front of the magnet and two gradients are oberved in front of the magnet edges. Besides, the magnetic pressure in the air gap is calculated and plotted on Fig.6 as well: the iso-pressure lines determine the seal contour, its size depends on the ferrofluid quantity. Indeed, the ferrofluid goes in the regions of high energy first (dark red ones). For an increasing volume of ferrofluid, the latter fills the regions of decreasing energy (from the red contours to the blue ones). So, for seals thicker than 0.5 mm, the seal expands along the whole magnet height. A smaller volume of ferrofluid would lead to the creation of two separate seals which would be quite thin and thus, to poor mechanical properties. This results from the shape of the magnet section: if it were rectangular along Z instead of square, two separate seals would appear too. The point is that the ferrofluid seeks the regions of both intense field gradient and high magnetic energy.

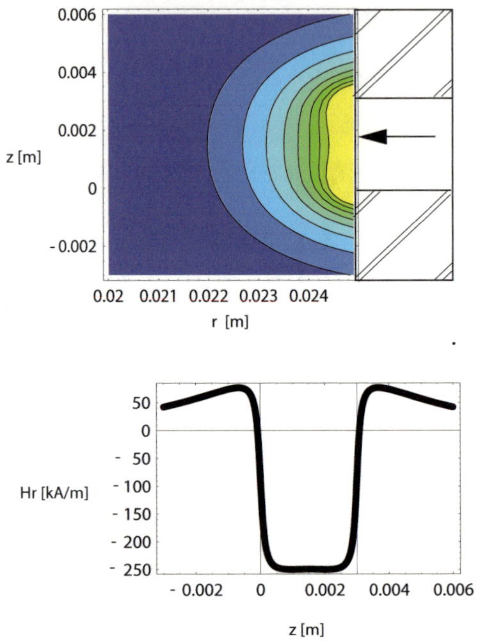

Fig. 6. Top right: upper and lower non-magnetic rings, middle ring permanent magnet radially polarized. Top left: magnetic pressure in front of the rings. Bottom: H_r along the Z axis at a 0.1 mm distance from the rings.

If the polarization direction of the ring magnet becomes axial, Fig. 7 shows that the magnetic pressure is at first sight rather similar to the previous one. Nevertheless, the seal shape differs, especially for large ferrofluid volumes. Moreover, the radial component of the magnetic field is no longer uniform in front of the magnet and presents instead a rather large gradient all over the magnet length and the non-magnetic rings.

5.3 Double magnet structures

The purpose is to describe how the seal shape and properties evolve when the magnetic structure becomes gradually more complicated but also maybe more efficient.
Then the structures considered are obtained by stacking two ring permanent magnets. The rings are identical in dimensions but are opposedly polarized, either radially as in Fig. 9 or axially as in Fig. 8. The magnetic field in both cases is evaluated by superposing the single magnet fields.
As a consequence, each radial magnet creates a region of uniform field in front of itself and the field directions are opposite. The field intensity in each uniform region is higher than in the single magnet structure because the leakage is decreased. Then, three field gradients exist, and the one that appears in front of the magnets' interface is twice as high as those at the edges. From the gradient point of view, Fig. 9 can be compared with Fig. 6, and the former will prove more useful because the gradient is steeper. The axial double structure creates progressive field gradients with no peculiar interest.

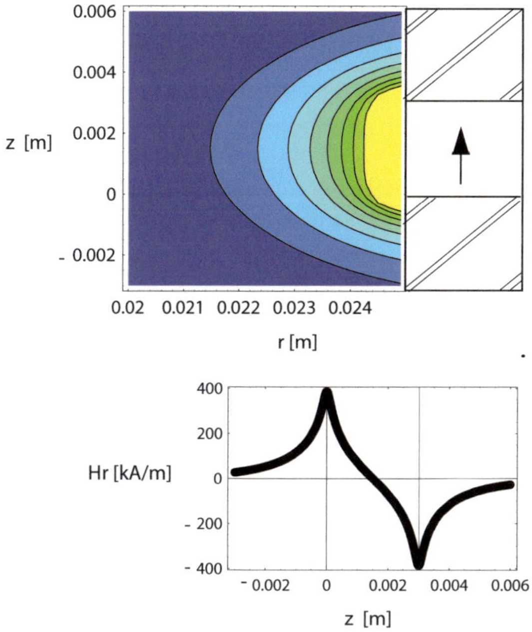

Fig. 7. Top right: upper and lower non-magnetic rings, middle ring permanent magnet axially polarized. Top left: magnetic pressure in front of the rings. Bottom: H_r along the Z axis at a 0.1 mm distance from the rings.

Moreover, the repartition of the magnetic energy density in the double magnet structure is not the superposition of the ones in the single magnet structures because the expression of the energy depends on the square of the field. Although the repartitions for radial and axial magnets seem alike at first sight, the radial structure is "more energetic" and its magnetic energy decreases slower at an increasing distance from the magnets. Nevertheless, the maximum energy density is in front of the magnets' interface and the ferrofluid seal will be located there. Eventually, the seal axial length in the single magnet structures is smaller than in the double magnet structures. Besides, the seal energy density is approximately doubled for the radially polarized magnets.

The evolution of the magnet shape can be observed when the axial dimension of the ring magnet is varied. For instance, Fig 10 shows the magnetic iso-pressure lines when both magnet heights are $h = 2$ mm, $h = 2.5$ mm , $h = 3$ mm, $h = 3.5$ mm, $h = 4$ mm and $h = 4.5$ mm respectively.

As expected, the magnetic field in the air gap increases when the magnet height increases. Figure 10 clearly shows that the longer the ring permanent magnet heights are, the stronger the magnetic field in the air gap is. However, as shown in Fig 10, the ferrofluid seal decreases in height when the ring permanent magnet heights increase. This implies that for a structure that requires a small ferrofluid seal with the greatest static capacity, the height of the ring permanent magnets must be greater than their radial widths.

5.4 Triple magnet structures

With the same reasoning as in previous section, the structures presented here are constituted of three stacked ring permanent magnets. Thus, the number of possible configurations increases. However, it isn't necessary to study all possibilities and the most interesting ones

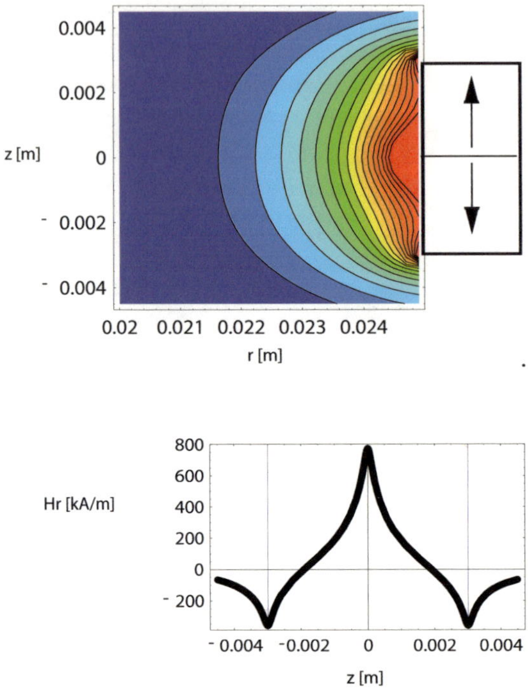

Fig. 8. Top right: Two ring permanent magnets with opposed axial polarization. Top left: magnetic pressure in front of the rings. Bottom: H_r along the Z axis at a 0.1 mm distance from the rings.

have been selected. Then, two main kinds of structures are brought out: structures with alternate polarizations and structures with rotating polarizations.

5.4.1 Alternate polarizations

The three rings are radially polarized: two have the same polarization direction, the third has the opposite direction and is located between both previous ones. Thus, this structure is the extension of the preceding double magnet structure, which can be generalized to even more ring magnets. Now, in the case of three ring magnets, the number of seals and their shape is closely related to the middle magnet axial height an to the ferrofluid total volume (Fig 11)
Indeed, the left top plot in Fig 11 shows that for three magnets of same dimensions and a small amount of ferrofluid, two seals are formed in front of the ring interfaces. Their axial dimension is rather small and they are very energetic. When the middle magnet height is decreased both seals get closer and join to form a single seal in front of the middle magnet. This seal is less energetic: it is normal as the middle magnet volume is decreased. Meanwhile, two secondary small seals appear at the extremities of the structure.
For a larger ferrofluid volume, a single seal is formed in front of the whole structure whatever the middle magnet dimensions and its energy is linked to the total magnet volume.

5.4.2 Rotating polarizations

The three ring magnets polarization directions are now alternately axial and radial and a 90 degrees rotation is observed from one magnet to its neighbor. Such a progressive rotation of the magnetic polarization is to put together with Halbach patterns (34).

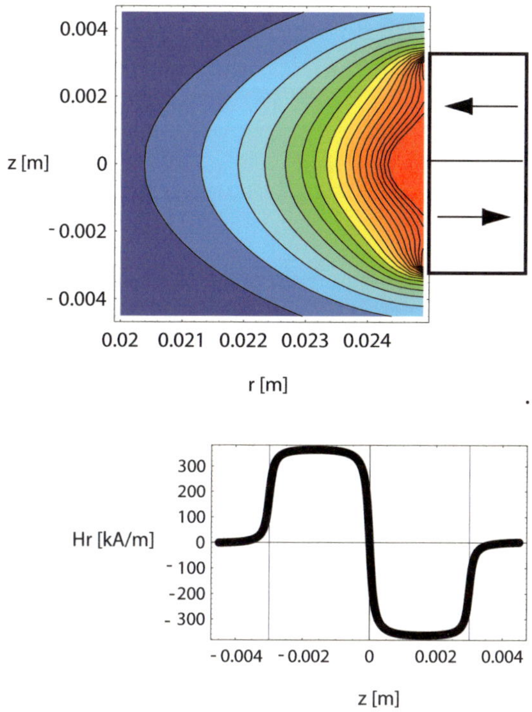

Fig. 9. Top right: Two ring permanent magnets with opposed radial polarization. Top left: magnetic pressure in front of the rings. Bottom: H_r along the Z axis at a 0.1 mm distance from the rings.

Two kinds of configurations are possible with three ring magnets: either two top and bottom axially and one middle radially polarized rings (Fig. 12) or the dual two top and bottom radially and one middle axially polarized rings (Fig. 13).
The energy density color plots show that two ferrofluid seals form in front of the magnets' interfaces. Their shape is rather similar in both structures and these seals are magnetically quite energetic. So, they will have "good" mechanical properties (such as a great radial stiffness for example). However, the magnetic field radial component proves different in each structure: it is fairly uniform in front of the middle radially polarized magnet whereas it varies with no particularly interesting properties in in front of the axially polarized one. As a consequence, the structure of Fig. 12 seams to be more useful for applications as the zone of uniform magnetic field can be optimized.
Indeed, the axial height of the middle magnet can be varied. For instance, the middle magnet is twice as high as each other magnet in Fig. 14 and half as small in Fig. 15. As a result, the magnetic field radial component is always rather uniform in front of the radially polarized magnet. So, the uniformity area increases with the height of the radially polarized magnet, but the field intensity decreases when the height of the radially polarized magnet is larger than the height of the axially polarized ones. Besides, when the height of the axially polarized magnets becomes too small the ferrofluid tends to expand over their whole axial length if the ferrofluid volume is sufficient. Thus, the seals can become quite large and they are well-fixed to the structure and have high mechanical performances because of their high magnetic energy and the steep field gradients.

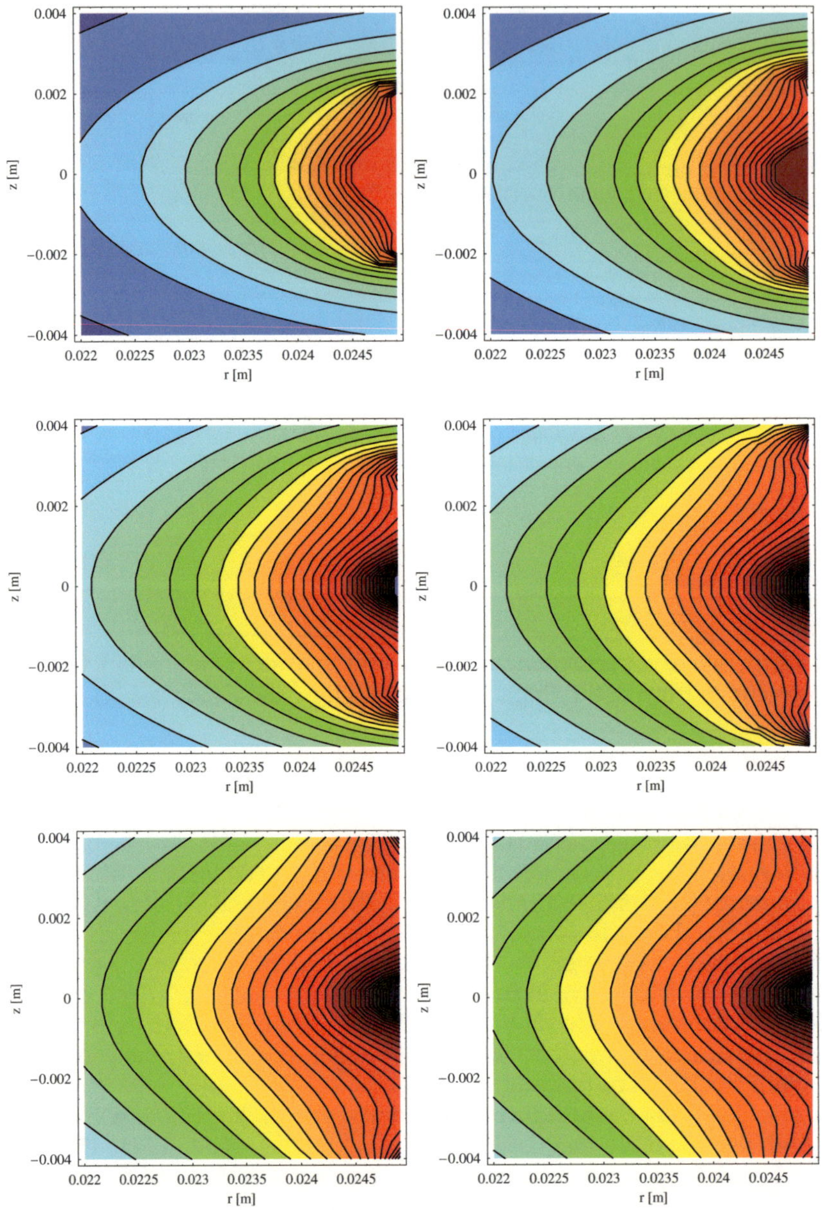

Fig. 10. Magnetic iso-pressure lines for increasing ring magnet heights; $h = 2$ mm (top left), $h = 2.5$ mm (top right), $h = 3$ mm (middle left), $h = 3.5$ mm (middle right), $h = 4$ mm (bottom left), $h = 4.5$ mm (bottom right).

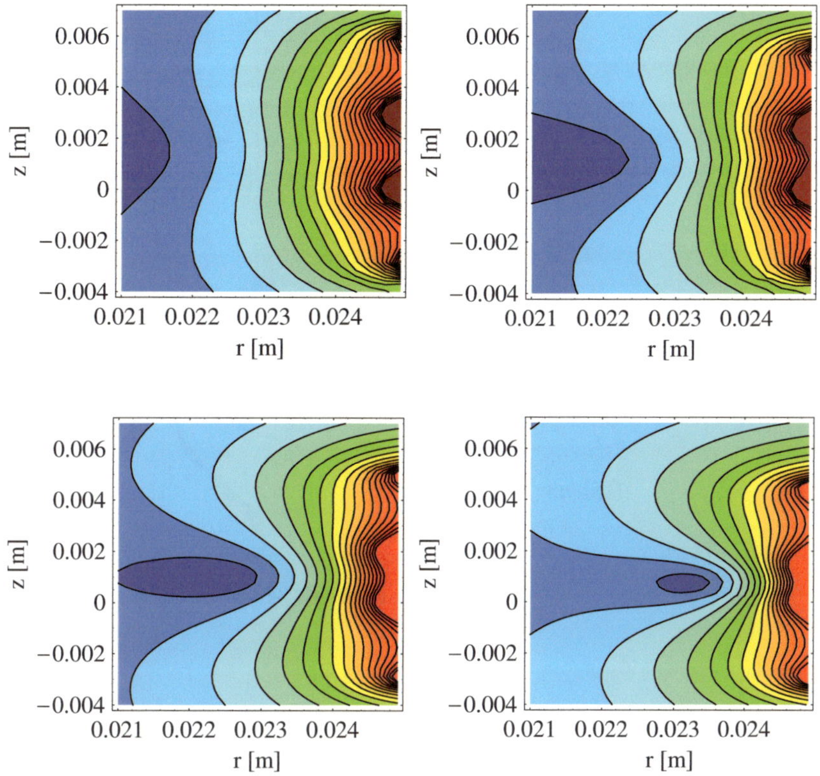

Fig. 11. Three magnet alternate structure. Magnetic iso-pressure lines in the air gap for a decreasing height of the middle magnet $h = 3$ mm, $h = 2.5$ mm, $h = 2$ mm, $h = 1.5$ mm.

Inversely, when the middle radially polarized magnet height becomes too small, both seals gather to form a single high energetic one which expands over the whole height of the middle magnet.

6. Mathematical description of the ferrofluid seal

Writing the whole mathematical equations describing all the ferrofluid properties doesn't lead to easily workable expressions. This exercise is still very complicated even if the only equations considered are the ones related to the magnetic pressure. Nevertheless, in the double magnet structure with radially polarized ring magnets, the seal shape can be described in a very good approximation by an equation of ellipse.
This allows some further characterization of the seal and especially its behavior when it gets crushed and works as a bearing.

6.1 Shape of the free ferrofluid seal

This section considers the shape of the seal when its boundary surface is totally free, so in absence of the inner moving part or for volumes small enough not to reach the inner part.

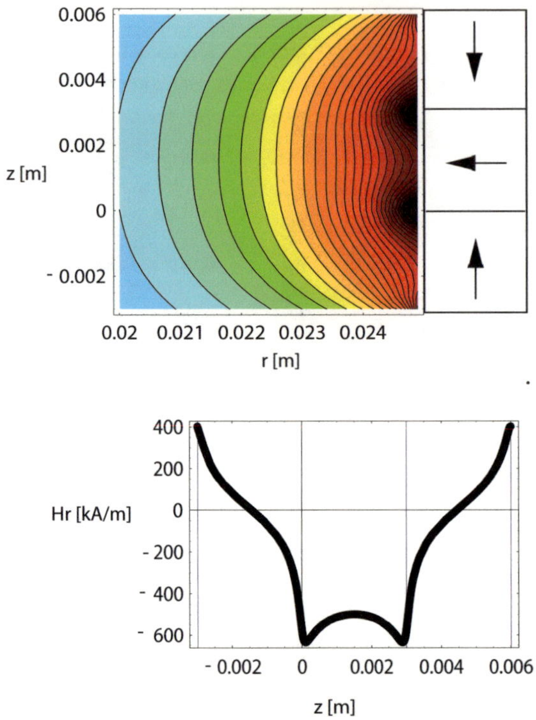

Fig. 12. Top right: axially polarized upper and lower rings, radially polarized middle ring. Top left: magnetic pressure in front of the rings. Bottom: H_r along the Z axis at a 0.1 mm distance from the rings.

Ellipse	$\mathbf{a_i}$	$\mathbf{b_i}$	$\mathbf{r_i}$	**error**
5%E	0.00025	0.000275	0.025	0.5%
10%E	0.00027	0.000297	0.025	0.9%
15%E	0.00029	0.000319	0.025	1.4%

Table 2. Parameters describing the free boundary ferrofluid seal shape.

For example, the contour of the ferrofluid seal in Fig 16 when its thickness is smaller than 0.4 mm can be written in terms of the following equation of an ellipse (21).

$$\frac{(r-r_i)^2}{a_i^2} + \frac{z^2}{b_i^2} = 1 \tag{21}$$

The parameter values are given in Table (2) when r is between 24.6 mm and 25 mm. Moreover, E is the total magnetic energy of the volume of ferrofluid located between 24.6 mm and 25 mm. Table (2) shows the proportion of energy located in the seal of considered dimensions. The error between the equations of ellipse and the real contour shape of the ferrofluid seal is also given.

When the ferrofluid volume increases and the seal goes further than $r = 24.6$ mm towards the axis, its shape changes and is no longer a portion of an ellipse. This gives the limits of our

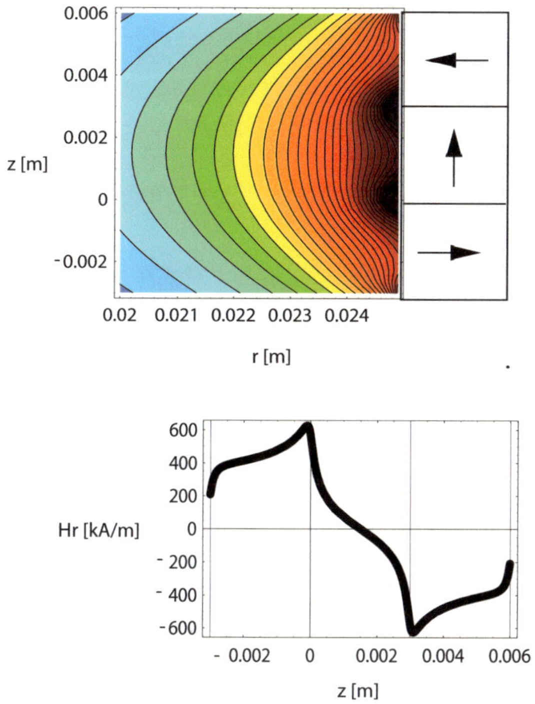

Fig. 13. Top right: radially polarized upper and lower rings, axially polarized middle ring. Top left: magnetic pressure in front of the rings. Bottom: H_r along the Z axis at a 0.1 mm distance from the rings.

modelling. However, it can be noted that the most energetic seals are the ones whose contour remain a portion of an ellipse.

6.2 Shape of the crushed ferrofluid seal

So, the model can be used further to describe the seal in presence of the inner moving part. As previously said, this part is a non magnetic cylinder forming an air gap with the permanent magnet structure. The device dimensions and the ferrofluid volume are chosen so that the seal plays its watertightening role. This means that the seal is crushed by the inner part (Fig.17). The point is that for adequate ferrofluid volumes, the ferrofluid seal contour can still be described in terms of an equation of ellipse.
If the same ferrofluid volume is considered as in the previous section, then adding the inner part results in the fact that the ferrofluid volume can't remain in the radial space and is driven away in the axial direction towards the free space of the air gap. As the geometry is axisymmetrical, the new seal contour is symmetrical and can be described as a truncated ellipse (Fig. 17).
The point is that when the seal shape changes because it is crushed its magnetic energy decreases. This is illustrated by the values of Table 3 which present the energy reduction when the inner cylinder radius grows of respectively 0, 1 mm, 0, 15 mm and 0, 2 mm. This is of importance as the seal properties or performances are directly related to its magnetic energy.
It can be noted that a 0.2 mm increase of the inner cylinder radius causes a 68% energy reduction.

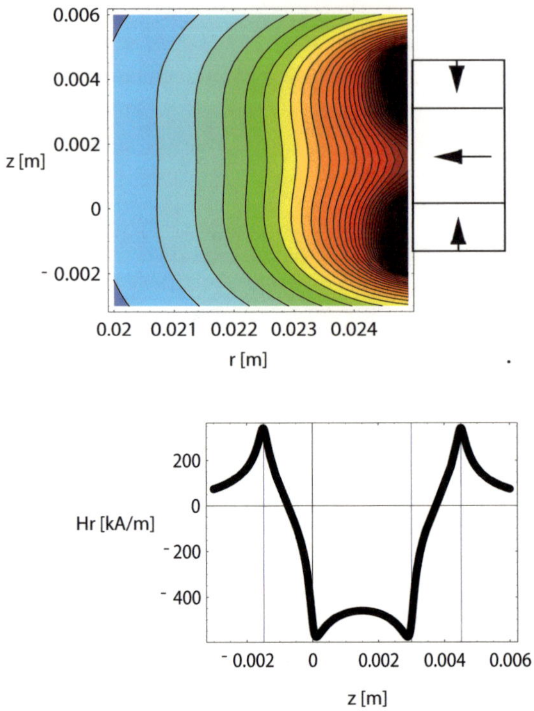

Fig. 14. Top right: axially polarized upper and lower rings, radially polarized middle ring, the axial magnet height is the half of the radial magnet one. Top left: magnetic pressure in front of the rings. Bottom: H_r along the Z axis at a 0.1 mm distance from the rings.

Radius increase	Energy reduction
0, 1 mm	13%
0, 15 mm	35%
0, 2 mm	68%

Table 3. The inner cylinder radius increase causes an energy reduction in the seal.

This modelling is useful to evaluate the mechanical properties and performances of ferrofluid seals.

7. Capacity of the ferrofluid seal

The static capacity of the ferrofluid seals is an important characteristic as it determines the maximal axial pressure they can undergo without losing their tightness property. It depends naturally on the chosen configuration and this issue will be discussed later on.

In some applications, the pressures on each side of the seal can be pretty different. Consequently, a pressure gradient appears: the ferrofluid seal is deformed and can be pierced (or blown) along the moving part for too high pressures. Therefore, the knowledge of the seal capacity is necessary or, inversely, the seal may be dimensionned to have a given capacity.

So, the configuration considered for the calculation corresponds to the case when a cylindrical air gap appears in the seal along the cylinder because of an applied pressure on one side of the seal (Fig.18).

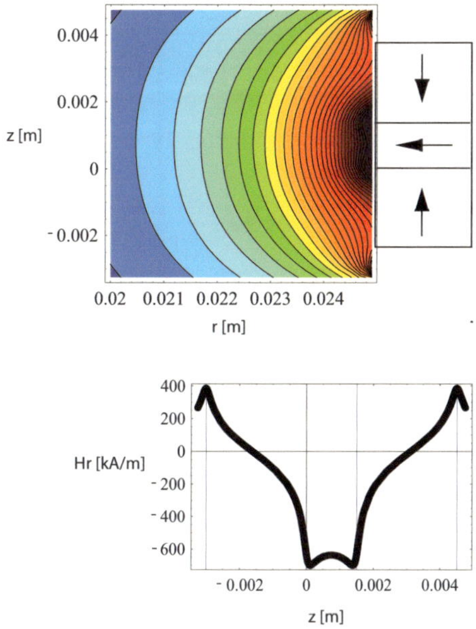

Fig. 15. Top right: axially polarized upper and lower rings, radially polarized middle ring, the axial magnet height is the double of the radial magnet one. Top left: magnetic pressure in front of the rings. Bottom: H_r along the Z axis at a 0.1 mm distance from the rings.

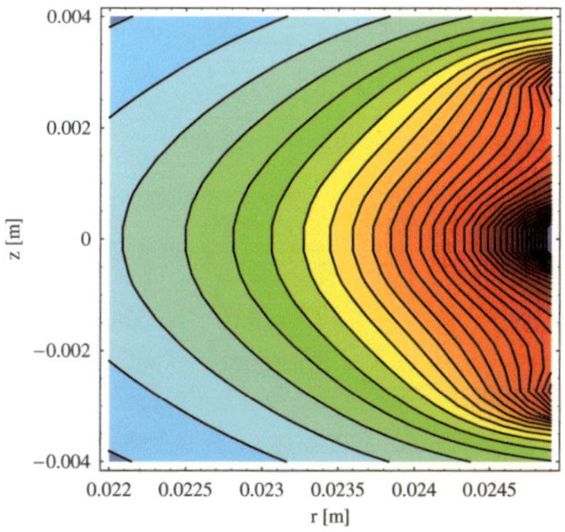

Fig. 16. Magnetic iso-pressure lines for $h = 3$ mm.

The ferrofluid seal capacity is determined in two steps. First, the potential energy of the ferrofluid seal is evaluated without any hole in the seal. The numerical integration of (24) leads to a first value of the potential energy, $E_m(1)$. Second, the potential energy, $E_m(2)$, of the ferrofluid seal with the hole is evaluated with another numerical integration. The energy difference, ΔE_m, corresponds to the pressure work $\delta W(P)$ and satisfies (22):

$$\Delta E_m = E_m(1) - E_m(2) = \delta W(P) = P_{lim} S d \tag{22}$$

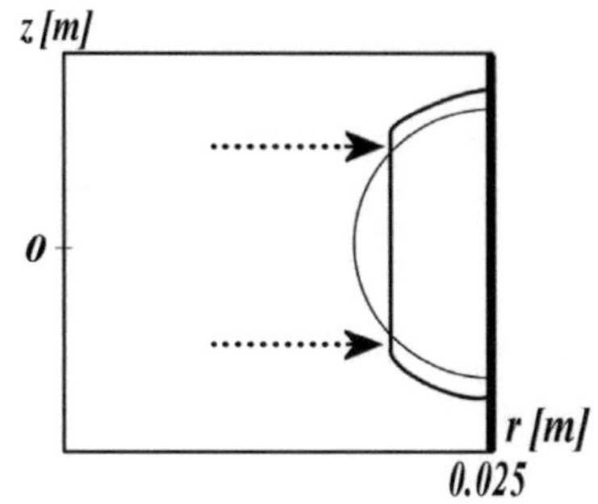

Fig. 17. Crushed ferrofluid seal.

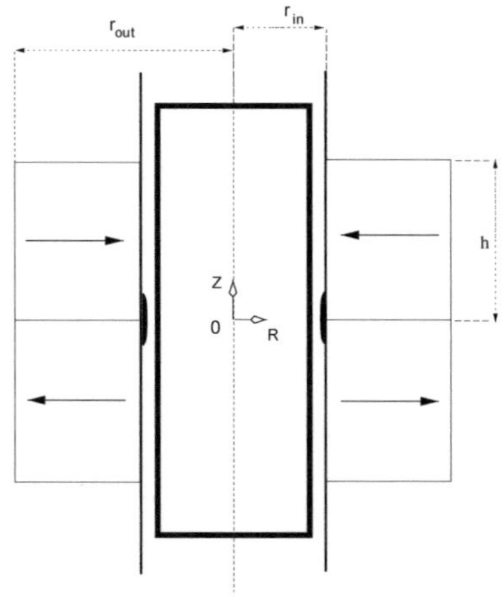

Fig. 18. Seal pierced along the inner cylinder.

Seal thickness	**Volume**	H_{lim}
$0,1$ mm	4.7 mm^3	700 000 A/m
$0,3$ mm	12 mm^3	600 000 A/m
$0,5$ mm	21 mm^3	450 000 A/m

Table 4. Seal volume and Magnetic field for a given ferrofluid seal thickness

where S is the surface of the air gap and d is the thickness of the hole. Consequently, the capacity P_{lim} is given by (23):

$$P_{lim} = \frac{\delta W(P)}{Sd} \tag{23}$$

Numerical values have been calculated with several ring inner radii and are shown on Fig.19 where the ferrofluid seal thickness is defined as the axial thickness of the ferrofluid in the air gap.

Figure 19 also illustrates that thin ferrofluid seals resist to higher pressure gradients than thick ones. Besides, Table 4 gives the ferrofluid volume corresponding to each seal thickness and the lowest magnetic field H_{lim} in the seal volume.

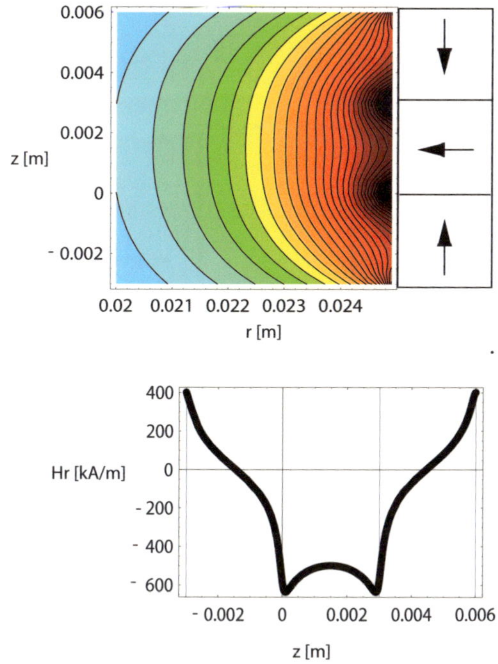

Fig. 19. The static capacity P_{lim}[Pa] of the ferrofluid seal depends on its thickness $ft[mm]$

8. Ferrofluid seals as bearings

One of the issues of the use of ferrofluid seals is their behavior as radial bearings. Indeed, when the inner moving part is no longer radially centered the ferrofluid seal exerts a pull back centering force and the corresponding radial stiffness can be evaluated.

8.1 Radial stiffness evaluation

The considered position of the inner moving part is illustrated both on Fig.20 and Fig.21.

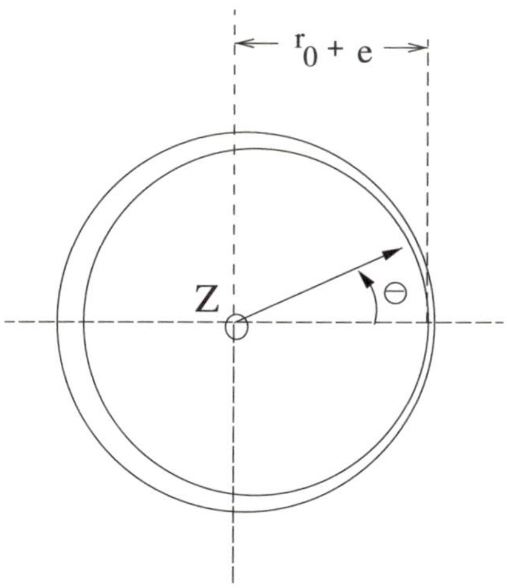

Fig. 20. Decentered inner cylinder: cross-section.

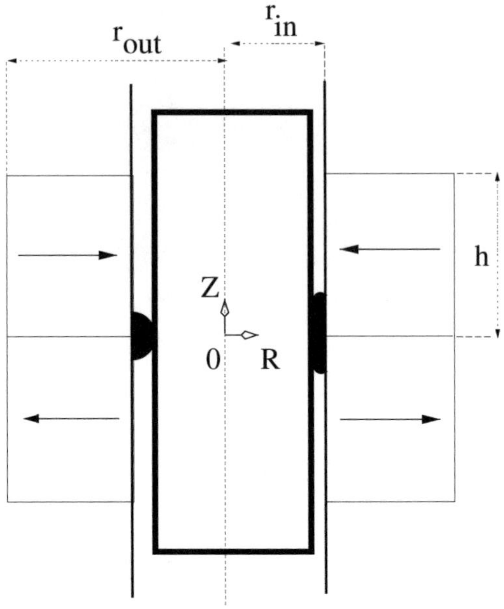

Fig. 21. Decentered inner cylinder: crushed ferrofluid.

They show that the air gap radial dimension depends on the angle θ. Moreover, the problem is now fully three-dimensional. Indeed, when the inner cylinder moves radially, the ferrofluid is more crushed on the narrow air gap side and it is driven away not only in the axial direction but also around the inner cylinder towards the regions of broader airgap. The new ferrofluid repartition is achieved according to energy considerations.
The evaluation of the radial stiffness is carried out in two steps.
First, the potential energy $E_m(1)$ is calculated when the non-magnetic cylinder is centered (Fig.1) with (24):

$$E_m = -\int\int\int_{(\Omega)} p_m(r,z)dV \tag{24}$$

Then, the potential energy $E_m(2)$ is calculated when the non-magnetic cylinder is decentered (Fig.21). In this second configuration, the limits of the integrals depend on the angle θ. Thus, the potential energy $E_m(2)$ is determined with (25):

$$E_m(2) = \int_0^{2\pi} \int_{r_0+e\cos(\theta)}^{r_{in}} \int_{-z_0}^{z_0} e_m(r,z)rdrd\theta dz \tag{25}$$

where r_0 and e are determined by the equation of the decentered circle (26):

$$r(\theta) = r_0 + e\cos(\theta) \tag{26}$$

where $r_0 = 24.7$ mm and $e = 0.1$ mm.
The radial force F_r is then calculated with (27):

$$F_r = \frac{E_m(1) - E_m(2)}{2\Delta r} \tag{27}$$

where $\Delta r = 1$ mm is the radial decentering of the inner cylinder.

Eventually, the radial stiffness k_r is determined by (28):

$$k_r = \frac{F_r}{\Delta r} \tag{28}$$

As a result, the numerical value of the radial stiffness is $k_r = 5.6$ N/mm for a ferrofluid seal of 0.3 mm thickness.

9. Comments and discussion

Thus far, this chapter described how ferrofluid seals are formed in magnetic structures and which shape and characteristics they have. Now, this section intends to comment and compare these structures with regard to the seals properties and performances in relation with the intended kind of application.
Indeed, the design depends on the goal and two major trends can be highlighted: the application consists in creating a seal for tightness purposes only or it intends to create a seal and a useful magnetic field, as in voice coil motors for example.

9.1 General purpose

In each case the seal mechanical properties are one of the issues, if not the only one.
Now, the structures presented can be very simple or more elaborate. Of course, the simplest ones are the single magnet structures, which lead to rather similar seals whatever the polarization direction. However, axially polarized ring magnets are cheaper and more easily available than radially polarized ones because the axial polarization is technically far easier to achieve. So, for simple and cheap solutions the choice should be a single axially polarized ring magnet.
Though, the mechanical robustness of the seal is linked to the magnetic pressure and the magnetic potential energy in the seal. Therefore, the high performance structures are not the simplest ones. Indeed, the previous sections show that multi-magnet structures create higher magnetic fields as well as larger field gradients to trap and fix the ferrofluid and that the thus formed seals are in areas of higher magnetic pressure.
The air gap dimensions can be chosen but must fulfill mechanical constraints: the movement of the inner part must be possible with the known mechanical clearances, the machining tolerances of the parts must be taken into account...Besides, the performances of a ferrofluid seal depend on the magnetic structure dimensions and on the seal thickness. For instance, the triple magnet structures of Fig. 15 or Fig. 12 lead to efficient seals thinner than the double magnet structures of Fig. 9 because the region of high magnetic pressure is radially thinner. Then, for given air gap dimensions, the type of magnetic structure is chosen with regard to the intended class of performances (low or high...) and then the ferrofluid volume is determined to achieve the optimal properties for this structure. And the ferrofluid shape depends on both the magnetic pressure and the ferrofluid volume.
Furthermore, "optimal properties" often means that the seal is robust and can resist rather high axial pressures, so, that its capacity is high. The preceding sections have shown that this is the case when the seal exerts an opposing force, related itself to energy variations in the ferrofluid seal. A great energy variation creates a high intensity force. Therefore, a high capacity is achieved when the iso-pressure lines are very close together in the axial direction: the energy variation will be large for axial displacements of the seal contour thus creating a considerable force. The preceding calculations and their illustrations show that thin seals are more energetic. Then, if a thin seal must achieve the tightness the air gap should be rather thin either. But the notion of "thin air gap" or "thin seal" is a relative one, defined in fact

with regard to the magnetic structure axial dimension. Indeed, in the taken examples a thin seal corresponds to a seal whose radial dimension is around the tenth of a ring magnet axial thickness. So, the seal can be the twentieth or the thirtieth of the axial total dimension for multi-magnets devices. And the axial dimension of the magnets is a tunable parameter.
As shown in Fig 10, the ferrofluid seal axial height decreases when the ring magnet height increases. This implies that for a structure that requires an axially narrow ferrofluid seal with a high static capacity, the height of the ring magnets must be larger than their radial width.
Moreover, the radial magnet dimension can also be tuned to achieve a high static capacity. Indeed, if the radial dimension is increased, the magnetic field intensity in the air gap is increased and so is the potential energy.
Thus, the design will generally be the result of trade-offs and the modelling presented will greatly help the device dimensioning and optimizing.
As the seal has a radial bearing behavior too, it must be characterized with regard to this function. So, the maximal radial decentering without creation of a hole in the seal for a given ferrofluid volume has to be determined. This can also be evaluated thanks to the presented modelling. For example, for a ferrofluid volume corresponding to a free boundary seal of 0.4 mm radial thickness and an inner moving part creating a 0.3 mm wide air gap, the maximal radial displacement of the moving part without tightness loss is 0.14 mm (ring magnets dimensions: $r_{in} = 25$ mm, $r_{out} = 28$ mm, $h = 3$ mm).

9.2 Special applications: transducers

Now, if the created magnetic field itself is of importance, like in actuators or sensors, the intensity and the uniformity of the field in the air gap are additionnal issues.
Indeed, voice coil motors require the creation of a force which is as exactly as possible the image of the driving current. This can be achieved if the magnetic field created by the permanent magnets and applied to the coil is uniform. Moreover, the efficiency is proportionnal to the field level.
These requirements are the same in transducers like seismometers, in which ferrofluid seals can be used as bearings and guiding systems. But, the field level will characterize the sensor sensitivity rather than an efficiency.
And they are even more important in special voice coil motors such as loudspeakers, in which the ferrofluid seal is used to ensure the airtightness, to transfer the heat from the moving part to the steady one, to work as a radial bearing and to replace the loudspeaker suspension thus contributing to the improvement of the loudspeaker linearity. Therefore, the structures presented in Figs. 5, 9 and 12 with radialy magnetized magnets and creating uniform magnetic field radial components are the most useful for loudspeaker appications. The triple magnet structures, though more complicated, offer the possibility of a multi-criteria optimization which will make the design easier and is enabled by the proposed analytical formulations and modelling.

10. Conclusion

Thus this chapter presents how ferrofluids can be used to form seals in various ironless magnetic structures. As preamble, the analytical evaluation of the magnetic field created by ring permanent magnets is given. Then, a simple analytical model is proposed to describe the static behavior of the ferrofluid seals: their shape, their static capacity, their working as bearings.
The originality of the considered structures lies in the fact that they are made of permanent magnet only, without any iron on the static part nor on the moving part which is a non magnetic cylinder.

Various magnetic structures are presented: some are very simple with a single ring magnet, some more complicated with three ring magnets. For each of them, the forming of one or more ferrofluid seal is decribed, with regard to the magnetic structure geometry and the ferrofluid volume. The properties and performances of the seals are also characterized and compared thanks to the concepts of magnetic pressure and magnetic potential energy. The modelling proves to be very useful and efficient to illustrate the magnet role and deduct design rules for ferrofluid seals with given mechanical characteristics.
Some information is also given for peculiar application such as voice coils motors in which the magnetic structure plays the double role of fixing the ferrofluid seal and creating a magnetic field with special properties such as uniformity.

11. References

[1] J. Bajkowski, J. Nachman, M. Shillor, and M. Sofonea, "A model for a magnetorheological damper," *Mathematical and computer modelling*, vol. 48, pp. 56–68, 2008.

[2] R. E. Rosensweig, *Ferrohydrodynamics*. Dover, 1997.

[3] O. Doutres, N. Dauchez, J. M. Genevaux, and G. Lemarquand, "On the use of a loudspeaker for measuring the viscoelastic properties of sound absorbing materials," *Journal of the Acoustical Society of America Express Letters*, vol. 124, no. 6, pp. EL335–EL340, 2008.

[4] X. Li, K. Yao, and Z. Liu, "Cfd study on the magnetic fluid delivering in the vessel in high-gradient magnetic fields," *Journal of Magnetism and Magnetic Materials*, vol. 320, pp. 1753–1758, 2008.

[5] G. S. Park and K. Seo, "New design of the magnetic fluid linear pump to reduce the discontinuities of the pumping forces," *IEEE Trans. Magn.*, vol. 40, pp. 916–919, 2004.

[6] I. Tarapov, "Movement of a magnetizable fluid in lubricating layer of a cylindrical bearing," *Magnetohydrodynamics*, vol. 8, no. 4, pp. 444–448, 1972.

[7] R. C. Shah and M. Bhat, "Ferrofluid squeeze film in a long bearing," *Tribology International*, vol. 37, pp. 441–446, 2004.

[8] S. Chen, Q. Zhang, H. Chong, T. Komatsu, and C. Kang, "Some design and prototyping issues on a 20 krpm hdd spindle motor with a ferro-fluid bearing system," *IEEE Trans. Magn.*, vol. 37, no. 2, pp. 805–809, 2001.

[9] Q. Zhang, S. Chen, S. Winoto, and E. Ong, "Design of high-speed magnetic fluid bearing spindle motor," *IEEE Trans. Magn.*, vol. 37, no. 4, pp. 2647–2650, 2001.

[10] P. Kuzhir, "Free boundary of lubricant film in ferrofluid journal bearings," *Tribology International*, vol. 41, pp. 256–268, 2008.

[11] M. Miwa, H. Harita, T. Nishigami, R. Kaneko, and H. Unozawa, "Frequency characteristics of stiffness and damping effect of a ferrofluid bearing," *Tribology Letter*, vol. 15, no. 2, pp. 97–105, 2003.

[12] W. Ochonski, "The attraction of ferrofluid bearings," *Mach. Des.*, vol. 77, no. 21, pp. 96–102, 2005.

[13] Z. Meng and Z. Jibin, "An analysis on the magnetic fluid seal capacity," *Journal of Magnetism and Magnetic Materials*, vol. 303, pp. e428–e431, 2006.

[14] J. Walker and J. Backmaster, "Ferrohydrodynamics thrust bearings," *Int. J. Eng. Sci.*, vol. 17, pp. 1171–1182, 1979.

[15] N. Tiperi, "Overall characteristics of bearings lubricated ferrofluids," *ASME J. Lubr. Technol.*, vol. 105, pp. 466–475, 1983.

[16] S. Miyake and S. Takahashi, "Sliding bearing lubricated with ferromagnetic fluid," *ASLE Trans.*, vol. 28, pp. 461–466, 1985.

[17] H. Chang, C. Chi, and P. Zhao, "A theoretical and experimental study of ferrofluid lubricated four-pocket journal bearing," *Journal of Magnetism and Magnetic Materials*, vol. 65, pp. 372–374, 1987.
[18] Y. Zhang, "Static characteristics of magnetized journal bearing lubricated with ferrofluids," *ASME J. Tribol.*, vol. 113, pp. 533–538, 1991.
[19] T. Osman, G. Nada, and Z. Safar, "Static and dynamic characteristics of magnetized journal bearings lubricated with ferrofluid," *Tribology International*, vol. 34, pp. 369–380, 2001.
[20] R. C. Shah and M. Bhat, "Anisotropic permeable porous facing abd slip velocity squeeze film in axially undefined journal bearing with ferrofluid lubricant," *Journal of Magnetism and Magnetic Materials*, vol. 279, pp. 224–230, 2004.
[21] F. Cunha and H. Couto, "A new boundary integral formulation to describe three-dimensional motions of interfaces between magnetic fluids," *Applied mathematics and computation*, vol. 199, pp. 70–83, 2008.
[22] R. E. Rosensweig, Y. Hirota, S. Tsuda, and K. Raj, "Study of audio speakers containing ferrofluid," *J. Phys. : Condens. Matter*, vol. 20, 2008.
[23] G. Lemarquand, "Ironless loudspeakers," *IEEE Trans. Magn.*, vol. 43, no. 8, pp. 3371–3374, 2007.
[24] R. Ravaud and G. Lemarquand, "Modelling an ironless loudspeaker by using three-dimensional analytical approaches," *Prog. Electromagn. Res.*, vol. 91, pp. 53–68, 2009.
[25] R. Ravaud, G. Lemarquand, V. Lemarquand, and C. Depollier, "Ironless loudspeakers with ferrofluid seals," *Archives of Acoustics*, vol. 33, no. 4S, pp. 3–10, 2008.
[26] R. Ravaud, G. Lemarquand, and V. Lemarquand, "Magnetic pressure and shape of ferrofluid seals in cylindrical structures," *J. Appl. Phys.*, vol. 106, no. 3, p. 34911, 2009.
[27] R. Ravaud, M. Pinho, G. Lemarquand, N. Dauchez, J. M. Genevaux, V. Lemarquand, and B. Brouard, "Radial stiffness of a ferrofluid seal," *IEEE Trans. Magn.*, vol. 45, no. 10, pp. 4388–4390, 2009.
[28] R. Ravaud, G. Lemarquand, and V. Lemarquand, "Mechanical properties of ferrofluid applications: centering effect and capacity of a seal," *Tribology International*, vol. 43, no. 1-2, pp. 76–82, 2009.
[29] A. Ivanov, S. Kantorovich, V. Mendelev, and E. Pyanzina, "Ferrofluid aggregation in chains under the influence of a magnetic field," *Journal of Magnetism and Magnetic Materials*, vol. 300, pp. e206–e209, 2006.
[30] S. I. Babic and C. Akyel, "Improvement in the analytical calculation of the magnetic field produced by permanent magnet rings," *Prog. Electromagn. Res. C*, vol. 5, pp. 71–82, 2008.
[31] R. Ravaud, G. Lemarquand, and V. Lemarquand, "Magnetic field created by tile permanent magnets," *IEEE Trans. Magn.*, vol. 45, no. 7, pp. 2920–2926, 2009.
[32] R. Ravaud and G. Lemarquand, "Comparison of the coulombian and amperian current models for calculating the magnetic field produced by arc-shaped permanent magnets radially magnetized," *Prog. Electromagn. Res.*, vol. 95, pp. 309–327, 2009.
[33] R. Ravaud, G. Lemarquand, V. Lemarquand, and C. Depollier, "The three exact components of the magnetic field created by a radially magnetized tile permanent magnet.," *Prog. Electromagn. Res.*, vol. 88, pp. 307–319, 2008.
[34] K. Halbach, "Design of permanent multiple magnets with oriented rec material," *Nucl. Inst. Meth.*, vol. 169, pp. 1–10, 1980.

Part 2

Tribology in Cutting

6

Characteristics of Abrasive Particles and Their Implications on Wear

Giuseppe Pintaude
Federal University of Technology – Paraná
Brazil

1. Introduction

Geometry, hardness and size can be considered the most important characteristics affecting the wear rate caused by abrasive particles. The combined effect of these characteristics on wear has not yet been modeled as a whole. One particular model, proposed by Rabinowicz et al. (1961), involves a single particle being considered as a pointed tool, so that the effect of the geometry can be determined.

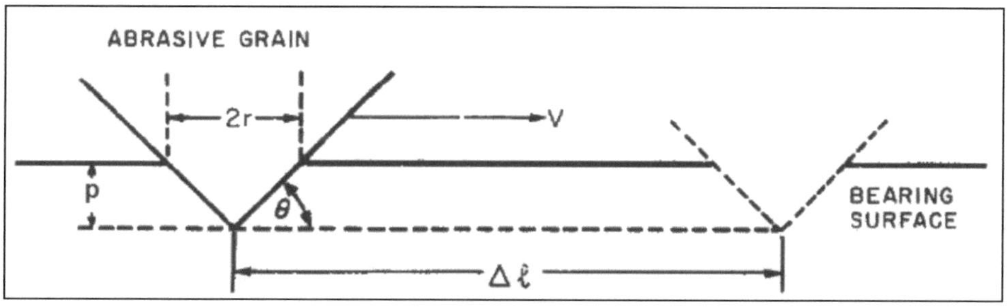

Fig. 1. Conical abrasive particle cutting a groove of depth p, and sliding for a distance Δl to produce a wear volume V (Rabinowicz et al., 1961).

The wear rate is expressed as follows:

$$dV / dl = \frac{tan\theta}{\pi}\frac{W}{H} \tag{1}$$

where $tan\theta$ is the average tangents of all roughness angles of the abrasive particles, H is the hardness of bearing surface and W is the total load.

In this case the particle is considered to be a solid body that neither degrades nor changes size and whose properties remain constant throughout the whole wear process. This hypothesis is not verified, especially for those situations where the hardness of abrasive is lower than the abraded material. This situation led to small wear rates, and it is usually classified as mild wear (Gates, 1998). The fragmentation of particles depends on their fracture toughness, and it is possible to verify in the literature data that the higher the hardness, the higher the fracture toughness (Fig. 2). Therefore, for softer abrasives the breakage is easier.

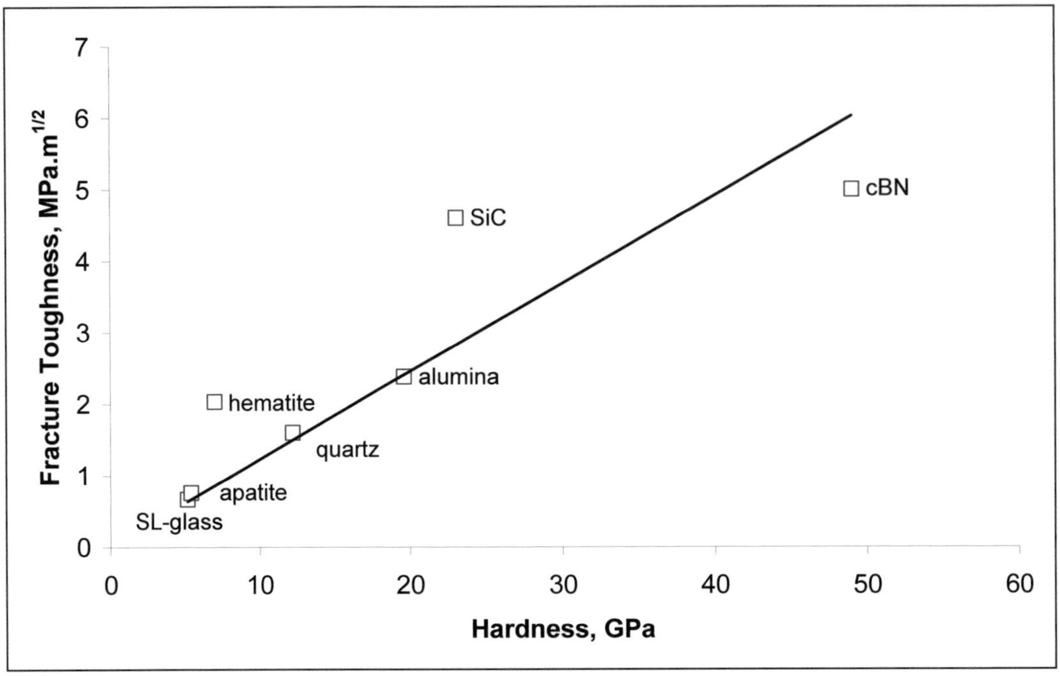

Fig. 2. Hardness and fracture toughness for usual abrasive materials. Data from Broz et al. (2006) for apatite, quartz and alumina; Rhee et al. (2001) for glass; Kaur et al. (2009) for SiC; Taniguchi et al. (1996) for cBN, and Tromans & Meech (2002) and Beste & Jacobson (2003) for hematite.

A likely effect of fragmentation is the reduction of size, another important characteristic of abrasive not incorporated in (1) (Rabinowicz et al., 1961). The well-known effect of particle size in wear rates was studied by Coronado & Sinatora (2009) for mottled cast iron with different contents of retained austenite (Fig. 3).

From Fig. 3 a high increase of mass loss with abrasive size can be observed, up to approximately 70 microns. After this, the mass losses do not increase at the same rate, and even in some cases, they do not change from a critical particle size. Many theories have been proposed to explain the effect of particle size in abrasive wear (Misra & Finnie, 1981), but a particularly important model, proposed by Sin et al. (1979) related this behavior to the abrasive characteristic considered by Rabinowicz et al. (1961): the particle geometry.

Sin et al. (1979) assumed that the abrasive particles are conical in shape with a hemispheric tip. With this hypothesis, they explained the effect of particle size on wear rates due to the changes in the tip radius. Therefore, the smaller the particle, the higher the tip radius, i.e., the small particles have rounder tips, and one can assume that they abrade a surface in the same manner of spherical particles.

Although this theory was criticized by Misra & Finnie (1981), there are many experimental evidences that support it. A good aim is to design the particle geometry that reproduces the practical results. For example, three numerical simulations (Jacobson et al., 1988, Jiang et al., 1998 and Fang et al., 2009) assumed the same geometry adopted by Sin et al. (1979) for the abrasive particles. In all cited investigations the particle size effect was very well-reproduced. Besides these studies, two experimental evidences can be mentioned also.

Firstly, Graham & Baul (1972) obtained for an aluminum alloy tested in a single pass scratch the results shown in Fig. 4.

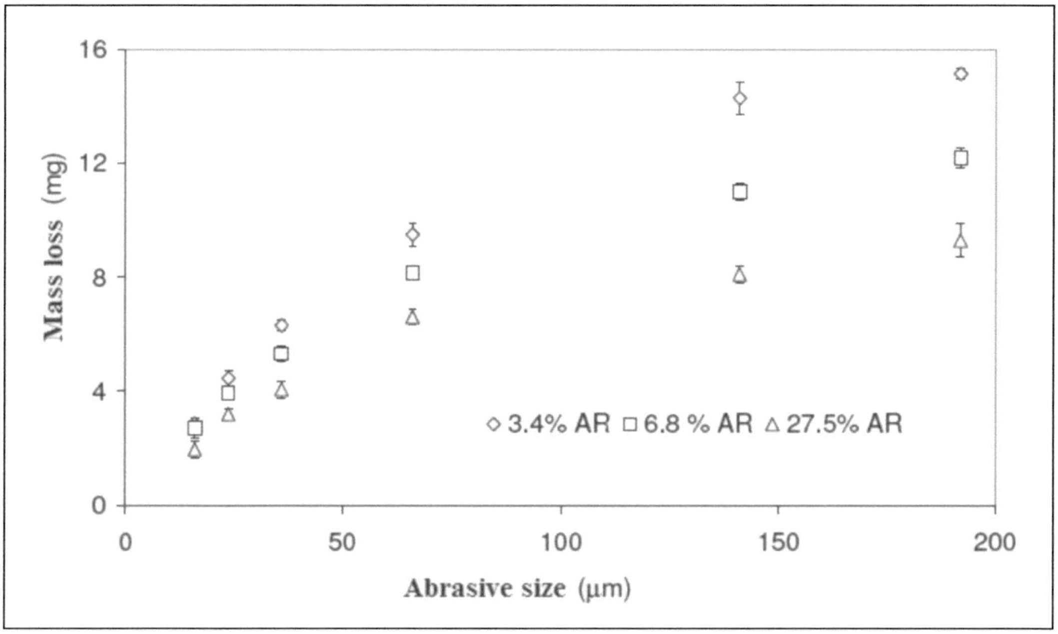

Fig. 3. Effect of abrasive size on mass loss of mottled cast iron with different retained austenite contents (Coronado & Sinatora, 2009).

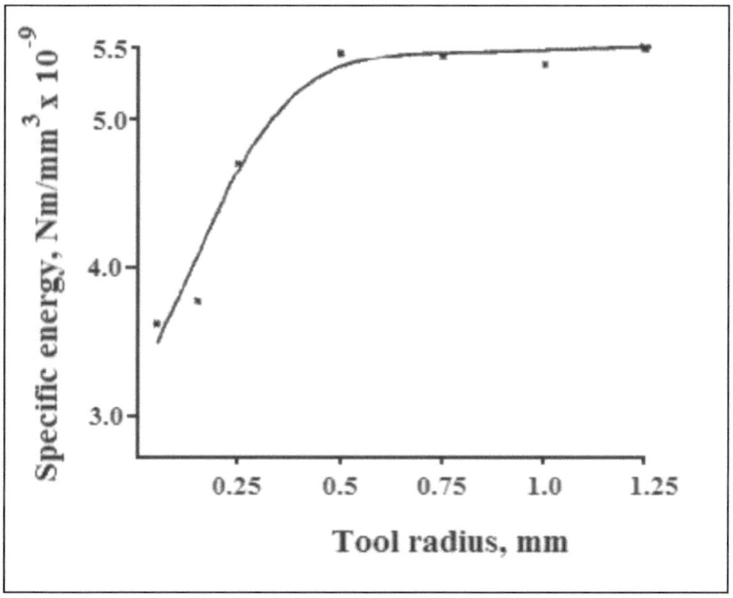

Fig. 4. Specific energy of cutting as a function of tool radius of spherical tip tool on the apex of a 90° cone (Graham & Baul, 1972).

Although these results are expressed in terms of specific energy as a function of tool radius, they can be correlated with results as those presented in Fig. 3. In the scratch test an increase in the tool radius means that the aluminum alloy experienced a contact with a blunt tool. Following Sin et al. (1979), small particles behave like blunt ones. In addition, the specific energy for cutting is high for small particles, as demonstrated in some investigations (Jacobson et al., 1988, Jiang et al., 1998).

The second example favorable to the bluntness theory was given by Gahlin & Jacobsson (1999). They produced controlled surfaces using micro mechanical etching technique, and different particle sizes were simulated by changing the packing density. Their main results are presented in Fig. 5, having tin as the worn material.

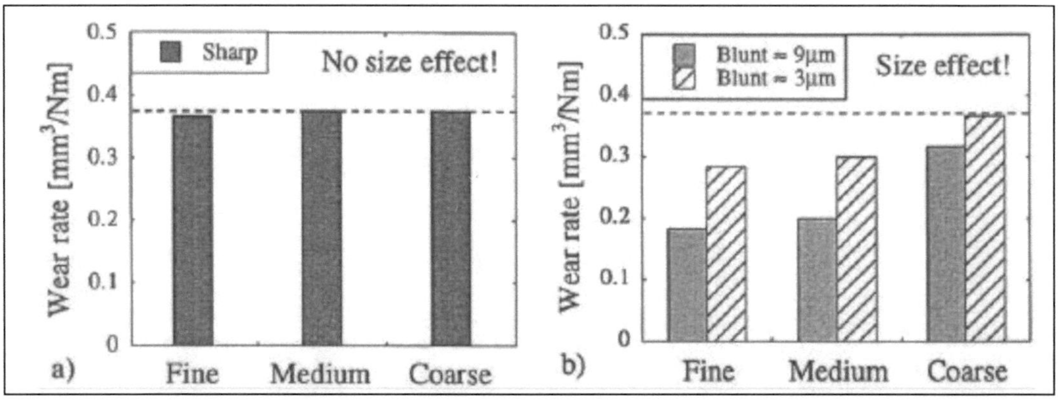

Fig. 5. The wear rate varying with packing density, considering: (a) sharp and (b) blunt tips (Gahlin & Jacobsson, 1999).

Fig. 5a shows that the size effect was not observed for the sharp tips. On the other hand, blunt tips exhibited a significant size effect. These results thus confirm that the relative bluntness theory is able to explain the effect of abrasive size in wear rates.

In the next item we will discuss possible approaches to characterize the abrasive wear, considering the difficulty to determine the level in which a particle loses its sharpness and the strong relationship between this and the mechanical properties of abrasive, such as hardness and fracture toughness, as previously described.

2. Approaches to characterize abrasive wear

Fig. 6 presents a sequence of events since a particle abrading a surface to the material removal step, resulting in a production of debris.

The wear debris characterization is still little employed in abrasion process (Stachowiak et al., 2008), so that this aspect will be not treated here.

The first approach to characterize the abrasive wear process can be made through the shape evaluation of abrasive particles. There are a lot of shape parameters, including the aspect ratio, roundness and fractal dimension (Kaye, 1998). Special attention to new developments to shape evaluation is given by Stachowiak and co-workers in a series of investigations. A remarkable definition is the spike parameter - quadratic fitting (SPQ) (Hamblin and Stachowiak, 1996), which is based on the localization of the centroid of the bi-dimensional section of particle and the circle in which radius is equal to the average radius

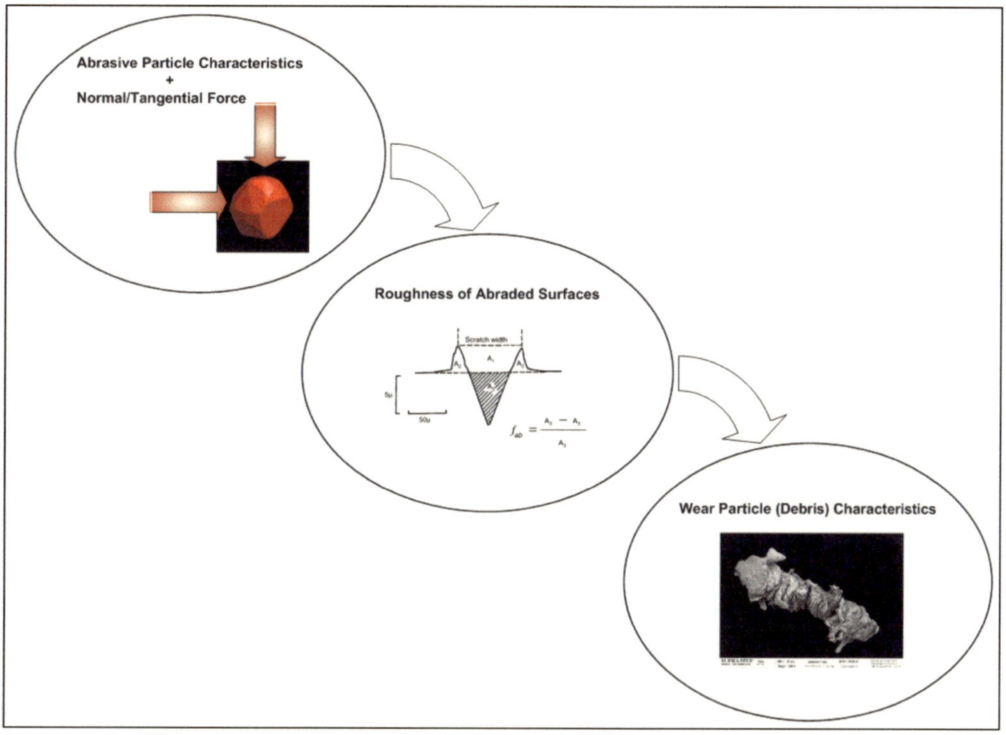

Fig. 6. Three ways to characterize the abrasive wear. Images adapted from De Pellegrin & Stachowiak (2005), Buttery & Archard (1971) and Stachowiak et al. (2008).

of particle (Fig. 7). The areas outside the circle are considered as interest regions, whereas the bulk is suppressed. The maximum local diameter is determined for each region outside the circle and this point is treated as spike vertices, M. The spike laterals, which are between the segments sp-mp and mp-ep are represented by polynomial quadratic functions. Differentiating the functions at point mp, led in the apex angle θ.

$$sv = \cos(\theta / 2) \tag{2}$$

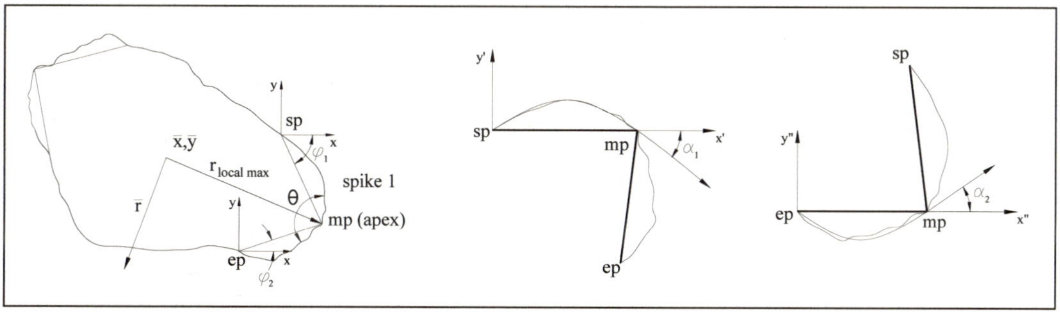

Fig. 7. Routine to calculate the spike parameter SPQ.

The spike parameter - quadratic fitting (SPQ) is thus obtained from the average value of valid spikes, where n is the number of found spikes:

$$SPQ = \frac{1}{n}\sum sv \tag{3}$$

The second possible form to characterize the abrasion process is to measure the surface roughness of an abraded material. Spurr (1981) proposed that the ploughing contribution to friction can be related to the roughness parameters, as follows:

$$\mu_P = \frac{8}{\pi}\frac{Ra}{\lambda a} \tag{4}$$

Taken into account the definition of $\lambda a\,(= 2\pi\, Ra\,/\,\Delta a)$, (4) can be rewritten as

$$\mu_P \cong 0.4\;\Delta a \tag{5}$$

where Δa is defined as the mean absolute profile slope over the assessment length.
Torrance (2002) defined the parameter Rqs as the rms slope of abrasive in contact with abraded material. In his description, both wear and friction coefficients are depended on this parameter. A simple definition provided by Torrance for this parameter is:

$$Rqs = p\sqrt{\frac{H_S}{f_n}} \tag{6}$$

where f_n is the normal force per grit, H_S is the surface hardness of metal, and p is the average scratch depth made by abrasives. If the geometry of abrasive was perfectly transferred to the worn surface, it would be possible to associate the Torrance's definition to the rms slope of the profile, Δq .

More recently, Da Silva & De Mello (2009) used the hybrid bi-dimensional parameter $\lambda q\,(= 2\pi\, Rq\,/\,\Delta q)$, which represents the root mean square wavelength of the profile, and considering that this parameter can be associated to the width of the scratches, they found a relationship between scratch widths and normal load using the single scratch experiments.
All definitions above mentioned made use of a roughness parameter of height and another that express the slope. The relationship between each other can be a good indicative of the contact area between abrasive and abraded surface. For example, McCool (1987) proposed an easily programmed method using the rms height and rms slope to estimate the contact area in a microcontact model.

3. Experimental

The wear and friction coefficients were presented elsewhere in Pintaude et al. (2003) and Pintaude et al. (2009), in tests performed under sliding abrasion using pin-on-disk apparatus. The tested materials were obtained from three groups:

- Low-carbon steel (composition similar to AISI 1006 steel).
- Bearing wiredrawn steel (AISI 52100).
- High-chromium cast iron.

The low-carbon steel was tested in the as-cast condition. The bearing steel (Q&T 52100 steel) and the high-chromium cast iron (HCCI) were heat-treated. The Vickers hardness of tested materials is presented in Table 1.

Material	HV_1, MPa
Low-carbon steel (1006 steel)	973
Bearing steel (Q&T 52100 steel)	4070
High-chromium cast iron (HCCI)	4600

Table 1. Vickers hardness of tested materials.

Pins of 3 mm diameter and 21 mm length were tested in a pin-on-disk apparatus, using abrasive papers coated disks. Conditioning of the pin surface before starting the test was made using a grit 600 corundum paper. The abrasives used in the tests were glass paper, either with 80 grit (0.20 mm average grain size) or 240 grit (0.06 mm of average grain size).
During tests the following variables were controlled: rotational and tangential velocities, number of revolutions, position of the pin in relation to the disc, sliding distance, tangential and normal forces, friction coefficient, room relative humidity and room temperature.
All tests were performed using 20 N of normal force, applied by deadweight. At least seven mass determinations were made for each test piece.
An analytical scale with 0.1 mg resolution was used to measure pin mass. A minimum mass variation of 1 mg was established for each test, in order to avoid the error of scale. In the case of the more wear resistant test pieces, up to 10 abrasive papers were used for each test.
With the measurement of mass losses and friction coefficients, it is possible to calculate the abrasion factor, f_{ab}, defined as the proportion of groove volume removed. The volume of material displaced is equal to the area A3 but, of this material, a volume equal to the area A2 is retained upon the surface as piled-up material. Thus, the volume of material removed (wear) is equal to (A3-A2) (see Fig. 8). An alternative used to determine f_{ab} is presented by Jacobson et al. (1988), defined in (7):

$$f_{ab} = K_A / \mu_P \tag{7}$$

where K_A is the wear coefficient and μ_P is the ploughing component of the friction coefficient. To find μ_P, the adhesive component of the friction coefficient is assumed as 0.2, so that this value is reduced from the whole friction coefficient values provided by Pintaude et al. (2003, 2009).
After the wear tests, the surface roughness of worn pins was determined, using 1.75 mm as sampling length, performed in PERTHOMETER v 6.2 equipment. The calculations of roughness parameters were made using MATLAB 5® software, 5.3 version. In this software, two filters were applied: one to remove the waviness of profile and another one to determine the mean line. After that, the profiles were divided into 7 cut-offs, of which the initial and final were discarded, and the remaining 5 were used to obtain the roughness parameters values. The routine of calculation was adequately tested in a standard block supplied with the roughness measuring equipment.
The glass particles were extracted from papers to analyze their shape factors. Twenty particles were randomly selected from two samples of glass, whose average size are 72 and 455 microns. These particles were removed from papers of #80 and #240 mesh, respectively. The images obtained in optical microscope were then submitted to threshold process in the Image-Pro Plus software to convert the grayscale to binary image (Fig. 9). The output binary images has values of 1 (white) for all pixels in the input image with luminance greater than defined level and 0 (black) for all other pixels. The transition level was based on the

histogram of images. The Image-Pro Plus software has routines to calculate the average values of roundness factor, fractal dimension and aspect ratio and these shape factors were determined for each group of glass particles.

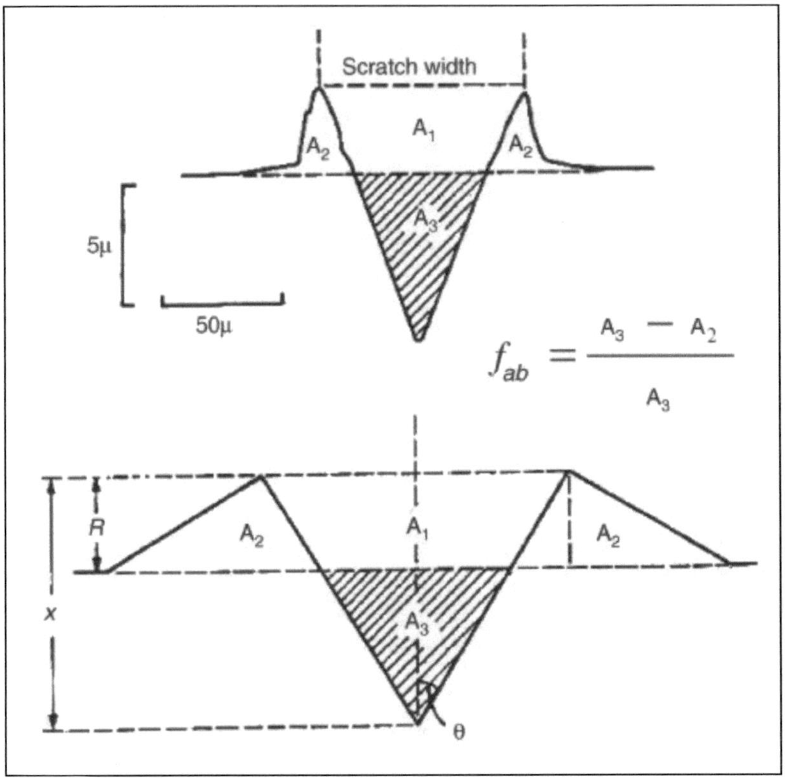

Fig. 8. Abrasion factor definition (Buttery & Archard, 1971).

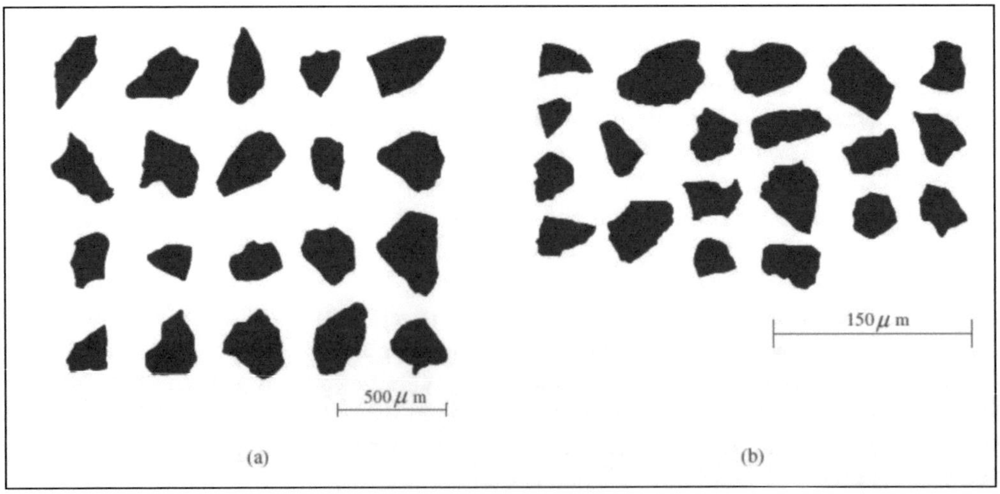

Fig. 9. Binary images of glass particles: (a) 72 microns and (b) 455 microns average size.

4. Results and discussion

Table 2 shows the shape factors determined for glass particles removed from #80 and #240 papers. Also, the wear rates promoted by these particles abrading quenched and tempered 52100 steel with 4.07 GPa Vickers hardness after sliding abrasion tests are provided.

Average size, microns	Wear rate (m^3/m)	Aspect ratio	1/Roundness	Fractal dimension
72	7.25E-12	1.5 ± 0.3	1.4 ± 0.1	1.06 ± 0.01
455	6.84E-11	1.5 ± 0.3	1.5 ± 0.1	1.07 ± 0.01

Table 2. Shape factors values for different glass particle sizes and the respective wear rates of 52100 steel caused by them

The wear rates were very much affected by the glass particle size. The increase from 72 to 455 microns caused an increase of one order of magnitude in wear rates. The values of shape factors presented in Table 2 do not corroborate the theory given by Sin et al. (1979), since there is no difference among them.

The insignificant effect of average size on shape factor was also demonstrated by Bozzi & De Mello (1999). When they tested silica grains against WC-12%Co thermal sprayed coating in three-body abrasion during 330 min, the average size of abrasive particles were reduced in 38.2%. This reduction did not occur in the same proportion for the roundness factor: only 2.9% of reduction was observed for the shape value. An important aspect of tests performed by Pintaude et al. (2009) and Bozzi & De Mello (1999) is that the hardness of abrasive is lower than that of worn material, resulting in a mild wear. In these cases, a possible explanation for the failure of a particle to penetrate another surface is that the geometry of the particle that is not sufficiently hard to produce a scratch on the other material must have undergone a change after its breakage. The particles indeed break, as has been shown in an earlier study (Pintaude et al., 2003). Thus, instead of having more points to cut with, the broken particle ends up becoming blunter, so that it cannot cut. However, the shape characterization did not prove this.

Another set of results was obtained by De Pellegrin & Stachowiak (2002) (Fig. 10), broader than those presented in Table 2 and by Bozzi & De Mello (1999). Again, no one can observe any variation of the shape factor (aspect ratio) with particle size.

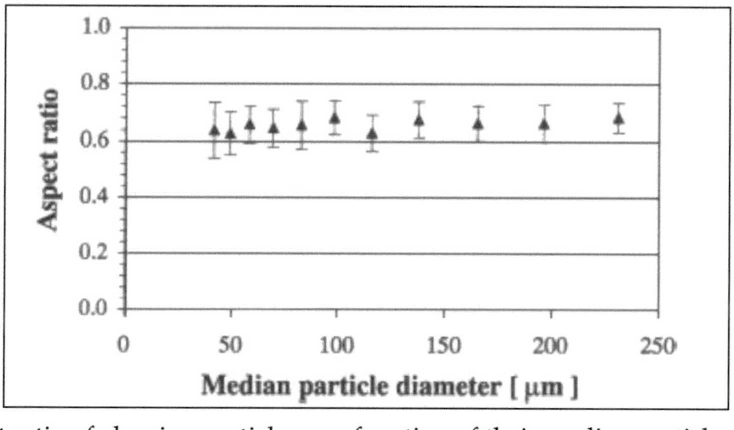

Fig. 10. Aspect ratio of alumina particles as a function of their median particle diameter (De Pellegrin & Stachowiak, 2002).

Although the presented results had been contrary to the bluntness theory, the same cannot be discharged due to an important reason. The shape factors determination should be considered as a bi-dimensional analysis and the action of abrasive during mechanical contact occurred in 3D dimension. Thus, De Pellegrin & Stachowiak (2002) pointed out that the presence of re-entrant features made a difference between the induced groove area and the calculated one from the particle projection. For this reason, we will test now some ideas about roughness characterization of abraded surfaces.

Table 3 presents the results of abrasion factors for 1006 and 52100 steels and for high-chromium cast iron abraded by glass papers. In addition, the root mean square wavelength values of abraded surfaces were also presented.

	λq , mm		f_{ab} ($\equiv K_A/\mu_P$)	
Worn material	#80 paper	#240 paper	#80 paper	#240 paper
1006 steel	23.2	16.3	0.0411	0.021
Q&T 52100 steel	39	26.9	0.106	0.037
HCCI	26.6	N.T.	0.00895	N.T.

Table 3. The root mean square wavelength of the profile and the estimated abrasion factor of three materials tested in sliding abrasion using glass as abrasive. N.T.: not tested.

The results presented in Table 3 show expected trends for steels, abraded in severe wear: the abrasion factor is higher for the hardest steel and lower as the abrasive particle size is reduced. In addition, the volume removed as debris to volume of micro-grooves of pins in repeated sliding determined by Hisakado et al. (1999) was in the same order of magnitude of those presented in Table 3 for tested steels.

Now, it is important to establish a possible relationship between the λq and f_{ab} values. Taken into account the results obtained for 1006 steel tested with #80 paper and for 52100 steel abraded by #240 glass particles one can conclude that the abrasion factors were similar, and at the same time, as well as the λq ones. Again, it is remarkable that the wear in these cases was severe, i.e., the hardness of abrasive is higher than the hardness of steels.

An important observation from λq results is that these values are more affected by Rq than the Δq values, i.e., the increase in particle size leads to an increase in the height profile, and the slope kept almost unmodified. This kind of result was already described by Hisakado & Suda (1999) in abrasive papers, when they measured the slope of SiC particles with different grain sizes (Table 4).

Abrasive papers	Average size, microns	Slope angle of abrasive grain	Rms roughness, microns
#100	125	47.2	68.2
#1000	16.3	54.1	10.7

Table 4. Topographical properties measured on abrasive papers constituted of SiC particles (Hisakado & Suda, 1999).

In order to reinforce the above discussion, a scheme given by Gahlin & Jacobson (1999) (Fig. 11) shows how the increase of particle size can mean a change only in the height roughness parameter with no variation in the slope of surface. In Fig. 11, D3 > D2 > D1, being D the diameter of particle, and H3 > H1, being H the total height imprint at surface.

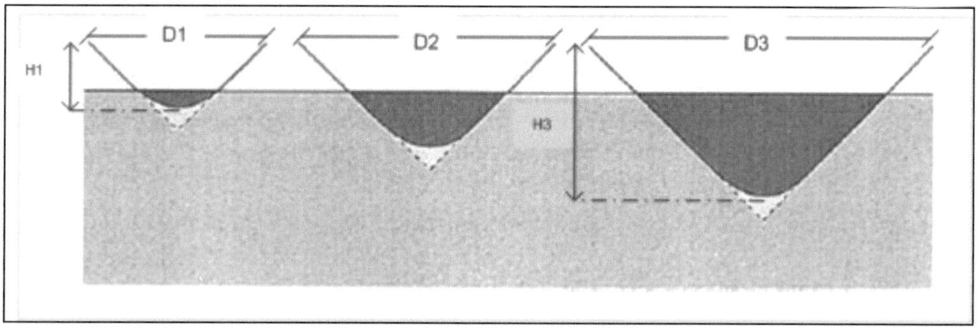

Fig. 11. Illustration showing a simultaneously increase of particle diameter and total height (adapted from Gahlin & Jacobson, 1999).

From the above analysis, we can conclude that the λq roughness parameter is a powerful variable to characterize an abraded surface, discriminating the effect of particle size under severe wear. In this situation, the abrasive characteristics are changed a little during the mechanical contact.

On the other hand, a very different situation occurred for HCCI. At present, this material was abraded under mild wear, and a severe fragmentation of glass particles was observed. The f_{ab} is very lower than that observed for 52100 steel (Tab. 3), despite the fact that their difference in hardness is not significant. In addition, any kind of correlation is possible to make with the λq value, as made for the steels between λq and f_{ab}.

Here, we identified a lack in the literature proposals to identify changes that happens during the contact between a soft abrasive and a hard abraded surface, even the bluntness theory proposed by Sin et al. (1979) has received good experimental evidences, as previously discussed.

We pay heed to other evidence in the literature to support it, since the relationship between static hardness tests and abrasion is always employed. Following the definitions provided by Buttery & Archard (1971) (Fig. 8), the fraction of material displaced is reduced as the severity of pile-up increases. The surface deformation, after the complete unloading, was evaluated by Alcalá et al. (2000) for spherical and Vickers geometry indenters, considering the static indentation process. The results obtained by these researches for work-hardened copper is presented in Fig. 12.

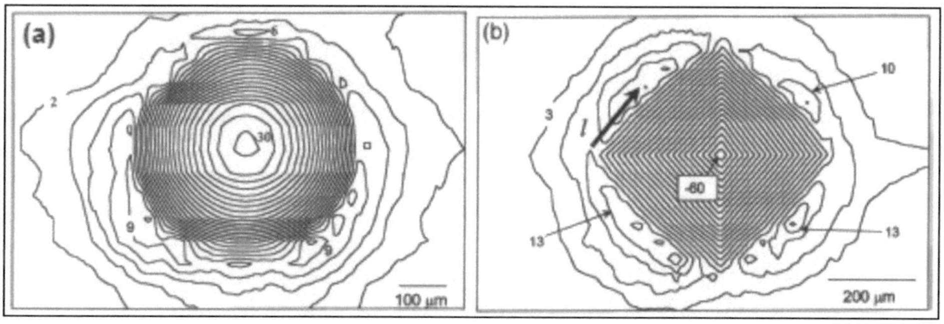

Fig. 12. Surface topography around spherical (a) and Vickers (b) indents for an indentation load of 160 N performed in a work-hardened copper. The dimensions are provided in microns (Alcalá et al., 2000).

The height dimensions at the center of indentation and at its ridges allow calculating the severity of pile-up (*s*), following (8). Thus, one can conclude that the spherical indentation gives rise to a larger severity of pile-up. It implies that f_{ab} produced by a spherical indenter should be smaller than that estimated for a pyramidal (angular). The small particles tested by Pintaude et al. (2009) produced low values of f_{ab}, confirming the possibility that they scratch the surface as spherical particles.

$$s = \frac{R}{x} \tag{8}$$

5. Conclusions and future trends

The measurement of shape factors using bi-dimensional technique is not useful to prove the theory put forward by Sin et al. (1979) used to explain the particle size effect in abrasive wear rates, although a series of experimental evidences support it. The main reason for this discrepancy is the 3D action of abrasives during the wear process, and a bi-dimensional characterization probably disregards the presence of re-entrant features of particles in this case.

For severe abrasion, when the hardness of abrasive is higher than the worn surface material, the use of roughness characterization by means of a hybrid parameter is a good way to discriminate the particle size effect, probably due to the undermost changes in the slope of particles, which have a high cutting capacity providing by a combination of their hardness and fracture toughness.

However, for mild abrasion, when the level of particles breakage is high, the surface characterization presented here is not yet enough to discriminate the size effects. Therefore, as a future trend we indicate the development of analytical tools able to detect the changes in abrasive sizes after their breakage, and the measurement of consequences of this process in their geometries.

6. References

Alcalá, J., Barone, A.C. & Anglada, M. (2000). The influence of plastic hardening on surface deformation modes around Vickers and spherical indents, *Acta Materialia*, Vol. 48, No. 13, pp. 3451-3464, ISSN: 1359-6454.

Beste, U. & Jacobson, S. (2003). Micro scale hardness distribution of rock types related to rock drill wear, *Wear*, Vol. 254, No. 11, pp. 1147-1154, ISSN: 0043-1648.

Bozzi, A.C. & De Mello, J.D.B. (1999). Wear resistance and wear mechanisms of WC–12%Co thermal sprayed coatings in three-body abrasion. *Wear*, Vol. 233–235, December, pp. 575–587.

Broz, M.E., Cook, R.F. & Whitney, D.L. (2006). Microhardness, Toughness, and Modulus of Mohs Scale Minerals, *American Mineralogist*, Vol. 91, No. 1, pp. 135–142, ISSN: 0003-004x.

Buttery, T.C. & Archard, J.F. (1970). Grinding and abrasive wear. *Proc. Inst. Mech. Eng.*, Vol. 185, pp. 537-552, ISSN: 0020-3483.

Coronado, J.J. & Sinatora, A. (2009). Particle size effect on abrasion resistance of mottled cast iron with different retained austenite contents, Wear, Vol. 267, No. 1-4, pp. 2077-2082.

Da Silva, W.M. & De Mello, J.D.B. (2009). Using parallel scratches to simulate abrasive wear. *Wear*, Vol. 267, No. 11, pp. 1987–1997.

De Pellegrin, D.V. & Stachowiak, G.W. (2005). Simulation of three-dimensional abrasive particles. *Wear*, Vol. 258, No. 1-4, pp. 208-216.

De Pellegrin, D.V. & Stachowiak, G.W. (2002). Assessing the role of particle shape and scale in abrasion using 'sharpness analysis' Part II. Technique evaluation. *Wear*, Vol. 253, No. 9-10, pp. 1026–1034.

Fang, L., Li, B., Zhao, J. & Sun, K. (2009). Computer simulation of the two-body abrasion process modeling the particle as a paraboloid of revolution. *Journal of Materials Processing Technology*, Vol. 209, No. 20, pp. 6124–6133, ISSN: 0924-0136.

Gahlin, R. & Jacobson, S. (1999). The particle size effect in abrasion studied by controlled abrasive surfaces. *Wear*, Vol. 224, No. 1, pp. 118-125.

Gates, J.D. (1998). Two-body and three-body abrasion: A critical discussion. *Wear*, Vol. 214, No. 1, pp. 139-146.

Graham, D. & Baul, R. M. (1972). An investigation into the mode of metal removal in the grinding process, *Wear*, Vol. 19, No. 3, pp. 301-314.

Hamblin, M.G. & Stachowiak, G.W. (1996). Description of abrasive particle shape and its relation to two-body abrasive wear. *Tribology Transactions*, Vol. 39, No. 4, pp. 803-810, ISSN: 1040-2004.

Hisakado, T. & Suda, H. (1999). Effects of asperity shape and summit height distributions on friction and wear characteristics. Wear, Vol. 225–229, Part 1, April, pp. 450–457.

Hisakado, T.; Tanaka, T. & Suda, H. (1999). Effect of abrasive particle size on fraction of debris removed from plowing volume in abrasive wear. *Wear*, Vol. 236, No. 1-2, December, pp. 24–33.

Jacobson, S., Wallen, P. & Hogmark, S. (1988). Fundamental aspects of abrasive wear studied by a new numerical simulation model. *Wear*, Vol. 123, No. 2, pp. 207-223.

Jiang, J., Sheng, F. & Ren, F. (1998). Modelling of two-body abrasive wear under multiple contact conditions. *Wear*, Vol. 217, No. 1, pp. 35-45.

Kaye, B.H. (1998). Particle shape characterization, In: *ASM Handbook Vol. 7 Powder Metal Technologies and Applications*, Lee, P.W. et al. (Ed.), 605-618, ASM International, ISBN 0-87170-387-4, Metals Park, OH

Kaur, S., Cutler, R.A. & Shetty, D.K. (2009). Short-Crack Fracture Toughness of Silicon Carbide. *J. Am. Ceram. Soc.*, Vol. 92, No. 1, pp. 179–185, ISSN: 1551-2916.

McCool, J.I. (1987). Relating profile instrument measurements to the functional performance of rough surfaces. *Journal of Tribology ASME*, Vol. 109, No. 2, pp. 264-270, ISSN: 0742-4787.

Misra, A. & Finnie, I. (1981). On the size effect in abrasive and erosive wear *Wear*, Vol. 65, No. 3, January, pp. 359-373.

Pintaude, G., Bernardes, F.G., Santos, M.M., Sinatora, A. & Albertin, E. (2009). Mild and severe wear of steels and cast irons in sliding abrasion. *Wear*, Vol. 267, No. 1-4, pp. 19-25.

Pintaude, G., Tanaka, D.K. & Sinatora, A. (2003). The effects of abrasive particle size on the sliding friction coefficient of steel using a spiral pin-on-disk apparatus. *Wear*, Vol. 255, No. 1-6, August-September, pp. 55-59.

Rabinowicz, E.; Dunn, L. A. & P. G. Russel, P.G. (1961). A study of abrasive wear under three-body conditions, *Wear*, Vol. 4, No. 5, pp. 345 - 355.

Rhee, Y-W., Kim, H-W., Deng, Y. & Lawn, B.R. (2001). Brittle fracture versus quasi plasticity in ceramics: a simple predictive index, *J. Am. Ceramic Soc.*, Vol. 84, No. 3, pp. 561-565.

Sin, H., Saka, N. & Suh, N.P. (1979). Abrasive wear mechanisms and the grit size effect. *Wear*, Vol. 55, No. 1, July, pp. 163-190.

Spurr, R.T. (1981). The abrasive wear of metals. *Wear*, Vol. 65, No. 3, pp. 315–324.

Stachowiak, G.P., Stachowiak, G.W. & Podsiadlo, P. (2008). Automated classification of wear particles based on their surface texture and shape features. *Tribology International*, Vol. 41, No. 1, January, pp. 34–43, ISSN: 0301-679X.

Taniguchi, T., Minoru Akaishi, M. & Yamaoka, S. (1996). Mechanical Properties of Polycrystalline Translucent Cubic Boron Nitride as Characterized by the Vickers Indentation Method. *J. Am. Ceramic Soc.*, Vol. 79, No. 2, pp. 547-549.

Torrance, A.A. (2002). The effect of grit size and asperity blunting on abrasive wear. *Wear*, Vol. 253, No. 7-8, pp. 813–819.

Tromans, D. & Meech, J.A. (2002). Fracture toughness and surface energies of minerals: theoretical estimates for oxides, sulphides, silicates and halides. *Minerals Engineering*, Vol. 15, No. 12, pp. 1027–1041, ISSN: 0892-6875.

7

Topographical Change of Engineering Surface due to Running-in of Rolling Contacts

R. Ismail[1], M. Tauviqirrahman[1], Jamari[2] and D.J. Schipper[1]
[1]*Laboratory for Surface Technology and Tribology, University of Twente*
[2]*Laboratory for Engineering Design and Tribology, University of Diponegoro*
[1]*The Netherlands*
[2]*Indonesia*

"If condition were wrong, piston rings could disappear within 24 h after start up, whereas after successful run-in piston ring life could be two years." (Summers-Smith, 1997)

1. Introduction

The above quotation indicates the important of running-in phase, which occurs at the beginning of the contact in a mechanical systems. Tribologist identifies that running-in takes place on the first stages of the practical mechanical system operation such as automotive engines, gears, camshaft and followers, and bearings. Kehrwald (1998) expressed the significance of running-in phase by predicting that an optimized running-in procedure has a potency to improve the life time of a mechanical system by 40% and more and to reduce the engine friction without any material modification.

The running-in phase is known as a transient phase where many parameters seek their stabilize form. During running-in, the system adjusts to reach a steady-state condition between contact pressure, surface roughness, interface layer, and the establishment of an effective lubricating film at the interface. These adjustments may cover surface conformity, oxide film formation, material transfer, lubricant reaction product, martensitic phase transformation, and subsurface microstructure reorientation (Hsu, et al., 2005). Next, the running-in phase is followed by a steady state phase which is defined as the condition of a given tribo-system in which the average dynamic coefficient of friction, specific wear rate, and other specific parameters have reached and maintained a relatively constant level (Blau, 1989).

Due to the complexity of the involved parameters, the discussion of running-in in this book chapter will be focused on the topographical change, contact stress and residual stress of an engineering surface which is caused by rolling contact of a smooth body over a rough surface. Specifically, the attention will be concentrated on the asperity of the rough surface. There are many applications of the rolling contact in mechanical components system, such as in bearing components, etc., therefore, the observation of the running-in of rolling contact becomes an interesting subject. The obvious examples are the contact of the thrust roller bearing and deep groove ball bearing where the running-in occurs on the rings. Its initial topography, friction, and lubrication regime change due to the contact with the balls on the first use of the bearing lifespan history.

This chapter is devided into six sub-chapters which the first sub-chapter deals with the significance of running-in as introduction. It is continued with the definition of "rolling contact" and "running-in" including with the types classification in sub-chapter 2 and 3, respectively. In sub-chapter 4, the model of running-in of rolling contact is studied by presenting an analytical model and numerical simulation using finite element analysis (FEA). A running-in model, derived analytically based on the static contact equation on the basis of ellipsoid deformation model (Jamari & Schipper, 2006) which is applied deterministically (Jamari & Schipper, 2008) on the real engineering surface, is proposed and verified with the experimental investigations. The topographical evolution from the initial to the final surface during running-in of rolling contact is presented. The numerical simulations of the two-dimensional FEA on the running-in of rolling contact are employed for capturing the plastic deformation, the stress and the residual stress. The localized deformations on the summit of the asperities and the transferred materials are discussed as well as the surface and subsurface stresses of the engineering surface during and after repeated rolling contacts. In sub-chapter 5, the experimental investigations, conducted by Jamari (2006) and Tasan et al. (2007), are explored to depics the topographical change of the engineering surface during running-in of rolling contact. With the semi-online measurement system, the topographical change is observed. The longitudinal and lateral change of the surface topography for several materials are presented. The last, concluding remarks close the chapter with some conclusions.

2. Rolling contacts

2.1 Definition of rolling contact

When two non-conformal contacting bodies are pressed together so that they touch in a point or a line contact and they are rotated relatively so that the contact point/line moves over the bodies, there are three possibilities (Kalker, 2000). First, the motion is defined as rolling contact if the velocities of the contacting point/line over the bodies are equal at each point along the tangent plane. Second, it is defined as sliding and the third is rolling with sliding motion.

According to Johnson (1985), a combination between rolling, sliding and spinning can be occured during the rolling of two contacting bodies, either for line contact or point contact. By considering the example of the line contact between body *1* and body *2*, as is shown in the Fig. 1, the rolling contact is defined as the relative angular velocity between the two bodies about an axis lying in the tangent plane. Sliding or slip is indentified as the relative velocity between the two bodies or surfaces at the contact point *O* in the tangent plane, whereas the spinning is the relative angular velocity between the two bodies about the common normal through *O*.

2.2 Types of rolling contact

Based on the contact area, the problems of rolling can be divided in three types (Kalker, 2000). (a) Problem in which the contact area is almost flat. The examples are a ball rolling over a plane; an offset printing press; and an automotive wheel rolling over a road. (b) Problems with non-conformal contact in the rolling direction plane and curved in the lateral. The examples are a railway wheel rolling over a rail and a ball rolling in a deep groove, as in ball bearings. (c) Problems in which the contact area is curved in the rolling direction, and conforming in the lateral direction where the example is a pin rolling in a hole.

In the case of rolling friction where the friction takes place on the rolling contact motion and produce the resistance to motion, Halling (1976) classified the rolling contact into: (a) Free rolling, (b) Rolling subjected to traction, (c) Rolling in conforming grooves and (d) Rolling around curves. Whenever rolling occurs, free rolling friction must occur, whereas (b), (c) and (d) occur separately or in combination, depending on the particular situation. The wheel of a car involves (a) and (b), in a radial ball bearing, (a), (b) and (c) are involved, whereas in a thrust ball bearing, (a), (b), (c) and (d) occur.

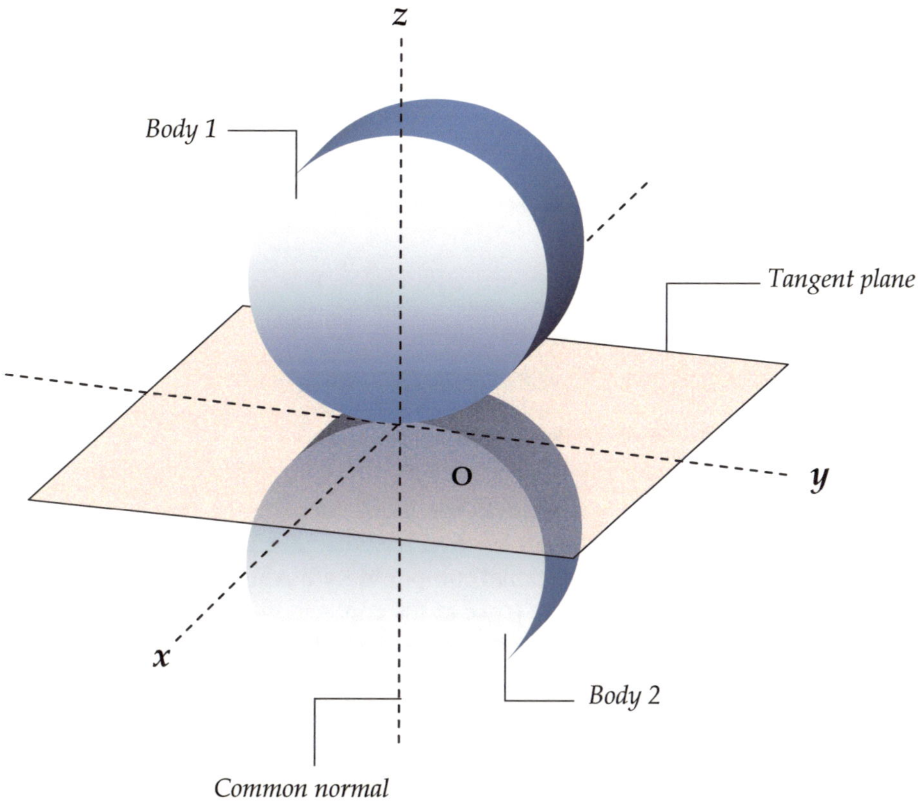

Fig. 1. Line contact between two non-conformal bodies, depicted on the coordinate system

Depending on the forces acting on the contacting bodies, rolling can be classified as free rolling and tractive rolling. Free rolling is used to describe a rolling motion in which there is no slip and the tangential force at the contacting point/line is zero. The term tractive rolling is used when the tangential force in the point/line of contact is not zero or a slip is exist.

3. Running-in

3.1 The definition of running-in

By definition, Summer-Smith (1994) describes running-in as: "The removal of high spots in the contacting surfaces by wear or plastic deformation under controlled conditions of running giving improved conformability and reduced risk of film breakdown during normal operation". While GOST (former USSR) Standard defines running-in as: "The change in the geometry of the sliding surfaces and in the physicomechanical properties of the surface layers of the material during the initial sliding period, which generally manifests itself, assuming constant external conditions, in a decrease in the frictional work, the temperature, and the wear rate" (Kraghelsky et al., 1982).

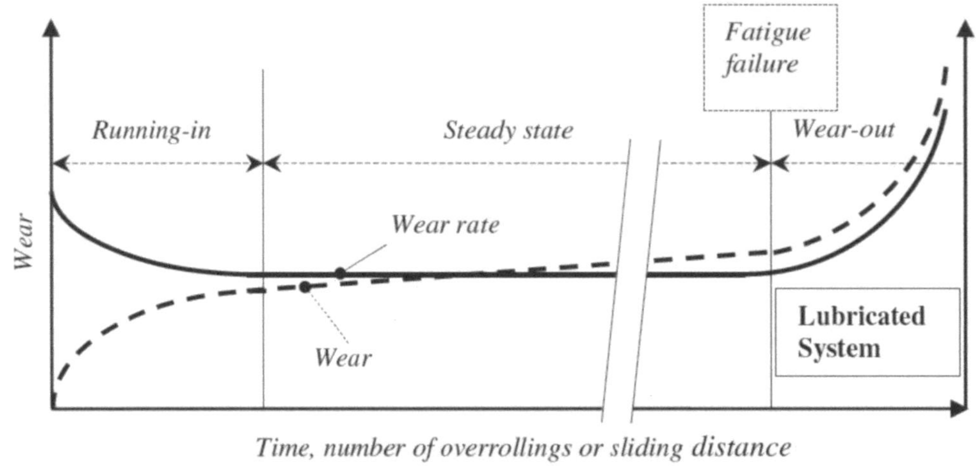

Fig. 2. Schematic representation of the wear behavior as a function of time, number of overrollings or sliding distance of a contact under constant operating conditions (Jamari, 2006)

Generally, the running-in, which is related to the terms breaking-in and wearing-in (Blau, 1989), has been connected to the process by which contacting machine parts improve in conformity, surface topography and frictional compatibility during the initial stage of use. It is focused on the interactions, which take place at the contacted interface on the macro scale and asperity scale, and involves the transition in the existing surface physical condition. For instance in gears contact of the transmission system, the tribologist observes the transition from the unworn to the worn state, from one surface roughness to another surface roughness, from one contact pressure to another contact pressure, from one frictional condition to another, etc. However, the physical change on the contacting surface in this phase, it also can be categorized as "physical damage" at the asperity level, is more beneficial instead of detrimental.

Lin and Cheng (1989) divided three types of wear-time behavior. Majority of the wear time curves observed is of type I, in which the wear rate is initially high and then decrease to a lower value. Wear of type II is more usually observed under dry conditions and the wear rate is constant in time, whilst wear rate of type III is increasing continuously with time. Jamari (2006) presented the wear-time curve which consists of three wear regimes: running-in, the steady state and accelerated wear/wear out as shown in Fig. 2. Each regime has a different wear behavior.

During running-in, the wear-time curve belongs to wear regime of type I. The surface of the material gets adjusted to the contact condition and the operating environment. Wear regime of type II usually takes place in the steady state wear process where the wear-time function is linear. In the wear out regime, the wear rate increases rapidly because of the fatigue wear which occurs on the upper layers of the loaded surface. The dynamic loading causes fatigue on the surface and results in larger material loss than small fragments associated with adhesive or abrasive wear mechanism. Breakdown of lubrication due to temperature increase, lubricant contaminant or environment factors are other causes of the increase of wear and wear rate in this regime (Lin & Cheng, 1989).

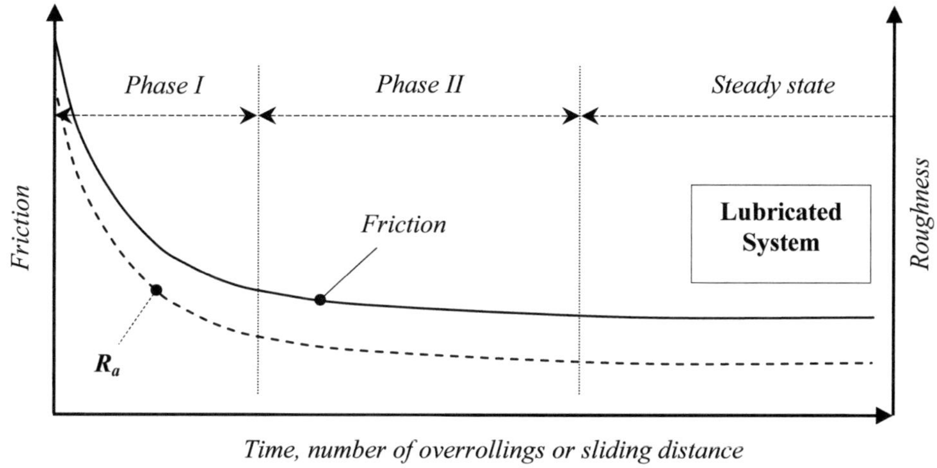

Fig. 3. The change of the coefficient of friction and a roughness as a function of time, number of overrollings or sliding distance of a contact under constant operating conditions (Jamari, 2006)

Figure 3 depicts the friction and roughness decrease as a function of time, the number of overrollings and/or sliding distance. In the running-in phase, the changes of coefficient of friction and roughness in surface topography is required to adjust or minimize energy flow, between moving surface (Whitehouse, 1980). Based on Fig. 3, phase I is indicated by the striking decrease of the surface roughness and the coefficient of friction. On phase II, the micro-hardness and the surface residual stress increases by work hardening and the changes in the geometry of the contact affects the contact behavior in repetitive contacts which leads only a slight decrease of the coefficient of friction and surface roughness. After a steady state condition is obtained, where there is no significant change in coefficient of friction, the full service condition can be applied appropriate with design specifications. The steady state phase is desirable for machine components to operate as long as possible.

3.2 The types of running-in

3.2.1 Based on the shape of the coefficient of friction

Blau (1981) started his work in determining the running-in behavior by collecting numerous examples of running-in experiments and conducting the laboratory experiments which resulted in sliding coefficient of friction versus time behavior graphs, in order to develop a physical realistic and useful running-in model. A survey of literature revealed eight common forms of coefficient of friction versus sliding time curves. Some of the possible occurrences and causes related to each type of friction curve were intensively discussed

(Blau, 1981). Each type is not uniquely ascribed to a single process or unique combination of processes, but rather must be analyzed in the context of the given tribosystem.

3.2.2 Based on the induced system

Blau (2005) divided the tribological transition of two types, namely induced and non-induced or natural transition. The induced transition is referred to when an operator applies a specified set of the first stage procedures in order to gain the desired surface condition after running-in of certain contacting components. For example, the induced running-in takes place when the new vehicle owner's drive the new car by following the manual book recommendation for the first 100 km.

Non-induced or natural running-in occurs as the system 'ages' without changing the operating contact conditions such as decreasing the load, velocity et cetera. The change of the friction and wear during the sliding contact of a reciprocating piston ring along the cylinder wall is a good illustration of the natural transition. The hydrodynamic or mixed film lubrication regime which is performed during the piston ring reaches its highest sliding velocity at the mid of the stroke. Then, the lubrication regime changes to the boundary film condition when the piston rings reach its lowest velocity at the bottom and top of the stroke. The different regime of lubrication during the piston stroke is realized by the engine designer but the fact that the wear is higher at the bottom or top of the stroke due to the lubrication regime is not intentionally arranged by the designer (Blau, 2005).

3.2.3 Based on the relative motion

Based on the relative motion as explained by Kalker (2000), there are three types of motion, namely rolling, sliding and rolling-sliding contact which generate the different mode in surface topographical change. Considering the surface topographical change during the running in period, there are two dominant mechanisms: plastic deformation and mild wear (Whitehouse, 1980). Shortly after the start of sliding, rolling or rolling-sliding contact between fresh and unworn solid surface, these mechanisms occur.

The rolling contact motion induces the plastic deformation at the higher asperities when the elastic limit is exceeded, as investigated experimentally by Jamari (2006) and Tasan et al. (2007). On the ball on disc system, the rolling contact generates the track groove on the disc rolling path which modifies the rough surface topography after a few cycles on the running-in phase. In this case, the plastic deformation mechanism due to normal loading is a key factor in truncating the higher asperities, decreasing the center line average roughness, R_a, and changing the surface topography (Jamari, 2006).

In the sliding contact, the change of the surface topography is commonly influenced by mild wear, considering several wear mechanisms such as abrasive, adhesive and oxidative. Many models, in predicting the surface topography change on the running-in of sliding contact, proposed with ignoring the plastic deformation (Jeng et al., 2004). Sugimura et al. (1987) pointed that the wear mechanism, i.e. abrasive wear, contributes to the surface topographical change of a Gaussian surface model during running-in of sliding contact. The work continued by Jeng et al. (2004) which introduced the translatory system of a general surface into a Gaussian model. Their works successfully predicted the run-in height distribution of a surface after running-in phase of a sliding contact system.

Running-in of rolling contact with slip, which indicates the rolling-sliding contact, promotes both plastic deformation and wear in modifying the surface topography. Wang et al. (2000)

investigated the change of surface roughness, R_a as a function of sliding/rolling ratio and normal load. The small amount of sliding at the surface increased the wear rate, minimized the time to steady state condition and resulted into a smoother surface than with pure rolling. The combination of the plastic deformation model and wear model in predicting the material removal during the transient running-in of the rolling-sliding contact is proposed by Akbarzadeh and Khonsari (2011). They combined the thermal desorption model, which is the major mechanism of adhesive wear, with the plastic deformation of the asperity in predicting the material removal in macro scale. They measured wear weight, wear depth, surface roughness and coefficient of friction of the two rollers which was rotated for several rolling speeds and slide to roll ratio. The increasing of rolling speed resulted a better protecting film in lubrication regime and reduced the wear weight and wear depth while the increase of the slide to roll ratio increased the sliding distance and generated lower wear rate. The thermal desorption model indicated that increasing of the sliding speed caused the molecules have less time to detach from the surface and therefore the wear volume rate decreased.

4. Running-in of rolling contact model

The models for predicting the surface topography change due to running-in, published in the literature, are mostly related with sliding contact. Started from Stout et al. (1977) and King and his co workers (1978), the topographical changes in running-in phase is predicted by considering the truncating functions of Gaussian surface to obtain the run-in height distribution. Sugimura et al. (1987) continued by proposing a sliding wear model for running-in process which considers the abrasive wear and the effect of wear particles. Due to its limitation of the model for the Gaussian surface, Jeng and co-workers (2004) have developed a model which describes the change of surface topography of general surfaces during running-in.

Other approaches have been applied by researchers for modeling running-in. Lin and Cheng (1989) and Hu et al. (1991) used a dynamic system approach, Shirong and Gouan (1999) used scale-independent fractal parameters, and Zhu et al. (2007) predicted the running-in process by the change of the fractal dimension of frictional signals. Liang et al. (1993) used a numerical approach based on the elastic contact stress distribution of a three-dimensional real rough surface while Liu et al. (2001) used an elastic-perfectly plastic contact model. In running-in of sliding contact, some parameters such as: load, sliding velocity, initial surface roughness, lubricant, and temperature have certain effects. Kumar et al. (2002) explained that with the increase of load, roughness and temperature will increase the running-in wear rate on the sliding contact.

However, based on the literature review, there are less publications discussed the running-in of rolling contact model, especially, dealt with the deterministic contact of rough surface. Most of the running-in models available in literature, is devoted to running-in with respect to wear during sliding motion. These models are designed to predict the change of the macroscopic wear volume or the standard deviation of the surface roughness rather than the change of the surface topography locally on the real engineering surface during the running-in process.

On the next section, an analytical and numerical model are described to propose another point of view in surface topographical change due to running-in of rolling contact. The discussion of the rolling contact motion at running-in phase is focused on the free rolling contact between rigid bodies over a flat rough surface and neglects the tangential force, slip and friction on the contacted bodies. The point contact is explored in the analytical running-in contact model and experiments while the line contact is observed in numerical model using the finite element analysis.

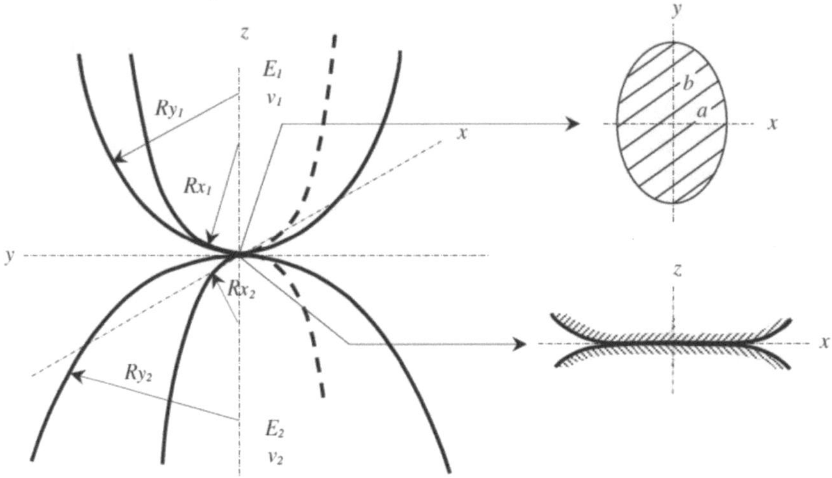

Fig. 4. Geometry of elliptical contact, after Jamari-Schipper (2006)

4.1 Analytical model

The change of surface topography due to plastic deformation of the non-induced running-in of a free rolling contact is presented in this model. On the basis of the elastic-plastic contact elliptical contact model developed by Jamari and Schipper (2006) and the use of the deterministic contact model of rough surfaces which has been explained extensively in Jamari and Schipper (2008), the surface topography changes during running-in of rolling contact is modeled.

Jamari and Schipper (2006) proposed an elastic-plastic contact model that has been validated experimentally and showed good agreement between the model and the experiment tests. In order to predict surface topography after running-in of the rolling contact, they modified the elastic-plastic model of Zhao et al. (2000) and used the elliptical contact situation to model the elastic-plastic contact between two asperities. Figure 4 illustrates the geometrical model of the elliptic contact where *a* and *b* express the semi-minor and semi-major of the elliptical contact area. The mean effective radius R_m is defined as:

$$\frac{1}{R_m} = \frac{1}{R_x} + \frac{1}{R_y} = \frac{1}{R_{x1}} + \frac{1}{R_{x2}} + \frac{1}{R_{y1}} + \frac{1}{R_{y2}} \tag{1}$$

R_x and R_y denote the effective radii of curvature in principal x and y direction; subscripts *1* and *2* indicate body *1* and body *2* respectively. The modification of the previous model leads the new equation of the elastic-plastic contact area A_{ep} and the elastic-plastic contact load P_{ep}, which is defined as follows:

$$A_{ep} = 2\pi R_m \omega \frac{\alpha}{\beta} + (2\pi\sqrt{R_x R_y \omega} - 2\pi R_m \omega \frac{\alpha\beta}{\gamma})\left[3\left(\frac{\omega - \omega_1}{\omega_2 - \omega_1}\right)^2 - 2\left(\frac{\omega - \omega_1}{\omega_2 - \omega_1}\right)^3\right] \tag{2}$$

$$P_{ep} = A_{ep}\left[c_h H - H\left(c_h - \frac{2}{3}K_v\right)\frac{\ln\omega_2 - \ln\omega}{\ln\omega_2 - \ln\omega_1}\right] \tag{3}$$

where ω is the interference of an asperity, subscripts *1* and *2* indicate body *1* and body *2* respectively, α and β are the dimensionless semi-axis of the contact ellipse in principal x and y direction respectively, γ is dimensionless interference parameter of elliptical contact, c_h is the hardness factor, H is the hardness of material and K_v is the maximum contact pressure factor related to Poisson's ratio v:

$$K_v = 0.4645 + 0.3141v + 0.1943v^2 \tag{4}$$

The change of the surface topography during running-in is analyzed deterministically and is concentrated on the pure rolling contact situation. Figure 5 shows the proposed model of the repeated contact model performed by Jamari (2006). Here, $h(x,y)$ is the initial surface topography. The surface topography will deformed to $h'(x,y)$ after running-in for a rolling contact. The elastic-plastic contact model in Eq. 2 and 3 are used to predict the $h'(x,y)$. The calculation steps are iterated for the number/distance of rolling contact.

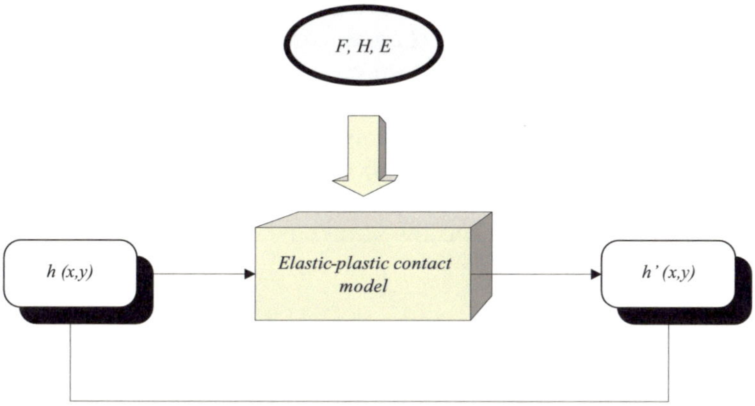

Fig. 5. The model of the surface topography changes due to running-in of a rolling contact proposed by Jamari (2006)

4.2 Finite element analysis of running-in of rolling contact

The next model of running-in of rolling contact is proposed numerically. In order to visualize the topographical change of the rough surface and observe the stress distribution during running-in phase, the two-dimensional finite element analysis (FEA) is conducted. A rigid cylinder was rolled over a rough surface in finite element software by considering the plain strain assumption. The free and frictionless rolling contact was assumed in this model. The cylinder was 4.76 mm in diameter while the asperity height on rough surface, Z_{as}, was 0.96 mm, the spherical tip on the summit of asperity, R_a, was 0.76 mm and the pitch of the rough surface, P was 1.5 mm. The dimensions of the rough surface control its wave length and amplitude. The model, simulation steps and validation, described respectively in this section, have been used in the previous FEA of rolling contact simulation (Ismail et al., 2010).

4.2.1 Material model and simulation steps

The rough surface was modeled as aluminum and was considered as elastic- plastic material with strain hardening behavior where the behavior of the stress versus strain curve was obtained from the tensile test conducted by Bhowmik (2007). The curve described the strain hardening effect of the elastic-plastic material model and had been verified with his experiments and finite element simulations. For the rough surface, the elastic modulus (E), yield stress (S_y), and Poisson's ratio (υ) were 70 GPa, 270 MPa, and 0.32, respectively whereas the cylinder was assumed to be a rigid body so that there is no deformation occured during rolling contact.

The four nodes element with plain strain model was specified on the rough surface with the refine mesh was applied on the top of the rough surface. As depicted in Fig. 6, the refinement was also set on one of the asperity (at the center position) from the asperity summit to the bulk material for better investigation on the contact stress and residual stress.

The simulation steps in FEA, as shown in Fig. 6, were conducted as follows: (a) the normal static contact was applied on the cylinder over the rough surface for an interference, ω; (b) by maintaining the vertical interference, the cylinder rolled over to the right direction incrementally until reached the end of the rough surface and the cylinder was moved up for unloading; (c) the rolling contact of the cylinder over the rough surface was repeated for three times in order to observe the transition of the running-in phase.

In order to compare the topographical change, contact stress and residual stress of the rolling contact, another simulation was carried out by conducting the repeated static contact of the rigid cylinder to the rough surface. The three steps on the repeated static contact were conducted as follows: (a) the normal static contact was applied on the cylinder over the rough surface with certain interference, ω; (b) unloading the contact load by moving the cylinder up to the origin position; and (c) repeating this loading and unloading contact for three cycles. The same interferences and the same model were used in the repeated rolling contact and repeated static contact.

4.2.2 FEM model verification

The critical interference, ω_c, and the critical contact width b_c proposed by Green (2005) for determining the yielding limit between the elastic and the elastic-plastic deformation of line contact were used to verify the model. These equations, derived by using the distortion energy yield criterion of maximum von Mises stress, are defined as:

$$\omega_C = R\left(\frac{CS_y}{E}\right)^2\left[2\ln\left(\frac{2E'}{CS_y}\right)-1\right] \tag{5}$$

$$b_C = \frac{2R(CS_y)}{E'} \tag{6}$$

where $C = 1.164 + 2.975\upsilon - 2.906\,\upsilon^2$ and E' is the equivalent elastic modulus. The analytical equations were compared with the results of the finite element simulation of the present model in predicting the yielding of the model.

Based on Eq (5) and (6), the critical interference and the critical contact width of the static contact between a rigid cyclinder versus a single asperity are 9.53×10^{-5} mm and 2.88×10^{-3} mm, respectively. A single asperity is employed as a representation of the rough surface.

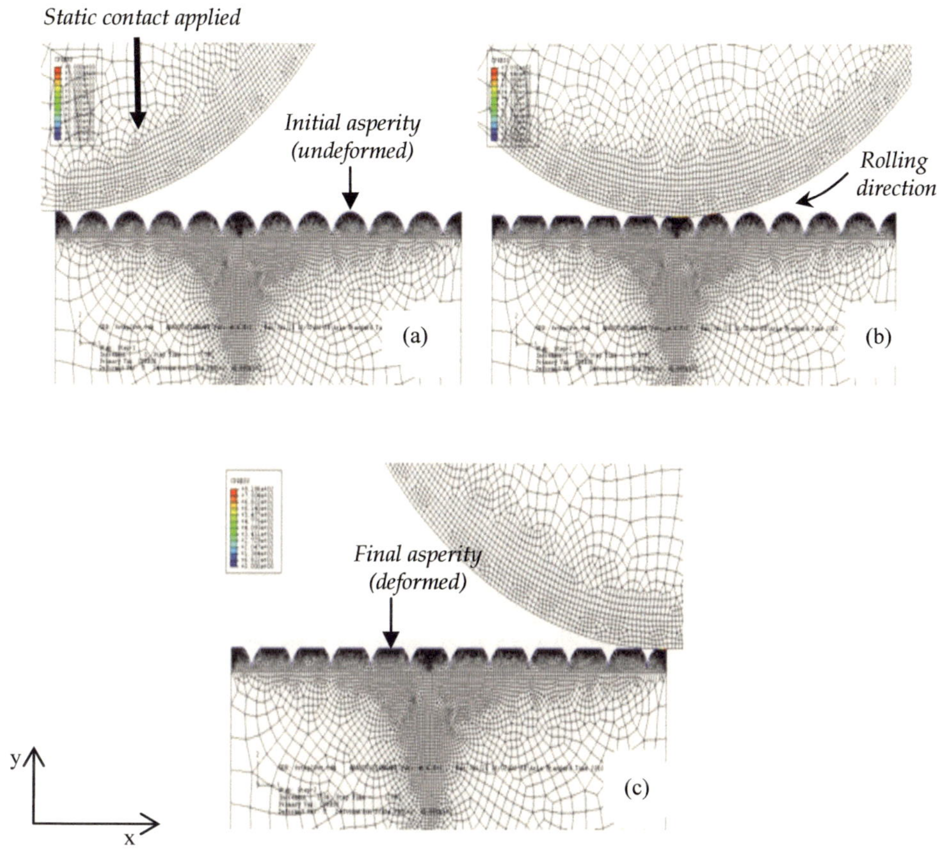

Fig. 6. The rolling contact simulation: (a) start from the static contact, (b) followed by rolling contact with maintaining the contact load, and (c) the roller reaches the end of the rough surface and is unloaded

When the calculated ω_c is applied on the present model of the finite element simulation, the measured contact width in FEA is 3 x 10^{-3} mm. The result implies the rational agreement between the analytical and numerical results where the deviation is 4.16 %. Then the von Mises yield stress criterion in the FEA was used to check the maximum stress. The obtained maximum yield stress indicates that the material starts to yield and the deformation is categorized in elastic-plastic regime. By comparing the von Mises stress of the obtained value in FEA (249.7 MPa) and the yield stress of the material model 270 MPa, a deviation of 7.52 % is found. The comparison of the analytical calculation and numerical simulation is listed in Table 1.

The previous critical interference and contact width are relatively small compared to the analytical calculation of the contact between the rigid cylinder with a flat surface. The analytical result of the critical interference and contact width in this case, where ω_c and b_c are 1.98 x 10^{-3} mm and 5.99 x 10^{-2} mm, respectively, are twenty times higher compared to the previous calculation. The accuracy of the analytical calculation of this model, compared to the results in FEA, increase significantly where the deviation of the contact width and the maximum von Mises stress are 2.93% and 0.13%, respectively. Table 1 shows the comparison between the prediction of the analytical model (Green, 2005) and the numerical simulation of single static contact occupying the present model. The table exhibits the transition of the elastic to elastic-plastic deformation with respect to the ω_c, the b_c and the maximum von Mises stress.

Contacting Bodies	ω_c analytic (mm)	b_c analytic (mm)	b_c numeric (mm)	Diff. of b_c (%)	Max stress (vM) numeric (MPa)	Diff. of max stress (%)
Cylinder vs singe asperity	9.53×10^{-5}	2.88×10^{-3}	3×10^{-3}	4.2	249.7	7.5
Cylinder vs flat surface	1.98×10^{-3}	5.99×10^{-2}	5.98×10^{-2}	0.1	262.1	2.9

Table 1. Comparison between the analytical model and numerical simulation for determining the transition of the elastic to elastic-plastic deformation for the static contact of the present FE model

The conclusion of the model verification is the present finite element simulation has a good agreement with the previous analytical model in predicting the critical interference, critical contact width and the maximum von Mises stress. However, the lower deviation, which is found in the contact between cylinder versus flat, argues that the Green's model has an opportunity to be derived by considering the contact between cylinders with the high ratio of the diameters such as the contact between cylinder and one asperity in this case. The modification of the new model is planned as the future work such that the prediction of the elastic-plastic deformation of two-dimensional rough surface can be done analytically.

4.2.3 Topographical change due to repeated rolling contact

In order to demonstrate the effect of the rolling contact to the asperity and the bulk material, the interferences in this simulation are set larger than the critical interference. Two interferences, $\omega_1 = 2 \times 10^{-2}$ mm and $\omega_2 = 4 \times 10^{-2}$ mm, are employed in this model to analyze the plastic deformation and the material transfer as well as the contact stress and the residual stress during and after the rolling contact. Respectively, these applied interferences are nearly 10 times and 20 times higher than critical interference of the contact model between the cylinder and flat surface.

Figure 6 shows the topographical evolution from the initial undeformed, contacted and the final deformed asperities due to plastic deformation of rolling contact. The asperity height on the rough surface is truncated and some materials are displaced after the rigid cylinder rolls over the rough surface. The observation is focused on a single asperity as a representative of the rough surface. Surface topographical changes of an asperity after several rolling contacts are presented in Fig. 7.

The change of the asperity height due to rolling contact can be seen on Fig. 7 (a) for ω_1 and Fig. 7 (b) for ω_2. The dashed and solid lines show the situation before and after the rolling contact deformation, respectively. The deformation was captured after the unloading of each cycle of the rolling contact. The figures exhibit that the first cycle causes the highest deformation and it is followed by only slight deformation at the second cycle. There is no significance difference on the asperity height after the second cycle which implies that, in this case, the steady state deformation is reached. The transition of the running-in of rolling contact occurs at the cycle 2.

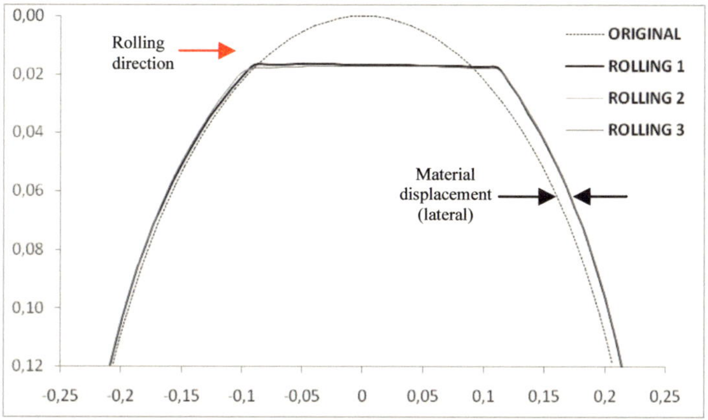

(a)

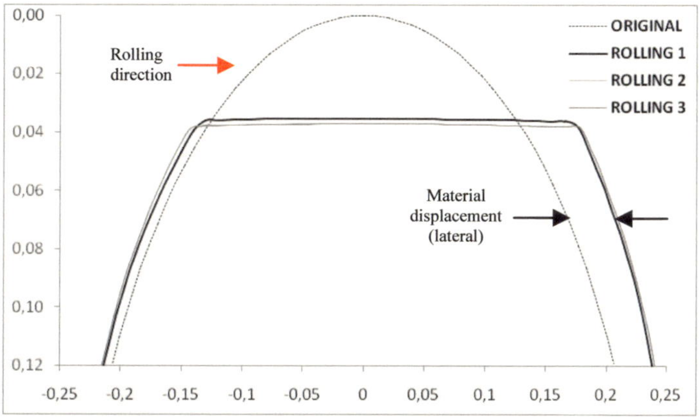

(b)

Fig. 7. Surface topographical change of a single asperity due to repeated rolling contact for: (a) $\omega_1 = 2 \times 10^{-2}$ mm and (b) $\omega_2 = 4 \times 10^{-2}$ mm

On their FEA of rolling cylinder of a deformable flat, Bijak-Zochowski and Marek (2007) also found that the steady state deformation was usually attained within the first two cycles for repeated rolling contact. The deformation for the next rolling has a small difference. The next discussion of the experimental running-in of rolling contacts, Jamari (2006) also reported that the censoring higher asperity of the rough surface is initially high on the first ten rolling cycle while Tasan et al. (2007) obviously resumed that the first rolling contact has a highest deformation and it is followed by the slight deformation for the next cycles.

The flattening of the asperities on the first rolling cycle means that the conformity of the contact increases during plastic deformation. With the increase of the contact conformity, the contact area gets wider, and the contact stress become more homogeneously distribute. The increasing conformity and contact area induce the stability of the plastic deformation on the asperities.

Considering the material displacement of the finite element simulation on Fig 7 (a) and (b), the material on the summit of the asperity is displaced laterally on the same direction of the rolling contact. The discussion of the material displacement is explored on the next section.

4.2.4 Stresses of rolling contact

The observation of the contact stress and the residual stress during and after the first cycle of repeated rolling contact for ω_2 is depicted on Fig. 8 (a) and (b), respectively. These figures are captured during and after the first rolling contact. Figure 8 (a) depicts the situation when the rigid cylinder is located at the centre of the contacted asperity, which is marked with the wide area of von Mises stress distribution while Fig. 8 (b) shows the residual stress after the first rolling contact is finished and unloaded.

The discussion of rolling contact stress is focused on the contacted asperity where the plastic deformation is easily noticed on the summit of the asperity and the maximum von Mises stress area reaches the surface. When the applied interference below its critical point and the deformation still behaves elastically, the highest von Mises stress area is located on the sub surface, few distances below the surface. The highest von Mises stress area moves to attain the surface as the applied interference is higher than the critical interference. This plastic deformation phenomenon was also discussed Jackson and Green (2005) which modelled the elastic-plastic static contact between the hemisphere and the rigid flat. Fig 8 (a) also depicts the plastic flow when the cylinder rolls over an asperity. The plasticity is formed on the right side of the asperity and produces a lateral material displacement as the same direction of the rolling motion.

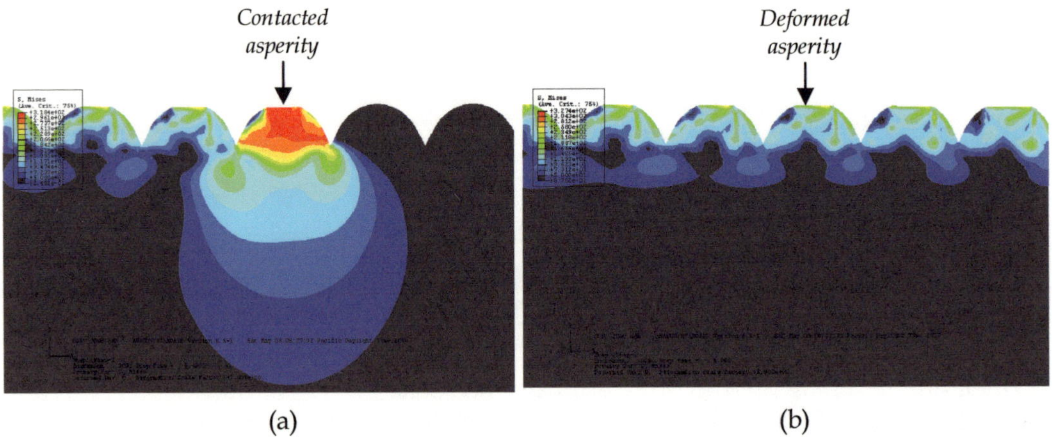

Fig. 8. The von Mises stress analysis for $\omega_2 = 4 \times 10^{-2}$ mm: (a) during the first rolling contact and (b) residual stress after the first rolling contact.

The von Mises stress of the contact stress and the residual stress are plotted as a function of the depth of the rough surface for the three repeated rolling contact, as seen in Figure 9 (a) and (b), respectively. In Fig. 9 (a), the behaviour of the stress distribution on the surface of the three repeated rolling contact only has a slight difference but at the subsurface of the asperity until the bulk material, the second and third rolling contact has a lower von Mises stress. The stress decreases after the first rolling contact and it is predicted that the material become harder due to strain hardening behavior. Kadin, et al. (2006) which studied the multiple loading-unloading of a spherical contact also reported the stability of plasticity distribution on the second loading-unloading contact. The linear strain hardening behavior was pointed as the causes of this phenomenon.

After the rolling cylinder is unloaded for each rolling contact, the stress distributions are captured again for analyzing the residual stress. Figure 8 (b) shows that the highest residual stress is found on the flattened asperity summit. Based on Fig. 9 (b) which depicts the residual stress as a funtion of the depth by using the von Mises criteria, the curves of three repeated rolling contact is nearly coincided after depth reaches 0.15 mm. The maximum of the residual stress, located on the asperity summit, from the first until the third rolling contact is increase from 179.7 MPa, 220.8 MPa and 248.8 MPa, respectively. The increase of the residual stress is a consequence of the strain hardening behavior on plastic the deformation. The material strain hardening on the summit of asperity has a protective effect in reducing further plastic deformation and induces a stability of the deformation after the first rolling contact.

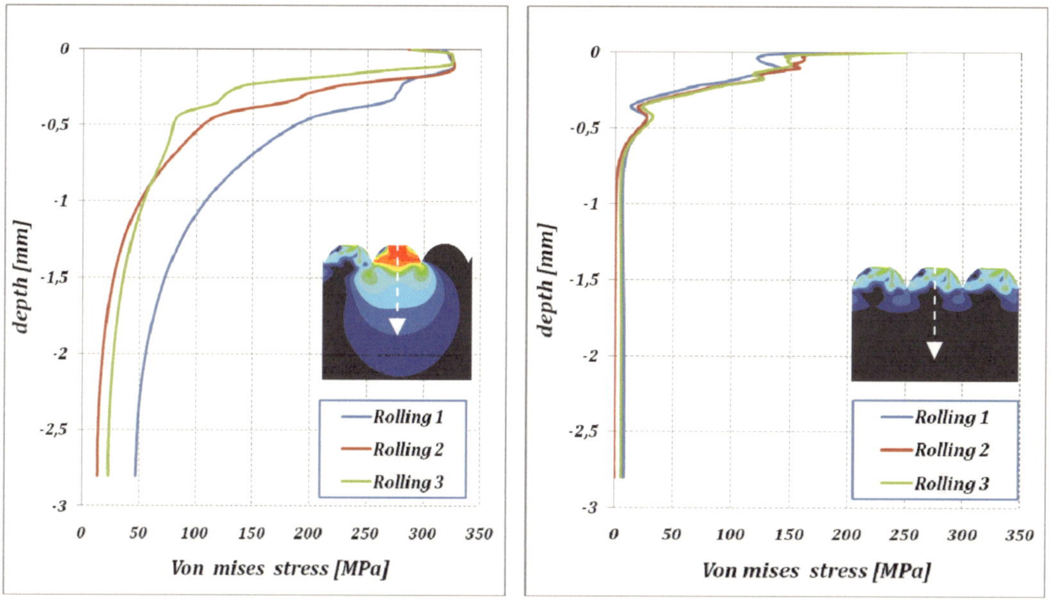

Fig. 9. von Mises Stress analysis (a) during rolling contact and (b) residual stress after the rolling contact finish.

The strength and expected life of mechanical components can be influenced by residual stress due to its effect on contact fatigue and wear (Bijak-Zochowski and Marek, 2007). Nelias and his co-workers (2006), assumed that the volume of material will detach from the surface after very few cycles. The detachment of the material occurs when the equivalent plastic strain found after unloading, located at the surface, exceeds a threshold value. Nelias et al. (2006) determined that the threshold value of the equivalent plastic strain is 0.2 %. In this present model, the maximum residual stress is found on the summit of the asperities which able to lead the detachments. The investigation are planed in the future for analizing the wear and the equiavalent plastic strain by combining the FEA model and Nelias et al. model (2006).

4.2.5 Comparison with the repeated static contact

Jamari (2006) reported three methods in repeated contact on the rough surface for observing the topographical change: (a) repeated static contact; (b) repeated moving contact; and (c) repeated rolling contact. The proposed elastic-plastic contact model of Jamari and Schipper (2006) could predict the three types of the repeated contact quite well. In the case of free or pure rolling contact, it does not contain a tangential force. Therefore, it is reasonable that this type of motion can be modeled by multiple-indentation of one body to another body without changing the indentation position.

A comparison between the repeated rolling contact and the repeated static contact is discussed to show the difference of the topographical change for the both mechanisms. The topographical change of a single asperity due to repeated static contact is depicted in Fig. 10. It shows the symmetrical material displacement on the both side of the asperity, whereas in the repeated rolling contact, the asymmetrical material displacement is found with the majority of the material is displaced at the same direction of the rolling, as seen in Fig 7 (a) and Fig 7 (b). The deformed asperity on the repeated static contact shows that the centre of the asperity summit has larger deformation than its edge whereas on the repeated rolling contact the deformation is almost flat from the edge and centre of the asperity.

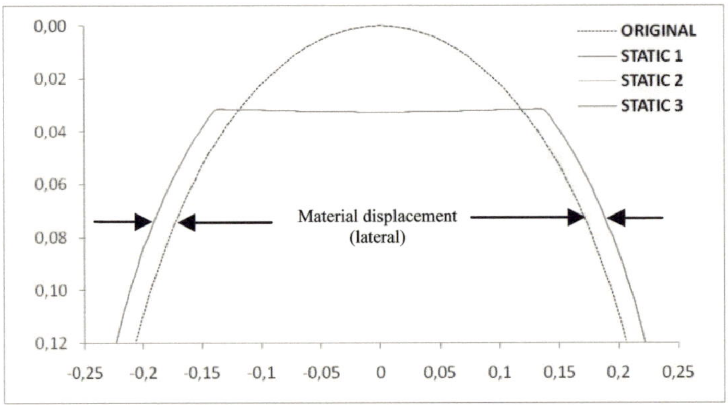

Fig. 10. The surface topographical change due to repeated static contact for a single asperity for $\omega_2 = 4 \times 10^{-2}$ mm

The von Mises stress distribution of the repeated static contact for ω_2 is shown in Fig. 11 (a) for the stresses during first loading and in Fig. 11 (b) for residual stress after the first unloading. It can be seen on the Fig. 11 (a) that the cylinder is in contact with three asperities and the highest contact stress takes place on the middle asperity. The centre of the middle asperity deformation due to repeated static contact is observed in Fig. 10. The concave shape

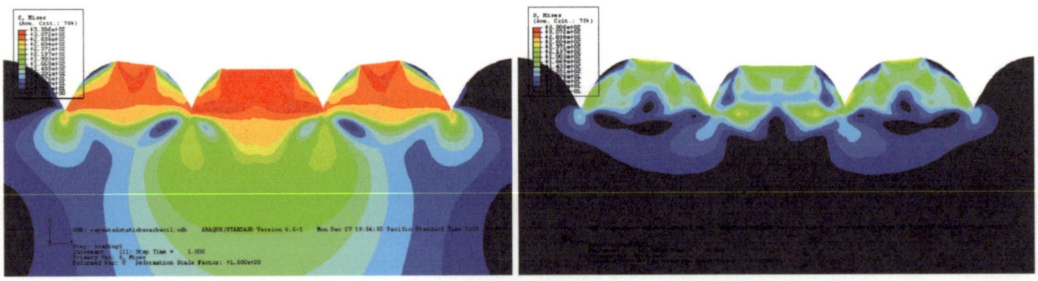

Fig. 11. The von Mises Stress analysis for $\omega_2 = 4 \times 10^{-2}$ mm: (a) during first static contact and (b) residual stress after the first static contact finish.

on the asperity surface is formed as the consequences of the highest contacted stress on the centre of the asperity. In Fig. 11 (b) the residual stress is found on the summit of the asperity and on the some area in bulk material, especially on the sharp valley of the rough surface. On the centre of the middle asperity, both of the contacted stress and residual stress on the repeated static contact perform symmetrical stress distribution.

5. Running-in of rolling contact experiments

Two running-in of rolling contact experiments of Jamari (2006) and Tasan et al. (2007) are presented in this section. There are two goals in discussing the experiments of running-in of rolling contact. First, the experiments are used to validate the running in model of Jamari (2006) which has been explored on the previous section. Second, the experiments are employed for investigating the change of the surface topography due to running-in of rolling contact at the lateral and longitudinal of rough surface direction.
The running-in of rolling contact experiments are conducted on the measurement setup where the details of the arrangement of the setup are presented in Fig. 12(a). The spherical indenter (ball specimen) is held by the clamping unit. The loading arm and the rotating table are positioned on the X-Y table such that the spherical indenter is located on one side of the disk, whilst the interference microscope is positioned on the other side of the disk and stands separately from the X-Y table. Figure 12(b) depicts the detail of ball holder and the position of the ball specimen.

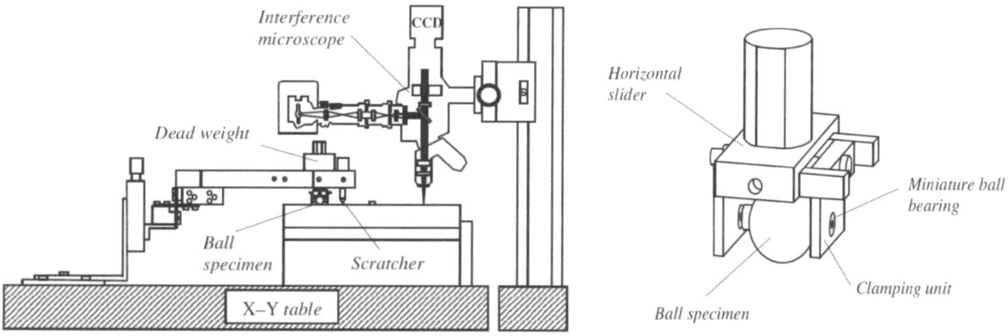

(a) (b)

Fig. 12. (a) The main view of semi on-line measurement of running-in rolling contact and (b) the rolling ball specimen and the holder (Jamari, 2006)

Silicon carbide ceramic balls SiC (H = 28 GPa, E = 430 GPa and v = 0.17) with a diameter of 6.35 mm were used as hard spherical indenters. The center line average roughness Ra of the ceramic ball of 0.01 mm was chosen to comply with the assumption of a perfectly smooth surface. Elastic-perfectly plastic aluminium (H = 0.24 GPa, E = 75.2 GPa and v = 0.34) and mild-steel (H = 3.55 GPa, E = 210 GPa and v = 0.3) were used for the rough flat surface specimens. The center line average roughness of the flat specimens varied from 0.7 to 2 mm.
Results of the rolling contact experiment, along with the model prediction for the aluminium and mild-steel surfaces are presented in Fig. 13 (a) and (b), repsectively. It can be seen that the change of the surface deformation for each number of rolling contact reaches

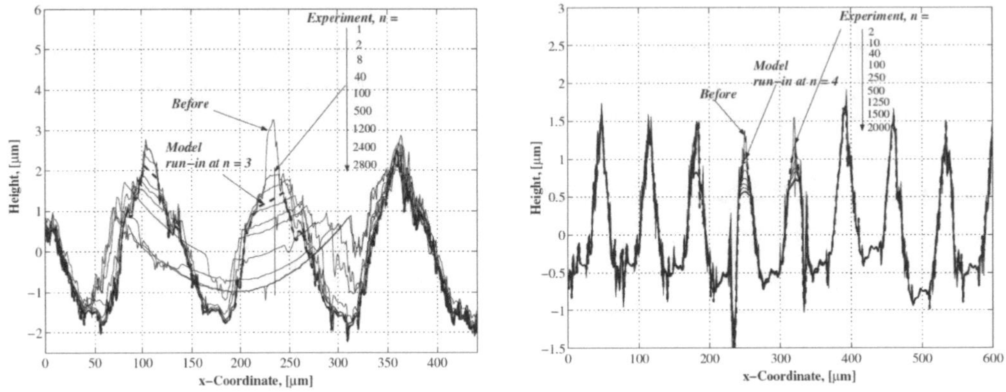

(a) (b)

Fig. 13. Profile of the matching and stitching result across the direction of rolling of (a) aluminium surface at y = 160 mm and (b) mild-steel surface at y = 158 mm (Jamari, 2006)

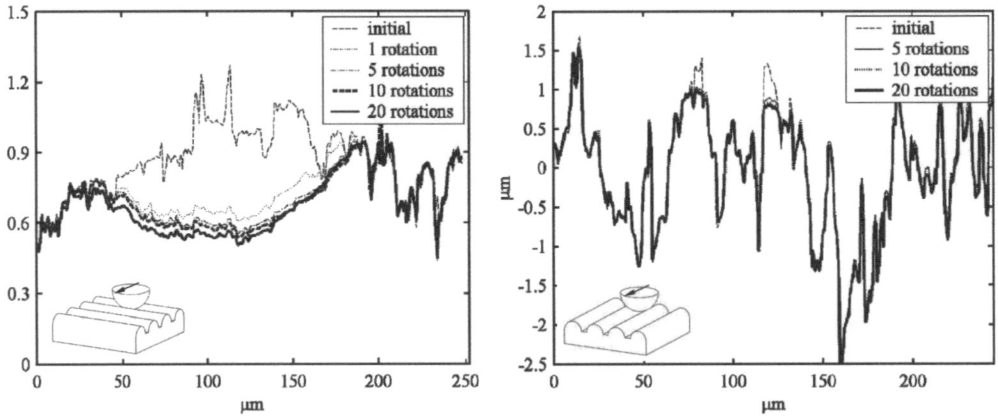

(a) (b)

Fig. 14. Profiles across the direction of rolling for (a) lateral and (b) longitudinal roughness as the function of the number of rolling cycles (Tasan, et al, 2007)

its highest gap on the first 10 cycles. There is only slight difference on the surface topographical change for each number of rolling contact after the first 10 cycles. Hsu et al. (2005) concluded that as the contact progresses to a nearly steady-state operation, the surface roughness of the tribocontacts becomes less variable. The model predicts the surface topographical change at cycle n = 3 for the aluminium surface and at cycle n = 4 for the mild-steel surface. The proposed model performs a rational agreement with the experiments for the mentioned number of rolling contact. However, on the set-up of the rolling contact experiment, the slight presence of slip is can not be avoided where its appearance can switch the pure rolling contact motion to the rolling contact with slip or rolling-sliding contact.

Compares to the Jamari's experiments, the investigations of Tasan et al (2007) on the topographical change after rolling contact between an SiC ball in contact with a rough mild steel disk (DIN 100MnCrW4) are presented. A semi on-line measurement setup was utilized to perform rolling contact experiments where the detail description can be found in (Tasan, 2005). Figure 14 depicts the topographical change of the surface on (a) the lateral direction and (b) the longitudinal direction due to running-in of rolling contact. On the lateral and longitudinal direction of this case, it is found that the rough surface deformed plastically on the contacted area at the first five cycle, where the largest deformation occurred on the first cycle, and then the deformation decreases and reaches its steady state. The experiments also exhibit the running-in phase of the rolling contact and followed by the steady state phase afterward.

In finite element simulation in Fig. 7(a), it also performs the lateral material displacement due to rolling contact experiments. The lateral displacement can be happen when there is some space for the material to be removed plastically. In repeated rolling contact, the displacement indicates that the plastic deformation is more dominant in topographical change on the asperity level rather than material removal due to wear mechanisms.

6. Concluding remarks

The studies of topographical change due to running-in of rolling contact are presented in this book chapter analytically, experimentally and numerically. A running-in model of rolling contact by considering the deterministic contact of the engineering surface has been proposed. The proposed model of running-in of rolling contact performs a good agreement with the experiments. Finite element simulations of two-dimensional rolling contact model contribute in illustrating the truncation on the asperity summit, the contact stress and the residual stress due to plastic deformation. The comparison of the repeated static contact and the repeated rolling contact leads un-similarity in material transfer direction and asperity deformation.

As the critical effect of the surface topography on the running-in stage for prolonging the lifespan of the contacted mechanical components, this investigation can contribute in predicting the initial, the change, and the final of the surface topography for a success running-in stage. Running-in plays an important role in plastic deformation, friction and wear of tribology systems during the steady-state period. Ignoring the running-in aspects means overlooking the important clues to the evolution of conjoint processes which leads to the final long-term steady-state friction and wear behavior.

7. References

Akbarzadeh, S. & Khonsari, M.M. (2011). Experimental and Theoretical Investigation of Running-in, *Tribology International*, Vol. 44, (February 2011), pp. 92-100, ISSN 0043-1648

Bhowmik, K. (2009). *Experimental and Finite Element Study of Elastic-Plastic Indentation of Rough Surfaces*, Master Thesis, Indian Institute of Science, Bangalore, India

Bijak-Zochowski, M. & Marek, P. (1997). Residual Stress in Some Elasto-Plastic Problems of Rolling Contact with Friction, *International Journal of Mechanic Science,* Vol. 39. No. 1, (January 1997), pp. 15-32, ISSN 0020-7403

Blau, P.J. (1981). Interpretations of the Friction and Wear Break-in Behaviour of Metal in Sliding Contact, *Wear,* Vol. 71, (September 1981), pp. 29 - 43, ISSN 0043-1648

Blau, P.J. (1989). *Friction and Wear Transitions of Materials: Break-in, Run-in, Wear-in,* Noyes Publications, ISBN 0-815-511965, Park Ridge, NJ, USA

Blau, P.J. (2005). On the Nature of Running-in, *Tribology International,* Vol. 38, (August 2005), pp. 1007 - 1012, ISSN 0043-1648

Halling, J. (1976) *Introduction to Tribology,* Wykeham Publication Ltd., ISBN 0387911286, London, UK

Green, I. (2005). Poisson Ratio Effects and Critical Values in Spherical and Cylindrical Hertzian Contacts, *International Journal of Applied Mechanics,* Vol. 10, No. 3, pp. 451-462

Hu, Y.Z.; Li, N. & Tonder, K. (1991). A Dynamic System Model for Lubricated Sliding Wear and Running-in, *ASME Journal of Tribology ,* Vol. 113, No. 3, (July 1991), pp. 499 - 505, ISSN 0742-4787

Hsu, S.M.; Munro, R.G.; Shen, M.C. & Gate, R.S. (2005). Boundary Lubricated Wear, In: *Wear-Materials, Mechanisms & Practice,* G.W. Stachowiak (Ed.), 37-69, John Wiley & Sons Inc., ISBN 978-0-470-01628-2, London, UK

Ismail, R.; Tauviqirrahman, M.; Jamari & D.J. Schipper. (2010). Elastic-Plastic Deformation of a Rough Surface Due To Repeated Rolling Contact, *Proceeding of International Conference on Physics and Its Applications,* Solo, Indonesia, pp. 49-52

Jackson R.L. & Green I. (2005). A Finite Element Study of Elasto-plastic Hemispherical Contact. *ASME Journal of Tribology,* Vol. 127, (April, 2005), pp. 343–54, ISSN 0742-4787

Jamari, J. (2006). *Running-in of Rolling Contacts.* PhD Thesis, University of Twente, Zutphen, ISBN: 90-365-2314-1, Enschede, The Netherlands

Jamari, J. & Schipper, D.J. (2006). An Elastic-Plastic Contact Model of Ellipsoid Bodies, *Tribology Letters,* Vol. 21, No. 3, (March, 2006), pp. 262-271, ISSN 1573-2711

Jamari, J. & Schipper, D.J. (2008). Deterministic Repeated Contact of Rough Surfaces, *Wear,* Vol. 264, (February 2008), pp. 349–358, ISSN 0043-1648

Jeng, Y.R. (1996). Impact of Plateaued Surfaces on Tribological Performance, *Tribology Transaction* Vol. 39, pp. 354 - 361, ISSN 1547-397X

Jeng, Y.R.; Lin, Z.W. & Shyu, S.H. (2004). Changes of Surface Topography during Running-in Process, *ASME Journal of Tribology,* Vol. 126, No. 3, pp. 620 - 625, ISSN 0742-4787

Johnson, K.L. (1985). *Contact Mechanics,* Cambridge University Press, ISBN 0-521-347963, Cambridge, England

Kadin, Y.; Kligerman, Y. & Etsion, I. (2006). Multiple Loading–Unloading of an Elastic–Plastic Spherical Contact, *International Journal of Solid and Structure,* Vol. 43, (November 2006), pp. 7119-7127, ISSN 0020-7683

Kalker, J.J. (2000). Linear Elasticity in Rolling Contact, In: *Rolling Contact Phenomena,* B. Jacobson & J.J. Kalker (Ed.), 1-16, Springer-Verlag Wien ISBN 3-211-83332-3, New York, USA

Kehrwald, B. (1998). *Einlauf Tribologischer Systeme*, Ph.D. Thesis, University of Karlsruhe, in German

King, T.G.; Watson, W. & Stout, K.J. (1978). Modelling the Micro-geometry of Lubricated Wear, *Proc. 4th Leeds-Lyon Symposium*, pp. 333 - 343, MEP, London, UK

Kraghelsky, V.; Dobychun, M.N. & Kombalov, V.S. (1982). *Friction and Wear Calculation Methods*, Pergamon Press, ISBN 0080254616, Oxford, UK

Kumar, R.; Prakash, B. & Sethuramiah, A. (2002). A Systematic Methodology to Characterize the Running-in and Steady-state Processes, *Wear,* Vol. 252, (March 2002), pp. 445 - 453, ISSN 0043-1648

Liang, X.; Kaiyuan, J.; Yongqing, J. & Darong, C. (1993). Variations in Contact Stress Distribution of Real Rough Surfaces During Running-in, *ASME Journal of Tribology,* Vol. 115, Vol. 4, (October 1993), pp. 602 - 606, ISSN 0742-4787

Lin, J.Y. & Cheng, H.S. (1989). An Analytical Model for Dynamic Wear, *ASME Journal of Tribology,* Vol. 111, Vol. 3, (July 1989), pp. 468 - 474, ISSN 0742-4787

Liu, Z.; Neville, A. & Reuben, R.L. (2001). Analyzing Elastic-Plastic Real Rough Surface Contact in Running-in, *Tribology Transactions,* Vol. 44, pp. 428 - 436, ISSN 1547-397X

Shirong, G. & Gouan, C. (1999). Fractal Prediction Models of Sliding Wear during the Running-in Process, *Wear,* Vol. 231, (July 1999), pp. 249 - 255, ISSN 0043-1648

Stout, K.J.; Whitehouse, D.J. & King, T.G. (1977). Analytical Techniques in Surface Topography and Their Application to a Running-In Experiment, *Wear,* Vol. 43, (May 1977), pp. 99 - 115, ISSN 0043-1648

Summer-Smith, J.D. (1994). *An Introductory Guide to Industrial Tribology,* Mechanical Engineering Publications Limited, ISBN 0-85298-896-6 London, UK

Summer-Smith, J.D. (1997). *A Tribology Casebook,* Mechanical Engineering Publications Limited, ISBN 0-85298-896-6 London, UK

Sugimura, J.; Kimura, Y. & Amino, K. (1987). Analysis of the topography changes due to wear–Geometry of the running-in process, *JSLE* , Vol. 31, No.11, pp. 813 - 820

Tasan, Y.C., (2005). *Measurement of Deformation in Rolling and Sliding Contacts,* PhD Thesis, University of Twente, Print Partners Ipskamp B.V., ISBN 90-365-2193-9, Enschede, The Netherlands

Tasan, Y.C.; de Rooij, M.B. & Schipper, D.J. (2007). Changes in the Micro-geometry of a Rolling Contact, *Tribology International,* Vol. 40, (April, 2007), pp. 672–679, ISSN 0043-1648

Wang, W.; Wong, P.L. & Zhang, Z. (2000). Experimental Study of the Real Time in Surface Roughness during Running-in for PEHL Contacts, *Wear,* Vol. 244, (August 2000), pp. 140 - 146, ISSN 0043-1648

Whitehouse, D.J., 1980, "The Effect of Surface Topography on Wear," in *Fundamentals of Tribology,* edited by Suh and Saka, The MIT Press, ISBN 978-0262191838, pp. 17-52, Massachusetts, USA

Zhao, Y.; Maietta, D.M. & Chang, L. (2000). An Asperity Microcontact Model Incorporating the Transition from Elastic Deformation to Fully Plastic Flow, *ASME Journal of Tribology,* Vol. 122, (January 2000), pp. 86 - 93, ISSN 0742-4787

Zhu, H.; Ge, S.; Cao, X. & Tang W. (2007). The change of fractal dimension of frictional signals in the running-in wear process, *Wear*, Vol. 263, (September, 2007), pp. 1502-1507, ISSN 0043-1648

8

Tribology in Water Jet Processes

Seiji Shimizu
Nihon University, College of Engineering
Japan

1. Introduction

In water jetting technology, a high-velocity stream of water is used for cleaning and cutting purposes. Modern water jetting technology has a history of more than forty years and has been used in a wide range of practical applications. These include machining and manufacturing, stone cutting, demolition work, surface preparation, rock and soil excavation, mining, agriculture, food treatment, and medical applications. In ordinary cutting processes using a water jet, water is usually pressurized up to 300 to 400 MPa by a high-pressure pump. Friction and wear between the cylinder and the piston are important problems that have significant influence on the efficiency, reliability, and lifetime of the high-pressure pump. Corrosion and erosion in valves and nozzles are serious problems that affect the reliability of the water jetting system. The primary material removal mechanism of pure water jets is erosion generated by water droplet impingement. Erosion by solid particle impingement is the material removal mechanism of abrasive water jet machining. Equipment for water jet processes and the mechanisms of cleaning and cutting by water jets are directly related to tribology. In this chapter, a brief review of water jetting technology related to tribology is presented. The review describes the history of water jetting technology, high-pressure pumps and water jet machining systems, various water jets used in water jet processes, and the material removal mechanisms of water jets.

2. History of water jetting technology

Water jetting technology has its origin in hydraulic mining, in which a high-speed water stream is used to break up soil and rock, and can be traced back to the end of the nineteenth century. Hydraulic mining was used in gold production in California from 1853 to 1886 (Summers, 1984). However, modern water jetting technology may be traced directly to the development of the high-pressure pump in the 1960s. According to Summers (1984), Franz et al. developed a prototype system for cutting material at pressures of up to 400 MPa and found that high-speed water jets could be used to cut through wood products at relatively high cutting speeds. Imanaka et al. (1972) developed a high-pressure pump in the early 1960s and conducted cutting tests on various materials at jetting pressures of up to 1,000 MPa. On the other hand, studies on the material removal mechanism of the impingement of water droplets began in the 1950s with studies on rain erosion of aircraft components. Springer (1976) published a comprehensive survey on erosion associated with liquid impact. The development of the high-pressure pump and the clarification of the material removal mechanism of water droplets has attracted growing interest in high-speed water jet

applications. In 1972, the British Hydromechanics Research Association (BHRA), currently BHR Group Ltd., held the 1st International Symposium on Jet Cutting Technology, and this symposium has been held biannually since its inception.
The main areas of research in water jetting technology during the 1970s were cutting by pure water jets and excavation of rock and soil connected with mining and civil engineering. In 1982, four papers on abrasive water jet cutting were presented at the 6th International Symposium on Jet Cutting Technology. In a conventional abrasive water jet system, an abrasive water jet is formed by entraining abrasive particles into a high-speed water jet stream in a tubular abrasive nozzle. This abrasive water jet is referred to as the abrasive water injection jet, the abrasive injection jet (AIJ), or simply the abrasive water jet. Since abrasive water jets can cut hard materials such as metals, ceramics, concrete, and rocks at practical cutting speeds, applications of water jetting technology in manufacturing, civil engineering, and construction have expanded rapidly. In 1986, Fairhurst et al. (1986) reported an abrasive water jet system of a different type, which is now referred to as the abrasive water suspension jet, the abrasive suspension jet (ASJ), or the abrasive slurry jet. The abrasive suspension jet has been shown to have a greater capacity for drilling and cutting than the conventional abrasive water jet (Brandt et al., 1994). However, there are a number of problems that remain to be solved before widespread application of the abrasive suspension jet system can be realized. For instance, the development of a reliable and long-life high-pressure slurry valve is necessary. Abrasive suspension jet systems are used in fire fighting and rescue operations (Holmstedt, 1999; Inoue et al., 2008) as well as semiconductor manufacturing (Jiang et al., 2005). Recently, a number of innovative water jetting technologies have been proposed, e.g., liquefied gas jets, ultra-high-pressure pumps that realize pressures exceeding 600 MPa, and a five-axis water jet machining system. Water jetting technology has made steady progress, and applications of water jetting technology continue to expand in various industries.

3. High-pressure pump and water jet machining system

Two types of pump are commonly used in the water jet industry, namely, the direct drive plunger pump and the intensifier pump. At pressures lower than 150 MPa, the direct drive plunger pump is the most commonly used type. In a typical triplex pump, three plungers are equally arranged about a crankshaft at 120-degree increments, and the plungers move backward and forward in the cylinders. At pressures higher than 150 MPa, the intensifier pump is used as a pumping unit. A typical circuit of an intensifier pump system is shown in Figure 1 (Ibuki et al., 1993). The intensifier is a reciprocating pump in which the piston assembly consists of a large-diameter piston and small-diameter plungers at both ends. Hydraulic oil pushes against the larger-diameter piston, and the smaller plunger generates higher pressure on the water contained in the cylinder. Since the typical area ratio of the plunger and the piston is 1:20, the intensification ratio of pressure is 20:1. When the oil pressure is 20 MPa, the resultant water pressure becomes 400 MPa. In the water jet machining industry, pressures of 300 to 400 MPa are commonly used. In order to obtain higher cutting speeds and reduce abrasive consumption, ultra-high-pressure water jet machining systems at working pressures of 400 to 700 MPa have been used in practical applications. A two-stage intensifier pump system is proposed in order to reduce dynamic loading and obtain longer system lifetimes for ultra-high-pressure water jet machining systems (Koerner et al., 2002).

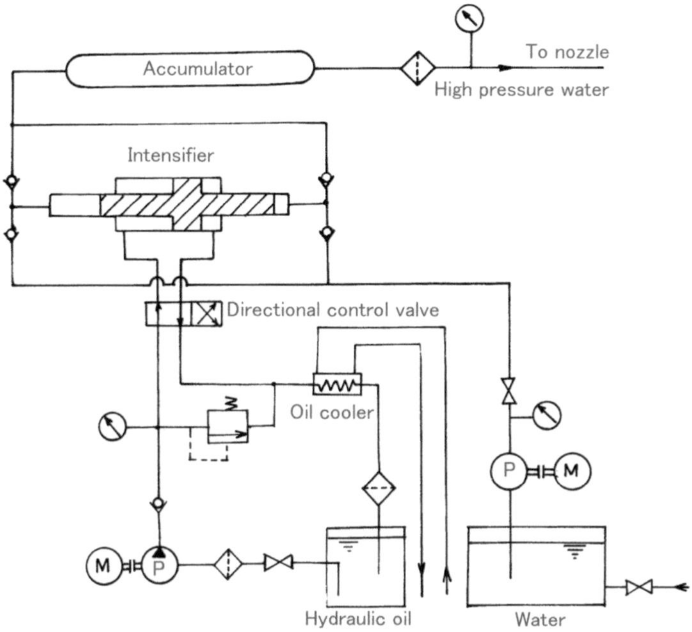

Fig. 1. Circuit of the intensifier pump (Ibuki et al., 1993)

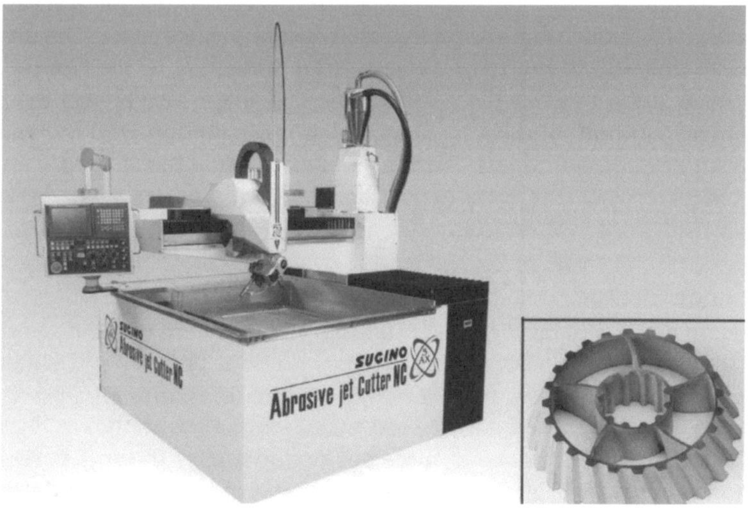

Fig. 2. Modern five-axis water jet machining system and sample part manufactured by the system (Sugino Machine Ltd., 2007)

XY tables are the most common forms of water jet machining motion equipments. These machines are used for two-dimensional cutting. With recent advances in control and motion technology, five-axis water jet machining systems have been used practically. An example of the latest five-axis water jet machining system and a sample part manufactured using this system are shown in Figure 2 (Sugino Machine Ltd., 2007). In addition to the X (back/forth), Y (left/right), and Z (up/down) axes, two degrees of freedom are added to the nozzle movement, namely, the angle from the perpendicular and rotation around the Z-axis. In the case of two-dimensional cutting, more accurately manufactured parts can be produced at higher cutting speeds by the five-axis machine. Three-dimensional complex shapes can be produced by the five-axis machine in single-pass operation.

4. Various water jets and material removal mechanisms

High-velocity-water jets used in water jet processes can be categorized according to the environment around the jet, the fluid medium, and the jetting regime, as shown in Figure 3 (Shimizu, 2003). Depending on the environment around the jet, water jets are classified as either water jets in air or submerged water jets. The submerged water jets are further classified as cavitating jets or non-cavitating jets, depending on the jetting and ambient pressures. In ordinary pressure levels of water jet processes, cavitation occurs in the region of high shear at the boundary between the jet and the surrounding water. Depending on the fluid medium, water jets are categorized as either pure water jets or abrasive water jets. In some cases of pure water jets, water-soluble polymeric additives are added in order to reduce the friction in the plumbing and improve the compactness of the jet issuing from the nozzle. Abrasive water jets are classified as either abrasive injection jets (AIJs) or abrasive suspension jets (ASJs) based on the generation mechanism and the phase composition. Abrasive injection jets are solid-air-liquid three-phase jet flows, and ASJs are solid-liquid two-phase jet flows. With respect to the jetting regime, water jets can be classified as continuous water jets or discontinuous water jets. Although all high-velocity water jets generate a discontinuous phase during impact, continuous water jets are considered to be water jets that are not broken up artificially by external mechanisms. The structures of the jets and the material removal mechanisms are differ according to the combination of these three factors. In addition to these three factors, nozzle shape also affects the flow structure of the jet. The most common nozzles have a circular cross section, and the water jets issued from such nozzles become round jets. Nozzles that can form a fan shaped water jet are often used to clean large areas. This type of nozzle is referred to as a fan jet nozzle. In the following, we consider only round jets.

4.1 Pure water jets

The most common water jets used in water jetting processes are continuous pure water jets in air issued from a nozzle having a circular cross section. This type of water jet is widely used in water jetting industries for cleaning, surface preparation, and cutting of soft materials. A schematic diagram of a high-seed water jet in air is shown in Figure 4 (Yanaida and Ohashi, 1980). The jet consists of three regions, namely, the initial region, the main region, and the final region. In the initial region, the stagnation pressure is considered to be the same as that at the nozzle exit, and the initial region length is determined from extrapolation of the decrease in the stagnation pressure. In the main region, the axial velocity of water is considered to be constant, irrespective of the axial distance from the nozzle exit. The breakup length exists in the main region. The continuous structure of the water jet disintegrates at the breakup length, and the jet becomes a droplet flow composed of water lumps and droplets surrounded by fine droplets. The velocities of the water lumps and relatively large droplets remain unchanged in the main region. In the final region, the decrease in the water droplets velocities becomes noticeable as the droplets break up into finer droplets.

Environment around water jets
- Water jets in air
- Submerged water jets
  - Non cavitating jets
  - Cavitating jets

Fluid mediums
- Pure water jets
- Abrasive water jets
  - AIJs
  - ASJs

Jetting regime
- Continuous jets
- Discontinuous jets

Fig. 3. Division of water jets (Shimizu, 2003)

When we consider a water jet impacting a solid material, two different pressures are considered to occur at the point of impact. If a continuous jet impinges on a solid material, then the jet stagnation pressure p_s is generated at the impact point. If a water lump or droplet impinges on a solid material, then the water hammer pressure p_w is generated at the instant of impact. The stagnation pressure p_s and the water hamper pressure p_w are given by the following equations:

$$p_s = \rho v^2/2 \tag{1}$$

$$p_w = \rho c\, v \tag{2}$$

where ρ is the density of water, c is the velocity of sound in water, and v is the velocity of water. Since the velocity of sound in water at normal temperature and atmospheric pressure is approximately 1,500 m/s, the water hammer pressure p_w is much larger than the stagnation pressure p_s under ordinary impact conditions.

The condition of impact, i.e., whether a stream of water impinges continuously or water lumps and droplets impinge intermittently, affects the pressure generated at the impact point. If the pressure generated at the impact point is larger than the strength of material such as the yield strength, material removal occurs at the impact point. Accordingly, material removal by a pure water jet is significantly affected by the flow structure of the jet. In general, sharp cutting by a pure water jet is conducted in the initial region, and massive material removal over a wide area is realized in and around the region in which jet breakup occurs.

Since the water hammer pressure is much larger than the stagnation pressure, discontinuous water jets generated by some external mechanisms have much larger destructive power than continuous water jets. Vijay and Foldyna (1994) developed a forced pulsed water jet nozzle containing a vibrating tip. The tip is forced to vibrate at high frequency by an ultrasonic piezoelectric or magnetostrictive transducer. Figure 5 (Yan, 2007) shows a photograph of the fully developed forced pulsed water jet taken with a Nd-YAG laser. Mushroom shaped water lumps are generated in the jet. The material removal capability by the forced pulsed water jet at the standoff distance, at which distinct water lumps are formed, is much larger than material removal capability of ordinary continuous water jets.

When a high-velocity water jet is injected in a submerged environment, a free shear layer is established around the jet, and this layer grows in thickness as the jet emerges from the nozzle. The shear layer quickly becomes unstable, breaks down into turbulent motion and

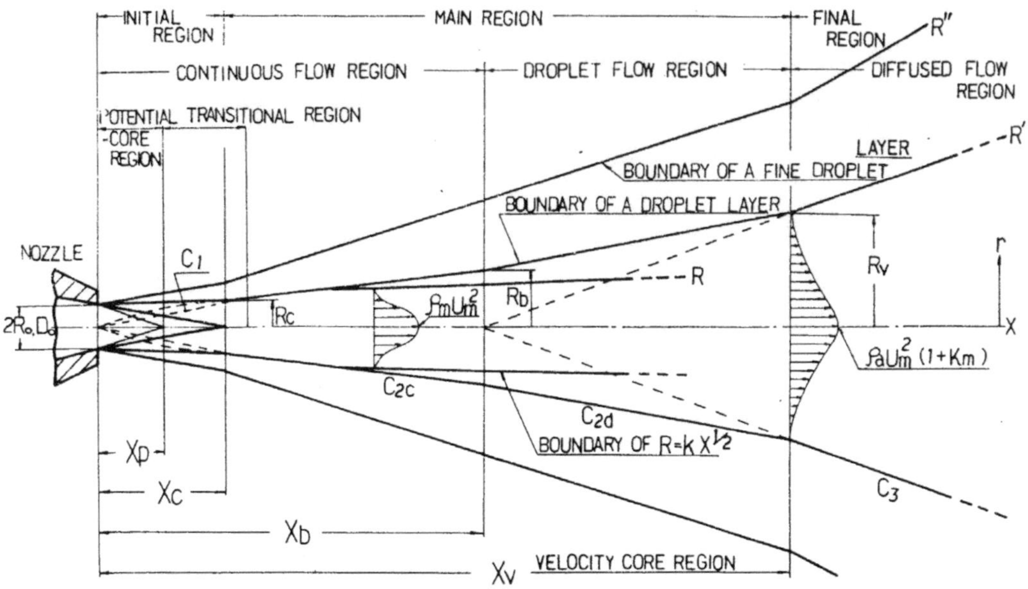

Fig. 4. Structure of a water jet in air (Yanaida and Ohashi, 1980)

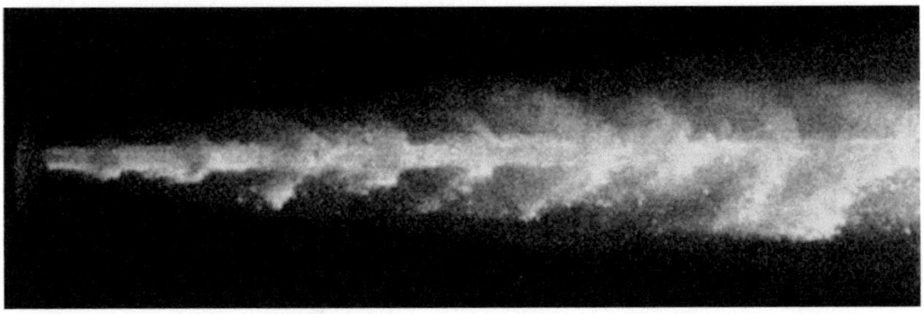

Fig. 5. Photograph of a fully developed forced pulsed water jet (Yan, 2007)

starts to spread, entraining surrounding water. In an ordinary submerged environment, cavitation occurs in the region of high shear at the boundary between the jet and the surrounding water. Cavitation around the jet suppresses the deceleration of the jet and produces cavitation erosion by the collapse of cavitation bubbles. Accordingly, cavitation around the jet significantly affects the material removal characteristics of submerged water jets. The parameter similitude in cavitation is defined by the cavitation number σ as follows:

$$\sigma = (p_a - p_v) \,/\, (p_i - p_a) \tag{3}$$

where p_a , p_i, and p_v are the ambient pressure, the injection pressure, and the vapor pressure of water, respectively. The cavitation number σ measures the resistance of the flow to cavitation. The lower the cavitation number, the more likely cavitation is to occur. If cavitation occurs, lowering the cavitation number will increase the extent of cavitation, i.e., the number and size of vapor bubbles will increase. Figure 6 shows an example of an instantaneous photograph of a cavitating jet at p_i = 69 MPa and σ = 0.006. The nozzle diameter is 0.5 mm. Since the jet is illuminated from behind, the cavitation clouds appear to be black. The cavitation clouds are continuous near the nozzle exit but separate and develop into lumps as they travel with the jet. Shimizu et al. (1998) conducted erosion tests using submerged water jets at injection pressures ranging from 49 to 118 MPa and cavitation numbers ranging from 0.006 to 0.022. Since the jet decelerates faster under a submerged environment, material removal by jet impingement is restricted in the region near the nozzle exit, as compared to jets in air. In addition to high-speed jet impingement, cavitation erosion is an additional material removal mechanism in the submerged environment. Cavitating jets are used for cleaning and shot-less peening (Soyama et al., 2002) in the water jetting industry.

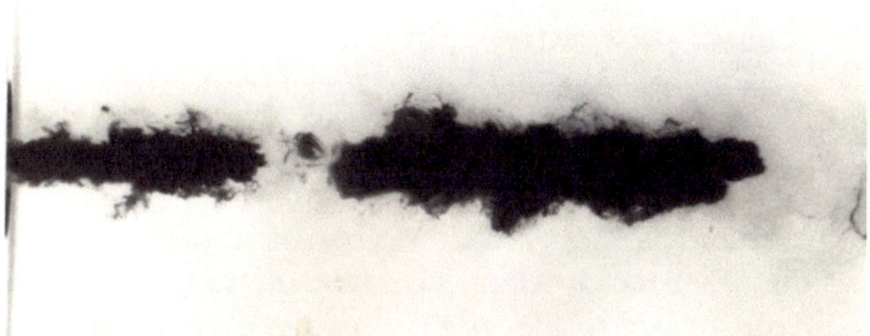

Fig. 6. Cavitating jet at p_i = 69 MPa and σ = 0.006 (flow direction is from left to right)

4.2 Abrasive jets

The material removal capability of abrasive water jets, in which abrasive particles are added to the water stream, is much larger than the material removal capability of the pure water jets. In an abrasive water jet, the stream of the water jet accelerates abrasive particles, which erode the material. The material removal capability of the water is slight in abrasive water jet processes. The impact of single solid particles is the basic material removal mechanism of abrasive water jets. Meng and Ludema (1995) defined four mechanisms by which solid particles separate material from a target surface, as shown in Figure 7 (Momber and Kovacevic, 1998). These mechanisms are cutting, fatigue, brittle facture, and melting, which generally do not work separately, but rather in combination. The importance of these mechanisms for a particular erosion process depends on several factors, such as the impact angle, the particle kinematic energy, the particle shape, the target material properties, and the environmental conditions.

Abrasive water jets can be classified as abrasive injection jets (AIJs) or abrasive suspension jets (ASJs), as stated earlier. Abrasive injection jets are formed using the nozzle head shown in Figure 8. A high-speed water jet is injected through the nozzle head. The diameter of the water jet nozzle is typically 0.2 to 0.4 mm. The high-speed water jet stream creates a vacuum, which draws abrasive particles into the mixing chamber along with air. The water jet stream accelerates the abrasive particles and air in the mixing tube, which is typically 0.5 to 1.5 mm in diameter. The cutting width of the AIJs depends on the diameter of the mixing tube and the standoff distance. For a mixing tube of 1.0 mm in diameter and the standoff distance of 3 to 5 mm, the cutting width is approximately 1.2 mm.

The three-phase jet flow discharged from the mixing tube consists of abrasive particles, water, and air. The material removal capability of the AIJ formed by a certain nozzle head (the dimensions and shape of the nozzle head are fixed) is affected by the pump pressure and the type and mass flow rate of abrasive. In general, the higher the pump pressure, the greater the material removal capability. When the abrasive flow rate is relatively small, the material removal capability increases with the abrasive mass flow rate, because the higher

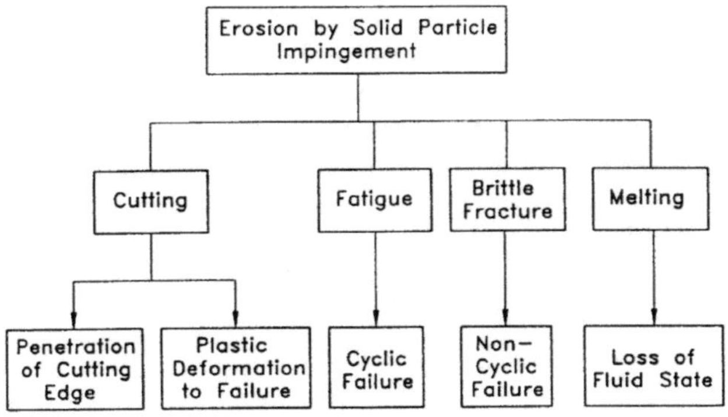

Fig. 7. Mechanisms of material removal by solid particle erosion (Momber and Kovacevic, 1998)

the abrasive mass flow rate, the higher the number of abrasive particles involved in the cutting processes. On the other hand, when too many abrasive particles are supplied to the nozzle head, the kinematic energy of the single abrasive particles tends to decrease because of the limited kinematic energy of the water jet. Thus, there exists an optimum abrasive mass flow rate. In addition, an uneven abrasive supply to the nozzle head can cause violent pulsation in AIJs. Shimizu et al. (2009) conducted high-speed observations of AIJs using high-speed video. Figure 9 shows a series of photographs of an AIJ issuing from the nozzle head at an injection pressure of 300 MPa and a time averaged abrasive mass flow rate of 600 g/min. The time interval between frames is 12.29 μs, and the flow direction is downward. Frame numbers are indicated at the top of each photograph. At frame number 1, the jet spreads radially just downstream of the mixing nozzle exit. As time proceeds, the hump of the jet develops into a large lump and moves downstream while growing in the stream-wise direction. As the lump leaves the mixing nozzle exit at frame number 10, another hump of the jet appears just downstream of the mixing tube exit. Observations of the flow conditions in the abrasive supply tube just upstream of the mixing chamber of the abrasive nozzle head were also conducted. Based on image analysis of the video, Shimizu et al. concluded that the pulsation of an AIJ at a frequency of less than 100 Hz is closely related to the fluctuation of the abrasive supply.

Wearing of the mixing tube is a serious problem in abrasive water jet machining. In the early days of abrasive water jet machining, the lifetime of a mixing tube constructed of standard tungsten carbide was only approximately five hours. However, advances in anti-wear materials technology have extended the lifetime of the mixing tube to 100 to 150 hours.

In contrast to the abrasive injection jets, abrasive suspension jets are solid-liquid two-phase jet flows. As shown in Figure 10, abrasive suspension jets are classified into two systems according to the generation mechanism (Brandt et al., 1994), namely, the bypass system and the direct pumping system. In the bypass system, part of the water flow is used to draw the abrasive material out of the storage vessel and to mix it back into the main water flow line. In the direct pumping system, the pre-mixed slurry charged in a pressure vessel is pressurized by high-pressure water. An isolator is used to prevent mixing of the slurry and the water.

In the case of the AIJ, the addition of abrasive particles increases the jet diameter and decreases the jet velocity. The velocity of the ASJ discharged from the nozzle is 0.90 to 0.95 times the theoretical jet velocity calculated by Bernoulli's equation assuming the loss in the nozzle to be zero (Shimizu, 1996). Moreover, a compact ASJ can be formed if a suitable

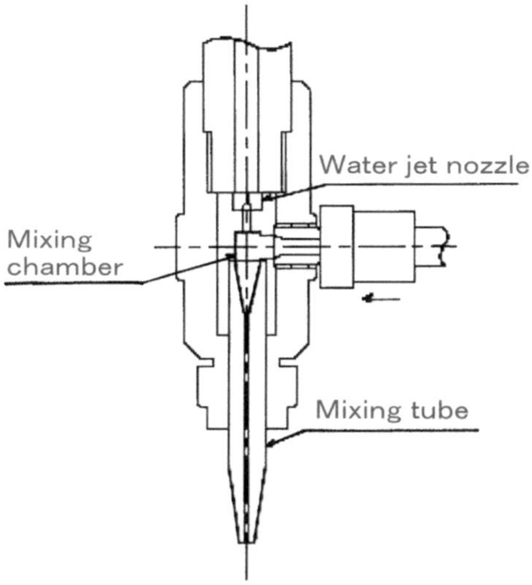

Fig. 8. Abrasive water jet nozzle head

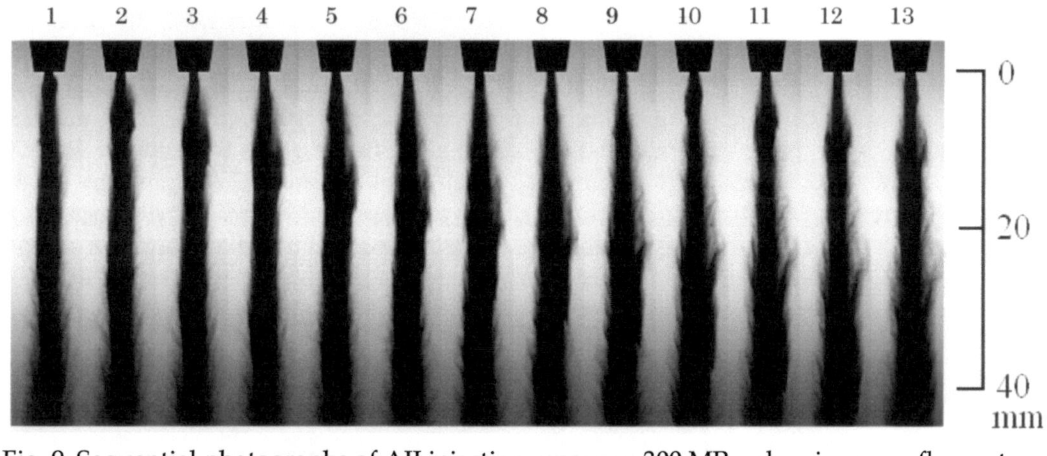

Fig. 9. Sequential photographs of AIJ,injection pressure: 300 MPa, abrasive mass flow rate: 600 g/min, abrasive: #80 garnet (Shimizu et al., 2009)

nozzle shape is adopted. It is also possible to form an ASJ with a very high abrasive concentration, such as 50 wt%. Accordingly, the abrasive suspension jet has a greater capability for drilling and cutting than the abrasive water injection jet. Brandt et al. (1994) compared the cutting performances of the ASJ and the AIJ under the same hydraulic power ranges and the same abrasive mass flow rate. They concluded that the ASJ cuts at least twice as deep as the AIJ at the same hydraulic power. A micro-abrasive suspension system with a nozzle diameter of 50 μm was also developed (Miller, 2002). Since a cutting width of 60 to 70 μm can be realized using such a system, applications in micro-machining and semiconductor industries are expected.

In the ASJ system, a convergent nozzle followed by a constant diameter straight passage (focusing section) of suitable length is generally used. Since abrasive-water slurry flows at high-speed in the nozzle, slurry erosion of the nozzle is a serious problem. Therefore, in

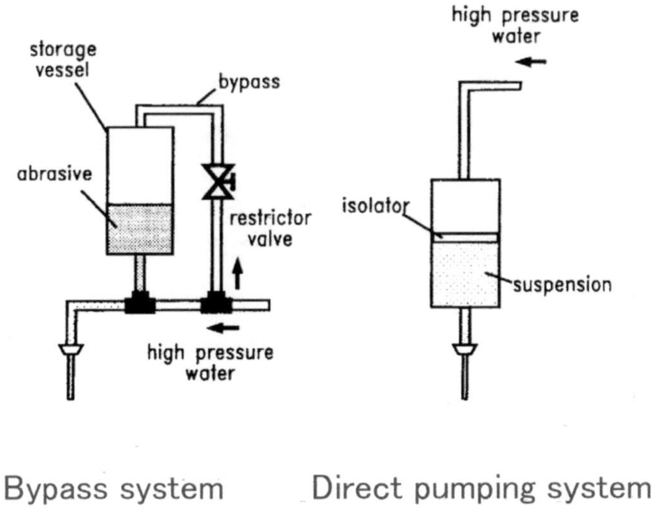

Fig. 10. Abrasive suspension systems (Brandt et al., 1994)

order to reduce nozzle wear, the outlet of the convergent section and the focusing section are constructed of wear resistance materials, such as sintered diamond. In order to investigate the effects of the wear of the nozzle focusing section on the material removal capability of the jet, an experimental nozzle was used to perform drilling tests (Shimizu et al., 1998). The outlet of the convergent section was constructed of sintered diamond, and the focusing section was constructed of cemented carbide. The drilling tests were conducted at a jetting pressure of 11.9 MPa with specimens of stainless steel and #220 aluminum oxide abrasive. Figure 11 shows the variation of drilling pit depth with standoff distance for a jetting duration of 60 s. The numbers in the figure are the order of the tests. The cross section of the nozzle after the drilling tests is shown in Figure 12. The total jetting duration was 780 s. The focusing section (indicated by the arrow) is worn, and the wear of the focusing section causes a serious reduction in drilling capability, as shown in Figure 11.

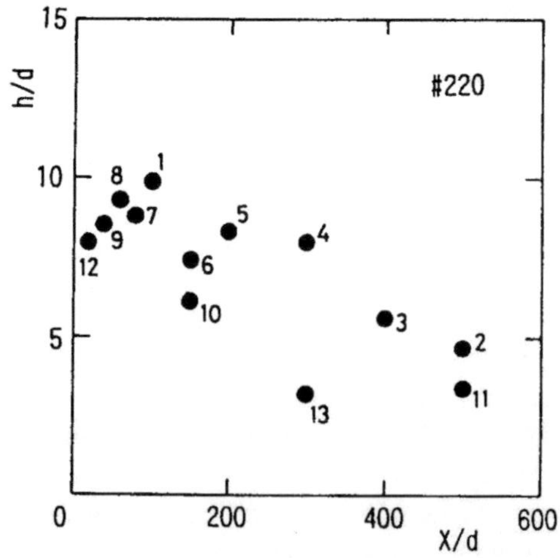

Fig. 11. Effect of nozzle wear on pit depth (Shimizu et al., 1998)

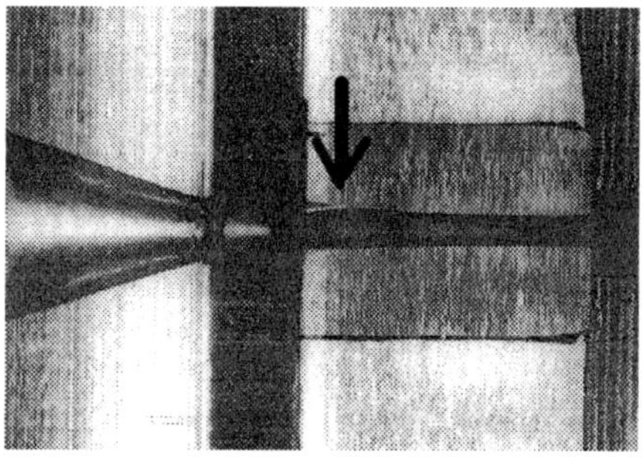

Fig. 12. Nozzle after drilling tests, jetting pressure: 11.9 MPa, abrasive of aluminum oxide mesh designation of #220 (Shimizu et al., 1998)

5. Conclusion

Friction and wear between the cylinder and the piston of high-pressure pumps used in the water jetting processes are important problems greatly influence the efficiency, reliability, and lifetime of the high-pressure pump. Corrosion and erosion in valves and nozzles are serious problems that affect the reliability of water jetting systems. Erosion by water droplet impingement is the material removal mechanism of pure water jets, and erosion by solid particle impingement is the material removal mechanism of abrasive water jet machining. Knowledge of tribology is indispensable in order to realize more reliable and more efficient water jet machining systems.

6. References

Brandt, C., Louis, H., Meier, G., & Tebbing, G. (1994), Abrasive Suspension Jets at Working Pressures up to 200 MPa, *Jet Cutting Technology*, Allen, N.G. Ed. pp.489-509, Mechanical Engineering Publications Limited, 0-85298-925-3, London

Faihurst, R.A., Heron, R.A., & Saunders, D.H. (1986), 'DIAJET' –A New Abrasive Water Jet Cutting Technique, *Proceedings of 8th International Symposium on Jet Cutting Technology*, pp.395-402, 0-947711-17-1, Durham, England, September, 1986, BHRA, Cranfield

Holmstedt, G. (1999), An Assessment of the Cutting Extinguisher Advantages and Limitations, *Technical Report from the Lund Institute of Technology*, Department of Fire Safty Engineering, Lund University

Ibuki, S., Nakaya, M, & Nishida, N. (1993), *Water Jet Technology Handbook*, The Water Jet Technology Society Japan Ed., pp.89-103, Maruzen Co., Ltd., 4-621-03901-6C3550, Tokyo

Imanaka, O., Fujino, S. Shinohara, K., & Kawate, Y. (1972), Experimental Study of Machining Characteristics by Liquid Jets of High Power Density up to 10^8 Wcm^{-2}, *Proceedings of the first International Symposium on Jet Cutting Technology*, pp.G3-25–G3-35, Coventry, England, April, 1972, BHRA, Cranfield

Inoue, F., Doi, S., Katakura, H., & Ichiryu, K. (2008), Development of water Jet Cutter System for Disaster Relief, *Water Jetting*, pp.87-93, BHR Group Limited, 978-1-85598-103-4, Cranfield

Jiang, S., Popescu, R., Mihai, C., & Tan, K. (2005), High Precision and High Power ASJ Singulations for Semiconductor Manufacturing, *Proceedings of 2005 WJTA American Waterjet Conference*, Hashish M. Ed., Papser 1A-3, Houston, Texas, August 2005, The WaterJet Technology Association, St. Louis, MO

Koerne, P., Hiller, W., & Werth, H. (2002), Design of reliable Pressure Intensifiers for Water-Jet Cutting at 4 to 7 kbar, *Water Jetting*, pp.123-132, BHR Group Limited, 1-85598-042-8, Cranfield

Meng, H.C. & Ludema, K.C. (1995), Wear Models and Predictive Equations: Their Form and Content, Wear 181-183, pp, 443-457

Miller, D.S. (2002), Micromachining with abrasive waterjets, *Water Jetting*, pp.59-73, BHR Group Limited, 1-85598-042-8, Cranfield

Momber, A.W. & Kovacevic, R. (1998), *Principles of Abrasive Water Jet Machining*, Springer, 3-540-76239-6, London

Shimizu, S. (1996), Effects of Nozzle Shape on Structure and Drilling Capability of Premixed Abrasive Water Jets, *Jetting Technology*, Gee, C. Ed., pp.13-26, Mechanical Engineering Publications Limited, 1-86058-011-4, London

Shimizu, S., Miyamoto, T., & Aihara, Y. (1998), Structure and Drilling Capability of Abrasive Water Suspension Jets, *Jetting Technology*, Louis, H. Ed. pp.109-117, Professional Engineering Publishing Ltd., 1-86058-140-4, London

Shimizu, S. (2002), High Velocity Water Jets in Air and Submerged Environments, *Proceedings 7th Pacific Rim International Conference on Water Jetting Technology*, Lee, C-I., Jeon S., and Song J-J. Eds. pp.37-45, Jejyu, Korea, September 2003, The Korean Society of Water Jet Technology, Seoul

Shimizu, S. , Ishikawa, T., Saito, A. & Peng, G. (2009), Pulsation of Abrasive Water-Jet, Proceedings of 2009 American WJTA Conference and Expo, Paper 2-H, Houston Texas, August 2009, Water Jet Technology Association

Soyama, H. Saito, K. & Saka, M. (2002), Improvement of Fatigue Strength of Aluminum Alloy by Cavitation Shotless Peening, *Transaction of the ASME, Journal of Engineering Materials Technology*, Vol. 124, No.2, pp.135-139.

Springer, G. S. (1976), *Erosion by Liquid Impact*, Scripta Publishing Co. 0-470-15108-0, Washington, D.C.

Sugino Machine Ltd. (2007), *Catalogue by Sugino Machine Ltd.*

Summers, D.A. (1995). *Waterjetting Technology*, E & FN Spon, 0-419-19660-9, Great Britain

Vijay, M.M. & Foldyna, J. (1994), Ultrasonically Modulated Pulsed Jets: Basic Study, *Jet Cutting Technology*, pp.15-35, Mechanical Engineering Publications Limited, 0-85298-925-3, London

Yan, W. (2007), Recent Development of Pulsed Waterjet Technology Opens New Markets and Expands Applications, *WJTA Jet News*, August 2007, WaterJet Technology Association, St. Louis

Yanaida. K. & Ohashi, A. (1980), Flow Characteristics of Water Jets in Air, *Proceedings of 5th International Symposium on Jet Cutting Technology*, Paper A3, pp.33-44, Hanover, June 1980, BHRA, Cranfield

9

The Elliptical Elastic-Plastic Microcontact Analysis

Jung Ching Chung
Department of Aircraft Engineering,
Air Force Institute of Technology
Taiwan ROC

1. Introduction

The elastic-plastic contact of a flat and an asperity which shape is a sphere or an ellipsoid is a fundamental problem in contact mechanics. It is applicable in tribological problems arising from the points of contact between two rough surfaces, such as gear teeth, cam and follower and micro-switches etc. Indeed, numerous works on the contact of rough surfaces were published so far (see review by Liu et al.). Many of these works are based on modeling the contact behavior of a single spherical asperity, which is then incorporated in a statistical model of multiple asperity contact. Based on the Hertz theory, the pioneering work on contact models of pure elastic sphere was developed by Greenwood and Williamson (GW) . The GW model used the solution of the frictionless contact of an elastic hemisphere and a rigid flat to model an entire contacting surface of asperities with a postulated Gaussian height distribution. The basic GW model had been extended to include such aspects as curved surfaces (by Greenwood and Tripp), two rough surfaces with misaligned asperities (by Greenwood and Tripp) and non-uniform radii of curvature of asperity peaks (by Hisakado). Abbott and Firestone introduced the basic plastic contact model, which was known as surface micro-geometry model. In this model the contact area of a rough surface is equal to the geometrical intersection of the original undeformed profile with the flat. Based on the experimental results, Pullen and Williamson proposed a volume conservation model for the fully plastic contact of a rough surface.

The works on the above two models are suitable for the pure elastic or pure plastic deformation of contacting spheres. In order to bridge the two extreme models, elastic and fully plastic, Chang et al. (CEB model) extended the GW model by developing an elastic-plastic contact model that incorporated the effect of volume conservation of a sphere tip above the critical interference. Numerical results obtained from the CEB model are compared with the other existing models. In the CEB model, there is no transition regime from the elastic deformation to the fully plastic deformation regime. These deficiencies triggered several modifications by other researchers. Zhao et al. (the ZMC model) used mathematical smoothing expressions to incorporate the transition of the contact load and contact area expression between the elastic and fully plastic deformation regions. Kogut and Etsion (KE model) performed a finite element analysis on the elastic-plastic contact of a deformable sphere and a rigid flat by using constitutive laws appropriate to any mode of deformations. It offered a general dimensionless relation for the contact load, contact area

and mean contact pressure as a function of contact interferences. Jackson and Green had done recently a similar work. In this work, it accounted for geometry and material effects, which were not accounted for in the KE model. Jackson et al. presented a finite element model of the residual stresses and strains that were formed after an elastoplastic hemispherical contact was unloaded. This work also defines an interference at which the maximum residual stress transitions from a location below the contact region and along the axis of symmetry to one near to the surface at the edge of the contact radius (within the pileup).
The aforementioned models deal with rough surfaces with isotropic contacts. However, rough surface may have asperities with various curvatures that the different ellipticity ratios of the micro-contacts formed. Bush et al. treated the stochastic contact summits of rough surfaces to be parabolic ellipsoids and applied the Hertzian solution for their deformations. McCool took account of the interaction between two neighboring asperities and modelled the elastic-plastic contact of isotropic and anisotropic solid bodies. Horng extended the CEB model to consider rough surfaces with elliptic contacts and determined the effects of effective radius ratio on the microcontact behavior. Jeng and Wang extend the Horng's work and the ZMC model to the elliptical contact situation by incorporating the elastic-plastic deformation effect of the anisotropy of the asperities. Chung and Lin used an elastic-plastic fractal model for analyzing the elliptic contact of anisotropic rough surfaces. Buczkowski and Kleiber concentrated their study on building an elasto-plastic statistical model of rough surfaces for which the joint stiffness could be determined in a general way. Lin and Lin used 3-D FEM to investigate the contact behavior of a deformable ellipsoid contacting with a rigid flat in the elastoplastic deformation regime. The above works provided results for the loaded condition case. Calculations of the stress distribution at the points of the compression region only under normal load within the ellipse of contact were dealt with in a number of works. The combined action of normal and tangential loads was also discussed in some works whose authors examined the stress conditions at points of an elastic semi-space. However, the above-mentioned works just discussed the distribution of stresses under the elliptical spot within the elastic deformation regime. The distribution of stresses within the elastoplastic deformation regime was still omitted. Chung presented a finite element model (FEM) of the equivalent von-Mises stress and displacements that were formed for the different ellipticity contact of an ellipsoid with a rigid flat.

2. Important

The present chapter is presented to investigate the contact behavior of a deformable ellipsoid contacting a rigid flat in the elastoplastic deformation regime. The material is modeled as elastic perfectly plastic and follows the von-Mises yield criterion. Because of geometrical symmetry, only one-eighth of an ellipsoid is needed in the present work for finite element analysis (FEA). Multi-size elements were adopted in the present FEA to significantly save computational time without losing precision. The inception of the elastoplastic deformation regime of an ellipsoid is determined using the theoretical model developed for the yielding of an elliptical contact area. k_e is defined as the ellipticity of the ellipsoid before contact, so the contact parameters shown in the elastoplastic deformation regime are evaluated by varying the k_e value. If the ellipticity (k) of an elliptical contact area is defined as the length ratio of the minor-axis to the major-axis, it is asymptotic to the k_e value when the interference is sufficiently increased, irrespective of the k_e value. The ellipticity (k) of an elliptical contact area varies with the k_e parameter. The k values evaluated at various dimensionless interferences and two $k_e$values (k_e=1/2 and k_e=1/5) are presented. Both interferences, corresponding to the inceptions of the elastoplastic and fully plastic deformation regimes, are determined as a function of the ellipticity of the ellipsoid (k_e).

The work also presents the equivalent von-Mises stress and displacements that are formed for the different ellipticities. According to the results of Johnson, Sackfield and Hills, the severest stress always occurs in the z-axis. In this work, we can get the following result: the smaller the ellipticity of the ellipsoid is, the larger the depth of the first yield point from the ellipsoid tip happens. The FEM produces contours for the normalized normal and radial displacement as functions of the different interference depths. The evolution of plastic region in the asperity tip for a sphere (k_e=1) and an ellipsoid with different ellipticities (k_e=1/2 and k_e=1/5) is shown with increasing interferences. It is interesting to note the behavior of the evolution of the plastic region in the ellipsoid tip for different ellipticities, k_e, is different. The developments of the plastic region on the contact surface are shown in more details. When the dimensionless contact pressure is up to 2.5, the uniform contact pressure distribution is almost prevailing in the entire contact area. It can be observed clearly that the normalized contact pressure ascends slowly from the center to the edge of the contact area for a sphere (k_e=1), almost has uniform distribution prevailing the entire contact area for an ellipsoid (k_e=1/2), and descends slowly from the center to the edge of the contact area for an ellipsoid (k_e=1/5). The differences in the microcontact parameters such as the contact pressure, the contact area, and contact load evaluated at various interferences and two k_e values are investigated.

The elastic-plastic fractal model of the elliptic asperity for analyzing the contact of rough surfaces is presented. Comparisons between the fractal model and the classical statistical model are discussed in this work. Four plasticity indices ($\psi = 0.5$, 1, 2, and 2.5) for the KE (Kogut and Etsion) model are chosen. The topothesy (G) and fractal dimension (D) values, which are corresponding to these four plasticity indices in the present model, will thus be determined.

3. Theoretical background

In the present chapter, Figure 1 shows that a deformable ellipsoid tip contacts with a rigid flat. The lengths of the semi-major axis of an ellipsoid and the semi-minor axis are assumed to be cR ($1 \le c < \infty$) and R, respectively. From the geometrical analysis, the radii of curvature at the tip of an ellipsoid, $R_{1x}(=c^2R)$ and $R_{1z}(=R)$, are obtained. the ellipticity of an ellipsoid is defined as k_e, and $k_e = (R_{1z} / R_{1x})^{1/2} = 1/c = R/cR$. For $c=1$, $k_e = 1$, corresponds to the spherical contact; for $c \to \infty$, $k_e = 0$, corresponds to the cylindrical contact. The simulations by FEM are carried out under the condition of a given interference δ applied to the microcontact formed at the tip of an ellipsoid. Because of geometrical symmetry, only one-eighth of an ellipsoid volume is needed in the finite element analysis (see Figure 2). At an interference, δ, an elliptical contact area is formed with a semi-major axis, a, and a semi-minor axis, b. The length ratio k is here defined as $k=b/a$, which is called the ellipticity of this elliptical contact area. The material of this ellipsoid is modeled as elastic perfectly plastic with identical behavior in tension and compression.

The contact area of an asperity here is elliptical in shape, having two semi-axis lengths, a and b ($b<a$), in the present study. The eccentricity of the contact ellipse (e) is

$$e = \left(1 - \frac{b^2}{a^2}\right)^{1/2} \tag{1}$$

Define the C' parameter as $C' \equiv (ab)^{1/2}$, this parameter has been derived by Johnson [25] as

$$C' = (ab)^{1/2} = \left(\frac{3FR_e}{4E^*}\right)^{1/3} \{F_1(e)\}^{1/3} \tag{2}$$

Where R_e denotes the effective radius of curvature of an asperity.

$$\mathrm{F}_1(e) = \frac{4}{\pi \mathrm{e}^2}\left(\frac{\mathrm{b}}{\mathrm{a}}\right)^{3/2} \left\{\left[\left(\frac{\mathrm{a}}{\mathrm{b}}\right)^2 \mathrm{E(e)} - K(e)\right]\left[K(e) - E(e)\right]\right\}^{1/2} \tag{3}$$

E^* in Eq.(1) denotes the effective Young's modulus of two solid contact bodies with the Young's moduli, E_1 and E_2, and the Poisson ratios, ν_1 and ν_2, respectively. It is stated as

$$\frac{1}{E^*} = \frac{1-\nu_1^2}{E_1} + \frac{1-\nu_2^2}{E_2} \tag{4}$$

Where F denotes the normal load of an asperity at the Hertz contact area. $K(e)$ and $E(e)$ in the formula of $F_1(e)$ denote the complete first and second elliptic integrals of argument (e), respectively. They are expressed by Johnson as

$$\mathrm{K(e)} = \int_0^{\pi/2} \frac{\mathrm{d}\theta}{\sqrt{1 - \mathrm{e}^2 \sin^2\theta}} \tag{5}$$

$$\mathrm{E(e)} = \int_0^{\pi/2} \sqrt{1 - \mathrm{e}^2 \sin^2\theta}\, \mathrm{d}\theta \tag{6}$$

The onset of the plastic yield of ductile materials usually occurs when the von Mises' shear strain-energy criterion reaches

$$J_2^* = (k')^2 = \frac{Y^2}{3} \tag{7}$$

Where J_2^* is the maximum value of the second invariant of the deviator stress tensor (J_2) at yielding and k' is the material yield stress in simple shear. The second invariant of the deviator stress tensor can be written as:

$$J_2 = \frac{1}{6}\left[(\sigma_1 - \sigma_2)^2 + (\sigma_2 - \sigma_3)^2 + (\sigma_3 - \sigma_1)^2\right] \tag{8}$$

Where σ_1, σ_2, and σ_3 are the three principal stresses. In the study of Sackfield and Hills, the stress distributions formed by the Hertz contact pressure acting on an elliptical contact surface were developed and it was shown that the severest stress always occurs on the z axis, and the maximum value of J_2 should occur at a certain point on this axis. The position of the point Z^* can be determined from the solution of the following equation:

$$\left.\frac{\partial J_2}{\partial Z}\right|_{Z=Z^*}=0 \tag{9}$$

Where $Z=z/a$ is the dimensionless z-coordinate, and Z^* denotes the Z parameter when J_2 has the maximum value at yielding.

The interference at the initial point of yielding is known as the critical interference, δ_y, which is derived analytically by using the von Mises yield criterion and given by Lin and Lin as

$$\delta_y=\left(\frac{9F_y^{\,2}}{16E^{*2}R_e}\right)^{\frac{1}{3}}\frac{2}{\pi}k^{\frac{1}{2}}\left[F_1(e)\right]^{-\frac{1}{3}}K(e) \tag{10}$$

Where

$$F_y=\frac{\pi^3R_e^{\,2}}{6E^{*2}}\left[F_1(e)\right]^2\left[K\left(k,\nu,Z^*\right)Y\right]^3 \tag{11}$$

The corresponding critical contact area is expressed as

$$a_c=\pi\left(\frac{3F_yR_e}{4E^*}\right)^{\frac{2}{3}}\left[F_1(e)\right]^{\frac{2}{3}} \tag{12}$$

Where Y is the yielding stress of the ellipsoid material. $K(k,\nu,Z^*)$ denotes the factor of the maximum contact pressure arising at yielding. This factor is expressed as a function of the ellipticity of the contact area, k, and the Poisson ratio of a material, ν. Z^* is the location of first yielding point on z-coordinate. The derivation of $K(k,\nu,Z^*)$ is shown in Lin and Lin's work. Ellipsoid deforms elastically as $\delta/\delta_y<1$. When $\delta/\delta_y>1$ the ellipsoid is in the elastoplastic deformation.

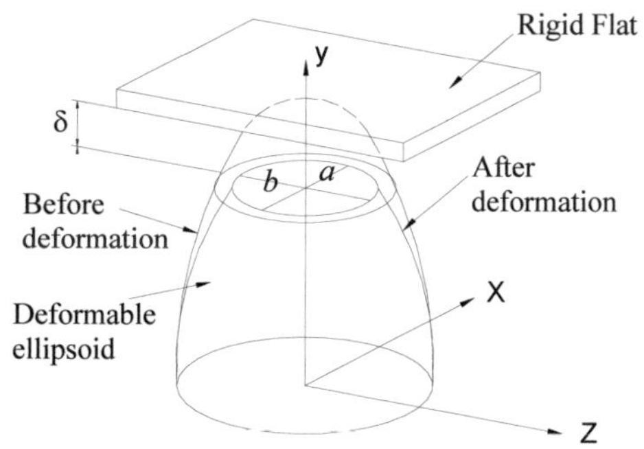

Fig. 1. The contact schematic diagram of a rigid flat with an ellipsoid.

4. Finite element model

In the present work, a commercial ANSYS-8.0 software package is applied to determine the elastoplastic regime arising at a deformable ellipsoid in contact with a rigid flat (see figure 2). There are two ways to simulate the contact problem. The first applies a force to the rigid body and then computes the resulting displacement. The second applies a displacement and then computes the resulting contact force. The present finite element solution is generated under a given interference δ applied to the contact area formed at the tip of an ellipsoid. This method is used because the resulting solution converges more rapidly than the former. In order to satisfy the geometrically symmetric condition and to assure that the nodes on the boundary of $y=0$ are far away from the contact area, the selection of one-eighth of the ellipsoidal volume as the simulation domain is made. Several option settings of ANSYS-8.0 software package have been made to reduce error in finite element calculations. The option of a static large displacement is adopted for the calculations in the elastoplastic and fully plastic regimes. The choice of the displacement style is based on the stress-strain (or load-displacement) behavior exhibited in each of these two deformation regimes. The ellipsoid is assumed to be an elastic-perfectly plastic material with identical behavior in tension and compression. This assumption was also made in the studies of Kogut and Etsion, and Jackson and Green. Frictionless and standard contact was also assumed as in their numerical simulations.

To increase the accuracy and efficiency of computation, one-eighth of an ellipsoid is used. Several mesh refinements have been performed to reduce the error in calculating von-Mises stress. For this investigation ANSYS element types 10-node, tetrahedron SOLID 92 element is selected for this nonlinear contact problem. Three sizes, 0.0005*R*, 0.0008*R*, and 0.001*R* (*R*: the semi-minor radius of ellipsoid), are the smallest element sides in the contact region set for ellipsoids with k_e=1, 1/2 and 1/5, respectively. In the present numerical model, the mesh size was refined according to its distance from the *y*-axis and the contact area of an ellipsoid. The fine mesh size of the volume element near the tip of the ellipsoid is varied in order to allow the ellipsoid's curvature to be captured and accurately simulated during deformations (see Figure 2). As to the region far away from the *y*-axis and the contact area, different coarser element size can be given in order to save computational time without sacrificing the precision of the solutions. As shown in Figure 2, constraints in the *x* directions and *z* direction are applied to the nodes on the *x*=0 plane and *z*=0 plane, respectively, while a constraint in *y* direction is applied on the base (the *y*=0 plane). The boundary condition may be valid for the modeling of asperity contacts for two reasons: (1) the asperities are actually connected to a much larger bulk material at the base and will be significantly restrained there, and (2) since the high stress region occurs near the contact, the boundary condition at the base of the ellipsoid will not greatly effect the solution because of Saint Venant's Principle.

In order to validate the model, mesh convergence must be satisfied. The mesh density is iteratively increased until the contact force and contact area differed by less than 1% between iterations. In the finite element analyses, the resulting meshes consist of at least 124572,125714, and 222913 elements correspond to ellipsoids with k_e=1, 1/2 and 1/5, respectively. These three node numbers are sufficient to obtain the numerical solutions with a high precision, compared with the theoretical solutions developed for the elastic deformation region. It is found that an excessive increase in the number of elements does not bring a significant improvement in the solution precision. The "contact wizard" in the software determines the relationship of the contact pair. The rigid flat is set as "Target", and the deformable ellipsoid is set as "Contact".

In addition to mesh convergence, the present work also compares well with the Hertz solution at interference below the critical interference. The numerical solutions for several contact parameters with different k_e values (k_e=1, k_e=1/2 and k_e=1/5) are listed in Table 1. The error between the theoretical and numerical solutions for all contact parameters is found to be always less than 1.5%. In the present study, the FKN (contact stiffness) value is varied in a range of 10 to 100 in the finite element analyses. Since precise solutions in all contact parameters are ensured, the accuracy of the solutions in the elastoplastic and fully plastic deformation regimes obtained by the present mesh scheme is ascertainable. Furthermore, because $k_e = 1$ represents a spherical contact, the present work compares with the results which are obtained by Jackson and Green and shows good agreement.

Since this contact problem is nonlinear and highly difficult to converge. An iterative scheme is used to solve for the solution, the minimum and maximum substeps are set in the range of 10 to 2000 such that (δ / δ_y)/(substep number) has a value varying in the range of 0.05~0.2. This is done to ensure the load increment is sufficiently small at each load step, thus improving the convergence behavior and minimizing the Newton-Raphson equilibrium iterations required.

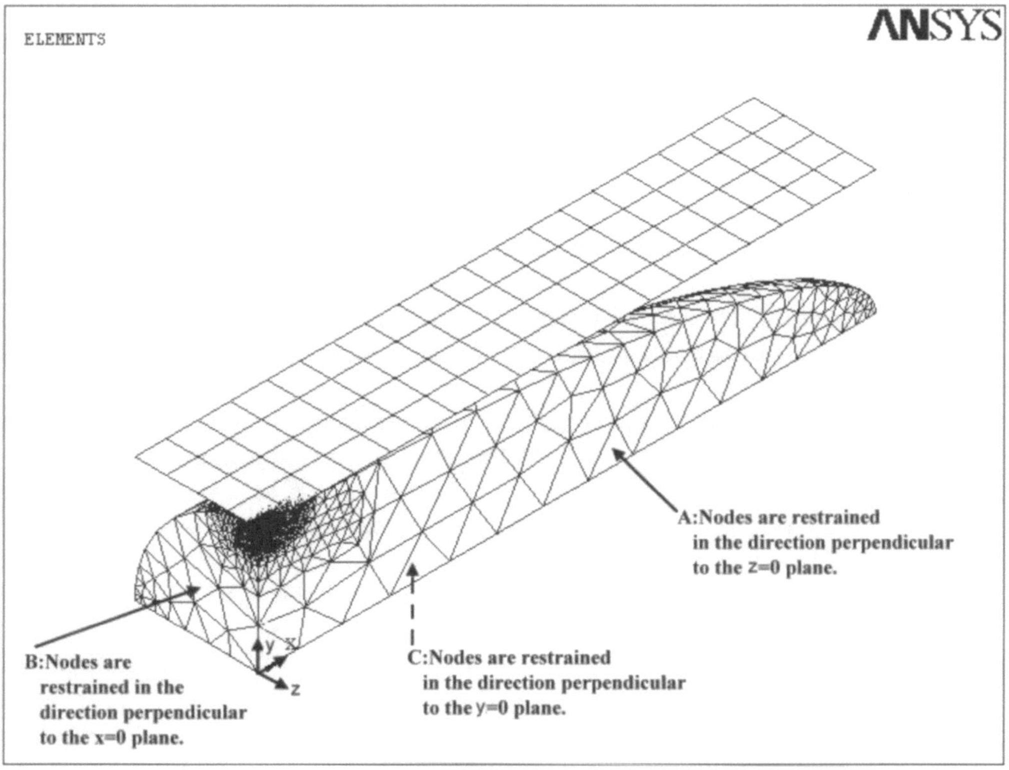

Fig. 2. The finite element analysis and the meshed model for simulations.

	k_e=1		k_e=1/2		k_e=1/5	
	Hertz solution	FEA solution	Hertz solution	FEA solution	Hertz solution	FEA solution
δ_y/δ_{y1}	1	0.99	1.85	1.83	3.21	3.19
S_{eqv}/Y	1	1	1	0.99	1	0.99
P_{max}/Y	1.61	1.63	1.65	1.62	1.73	1.74

Table 1. The comparison of numerical evaluated results at the critical interference with the theoretical solutions in the elastic deformation case

5. Results and discussion

The results are presented for a range of interferences, δ, which are normalized by each corresponding critical interference, δ_y, from 1 to 120 for a sphere, 1 to 100 for an ellipsoid (k_e =1/2) and 1 to 70 for an ellipsoid (k_e =1/5). The factor of the maximum contact pressure arising at yielding criterion, K, as shown in Eq. 7, is expressed as a function of $f\left(k,\nu,Z^*\right)$. The used material properties are for a steel material and present as $\nu = 0.3$, E=2.07x10^{11}Pa, Y=7x10^{8}Pa, and R=10^{-4}m. These material properties allow for effective modeling of all the elastoplastic contact regimes in the FEM simulation. The force convergence tolerance is 0.01 for the nonlinear solutions. Once the mesh is generated, computation takes from 1 hour for small interference to 50 hours for large interference by using an IBM p690 computer.

Figure 3-a shows the first yield point in the ellipsoid tip for (a) sphere (k_e =1), (b) ellipsoid (k_e =1/2), (c) ellipsoid (k_e =1/5) while the deformation equals to each critical interference δ_y. It is found that the first yield point happens in the larger depth from the ellipsoid tip for the smaller ellipticity of an ellipsoid, where the depth of the first yielding point is calculated as the distance between the top of the tip of an ellipsoid and the locations of the first yielding point. The smaller the ellipticity of the ellipsoid is, the larger the depth from the ellipsoid tip happens. While the first yielding depth values are normalized by the depth of ellipticity k_e =1, the values corresponding to k_e =1, k_e =1/2, k_e =1/5 are 1, 1.5 and 1.8, respectively.

Figure 3-b shows the comparisons of the critical interference δ_y and the location of the first yielding point with the ellipticity of a contact area *k*. The critical interference, δ_y, is significantly increased when the ellipticity of a contact area, *k*, is reduced. The depth of the first yielding point is deeper as the ellipticity of a contact area becomes smaller. The ellipticity of a contact area, k, is actually governed by the ellipticity of the ellipsoid. The ellipticities of a contact area formed at the yielding point for k_e =1, k_e =1/2, k_e =1/5 are *k*=1, 1/2.5 and 1/7.95 respectively. If the ellipticity of a contact area is smaller, it will correspond to the smaller ellipticity of an ellipsoid. The described phenomena can be seen on figure 3-a.

Figure 4 presents the evolution of the plastic region inside the ellipsoid tip for (a) sphere (k_e =1),(b) ellipsoid (k_e =1/2) and (c) ellipsoid (k_e =1/5) while $1 \le \delta / \delta_y \le 30$. Connecting the nodes with the equivalent von-Mises stress equals yield stress, *Y*, which is recorded by the commercial finite element program ANSYS 8.0, draws the elastic-plastic boundary line.

The behavior of the evolution of the plastic region in the ellipsoid tip for different ellipticities of an ellipsoid, k_e, is different. The development of plastic region on the contact surface is shown in more details in figure 7.

The evolution of the plastic region inside the ellipsoid tip at larger normalized interference, $40 \le \delta / \delta_y \le 120$, is shown in figure 5. As the interference increases above $\delta / \delta_y = 80$, the normal penetration of the plastic region is coincided at about 0.805x10^{-4}m as shown in figure 5(a) for a sphere tip (k_e =1). The above phenomena doesn't happen in the ellipsoid tip for ellipticity k_e =1/2 and k_e =1/5 at larger interference. The shapes of the plastic region in the ellipsoid tip for different ellipticities are also different.

Figure. 6 presents the three dimension contour plots of the equivalent von-Mises stress on the contact surface for (a) sphere (k_e =1), (b) ellipsoid (k_e =1/2) and (c) ellipsoid (k_e =1/5) at interference $\delta / \delta_y = 10$. At this interference the plastic region reaches the contact surface for both a sphere and an ellipsoid, which is shown in more details in figure 7. At this point an elastic core remains locked between the plastic region and the surface for a sphere (k_e =1) and an ellipsoid (k_e =1/2). It is interesting to note that the center of contact surface for an ellipsoid (k_e =1/5) has reached the plastic deformation. The plastic region reaches on both of the center area and an elliptical annular area on the contact surface for an ellipsoid (k_e =1/5). The evolution of the plastic region on the contact surface will behave in a different way for a sphere and both for an ellipsoid tip. At $\delta / \delta_y = 10$, the boundary of the plastic region that reaches the contact surface, which is obtained from curve fitting of the finite element analysis numerical results is plotted as figure 7. The lengths of semi-minor contact axis that are normalized by the critical contact radius of a sphere are about 2.4, 3.2 and 4.4 for the sphere (k_e =1), ellipsoid (k_e =1/2) and ellipsoid (k_e =1/5), respectively. The lengths of semi-major contact axis that are normalized by the critical contact radius of a sphere are about 2.4, 6.8 and 23.5 for the sphere (k_e =1), ellipsoid (k_e =1/2) and ellipsoid (k_e =1/5), respectively. The curve fitting length on semi-minor contact axis compared to the curve fitting length on semi-major contact axis is 1/2.13 and 1/5.34 for the ellipsoid (k_e =1/2) and ellipsoid (k_e =1/5), respectively. The above comparison values for the ellipsoid (k_e =1/2) and ellipsoid (k_e =1/5) aren't equal to the ellipticity of an ellipsoid.

Figure 8 presents the evolution of the plastic region on the contact surface for a sphere (k_e =1) while $10 \le \delta / \delta_y \le 120$. The boundary of the plastic region on the contact surface obtained from curve fitting of the finite element analysis numerical results of nodes is plotted as figure 8. At about $\delta / \delta_y = 6$ an annular plastic region first reaches the contact surface of a sphere. It is clear to see that an elastic core is locked by the annular plastic region. As the interference increases, the elastic core gradually shrinks and the annular plastic region will increase both to the center and outer line of the contact area. Finally, the elastic core disappears and the plastic region will dominate the most part of the contact area except for the outer annular area surrounded by the elastic region as shown in figure 8. The same conclusions have been obtained in the Kogut and Etsion's studies and are also seen by Jackson and Green.

Figure 9 presents the evolution of the plastic region on the contact surface of an ellipsoid (k_e =1/2) for increasing interference values up to $\delta / \delta_y = 90$. Up to $\delta / \delta_y = 10$ the elastic region dominate the contact surface. At $\delta / \delta_y = 10$ the plastic region first reaches the contact surface and forms an annular plastic region as shown in Figure 6 and figure 9. For $10 \le \delta / \delta_y \le 30$, the annular plastic region disappears dramatically and an elliptical plastic region appears on the center of contact surface. As the interference increases thereafter, the plastic region expands from center toward the edge of the contact surface. Even at this interference $\delta / \delta_y = 90$, the plastic region still doesn't coincide with the boundary edge of the contact area. But as figure 15(b) shows, the P/Y value is asymptotic to a constant value at this interference $\delta / \delta_y = 100$. Obviously, the interference $\delta / \delta_y = 90$ is near the inception of the fully plastic deformation for k_e =1/2.

Figure 10 presents the evolution of the plastic region on the contact surface of an ellipsoid (k_e =1/5) for increasing interference values up to $\delta / \delta_y = 70$. Up to $\delta / \delta_y = 10$ the elastic region dominates the contact surface. At $\delta / \delta_y = 10$ the plastic region first reaches the contact surface and forms an annular plastic region as shown in figure 6 and figure 10. In addition to the annular plastic region, the center of the contact surface also forms a plastic subregion. For $10 \le \delta / \delta_y \le 30$ the annular plastic region disappears dramatically and an elliptical plastic region on the center of contact surface extends the original dominated area. As the interference increases thereafter, the plastic region expands from center toward the edge of the contact surface. Even at interference $\delta / \delta_y = 70$, the plastic region still doesn't coincide with the boundary edge of the contact area. It can be seen the plastic region extends toward the direction of the major contact axis as the interference increases. But as figure 15(c) shows, the P/Y value is asymptotic to a constant value at this interference $\delta / \delta_y = 70$. Obviously, the interference $\delta / \delta_y = 70$ is near the inception of the fully plastic deformation for k_e =1/5.

The normal and radial surface displacements of the nodes on the ellipsoid surface are monitored in order to investigate the deformation of an ellipsoid. As shown in figure 11~14 the normal and radial directions (including semi-major and semi-minor contact axis) correspond to the y- and x-, z-axis, respectively. figure 11 and 12 show the surface displacement, Uy/ δ_{y1}, in the normal direction for the sphere (k_e =1), ellipsoid (k_e =1/2) and ellipsoid (k_e =1/5) vs. the normalized semi-major axis and semi-minor axis. These plots show the evolution of the surface normal deformation with increasing normalized interferences, δ / δ_y . As expected, the normal deformation on contact surface increases with increasing the normalized interference depth. The boundary between the contact region and the free surface boundary of the ellipsoid can be seen on the line edge in the plots. The slope of the normal displacement on the semi-minor axis is larger than the slope on the semi-major axis. As figure 11 and 12 show, both of the slopes of the normal displacement on the major and minor axis directions for different ellipticities satisfy: (k_e=1) > (k_e=1/2) > (k_e=1/5). The slope near the center and edge of contact surface becomes flat. Figure 13 and 14 show the surface displacement, U_x/ δ_{y1}, in the x radial direction vs. the normalized semi-major axis and the surface displacement, U_z/ δ_{y1} vs. the semi-minor axis for the sphere (k_e=1), ellipsoid (k_e=1/2) and ellipsoid (k_e=1/5). In the smaller interference depths, the surface displaces radially in mostly the negative direction for a sphere and an ellipsoid. The ellipsoid has the same compression behavior like a sphere. In other word, their contact areas are smaller than the geometrical intersection of the original undeformed profile with the flat in the smaller interference depths. This is because at the smaller normalized interferences, most of the materials in the sphere and ellipsoid are deforming elastically and are allowed to compress. As the interference is larger and significantly increases past the critical deformation, the material of the contact region displaces outward into the positive x direction and negative z direction. It can be seen in figure 14 that the radial displacement isn't positive until the normalized interference, $\delta / \delta_y = 90$, for a sphere, $\delta / \delta_y = 50$ for an ellipsoid (k_e=1/2) and past $\delta / \delta_y = 50$ for an ellipsoid (k_e=1/5). This bulginess occurs because the yielding material flows plastically and abides by the outlying material of the contact area. As shown in figure 14, the normalized radial displacement, U_x / δ_{y1}, become flat near the edge of the semi-major axis of contact area for an ellipsoid (k_e=1/5). When compared the boundary of plastic contact area in figure 10, their locations of semi-major contact axis are the same.

Figure 15 presents the normalized contact pressure, P/Y, profile vs the normalized semi-major contact axis, a/a_{c1}, and semi-minor contact axis, b/a_{c1}, for (a) sphere (k_e =1) at $\delta / \delta_y = 120$ (b) ellipsoid (k_e =1/2) at $\delta / \delta_y = 100$, (c) ellipsoid (k_e =1/5) at $\delta / \delta_y = 70$. When the normalized interferences, δ / δ_y, equal to 120, 100 and 70, for (a) sphere (k_e=1), (b) ellipsoid (k_e=1/2) and ellipsoid (k_e=1/5), respectively, the uniform contact pressure distribution is almost prevailing in the entire contact area, in which the dimensionless contact pressure is up to 2.5. The determination of the inception of the fully plastic deformation regime is based on the observed phenomenon that the normalized contact pressure and the normalized equivalent von-Mises stresses formed at the contact area have a uniform distribution. If interference is increased further, these uniform normalized contact pressures are found to be unchanged. It can be observed clearly that the normalized contact pressure ascends slowly from the center to the edge of the contact area for a sphere (k_e =1), the normalized contact pressure almost has a uniform distribution prevailing the entire contact area for an ellipsoid (k_e =1/2), and the normalized contact pressure descends slowly from the center to the edge of the contact area for an ellipsoid (k_e =1/5).

Figure 16 shows the variations of the contact-area ellipticity (k) in the elastoplastic deformation regime with the ellipticity of ellipsoid (k_e) and the dimensionless interference, δ / δ_y. The radii of curvature at the tip of the ellipsoid, R_{1x} and R_{1y}, are varied with the contact deformation. The effect of changing the radii of curvature due to contact deformations has been included in the evaluation of the k value. Data marked by the "Δ" and "□" symbols is obtained by assuming $k_e = 1/2$ and $k_e = 1/5$, respectively.

Both data sets show that the ellipticity of contact area, k, is increased by increasing the dimensionless interference (δ / δ_y). The data is asymptotic to a constant value equal to the k_e value associated with it if δ / δ_y is sufficiently large. It should be mentioned that the ellipticity of the contact area in the elastic deformation regime is always a constant value, which is equal to the datum shown in Fig. 16 at δ / δ_y =1. Therefore, the k value of an elliptical contact area is a variable when operating in the elastoplastic deformation regime.

The dimensionless contact areas (A^*) in the elastoplastic deformation regime varying with the dimensionless interference (δ / δ_{sy}) and the ellipticity of the ellipsoid (k_e) are shown in figure 17. The data for each k_e value can be expressed in a linear form in the log A^* -log (δ / δ_{sy}) plot. With the same dimensionless interference δ / δ_{sy}, the dimensionless contact area (A^*) is lowered by decreasing the ellipticity of the ellipsoid (k_e).

Figure 18 shows for the dimensionless contact load (F^*) results needed in the applications of different interferences (δ / δ_{sy}) and ellipticity values of an ellipsoid (k_e). Each of these curves shows a nonlinear relationship between F^* and δ / δ_{sy}. However, the effect of k_e on F^* is exactly opposite to that shown in the dimensionless contact area.

Kogut and Etsion used the plasticity index $\psi=\left(\delta_y/\sigma_s\right)^{-1/2}$ in the evaluations of the deformations arising at the contacts of rough surfaces, where σ_s denotes standard deviation of asperity heights. However, this plasticity index doesn't appear in the fractal analyses. In the Chung and Lin's study, four plasticity indices ($\psi=0.5,\ 1,\ 2,$ and 2.5) for the KE model are chosen. The dimensionless topothesy, $\overline{G}$, and fractal dimension, D, values applied are thus needed to determine on the base of these four plasticity indices. The plasticity index introduced by Greenwood and Williamson can be further expressed as $\psi=\left(\delta_y/\sigma_s\right)^{-1/2}=\left(\delta_y/\sigma\right)^{-1/2}\left(\sigma/\sigma_s\right)^{-1/2}$. In the study of Chung and Lin, $\left(\delta_y/\sigma\right)^{-1/2}$ is given as $\left(\dfrac{\delta_y}{\sigma}\right)^{-1/2}=\left(\dfrac{E^*}{K'H}\right)\sqrt{\dfrac{\sigma}{R_e(a_c)}}\left(\dfrac{\pi}{2}Q(\gamma)\right)^{-1/2}=\left(\dfrac{E^*}{K'H}\right)\sqrt{\dfrac{1}{R_e(a_c)}}\left(\dfrac{\pi}{2}Q(\gamma)\right)^{-1/2}$. In the study of McCool, the relationship between σ and σ_s is given as: $\left(\dfrac{\sigma_s}{\sigma}\right)^2=1-\dfrac{3.717\times10^{-4}}{(\eta R\sigma)^2}$, where the $\eta R\sigma$ values corresponding to $\psi=0.5,\ 1,\ 2,$ and 2.5 are 0.0302, 0.0414, 0.0541, and 0.0601, respectively. Therefore, $\left(\dfrac{\sigma_s}{\sigma}\right)_{\psi=0.5}=0.7697$, $\left(\dfrac{\sigma_s}{\sigma}\right)_{\psi=1}=0.8849$, $\left(\dfrac{\sigma_s}{\sigma}\right)_{\psi=2.0}=0.9343$, $\left(\dfrac{\sigma_s}{\sigma}\right)_{\psi=2.5}=0.9472$. By the expression $\psi=\left(\delta_y/\sigma\right)^{-1/2}\left(\sigma/\sigma_s\right)^{-1/2}$, the $\left(\delta_y/\sigma\right)^{-1/2}$ values corresponding to $\psi=0.5,\ 1,\ 2,$ and 2.5 are, 0.5699, 1.063, 2.069 and 2.5687, respectively. The $\overline{G}$ and D values corresponding to these four ψ values are thus shown in figure 19. In figure 19, the ellipticity $k=1$. The results of the dimensionless contact load predicted by the Chung and Lin's model are quite close to those predicted by the KE model, irrespective of the plasticity index values.

6. Conclusions

The work presents a finite element model (FEM) of the equivalent von-Mises stress and displacements that are formed for an elastoplastic ellipsoid is loaded. The material is modeled as elastic perfectly plastic and follows the von-Mises yield criterion. One-eighth of an ellipsoid in contact with a rigid flat is used to calculate the von-Mises stresses and deformations. A three-dimensional, 10-node, tetrahedron SOLID 92 element was selected for the nonlinear contact problem. Three sizes, 0.0005*R*, 0.0008*R*, and 0.001*R* (*R*: the semi-minor radius of ellipsoid), are the smallest element sides in the contact region set for a sphere with k_e =1 and ellipsoids with k_e =1/2 and 1/5, respectively

The FEM produce the evolution contour plots of the von-Mises stress with the different interferences. It is found that the first yield point happens in the larger depth from the ellipsoid tip for the smaller ellipticity of an ellipsoid. The smaller the ellipticity of the ellipsoid is, the larger the depth from the ellipsoid tip happens. While the first yielding depth values are normalized by the depth of an ellipticity k_e =1, the values corresponding to k_e =1, k_e =1/2, k_e =1/5 are 1, 1.5 and 1.8, respectively.

At a normalized interference, $\delta / \delta_y = 10$, the plastic region reaches the contact surface for a sphere and an ellipsoid, which is shown in more details in figure 7. At this point an elastic core remains locked between the plastic region and the surface for a sphere (k_e =1) and an ellipsoid (k_e =1/2). It is interesting to note that the center of contact surface for an ellipsoid (k_e =1/5) has reached the plastic deformation. The plastic region reaches on both of the center area and an elliptical annular area on the contact surface for an ellipsoid (k_e =1/5).
The normal deformation on the contact surface increases with increasing the normalized interference depth. The boundary between the contact region and the free surface boundary of an ellipsoid can be seen on the line edge in the plots. The slope of normal displacement on the semi-minor axis is larger than the slope on the semi-major axis.
The surface displaces radially in mostly the negative direction for a sphere and an ellipsoid. The ellipsoid has the same compression behavior like a sphere. In other word, their contact areas will shrink compared to the original cutting areas without deformation in the smaller interference depths. As the interference is larger and significantly increases past the critical deformation, the material in contact region displaces outward into the positive x direction and negative z direction.
When the normalized interferences, δ / δ_y, equal to 120, 100 and 70, for (a) sphere (k_e =1), (b) ellipsoid (k_e =1/2) and ellipsoid (k_e =1/5), respectively, the uniform contact pressure distribution is almost prevailing in the entire contact area, in which the dimensionless contact pressure is up to 2.5. It can be observed clearly that the normalized contact pressure ascends slowly from the center to the edge of the contact area for a sphere (k_e =1), the normalized contact pressure almost has a uniform distribution prevailing the entire contact area for an ellipsoid (k_e =1/2), and the normalized contact pressure descends slowly from the center to the edge of the contact area for an ellipsoid (k_e =1/5).

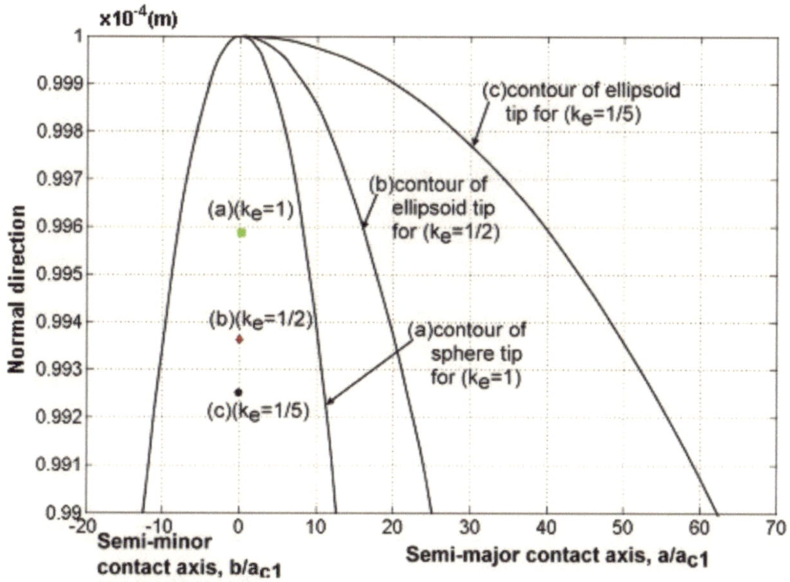

Fig. 3-a. The first yielding point happens in the ellipsoid for (a) sphere (k_e =1) (b) ellipsoid (k_e =1/2) (c) ellipsoid (k_e =1/5)

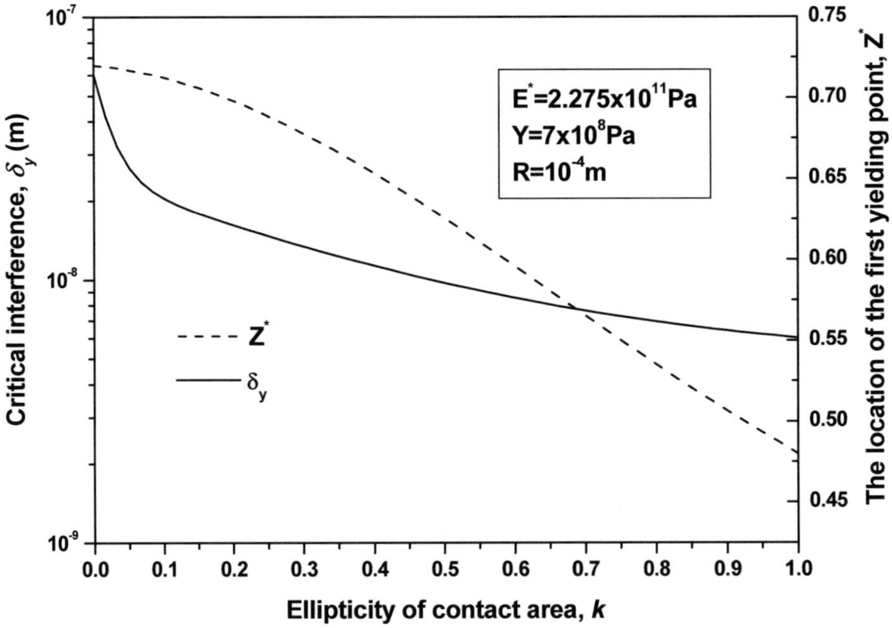

Fig. 3-b. Comparisons of the critical interference and the location of the first yielding point with the ellipticity of a contact area, *k*

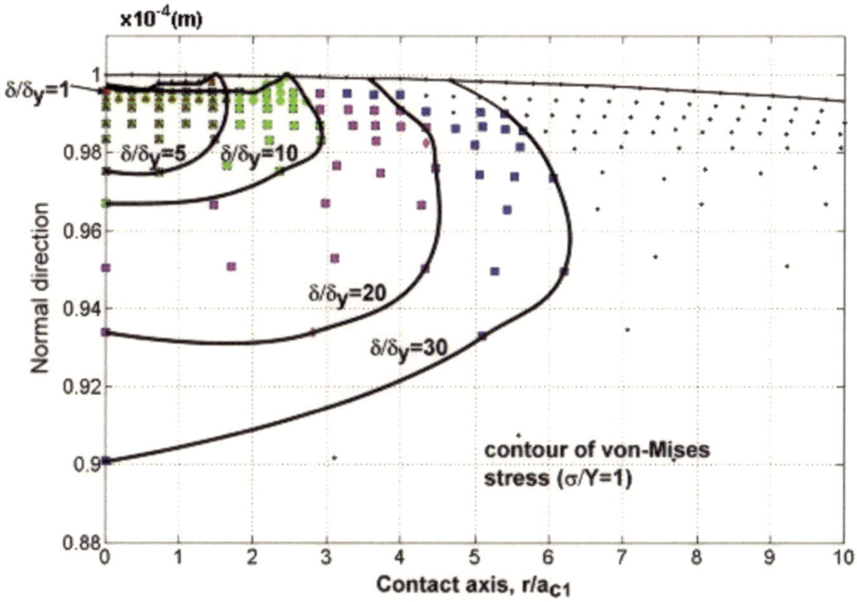
x10^-4(m)
δ/δy=1
δ/δy=5
δ/δy=10
δ/δy=20
δ/δy=30
Normal direction
contour of von-Mises stress (σ/Y=1)
Contact axis, r/ac1

(a)

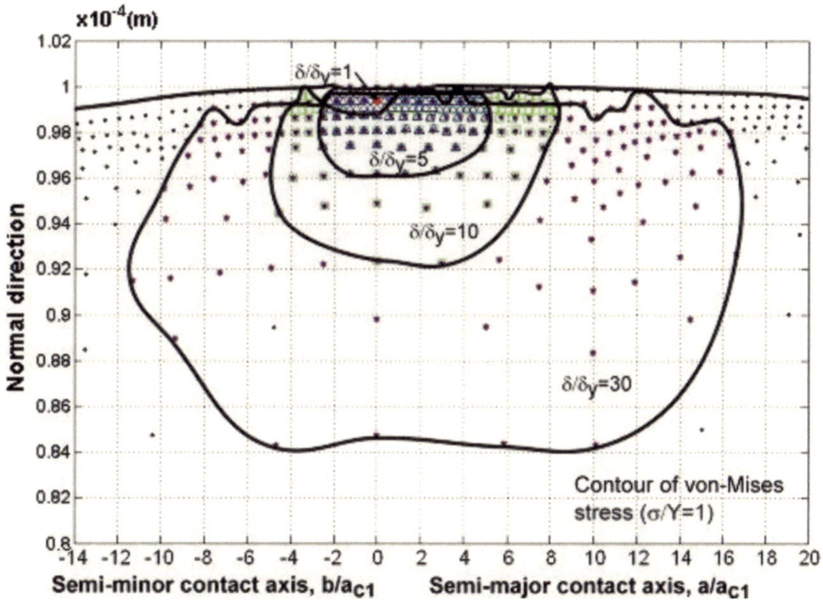
x10^-4(m)
δ/δy=1
δ/δy=5
δ/δy=10
δ/δy=30
Normal direction
Contour of von-Mises stress (σ/Y=1)
Semi-minor contact axis, b/ac1
Semi-major contact axis, a/ac1

(b)

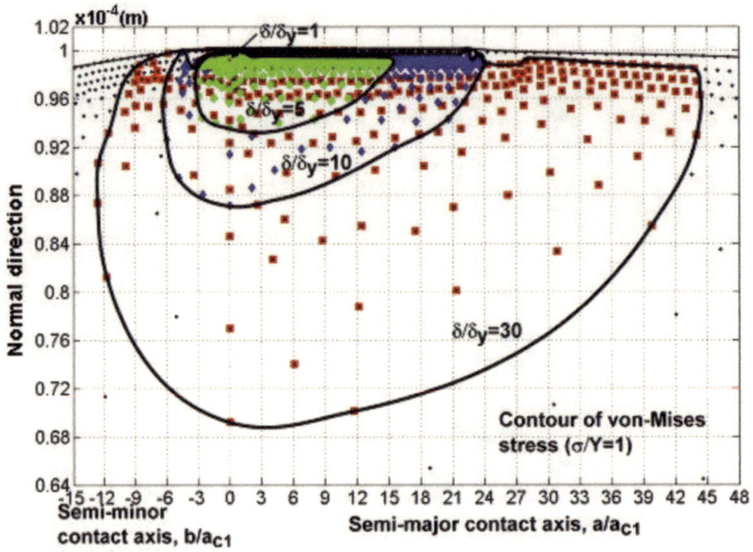

(c)

Fig. 4. Evolution of the plastic region in the ellipsoid tip for (a) sphere (k_e =1), (b)ellipsoid(k_e =1/2) and (c) ellipsoid(k_e =1/5) while $1 \le \delta / \delta_y \le 30$

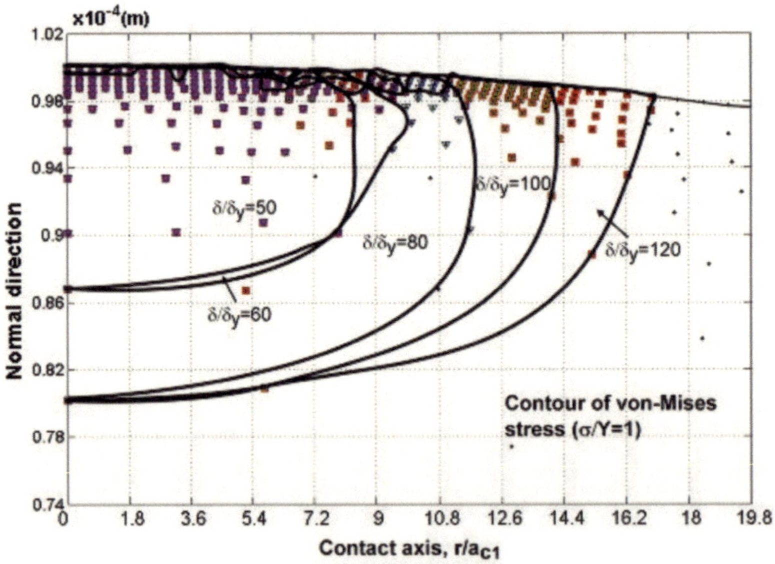

(a)

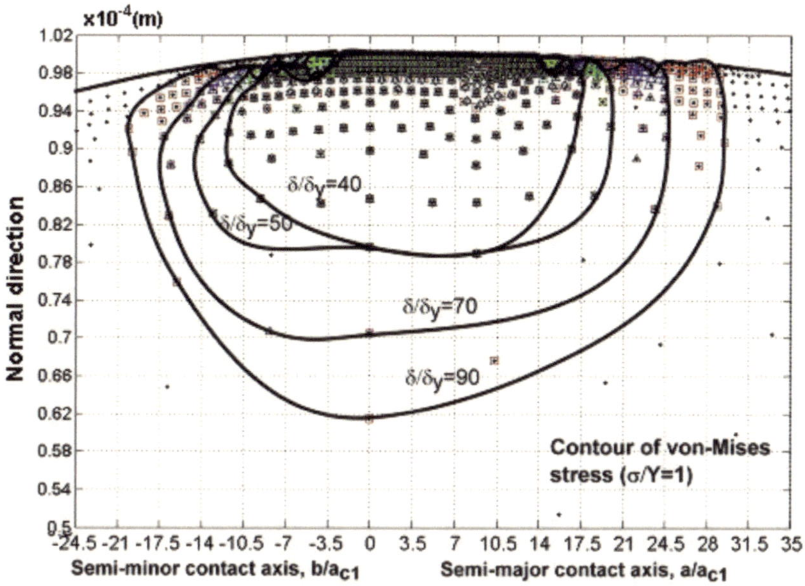

(b)

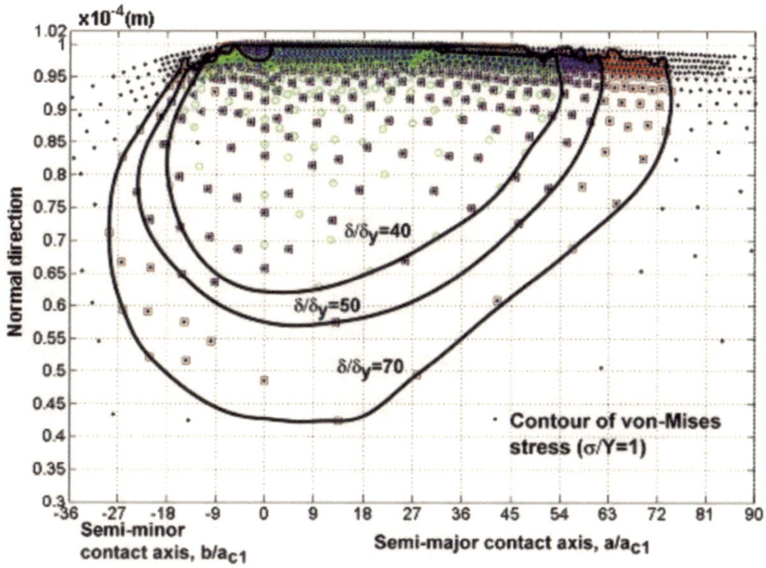

(c)

Fig. 5. Evolution of the plastic region in the ellipsoid tip for (a) sphere (k_e =1), (b)ellipsoid(k_e =1/2) and (c) ellipsoid(k_e =1/5) while $40 \le \delta / \delta_y \le 120$

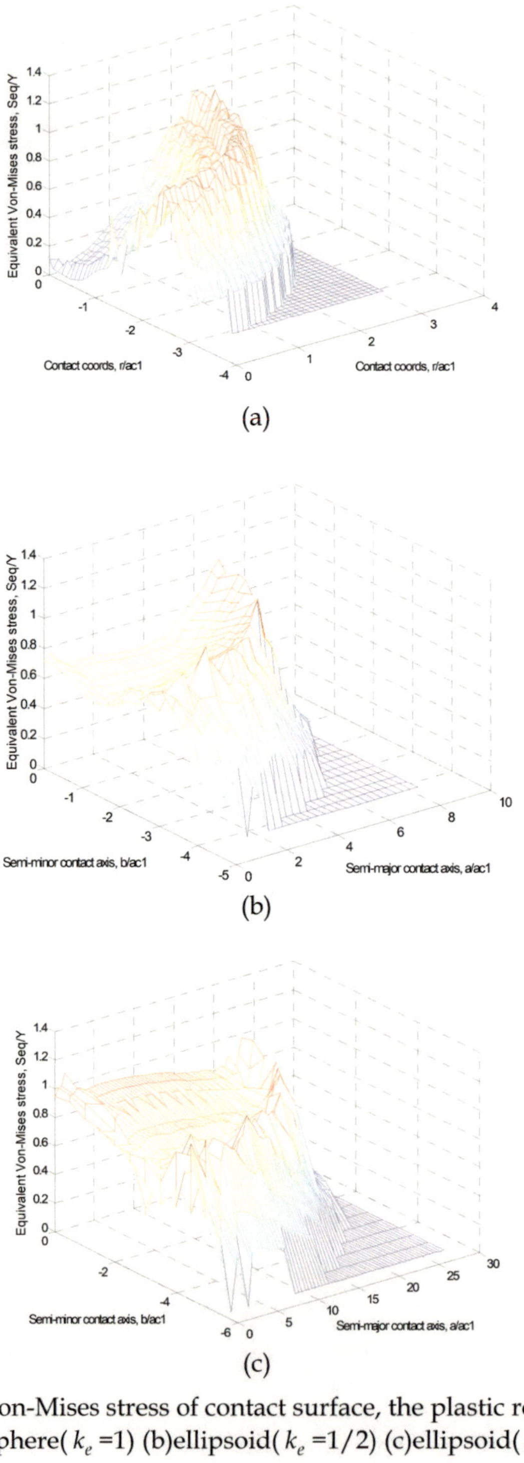

Fig. 6. The equivalent von-Mises stress of contact surface, the plastic region first reaches the contact surface, for (a)sphere(k_e =1) (b)ellipsoid(k_e =1/2) (c)ellipsoid(k_e =1/5) at interference $\delta / \delta_y = 10$

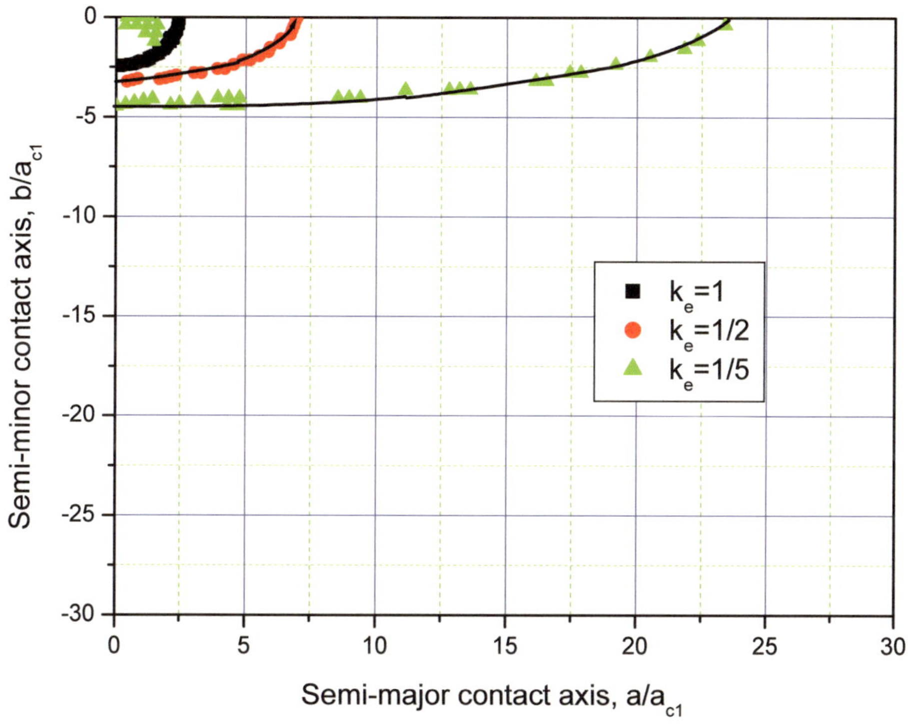

Fig. 7. The Semi-major and minor contact axis of contact surface, the plastic region reaches the contact surface at $\delta / \delta_y = 10$ for sphere (k_e =1), ellipsoid (k_e =1/2), ellipsoid (k_e =1/5).

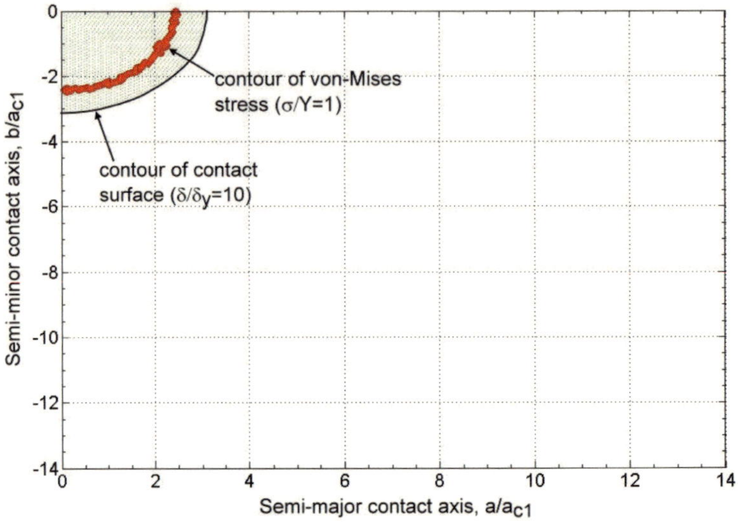
contour of von-Mises stress (σ/Y=1)
contour of contact surface (δ/δy=10)
Semi-minor contact axis, b/ac1
Semi-major contact axis, a/ac1

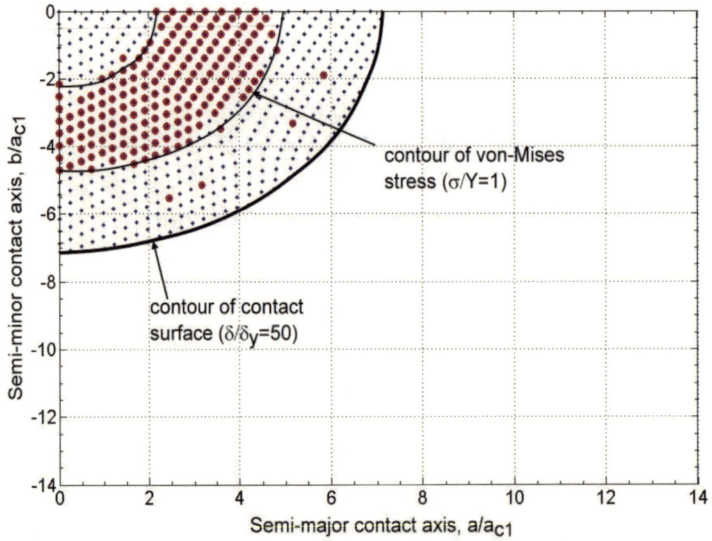
contour of von-Mises stress (σ/Y=1)
contour of contact surface (δ/δy=50)
Semi-minor contact axis, b/ac1
Semi-major contact axis, a/ac1

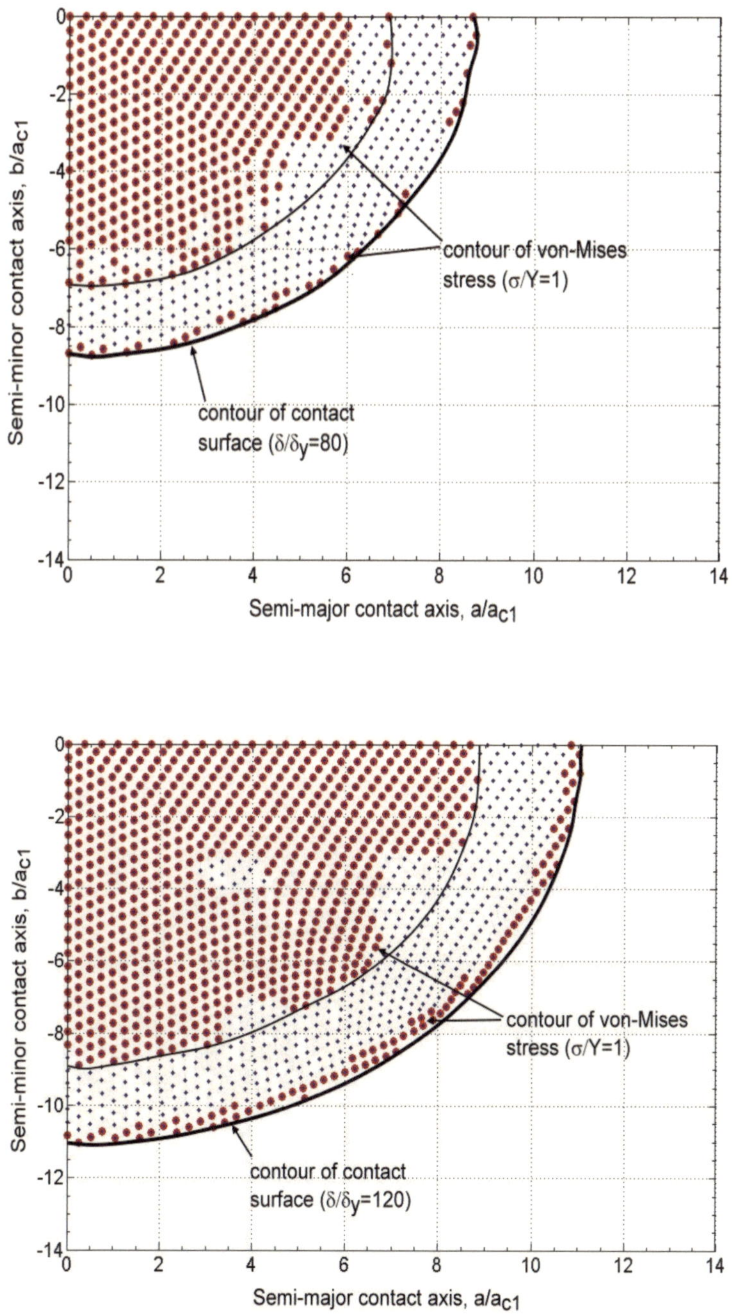

Fig. 8. Evolution of the plastic region on the contact surface for sphere (k_e=1) while $10 \le \delta / \delta_y \le 120$

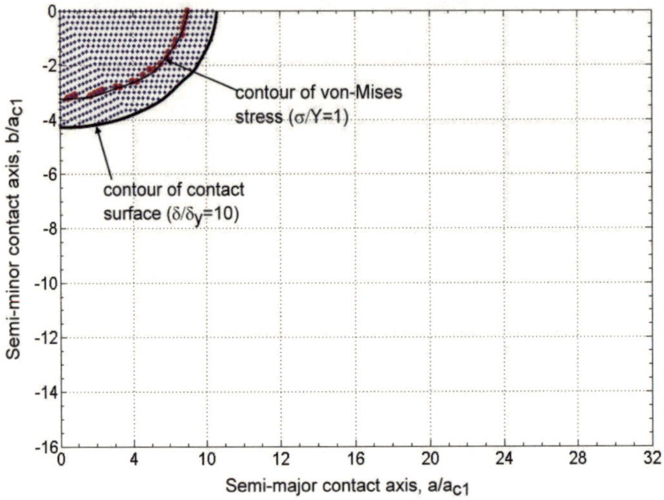

contour of von-Mises stress (σ/Y=1)
contour of contact surface (δ/δy=10)
Semi-minor contact axis, b/ac1
Semi-major contact axis, a/ac1
0
-2
-4
-6
-8
-10
-12
-14
-16
0
4
10
12
16
20
24
28
32

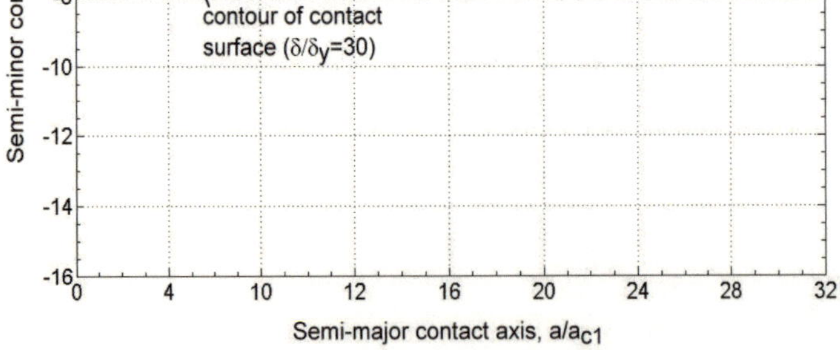

contour of von-Mises stress (σ/Y=1)
contour of contact surface (δ/δy=30)
Semi-minor contact axis, b/ac1
Semi-major contact axis, a/ac1
0
-2
-4
-6
-8
-10
-12
-14
-16
0
4
10
12
16
20
24
28
32

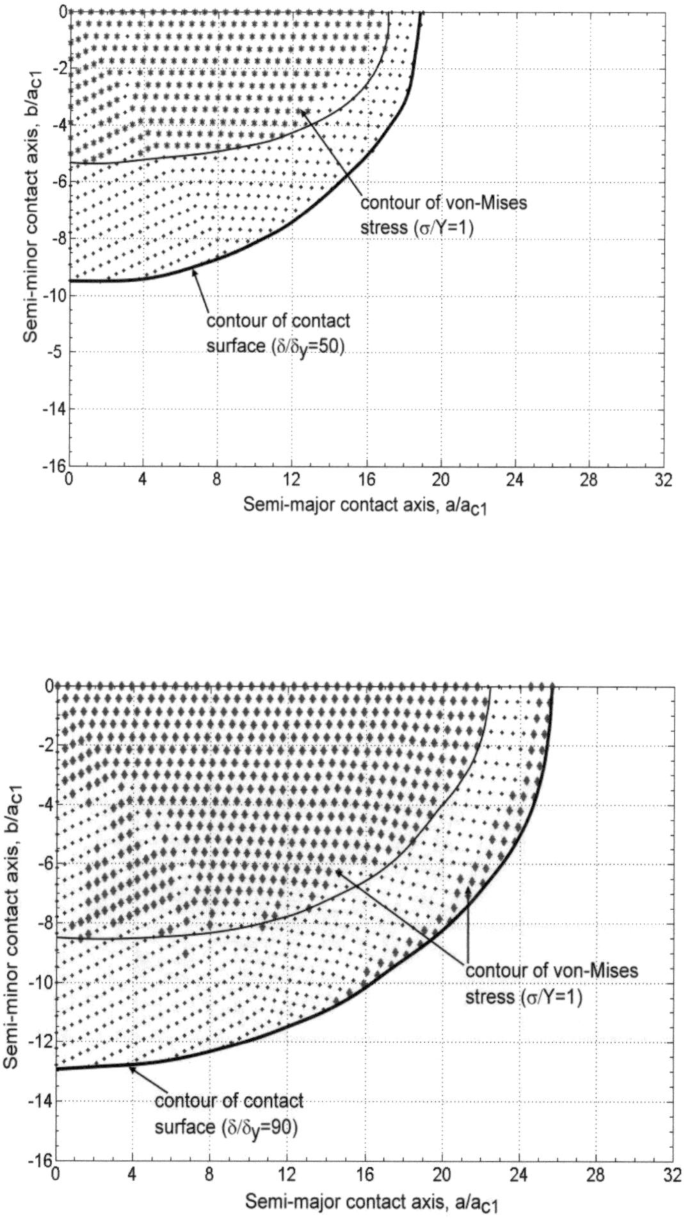

Fig. 9. Evolution of the plastic region on the contact surface of ellipsoid (k_e=1/2) for $10 \le \delta / \delta_y \le 90$

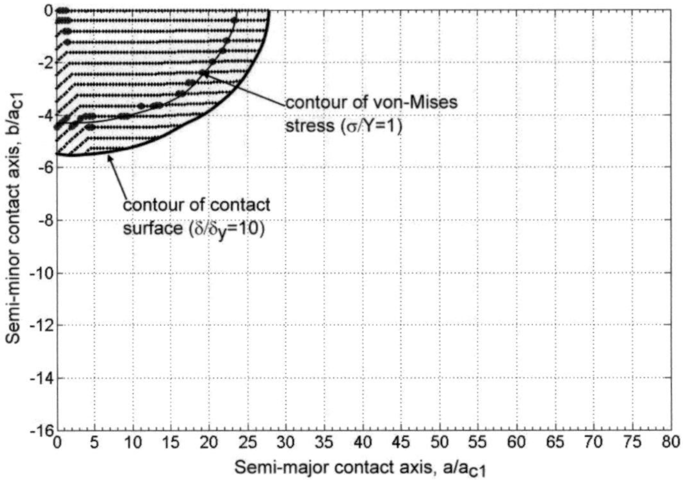
contour of von-Mises stress (σ/Y=1)
contour of contact surface (δ/δy=10)
Semi-minor contact axis, b/ac1
Semi-major contact axis, a/ac1
0 -2 -4 -6 -8 -10 -12 -14 -16
0 5 10 15 20 25 30 35 40 45 50 55 60 65 70 75 80

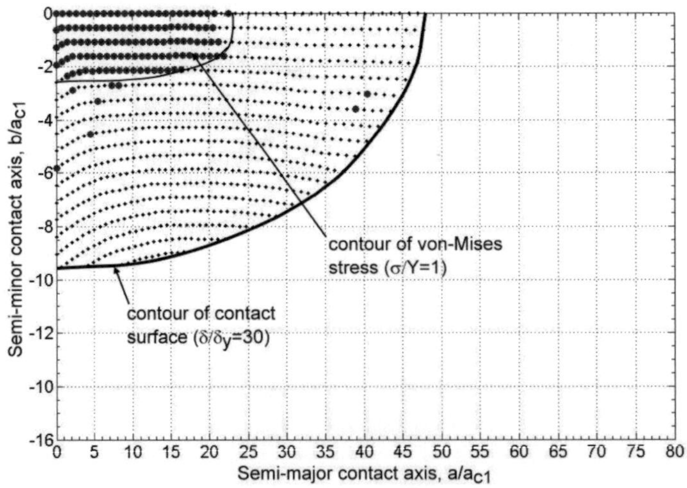
contour of von-Mises stress (σ/Y=1)
contour of contact surface (δ/δy=30)
Semi-minor contact axis, b/ac1
Semi-major contact axis, a/ac1
0 -2 -4 -6 -8 -10 -12 -10 -16
0 5 10 15 20 25 30 35 40 45 50 55 60 65 70 75 80

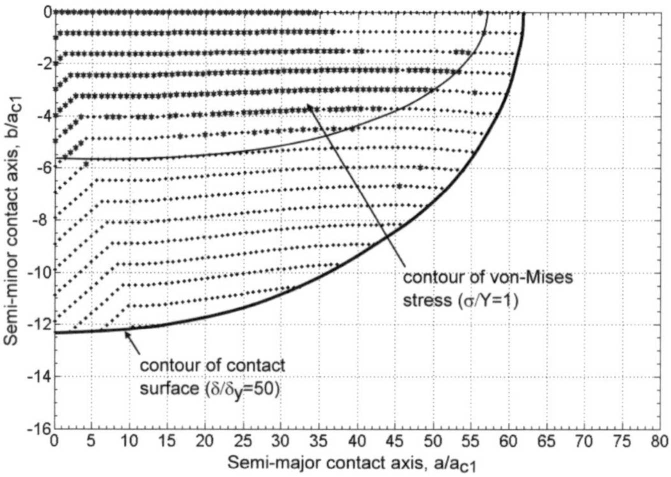

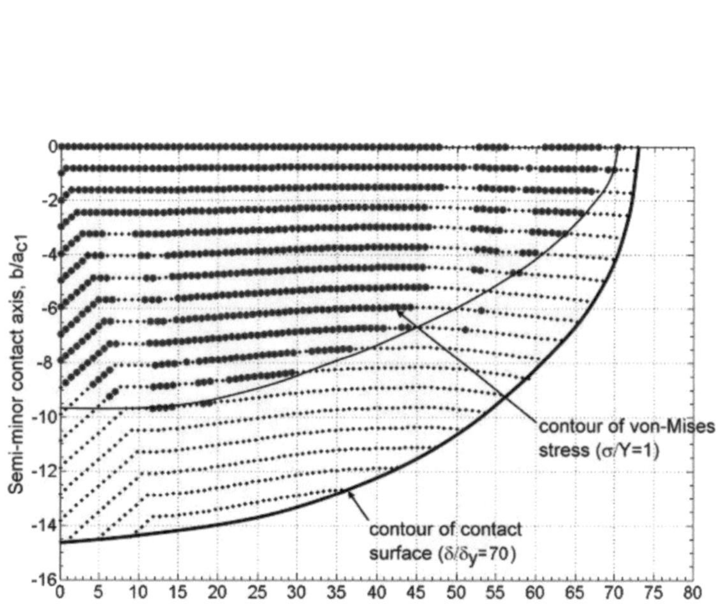

Fig. 10. Evolution of the plastic region on the contact surface of ellipsoid (k_e=1/5) for $10 \le \delta / \delta_y \le 70$.

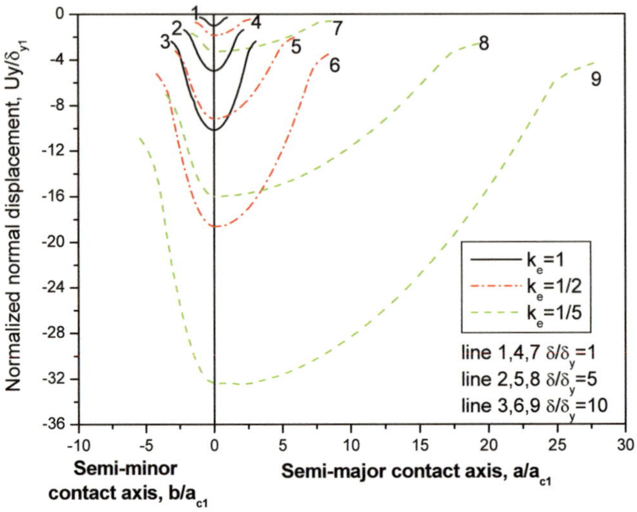

Fig. 11. The normalized normal displacement, $U_y/\ \delta_{y1}$, vs the Semi-major contact axis, a, and Semi-minor contact axis, b, for sphere (k_e =1), ellipsoid (k_e =1/2) , ellipsoid (k_e =1/5) in smaller interference depths.

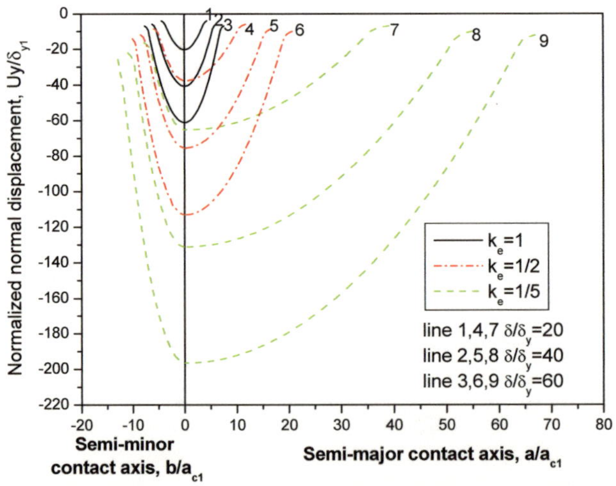

Fig. 12. The normalized normal displacement $U_y/\ \delta_{y1}$ vs the Semi-major contact axis, a, and Semi-minor contact axis, b, for sphere (k_e =1), ellipsoid (k_e =1/2), ellipsoid (k_e =1/5) in larger interference depths.

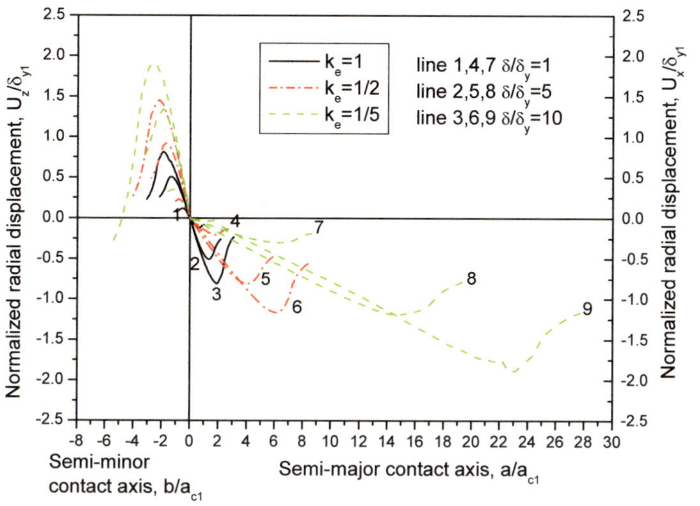

Fig. 13. The normalized radial displacement $U_x/\ \delta_{y1}$ vs the Semi-major contact axis, a, and $U_z/\ \delta_{y1}$ vs Semi-minor contact axis, b, for sphere (k_e =1), ellipsoid (k_e =1/2), ellipsoid (k_e =1/5) in smaller interference depths.

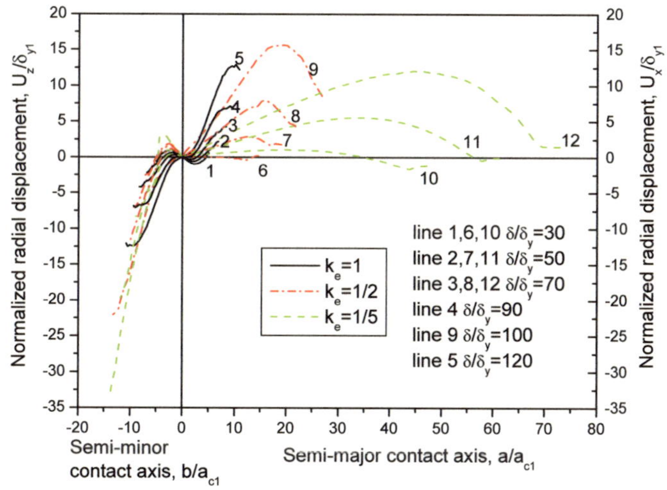

Fig. 14. The normalized radial displacement $U_x/\ \delta_{y1}$ vs the Semi-major contact axis, a, and $U_z/\ \delta_{y1}$ vs Semi-minor contact axis, b, for sphere (k_e =1), ellipsoid (k_e =1/2), ellipsoid (k_e =1/5) in larger interference depths.

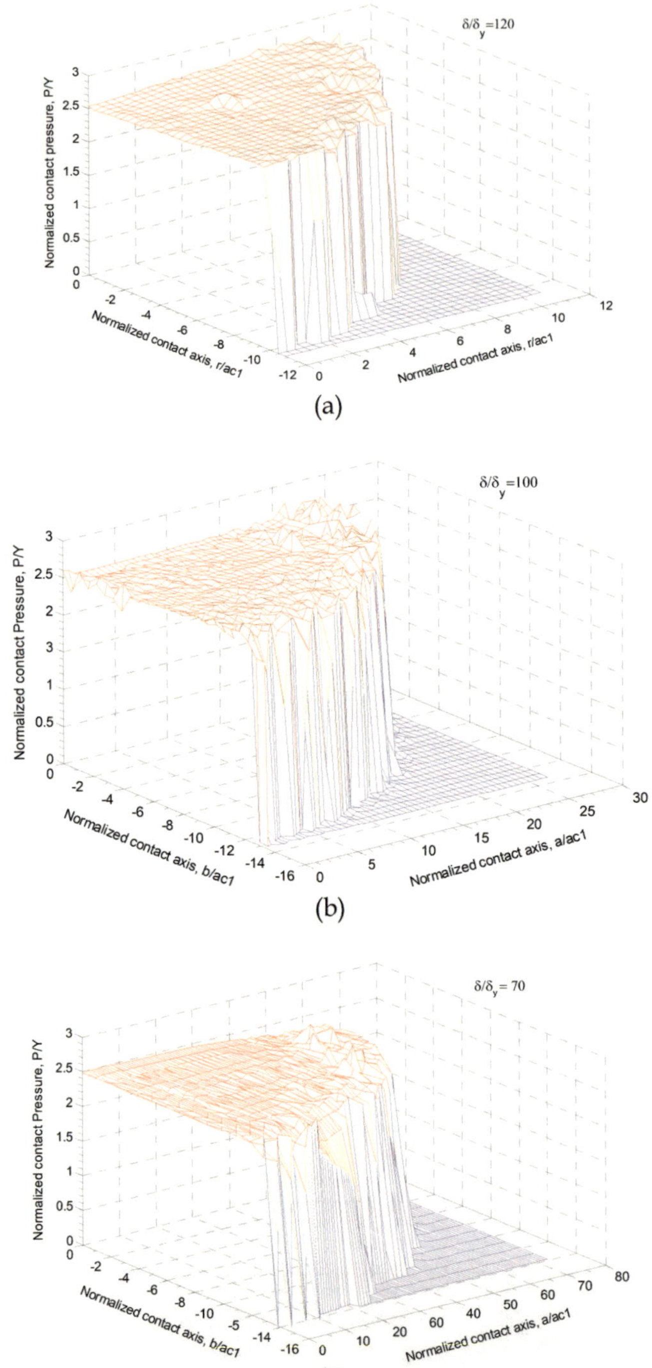

Fig. 15. The normalized contact pressure, *P/Y* vs Normalized contact axis, a/a_{c1}, and Semi-minor contact axis, b/a_{c1}, for (a) sphere (k_e =1) at $\delta / \delta_y = 120$ (b)ellipsoid (k_e =1/2) at $\delta / \delta_y = 100$, (c)ellipsoid (k_e =1/5) at $\delta / \delta_y = 70$.

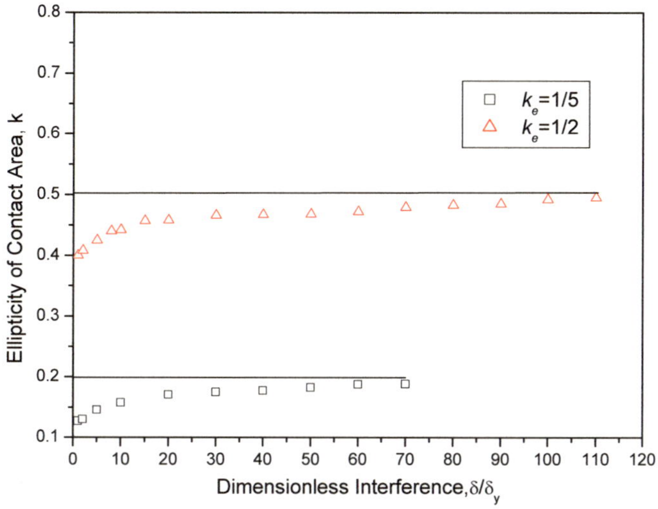

Fig. 16. Variations of the ellipticity of contact area with the dimensionless interference in the elastoplastic deformation regime.

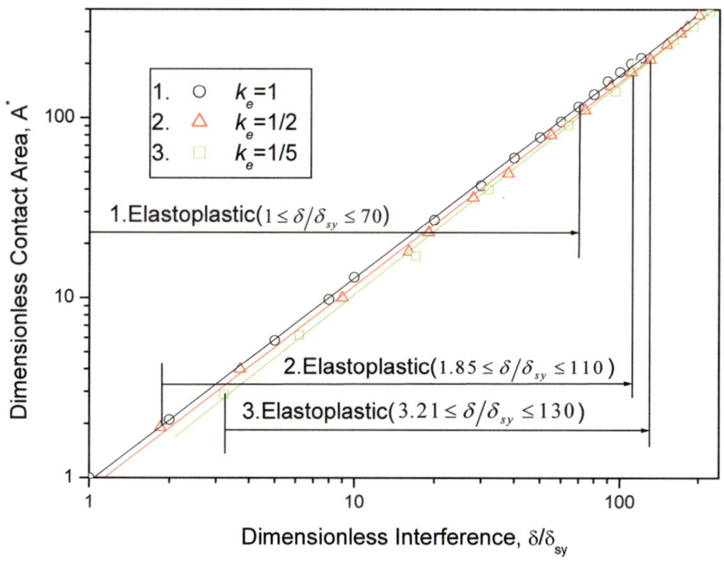

Fig. 17. Variations of the dimensionless contact area with the dimensionless interference.

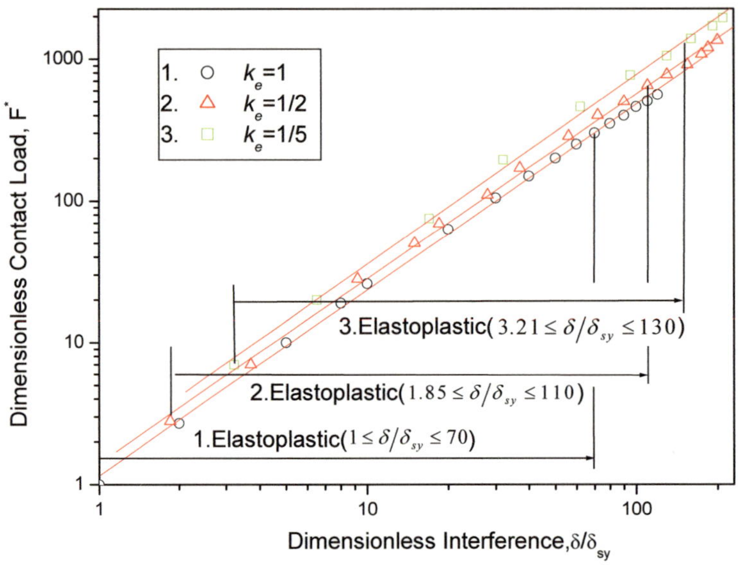

Fig. 18. Variations of the dimensionless contact load with the dimensionless interference.

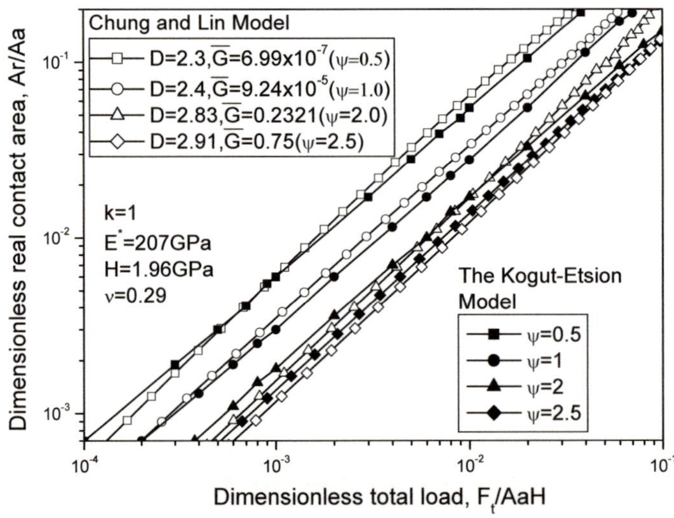

Fig. 19. Variations of the dimensionless real contact area with the dimensionless total load.

7. Reference

Abbott, E. J. & Firestone, F. A. (1933). Specifying Surface Quality-A Method Based on Accurate Measurement and Comparison, *Mech. Eng. (Am. Soc. Mech. Eng.)*, 55, pp. 569-572.

Belyaev N. M. (1957). Theory of Elasticity and Plasticity, *Moscow*.

Bush, A. W.; Gibson, R. D. & Keogh, G. D. (1979). Strong Anisotropic Rough Surface, *ASME J. Tribol.*, 101, pp. 15-20.

Bryant M. D. & Keer L. M. (1982). Rough Contact Between Elastically and Geometrically Identical Curved Bodies, *ASME, J. Appl. Mech.*, 49, pp. 345-352.

Buczkowski R. & Kleiber M. (2006). Elasto-plastic statistical model of strongly anisotropic rough surfaces for finite element 3D-contact analysis, *Comput. Methods Appl. Mech. Engrg.*, 195, pp. 5141–5161

Chang, W. R.; Etsion, I. & Bogy, D. B. (1987). An Elastic-Plastic Model for the Contact of Rough Surfaces, *ASME J. Tribol.*, 109, pp. 257-263.

Chung, J. C. & Lin J. F. (2004). Fractal Model Developed for Elliptic Elastic-Plastic Asperity Microcontacts of Rough Surfaces, *ASME J. Tribol.*, 126, pp. 82-88.

Chung, J. C. (2010). Elastic-Plastic Contact Analysis of an Ellipsoid and a Rigid Flat, *Tribology International*, 43, pp. 491-502

Greenwood, J. A. & Williamson, J. B. P. (1966). Contact of Nominally Flat Surfaces, *Proc. R. Soc. London*, Ser. A, 295, pp. 300-319.

Greenwood, J. A. & Tripp, J. H. (1967). The Elastic Contact of Rough Spheres, *ASME J. of Appl. Mech.*, Vol. 34, pp. 153-159.

Greenwood, J. A. & Tripp, J. H. (1970-71). The Contact of Two Nominally Flat Rough Surfaces, *Proc. Instn. Mech. Engrs.*, Vol. 185, pp. 625-633

Hisakado, T. (1974). Effects of Surface Roughness on Contact Between Solid Surfaces, *Wear*, Vol. 28, pp. 217-234.

Horng, J. H. (1998). An Elliptic Elastic-Plastic Asperity Microcontact Model for Rough Surface, *ASME J. Tribol.*, 120, pp. 82-88.

Johnson, K. L. (1985). Contact Mechanics, *Cambridge University Press*, Cambridge.

Jeng, Y. R. & Wang P. Y. (2003). An Elliptical Microcontact Model considering Elastic, Elastoplastic, and Plastic Deformation, *ASME J. Tribol.*, 125, pp. 232-240.

Jackson, R. L. & Green I. (2005a). A Finite Element Study of Elasto-Plastic Hemispherical Contact Against a Rigid Flat, *ASME J. Tribol.*, 127, pp. 343-354.

Jackson, R. L.; Chusoipin I. & Green I. (2005b). A Finite Element Study of the Residual Stress and Deformation in Hemispherical Contacts, *ASME J. Tribol.*, 127, pp. 484-493.

Kogut, L. & Etsion, I. (2002). Elastic-Plastic Contact Analysis of a Sphere and a Rigid Flat, *ASME, J. Appl. Mech.*, 69(5), pp. 657-662.

Liu, G.; Wang, Q. J. & Lin, C. (1999). A Survey of Current Models for Simulating the Contact between Rough Surfaces, *Tribol. Trans.*, 42, pp. 581-591.

Lin L. P., & Lin J. F. (2007). An Elliptical Elastic-Plastic Microcontact Model Developed for an Ellipsoid in Contact With a Smooth Rigid Flat, *ASME J. Tribol.*, 129, pp. 772-782.

Mindlin R. D. (1949). Compliance of Elastic Bodies in Contact, *ASME, J. Appl. Mech.*, 7, pp. 259

McCool, J. I. (1986). Comparison of Model for Contact of Rough Surfaces, *Wear*, Vol. 107, pp. 37-60.

Pullen, J. & Williamson, J. B. P. (1972). On the Plastic Contact of Rough Surfaces, *Proc. Roy. Soc. (London)*, A 327, pp. 159-173.

Zhao, Y.; Maletta, D. M., & Chang, L. (2000). An Asperity Microcontact Model Incorporating the Transition From Elastic Deformation to Fully Plastic Flow, *ASME J. Tribol.*, 122, pp. 86-93.

Sackfield, A. & Hills, D.A. (1983). Some Useful Results in the tangentially loaded Hertz Contact Problem, *J. of Strain Analysis*, 18, pp. 107-110.

10

Methods of Choosing High-Strengthened and Wear-Resistant Steels on a Complex of Mechanical Characteristics

Georgy Sorokin and Vladimir Malyshev
Gubkin Russian State University of Oil and Gas
Russia

1. Introduction

Tribology, as the science, has passed a long and complicated path of development, but still has not received that stage of completeness which guesses the decision of engineering tasks connected with increase of wear resistance of machines and instruments' parts in factory practice. In a large array of works on different aspects of tribology published for the last half century there are not enough investigations about the role of metal science in a nature of wear. It is characteristic specially for knots of machines working under abrasive affect conditions that cause an intensive mechanical wear and loss of life by executive links (Kragelsky, 1965; Beckman & Kleis, 1983).

A role of mechanical characteristics and aspects of metal science began to study in tribology much later (Rabinowicz, 1965; Tribology handbook, 1973). For this reason, the providing wear resistance of machines parts was reached, primarily, by possibilities of the experienced designers' specialists trying to exclude their breaking and deformation in conditions of small-cycled and a long-lived loading of working links based on known methods of toughness computation.

In accordance with designer's ideas of development and machines creation with higher operational characteristics, there was an apparent necessity for more detailed study of outwearing nature, especially in conditions of abrasive affect, as one of the basic reasons of equipments refusal. Specially, it concerns the work of oil-industry machines and drilling equipment, ore-mining, coal-extracting, ore- grinding, agricultural, building and other equipments (Richardson, 1967; Wellinger, 1963). Thus, the independent direction was discovered in tribology - the investigation of mechanical wear nature at the different acts variants of external forces and abrasives: at the sliding friction, at the rolling friction, at the blow over an abrasive, in the stream of abrasive particles, in the not fastened abrasive mass, etc.

The final goal of these investigations was the search of criteria tie of wear for steels and alloys with their standard mechanical characteristics, with regimes of heat treatment and structure, with the purpose of technological possibilities revealing in industrial conditions to control the processes capable to influence positively on the wear resistance increase of machines' parts under mechanical wear conditions.

In the chapter given, the basic dependences describing this complex process are reviewed and the recommendations connected to the methodology of its study and the definitions of criteria for an estimation of wear resistance of materials in similar conditions are marked.

2. Materials and methods of investigations

Mechanical characteristics of steels defined by standard methods on which basis are carried out calculations of machine details, are not connected with their design features and practically do not change within time of equipment exploitation. Unlike these characteristics the wear resistance is being defined not only by initial properties of tested material in interaction with which occurs the outwearing at exploitation, and also by character of uploading, especially by temperature in a friction zone. Dependence of one material's wear resistance from conditions of wear and properties of another material contacting with him complicates an estimation of actual wear and a choice of methods for its definition.

The development of materials trial methods on outwearing is caused by necessity of reliable choice of wear-resistant materials for the purpose of resource increase of machines and mechanisms.

The basic investigations of mechanical wear nature were conducted by sliding friction over monolithic abrasive as one of the wide-spread kinds of wear rendering the most negative influence on work resource of equipment in numerous branches of machine industry. For this purpose, the original laboratory machine (Fig. 1) for conducting the wear trials of any materials by sliding friction over monolithic abrasive wheel was manufactured.

The methodical feature and difference of this machine from those that were used earlier is that the cylindrical sample is moving radially by its lower face on rotary abrasive wheel plain and is rotating in addition around of own axle. This is stipulated to eliminate the passage of sample on the friction surface "track in track" and thus to avoid the "blocking" of working surface of abrasive wheel.

Technical characteristics of laboratory machine are as follows:

Diameter of a sample (mm) 10
Length of a sample (mm) 25-30
Load on a sample (N) . . . . up to 1000
Abrasive . . . . Grinding wheel 350 x 70 x 40
a green silicon carbide SiC, graininess ≤0.070 mm, HV = 32 GPa
Rotating speed of a wheel (rad/s) 3.2
Radial submission of a sample on one turn-over of a wheel (mm) 4.3

Symbols

WR wear resistance (g^{-1}) Δm mass wear (g) σ_b ultimate strength (MPa) $\sigma_{0.2}$ conventional yield limit (MPa) ψ relative reduction of area (%) δ relative elongation (%) τ_{sh} shear strength (MPa)	*HRC* Rockwell hardness *KCV* impact strength (MJ/m^2) σ_{-1} endurance limit (MPa) ρ resistivity (Ω m) K_1 coefficient of heat resistance at the furnace heat K_2 coefficient of heat resistance at the heat-up from friction a_H coefficient of impact strength (kg m/cm^2)

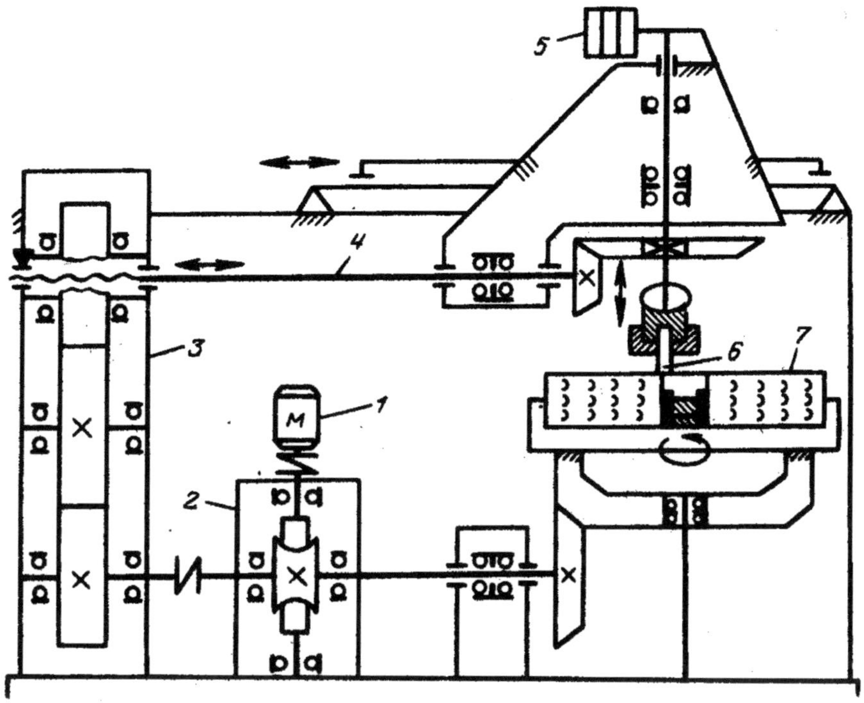

Fig. 1. A kinematics schema of original laboratory machine for materials trials on abrasive wear at the sliding friction: 1-electric motor; 2-worm reducer; 3-reducer; 4-feed screw; 5-weights; 6-sample; 7-abrasive wheel.

Such scheme of a trial ensures the higher convergence of tests data from experience to experience. The loading of sample was carried out by a lever with a weight. The outwearing path of sample on the abrasive wheel is 2.53 m for one-time pass. The velocity of samples slide over the abrasive wheel per tour of test was being changed from 0.1 up to 0.28 m/s. The unit load was selected 1.27 MPa experimentally that allowed to avoid a heat-up of friction surface at the trial. The wear was defined on a loss of samples mass Δm per tour of trial, i.e. for friction path 2.53 m. For comparative estimation of wear resistance of various steels the absolute parameter - the value return to mass wear - «$WR = 1/\Delta m$, g^{-1}» was chosen (Sorokin, 1991). Such indicator of wear resistance is most universal at comparison of this characteristic of steels tested in various conditions. The plots of dependences were built out of tests results as mean of minimum 5-6 experiences. The supplementary rotating of sample around own axle not only eliminates the directional roughness of samples friction surface, but also restores the cutting ability of the abrasive wheel as a result of gradual breaking down of its friction surface.

The advantage of this laboratory machine is the capability of trials conduction with chilling by any liquid environments, at the dry friction also and at the outwearing of the metal over the metal. In this case, the abrasive wheel is being substituted by the metal disk.

The abrasive outwearing is mechanical and represents the removing of metal from friction surface at the complex uploading. The removal of metallic particles at the outwearing is a destruction version by its nature, therefore it is quite lawful the using for it a classical concepts about toughness. In this connection it is methodically expedient to consider the role of all standard mechanical characteristics of steels, because other criteria of an estimation of steels' wear resistance are not present.

Regular investigations of wear resistance interrelation of hardened steels with all standard mechanical characteristics have been carried out. The steels of different structural classes with various levels of mechanical characteristics were selected for this goal: pearlitic class of average and high toughness, carbidic, austenitic and maraging classes. The trials have been complicated by using some other laboratory installations (for example Fig.2): along with tests at the sliding friction some trials were conducted at the blow over an abrasive and at the friction of metal surfaces without abrasive.

The basis of test method on this installation (Fig. 2) consists in outwearing of cylindrical samples by consecutive repeated blows on a layer of not fastened abrasive of the certain thickness located on a flat anvil. Installation is supplied by the adaptation allowing the regulation of abrasive layer thickness on the anvil and by the device for anvil moving after each cycle of trial. Energy of individual blow was being defined as product of weights placed on flat die on height of free fall (50 mm). Change of blow energy was possible in limits from 2.5 to 30 J. Frequency of blows were being changed from 60 to 120 min^{-1}.

Use of various installations at trials has allowed comparing influence of various schemes and conditions of mechanical outwearing on criteria of steels' wear resistance estimation.

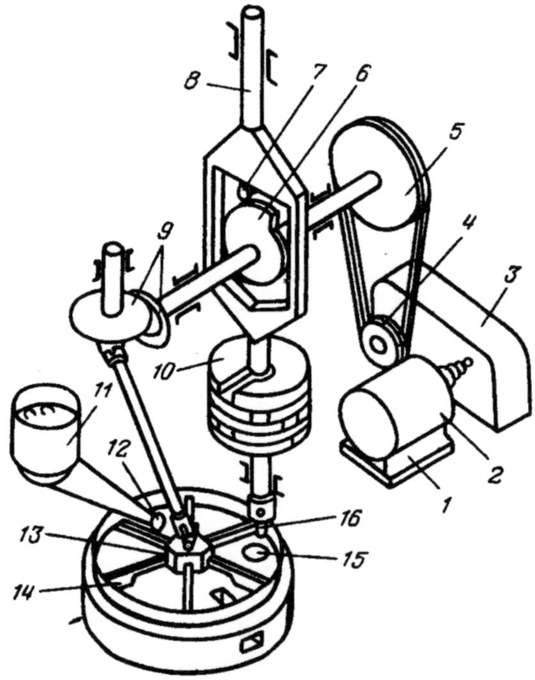

Fig. 2. Laboratory installation for wear trials at the blow on a not fastened abrasive: 1 - welding frame; 2- electric motor; 3 - reducer; 4,5 - pulleys of belt drive; 6 - cam; 7 - roller; 8 - spindle-flat die; 9 -bevel gearing; 10 - weights; 11 - hopper; 12 - batcher; 13 -rotated disk; 14 - brushes; 15 - anvil with abrasive; 16 - sample.

Apart from steels of different structural classes for which the chemical composition and mechanical characteristics are instituted by national standards (GOST) (Machine building Materials, 1980), the mechanical characteristics and wear resistance of experimental steels conditionally marked as D4, D5, D6 and D7 and created in different time under orders of petroleum industry were studied (Vinogradov, 1989). The elemental chemical composition of steels of different structural classes used in trials is given in Table 1.

Grade of steel	Content of chemical elements, %										
	C	*Si*	*Mn*	*Cr*	*Ni*	*Mo*	*V*	*S u P*	*Co*	*W*	*Ti*
95X18	1.0	≤0.8	≤0.7	18	-	-	-	≤0.03	-	-	-
110Г13Л	1.1	-	13	1	1	-	-	-	-	-	-
H18K9M5T	-	-	-	-	18	5	-	-	9	-	1
P18	0.8	≤0.4	≤0.4	4.2	≤0.4	0.3	1.2	≤0.03	-	18	-
X12M	1.55	0.25	0.35	12	-	0.5	0.25	≤0.03	-	-	-
40X13	0.4	0.30	0.65	1.3	≤0.4	-	-	≤0.04	-	-	-
40X	0.4	0.28	0.55	0.9	≤0.4	-	-	≤0.04	-	-	-
У8	0.8	0.25	0.45	0.20	0.15	-	-	≤0.03	-	-	-
У10	1.0	0.20	0.25	0.20	0.15	-	-	≤0.02	-	-	-
45	0.45	0.28	0.70	0.25	0.25	-	-	≤0.04	-	-	-
40	0.40	0.30	0.70	0.25	0.25	-	-	≤0.04	-	-	-
20	0.20	0.30	0.50	0.25	0.25	-	-	≤0.04	-	-	-
D4	0.39	0.28	0.54	0.4	1.1	-	-	-	-	-	-
D6	0.58	0.26	0.55	0.8	1.2	-	-	-	-	-	-
D7	0.7	0.25	0.42	0.6	1.5	-	0.22	-	-	-	-
D5	0.47	0.27	0.69	1	1.4	0.18	0.25	≤0.02	0.25	0.25	0.25

Note: *Fe* – the rest

Table 1. Chemical composition of tested steels

3. Results of investigations

The purpose of investigations on the first stage was the definition of functional bond of steels' wear resistance at the mechanical (abrasive) outwearing with their standard mechanical characteristics: ultimate strength σ_b, conventional yield limit $\sigma_{0.2}$, endurance limit σ_{-1}, Rockwell hardness *HRC*, relative elongation δ, relative reduction of area ψ and impact strength *KCV*.

3.1 Interrelation of wear resistance with indexes of steels' mechanical properties

At the analyses of correlation of each mechanical characteristics separately, "wear resistance-property", the enough defined tendencies are discovered: with increasing of strength characteristics (σ_b, $\sigma_{0.2}$, *HRC*) the wear resistance of steels grows, and the characteristics of plasticity and viscosity (δ, ψ, *KCV*) reduce the wear resistance with their increasing. The similar dependence is characteristic for all mechanical properties (Sorokin, 2000).

Mechanical characteristics depend, first of all, from class of steel and its structural features: it means here the type of steels' structure, the ability of structure to hardening at the heat treatment and its propensity to unhardening under thermal influence. If to combine graphics changes of mechanical characteristics of hardened steels of different structural classes depending on tempering temperature, it is possible to reveal characteristic tendencies in change of properties and their numerical values. There have been compared, first of all, the characteristics of toughness group - hardness, ultimate strength and conventional yield limit, and also the characteristics of plasticity - relative reduction of area.

3.1.1 Steels hardness change of various structural classes from tempering temperature

The hardness of hardened steels of various structural classes changes in a wide interval of numerical values at the rise of tempering temperature (Fig. 3). The law of hardness change is ambiguous: at the rise of tempering temperature the hardness can be constant - for steels of austenitic class, sharply decrease - for steels of pearlitic class and increase - for steels of carbidic class. Hardness of austenitic steel 110Г13Л is low - 18 HRC, but in the range of tempering temperatures 0-600 ⁰C it is constant. It can be explained by absence of structural transformations in this steel at tempering, and consequently, unhardening. Steels hardness of pearlitic class (20, 45, 40X, У10, D7) after hardening is various: the minimal hardness (35 HRC) has the steel 20 and the maximal hardness (65 HRC) has an experimental steel D7. At the rise of tempering temperature the hardness of these steels is decreasing: at tempering temperature 600 ⁰C the hardness for D7 is equal 38 HRC, and for steel 20 is equal 15 HRC. Steel hardness of carbidic class P18 directly after hardening is approximately 62 HRC; at the rise of tempering temperature the hardness of this steel not only does not decrease, but increases at tempering temperature 600 ⁰C until 65 HRC. The law of hardness change at the tempering of hardened steels of martensitic class 95X18, maraging class H18K9M5T and ledeburitic class X12M essentially differs from the law of steels hardness change of pearlitic and carbidic classes.

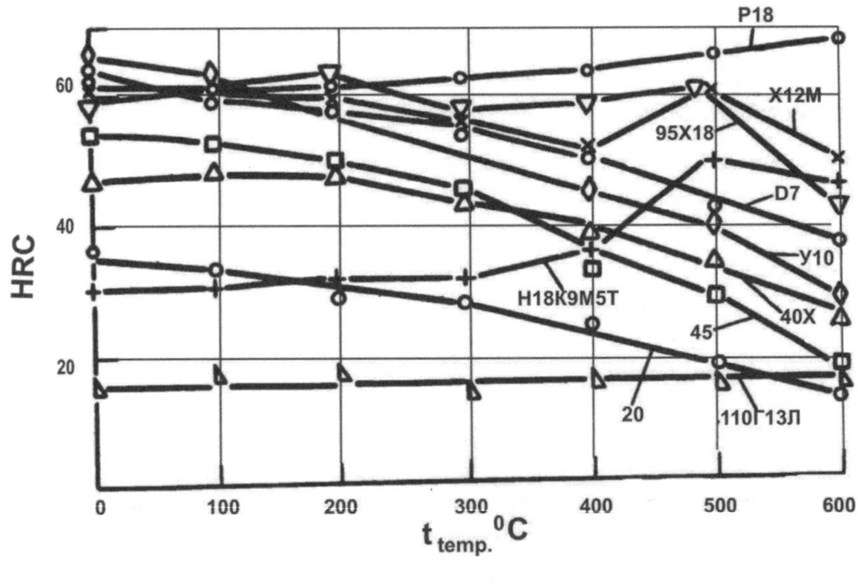

Fig. 3. Dependence of steels hardness change of various structural classes from tempering temperature

Steel's initial hardness of maraging class H18K9M5T (30 HRC) remains until tempering temperature 300 ⁰C; after this it starts to increase until 44 HRC at 500 ⁰C and is stabilizing at this level up to 600 ⁰C. Hardness of steel 95X18 decreases a little at the rise of tempering temperature until 400 ⁰C, then increases at 500 ⁰C, and decreases again (to 48 HRC) at 600 ⁰C. Hardness of steel X12M at tempering temperature until 500 ⁰C is constant and high enough, its heating up to 600⁰C reduces this value to 50 HRC.

Thus, the area of hardness change is in a range from 18 up to 62 HRC at tempering of hardened steels of basic structural classes in the range of temperatures from 0 to 600 ⁰C. The lower level of this area is limited by hardness of austenitic steel 110Г13Л and upper level - by hardness of carbidic steel P18. By comparison of steels hardness of various classes in the conditions of tempering becomes obvious, that for the hardened steels of pearlitic class it is characteristic a strong unstrengthen at heating; by this index they cannot be attributed to group of wear-resistant steels. For work in the conditions of heats when force uploading is accompanied by mechanical outwearing, the best steel with structural stability and hardness is the steel of carbidic class P18.

3.1.2 Change of ultimate strength for steels of various structural classes from tempering temperature

The ultimate strength was compared for the same hardened steels in the same interval of tempering temperatures. Polarization of this mechanical characteristic depending on tempering temperature (Fig. 4) is even more, than for hardness.

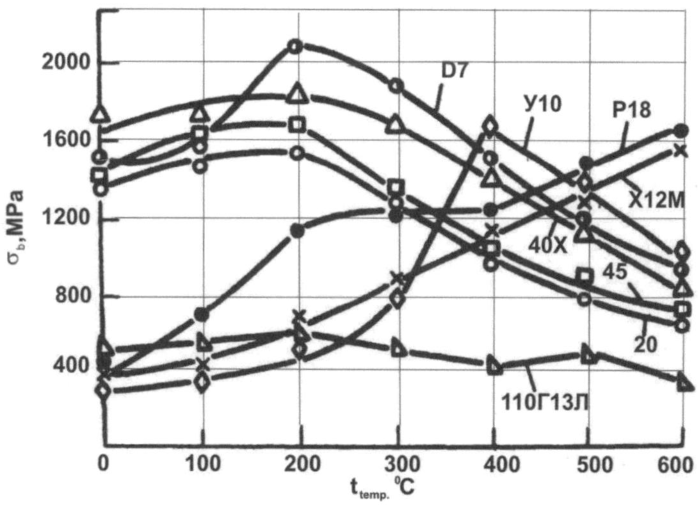

Fig. 4. Change of ultimate strength for steels of various structural classes from tempering temperature

The value of ultimate strength is stable in a wide interval of tempering temperatures for austenitic steel 110Г13Л and is minimal in relation to other steels - nearby 400 MPa. The ultimate strength of steels pearlitic class 20, 45, D7 changes under one law: it is increasing a little at tempering temperature 200 ⁰C and then decreasing monotonous. The maximum of ultimate strength is fixed for steel D7 at tempering temperature 200 ⁰C - 2200 MPa; after high tempering this value decreases approximately in 2 times (up to 1000 MPa). The ultimate strength of steel X12M almost linearly increases from 400 to 1860 MPa at rising of tempering temperature. The ultimate strength of steel P18 increases stably in process of rising tempering temperature and has a maximum at 600 ^{0}C. The analysis of these dependences shows that for conditions of static uploading the steels of pearlitic class have appreciable advantages before steels of other classes on level of ultimate strength, but stability of its maximum values is limited by an interval of tempering temperatures 100-300 ^{0}C.

3.1.3 Change of relative reduction of area for steels of various classes from tempering temperature

Relative reduction of area ψ for steels 20, 45, 40X, У10 is increasing at rising of tempering temperature, but for steels 110 Г13Л and X12M this characteristic does not change practically (Fig. 5).

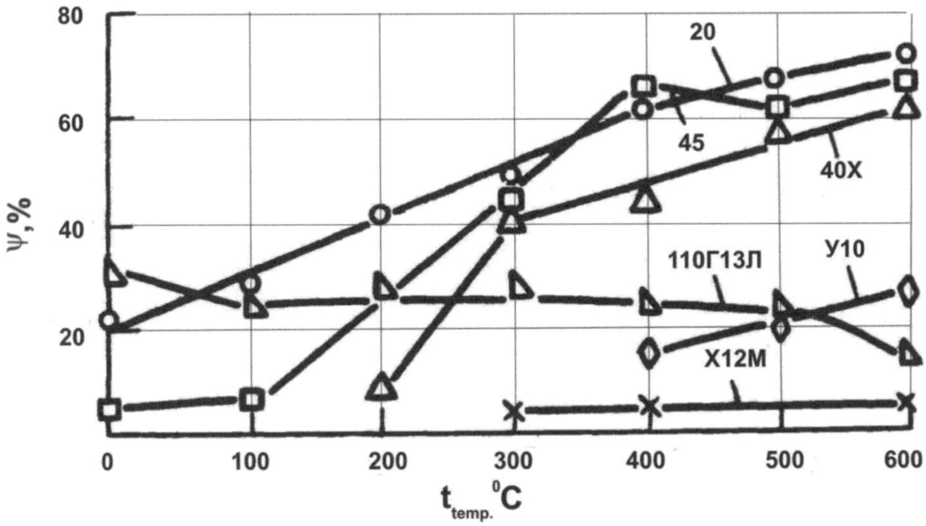

Fig. 5. Change of relative reduction of area for steels of various structural classes from tempering temperature

Relative reduction of area ψ and relative elongation δ vary practically under one law. Thus, relative reduction of area of the steels majority is maximum at high tempering (600 ^{0}C).

3.1.4 Dependence of steels' wear resistance from one parameter of mechanical properties

The steels' wear resistance may be defined for some external uploading conditions on one of the parameters (Fig.6) (Sorokin, 2000), for example,

- at a blow over a not fastened abrasive - the shear strength (τ_{sh}),
- at an erosive outwearing when the angle of attack is equal 90^0 - the relative elongation (δ),
- at a blow over a metal without abrasive - the endurance limit (σ_{-1}),
- at an abrasive outwearing of surface hardening alloys - the resistivity (ρ).

Thus, there are some external forces conditions of abrasive affecting or of blow of metal over metal, when one of mechanical properties can be selected as criterion of wear-resistant steels for defined work conditions. However, for more other cases of work conditions it is very difficult to find reliable criteria of steels wear resistance. The subsequent separated investigations of interrelation of steels wear resistance with all standard mechanical characteristics has allowed concluding that neither of them cannot serve as criterion for estimation of wear resistance, because they are not connected with wear resistance by univocal dependence. For revealing of more generalized dependence of steels wear resistance and their mechanical characteristics it was necessary to conduct the whole cycle of investigations.

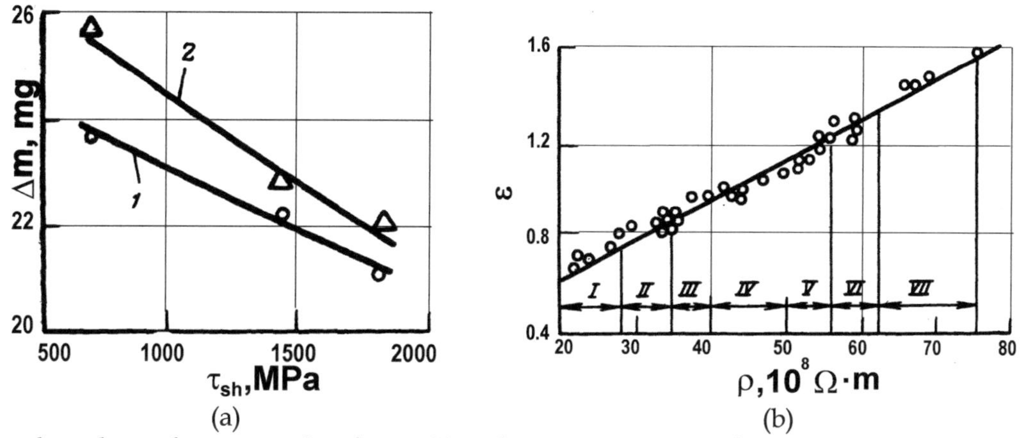

a - dependence of mass wear Δm of austenitic and martensitic structure from shear strength τ_{sh} at blow-abrasive wear and energy of blow accordingly, J: 1 - 5; 2 - 10;
b - dependence of relative wear resistance ε of surface hardening layer of system Fe-C-Mn from their resistivity ρ : I - ferrite + pearlit; II - pearlit + cementit; III - martensit; IV - austenit + disintegration products; V - martensit + carbides; VI - austenit + carbides; VII - austenit+martensit.

Fig. 6. Examples of unequivocal dependence of wear resistance parameters and one of physical and mechanical characteristics of steels:

3.1.5 The law of change conformity of toughness characteristics and wear resistance of steels from tempering temperature

The analysis of pairs ties of type "wear resistance - one of steels characteristics" gives the basis for assuming that the resistance to abrasive outwearing is more complicated by the character of forces interaction into friction surfaces, than resistance to introduction of indentor at hardness definition or resistance to tension at toughness characteristics definition - ultimate strength, conventional yield limit, relative elongation etc.
For more detailed analyses of cause of this dependence the correlations of wear resistance with steels mechanical characteristics of all structural classes were studied.
If abrasive wear is considered as mechanical destruction it is necessary to recognize its toughness basis. So, the interrelation between wear resistance and other mechanical characteristics for steels of different classes (Fig. 7) is received.
Character of toughness parameters change and wear resistance is identical: the decreasing at the rising of tempering temperature. As the standard for comparison the steel 45 is accepted; its relative wear resistance is accepted for unit. In each class of steels the tendency of change of toughness and plasticity characteristics are not identical at the tempering in the conditions of heating:

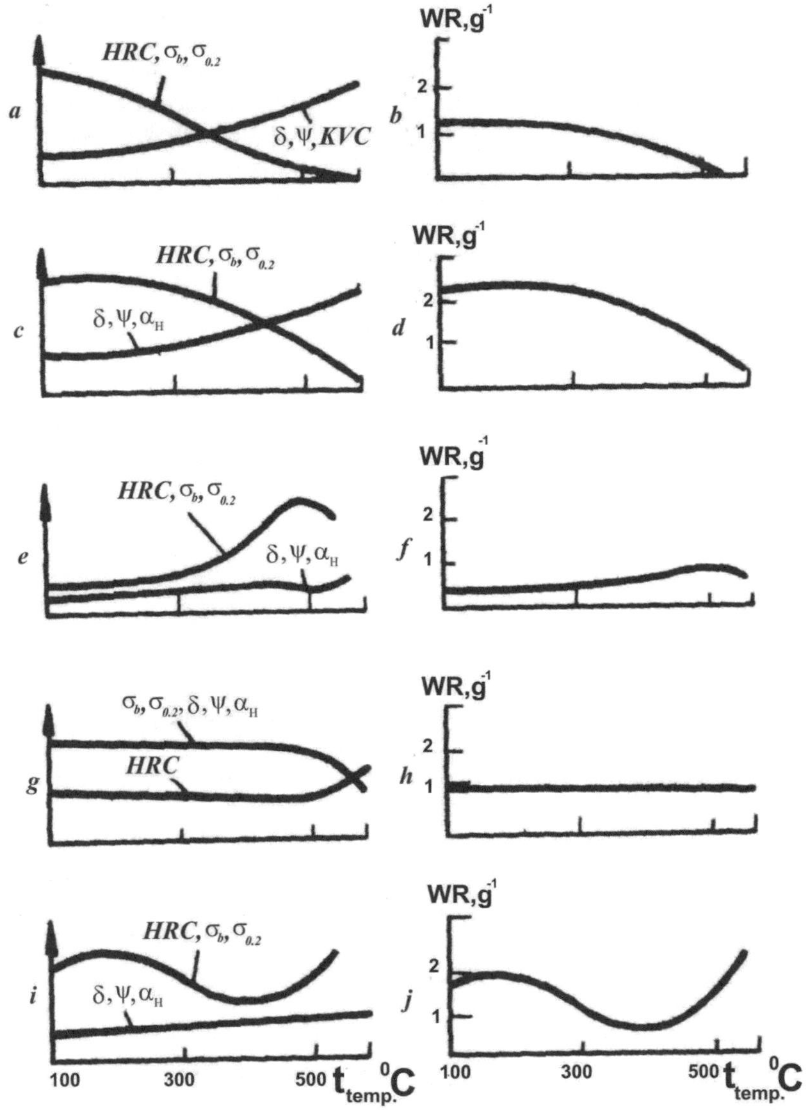

Fig. 7. Curves changes of toughness characteristics *(a,c,e,g,i)* and wear resistance (*b,d,f,h,j*) for steels of various structural classes from tempering temperature: *a,b* –steel 45 of pearlitic class; *c,d* - 95X18 of martensitic class; *e,f* –H18K9M5T of maraging class; *g,h* - 110Г13Л of austenitic class; *i,j* - P18 of carbidic class

For steels of pearlitic class at the rising of tempering temperature the toughness parameters are decreasing, and the plasticity characteristics are increasing;
For martensitic class steels is the same tendency, like for pearlitic class steels, but decrease of toughness characteristics and increase of plasticity characteristics are displaced into area for higher tempering temperature;
For maraging steels in process of rise of tempering temperature until 500 ^{0}C the toughness parameters increase at preservation of high plasticity;

For austenitic class steels at the rising of tempering temperature until 400 ^{0}C the toughness and plasticity characteristics do not change; the further rising of tempering temperature leads to decreasing of ultimate strength and plasticity characteristics; the hardness of steels is being raised a little.

For steels of carbidic class in rise process of tempering temperature the toughness characteristics are decreasing at first, and at the tempering temperature above than 400 ^{0}C start to increase; the plasticity characteristics do not change almost.

For the first time was established the conformity between changes of toughness characteristics and wear resistance depending on tempering temperature for steels of each class.

The results of tribological investigations have allowed to determine the law of conformity between variations of toughness characteristics (σ_b, $\sigma_{0.2}$, *HRC*) and wear resistance at different temperatures of tempering for hardened steels of all structural classes (Sorokin et al., 1991). These data have allowed concluding that in a nature of mechanical wear, the toughness ground lays, but the mechanism of these processes is more complicated.

The wear resistance estimation of several steels grades of different structural classes by the one characteristic of mechanical properties reveals the complicated dependence (Fig. 8). Its feature is that the different wear resistance corresponds to one value of any mechanical steels characteristics of different structural classes.

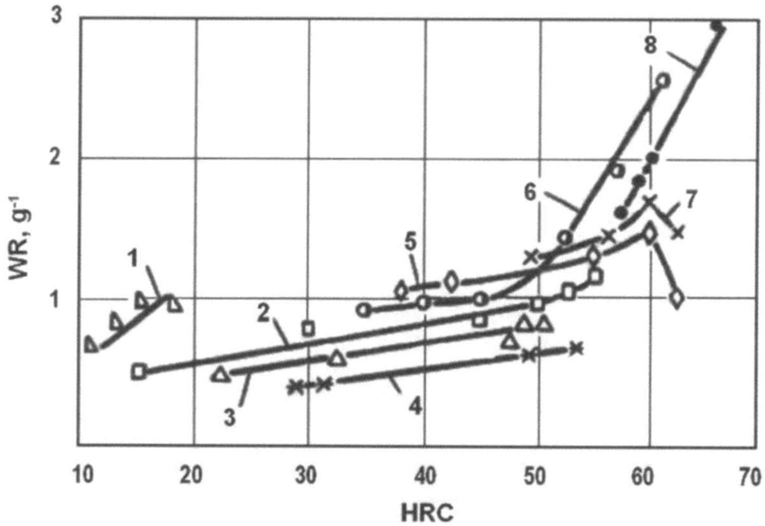

Fig. 8. Dependence of steels wear resistance *WR* from hardness *HRC*: 1—110Г13Л, 2—45 (BS En8), 3—40 (BS En8), 4-H18K9M5T, 5—У10 (tool steel), 6—D7, 7—X12M, 8—P18.

There was a basis to consider that at the mechanical outwearing only one of toughness characteristics (σ_b, $\sigma_{0.2}$, *HRC*) cannot be the full criterion of steels' wear resistance, because on the final process of forming and separating the corpuscles of wear from a friction surface, apart from strength properties, other mechanical characteristics exercise influence also.

This supposition was confirmed by analyses of steels plasticity characteristics correlations (δ, ψ, *KCV*) with their toughness characteristics.

It became apparent that the advantage of steels' wear resistance at the equal toughness is connected to a higher plasticity. There was a necessity to demonstrate these reasons experimentally.

3.2 The elaboration of wear resistance definition method

Such a problem was decided with applying a new wear resistance definition method which is taking into account simultaneously two properties "the toughness and the plasticity" (Fig. 9).

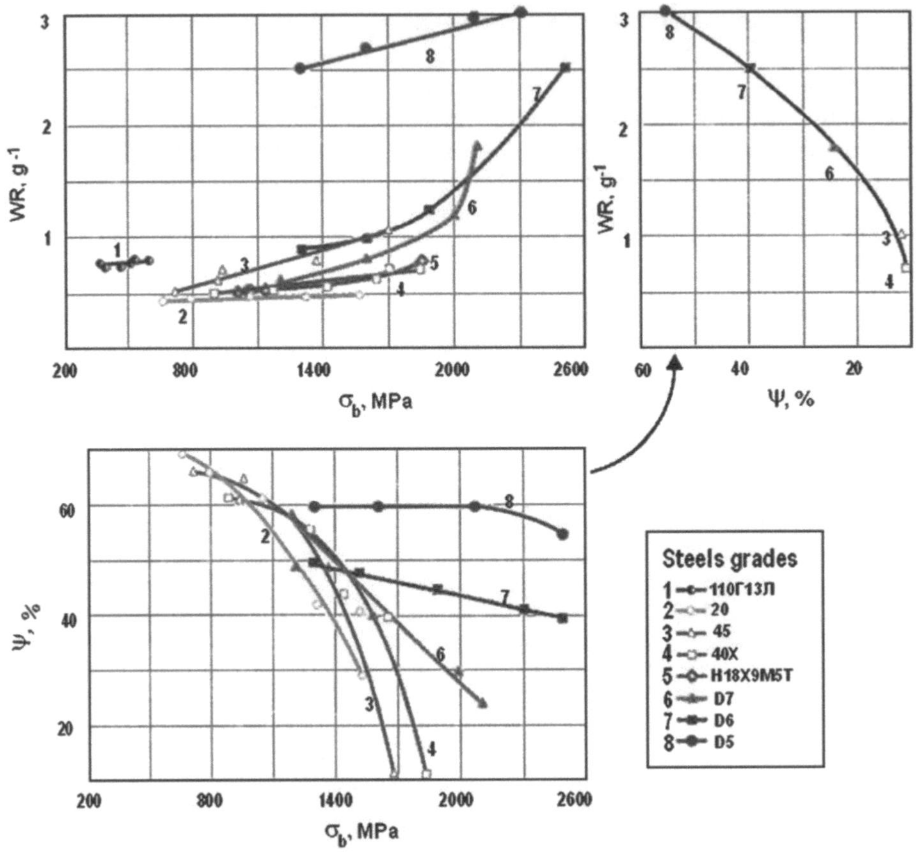

Fig. 9. Dependence of steels' wear resistance *WR* from ultimate strength σ_b and relative reduction of area ψ: 1 – 110Г13Л, 2 - 20, 3 - 45, 4 - 40X, 5 - H18K9M5T, 6 - D7, 7 - D6, 8 - D5.

The essence of this method consists in combination of two functional dependences: "wear resistance - toughness" and "toughness-relative reduction of area". Then, out of these dependences data, the final parameter in coordinates "wear resistance-relative reduction of area" is being defined. This method convincingly has confirmed that in a nature of mechanical wear at sliding friction over an abrasive the leading role belongs to steels' toughness, but the level of strength properties is more significant with higher plasticity.

All standard mechanical characteristics such as σ_b, $\sigma_{0.2}$, *HRC* enter into group of toughness. It is a dignity of this method because the selection of wear-resistant steels in factories conditions is being simplified. For this purpose it is enough to have one of three known characteristics.

The relative reduction of area is enough to have as an index of plasticity. The shape of handling and constructing the graphic dependences can be simplified, without representing a tie of relative reduction of area with toughness characteristic, and can be restricted by the dependence "wear resistance-plasticity" only (Fig. 10).

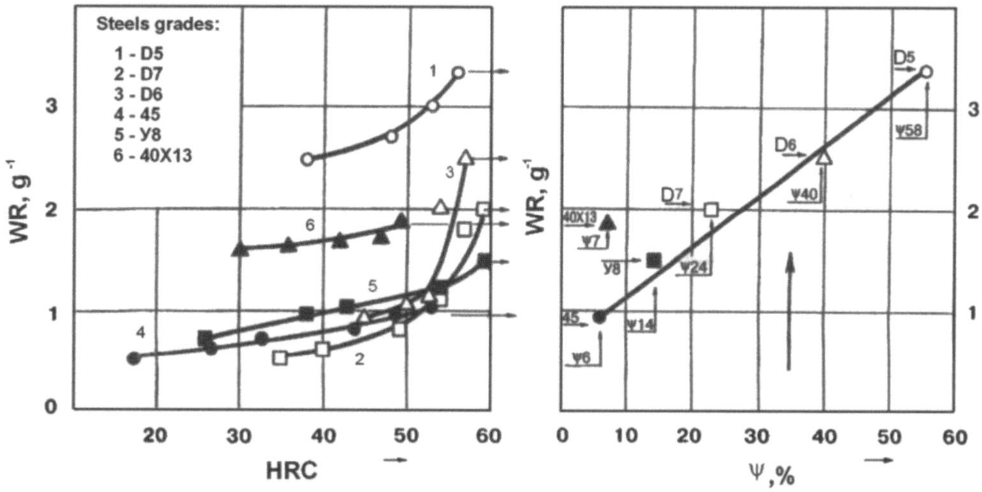

Fig. 10. Dependence of steels' wear resistance *WR* from hardness *HRC* and relative reduction of area ψ: 1 - D5, 2 - D7, 3 - D6, 4 - 40, 5 - Y8 (tool steel);6 - 40X13.

3.3 Methods of steels' wear resistance ranking

We also used other methods for ranking of steels' wear resistance. In this case, the combinations of two characteristics were applied: product of hardness on relative reduction of area ($HRC \cdot \psi$) versus ultimate strength (Fig.11) and product of ultimate strength on relative reduction of area ($\sigma_b \cdot \psi$) versus hardness (Fig.12).

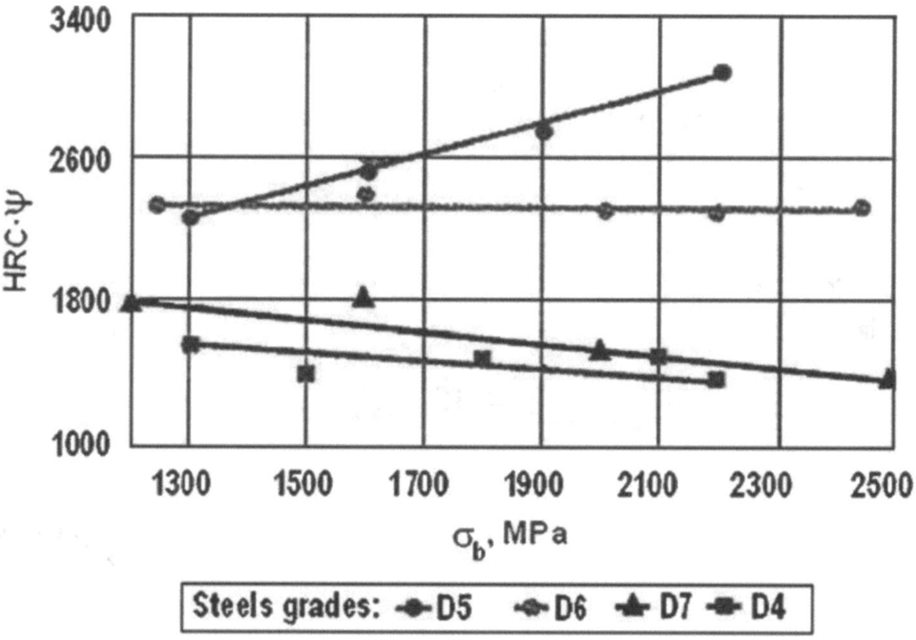

Fig. 11. Correlation of product of hardness on relative reduction of area ($HRC \cdot \psi$) from ultimate strength σ_b.

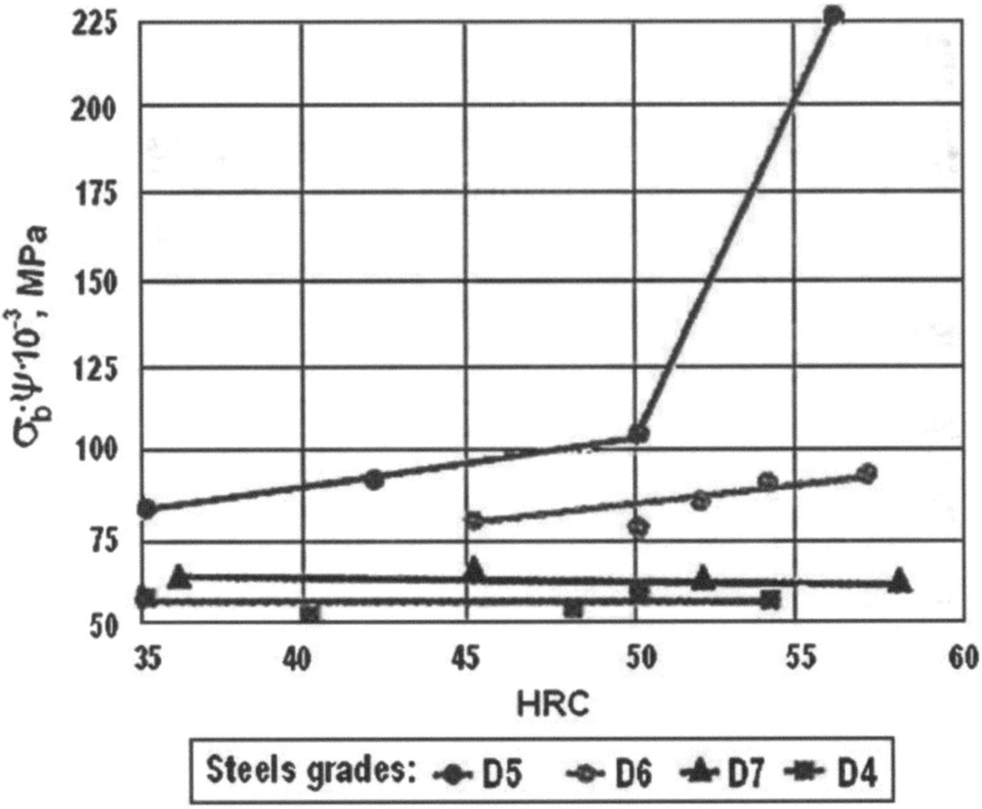

Fig. 12. Correlation of product of ultimate strength on relative reduction of area ($\sigma_b \cdot \psi$) from hardness HRC.

The points recieved on these plots represent the outcomes of experiments for five to six samples of each tested steels. All steels on these dependences can be divided in three groups: steels for which with growth of ultimate strength or hardness the parameters of wear resistance ($\sigma_b \cdot \psi$) or ($HRC \cdot \psi$) were being diminished, remained constant or increased. Thus, the principle of selection of wear-resistant steels out of these dependences is as follows: it is necessary to recommend for industrial production such steels for which the parameters of wear resistance ($\sigma_b \cdot \psi$) or ($HRC \cdot \psi$) are maximal and tends of growth with increase of second (pair) characteristic.

3.4 The influence of carbon content in steels on their wear resistance

With a purpose of studying the influence of carbon content in steels on their wear resistance, several steels with miscellaneous carbon content (from 0.2% up to 1.2%) at the equal hardness, 30, 40, 50, 55, 60 *HRC*, were selected for experiments. The dependences of steels' wear resistance from carbon content at different levels of hardness are shown on Fig. 13. The carbon content renders the direct influence on structure of steel forming and consequently, its mechanical characteristics and, first of all, the hardness. Our investigation's outcomes of structural stability influence on wear resistance of steels have allowed more widely considering this problem. The selected steels were tested after hardening and tempering at different temperatures to receive all possible structural statuses. At the low level of hardness (up to 30 *HRC*), changing of carbon content in a wide interval (0.2-1.0%) practically does not influence wear resistance. At the hardness 40 - 50 *HRC* the wear resistance of steels is being increased proportionally carbon content in them. At *HRC* 60 the carbon content in steels from 0.6% up to 1.2% cause a sharp falling of wear resistance.

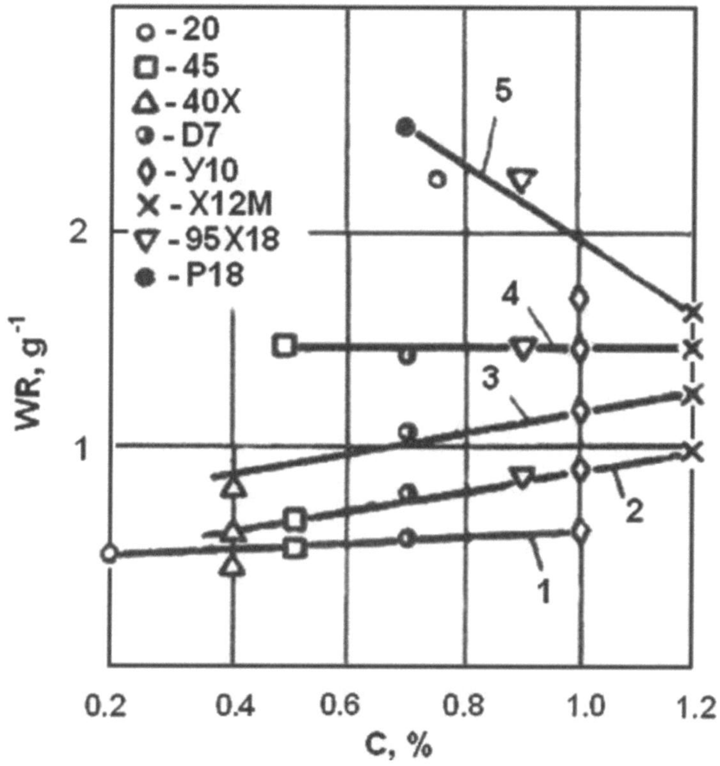

Fig. 13. Dependence of steels wear resistance *WR* from carbon content *C* at the different levels of hardness (*HRC*): 1 - 30, 2 - 40, 3 - 50, 4 - 55, 5 - 60.

4. Discussion

The investigations of last years have set more complicated task to which all tribologists are aspired. This task consisted in the elaboration of methods for ranking steels wear resistance without their trials on wear.

The logic of reasoning has specified the methodical necessity to use as criterion of outwearing simultaneously two characteristics of mechanical properties - the toughness (σ_b) and the plasticity (ψ). So, it was proposed that the product of ultimate strength on relative reduction of area ($\sigma_b \cdot \psi$) can be used as the complex criterion for an estimation of steels wear resistance at the mechanical outwearing.

The advantage of this method is in the kept dimensionality of toughness (MPa) which is "strengthened" by the influence of plasticity. This criterion takes into consideration the nature of dependences shown on Figs. 9 and 10. Besides, the criterion ($\sigma_b \cdot \psi$) in a certain measure reflects the power consumption of steel, because it considers actually two indexes: static toughness and plasticity.

The possibility of using this criterion was exhibited under different conditions of external forces influence. It was enough reliable at an estimation of steels wear resistance in conditions of sliding and rolling friction over an abrasive, of erosive wear with angles of attack less than 90^0 (Sorokin et al., 1991). The product of ultimate strength on relative reduction of area ($\sigma_b \cdot \psi$) has appeared universal, allowing to explain not only distinction of steels wear resistance at an equal value of one toughness characteristics, but the difference of endurance strength at an equal value of ultimate strength, also (Fig. 14) (Sorokin & Malyshev, 2008).

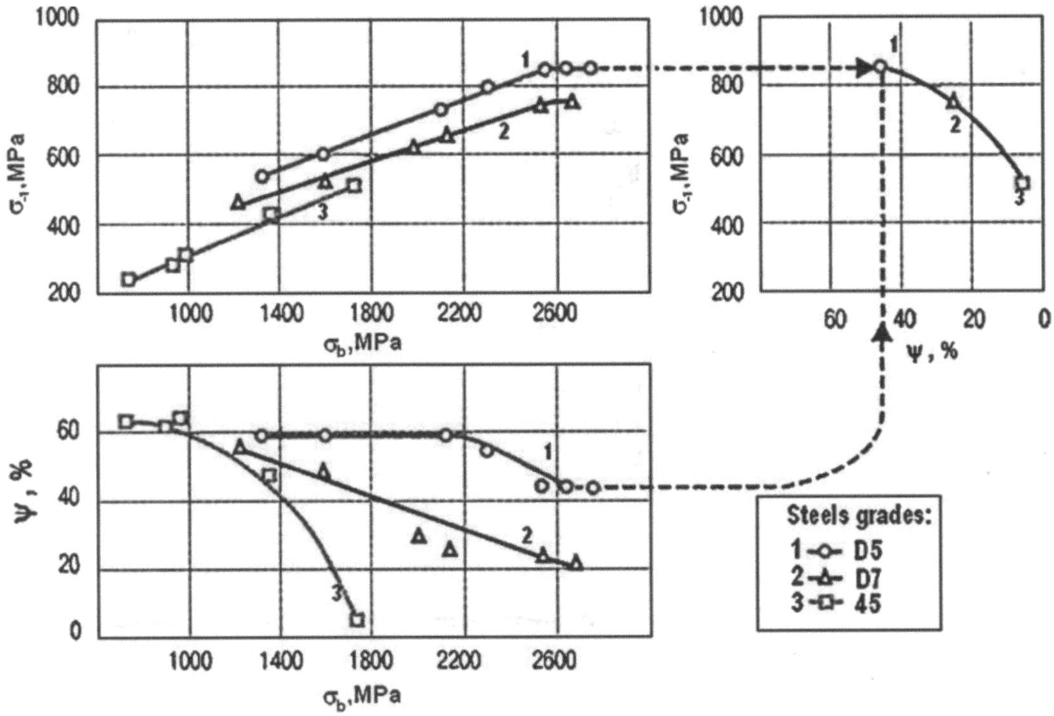

Fig. 14. Dependence of endurance strength σ_{-1} from ultimate strength σ_b and relative reduction of area ψ.

The ranking of steels wear resistance of miscellaneous structural classes was held using obtained criteria and taking into account their structural stability under thermal affect conditions.

There were used the coefficients of structural stability (thermo stability) which were taking into account the destruction of original structure at the heating in furnace K_1 and as a result of heating by friction K_2. The coefficient K_1 was being defined as hardness ratio of steels after their hardening and tempering at defined temperature (100 ^{0}C) to steels hardness after hardening, but without tempering. The coefficient K_2 was being defined as a ratio of steels wear resistance during defined time to its initial wear resistance after a trial within 3 min.

All steels on their thermo stability at the outwearing may be subdivided to as self-hardening, self-softening and stable steels. Coefficient of thermo stability (K_1) may change from 0.5 up to 1.5 (Sorokin, 2000). The maximal steels wear resistance of different structural classes was distributed as follows:

Structural class of steel	Criterion,($\sigma_b \cdot \psi$) (MPa)	Wear resistance, *WR* (g^{-1})
Pearlitic experimental steel	140250	2.5
Martensitic steel 95X18	14400	2.0
Austenitic steel 110Г13Л	9156	0.89
Maraging steel H18K9M5T	-	0.57
Carbidic high-speed steel P18	-	3.2

It is necessary to mark that the warmly-resistant steels unconditionally have advantage before other steels in their capacity to keep an initial wear resistance under high thermal conditions. But their large shortcoming consists in a low magnitude of complex criterion ($\sigma_b \cdot \psi$) that characterizes a capability to perceive the high external forces of uploading on the executive links of mechanisms. In particular, the value of complex criterion ($\sigma_b \cdot \psi$) for the warmly-resistant austenitic steel 110Г13Л has 10 times less magnitude, than for steel of the pearlitic class D5. The carbidic steel (P18) has a high wear resistance under thermal conditions, but as steel for machine building it can't be applied because has a low value of plasticity. This feature should be taken in consideration by designers at the creation of machines. The complex criterion $\sigma_b \cdot \psi$ is suitable for an estimation of steels wear resistance not only in the conditions of sliding friction over an abrasive, but also in the conditions of rolling friction that Fig. 15 is visually illustrates.

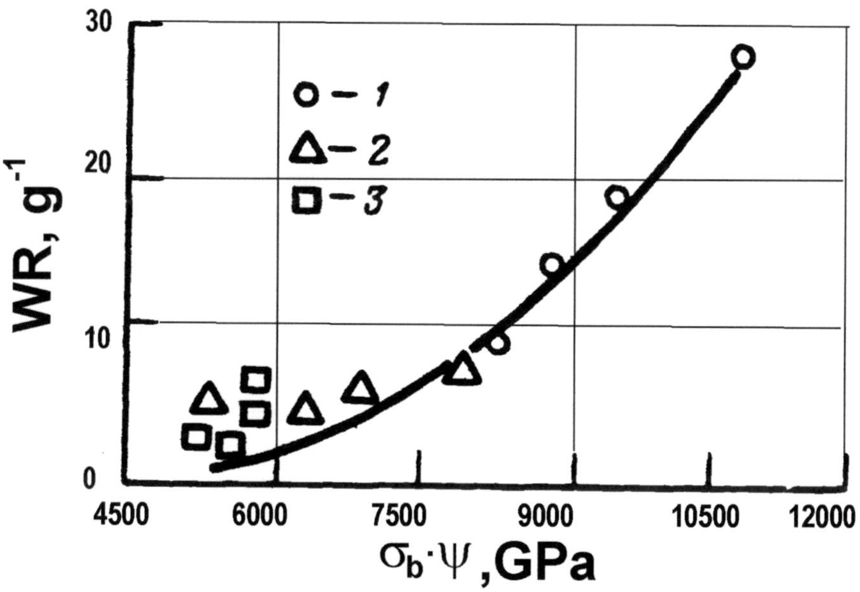

Fig. 15. Dependence of wear resistance *WR* of steels at rolling friction on an abrasive from criterion $\sigma_b \cdot \psi$: 1 - D5; 2 - D7; 3 - steel 55CM5ФA

The carbon content in steels influences their wear resistance in that measure in which are increasing the toughness characteristics under condition of providing the indispensable reserve of structure plasticity.

When this condition is not being observed - the wear resistance diminishes because of rising fragility of structure causing a crumbling in micro volumes of a friction surface. The analysis of obtained data shows that for the steels wear resistance estimation it is necessary to take into consideration their chemical composition.

The obtained new information has allowed to discover the law of abrasive outwearing and to show the influence of mechanical characteristics and their combinations on steels wear resistance of miscellaneous structural classes, to formulate the mechanism and criteria of this kind of outwearing from positions of metal science and toughness of metals. At the analysis of detected dependences became apparent that in a nature of mechanical (abrasive) outwearing lies a strengthened ground which allows simplifying the criteria connected to an estimation of steels wear resistance using only a well-known in industrial conditions the standard characteristics of mechanical properties and their combinations.

Out of obtained results data of investigations became indisputable the significance of plasticity reserve in steels with an obligatory high value of all characteristics of toughness (σ_b, $\sigma_{0.2}$, *HRC*) (Sorokin,2000; Gokhfeld, 1996; Kimura, 1975).

Thus, a toughness "reinforced" of indispensable plasticity, is a basic component in understanding mechanical wear nature under complicated external forces affect conditions, where simultaneously with one-time contact of a single abrasive particle can take place the low-cycle fatigue from the repetitive multiple acts of such affect.

This feature can be explained by the mechanism of interplay of single abrasive particle with wear surface: at the friction of solid particle on the steel surface is occur simultaneously its intrusion with defined effort and consequent migration. This, at the result, shapes the "products" of wear and leaves on the contact surface the risks or crushing oriented in direction of particles' moving. The intrusion of a corpuscle in metal and its migration meets a complex resistance in which ones participate the characteristics of toughness and plasticity. The combination of these characteristics can be various, but the positive effect will be in the events when the selected combination of mechanical characteristics ensures indispensable resistance:

- to an intrusion - this function executes the hardness *HRC;*
- to a tension, shear, crushing - this is ensured with high values of ultimate strength, yield limit in combination with relative elongation, relative reduction of area or difference (σ_b - $\sigma_{0.2}$).

Namely, such purpose was pursued by the selection of reviewed above mechanical characteristics and their combinations.

Definition of estimation criteria of materials wear resistance and, first of all, steels are one of the major problems of tribology development in the near future. The successful decision of this problem will open wide prospect of a choice and creation of wear-resistant materials.

There is necessary to notice that tribological toughness of materials is a complicated concept and completely is not discovered; it will be gradually specified in process of accumulation of new experimental data. This new characteristic will be connected with studying of new aspects, and first of all, metal science and classical laws of strength. It means the behavior of steels of different structural classes in difficult conditions of force uploading and temperature influence. New data, certainly, will allow expanding representations about the mechanism of mechanical outwearing. But already today it is possible to assert that hardness of materials, especially steels, will be the base characteristic of wear resistance at mechanical wear. Hardness as the measure of resistance of material to introduction in its surface of a solid body defines a possibility of development of the basic stage of mechanical outwearing - the introductions of a solid particle into another surface. If particle introduction occurs, there is possible the following stage of wear process consisting in micro cutting or plastic deformation at moving in the conditions of sliding. If initial introduction is absent, the second stage, i.e. actually the outwearing or damage of a contact surface, becomes impossible. In this case the particle is not capable to damage a surface. It means that a wear is absent and the wear resistance as an inversely value of wear aspires to infinity, i.e. it is become unwear situation that at the mechanical outwearing happens extremely seldom.

Thus, there are possible two ways of wear resistance increase at mechanical outwearing: the creation of structures with high initial hardness and thermo-resistance in all volume of detail or in superficial layer only. These structures shielding the surface from introduction of abrasive particles are capable, if not in whole, but significant reducing a wear. However, on the modern stage of metal science development such structures are practically not present - all steels are exposed to mechanical outwearing.

Quite probably that the creation of new steels structures considerably surpassing in hardness the existing steels (60-62 *HRC*) can cardinally decide the problem of wear resistance rising. But, the difficulty consists in that that at the rise of toughness characteristics and hardness the plasticity characteristics reduce very sharply and it lead to undesirable fragile destruction. Analysis of works on the creation of high-strengthened steels for the last quarter of the century confirms all complexity of realization this direction in metal science. The repeated increase of limits of ultimate strength, fluidity, endurance, shear strength resistance and hardness was not possible to reach as at us in Russia and abroad.

The second way of wear resistance increase of machine details supposes the creation in superficial layer the structure of high hardness on small depth from friction or blow zone. This way has more perspectives from tribology positions because it does not demand high strength of steel structure in whole volume of detail. Increase of superficial hardness in some cases influences significant positively on the wear resistance because the relation hardness law of abrasive and metal at the high hardness of metal provides sharply increase of wear resistance.

Results of the analysis of an extensive experimental data show that for providing the best indexes of wear resistance at mechanical outwearing it is necessary to combine three components: high static toughness, hardness and plasticity. Only the combination of such characteristics provides the best results regarding the wear resistance increase.

The problem of steels wear resistance rise is a major task of technical progress in machine industry. The path to successful decision of this task is the creation of high strengthened steels that not only have the separate high characteristics of mechanical properties, but also the exceeding from other steels by high values of combinations of these separate characteristics, such as ($\sigma_b \cdot \psi$), ($HRC \cdot \psi$), (σ_b - $\sigma_{0.2}$) and high thermo stability. The difference of ultimate strength and yield limit (σ_b - $\sigma_{0.2}$), the so-called - "barrier effect" (Alekhin,1983), is almost linearly connected to ultimate strength and positively influences on wear resistance of steels. Out of our data (Sorokin et al., 1991), the maximal ultimate strength is fixed, when (σ_b - $\sigma_{0.2}$) is in a spacing 500 - 700 MPa. The steel D5 has the best indexes of the wear resistance and endurance strength from among experienced steels that is being provided by their higher combination of all mechanical characteristics (see Table 2).

Grade of steel	σ_b, MPa	$\sigma_{0.2}$, MPa	*HRC*	δ, %	ψ, %	*KCV*, MJ/м²	$\sigma_b \cdot \psi$, MPa	$HRC \cdot \psi$	σ_b-$\sigma_{0.2}$, MPa	*WR*, g^{-1}
D4	2000	1900	53	8	35	0.34	70000	1855	100	1.55
D5	2550	1850	56	12	55	0.55	140250	3080	700	1.94
D6	2500	2000	57	8	40	0.30	100000	2280	500	1.82
D7	2100	2100	57	7.5	33	0.32	69300	1881	0	1.75
H18K9M5T	1070	900	31	7	-	-	-	-	170	0.54
95X18	1800	1700	61	8	8	0.15	14400	488	100	2.4
110Г13Л	327	327	16	37	28	1.53	9156	448	0	0.79
X12M	730	730*	59	-	-	0.016	-	-	0	1.69
40X13	1850	1520	48	7	8	0.18	14800	384	330	0.73
P18	1150	1150*	58	-	-	0.015	-	-	0	1.51
45	1700	1700*	49	8	8	0.39	13600	392	0	0.98

Note: *The yield limit of samples that were being destroyed at the test fragile is accepted equal to ultimate strength.

Table 2. Steels' mechanical properties at the tempering temperature 200 ^{0}C and wear resistance *WR*

Thus, the long-term problem connected to a searching of reliable criteria of an estimation of steels wear resistance, without the necessity of labor-intensive and not always realized wear trials in industrial conditions has received its further development. This task in defined aspects is finished up to an engineering decision and can be used in designer's practice for choosing the wear-resistant steels and alloys. The tendered methods allow not only to produce an estimation of suitable steels for different conditions of wear and external forces of uploading, but also to orient metallurgists to melting new steels with quite defined mechanical properties and their combinations.

The perspectives of wear resistance rise of steels and alloys is necessary to bind not only with toughness increase of materials matrix structure, but also with searching of methods for strengthening surface parts of machines, as one of the possible ways to increase the work resource of machines. In this event, the different methods of forming wear-resistant surface layers are of interest. There are different coatings obtained by laser strengthening (Gnanamuthu D.S.,1979), by spark doping and also ceramic coatings formed by microarc oxidation method (Malyshev & Sorokin, 1996).

It is necessary to recognize that tribology as an effective facility for increase of work resource of machines by means of protecting them from wear could give for factory practice much more for machinery production of different assignment if it have been founded, first of all, on the basis of metal science investigations, especially under mechanical wear conditions.

5. Conclusion

1. The mechanical outwearing of steels under abrasive affect conditions is a variety of breaking down of metals in their nature . Its basic difference is conditioned by the

scaling factor by means of formation and removing of wear corpuscles in micro volumes of friction surfaces.
2. The functional tie of steels wear resistance with toughness characteristics at the indispensable reserve of plasticity is the confirmation of this hypothesis. The toughness characteristic intensified of necessary plasticity is the main component in understanding of mechanical outwearing nature under complex external forces affects conditions.
3. The method reviewed in this chapter allows in the practice of designing machinery to select the more wear-resistant steels without necessity of much labor-intensive and not always accessible in industrial conditions the wear trials using only standard mechanical characteristics - the indexes of toughness and plasticity.
4. It is possible to assert definitely that in the general problem of wear resistance increase of machines the metal science role is exclusively great: in any specific target of resource increase of machines details two third of volume of possible measures will always make aspects of metal science or if to speak more widely - materials of different structural systems.

6. References

Alekhin V.P. (1983).Physics of strength and plasticity of materials surface layers . Nauka, Moscow. 280p. (in Russian).

Beckman G. & Kleis Y. (1983). Abtragverschleiss von metallen. VEB Deutscher Verlag fuer Grundstoff Industrie, Leipzig. 200s.

Gnanamuthu D.S. (1979).Laser surface treatment. In: Metzbower E.A.,editor. Application of lasers in material processing. DC: American Society for Metals, Washington.

Gokhfeld D.A., Getsov L.B., Kononov K.M., et al. (1996). Mechanical properties of steels and alloys at non-stationary loading. Handbook. Ural branch of Russian Academy of Sciences, ISBN 5-7691-0570-4, Ekaterinburg. 408p (in Russian).

Kimura Y. (1975).An interpretation of wear as fatigue process. JSLE-ASLE.In: Proceedings of the international lubrication conference, Tokyo. p. 89-95.

Kragelsky I.V. (1965).Reibung und verschleiss.: VEB Verlag Technik, Berlin. 423s.

Machine building Materials. (1980).In: Raskatov VM, editor. Short Hand-book. Mashinostroenie, Moscow. 511p (in Russian).

Malyshev V.N. & Sorokin G.M. (1996).Criteria of wear for coatings formed by microarc oxidation method. Friction and Wear,V.17, № 5, p. 653-657 (in Russian).

Metals handbook. (1990).Vol. 1: properties and selection: irons, steels and high performance alloys. ASM International, Metals Park, OH.1300p

Tribology handbook. (1973). Neale M.J. editor. Butterworths. London

Rabinowicz E. (1965). Friction and wear of materials. Wiley, New York. 244p.

Richardson R.CD. (1967).The wear of metals by hard abrasives. Wear, 10: 4.

Sorokin G.M., Grigoryev S.P., Saphonov B.P. (1991).Wear by abrasives: methodology and results of investigations. Tribology International; 24(1): p.3-9.

Sorokin G.M. (2000).The tribology of steels and alloys. Nedra, Moscow. 315p. (in Russian).

Sorokin G.M. & Malyshev V.N. (2008). Steels wear resistance definition method by their standard mechanical characteristics.Tribology International, 41(5): p.515-523.

Vinogradov V.N., Sorokin G.M., Bobrov S.N.. (1989). Special high-strength machine-building steel D5. Metal Science and Heat Treatment, № 2, p. 53-61 (in Russian).

Wellinger K. & Uetz H. (1963).Vershleiss duerch koernige mineralische Stoffe.Aufberichtungs-Technik, Jg.4, № 8, S.319.

11

A Comparison of the Direct Compression Characteristics of *Andrographis paniculata*, *Eurycoma longifolia* Jack, and *Orthosiphon stamineus* Extracts for Tablet Development

Yus Aniza Yusof[1], Aziana Azlin Abdul Hamid[1], So'bah Ahmad[1], Norawanis Abdul Razak[1], Chin Nyuk Ling[1] and Suhaila Mohamed[2]

[1]*Department of Process and Food Engineering, Faculty of Engineering*
[2]*Department of Food Service and Management, Faculty of Food Science and Technology*
Universiti Putra Malaysia, 43400 UPM, Serdang, Selangor
Malaysia

1. Introduction

Tropical herb extracts, such as *Andrographis paniculata, Eurycoma longifolia* Jack and *Orthosiphon stamineus,* which have enormous beneficial health effects need to be made available in a convenient and affordable tablet form. Information on the powder extract characteristics during compression is lacking and would be useful in investigations of powder deformability for optimizing operating conditions. The current market value of herbal products for traditional and complementary medicine is estimated to be between USD 40 to 100 billion with an average growth rate of 15 to 20% each year (Abdul Aziz et al., 2004).

Andrographis paniculata (locally known as 'hempedu bumi' or 'akar cerita') has analgesic, anti-malaria, anti-inflammatory, anti-neoplastic, anti-ulcerogenic, anti-bacterial, febrifuge, anti-platelet, anti-diarrhoeal anti-thrombotic, anti-diabetic, and anti-hypertensive properties (Ahmad and Asmawi, 1993). The roots of *Eurycoma longifolia* Jack are traditionally used for curing fever, mouth ulcers, and intestinal worms, and as a general tonic after childbirth, and are believed to be effective for treating hypertension and malaria. Studies have shown that *Eurycoma longifolia* Jack possesses sexual enhancement and aphrodisiac properties (Sambandan et al., 2006). *Orthosiphon stamineus,* or 'misai kucing', has also been scientifically proven to have anti-hypertensive, anti-inflammatory, and diuretic properties, and has also been proven to be beneficial for urinary tract ailments, renal stones, arteriosclerosis, nephritis, diabetes mellitus, hypertension, rheumatism, tonsillitis, and menstrual disorders (Jagarath and Ng, 2000). Currently, *Orthosiphon stamineus* is categorized as one of the herbs with a great potential for commercialization in Malaysia due to its easy cultivation and its validated health benefits, particularly in treating ailments associated with the kidney and urinary system (Jagarath and Ng, 2000).

Most of the research conducted on these herbs so far has been limited to their chemical and medicinal properties. This work attempted to obtain scientific and engineering data to determine the technical feasibility and economic viability of producing solid oral dosage tablets of these herb extracts via direct compression. Takeuchi et al. (2004) used this method to study the compression properties of several pharmaceutical materials such as potassium chloride, pre-gelatinized starch (PCS), cornstarch, low-substituted hydroxypropylcellulose (L-HPC), crystalline lactose, and ascorbic acid by measuring the die wall force. The direct compression method is advantageous due to its simplicity of combining blended ingredients and pressing them into a tablet without having to change the formulation or ingredients and, simultaneously, the forces generated during compression may be used to understand the roll compaction mechanisms (Yusof et al., 2005). The study and application of the powder compression process has continued to develop in line with the needs of the pharmaceutical (Varthalis and Pilpel, 1976; Mohammed et al., 2006; Yamamoto et al., 2009), food (Yusof et al., 2005), and herbal industries (Eggelkraut-Gottanka et al., 2002; Schiller et al., 2003).

The characteristics and performance of tropical herbal powder extracts during direct compression are neither well documented nor understood. This chapter discusses the properties of such herbal extract powders from *Andrographis paniculata, Eurycoma longifolia* Jack and *Orthosiphon stamineus* from a freeze dried process during direct compression, portraying them using the Heckel (1961), Kawakita and Lüdde (1970/71), and Walker (1923) models.

2. Materials and methods

2.1 Herbs and powder evaluation

Andrographis paniculata, Eurycoma longifolia Jack and *Orthosiphon stamineus* freeze-dried extract powders were supplied by Phytes Biotek Sendirian Berhad, Malaysia. The herbal powders were evaluated in terms of particle size, moisture content, and density before the tableting process. Particle size and particle size distribution for all samples were measured by a Malvern Mastersizer 2000 instrument (Malvern Instruments Ltd, Worcestershire, UK). Approximately 5 ml of powder was used for each measurement and particle size distributions were recorded. Each sample was measured three times and the results are shown in Table 1. The moisture content of the herb powder extracts was determined gravimetrically using a digital Moisture Analyzer (Ohaus Corporation, Pine Brook, NJ, USA), set at 105 °C, with approximately 5 g of the extract powder. The instrument was allowed to cool between tests for each powder. A gas pycnometer (Quantachorome Corp., Boynton Beach, Fl, USA) was used to measure the density and volume of the accurately

Material	Particle size (μm)	Moisture content (%)	Density (kg/m³)			Carr index (1965) (%)	Hausner ratio (1967) (HR)
			Bulk	Tap	True		
Orthosiphon stamineus	16.4±0.124	5.52±0.100	589.9±1.00	893.4±1.00	1628.5±5.10	33.96±0.250	1.51±0.006
Eurycoma longifolia Jack	26.0±0.395	3.59±0.238	438.7±0.625	645.1±0.675	1304.3±0.400	32.00±0.089	1.47±0.002
Andrographis paniculata	15.6±3.446	4.64±0.400	589.8±0.370	879.7±0.500	1612.6±0.001	32.96±0.001	1.49±0.001

Table 1. Material properties of herb extract powders

weighed powders. The bulk density, which is the ratio of the sample weight to the volume of the powdered sample, was determined by placing 30 ml of powder into a weighed 50 ml glass cylinder. For the tap density, a cylinder filled with the powder was tapped until a constant volume was reached using an Automated Tap Density Tester Model ETD-1020, (Electrolab, Mumbai, India), at 500 taps per minute. All measurement data are presented as means and standard deviations from triplicate measurements.

2.2 Tablet compression, testing and evaluation

A 13 mm cylindrical-uniaxial-stainless steel die (Ranning Enterprise, Selangor) was used to study the compression characteristics of the powdered herb extracts. The uniaxial die was attached to the upper moving crosshead (5 mm min^{-1}) of the Instron Testing Machine (Instron Corp., Canton MA). The maximum allowable force of the machine was 10 ± 0.1 kN. Samples comprising 0.5 g and 1.0 ± 0.01 g of powder were manually placed into the die using a plastic funnel. The die was cleaned with acetone prior to compression of the powders. The compressions were performed at room temperature between 23 °C and 26 °C with the humidity between 37 and 42% RH. A schematic diagram of the uniaxial die compression is shown in Figure 1.

All of the tablets were numbered and wrapped in polyethylene bags placed over silica gel and stored in desiccators in the same laboratory where they were produced. The tablets were kept for tensile strength testing for at least 24 hours after compaction to allow consistent elastic recovery and hardening. Tablet diameter, height, mass, and volume were measured to obtain the density before and after compression. These procedures were adopted to observe signs of plastic recovery and hardening, and consequently to prevent false low-yield values (Odeku et al., 2005). This information was used to determine the essential parameters for tensile strength and for data validation using the Heckel, Kawakita and Lüdde, and Walker models.

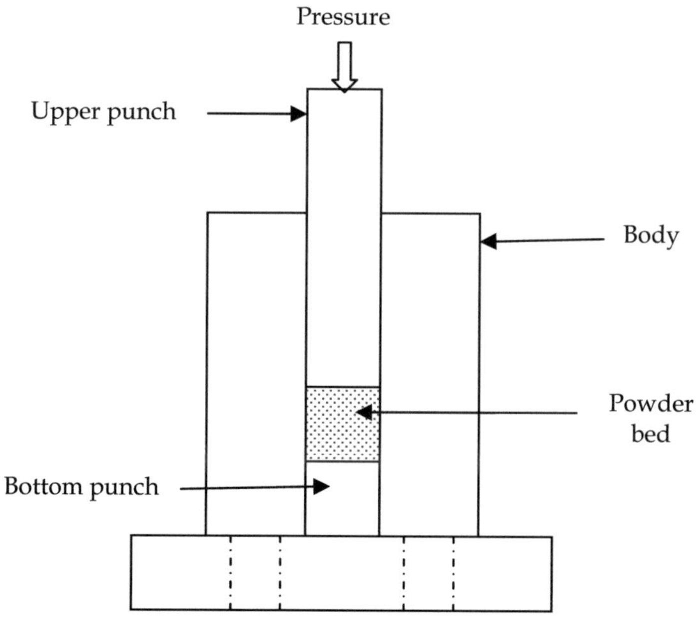

Fig. 1. A schematic diagram of a compression unit

The tensile strength test was carried out via a diametrical compression test and was then calculated (Fell and Newton, 1970). Forces at the top punch, F_t, were measured using a load cell (10 ± 1 kN). The stress was calculated by dividing the respective forces with the cross-sectional area of the die as described in equation (1) as follows:

$$\sigma_s = \frac{2F_t}{\pi d} \tag{1}$$

where F_t is the tensile force (N) , d is the diameter of the tablet (m) and t is the height of the tablet (m).

2.3 Models

2.3.1 The Heckel model

The volume reduction mechanism under the compression force was determined using the Heckel model for relating the relative density of the powder bed during compression to the applied pressure (Heckel, 1961). It is used with the assumption that powder compression follows a first order kinetics with the interparticulate pores as the reactants and densification of the powder as the product (Heckel, 1961), following the equation:

$$\ln \frac{1}{1 - \rho_f} = kP + A \tag{2}$$

where ρ_f is the relative density of the powder bed at compression pressure P.

2.3.2 The Kawakita and Lüdde model

The Kawakita and Lüdde (1970/1971) model expresses the relationship between the degrees of volume reduction of the powder bed under compression:

$$\frac{P}{C} = \frac{P}{a} + \frac{1}{ab} \tag{3}$$

and,

$$C = \frac{(V_o - V)}{V_o}$$

where P is the compression pressure, constant a is equal to the total volume reduction for the powder bed and constant b is inversely related to the yield strength of the particles (Zhang et al., 2003), C is the degree of volume reduction, V_o is the initial volume of the powder bed and V is the powder volume after compression or volume of the tablets.

2.3.3 The Walker model

The compressibility of the powders and the expressed volume changes of the powder were analysed using the Walker model (Walker, 1923):

$$V_R = W \, lnP + B \tag{4}$$

and

$$V_R = \frac{V}{V_0}$$

where V is the volume of compressed powder at compression pressure P and V_o is the volume of powder before compression. Walker's constants W and B describe the compressibility and constant model, respectively.

3. Results and discussion

3.1 Tensile strength

Figures 2 (a-b) depicts the increase in tensile strength of the tablets with increasing compression pressures for both 0.5 g and 1.0 g quantities of feed powder. Clear linear correlations between the compression pressure and tensile strength of the extract powders in both figures indicate a similar trend of compression characteristic. The *Eurycoma longifolia* Jack extract powder produced the strongest tablets, followed by *Andrographis paniculata* and lastly *Orthosiphon stamineus*, for both quantities of feed powder. Particle size and moisture content were found to influence the tensile strength properties of the tablets. Table 1 shows that although *Eurycoma longifolia* Jack had the largest particle size, it exhibited the highest tensile strength compared to the other powders. Previous studies (Fichtner et al., 2005) showed that small particles do not necessary give strong tablets. The amount of water associated with a solid at a particular relative humidity and temperature depends on its chemical affinity for the solid and the number of available sites of interaction, the surface area, and the nature of the material (Nokhodchi, 2000). The tablets formed from *Orthosiphon stamineus* with the highest moisture content showed low tensile strength and exhibited capping. This was probably because during compression, air was evacuated, which pushed the fine powders with inter-particular forces and contacts due to moisture, (Sebhatu et al., 1997). Additionally for spray-dried lactose tablets of the highest moisture content (6.1%) exhibited low tensile strength (6.47±0.24 MPa). Tablets with a low tensile strength thus show a tendency to cap or laminate (Sebhatu et al., 1997).

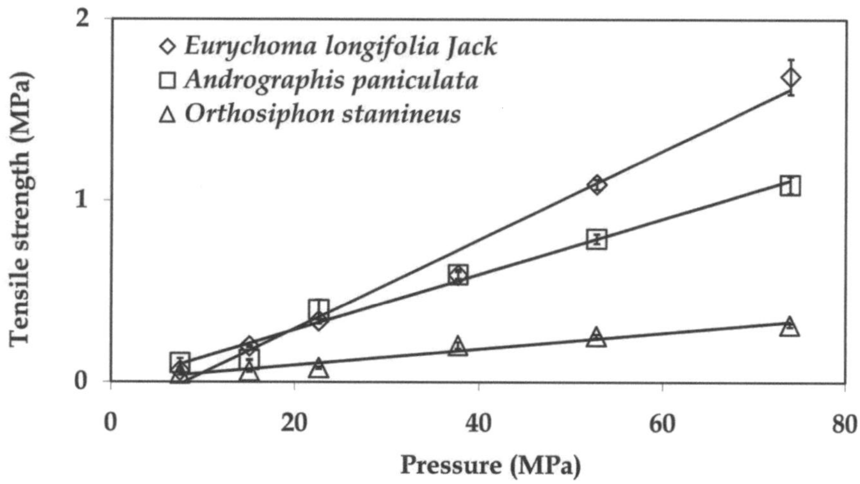

Fig. 2. (a) The tensile strength of tablets prepared from *Eurycoma longifolia Jack*, *Andrographis paniculata* and *Orthosiphon stamineus* at 0.5 g of feed powders.

These data can be supported with further measurement the Hausner ratio and the Carr index (Carr, 1965; Hausner, 1967). The Hausner ratio and the Carr index gave the lowest values of

Eurycoma longifolia Jack extract powder. The Carr index values (above 25%) and Hausner ratio values (greater than 1.4) for all of the powders indicated poor and difficult to achieve flow behaviours (Carr, 1965; Hausner, 1967). This inferred that the powders were easily compressible and may form strong coherent junctions between the particles (Yusof et al., 2005).

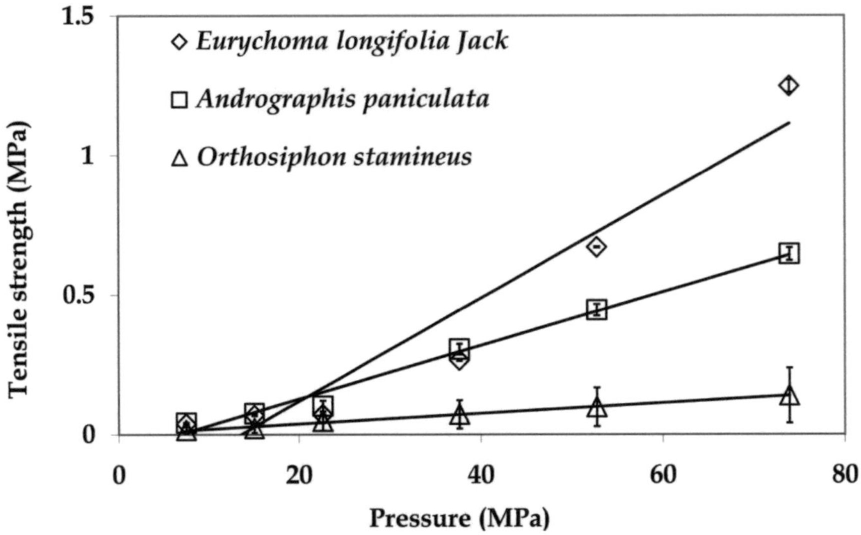

Fig. 2. (b) The tensile strength of tablets prepared from *Eurycoma longifolia Jack*, *Andrographis paniculata* and *Orthosiphon stamineus* at 1.0 g of feed powders.

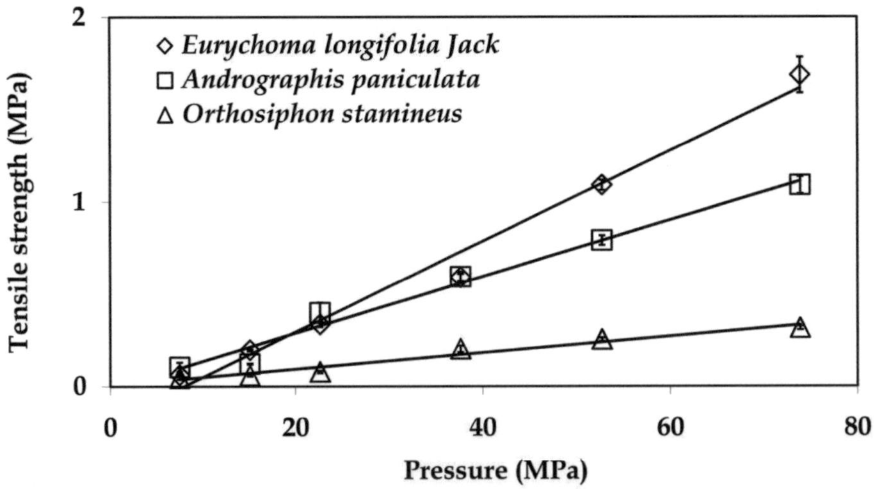

Fig. 3. (a) Heckel plots of tablets prepared from *Eurycoma longifolia Jack*, *Andrographis paniculata* and *Orthosiphon stamineus* for 0.5 g of feed powders.

3.2 Model validations

3.2.1 The Heckel model

Figures 3 (a-b) shows the compression of feed powders of 0.5 g and 1.0 g quantities using the Heckel model to describe the compaction characteristics of the extract powders. A higher ρ_f value indicates that there will be a higher volume reduction of the powder due to particle rearrangement (Adapa et al., 2005). Constant k is the measure of the plasticity of a compressed material. A larger k value indicates the onset of plastic deformation at relatively low pressures, thus, powders are more compressible (Adapa et al., 2005). Constant A is related to the die filling and particle rearrangement before deformation and bonding of the discrete particles. The Heckel plot enables bonding mechanism interpretation during compression (Zhang et al., 2003).

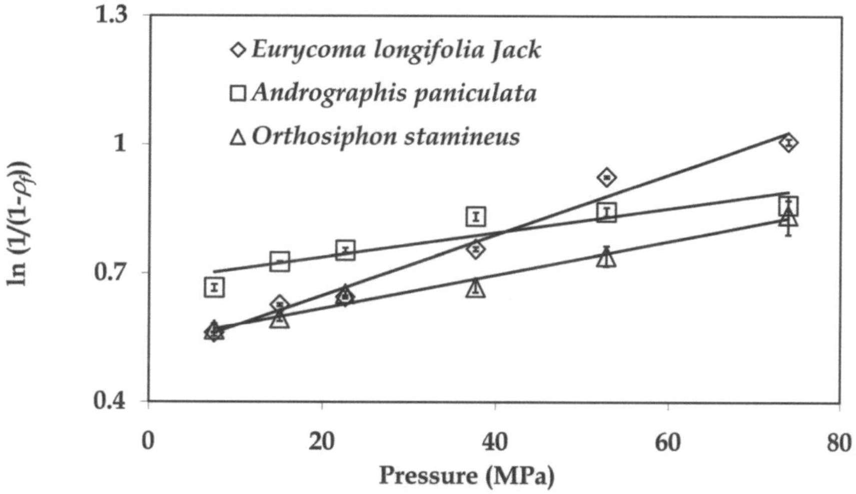

Fig. 3. (b) Heckel plots of tablets prepared from *Eurycoma longifolia Jack*, *Andrographis paniculata* and *Orthosiphon stamineus* for 1.0 g of feed powders.

The slopes of the Heckel plots positively correlated with plastic deformation (Heckel, 1961; Ramakrisnan et al., 1997). The trends of these graphs are similar to those of Ramakrisnan et al., (1997) for a compression study on ceramic powders. The constant values of the Heckel plots of the herb extract powders are displayed in Table 2. The reciprocal of slope k is the yield pressure (Korhonen et al., 2002). The *Eurycoma longifolia* Jack extract powder showed the highest compression values for both 0.5 g and 1.0 g amounts of feed powders. The yield pressure $\left(\frac{1}{k}\right)$ needed to induce plastic deformation was lowest for *Eurycoma longifolia* Jack extract powder. This indicates that *Eurycoma longifolia* Jack extract powder is the most compressible and easiest to deform into tablets, thus explaining the highest tensile strength value compared with the other two extract powders. The plasticity and tensile strength, however, decreased with increasing feed powder quantities for all extract powders as higher yield pressures were required to compress the extract powders (Figure 2(a)). Generally, plasticity was the highest for *Eurycoma longifolia* Jack, followed by *Andrographis paniculata* and lastly *Orthosiphon stamineus*, for both feed powder amounts.

Material	Mass (g)	Constants			R^2 value
		k	*1/k*	*A*	
Andrographis paniculata	0.5	0.005±0.000	200±9.713	0.798±0.018	0.987
Eurycoma longifolia Jack		0.009±0.000	111.111±0.000	0.638±0.004	0.970
Orthosiphon stamineus		0.004±0.001	238.095±0.001	0.643±0.016	0.909
Andrographis paniculata	1.0	0.003±0.000	357.143±38.263	0.660±0.000	0.890
Eurycoma longifolia Jack		0.007±0.000	142.857±0.000	0.518±0.003	0.980
Orthosiphon stamineus		0.003±0.001	303.030±0.001	0.553±0.036	0.969

Table 2. The Heckel model

3.2.2 The Kawakita and Lüdde model

The Kawakita and Lüdde equation is widely applicable for metallic and medical powders in the fields of powder metallurgy and pharmaceutics (Kawakita and Lüdde, 1970/71). Following this equation, the particles which are subjected to a compressive force in a confined space are viewed as a system in equilibrium at all stages of compression, so that the products of the pressure term and the volume term are constants.

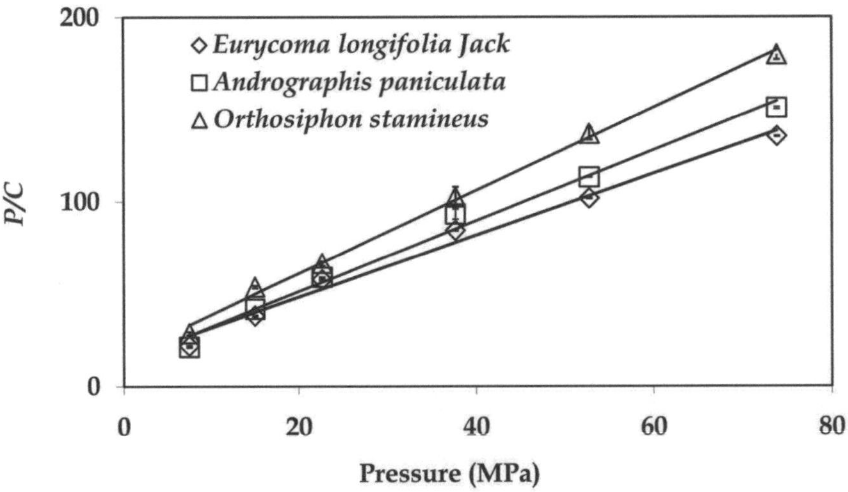

Fig. 4. (a) Kawakita and Lüdde plots of tablets prepared from *Eurycoma longifolia Jack*, *Andrographis paniculata* and *Orthosiphon stamineus* for 0.5 g of feed powders.

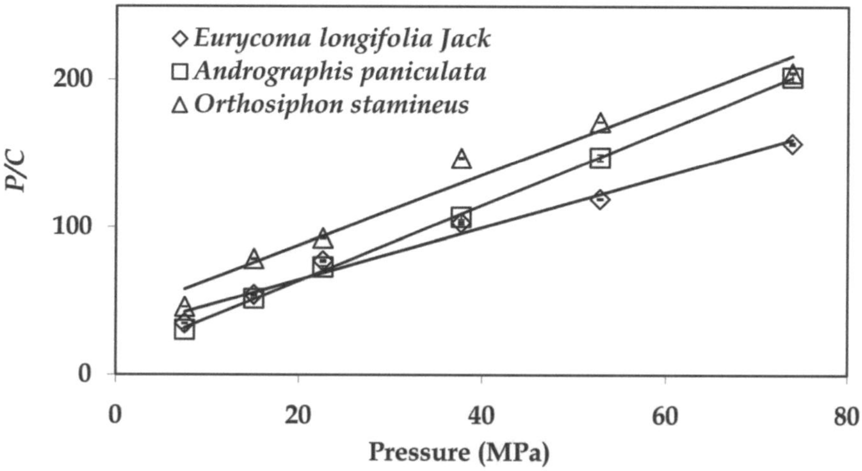

Fig. 4. (b) Kawakita and Lüdde plots of tablets prepared from *Eurycoma longifolia Jack*, *Andrographis paniculata* and *Orthosiphon stamineus* for 1.0 g of feed powders.

Figures 4 (a-b) shows that the linear trends of the Kawakita and Lüdde plots for 0.5 g and 1.0 g of herbal extract feed powders were similar to the compression characteristics of cellulose polymers (Shivanand and Sprockel, 1992), and the trends reported by Ahmad (2007) for *Eurycoma longifolia* Jack and *Andrographis paniculata* ground powders at similar pressures and feed quantities.

Based on the constant *a*, the *Eurycoma longifolia* Jack extract powder exhibited the highest volume reduction of the powder bed followed by *Andrographis paniculata* and *Orthosiphon stamineus* extract powders for 0.5 g of feed powders (Table 3). However, the difference between the *a* values of *Andrographis paniculata* and *Orthosiphon stamineus* was only 0.043, which was quite close. Likewise, this trend was similar for the 1.0 g of feed powders, except that they exhibited lower *a* values compared to the 0.5 g of feed powders. This shows that the ease of volume reduction is lower with an increase in the amount of feed powder. Thus, a smaller amount of powder can form a 'good' and coherent tablet, whilst it is more difficult to compress a greater amount of powder at the same pressure.

Material	Feed powder (g)	Constants		R^2 value
		a	*b*	
Andrographis paniculata	0.5	0.496±0.003	0.206±0.023	0.997
Eurycoma longifolia Jack		0.592±0.002	0.116±0.001	0.986
Orthosiphon stamineus		0.453±0.005	0.120±0.017	0.998
Andrographis paniculata	1.0	0.393±0.012	0.169±0.016	0.850
Eurycoma longifolia Jack		0.568±0.003	0.062±0.001	0.980
Orthosiphon stamineus		0.386±0.066	0.072±0.041	0.980

Table 3. The Kawakita and Lüdde model

This trend is similar to the trend reported by Ahmad (2007) for the compression of ground *Eurycoma longifolia* Jack at similar compression pressures and amounts of feed powders. The *a* value for 0.5 g of ground *Eurycoma longifolia* Jack root was 0.77 whereas at 1.0 g it was 0.69 (Ahmad, 2007).

Eurycoma longifolia Jack extract powder exhibited the lowest *b* values followed by *Orthosiphon stamineus* and *Andrographis paniculata* for both 0.5 g and 1.0 g of feed powders. However, these values were inversely related to the yield strength of the particles (Nordström et al., 2008) whereby *Eurycoma longifolia* Jack extract powder showed the highest tensile strength followed by *Orthosiphon stamineus* and *Andrographis paniculata* for both feed powder quantities (Figures 2 (a-b)), thus explaining the increase in yield strength with increasing volume reduction of the powder bed in the tablet formation.

3.2.3 The Walker model

The Walker (1923) model is based on the assumption that the rate of change of compression pressure with respect to volume is proportional to the compression pressure. Figures 5 (a-b) depicts the decrease in the volume of tablets made from the three different extract powders with increased compression pressure at various quantities of feed powder. A larger value for the slope *W* is related to a greater amount of density in the material (Walker, 1923). The Walker constants *W* and *B* for each extract powder of the herbs are listed in Table 4. The densities decreased in the following order for 0.5 g of feed powder: *Eurycoma longifolia* Jack, *Andrographis paniculata* and *Orthosiphon stamineus* extract powders. For the 1.0 g of feed powder, the densities decreased from *Eurycoma longifolia* Jack, to *Orthosiphon stamineus*, and

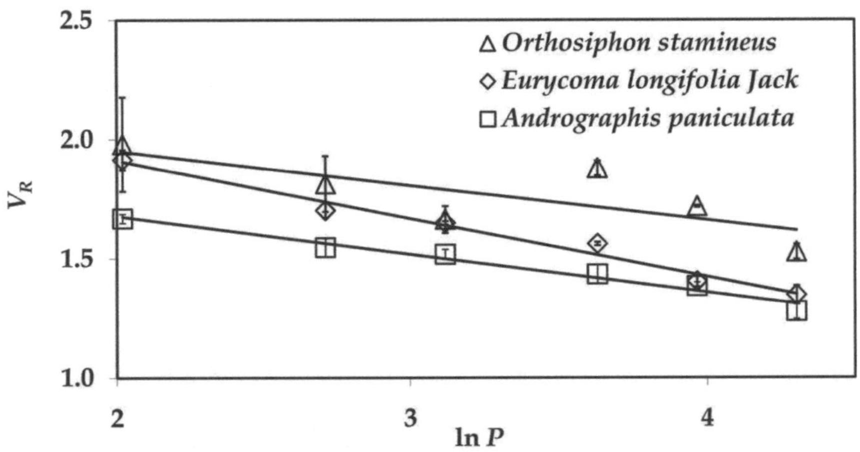

Fig. 5. (a) Walker plots of tablets prepared from *Eurycoma longifolia Jack*, *Andrographis paniculata* and *Orthosiphon stamineus* for 0.5 g of feed powders.

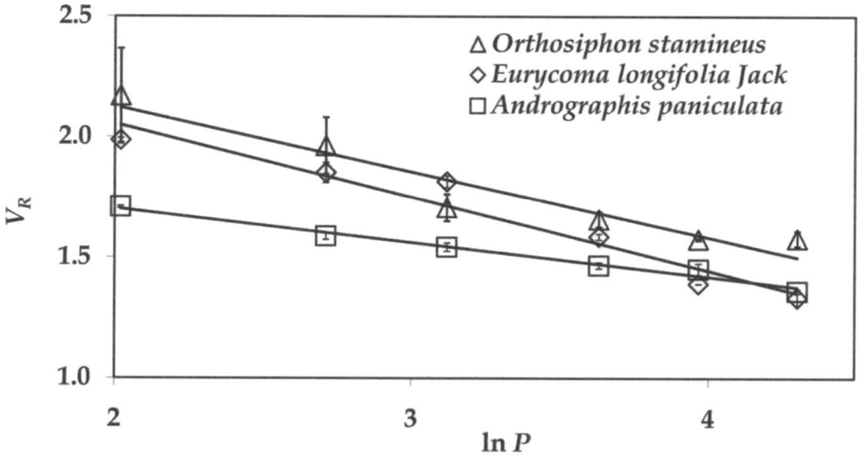

Fig. 5. (b) Walker plots of tablets prepared from *Eurycoma longifolia Jack*, *Andrographis paniculata* and *Orthosiphon stamineus* for 1.0 g of feed powders.

lastly the *Andrographis paniculata* extract powder. The high density gave the high value of the tensile strength, which was related to the reduction in the void space between particles in the powders during tablet formation with respect to the values of the slopes, which decreased as the tensile strength increased.

Material	Feed powder (g)	Constants		R^2 value
		W	*B*	
Andrographis paniculata	0.5	-0.16±0.001	2.00±0.006	0.971
Eurycoma longifolia Jack		-0.24±0.001	2.33±0.056	0.984
Orthosiphon stamineus		-0.14±0.056	2.24±0.243	0.957
Andrographis paniculata	1.0	-0.14±0.002	1.99±0.000	0.948
Eurycoma longifolia Jack		-0.31±0.001	2.66±0.053	0.992
Orthosiphon stamineus		-0.27±0.105	2.67±0.421	0.928

Table 4. The Walker model

4. Conclusion

This study on direct compression characteristics of selected Malaysian herb extract powders helped to deduce and understand some of the important principles of tablet development. The *Eurycoma longifolia* Jack extract powder was the easiest of the three herb powders to compress, and it underwent significant particle rearrangement at low compression pressures, resulting in low values of yield pressure. The compression characteristics of the *Eurycoma longifolia* Jack powder were consistent when validated with all of the models used. Another significant finding showed that the characteristics of 0.5 g of feed powder are better than for 1.0 g of feed powder, as proven from the tensile strength test; hence a more coherent tablet can be obtained. Thus, herbal parameters are superior when screening extract powders with the desired properties, such as plastic deformation. This study also validated the use of Heckel, Kawakita and Lüdde, and Walker model parameters as acceptable predictors for evaluating extract powder compression characteristics.

5. Acknowledgements

This work was supported by a research grant from the Ministry of Higher Education Malaysia, Fundamental Research Grant Scheme with project number: 5523035 and Universiti Putra Malaysia (UPM) Research University Grant Scheme with project number: 91838. Some of the authors were sponsored by a Graduate Research Fellowship from the UPM.

6. References

Abdul Aziz, R., Kumaresan, S., Mat Taher, Z. & Chwan Yee, F.D. (2004). Phytochemical Processing: The Next Emerging Field In Chemical Engineering: Aspects and Opportunities (available from http%3a//kolmetz.com/pdf/Foo/IJKM_Phytochemical.pdf - Accessed 20/10/2008).

Adapa, P.K., Tabil, L.G., Schoenau, G.J., Crerar, B., Sokhansanj, S. & Canada (2005). Compression characteristics of fractionated alfafa grinds. *International Journal of Powder Handling and Processing* 14(4), 252-259, ISSN: 09347348.

Ahmad, S. (2007). Mechanical Granulation by tabletting of Eurycoma longifolia jack, B. Eng. Hons. (Process and Food) Thesis, Universiti Putra Malaysia.

Ahmad, M., and Asmawi, M.Z. (1993). Int. Conf. on the use of Traditional Med. & Other Natural Products in Health Care, USM. (available from http://www.imr.gov.my/org/hmrc_r1.htm - Accessed 11/11/2008).

Carr, R.L. (1965).Evaluating flow properties of powders, *Chemical Engineering. 72*, 163-167 ISSN 1385-8947.

Eggelkraut-Gottanka, A.G., Abu Abed, S., Müller, W. & Schmidt, P.C. (2002). Roller compaction and tabletting of St. John's Wort plant dry extract using a gap width and force controlled roller compactor I. Study of roller compaction variables on granule and tablet properties by a 33 factorial design. *Pharmaceutical Development Technology*, 7, 447-45, ISSN: 1083-7450.

Fell, J.T. & Newton J.M. (1970). Determination of Tablet Strength by Diametrical-Compression Test. *International Journal of Pharmaceutics*, 59, 688-691, ISSN: 0378-5173.

Fichtnera, F., Rasmuson, A. & Alderborn. G. (2005). Particle size distribution and evolution in tablet structure during and after compaction. *International Journal of Pharmaceutics* 292, 211–225, ISSN: 0378-5173.

Hausner H.H. (1967). Friction conditions in a mass of metal powder, *International Journal of Powder Metallurgy*, 3 (4), 7-13, ISSN: 0888-7462.

Heckel, R.W. (1961). An analysis of powder compaction phenomena. *Transaction of Metallurgy Society, AIME*, 221, 671-675, ISSN 0543-5722. 221.

Jaganath, I.B. & Ng, L.T. (2000). *Herbs. In The green pharmacy of Malaysia*, Jaganath, I.B., *Vinpress Sdn. Bhd and Malaysian Agricultural Research and Development Institute*, ISBN No.967-81-0281-1, Kuala Lumpur, Malaysia.

Kawakita, K. & Lüdde, K.H. (1970/71). Some considerations on powder compression Equations. *Powder Technology*, 4, 61-68, ISSN 0032-5910.

Korhonen, O., Pohja, S., Peltonen, S., Suihko, E., Vidgren, M., Paronen, P. & Ketolainen, J. (2002). Effects of Physical Properties for Starch Acetate Powders on Tableting.

Pharmaceutics Science Technology AAPS., 34, ISSN 1530-9932. (available from http://www .apppspharmscitech. org - Accessed 21/6/2008).

Mohammed, H., Briscoe, B.J. & Pitt, K.G. (2006). The intrinsic nature and coherence of compacted pure pharmaceutical tablets. *Powder Technology,* 165, 11-21, ISSN 0032-5910.

Nokhodchi, A. (2000). An overview of the effect of moisture on compaction and compression. *Pharmaceutical Technology.* (available www.pharmtech.com - Accessed 3/7/2007).

Nokhodchi A. (2005). An Overview of The Effect of Moisture on Compaction and Compression. *Pharmaceutical Technology* (available from www.pharmatech.com - Accessed 20/6/2009).

Nordström, J., Welch, K., Frenning, G. & Alderborn, G. (2008). On the physical interpretation of the Kawakita and Adams parameters derived from confined compression of granular solids. *Powder Technology* 182, 424–435 , ISSN 0032-5910.

Odeku, O. A., Awe, O.O., Popoola, B., Odeniyi, M. A. & Itiola, O. A. (2005). Compression and Mechanical Properties of Tablet Formulations Containing Corn, Sweet Potato, and Cocoyam Starches as Binders. *Pharmaceutical Technology,* 29 (4): 82-90, ISSN: 1543-2521.

Ramakrishnan, K.N., Nagarajan, R., RamaRao, G.V. & Venkadesan, S., 1997. A Compaction Study on Ceramic Powders. *Elsevier Material Letters,* 33, 191-194, ISSN 0167-577X.

Sambandan, T., G., Rha, C. K., Abdul Kadir, A., Aminudin, N & Mohammed Saad, J. (2006). Bioactive Fraction of *Eurycoma Longifolia. United States Patent* 0087493 A1.

Schiller, M., von der Heydt, H., März, F. & Schmidt, P.C. (2003). Enhanced processing properties and stability of film-coated tablets prepared from roller –compacted and ion-exchanged Eschscholtzia Californica Cham. Dry Extract, *STP Pharmaceutical Science,* 13, 111-117, ISSN 1157-1489.

Sebhatu, T., Ahlneck, C. & Alderborn, G. (1997). The effect of moisture content on the compression and bond-formation properties of. amorphous lactose particles. *International Journal of Pharmaceutics* 146, 101-114, ISSN: 0378-5173.

Shivanand, P. & Sprockel, O.L. (1992). Compaction behavior of cellulose polymers. *Powder Technology,* 69, 177-184, ISSN 0032-5910.

Takeuchi, H., Nagira, S., Hiromitsu Y. & Kawashima, Y. (2004). Die wall pressure measurement for evaluation of compaction property of pharmaceutical materials, *International Journal of Pharmaceutics,* 274, 131–138, ISSN: 0378-5173.

Varthalis, S. & Pilpel, N. (1976). Anomalies in some properties of powder mixtures. *Journal of Pharmacy and Pharmacology,* 28, 415-419, ISSN: 0022-3573.

Walker, E.E. (1923). The Properties of powders- Part VI: The compressibility of powders; *Transactions of the Faraday Society,* 19, 73-82, ISSN: 0014-7672.

Yamamoto, Y., Fujii, M., Watanabe, K., Tsukamoto, M., Shibata, Y., Kondoh, M. & Watanabe, Y. (2009). Effect of powder characteristics on oral tablet disintegration. *International Journal of Pharmaceutics* 365, 116–120, ISSN: 0378-5173.

Yusof, Y. A., Smith, A. C., & Briscoe, B. J. (2005). Roll compaction of maize powder. *Chemical Engineering Science,* 60 (14), 3919-3931, ISSN 0009-2509.

Zhang Y., Law Y. & Chakrabarti, S. (2003). Physical Properties and Compact Analysis of commonly Used Direct Compression. *Pharmaceutical Science Technology,* 4(4), 62, ISSN 1530-9932. (available from http://www .apppspharmscitech. org - Accessed 21/6/2008).

Part 3

Tribology and Low Friction

12

Frictional Property of Flexible Element

Keiji Imado
Oita University
Japan

1. Introduction

In the calculation of frictional force of a flexible element such as a belt, rope or cable wrapped around the cylinder, the famous Euler's belt formula (Hashimoto, 2006) or simply known as the belt friction equation (Joseph F. Shelley, 1990) is used. The formula is useful for designing a belt drive or band brake (J. A. Williams, 1994). On the other hand, a belt or rope is conveniently used to tighten a luggage to a carrier or lift up the luggage from the carrier. In that case, for the sake of adjusting the belt length and keeping an appropriate tension during transportation, various kinds of belt buckles are used. These belt buckles have been devised empirically and there was no theory about why it can fix the belt. The first purpose of this chapter is to present the theory of belt buckle clearly by considering the self-locking mechanism generated by wrapping the belt on the belt. Making use of the belt tension for a locking mechanism, a belt buckle with no locking mechanism can be made. The principle and some basic property of this new belt buckle are also shown.

The self-locking of belt may occur even in the case where a belt is wrapped on an axis two or more times. The second purpose of this chapter is to present the frictional property of belt wrapped on an axis two and three times through deriving the formulas corresponding to an each condition. Making use of this self-locking property of belt, a belt-type one-way clutch can be made (Imado, 2010). The principle and fundamental property of this new clutch are described.

As the last part of this chapter, the frictional property of flexible element wrapped on a hard body with any contour is discussed. The frictional force can be calculated by the curvilinear integral of the curvature with respect to line element along the contact curve.

2. Theory of belt buckle

Notation

C	Magnification factor of belt tension
F	Frictional force, N
$F_{ij} = F_{ji}$	Frictional force between point P_i and P_j, N
L	Distance between two cylinder centers, m
N	Normal force of belt to surface, N
$N_{ij} = N_{ji}$	Normal force of belt between point P_i and P_j, N
P_i	Boundary of contact angle
R	Radius of main cylinder, m

T_i	Tension of belt in i'th interval, N
r	Radius of accompanied cylinder, m
μ	Coefficient of friction for belt-cylinder contact
μ_b	Coefficient of friction for belt-belt contact
θ	Angle
θ_i	Angle of point P_i
$\theta_{ij} = \theta_{ji}$	Contact angle between P_i and P_j

2.1 Friction of belt in belt buckle

Figure 1 (a) shows a cross sectional view of a belt buckle and a belt wrapped around the two cylindrical surfaces. T_1 and T_4 ($T_1 > T_4$) are tensions of the belt at both ends. There is a double-layered part where the belt is wrapped over the belt. Figure 1 (b) shows the enlarged view around the main axis. For simplicity, the thickness of the belt was neglected. According to the theory of belt friction, following equations are known for belt tensions of T_1, T_2 and T_3 (Joseph F. Shelley, 1990).

$$T_1 = e^{\mu_b \theta_{12}} T_2, \quad T_2 = e^{\mu \theta_{34}} T_3 \tag{1}$$

T_4' and T_4'' are of inner belt tension at P_1 and P_2 respectively. The normal force to a small element of the inner belt at angle θ is denoted as dN_b, which can be written as

$$dN_b = e^{\mu_b (\theta - \theta_2)} T_2 d\theta \tag{2}$$

Making use of T_4' and T_4'', the normal forces of inner belt for an each section are expressed as

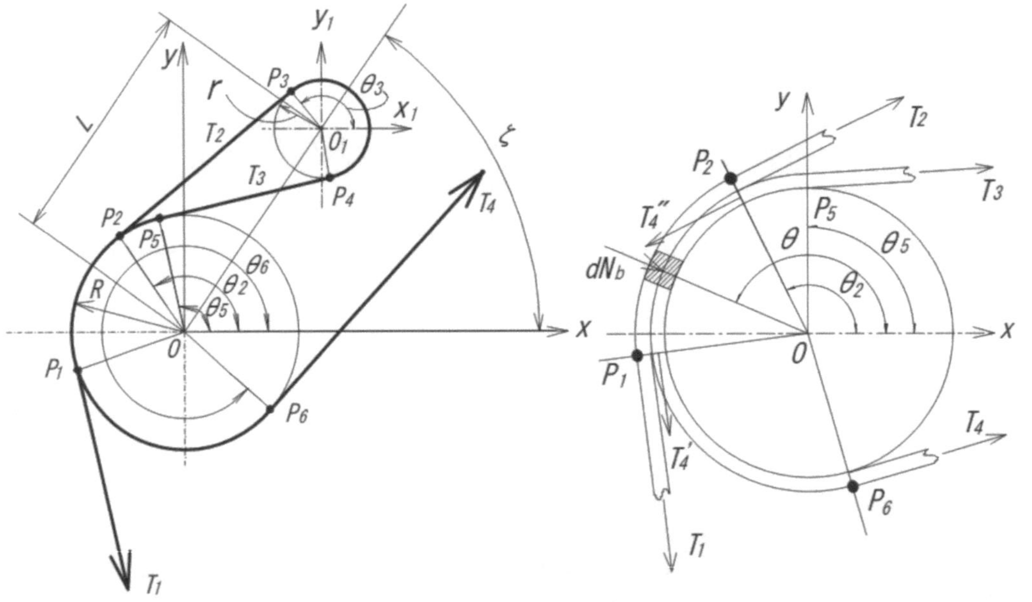

(a) Belt buckle (b) Enlarged view

Fig. 1. Mechanical model of belt buckle and enlarged veiw around main axis

$$\left.\begin{aligned} dN_{25} &= e^{\mu(\theta_2-\theta)}T_4''d\theta \\ dN_{12} &= e^{\mu(\theta_1-\theta)}T_4'd\theta \\ dN_{16} &= e^{\mu(\theta_6-\theta)}T_4\,d\theta \end{aligned}\right\} \quad (3)$$

The frictional force between P_1 and P_6 is

$$F_{16} = \int_{\theta_1}^{\theta_6} \mu dN_{16} = (e^{\mu\theta_{16}} - 1)T_4 \quad (4)$$

The inner belt tension T_4' is the sum of the frictional force F_{16} and the belt tension T_4.

$$T_4' = T_4 + F_{16} = e^{\mu\theta_{16}}T_4 \quad (5)$$

The frictional force F_{12} acting on the inner belt is composed of two forces denoted as F_{12in} and F_{12out}. The frictional force F_{12in} is acting on the cylindrical surface, which is generated by the normal forces dN_b and dN_{12}. The normal force dN_b is exerted from the outer belt. The other normal force dN_{12} is generated by the inner belt tension. So, F_{12in} is given by

$$F_{12in} = \int_{\theta_2}^{\theta_1} \mu\, dN_b + \int_{\theta_2}^{\theta_1} \mu\, dN_{12} = (e^{\mu_b\theta_{12}} - 1)\frac{\mu T_2}{\mu_b} + (e^{\mu\theta_{12}} - 1)T_4' \quad (6)$$

Making use of Eq. (2), the frictional force F_{12out} acting on the belt-belt boundary can be written as

$$F_{12out} = \int_{\theta_2}^{\theta_1} \mu_b\, dN_b = (e^{\mu_b\theta_{12}} - 1)T_2 \quad (7)$$

The frictional force F_{12} is the sum of Eqs. (6) and (7).

$$F_{12} = (e^{\mu_b\theta_{12}} - 1)\left(1 + \frac{\mu}{\mu_b}\right)T_2 + (e^{\mu\theta_{12}} - 1)T_4' \quad (8)$$

As the belt tension T_4'' is the sum of F_{12} and T_4', making use of Eq. (5) and (8), T_4'' can be written as

$$T_4'' = F_{12} + T_4' = e^{\mu\theta_{26}}T_4 + (e^{\mu_b\theta_{12}} - 1)\left(1 + \frac{\mu}{\mu_b}\right)T_2 \quad (9)$$

Making use of Eq. (3), the frictional force F_{25} can be written as

$$F_{25} = \int_{\theta_5}^{\theta_2} \mu\, dN_{25} = (e^{\mu\theta_{25}} - 1)T_4'' \quad (10)$$

As the belt tension T_3 is the sum of F_{25} and T_4'', making use of Eqs. (9) and (10), T_3 can be expressed as

$$T_3 = F_{25} + T_4'' = e^{\mu\theta_{25}}\left\{e^{\mu\theta_{26}}T_4 + (e^{\mu_b\theta_{12}} - 1)\left(1 + \frac{\mu}{\mu_b}\right)T_2\right\} \quad (11)$$

Substituting Eq. (1) into Eq. (11) to eliminate T_2 gives

$$T_3 = \frac{e^{\mu\theta_{56}}}{1 - e^{\mu(\theta_{34}+\theta_{25})}(e^{\mu_b\theta_{12}} - 1)(1 + \mu/\mu_b)} T_4 \tag{12}$$

Substituting Eq. (1) into Eq. (12) to get the relation between T_1 and T_4 gives

$$T_1 = \frac{e^{\mu_b\theta_{12}} e^{\mu(\theta_{34}+\theta_{56})}}{1 - e^{\mu(\theta_{34}+\theta_{25})}(e^{\mu_b\theta_{12}} - 1)(1 + \mu/\mu_b)} T_4 \tag{13}$$

In the same manner from Eq. (1) to Eq. (13), in the case of $T_1 < T_4$, corresponding relation of Eq. (13) yields as

$$T_4 = \left\{ e^{\mu(\theta_{34}+\theta_{56})} e^{\mu_b\theta_{12}} + e^{\mu\theta_{16}} (e^{\mu_b\theta_{12}} - 1)\left(1 + \frac{\mu}{\mu_b}\right) \right\} T_1 \tag{14}$$

2.2 Property of formulas of belt buckle

The validity of Eqs. (13) and (14) might be checked by supposing an extreme case of either μ=0 or μ_b=0. Substituting μ=0 into Eq. (13) gives

$$T_1 = \frac{e^{\mu_b\theta_{12}}}{2 - e^{\mu_b\theta_{12}}} T_4 \tag{15}$$

Next, substituting μ_b=0 into Eq. (13) gives

$$T_1 = \frac{e^{\mu(\theta_{34}+\theta_{56})}}{1 - \theta_{12}\,\mu\, e^{\mu(\theta_{34}+\theta_{25})}} T_4 = C\, e^{\mu(\theta_{34}+\theta_{56})} T_4 \tag{16}$$

Substituting μ_b=0 into Eq. (15) or substituting μ=0 into Eq. (16) gives T_1=T_4. Substituting $\theta_{12} = 0$ into Eq. (13) to remove the double-layered segment on the ratio of belt tension yields the conventional equation of belt friction.

$$T_1 = e^{\mu(\theta_{34}+\theta_{56})} T_4 \tag{17}$$

Equation (17) is also obtained by substituting $\theta_{12} = 0$ into Eq. (16). This means that the ratio of belt tension is magnified by the factor C

$$C = \frac{1}{1 - \theta_{12}\,\mu\, e^{\mu(\theta_{34}+\theta_{25})}} \tag{18}$$

due to the double-layered segment even in the case of μ_b=0. As far as these inspections are concerned, there is no contradiction in Eq. (13). As Eqs. (13), (15) and (16) are of fractions, the factor of T_4 might become infinity meaning T_4/T_1=0. This fact virtually implies the occurrence of self-locking. Figure 2 shows the relation of μ_b and θ_{12} satisfying $e^{\mu_b\theta_{12}} = 2$ in Eq. (15). Self-locking occurs in the region above this curve where $e^{\mu_b\theta_{12}} > 2$. On the other hand, in the region below this curve, self-locking does not occur. In the case of μ=0, the equilibrium of moment of belt tension about O in Fig. 1 gives

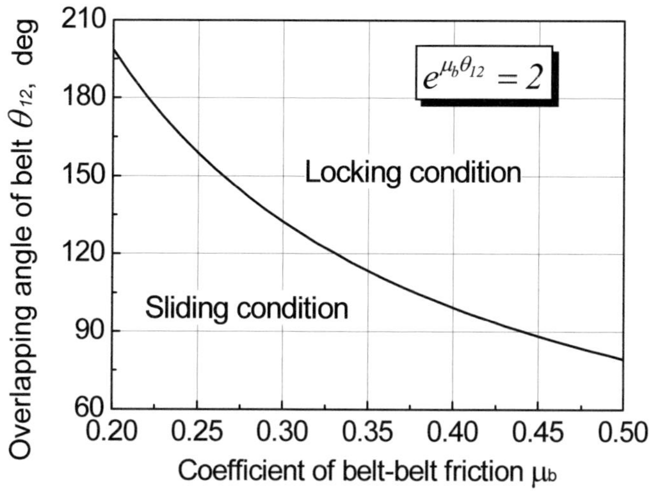

Fig. 2. Boundary curve between self-locking condition and sliding condition

$$T_4 = 2T_2 - T_1 \tag{19}$$

In the locking state with μ=0, T_4=0 so that T_1=2T_2=2T_3. It means that belt tension T_1 is halved to T_2 by the belt-belt friction.

As the angle of double-layered segment θ_{12} is determined by the geometry of the buckle, some calculations were carried out to know the properties of Eq. (13) and Eq. (15) providing r/L=R/L=1/4. The direction of belt tension T_1 and T_4 were assumed to be the same direction for simplicity. Results are shown in Figs. 3 and 4. Figure 3 corresponds to the Eq. (15) where

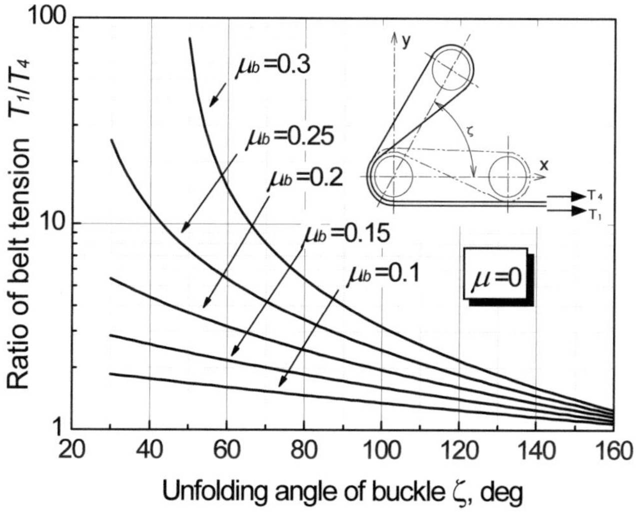

Fig. 3. Change of belt tension ratio with unfolding angle ζ in the case of μ=0. Belt tension ratio increases greatly with an increment of the coefficient of friction μ_b especially in the vicinity of locking condition. It is very sensitive to angle ζ.

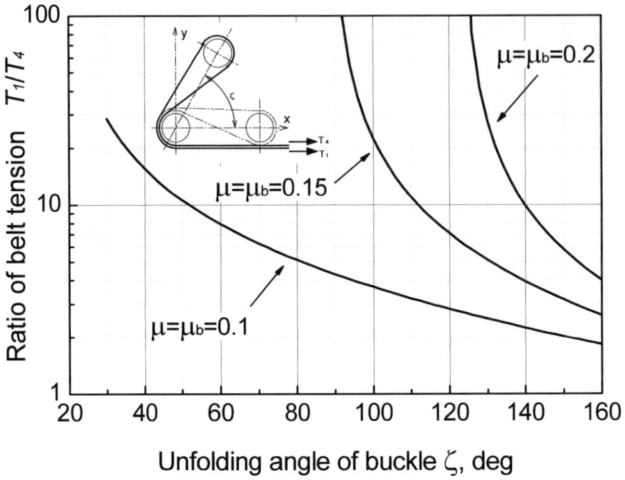

Fig. 4. Change of belt tension ratio with unfolding angle ζ in the case of $\mu=\mu_b$. Belt tension ratio increases greatly with an increment of the coefficient of friction.

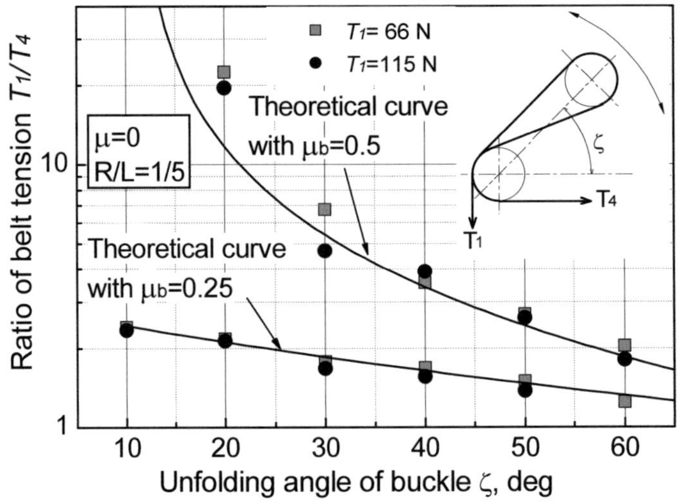

Fig. 5. Change of belt tension ratio with unfolding angle ζ in the case of μ=0. The ratio of belt tension changes according to Eqs. (13) or (15).

the coefficient of friction is μ=0. The ratio of belt tension increases with an increment of the coefficient of friction μ_b. It increases greatly when it approaches the locking condition. Figure 4 shows some results obtained by Eq. (13) providing $\mu=\mu_b$. The ratio of belt tension becomes far bigger than the that of Fig. 3.

Some experiments were carried out to verify the validity of Eq. (15) by wrapping a belt around the outer rings of rolling bearings to realize the condition of μ=0. Belt tension T_1 was applied by the weight. Belt tension T_4 was measured by the force gauge. Figure 5 shows the results. Experimental data are almost on the theoretical curves. As predicted by the Eq. (15), self-locking was confirmed for the belt with μ_b=0.5 in the region of ζ<10° where $e^{\mu_b\theta_{12}} > 2$.

2.3 Calculation of arm torque

Figure 6 (a) shows the mechanical model of belt buckle (Imado, 2008 a). Figure 6 (b) shows a three-dimensional model of the buckle. The arm of the buckle rotates around the point O_2. The angle of arm is denoted by α. The intersection angle of the line O-O_2 and O_2-O_1 is denoted by β. From geometrical consideration, the angle β is given by

$$\beta = \pi + \alpha - \phi \tag{20}$$

Applying the cosine theorem to the triangle OO_1O_2, length L is given by

$$L = L_2\sqrt{1+\kappa^2 + 2\kappa\cos(\alpha - \phi)}\ , \text{ where } \kappa = L_1 / L_2 \tag{21}$$

The symbol ζ denotes the angle of line O-O_1.

$$\zeta = \phi - \beta_1 \tag{22}$$

Applying the cosine theorem and sine theorem to the triangle OO_1O_2 gives

$$\cos\beta_1 = \frac{L^2 + L_1^2 - L_2^2}{2LL_1}, \quad \sin\beta_1 = \frac{L_2}{L}\sin(\phi - \alpha) \tag{23}$$

Substituting Eq. (21) into Eq. (23) and substituting Eq. (23) into Eq. (22) gives

$$\zeta = \phi - \tan^{-1}\left\{\frac{1}{\kappa + \cos(\alpha - \phi)}\sin(\phi - \alpha)\right\} \tag{24}$$

ζ, the angle of center line O-O_1, can be calculated from the arm angle α by Eq. (24). Note the angle ζ is equal to α when L_1 becomes 0.

The moment of the arm about point O_2 due to belt tensions T_2 and T_3 is expressed by

$$M = c_2T_2 + c_3T_3 \tag{25}$$

where c_2 and c_3 are geometrical variables that can be calculated from the position of contact boundaries P_2, P_3, P_4 and P_5. Dividing the arm torque M with RT_1, torque due to belt tension T_1 about point O, gives non-dimensional moment N.

$$N = \frac{1}{R}\left(c_2\frac{T_2}{T_1} + c_3\frac{T_3}{T_1}\right) \tag{26}$$

Making use of Eq. (1), the fractions of belt tension in Eq. (26) can be calculated by

$$\frac{T_2}{T_1} = \frac{1}{e^{\mu_b\theta_{12}}}, \quad \frac{T_3}{T_1} = \frac{1}{e^{\mu_b\theta_{12}}e^{\mu\theta_{34}}} \tag{27}$$

Figure 7 shows some examples of non-dimensional torque N. For simplicity, the coefficients of friction were taken to be $\mu=\mu_b$. The non-dimensional torque N decreases to be negative value with decrement of arm angle α. It means an occurrence of directional change in arm torque. This negative torque acts so as to hold the arm angle in a locking state without any locking mechanism. The angle where arm torque N becomes 0 is denoted by α_C. It depends on the geometry of buckle and the coefficients of friction μ and μ_b. Making use of Eqs. (13) and (24), the fraction of belt tension can be calculated. Figure 8 shows some results. The fraction of belt tension, T_4/T_1, decreases with arm angle α. It becomes 0 at $\alpha = \alpha_L$. According to Eq. (13), the fraction of belt tension T_4/T_1 becomes negative when arm angle α becomes less than α_L, $\alpha < \alpha_L$. The physical meaning of negative value in the fraction of belt tension is that the belt tension T_4 should be compressive so as to satisfy the equilibrium condition of the force. But a belt cannot bear compressive force so that negative value in the fraction of belt tension is actually unrealistic. It means the belt was locked with the buckle. The angle α_L becomes larger with an increment of the coefficients of friction. As the coefficient of friction is generally greater than 0.15, the locking condition is easily satisfied. Once the locking condition is satisfied, the belt is dragged into the buckle with a decrement of arm angle α. Then the belt tension becomes greater.

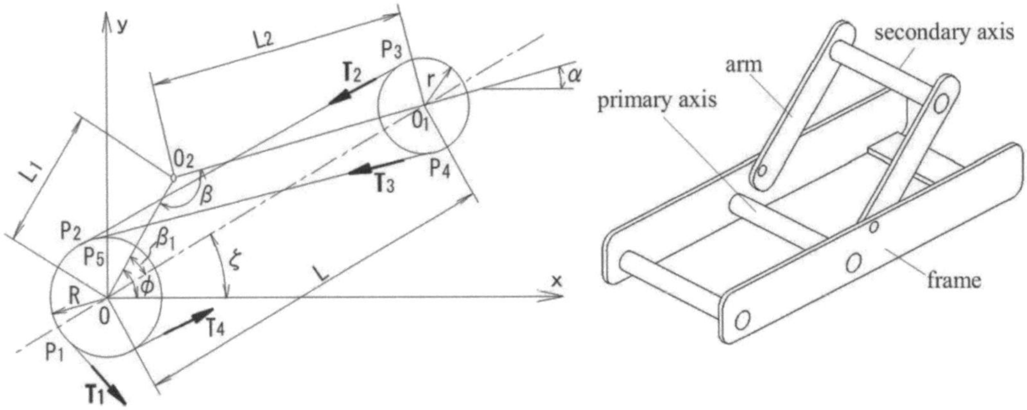

Fig. 6. Mechanical model of belt buckle to calculate arm torque and 3D model

3. Theory of belt friction in over-wrapped condition

3.1 Friction of belt wrapped two times around an axis

Figure 9 shows a mechanical model (Imado, 2008 b). The point P_i (i=1, 2, 3) is a boundary of contact and T_i (i=1, 2, 3, 4) is tension of the belt. Symbol θ_i denotes the angle of point P_i. The belt is over-wrapped around the belt in the range from P_1 to P_2 denoted by θ_1. The axis x is taken so as to pass through the point P_2, which is an end of the belt. T_1 is bigger than T_4. T_4 is an imaginary belt tension. There is no contact from P_2 to P_3 due to the thickness of the belt-end. According to the theory of belt friction (Joseph F. Shelley, 1990), analysis starts with the conventional equation.

$$T_1 = e^{\mu_b \theta_1} T_2 = e^{\mu_b \theta_1} T_3 \tag{28}$$

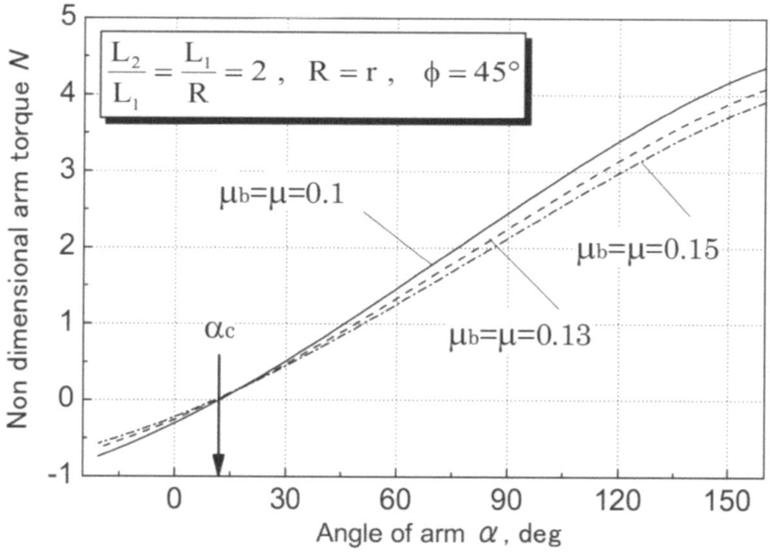

Fig. 7. Non-dimensional arm toruque N decreases with arm angle α

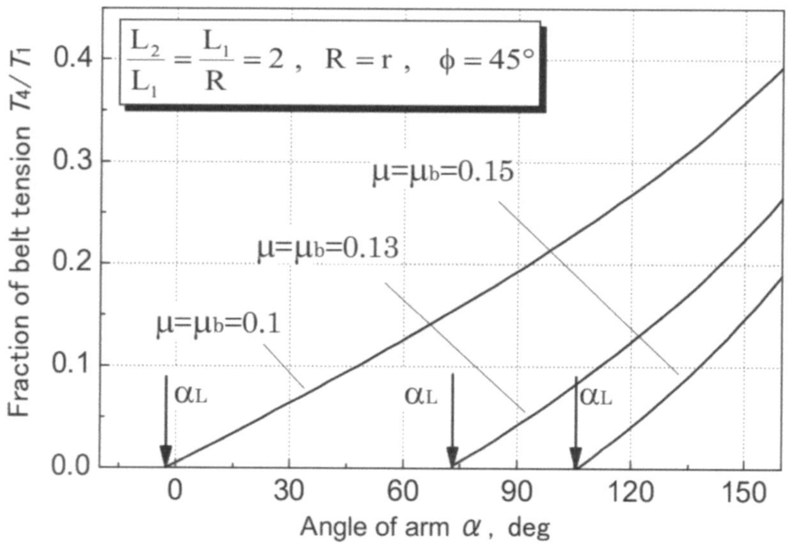

Fig. 8. Fraction of belt tension T_4/T_1 decreases with arm angle α

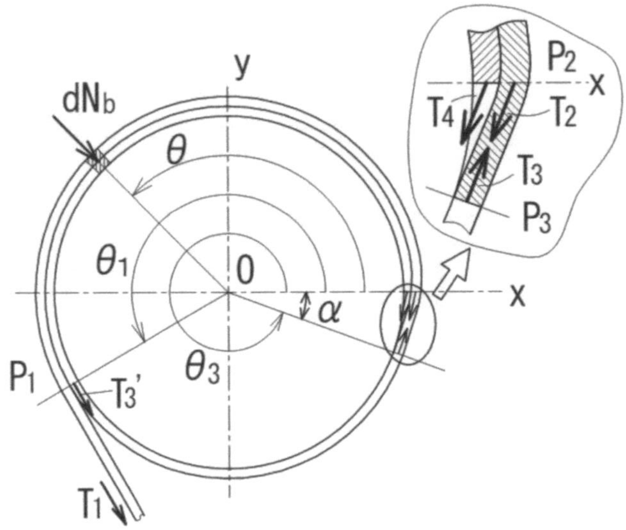

Fig. 9. Mechanical model of belt wrapped two times around an axis

The belt tension T_2 or T_3 can be expressed by the belt tension T_3', where T_3' is inner belt tension at the point P_1 as shown in Fig. 9.

$$T_2 = T_3 = e^{\mu(\theta_3 - \theta_1)} T_3' \tag{29}$$

Making use of Eqs. (28) and (29), T_1 can be expressed as

$$T_1 = e^{\{\mu_b \theta_1 + \mu(\theta_3 - \theta_1)\}} T_3' \tag{30}$$

The inner belt is normally pressed onto the cylinder by the outer belt. The normal force to a small segment of the inner belt at angle θ denoted by dN_b is

$$dN_b = e^{\mu_b \theta} T_2 d\theta \tag{31}$$

On the other hand, the normal force is also generated by inner belt tension itself. The normal force exerted on the cylinder between P_i and P_j is denoted by N_{ij}. Normal force acting to a small segment of the cylinder at angle θ is given by

$$dN_{21} = e^{\mu\theta} T_4 d\theta \tag{32}$$

Then, making use of Eqs. (31) and (32), the frictional force between the inner belt and cylinder denoted by F_{12in} is given by

$$F_{12in} = \int_0^{\theta_1} \mu dN_b + \int_0^{\theta_1} \mu dN_{21} = (e^{\mu_b \theta_1} - 1)\frac{\mu T_2}{\mu_b} + (e^{\mu\theta_1} - 1)T_4 \tag{33}$$

Denoting the radius of cylinder by r and neglecting the thickness of the belt, the equilibrium equation of moment of the cylinder is

$$T_1 r = (F_{12in} + F_{13} + T_4) r \tag{34}$$

Here, the frictional force F_{13} exerted on the surface between P_1 and P_3 is given by

$$F_{13} = \mu \int_{\theta_1}^{\theta_3} e^{\mu(\theta-\theta_1)} T_3 ' d\theta = \{e^{\mu(\theta_3-\theta_1)} - 1\} T_3 ' \tag{35}$$

Substituting Eqs. (33) and (35) into Eq. (34) gives

$$T_1 = (e^{\mu_b \theta_1} - 1)\frac{\mu T_2}{\mu_b} + \{e^{\mu(\theta_3-\theta_1)} - 1\} T_3 ' + e^{\mu\theta_1} T_4 \tag{36}$$

Substituting T_2 and T_3' in Eq. (36) as functions of T_1 by making use of Eqs. (28) and (30) gives

$$T_1 = \frac{e^{\theta_1(\mu+\mu_b)}}{(1-e^{\mu_b\theta_1})\left(\frac{\mu}{\mu_b} - 1\right) + e^{-\mu(\theta_3-\theta_1)}} T_4 \tag{37}$$

This is the targeted equation that expresses the relation between T_1 and T_4.
Equation (37) can be checked by supposing an extreme case of either μ=0 or μ_b=0. Substituting μ=0 into Eq. (37) gives T_1=T_4 as a matter of course. Substituting of μ_b=0 into Eq. (37) requires limiting operation.

$$\lim_{\mu_b \to 0} (1 - e^{\mu_b \theta_1}) \frac{\mu}{\mu_b} = -\mu\theta_1 \tag{38}$$

Making use of Eq. (38), Eq. (37) becomes Eq. (39) for the case of μ_b=0.

$$T_1 = \frac{e^{\mu\theta_1}}{-\mu\theta_1 + e^{-\mu(\theta_3-\theta_1)}} T_4 \tag{39}$$

Equation (39) implies the belt may be locked firmly around an axis when the denominator of the fraction in Eq. (39) becomes 0. Substituting μ=0 into Eq. (39) gives T_1=T_4 again as a matter of course.
Substituting μ=μ_b into Eq. (37) gives

$$T_1 = e^{\mu(\theta_1+\theta_3)} T_4 \tag{40}$$

Equation (40) is exactly the same form as the Euler's belt formula though was derived from the expression that took an effect of over-wrapping of belt into account. Equation (40) implies that the belt cannot be locked on the cylinder as far as the wrapping angle is finite.
Letting θ_1=0 in Eq. (37) to eliminate the over-wrapping part gives

$$T_1 = e^{\mu\theta_3} T_4 \tag{41}$$

This is the well-known Euler's belt formula. So the Euler's belt formula was proved to be included as a special case in Eq. (37). Equation (41) can also be obtained from Eqs. (39) and (40).
Next, let's consider some locking conditions. According to Eq. (37), the belt tension ratio T_4 / T_1 can be expressed as

$$\frac{T_4}{T_1}=\frac{(1-e^{\mu_b\theta_1})\left(\dfrac{\mu}{\mu_b}-1\right)+e^{-\mu(\theta_3-\theta_1)}}{e^{\theta_1(\mu+\mu_b)}}=\frac{\Gamma}{e^{\theta_1(\mu+\mu_b)}} \tag{42}$$

The locking condition is satisfied when the numerator of Eq. (42) becomes 0 meaning $T_4=0$. So, the discriminant of locking condition can be expressed as

$$\Gamma=(1-e^{\kappa\mu\theta_1})\left(\frac{1}{\kappa}-1\right)+e^{-\mu(\theta_3-\theta_1)} \tag{43}$$

Locking condition is satisfied in the case of $\Gamma\leq 0$. Critical point is $\Gamma=0$. Here, κ denotes a ratio of the coefficient of friction.

$$\kappa=\mu_b/\mu \tag{44}$$

As $e^{\kappa\mu\theta_1}\geq 1$ and $e^{-\mu(\theta_3-\theta_1)}>0$, κ should be less than unity to make the value of locking discriminant of Eq. (43) be $\Gamma<0$. As can be seen in Fig. 9, the angle θ_3 is smaller than 2π due to the thickness of the belt. From geometrical consideration in Fig. 9, following equation is obtained.

$$\cos\alpha=\frac{r}{r+t}\approx 1-\frac{t}{r} \tag{45}$$

Here, t is thickness of the belt and r is a radius of the cylinder. When angle α is small, the angle α can be roughly estimated by

$$\alpha\approx\sqrt{2t/r} \tag{46}$$

Supposing the angle of non-contact is $\alpha=15°$, the corresponding critical locking condition can be evaluated by solving Eq. (43). Figure 10 shows some solutions. The critical angle of belt locking θ_1 decreases with an increment of the coefficient of friction μ. Provided the coefficient of friction is constant, the critical angle of belt locking θ_1 increases with an increment of κ. This fact means that the belt is likely to lock with a decrement of κ. So the smaller coefficient of friction μ_b is preferable for self-locking. The limiting condition for the belt locking is $\kappa=0$ or $\mu_b=0$.

Figure 11 illustrates the effect of κ on the fraction of belt tension T_4/T_1 for the case of $\mu=0.3$ and $\theta_3=345°$. Making use of Eq. (41), the convergence point is calculated. It is $T_4/T_1=\exp(-\mu\theta_3)\approx 0.164$. It is clear that the fraction of belt tension T_4/T_1 is greatly influenced by the magnitude of κ, μ_b/μ. The belt tension ratio T_4/T_1 decreases with an increment of over-wrapping angle θ_1 except for the case of $\kappa=1.4$. When $\kappa\geq 1$, the fraction of belt tension is always positive, so that the self-locking never occurs. Provided $\theta_1=360°$, $\theta_3=345°$ and $\mu=0.3$, the critical ratio of the coefficient of friction κ_c for the self-locking with two times over-wrapping condition was calculated by using the discriminant Eq. (43). It was $\kappa_c=0.735$. The corresponding line was plotted with a dashed line in Fig. 11. The magnitude of κ should be smaller than κ_c to cause the self-locking.

Figure 12 shows a method by which the coefficient of friction between the belt and belt can be reduced so as to satisfy the self-locking condition. When a polyethylene film was wrapped together with belt, an occurrence of self-locking was confirmed. But self-locking never occurred without polyethylene film.

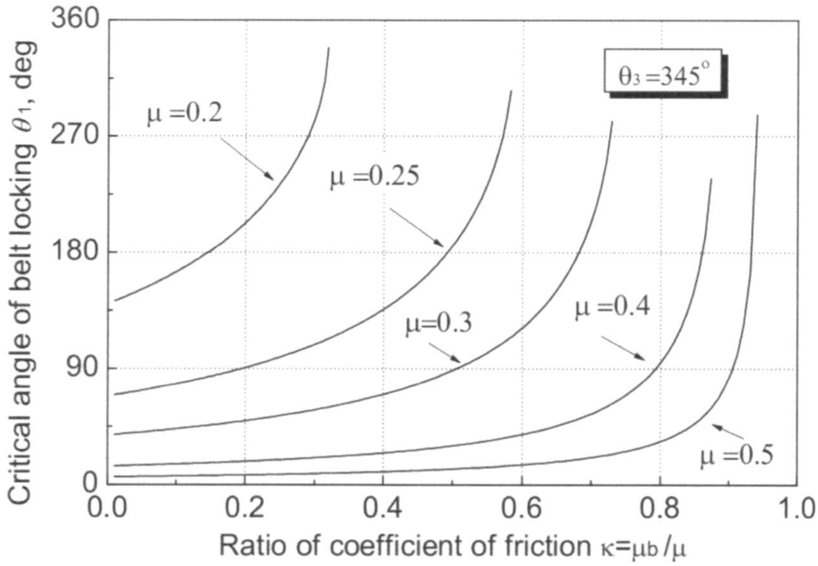

Fig. 10. Change of critical over-wrapping angle θ_1 for self-locking with ratio of the coefficients of friction κ.

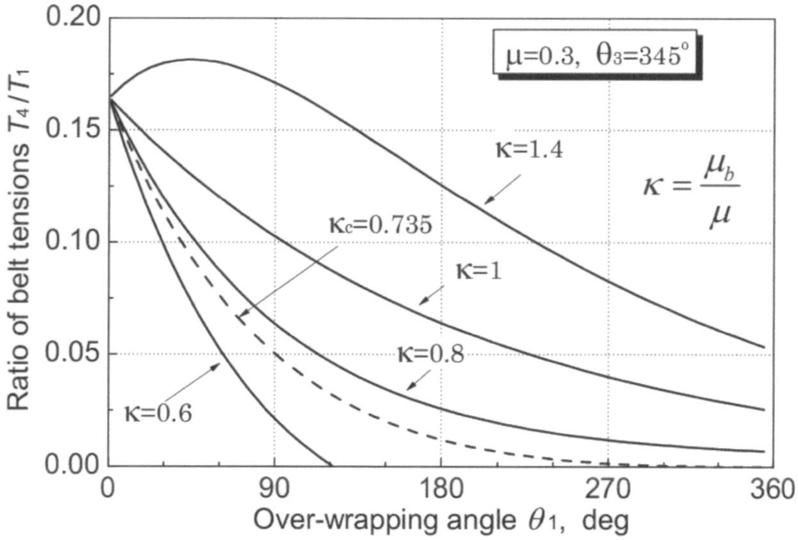

Fig. 11. Fraction of belt tension T_4 / T_1 decreases rapidly with increment of over-wrapping angle θ_1 for the case of smaller κ.

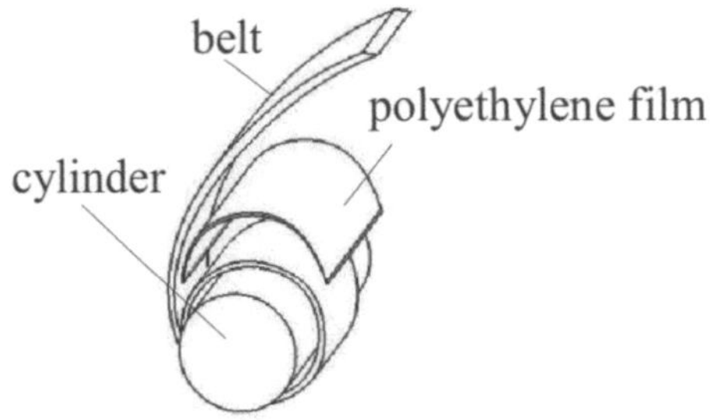

Fig. 12. Polyethylene film was wrapped together with belt to reduce the coefficient of friction μ_b. Self-locking was recognized in experiment with polyethylene film. But it never occurred without polyethylene film.

3.2 Friction of belt wrapped three times around axis

A belt can be wrapped more than two times around an axis. Let us consider the case where a belt is wrapped three times around an axis as shown in Fig. 13. The point P_i (i=1, 2, 3) is a boundary of contact. Tension of belt is denoted by T_i (i=1, 2, 3, 4) or T_i' and $T_1>T_4$. There are two kinds of the coefficients of friction μ and μ_b. μ_b is the coefficient of friction between belt and belt. The belt does not in contact with the axis from the point P_2 to P_3 due to the thickness of belt-end. In order to consider the equation of belt friction, the belt is divided into 5 sections from outside to inside as a, b, c, d and e in terms of frictional force as shown in Fig. 14. The frictional force working on an each section is expressed by either F_{si} or F_{so} , where the first subscript s means the name of section and the second subscript i means inside and o means outside respectively. Note that F_{si} works clockwisely and F_{so} works in a counter-clockwise direction. Considering the equilibrium of the force in an each section, following equations are obtained.

$$T_1 = F_{ai} + T_2 \tag{47}$$

$$T_3 = T_2 = F_{bi} + T_1' \tag{48}$$

$$T_1' = F_{ci} - F_{co} + T_2' \tag{49}$$

$$T_3' = T_2' = F_{di} - F_{do} + T_1'' \tag{50}$$

$$T_1'' = F_{ei} - F_{eo} + T_4 \tag{51}$$

Denothing the normal force from the section a to c by N_{ac}, the normal force acting to a small segment at angle θ is given by

$$dN_{ac} = e^{\mu_b \theta} T_2 d\theta \tag{52}$$

Frictional force F_{ai} is calculated by integrating Eq. (52).

$$F_{ai} = \int_{\theta=0}^{\theta_1} \mu_b dN_{ac} = \left(e^{\mu_b \theta_1} - 1\right) T_2 \tag{53}$$

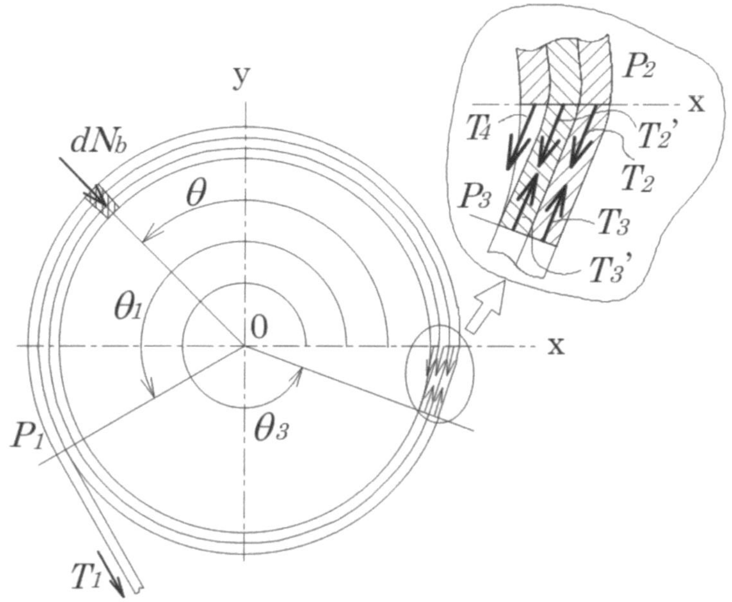

Fig. 13. Mechanical model of belt wrapped three times around an axis

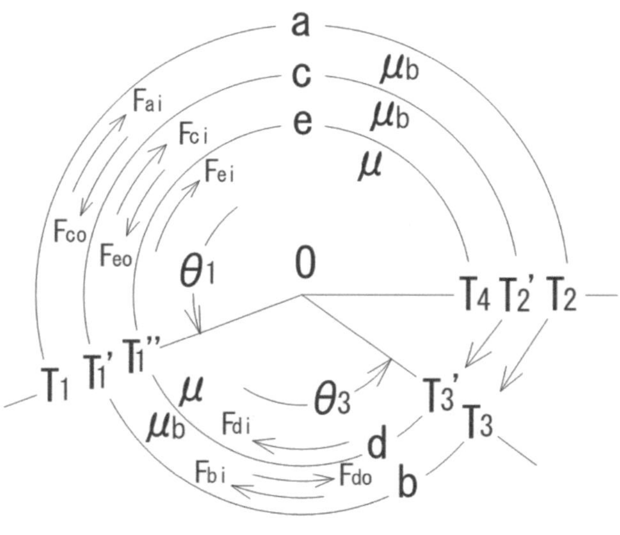

Fig. 14. Mechanical model with frictional force direction and the coefficients of friction corresponding to an each section.

In the same manner, infinitesimal normal force from the section b belt to d belt is given by

$$dN_{bd} = e^{\mu_b(\theta-\theta_1)} T_1' d\theta \tag{54}$$

Frictional force F_{bi} is calculated by integrating Eq. (54).

$$F_{bi} = \int_{\theta_1}^{\theta_3} \mu_b dN_{bd} = \int_{\theta_1}^{\theta_3} \mu_b e^{\mu_b(\theta-\theta_1)} T_1' d\theta = \left(e^{\mu_b(\theta_3-\theta_1)} - 1\right) T_1' \tag{55}$$

Making use of Eq. (52), infinitesimal normal force from section c belt to e belt is given by

$$dN_{ce} = e^{\mu_b\theta} T_2' d\theta + dN_{ac} = e^{\mu_b\theta} T_2' d\theta + e^{\mu_b\theta} T_2 d\theta \tag{56}$$

Frictional force F_{ci} is calculated by integrating Eq. (56).

$$F_{ci} = \int_0^{\theta_1} \mu_b dN_{ce} = \int_0^{\theta_1} \mu_b e^{\mu_b\theta} (T_2' + T_2) d\theta = \left(e^{\mu_b\theta_1} - 1\right)(T_2' + T_2) \tag{57}$$

Making use of Eq. (54), infinitesimal normal force from section d belt to the axis is given by

$$dN_d = e^{\mu(\theta-\theta_1)} T_1'' d\theta + dN_{bd} = e^{\mu(\theta-\theta_1)} T_1'' d\theta + e^{\mu_b(\theta-\theta_1)} T_1' d\theta \tag{58}$$

Frictional force F_{di} is calculated by integrating Eq. (58).

$$F_{di} = \int_{\theta_1}^{\theta_3} \mu dN_d = \int_{\theta_1}^{\theta_3} \mu\left(e^{\mu(\theta-\theta_1)} T_1'' + e^{\mu_b(\theta-\theta_1)} T_1'\right) d\theta = \left(e^{\mu(\theta_3-\theta_1)} - 1\right) T_1'' + \left(e^{\mu_b(\theta_3-\theta_1)} - 1\right) \frac{\mu}{\mu_b} T_1' \tag{59}$$

Making use of Eq. (56), infinitesimal normal force from section e belt to the axis is given by

$$dN_e = e^{\mu\theta} T_4 d\theta + dN_{ce} = e^{\mu\theta} T_4 d\theta + e^{\mu_b\theta} T_2' d\theta + e^{\mu_b\theta} T_2 d\theta \tag{60}$$

Then, the frictional force F_{ei} is given by

$$F_{ei} = \int_0^{\theta_1} \mu dN_e = \int_0^{\theta_1} \mu\left(e^{\mu\theta} T_4 + e^{\mu_b\theta} T_2' + e^{\mu_b\theta} T_2\right) d\theta = \left(e^{\mu\theta_1} - 1\right) T_4 + \left(e^{\mu_b\theta_1} - 1\right)(T_2 + T_2') \frac{\mu}{\mu_b} \tag{61}$$

Neglecting the thickness of the belt, the equilibrium requirement of the moment gives

$$T_1 = F_{di} + F_{ei} + T_4 \tag{62}$$

Substituting Eqs. (59) and (61) into Eq. (62) gives

$$T_1 = \left(e^{\mu(\theta_3-\theta_1)} - 1\right) T_1'' + \left(e^{\mu_b(\theta_3-\theta_1)} - 1\right) \frac{\mu}{\mu_b} T_1' + \left(e^{\mu_b\theta_1} - 1\right)(T_2 + T_2') \frac{\mu}{\mu_b} + e^{\mu\theta_1} T_4 \tag{63}$$

The belt tensions T_1', T_1'', T_2 and T_2' in Eq. (63) should be expressed by the function of T_4. From the law of action and reaction,

$$F_{co} = F_{ai}, \quad F_{eo} = F_{ci}, \quad F_{do} = F_{bi} \tag{64}$$

Substituting Eqs. (53), (55), (57), (59) and (61) into Eqs. (47) to (51) give

$$T_1 = F_{ai} + T_2 = e^{\mu_b\theta_1} T_2 \tag{65}$$

$$T_2 = T_3 = F_{bi} + T_1' = e^{\mu_b(\theta_3-\theta_1)} T_1' \tag{66}$$

$$T_1' = F_{ci} - F_{co} + T_2' = \left(e^{\mu_b\theta_1} - 1\right)(T_2' + T_2) - \left(e^{\mu_b\theta_1} - 1\right) T_2 + T_2' = e^{\mu_b\theta_1} T_2' \tag{67}$$

$$T_3' = T_2' = F_{di} - F_{do} + T_1'' = F_{di} - F_{bi} + T_1'' = e^{\mu(\theta_3-\theta_1)} T_1'' + \left(e^{\mu_b(\theta_3-\theta_1)} - 1\right)\left(\frac{\mu}{\mu_b} - 1\right) T_1' \tag{68}$$

$$T_1'' = F_{ei} - F_{eo} + T_4 = \left(e^{\mu_b\theta_1} - 1\right)\left(\frac{\mu}{\mu_b} - 1\right)(T_2 + T_2') + e^{\mu\theta_1} T_4 \tag{69}$$

Making use of Eqs. (65), (66) and (67) gives,

$$T_1 = e^{\mu_b(\theta_1+\theta_3)} T_2' = e^{\mu_b(\theta_1+\theta_3)} T_3' \tag{70}$$

Substituting Eq. (68) into Eq. (70) and making use of Eq. (67) gives

$$T_1 = e^{\mu_b(\theta_1+\theta_3)} \left\{ e^{\mu(\theta_3-\theta_1)} T_1'' + \left(e^{\mu_b(\theta_3-\theta_1)} - 1\right)\left(\frac{\mu}{\mu_b} - 1\right) e^{\mu_b\theta_1} T_2' \right\} \tag{71}$$

Making use of Eqs. (65), (66) and (67) gives

$$T_2' = \frac{T_1}{e^{\mu_b(\theta_1+\theta_3)}} \tag{72}$$

Substituting Eq. (72) into Eq. (71) gives

$$T_1 = e^{\mu_b(\theta_1+\theta_3)+\mu(\theta_3-\theta_1)} T_1'' + e^{\mu_b\theta_1}\left(e^{\mu_b(\theta_3-\theta_1)} - 1\right)\left(\frac{\mu}{\mu_b} - 1\right) T_1 \tag{73}$$

Rearranging Eq. (73) gives,

$$T_1'' = \frac{1 - \left(e^{\mu_b\theta_3} - e^{\mu_b\theta_1}\right)\left(\frac{\mu}{\mu_b} - 1\right)}{e^{\mu_b(\theta_1+\theta_3)+\mu(\theta_3-\theta_1)}} T_1 = A T_1 \tag{74}$$

Making use of Eqs. (65) and (72) gives

$$T_2 + T_2' = \frac{e^{\mu_b\theta_3} + 1}{e^{\mu_b(\theta_1+\theta_3)}} T_1 \tag{75}$$

Substituting Eq. (75) into Eq. (69) gives

$$T_1'' = \left(\frac{\mu}{\mu_b} - 1\right) \frac{\left(e^{\mu_b\theta_3} + 1\right)\left(e^{\mu_b\theta_1} - 1\right)}{e^{\mu_b(\theta_1+\theta_3)}} T_1 + e^{\mu\theta_1} T_4 = B T_1 + e^{\mu\theta_1} T_4 \tag{76}$$

Substituting Eq. (76) into the left hand side of Eq. (74) gives,

$$T_1'' = \left(\frac{\mu}{\mu_b} - 1\right)\frac{\left(e^{\mu_b\theta_3}+1\right)\left(e^{\mu_b\theta_1}-1\right)}{e^{\mu_b(\theta_1+\theta_3)}}T_1 + e^{\mu\theta_1}T_4 = BT_1 + e^{\mu\theta_1}T_4$$
$$= \frac{1-\left(e^{\mu_b\theta_3}-e^{\mu_b\theta_1}\right)\left(\frac{\mu}{\mu_b}-1\right)}{e^{\mu_b(\theta_1+\theta_3)+\mu(\theta_3-\theta_1)}}T_1 = AT_1 \tag{77}$$

Equation (77) can be written in the form of

$$T_1 = \frac{e^{\mu\theta_1}}{A-B}T_4 \tag{78}$$

where

$$A = \frac{1-\left(e^{\mu_b\theta_3}-e^{\mu_b\theta_1}\right)\left(\frac{\mu}{\mu_b}-1\right)}{e^{\mu_b(\theta_1+\theta_3)+\mu(\theta_3-\theta_1)}}, \quad B = \left(\frac{\mu}{\mu_b}-1\right)\frac{\left(e^{\mu_b\theta_3}+1\right)\left(e^{\mu_b\theta_1}-1\right)}{e^{\mu_b(\theta_1+\theta_3)}} \tag{79}$$

Eqs. (78) and (79) are the targeted equations that express the relation between T_1 and T_4 in the case of a belt wrapped three times around an axis.

3.3 Characteristics of belt friction equation with three times wrapping around axis

The equation derived in the previous section seems complex. It can be checked by assuming some extreme cases such as μ=0, μ_b=0 and μ=μ_b. In the case of μ=0, Eq. (79) becomes,

$$A = \frac{1+\left(e^{\mu_b\theta_3}-e^{\mu_b\theta_1}\right)}{e^{\mu_b(\theta_1+\theta_3)}}, \quad B = -\frac{\left(e^{\mu_b\theta_3}+1\right)\left(e^{\mu_b\theta_1}-1\right)}{e^{\mu_b(\theta_1+\theta_3)}} \tag{80}$$

then

$$A - B = \frac{e^{\mu_b(\theta_1+\theta_3)}}{e^{\mu_b(\theta_1+\theta_3)}} = 1 \tag{81}$$

Substituting Eq. (81) and μ=0 into Eq. (78) gives T_1=T_4.
In the case of μ_b=0, limiting operations are required. For the term A in Eq. (79),

$$\lim_{\mu_b\to 0}\frac{\mu}{\mu_b}\left(e^{\mu_b\theta_3}-e^{\mu_b\theta_1}\right) = \mu\left(\theta_3-\theta_1\right) \tag{82}$$

For the term B in Eq. (79),

$$\lim_{\mu_b\to 0}\frac{\mu}{\mu_b}\left(e^{\mu_b\theta_3}+1\right)\left(e^{\mu_b\theta_1}-1\right) = 2\mu\theta_1 \tag{83}$$

Then Eq. (79) becomes,

$$A = \frac{1-\mu\left(\theta_3-\theta_1\right)}{e^{\mu(\theta_3-\theta_1)}}, \quad B = 2\mu\theta_1 \tag{84}$$

Substituting Eq. (84) into (78) gives

$$T_1 = \frac{e^{\mu\theta_3}}{1-\mu\left(\theta_3-\theta_1+2\theta_1 e^{\mu(\theta_3-\theta_1)}\right)} T_4 \tag{85}$$

In order to consider the smallest wrapping angle of three times wrapping, substituting $\theta_1=0$ into Eq. (85) gives,

$$T_1 = \frac{e^{\mu\theta_3}}{1-\mu\theta_3} T_4 \tag{86}$$

On the other hand, substituting $\theta_1=\theta_3$ into Eq. (85) gives,

$$T_1 = \frac{e^{\mu\theta_3}}{1-2\mu\theta_3} T_4 \tag{87}$$

Equation (87) shows the relation of belt tension with the largest wrapping angle of three times wrapping. The locking condition is satisfied when the denominator of Eqs. (86) and (87) become 0, so that in the case of $\theta_1=\theta_3$, only 1/2 of the coefficient of friction is required for self locking in compared with the case of $\theta_1=0$.

In the case of $\mu_b=\mu$, Eq. (79) becomes,

$$A = \frac{1}{e^{2\mu\theta_3}}, \quad B = 0 \tag{88}$$

so that Eq. (78) becomes,

$$T_1 = \frac{e^{\mu\theta_1}}{A-B} T_4 = e^{\mu(\theta_1+2\theta_3)} T_4 \tag{89}$$

Substituting $\theta_1=0$ into Eq. (89) gives,

$$T_1 = e^{2\mu\theta_3} T_4 \tag{90}$$

Substituting $\theta_1=\theta_3$ into Eq. (89) gives,

$$T_1 = e^{3\mu\theta_3} T_4 \tag{91}$$

Note the magnitude of the wrapping angle of Eqs. (90) and (91). They are exactly the same form as the Euler's belt formula though they were derived considering the effect of over-wrapping of belt on belt friction.

Next, Substituting $\theta_1=0$ into Eq. (79) provided the boundary of two and three times over-wrapping of belt gives,

$$A = \frac{1+\left(1-e^{\mu_b\theta_3}\right)\left(\dfrac{\mu}{\mu_b}-1\right)}{e^{\theta_3(\mu+\mu_b)}}, \quad B = 0 \tag{92}$$

then Eq. (78) becomes

$$T_1 = \frac{1}{A-B}T_4 = \frac{e^{\theta_3(\mu+\mu_b)}}{(1-e^{\mu_b\theta_3})\left(\dfrac{\mu}{\mu_b}-1\right)+1}T_4 \tag{93}$$

On the other hand, substituting $\theta_1=\theta_3$ into Eq. (37) in the section 3.1 that was the equation for two times over-wrapping conditions gives,

$$T_1 = \frac{e^{\theta_3(\mu+\mu_b)}}{(1-e^{\mu_b\theta_3})\left(\dfrac{\mu}{\mu_b}-1\right)+1}T_4 \tag{94}$$

Equation (93) is completely corresponding to Eq. (94) so that both equations are continuous. Figures 15 and 16 show some calculated results by using Eqs. (37), (78) and (79). Figure 15 is of μ=0.25 and θ_3=350°. With an increment of κ, namely with an increment of μ_b, the ratio of belt tension T_4 / T_1 increases. Self-locking occurs with wrap angle less than 720° in the case of κ=0.5 and 0.6, so that they were calculated by Eq. (37). On the other hand, in the case of κ=0.7, 0.8 and 0.9, the wrap angle less than 720° is not enough for self-locking to occur. They requires wrap angle greater than 720° so that they were calculated by Eqs. (78) and (79). Figure 16 is of μ=0.2 and θ_3=350°. All of them require wrap angle greater than 720° to enter the self-locking condition.
The threshold of self-locking for three times wrapped belt is obtained by equating A to B in Eq. (79).

$$\left(\frac{\mu}{\mu_b}-1\right)\left\{\left(e^{\mu_b\theta_3}+1\right)\left(e^{\mu_b\theta_1}-1\right)e^{\mu(\theta_3-\theta_1)}+e^{\mu_b\theta_3}-e^{\mu_b\theta_1}\right\}=1 \tag{95}$$

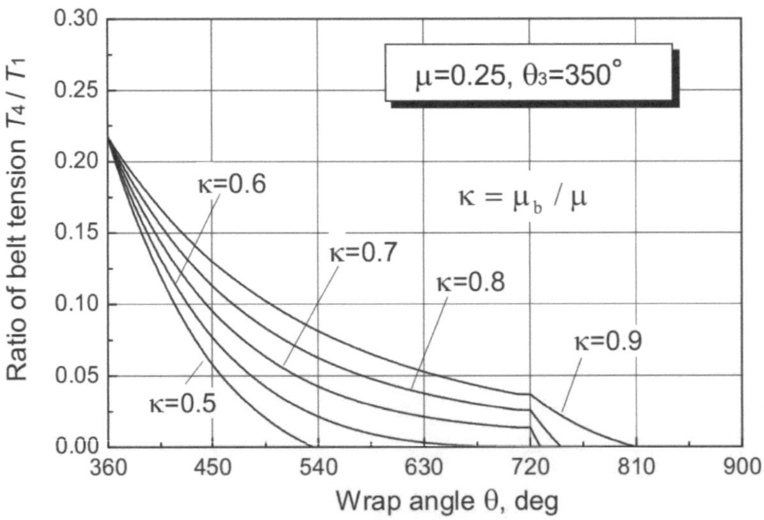

Fig. 15. Change of belt tension ratio with wrap angle

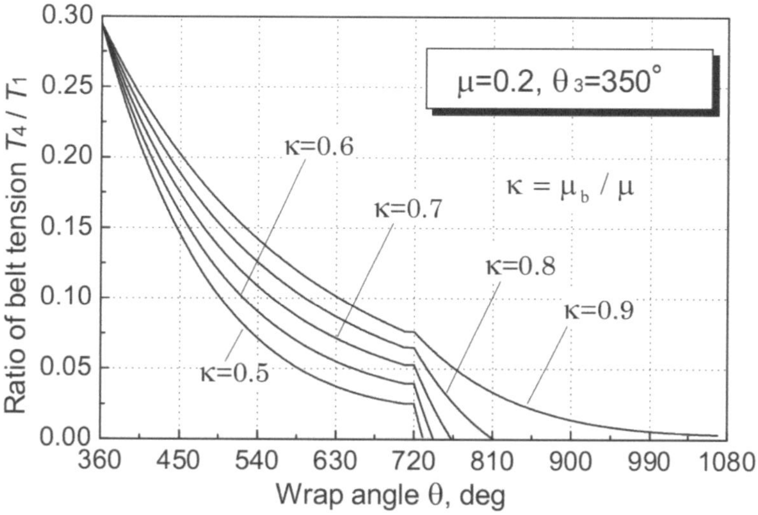

Fig. 16. Change of belt tension ratio with wrap angle

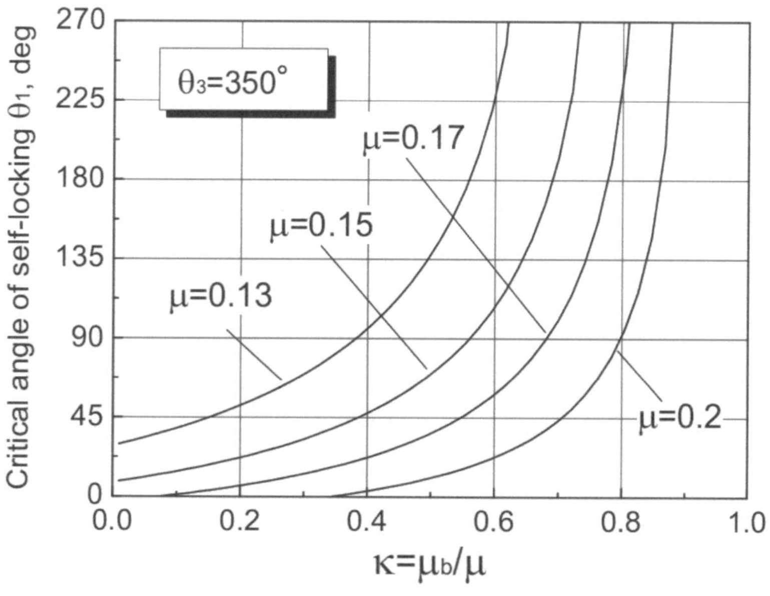

Fig. 17. Change of critical angle for self-locking with ratio of the coefficients of friction

Equation (95) is the discriminant of the self-locking condition for three times wrapped belt. When the coefficients of friction μ, μ_b and the angle θ_3 are given, the magnitude of critical angle θ_1 necessary for self-locking is calculated by solving Eq. (95). Figure 17 shows some solutions of Eq. (95) with angle θ_3=350°. If an angle θ_3 and the coefficient of frictions μ and μ_b are given, self-locking occurs with the wrap angle θ_1 over the corresponding curve. But it does not occur with wrap angle θ_1 under the corresponding curve. According to Fig. 17, it is clearly seen that wrap angle θ_1 becomes larger with an increment of κ. It also becomes larger with a decrement of the coefficient of friction μ. Provided κ is small enough, it is noticeable that the self-locking occurs theoretically even with these small coefficients of friction.

4. Novel clutch utilizing self-locking property of belt

Paying attention to the self-locking property of belt as described in the previous section, a novel clutch mechanism can be developed (Imado et al., 2010). Figure 18 shows a simplified three-dimensional image of the novel clutch. Figure 19 shows a cross sectional view of the clutch. Rotational torque is transmitted from the power ring to the inner axis by the belt. In declutching condition, a belt is only rotating with the power ring. Due to the centrifugal force or some restitutive property of belt, the belt is pressed against the internal face of the power ring. To transmit the rotation of power ring to the internal axis, the sleeve on the inner axis is slid along the axis to push the end face of the trigger pin that is attached at the end of the belt and rotating with the power ring. As the sleeve is rotating with the same angular speed of the inner axis, the frictional force to the trigger pin drags the belt so as to coil around the inner axis. The trigger pin works as a synchronizer. As soon as the belt comes in contact with the axis, the belt coils automatically around the axis by the frictional force between the belt and axis. Then due to the self-locking property of belt, the rotation of the power ring is transmitted to the inner axis without any slip as far as self-locking

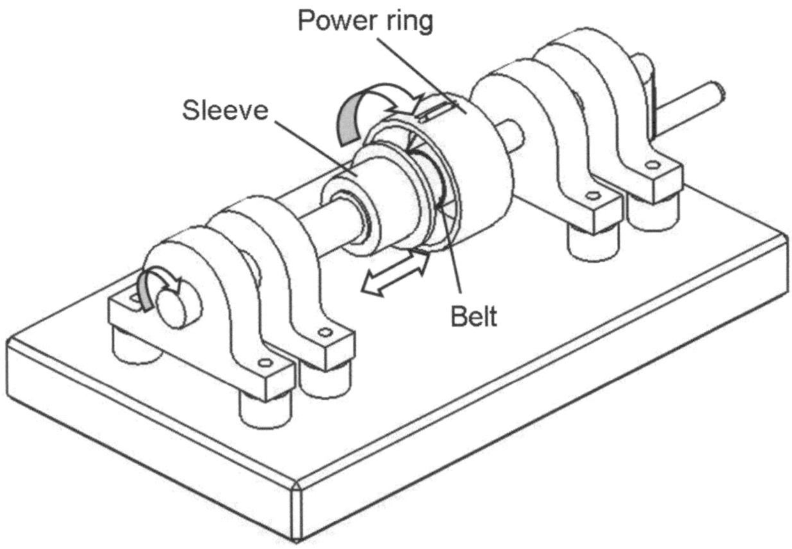

Fig. 18. Three-dimensional image of novel clutch. Rotational torque is transmitted from power ring to inner axis by self-locking belt.

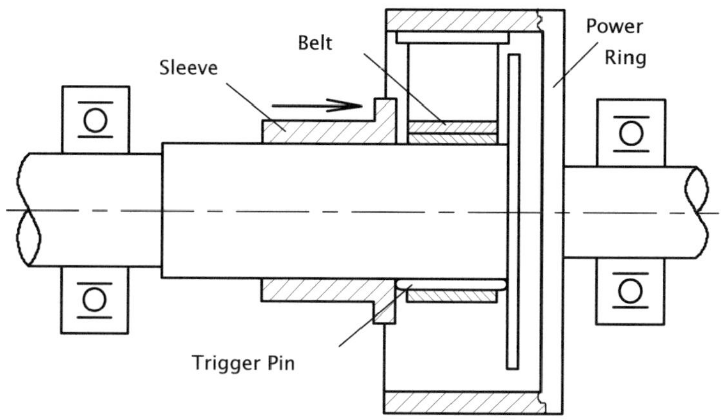

Fig. 19. Cross section of novel clutch

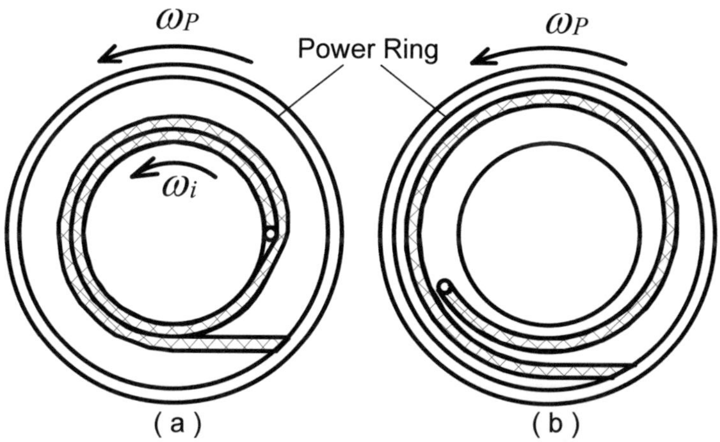

Fig. 20. Frontal views of main part of belt-type clutch in (a) locked-up condition and (b) unlocked condition

condition is satisfied. As long as driving torque is applied, the self-locking state is maintained. Semi-locking state can be realized by adjusting the over-wrapping angle of the belt. When the rotational speed of the power ring becomes smaller than that of the inner axis, the rotation of the inner axis uncoils the belt so that declutching occurs automatically.
Figure 20 shows frontal views of the main part of the clutch in a state of locked-up condition and unlocked condition respectively. From the mechanical point of view, an accurate centering operation is required in assembling individual rotational machine components. Because the torque is transmitted through a flexible belt, this delicate centering operation is not so strictly required for this novel clutch. The belt-type clutch works even in the case where a power ring and an inner axes are either slightly off-centered or inclined with each other.
Figure 21 shows prototype clutch. Brake torque can be applied by the belt brake. It was confirmed experimentally that rotational torque could be transmitted without any slip

Fig. 21. Photograph of belt-type clutch

where there was an eccentricity. A steel belt with 12 mm wide and 0.12 mm in thickness was used in the prototype clutch. In order to reduce the coefficient of friction between belt and belt μ_b, a small amount of grease of molybdenum disulfide, MoS_2, was spread between the belt and belt. Test condition was summarized in Table 1. According to Eq. (43), the critical wrap angle θ_1 of the clutch in Fig. 9 was 105° as shown in Table 1. Considering unsteadiness of the coefficients of friction, two kinds of experiments were carried out. One was of θ_1=90°, the other was of θ_1=120°. Then, self-locking occurred in the case of wrap angle θ_1=120°. On the other hand, self-locking never occurred in the case of θ_1=90°. As far as this experimental result was concerned, the validity of Eq. (43) was verified.

Item	
Rotational speed	60 rpm
Diameter of power ring	89 mm
Diameter of inner axis	30 mm
Width of steel belt	12 mm
Thickness of steel belt	0.12 mm
Coefficient of friction μ	0.25
Coefficient of friction with MoS_2 grease μ_b	0.074
Maximum center offset	8.2 mm
Critical over-wrap angle θ_1	105°

Table 1. Dimensions of clutch and the coefficients of friction

5. Generalization of belt/rope friction formula

The belt formula written in a text, it is usually explained by a figure illustrating a flexible element partially wrapped on a cylindrical surface. But actually there are many kinds of surface. So far, frictional force calculation of a flexible element to these surfaces has not been clearly explained in a text. In this section, the friction of flexible element in the generalized condition is studied. Fige 22 shows a belt wrapped around an arbitrary surface. The equilibrium equation of the force acting to an infinitesimal line element ds is (Hashimoto, 2006)

$$\mu d\theta = \frac{dT}{T} \tag{96}$$

Let denote the curvature and the radius of curvature by κ and ρ respectively. The small wrap angle $d\theta$ can be written as

$$d\theta = \frac{1}{\rho} ds = \kappa ds \tag{97}$$

Substituting Eq. (97) into Eq. (96) gives

$$\mu \kappa ds = \frac{dT}{T} \tag{98}$$

Equation (98) means that the friction of a flexible element on a generalized curve can be evaluated by line integral of the curvature with respect to curvilinear length s.

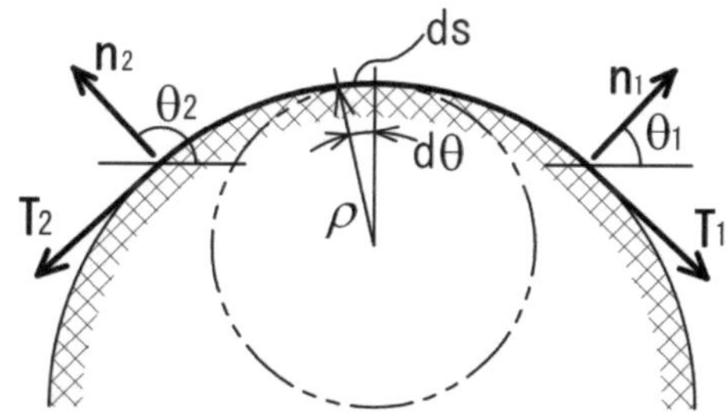

Fig. 22. Flexible element wrapped around body of arbitrary profile

Now, a position vector $\mathbf{r}$ of a curve C in parametric expression with t is

$$\mathbf{r}(t) = x(t)\mathbf{i} + y(t)\mathbf{j} + z(t)\mathbf{k} \tag{99}$$

Differential coefficient with respect to the parameter t is expressed as $\dot{\mathbf{r}}$ or $\dot{x}$. On the other hand, the differential coefficient with respect to curvilinear length s is expressed as $\mathbf{r}'$ or x'. The unit tangential vector $\mathbf{u}$, curvature κ and a line element ds of the curve C are (Yano & Ishihara, 1964)

$$\mathbf{u} = \frac{\dot{\mathbf{r}}}{|\dot{\mathbf{r}}|} = \frac{\dot{x}\mathbf{i} + \dot{y}\mathbf{j} + \dot{z}\mathbf{k}}{\sqrt{\dot{x}^2 + \dot{y}^2 + \dot{z}^2}} \tag{100}$$

$$\kappa = \frac{\sqrt{(\dot{\mathbf{r}} \cdot \dot{\mathbf{r}})(\ddot{\mathbf{r}} \cdot \ddot{\mathbf{r}}) - (\dot{\mathbf{r}} \cdot \ddot{\mathbf{r}})^2}}{(\dot{\mathbf{r}} \cdot \dot{\mathbf{r}})^{3/2}}, \quad ds = |\dot{\mathbf{r}}| dt \tag{101}$$

For a plane curve of z=0 in Eq. (99), substituting Eq. (99) into Eq. (101) gives

$$\kappa = \frac{\dot{x}\ddot{y} - \ddot{x}\dot{y}}{(\dot{x}^2 + \dot{y}^2)^{3/2}}, \quad ds = \sqrt{\dot{x}^2 + \dot{y}^2}\,dt \tag{102}$$

Substituting Eq. (102) into the left side of Eq. (98) gives

$$\mu\kappa ds = \mu \frac{\dot{x}\ddot{y} - \ddot{x}\dot{y}}{\dot{x}^2 + \dot{y}^2} dt \tag{103}$$

The unit principal normal vector **m** of the curve C is given by the formula (Yano & Ishihara, 1964)

$$\mathbf{m} = \frac{\mathbf{u}'}{|\mathbf{u}'|} = \frac{d\mathbf{u}}{dt}\frac{dt}{ds} / |\mathbf{u}'| \tag{104}$$

Making use of Eq. (100) gives

$$\mathbf{u}' = \frac{1}{(\dot{x}^2 + \dot{y}^2)^2} \{\dot{y}(\ddot{x}\dot{y} - \dot{x}\ddot{y})\mathbf{i} + \dot{x}(\dot{x}\ddot{y} - \ddot{x}\dot{y})\mathbf{j}\} \tag{105}$$

$$|\mathbf{u}'| = \frac{\dot{x}\ddot{y} - \ddot{x}\dot{y}}{(\dot{x}^2 + \dot{y}^2)^2} \sqrt{\dot{x}^2 + \dot{y}^2} = \kappa \tag{106}$$

Substituting Eqs. (105) and (106) into Eq. (104) gives

$$\mathbf{m} = \frac{-\dot{y}\,\mathbf{i} + \dot{x}\,\mathbf{j}}{\sqrt{\dot{x}^2 + \dot{y}^2}} \tag{107}$$

Here, the direction of the vector **m** is toward the center of curvature. Then, an outward normal vector **n** can be defined as

$$\mathbf{n} = -\mathbf{m} = \frac{\dot{y}\,\mathbf{i} - \dot{x}\,\mathbf{j}}{\sqrt{\dot{x}^2 + \dot{y}^2}} \tag{108}$$

The direction of the normal vector **n** is denoted by θ

$$\theta = \tan^{-1}\left(\frac{-\dot{x}}{\dot{y}}\right) \tag{109}$$

Differentiating Eq. (109) with respect to t gives

$$\dot{\theta} = \frac{\dot{x}\ddot{y} - \ddot{x}\dot{y}}{\dot{x}^2 + \dot{y}^2} \tag{110}$$

Comparing Eq. (110) with Eq. (103) gives

$$\mu\kappa ds = \mu \frac{\dot{x}\ddot{y} - \ddot{x}\dot{y}}{\dot{x}^2 + \dot{y}^2} dt = \mu\dot{\theta} dt \tag{111}$$

Hence, making use of Eq. (111), integration of Eq. (98) becomes

$$\int \mu \kappa ds = \int \mu \dot{\theta} dt = \mu(\theta_2 - \theta_1) = \log(T_2 / T_1) \tag{112}$$

Equation (112) means that fraction of belt tension is determined by angular difference of the outward normal vectors at the contact boundaries and is unrelated to the intermediate profile. Equation (98) might be applied to the three dimensional problems.

As an example, let's consider a rope spirally wrapped around a cylinder with radius a. The parametric expression of a spiral with parameter t is (Yano & Ishihara, 1964)

$$x = a\cos t, \quad y = a\sin t, \quad z = bt \tag{113}$$

Substituting Eq. (113) into Eqs. (99) and (101) gives

$$\kappa = a / \left(a^2 + b^2\right), \; ds = \sqrt{a^2 + b^2}\, dt \tag{114}$$

Substituting Eq. (114) into Eq. (98) and integrating with respect to the parameter t from t_1 to t_2 gives

$$\frac{T_2}{T_1} = \exp\left\{ \mu \frac{1}{\sqrt{1 + (b/a)^2}} (t_2 - t_1) \right\} \tag{115}$$

The member t_2-t_1 in Eq. (115) is usually a wrap angle for a plane problem. But it is not wrap angle in the three dimensional problem. In the case of b=0, Eq. (115) becomes well known Euler's belt formula. On the other hand, when b becomes infinity, Eq. (115) yields T_1=T_2. Hence, for a three-dimensional problem, the frictional force of a rope is influenced on a way of wrapping. Figure 23 (a) shows some results of calculation of Eq. (115) provided t_1=0 and t_2=$2n\pi$. Tension ratio T_2/T_1 decreases with an increment of the fraction of b/a.

Let's consider another example of a modified spiral defined by Eq. (116).

$$x = a\cos t, \quad y = a\sin t, \quad z = \left(1 - \frac{t}{4n\pi}\right) bt \tag{116}$$

According to Eq. (116), it can be seen that the velocity component in z direction decreases linearly with parameter t and becomes 0 at t=$2n\pi$. The components of Eq. (101) for the curve of Eq. (116) are

$$\dot{\mathbf{r}} \cdot \dot{\mathbf{r}} = a^2 + \left(1 - \frac{t}{2n\pi}\right)^2 b^2, \quad \ddot{\mathbf{r}} \cdot \ddot{\mathbf{r}} = a^2 + \left(\frac{b}{2n\pi}\right)^2, \quad \dot{\mathbf{r}} \cdot \ddot{\mathbf{r}} = \frac{b^2 (t - 2n\pi)}{4n^2\pi^2} \tag{117}$$

Substituting Eq. (117) into Eq. (101) gives

$$\kappa = \frac{a\sqrt{4a^2n^2\pi^2 + b^2\left(1 + 4n^2\pi^2 - 4n\pi t + t^2\right)}}{2n\pi\left\{a^2 + \left(1 - \frac{t}{2n\pi}\right)^2 b^2\right\}^{\frac{3}{2}}}, \quad ds = \sqrt{a^2 + \left(1 - \frac{t}{2n\pi}\right)^2 b^2}\, dt \tag{118}$$

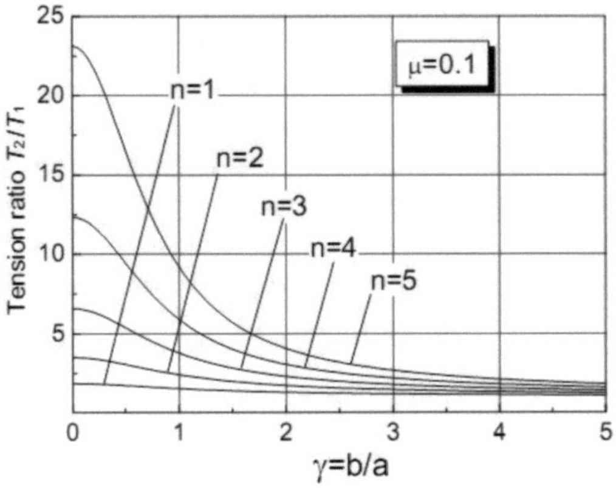

(a) Spiral

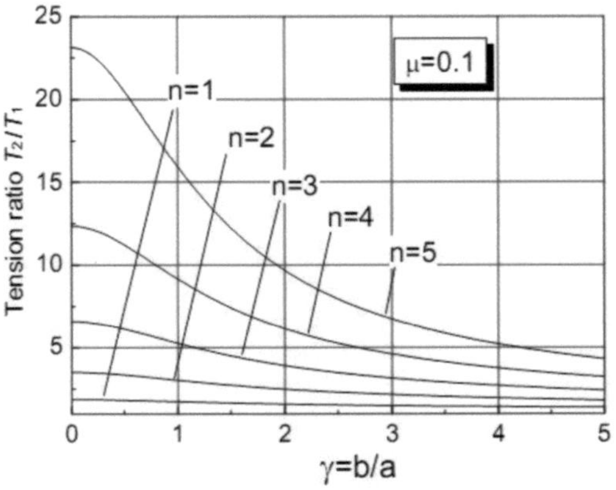

(b) Modified spiral

Fig. 23. Change of tension ratio T_2/T_1 with coefficient ratio b/a of spiral

Substituting Eq. (118) into Eq. (98) and integrating with respect to parameter t from t_1=0 to $t_2=2n\pi$ gives

$$\log\left(\frac{T_2}{T_1}\right)=\mu\left\{\frac{1}{\beta\gamma}\log\left(\frac{\sqrt{1+\beta^2\gamma^2}}{\sqrt{1+\beta^2\gamma^2+\gamma^2}-\gamma}\right)+\tan^{-1}\left(\frac{\beta\gamma^2}{\sqrt{1+\beta^2\gamma^2+\gamma^2}}\right)\right\} \tag{119}$$

where $\gamma=b/a,\quad \beta=1/(2n\pi)$

Considering the case of γ=0, namely b=0 of Eq. (119) requires limiting operation.

$$\lim_{\gamma\to 0}\frac{\mu}{\beta\gamma}\log\left(\frac{\sqrt{1+\beta^2\gamma^2}}{\sqrt{1+\beta^2\gamma^2+\gamma^2}-\gamma}\right)=\frac{\mu}{\beta}=\mu 2n\pi \tag{120}$$

Hence, the result of plane problem is included as a special case of γ=0 in Eq. (119). Figure 23 (b) shows some results of calculation of Eq. (119). The tension ratio of T_2/T_1 in Fig. 23 (b) becomes larger than that of the corresponding value of Fig. 23 (a).

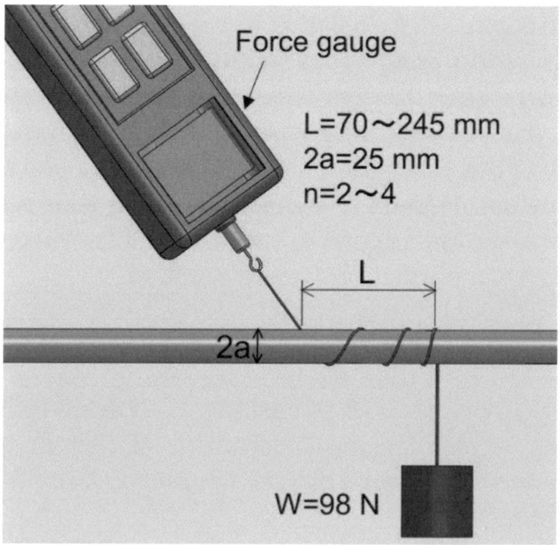

Fig. 24. Experimental method to evaluate the coefficient of friction of string wrapped spirally around cylinder

n	L, mm	b/a	μ
2	70	0.8913	0.143
2	140	1.7825	0.144
3	110	0.9337	0.142
3	250	2.1221	0.134
4	245	1.5597	0.151

Table 2. Summary of experimental result to evaluate the coefficient of friction of string wrapped spirally around cylinder

In order to confirm the validity of Eq. (119), simple experiments were carried out. Figure 24 shows experimental method. The diameter of the pipe was 25 mm. The string of 1.8 mm in diameter was wrapped around the pipe in a way according to Eq. (116) by using a steel scale. The weight of 98 N was hung at the end of the string. The other end of the string was connected to the force gauge that was fixed firmly to the stay. The weight was lifted up by hand at first. Then the weight was released quietly and string tension was measured by the force gauge. As the parameter t in Eq. (116) was taken from t=0 at z=0 to $t=2n\pi$ at $z=L$, the constant b in Eq. (116) can be calculated by $b=L/(n\pi)$. The coefficient of friction was calculated by Eq. (119). Experimental results are summarized in Table 2. Because almost same values were obtained regardless of the test condition, the validity of Eq. (119) was confirmed.

6. Closure

Frictional property of a flexible element was considered in this chapter. The theory of belt buckle has been clarified by considering an effect of over-wrapping of belt on belt friction. Frictional fixation of the belt buckle is caused by self-locking property of belt friction. Self-locking occurs even in the case where a belt is wrapped around an axis two or more times. Two conditions are required to bring about self-locking. One is smaller coefficient of belt-belt friction than that of belt-axis friction. The other is larger wrap angle than the critical wrap angle. Utilizing the self-locking property of belt, a novel one-way clutch was developed. The problem of this clutch is how to get the smaller and stable coefficient of belt-belt friction for long time use. Friction of a flexible element wrapped around a generalized profile was studied. However, the friction of twisted flexible element in a thread, rope and wire has not been clarified yet. Further research is required.

7. References

Hashimoto H., (2006). *Tribology*, Morikita publishing, ISBN 4-627-66591-1, Tokyo

Imado K., (2007). Study of Self-locking Mechanism of Belt Friction, *Proceedings of the STLE/ASME International Joint Tribology Conference*, ISBN 0-7918-3811-0, San Diego October 2007, ASME

Imado K., (2008 a). Study of Belt Buckle, *Proceedings of the JAST Tribology Conference*, pp.139-140, ISSN 0919-6005, Tokyo, May 2008

Imado K., (2008 b). Study of Belt Friction in Over-Wrapped Condition, *Tribology Online*, Vol.3, No.2, pp.76-79, ISSN 1881-2198

Imado K., Tominaga H., et al. (2010). Development of novel clutch utilizing self-locking mechanisms of belt. *Triloboy International*, 43. pp.1127-1131, ISSN 0301-679X

J. A. Williams (1994). *Engineering Tribology*, Oxford University Press, ISBN 0-19-856503-8, New York

Joseph F. Shelley (1990). *Vector Mechanics for Engineers*, McGraw-Hill, ISBN 0-07-056835-9, New York

Yano K. & Ishihara S., (1964). *Vector Analysis*, Shokabo, 3341-01060-3067, Tokyo

13

Surface Friction Properties of Fabrics and Human Skin

Mari Inoue
Graduate School of Human Development and Environment,
Kobe University, Hyogo, 657-8501,
Japan

1. Introduction

We will select and decide to buy our clothes not only by looking at the design and colour of the clothes, but also by handling the cloth. And for the people which their skin has any trouble, the surface friction property of fabrics is important. It is known that the fabric handle judged by hand is affected by the mechanical properties, surface property and the thermal and water transfer properties of the fabrics. The objective evaluation equations are developed by Kawabata and Niwa [1].

Figure 1 shows the factors concerning for the performance of clothing. The factors of the properties of clothing are the structure of clothing and the properties of fabrics. And the factors of the properties of fabrics are the structure of the fabrics and the properties of yarn, and the factors of the properties of yarn are the structure of the yarns and the properties of fiber.

In the objective evaluation equations of hand value, especially, *NUMERI* and *FUKURAMI*, the effects of surface properties is so large. In this study, objectives are to be remarkable about three points. At first, the friction properties of fabrics which differ from the kinds of fiber, yarn counts, and yarn density, secondly, the friction properties of the human skin and next, the friction properties between human skin and the fabrics are experienced.

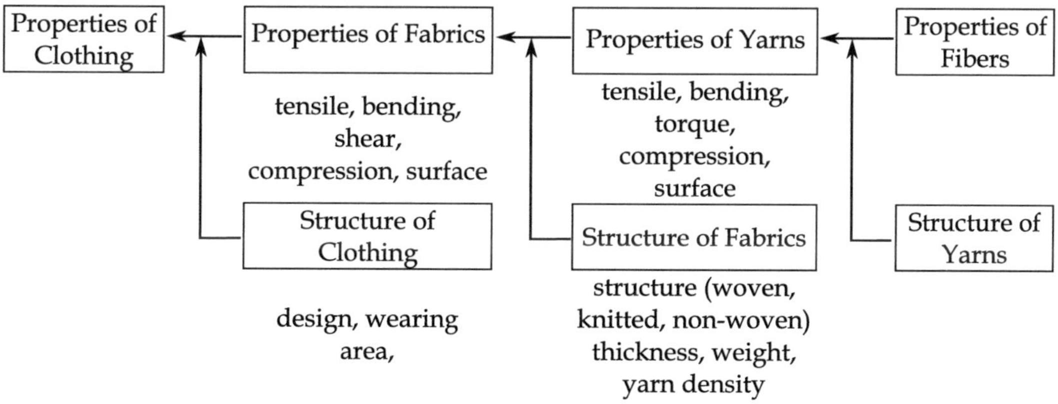

Fig. 1. The factors for properties of clothing

2. Experimental

2.1 Surface friction properties of fabrics

2.1.1 Measuring method

The surface friction properties of fabrics are measured by KES-SE surface friction tester as shown in Figure 2. Figure 3 shows the friction contactor. It consists of the twenty steel wires of which the diameter is 0.5 mm and the fingerprint is simulated. The contact area is 10mm x 10mm, and the contact load is 0.5N. The scan speed of the tester is 1 mm/sec. Measuring characteristics values are coefficients of the surface friction, *MIU* and the standard deviation of *MIU*, *MMD*. This tester is used in all experiments.

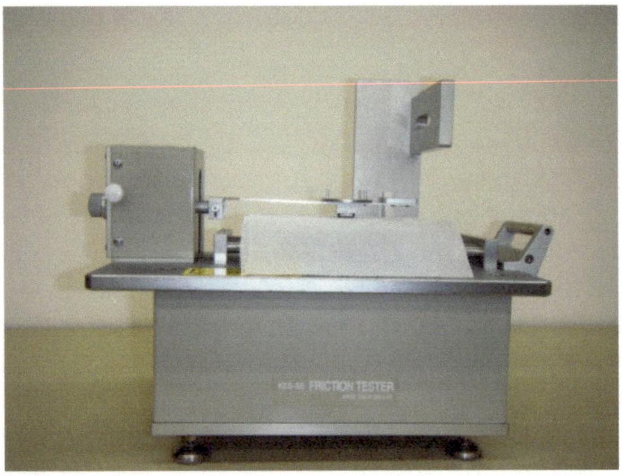

Fig. 2. KES-SE surface friction tester

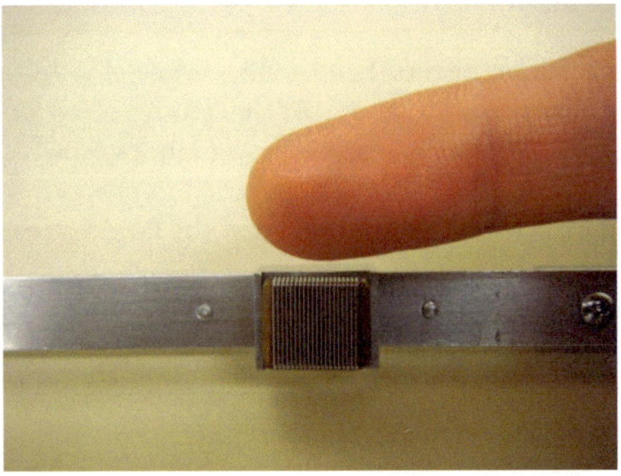

Fig. 3. Friction contactor

2.1.2 Samples

The properties of the fabrics are affected by the yarn properties and the structure of the fabrics. And the yarn properties are affected by the properties of fibers and the structure of the yarns. In these experiments, the samples are composed of different fibers as shown in Table 1. Another samples are shown in Table 2. Yarn counts of these samples are same, but yarn density is different in these groups.

symbol	Fiber		Yarn	Yarn counts tex(=×10^{-5}N/m)	
			structure	warp	weft
SC	Natural	cotton	staple	14.8	14.8
SL	fiber	linen	staple	7.4	7.4
SW		wool	staple	14.1	12.3
SS		silk	staple	8.4	8.4
FN	Synethetic	nylon	filament	7.8	7.8
FP	fiber	polyester	filament	5.6	8.3
SA		acrylic	staple	11.4	11.4

Table 1. Samples for fabric consisted of various fibers

symbol	Fiber	Yarn counts	Yarn density	
		tex(=×10^{-5}N/m)	ends/cm	picks/cm
C1	cotton	14.8	43.0	30.4
C2	(staple)	14.8	34.6	30.0
C3		14.8	43.0	20.2
C4		14.8	33.2	20.0
C5	cotton	7.4	47.0	39.0
C6	(staple)	7.4	46.2	30.0
C7		7.4	33.6	30.4
C8		7.4	45.8	20.4
P1	polyester	16.7	38.7	40.1
P2	(filament)	16.7	37.3	35.5
P3		16.7	36.3	31.7
P4		16.7	36.1	27.5

Table 2. Samples for fabric which are different density

2.2 Surface friction properties of human skin

Surface friction properties, *MIU* and *MMD* of human skin of twenty-six subjects in their twenties are measured by KES-SE. in Figure 2. Figure 4 shows the measurement of human skin and the figure 5 shows the example of the measurement result of the surface friction. And moisture regain of the skin also is measured as shown in figure 6.

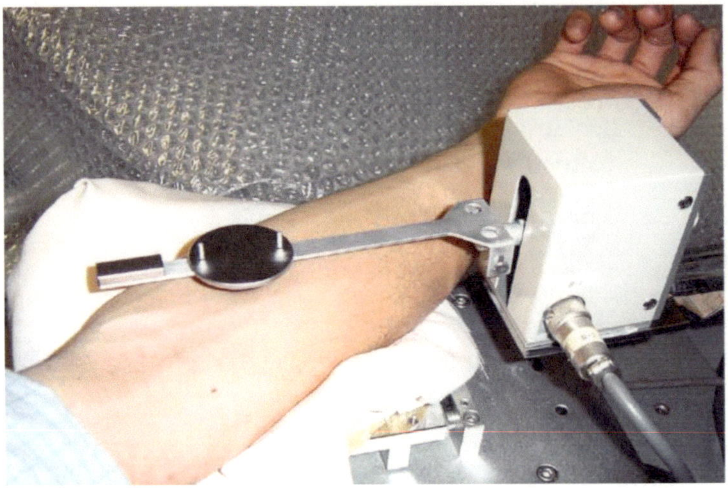

Fig. 4. Measurement of surface friction properties of human skin

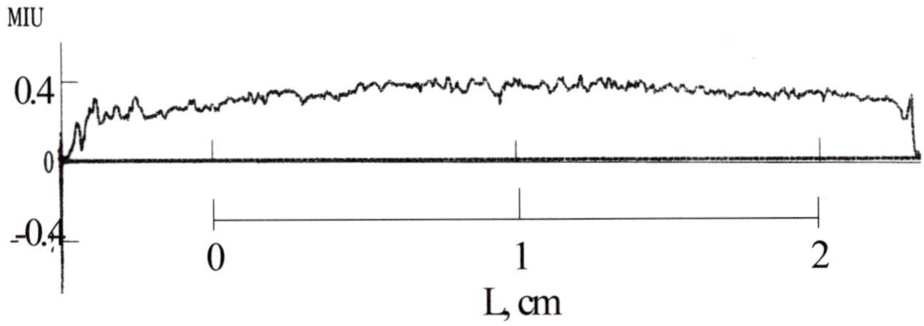

Fig. 5. The example of the measurement result of the surface friction

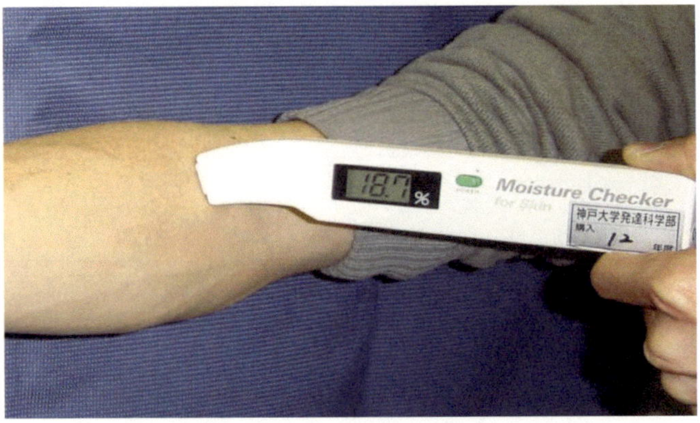

Fig. 6. The measurement of moisture regain of human skin

2.3 Friction properties between Human skin and fabric

Friction properties, that is, coefficients of the surface friction, *MIU* and the standard deviation, *MMD* of human skin of twenty-six subjects in their twenties are measured by KES-SE using contactor with fabrics between Human skin and fabric. Figure 7 shows the contactor.

The mounted fabrics are two knitted fabrics and two woven fabrics. The *MIU* and *MMD* of each fabric are shown in Table 3. *MIU*s of K2 and W2 are larger than K1 and W1, respectively.

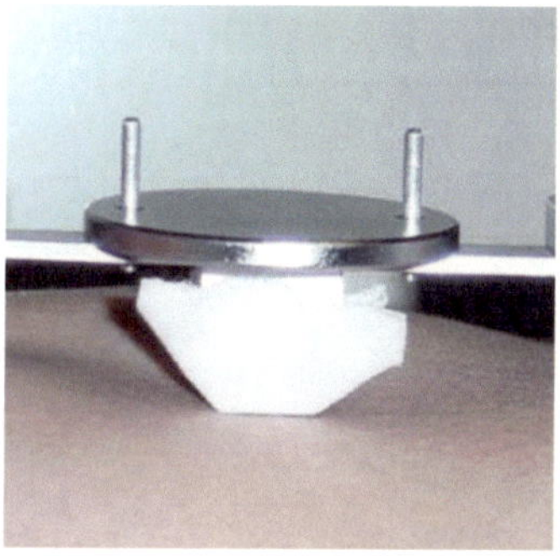

Fig. 7. Surface contactor mounted with fabric

sample	structure	fiber	*MIU* Ave.	*MIU* SD	*MMD* Ave.	*MMD* SD	thickness mm	weight mg/cm^2
K1	rib knitted	cotton 100%	0.163	0.016	0.0070	0.0016	0.78	21.6
K2	plain knitted	cotton 100%	0.273	0.037	0.0115	0.0015	2.41	32.0
W1	plain woven	cotton/PET 50/50%	0.131	0.002	0.0172	0.0051	0.34	11.0
W2	twill woven	cotton100%	0.227	0.007	0.0084	0.0012	1.49	21.3

Table 3. *MIU* and *MMD* of fabrics using friction experiments with human skin

3. Results and discussion

3.1 Surface friction properties of fabrics

Table 4 shows the *MIU* and *MMD* of specimen which is composed of different fiber. *MIU* of sample FN (nylon filament) shows the lowest value and the *MIU* and *MMD* of sample SW (wool staple) show the highest values. The tendency is that *MIU* and *MMD* of filament fiber are lower than staple fiber. But it's not remarkable.

The relationship between product of yarn density in the warp and weft direction and the *MIU* or *MMD* shows in Figure 8. In the case of staple yarn, the tendency is not remarkable, but it is remarkable that the higher density shows the higher *MIU* and *MMD* in the case of filament yarns.

symbol	*MIU* μm	*MMD* μm
SC	0.161	0.0104
SL	0.127	0.0149
SW	0.169	0.0154
SS	0.141	0.0148
FN	0.102	0.0145
FP	0.130	0.0125
SA	0.205	0.0099

Table 4. *MIU* and *MMD* of specimen composed of different fiber

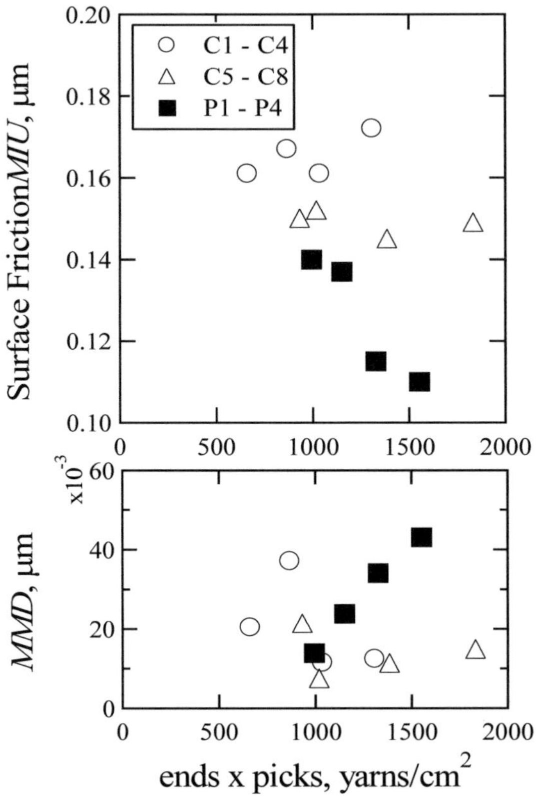

Fig. 8. The relationship between product of yarn density and *MIU* and *MMD*

3.2 Surface friction properties of human skin

Surface friction properties, that is, coefficients of the surface friction, *MIU* and the standard deviation, *MMD* of human skin of twenty-six subjects in their twenties are shown in Table 4. There is no difference between male and female, but there is large difference among individuals because of the large standard deviation.

Figure 9 shows the relationships between moisture regain and *MMD* of all subjects in 25 degree C and 65%RH. It does not show the remarkable tendency, but the it is consider that the larger moisture regain, the larger *MMD* it is.

Figure 10 shows the examples of coefficients of surface friction of skin versus moisture regain of skin in the same person. The coefficients of surface friction have not only the large difference among individuals, but also the difference of moisture regain. Therefore, it is consider that there are the differences between season or rhythm of one day.

	number	*MIU*		*MMD*		Moisture regain, %	
		Ave.	SD	Ave.	SD	Ave.	SD
male	13	0.405	0.220	0.0193	0.0136	32.3	4.5
female	13	0.430	0.144	0.0111	0.0065	29.6	3.2
all	26	0.419	0.187	0.0148	0.0114	30.8	4.2

Table 4. *MIU*, *MMD* and moisture regain of human skin

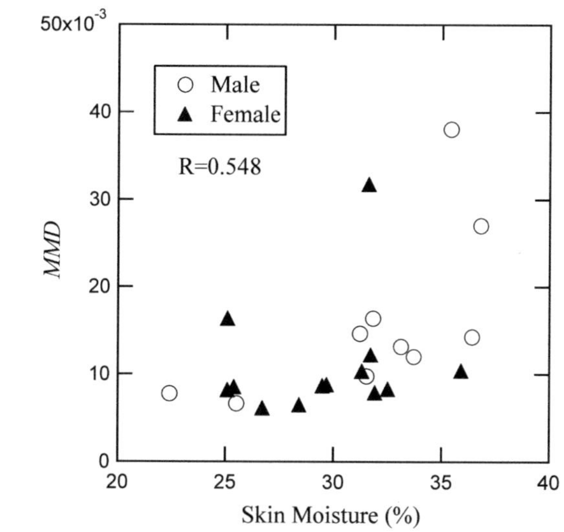

Fig. 9. The relationships between moisture regain and *MMD* of all subjects in 25 degree C and 65%RH

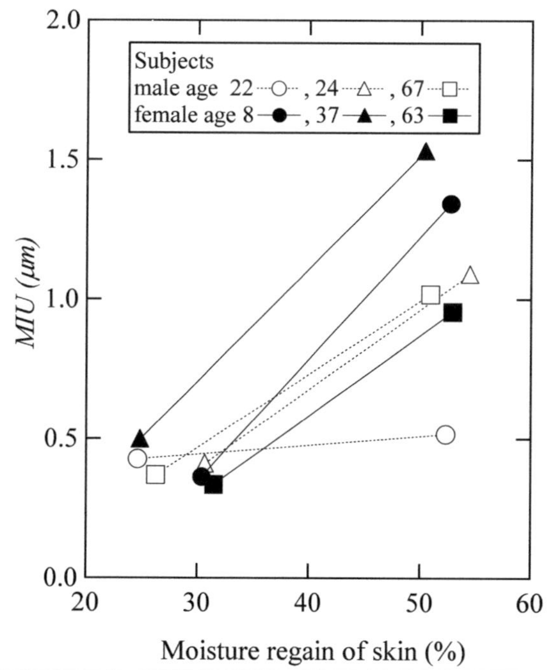

Fig. 10. The relationship between moisture regain and *MIU* of human skin

3.3 Friction properties between Human skin and fabric

Figure 11 shows the examples of *MIU* which the change of *MIU* is the largest one of twenty-six subjects. From these results, it is concluded that the *MIU* between human skin and fabric does not relate to the *MIU* of fabric, but moisture regain of skin.

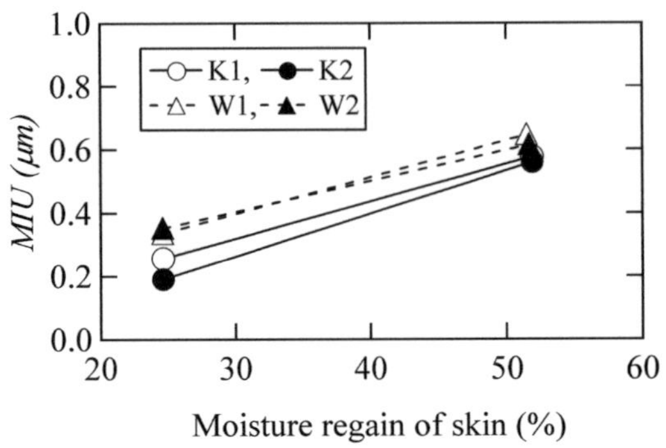

Fig. 11. The relationship between moisture regain and *MIU* of human skin/fabric

4. Conclusion

The hand of fabric used as clothing materials, the surface friction properties of skin and the friction between clothing materials and skin were measured. As the results, the tendency was that *MIU* and *MMD* of filament fiber were lower than staple fiber. And it was remarkable that the higher density showed the higher *MIU* and *MMD* in the case of filament yarns. Friction between human skin and fabrics were measured, and the effects of the moisture regain of human skin and the friction of fabrics were shown from the results. Our group will develop the new apparatus which the width of the part of contactor are wider one at present. On the basis of the results of this study, we would like to develop the apparatus which are close to human sense for friction properties.

5. References

[1] Sueo Kawabata, *"The standardization and analysis of hand evaluation (second edition)"*, The Hand Evaluation and Standardization Committee and The Textile Machinery Society of Japan, 1980
[2] Harumi Morooka and Masako Niwa, *Jpn. Res. Assn. Text. End-uses*, Vol.29, No.11, 486-493, 1988
[3]A.J.P.Martin, *J. Society of Dyers and Colourists*, Vol.60, 325-328, 1944
[4] P.Grosberg, *J. Text. Inst.*, Vol.46, T233-246, 1955
[5] B.Lincoln, *J. Text. Inst.*, Vol.45, T92-107, 1954
[6] H.G.Howell, *J. Text. Inst.*, Vol.44, T359-362, 1953
[7] C.Rubenstein, *J. Text. Inst.*, Vol.49, T13-32, 1958
[8] C.Rubenstein, *J. Text. Inst.*, Vol.49, T181-191, 1958
[9] E.J.Kaliski, *Text. Res. J.*, Vol.28, 325-329, 1958
[10] M. Nakao, *J. Text. Mach Soc. Jpn*, Vol.17, 293-297, 1964
[11] Y. Miura, *J. Seni Gakkai*, Vol.10 558-563, 1954
[12] K. Hirata, M.Yoshida and A.Hanawa, *Jpn, Res. Assn. Text. End-uses*, Vol.15, 47-53, 1974
[13] M.Nakura and N.Imoto, *Jpn, Res. Assn. Text. End-uses*, Vol.18, 74-78, 1977
[14] S.Kobayashi, *Jpn, Res. Assn. Text. End-uses*, Vol.8, 264-270, 1967
[15] S.Kobayashi, *Jpn, Res. Assn. Text. End-uses*, Vol.7, 290-296, 1966
[16] H.L.Roader, *J. Text, Inst*, Vol.44, T247-265, 1953
[17] B.Olofsson and N.Gralen, *Text. Res. J.*, Vol.20, 467-476, 1950
[18] M.Osawa and K.Namiki, J. *Text. Mach. Soc. Jpn*, Vol.19, T7-16, 1966
[19] M.Osawa, K.Namiki and H.Odaka, *J. Text. Mach. Soc. Jpn*, Vol.22, T31-38, 1969

14

Investigation of Road Surface Texture Wavelengths

Chengyi Huang and Shunqi Mei
Department of EME, Wuhan Textile University, Wuhan
P. R. China

1. Introduction

It is generally realized that pavement texture plays a vital role in the development of both pavement friction and tire wear. For the past several decades, pavement texture measurements and modeling analysis have attracted considerable interest of many researchers. Pavement profiles usually present many of the statistical properties of random signals, it is very difficult to distinguish the different surfaces through texture analyses. Based on ASTM E 867, pavement texture can be grouped into two classes micro- and macro-texture in terms of the deviations of pavement surface with characteristic dimensions of wavelength and amplitude. Pavement macrotexture has a substantial influence on the friction between tire and road surfaces, especially at high speeds and in wet pavement conditions. Kokkalis(1998) has shown a relationship between wet pavement accident rate and pavement macrotexture. As expected, the accident rate is reduced as macrotexture increases. Gunaratne et al. (1996) used an electro-mechanical profilometer to record the surface profiles made of asphalt and concrete. The data were later modeled using Auto Regressive (AR) models, where a Fast Fourior Transform (FFT) technique was used to graphically regenerate the pavement surface. Since the order of the models used in these studies was very low (AR(3)), they were able to only model macrotexture and could not capture the characteristics of microtexture. Fülöp et al. (2000) investigated the relationship between International Friction Index (IFI) and skid resistance and between IFI and surface macrotexture. It was found that the macrotexture relates to the hysteresis effects in the tire tread rubber and absorbs some of the kinetic energy of the vehicle. Hence, they concluded that macrotexture has a direct effect on skid resistance.

Today with the advance of measurement technology, by means of a sensor-measured texture meter, profile heights related to both microtexture and macrotexture can be obtained easily. Researchers have focused on the effect of microtexture on friction between the tire and the road surfaces. The investigations by Kokkalis (1998) classified the microtexture and macrotexture as the first and second order of pavement surface irregularities, respectively. Rohde (1976) demonstrated the importance of microtexture pattern as well as its amplitude on the load-carrying capability and the descent time of the tread element. Taneerananon and Yandell (1981) developed a model to simulate a rigid tread element sinking onto a cover of a road surface having microtexture and studied the effect of microtexture roughness on the braking force coefficient. They found that this effect becomes more important when the pavement surface is wet. Persson and Tosatti (2000) presented a comprehensive treatment of the hysteric contribution to the friction for viscoelastic solids sliding on hard substrates with different types of (idealized) surface roughness. They discussed qualitatively how the resulting friction force depends on the nature of the surface roughness. It was found that, when rubber is slowly sliding on the surface, at velocity less than 1cm/s (as in the case to ABS-braking of automotive tires on dry and wet road surface), the rubber will deform and fill out the nanoscale cavities associated with the short-ranged surface roughness and this gives an additional contribution to the sliding friction.

With increase in number of vehicles and increase in speed limits and the subsequent traffic fatalities, tire-road friction estimation has become an important research issue with Department of Transportation (DOT). In particular, researchers have paid more attention to the investigation of elevation road surface texture as a function of Average Daily Traffic (ADT). In the first part of this article, to further understand the features of polishing process on pavement surfaces, experimental texture measurements and Data Dependent Systems (DDS) approach were utilized to model and analyze the elevation profiles collected from polished and unpolished aggregate surfaces of Aggregate Wear Index (AWI) wear track. A key problem in texture measurement was how to determine sampling step sizes so as to reveal the properties of tire polishing process. Three step sizes were adopted to measure the aggregate surfaces. The DDS approach was then used to model and analyze those elevation profiles collected from polished and unpolished AWI wear track surface. It was found that the DDS approach was able to capture both the characteristics of the evolved macrotexture and microtexture and the polishing effect on the aggregate surfaces is found to reduce the microtexture roughness significantly. The second part in this article is to exhibit a texture analysis from several bituminous pavement surfaces obtained from Michigan, USA. Since traffic abrades the pavement surface, exposing aggregates and makes aggregates worn and polished, the polishing properties of coarse aggregates play a significant role in determining skid resistance. Therefore, 1 micron step size scan was used to collect the elevation profile from exposed aggregates and 45 micron step size scan was arranged to collect data from texture surface on each core surface, respectively. DDS approach was utilized to model and analyze the data for both 1 micron and 45 micron step size scans. The characteristics of both microtexture and macrotexture were derived by applying different criteria to DDS modeling analysis and they were correlated to the British Pendulum Tester numbers (BPNs) Laboratory Friction Tester values (LBF) and obtained on the same core. A good correlation was found from some mixed type of pavements.

2. Surface texture measurements

In order to simplify the analyses of road surfaces, aggregate surface textures on AWI wear track were investigated first. Figure 1 shows several polished aggregate on a portion of the AWI wear track obtained from Michigan Department of Transportation (MDOT). Since 1971, MDOT has been using a laboratory wear track to quantify the tendency of individual coarse aggregate sources to polish under the action of traffic (Dewey, et. al., 2001). The wear track consists of a pair of diametrically opposite wheels each attached to a common center pivot point. An electric motor is used to apply a driving force to the wheels through the center pivot point. The aggregate test specimens used on the wear track are trapezoidal in shape. Uniformly graded aggregates are placed in a layer directly against the mold and then covered by portland cement mortar. When 16 of the test specimens are placed end to end, they form a circular path about 2.13 meter in diameter. The surface of the wear track is consisted of limestone aggregate (from Port Inland, MI) of around 10mm size.

Fig. 1. Polished AWI wear track surface

The purpose of this section is to characterize the macrotexture and the microtexture present on both the polished (smooth) and the rough (unpolished or original) AWI aggregate surface. A laser profilometer was used to collect the elevation profiles on the surfaces. The profilometer can scan a 50.8×50.8 mm square area on any given sample surface. However, due to the restriction on the number of data points that can be effectively used in the subsequent DDS analysis, a maximum of 1024 points were collected for each scan length. Therefore, higher resolution scans were used for short scan lengths and vice versa. For example, if one micron step size is adopted to scan a surface, then the maximum scan length allowed is around one millimeter, in which the scan included 1024 data points.

Since the optimum step size for a given pavement is not known a priori, a number of step sizes (from 1 micron in Do, et. al.,(2000) to 20 millimeters in Perera, et. al., (1999)) have been chosen to measure road surface irregularities. Most of the texture measurements were characterized by Mean Texture Depth (MTD) (Gunaratne, et. al., (1996)) or Root Mean Square (RMS) of texture profile (Fülöp, et. al., 2000). Those measurement analyses seemed to have a good relationship with other road surface friction tests. In this paper, three step sizes, 1micron, 30micron and 45micron, were chosen to scan both the smooth and rough aggregate surfaces spanning 1mm, 7mm and 45mm, respectively. Typically, the 1μm and 30μm scans were limited to one aggregate surface and hence can provide the microtexture present on the individual aggregate, whereas the 45μm scans sampled several aggregates and the spaces in between, and therefore, were able to capture the features of both macrotextural and microtextural features of the wear surface. In addition, the 30micron scans can also provide a criterion for distinguishing between polished and unpolished aggregate surfaces for large scan step size. A total 10 scans were collected for each step size, 5 from polished surfaces and 5 from unpolished surfaces. Each scan data was imported into a DDS program so that parameters of the model such as frequency, wavelength, damping ratio and variance contribution could be determined. Comparisons of the model parameters from the polished and unpolished scans can reveal the differences between them.

3. Data Dependent System (DDS) methodology

DDS approach is commonly used for time series analysis of sequentially sampled data. The methodology provides an effective approach to model such series in a statistically optimal manner. The elevation profile collected by the laser profilometer is essentially a uniformly sampled time series or space series data. The DDS modeling of the texture of the aggregate surface is aimed at a complete frequency or wavelength decomposition of the surface.
The DDS approach for modeling the elevation profiles utilizes the Autoregressive Moving Average model, represented as ARMA(2n,2n-1) (Pandit and Wu, 1983) and is given by

$$X_t = \varphi_1 X_{t-1} + \varphi_2 X_{t-2} + \cdots + \varphi_{2n} X_{t-2n} + a_t - \theta_1 a_{t-1} - \theta_2 a_{t-2} - \cdots - \theta_{2n-1} a_{t-2n+1} \tag{1}$$

where the variable X_t denotes the "state" of a system at time t, i.e., the profile height in this analysis. The adequacy of the model implies that a single state X_t completely characterizes the behavior of the system by expressing the dependence of the present state, i.e., the current profile height X_t on past states $X_{t-1}, X_{t-2}, \ldots, X_{t-2n}$. The remainder a_t's are independent or uncorrelated random variables with zero mean and are often called as white noise. The order n of the model is increased until an adequate model is found, which will be explained later. In Eq. (1), the φ_i's are autoregressive parameters.
If the ARMA(2n, 2n-1) model is adequate, the roots λ_i (i=1, 2, 3,…, 2n) can be found from the characteristic equation

$$\lambda^{2n} - \varphi_1 \lambda^{2n-1} - \varphi_2 \lambda^{2n-2} \cdots - \varphi_{2n} = 0 \tag{2}$$

where a real root provides a decaying exponential dynamic mode and a complex conjugate pair of roots provide a decaying (damped or undamped) sinusoidal mode with certain decay rate and frequency or wavelength. Using the backshift operator $BX_t = X_{t-1}$,

$$X_t = \frac{1 - \theta_1 B - \theta_2 B^2 - \cdots - \theta_{2n-1} B^{2n-1}}{(1-\lambda_1 B)(1-\lambda_2 B)\cdots(1-\lambda_{i-1} B)(1-\lambda_i B)(1-\lambda_{i+1} B)\cdots(1-\lambda_{2n} B)} a_t = \sum_{j=0}^{\infty} G_j a_{t-j} \tag{3}$$

where

$$G_j = g_1 \lambda_1^j + g_2 \lambda_2^j + \cdots g_{2n} \lambda_{2n}^j \tag{4}$$

is called as Green's function and the coefficients corresponding to the root λ_i are given by

$$g_i = \frac{\lambda_i^{2n-1} - \theta_1 \lambda_i^{2n-2} - \cdots - \theta_{2n-1}}{(\lambda_i - \lambda_1)(\lambda_i - \lambda_2)\cdots(\lambda_i - \lambda_{i-1})(\lambda_i - \lambda_{i+1})\cdots(\lambda_i - \lambda_{2n})} \qquad i = 1,2,3,\cdots 2n \tag{5}$$

The g_i terms simply scale the magnitude of the response from the ith mode and can also introduce a phase shift when that mode is sinusoidal . To better clarify the role of complex conjugate pairs of roots, each λ_i, λ_i^* and associated g_i, g_i^* can be expressed in the form of

$$g_i \lambda_i^j + g_i^* \lambda_i^{j*} = 2|g_i||\lambda_i|^j \cos(\omega_i j + \beta_i) \tag{6}$$

where the damped frequency ω_i and phase shift β_i come from the root λ_i and the corresponding scaling factor g_i respectively (Pandit and Wu, 2001). The damped frequency can further be expressed in terms of the damping ratio ζ and natural frequency ω_n as

$$\omega_i = \omega_n \sqrt{1-\zeta^2} = \cos^{-1} \frac{\mathrm{Re}(\lambda_i)}{|\lambda_i|} \tag{7}$$

where the damped angular frequency ω_i and the natural frequency ω_n are expressed as angle per sampling interval, and can be converted into cycles per second (Hz) by dividing 2π or can be converted into wavelength by using the constant speed of the profilometer. For a real root, the break or pseudo-frequency defined by the half power point in the spectral domaine.

Once the model has been fitted to the corresponding elevation profile data, the variance can be written in terms of the roots as

$$\gamma_0 = Variance(X_t) = E(X_t^2) = d_1 + d_2 + \cdots + d_{2n} \tag{8}$$

where $$d_i = \sigma_a^2 \sum_{j=1}^{2n} \frac{g_i g_j}{1-\lambda_i \lambda_j}, \qquad i = 1,\ 2, \cdots,\ 2n \tag{9}$$

Thus, the power of a particular root, that is its contribution to the variance γ_0, is represented by the corresponding d_i.

The choice ARMA(2n,2n-1) sequence is mainly based on the configuration of the characteristic roots λ_i. Since the autoregressive parameters φ_i's are always real, the complex roots can occur only in conjugate pairs. For example, for an ARMA(2,1) model, we have

$$(1-\varphi_1 B - \varphi_2 B^2) = (1-\lambda_1 B)(1-\lambda_2 B)$$

$$\lambda_1,\ \lambda_2 = \frac{\varphi_1}{2} \pm \frac{\sqrt{\varphi_1^2 + 4\varphi_2}}{2}$$

and $$\kappa_1 = \lambda_1 + \lambda_2, \qquad \varphi_2 = -\lambda_1 \lambda_2 \tag{10}$$

If $\varphi_1^2 + 4\varphi_2 < 0$, then the roots λ_1 and λ_2 must be a complex conjugate pair. Therefore, if we increase the order by one, allowing odd autoregressive orders, one of the roots will be forced to be real. Another reason is that increasing the autoregressive order in steps of two is more economical than in step of one. One fits only half the number of models compared to the increase by step of one.

Using the above formulation, the experimentally obtained elevation profiles for each scan were modeled. The critical issue in modeling is to identify the correct model order $2n$, that completely captures the trends (or correlations) in the experimental data. To achieve this, the model order is continuously increased until the adequate order of the model is determined based on three criteria (Pandit and Wu, 2001): (1) Verify the independence of the residuals (the a_t's) of the fitted model by using the autocorrelations of the residuals, i.e., the chosen model is deemed to completely charaterize the data if the unified correlations (sample correlation divided by its standard deviation) are less than two which correspond to 95% probability in a normal distribution; (2) Once the data have been characterized completely, the residual sum of squares (RSS) is made as low as possible by introducing an F-test parameter that relates the RSS from the current model order $2n$ to the previous model order ($2n$-1) in the computer program. The F-test parameter value is smaller value than the one from an F-table corresponds to a statistically insignificant reduction in RSS; (3) The adequate model should capture an obviously known physical frequency, such as the one corresponding to the size of aggregate on the surface.

4. Analysis of polished and unpolished aggregate surface profiles

4.1 One micron step size scan

Figures 2a and 2b present two typical elevation profiles collected at 1micron step size from polished and unpolished surfaces of AWI wear track, respectively. Clearly, the vertical scale in these two plots indicates that the magnitudes of the elevations are significantly different in both the data, and hence the variance (averaged square deviation from the mean) is essentially higher for the unpolished surface compared to that on the polished surface. The unpolished scan also appears to have a more complicated profile than the polished scan. This is an important physical characteristic that will be utilized in interpreting the model order in the following DDS analyses.

The data for each scan from the polished and unpolished surfaces was modeled by the DDS program. The starting model for every scan was ARMA(2,1) and the model order was increased in steps of 2, until the adequate model that satisfies the three criteria mentioned above was found. Table 1 and Table 2 present the modeling results for the two scans in Figures 2a and 2b respectively, with adequate models ARMA(12,11) (for 01a polished profile) and ARMA(22,21) (for 011 unpolished profile), respectively. Note that since unpolished scan is generally more complicated than polished one, the adequate model for unpolished scan usually has a higher order compared to that of the polished surface. In these tables, the frequency refers to number of cycles per millimeter. The wavelength is the inverse of this spatial frequency. The damping ratio indicates how well a given wavelength component of the profile repeats at that frequency in the scan. For example, a damping ratio of zero indicates a perfect sinusoidal wave extending for infinite time or length. The maximum damping ratio tending to unity implies that the wavelength component does not repeat at all. In Figure 2a, there exits a dominant peak that shows up at half shape of a wave crest at the end. Generally, the dominant peak has the largest height and the largest wavelength compared to other wave crests or wave troughs, may not repeat in the same elevation profile and will show up as a real root with very large wavelength in DDS analysis. The DDS analysis can capture these features effectively. For 1mm scan, this dominant peak also provides a way to distinguish the difference between polished and unpolished surfaces in the DDS analysis. These dominant peaks are indicated by bold in the Tables. In Table 1, the dominant wavelength is 0.433839mm and the corresponding variance contribution is 2.01E-4 mm^2, which is less than the dominant contribution of 1.42E-3 mm^2 from the unpolished scan in Table 2. All other wavelengths given in these tables are significantly smaller with low variance contribution and typically have much smaller damping ratio indicating that these wavelengths repeat over a long period time. Thus, the 1mm scans capture the microntextural features effectively.

Table 3 presents the modeling results from 10 scans. It is clear that both the variances and the dominant variance contributions for unpolished surfaces are consistently larger compared to those of the polished surfaces. Comparison of the dominant wavelengths

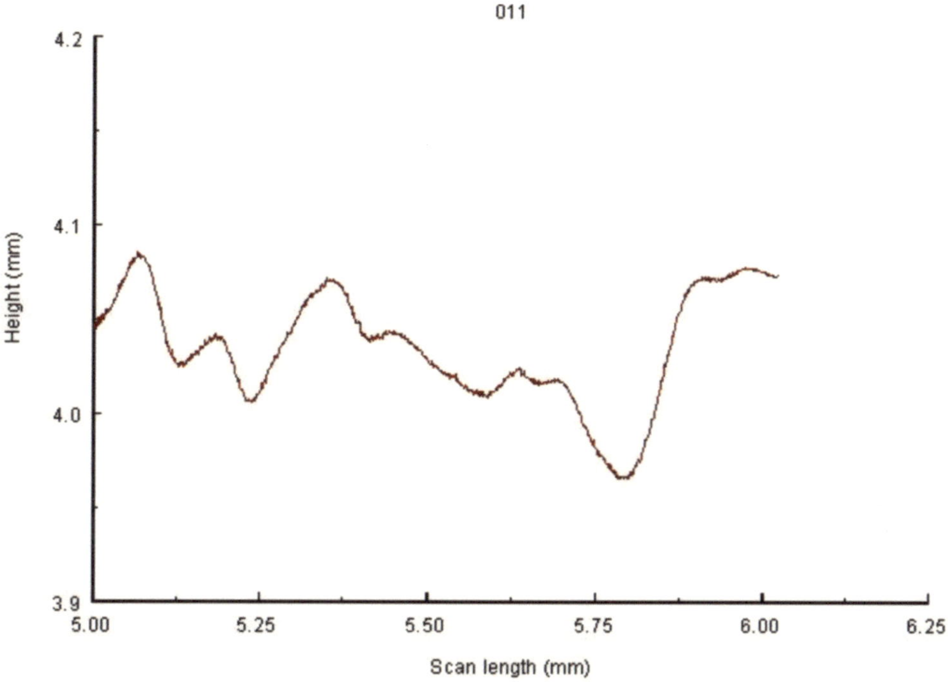

Fig. 2a. 1 micron scan from polished aggregate surface

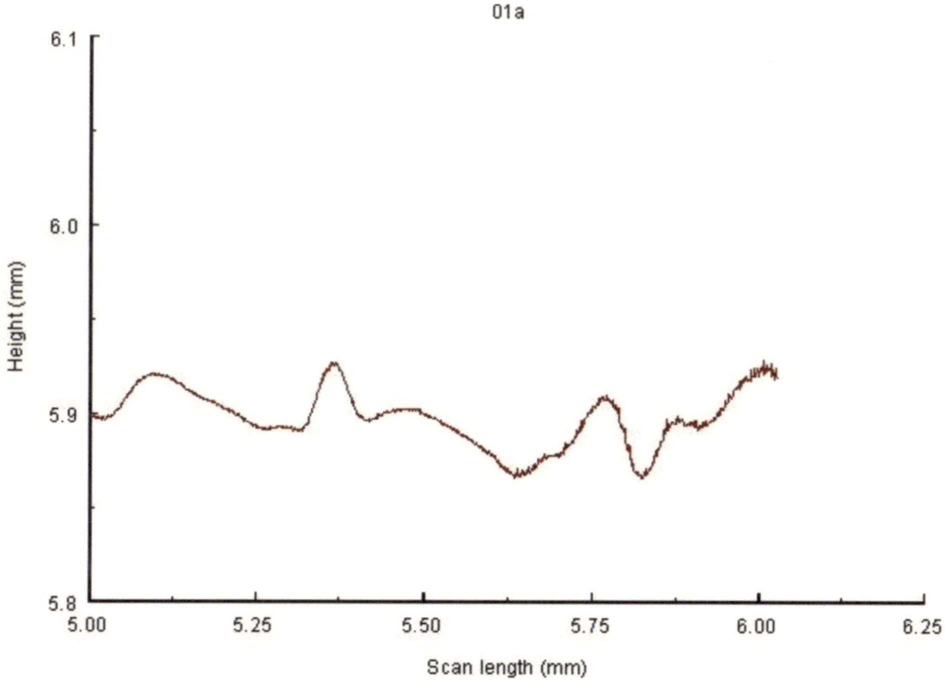

Fig. 2b. 1 micron scan from unpolished aggregate surface

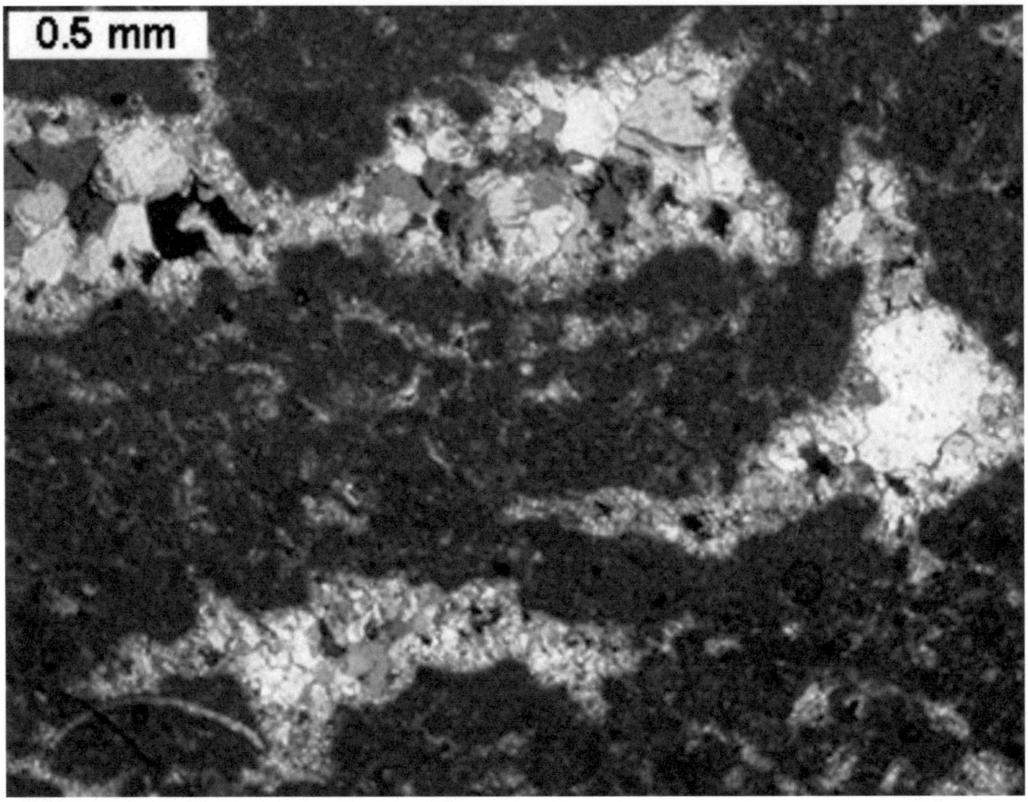

Fig. 3. Microstructure of Port Inland aggregate

obtained from the polished and unpolished surfaces reveals that these wavelengths do not present any trend implying that the tire polishing did not change the dominant wavelengths but only the overall variance contribution. Further analysis of Port Inland aggregate reveals that the dominant wavelengths have a strong relationship with the grain size of the aggregate. The grain size of the Port Inland aggregate used in this AWI wear track is usually in the range of 100micron to 500micron and this range seems to agree well with the dominant wavelengths in Table 3. Figure 3 presents the microstructure of Port Inland aggregate (The dark area is composed of algae lumps and they have a very fine grain size around of 2-10 micron, the white area is composed of calcite crystals with grain size around 20-500 micron). It is worth mentioning that for other shorter wavelengths, no trend can be found from the corresponding contributions when comparing between polished and unpolished aggregate surfaces is made. Therefore, the dominant wavelengths are the minimum wavelengths that can provide a trend between the polished and the unpolished aggregate surfaces in 1mm scans.

4.2 Thirty micron step size scan

In order to explore the effect of other larger wavelengths on the roughness of aggregate surfaces, 30 micron step size scans were collected from smooth and rough aggregate surfaces (5 scans from polished, 5 scans from unpolished). Because of the limited size of each aggregate, the 30 micron step size scans were composed of only 234 sampling points

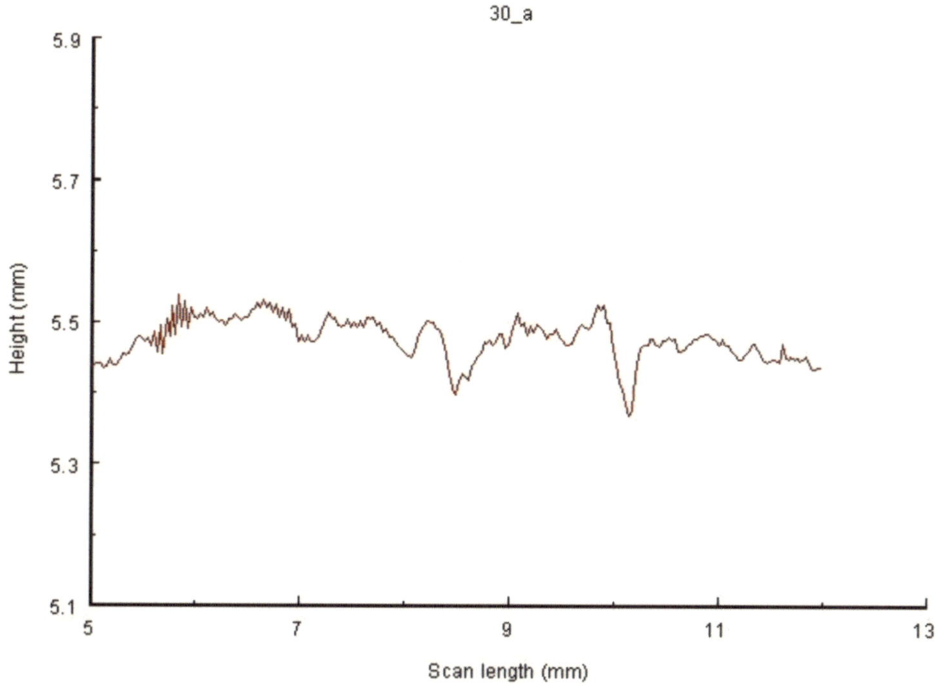

Fig. 4a. 30 micron scan from polished aggregate surface

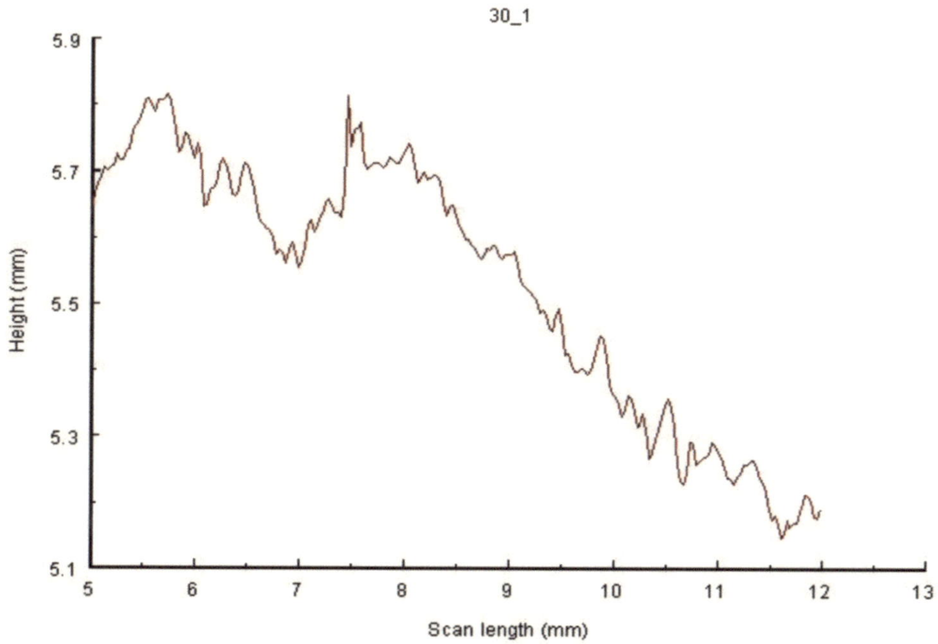

Fig. 4b. 30 micron scan from unpolished aggregate surface

extending over 7.02mm in length and were limited to one aggregate surface. Figures 4a and 4b present two typical elevation profiles collected from polished and unpolished surfaces. Similar to the 1micron step size scans, the vertical scale in these two plots indicates that the magnitudes of the elevations are significantly different and hence the variance of unpolished scan is essentially higher than that of the typical polished scan. However, since aggregate surface is not a flat plane naturally, even on the fully polished surface there still exist some significant irregularities where the tire rubber neither could contact aggregate surface entirely nor polish to reduce the height of irregularities. Figure 4a presents the situation where there are two large wave troughs that are hardly affected by the tire polishing action. These kinds of troughs may become the dominant wavelengths and affect the variance of elevation profiles significantly.

Another feature that affects the value of variance contributions may come from a slope in an elevation profile. In Table 4 and 5, the values of variance seem to present a trend between polished and unpolished surfaces. However, in Table 5, there are two real roots, one at 1.293mm and the others at 82.44mm wavelength. The dominant wavelength of 82.44mm is 11.7 times the scan length. This indicates that there is a slope in the overall elevation profile (Figure 4b) which has been effectively captured by the DDS program. Even in a 1mm scan, if a slope exits for the overall scan, a dominant root will reflect that feature. However, by carefully selecting a small flat region, likelihood of such a slope in a 1mm scan can be minimized. This can not be done in a 30mm scan, therefore, the value of variance does not present a criterion to distinguish the polished and unpolished surfaces. This can also be confirmed by the variance values in Table 6, where the variances are of comparable order between polished and unpolished aggregate surfaces. Therefore, the criterion used in 1micron step size scans to distinguish the difference between rough and smooth scans is not applicable to 30micron step size scans.

Usually, if the damping ratio is less than 10%, Eq. (7) shows that the frequency is nearly undamped and the component repeats regularly. Its contribution to the variance is given by Eq.(8) as d_i for a real root λ_i and $d_i + d_{i+1}$ for a complex conjugate pair λ_i, λ_{i+1}. Since a 30micron step size scan is 7 times the scan length of 1micron step size, it is possible that the dominant wavelength in 1micron step size scan may also appear several times in 30micron step size scans. Therefore, a new criterion based on damping ratio is introduced in 30micron step size scan analyses, i. e., contributions from all the wavelengths that have a damping ratio less than 10%, (see the column "criterion" in Table 4 and 5) are summed up to obtain a partial contribution (see the column "partial" in Table 4 and 5). Table 6 presents those partial contributions obtained from 10 such scans as well as the variation ranges of corresponding wavelengths. Since most of the wavelengths are less than 0.5mm, the texture can be depicted as 'microtexture' and the associated partial contributions physically describe the averaged squared microtexture roughness. In Table 6, there are two 'negative' partial contributions; the negative signs imply that these contributions have phase opposite to those with positive contribution. Comparison of the partial contributions between the polished and unpolished surfaces clearly indicates that polished aggregate line scans have significantly lower 'partial contributions' than those on the unpolished scans. This means that polishing wears away the micro-roughness present on the original aggregate surfaces. It is also interesting to note that the microtexture wavelengths satisfying the damping ratio criterion in 30micron step size are close to the dominant wavelengths in 1micron step size. Thus, both 1micron and 30micron step size scans capture the microtextural features effectively. Another interesting feature is that the microtexture wavelengths in the polished scan have a much smaller range and smaller wavelengths than in the unpolished scans. This means that the polishing effect is to either chip away or break the larger grains from the original surface due to the traffic.

4.3 Forty-five micron step size scan

In the above 1micron and 30micron step size analyses, the characteristics related to macrotexture have not been found. The reason is that 1micron or 30micron scan lengths are too short and span only one aggregate. Therefore, 45micron step size scan was adopted which spans over 45mm length encompassing several aggregates. Figure 5a and 5b present two typical elevation profiles collected at 45micron step size from polished and unpolished surfaces, respectively. It is clear that there are several profiles of aggregate in the two plots and they all have an average size of around 10 millimeters. Comparison of two elevation profiles reveals that the scans obtained from unpolished aggregate surfaces appear to be rougher than those from the polished surfaces. Also, due to the inherent irregularities present on any aggregate surface, they do not get polished uniformly. Hence every scan includes some portion of unpolished surface. This will increase the complexity of DDS model analysis.

It is required to choose every scan line carefully so as to reduce the rough portions included in any given scan on a polished surface.

Table 7 and Table 8 present the modeling results corresponding to Figure 5a and 5b, respectively. Note that the model order is significantly smaller for polished scans compared to that of unpolished scans. The dominant contributions correspond to the largest wavelengths in these tables are 10.4 mm and 6.64 mm, respectively resulting from real roots.

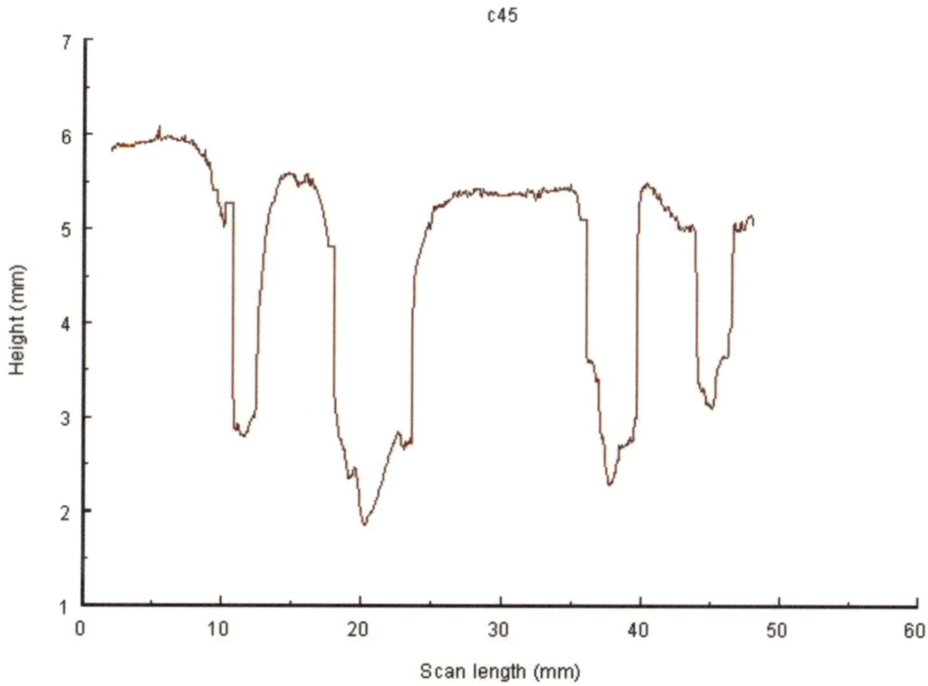

Fig. 5a. 45 micron scan from polished aggregate surface

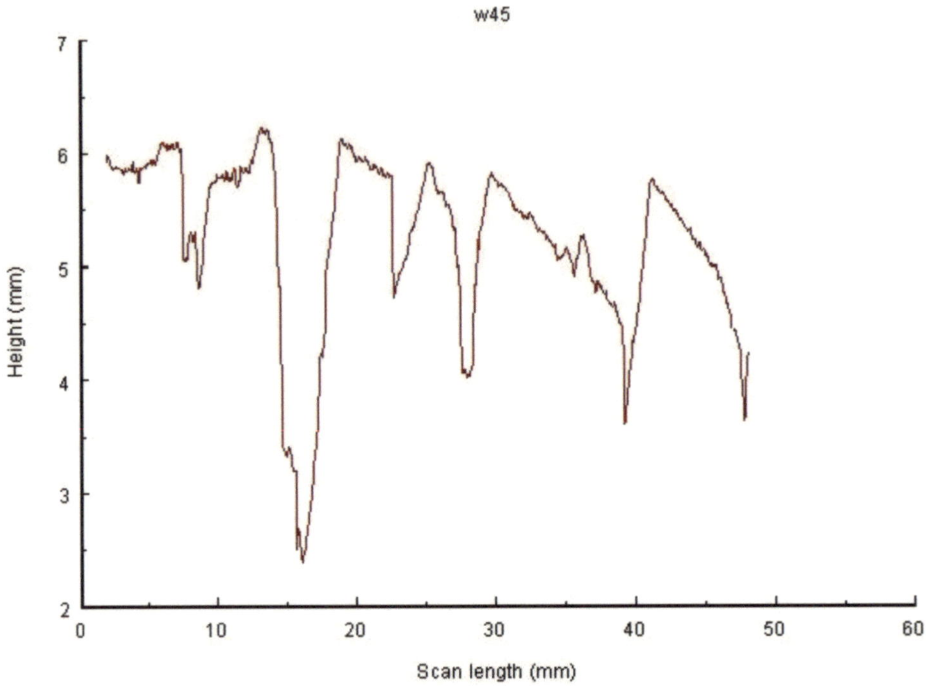

Fig. 5b. 45 micron scan from unpolished aggregate surface

These two wavelengths indicate physically the average size of the aggregates appearing in Figure 5. Considering the two tables, except for the largest wavelengths, the other wavelengths appear to be in the range of microtexture varying from 0.1mm to 0.5mm. It is once again verified that polishing the aggregate surface dose not change the wavelengths in the microtexture. To determinate contributions, the same criterion based on damping ratio as depicted in 30micron step size scans was used to distinguish the contributions between smooth and rough aggregate surfaces. It can be found from the tables that, when the value of 'damping ratio' is less than 0.1, the corresponding wavelengths are always less than 0.5mm, and these wavelengths agree well with the dominant wavelengths in 1micron step size scans or the grain size on aggregate surfaces. Therefore the characteristics of the microtexture can be captured by prescribing the small damping ratio criterion and the corresponding contribution describes the microtexture roughness. Comparing the two tables, it is clear that the polished surface has a smaller partial contribution (of the order of around 10^{-7}) and the unpolished one has a higher partial contribution (of the order of around 10^{-4}). Therefore, it appears that polishing reduces the microtexture roughness, but does not change the microtexture wavelengths significantly.

Table 9 summarizes the modeling results from 10 scans. It is clear that the variance does not show any trend between polished and unpolished surfaces. This is expected because of the large contributions that arise from the aggregate surfaces and spaces (troughs) in between the aggregates in addition to any contribution that may arise from any slope of the overall surface. These features completely mask the minor contributions arising from the polished regions. However, the partial contributions that are based on 10% damping ratio criterion reveal a clear trend between polished and unpolished surfaces, where the unpolished surface scans have contributions of the order 10^{-3} to 10^{-4} and the polished surface scans have contributions of order 10^{-4} to 10^{-7}. However, the range of microtexture wavelengths as well as their overall values is significantly smaller on the polished surfaces indicating that the polishing effect due to traffic is to either chip away the large particles (or grains) or break them into smaller fragments.

As mentioned in the introduction, most of the previous publications in the literatures have focused on obtaining a single number either in terms of MTD or RMS to distinguish between polished and unpolished surfaces. Although, such a size single number is desirable for simplicity, it can not provide a more in-depth description of the micro- as well as macro-textural features that are characteristics of any polishing process. Moreover, comprehensive descriptions of the evolving wavelengths are desirable for more precise correlations with any other laboratory of field measurements such as British Pendulum Number (BPN) or field friction number (FFN) on pavements. Characterization of the evolving micro-texture is also essential for identification of the wear mechanisms on a given aggregate surface due to the type or the level traffic, type of mix-designs and age of pavement, the weather conditions over a period of time or a combination of the above.

However, the aim of the current paper is to introduce a methodology that can be effectively used on aggregate surfaces to capture the induced micro- as well as the macro-textural features due to tire polishing. Similar method can be easily extended to analyze the induced the topological features on the on the pavement surfaces. Pavement surface texture can be thought of as a combination of microtexture and macrotexture. Friction between tire and road surface is strongly dependent on surface texture. Microtexture, in particular, plays a significant role in this interaction. In the previous sections, experimental texture measurements and DDS modeling methodology were introduced to analyze the polished and unpolished aggregate surfaces on AWI wear track. The elevation profiles collected from AWI wear track surface by a laser sensor were modeled and analyzed by use DDS methodology. Comparison of the modeling results leads to the following conclusions which are applicable to road pavement surface analyses.

1. Microtexture with a wavelength in the range of 0.1~0.5mm (which corresponds to the grain size of Port Inland aggregate) has a significant effect on the roughness of the aggregate surfaces. All the DDS modeling results for 1micron, 30micron and 45micron step size scans present the difference of microtexture between polished and unpolished aggregate surfaces.
2. For different step size scans, in order to obtain a consistent modeling results, different criteria have been adopted. For short scans, the dominant wavelength should be considered (when there is no overall slope in the elevation profile), but for longer scans the damping ratio criterion should be used.
3. Compared with other step size scans, 45 micron step size scan can capture both the macrotexture and the microtexture characteristics and can provide more texture information than other two scan sizes.
4. Since the unpolished surface is much more complex than the polished one, the modeling of the former surface requires a higher order of ARMA model than the latter.
5. The damping ratio criterion used in this paper is useful in microtexture analysis, especially for larger scan lengths. It will provide with the microtexture information, such as wavelengths and roughness resulting from DDS analysis.
6. Based on DDS analysis, the polishing effect due to traffic is to reduce the microtexture roughness of aggregate surfaces as well as the range of microtexture.

5. Investigation of road surface texture wavelengths

Pavement surface characteristics and traffic conditions are generally recognized as the major contributors to pavement friction. Statistic on pavement accidents indicates that each year approximately 15% of accidents that result in injury or fatality occurred during wet weather conditions. Some of these accidents resulted from loss of friction at the tire-pavement interface. The tire-pavement interaction has two contributions arising from adhesion and hysteric components. The adhesion term is interpreted as a thermally-activated molecular stick-slip action, which is described by a stationary stochastic process. The stationary stochastic process consists in the formation and breakage of adhesive linking chains that bind the rubber body to the textured surface (Schallamach, 1963; Chernyak and Leonow, 1986). The hysteric component results from the internal friction and subsequent dissipation of energy during cyclic deformation of sliding rubber arising from the asperities of rough substrate which exerting oscillating forces on the rubber surface. Kummer (1966) proposed a model for rubber friction that considered the above two components of the friction: the adhesion component based on microtexture and the hysteresis component based on the macrotexture. Persson (1998) qualitatively presented that the hysteric contribution was associated with the long wavelength roughness of substrate. The interaction between tire and substrate deformed the rubber so that it could 'follow' the short wavelength roughness of the substrate when the rubber slid at a low velocity. This deformation would provide an additional adhesion contribution to the friction. However, recent investigations indicated that for different type of tires, for example smooth and ribbed tires which are widely adopted for friction measurements on road surface, the friction measured on the same road surface will be significantly different. Generally the ribbed tire will present a higher friction and the smooth tire will result in a lower friction. Yandell and Sawyer (1994) pointed out that there was seldom agreement between any two different devices that measured friction on the same surfaces. For example, a runway friction tester and a pavement friction tester yielded an R-squared value of 0.02 for readings on about 25 wet open-graded textured roads using ribbed tires and an R-squared value of 0.75 on a large variety of wet asphalt surfaces using a smooth tire. Whitehurst (1978) also showed a 30 percent variation among seven different ASTM skid trailers reading identical surfaces. There are many reasons for this poor agreement, among which are the vagaries of tread rubber behavior. Due to the complexity of analysis in the tire characteristics of every vehicle to determine the friction between tire and road surface, it is imperative that pavement surfaces should be designed and constructed to provide adequate friction to minimize the accident rate as a result of frictional deficiencies.

Pavement texture is a feature of the road surface that ultimately determines most tire/road interactions, including wet friction, noise, splash and spray, rolling resistance and tire wear. The characteristics of pavement texture that affect tire and pavement interactions are arbitrarily categorized as microtexture, consisting of wavelengths (characteristic dimensions) of 1 μm to 0.5 mm, and macrotexture, consisting of wavelengths of 0.5 mm to 50 mm. It has been demonstrated that at low slip speeds the effect of microtexture dominates the friction measurement, whereas at high slip speeds the effect of macrotexture becomes important. Therefore, if both microtexture and macrotexture are maintained at high levels, they can provide sufficient resistance to skidding. A recent European study reports that increased macrotexture reduces total accidents, under both wet and dry conditions. Kokkalis (1998) also presented a relationship between wet pavement accident rate and pavement macrotexture. As expected, when macrotexture increases, the accident rate is reduced. Fülöp et al. (2000) investigated the relationship between International Friction Index (IFI) and skid resistance and the relationship between IFI and surface macrotexture. They developed a relation that the IFI threshold value of friction is a function of the macrotexture parameter. The theoretical estimations corresponded well to their experimental results.

On the other hand, more and more researchers believe that microtexture has a substantial effect on skid resistance and various methods are proposed to evaluate microtexture. Forster (1994) conducted a study to investigate the quantitative role played by small-scale surface texture (microtexture) in determining the skid resistance of a pavement. A non contact image analysis system was used to measure the microtexture profiles on a series of pavement cores. The measurements of microtexture were correlated to the British Portable Tester numbers (BPNs) obtained on the same cores. A linear regression fit of these data based on 87 cores yielded a correlation coefficient (R-squared value) of 0.68. Do et. al. (2000) adopted the ideas from Fahl (1982) who emphasized that large profile peaks and valleys play an important role in functional applications and developed a 'theta angle' to measure the microtexture of road surface. The 'theta angle' was derived from the two consecutive peaks and the horizontal between every segment on road surface profile. The theta angle distribution was used to characterize the microtexture roughness. Correlation between theta values and friction gave a correlation coefficient R-squared value of 0.8 for 24 data points. Rohde (1976) demonstrated the importance of microtexture pattern as well as its amplitude on the load-carrying capability and the descent time of the tread element. Persson and Tosatti (2000) presented a comprehensive treatment of the hysteric contribution to the friction for viscoelastic solids sliding on hard substrates with different types of (idealized) surface roughness. They found that, when rubber is slowly sliding on the surface, at velocity less than 1cm/s (as in the case to ABS-braking of automotive tires on dry and wet road surface), the rubber will deform and fill out the nanoscale cavities associated with the short-ranged surface roughness and this gives an additional contribution to the sliding friction. Huang (2010) utilized Data Dependent Systems (DDS) approach to model and analyze the elevation profiles collected from polished and unpolished aggregate surfaces of Aggregate Wear Index (AWI) wear track. It was revealed that the microtexture roughness of aggregate surfaces was influenced significantly by tire polishing effect and the DDS approach was able to capture both the characteristics of microtexture and macrotexture.

Due to the vital role of pavement texture in both pavement friction and tire wear, Michigan Department of Transportation (MDOT) initiated a research program of pavement texture analysis in the Center of Transportation Materials Research at Michigan Technological University. As the second part of this research program, the current work presents a texture analysis from several bituminous pavement surfaces obtained from Michigan. A total 212 road surface cores from 29 sites on suburban and rural lanes were obtained. From each site, samples of 6 inch diameter were cored from shoulder, both wheel paths and between wheel paths. A laser profilometer was used to collect elevation profiles on each core. The Data Dependent Systems (DDS) methodology (Pandit, 1991) was introduced to model and analyze the elevation profiles. Similar to the previous investigations on the aggregate surfaces of AWI wear track, 1 micron and 45 micron step sizes were chosen to measure the texture. A total of 1,024 readings were taken per individual line scan along the traffic direction. For bituminous pavement, skid resistance gradually decreases by the polishing action of traffic. A generally accepted explanation concerning the reduction process is that traffic abrades the pavement surface, exposing the aggregate and makes aggregate worn and polished which, in turn, reduces skid resistance. Therefore, polishing properties of coarse aggregates play a significant role in determining skid resistance. On the other hand, the 4th section in this article utilized the DDS methodology analyzing the tire polishing characteristics of aggregate surface on AWI wear track and found that the tire polishing process will mainly affect the microtexture of aggregate surfaces. Based on the above analyses, 1 micron step size scan was used to collect the elevation profile from exposed aggregate on each core surface. To collect the data randomly on any given core, a five by five grid of points was applied to an area of 25 cm^2 on each core surface. This eliminates any bias in the location of region from where the data was collected. The 1 mm long scan was collected closest to each grid point and a total 25 scans were obtained on each core. For the 45 micron step size, since DDS modeling can capture both the characteristics of macro- and micro-texture, the texture data collection included a series of twelve - 46 millimeter long linear traverses collected from an area of 100 cm^2 on each core surface for a total traverse length of approximately 552 mm. The characteristics of both microtexture and macrotexture reflected surface feature of pavement core were correlated to the British Pendulum Tester numbers (BPNs) and the Laboratory Friction Tester values (LBF) obtained on the same core.

6. Data analysis and correlation of DDS analysis with BPN value

Based on the DDS theory, the collected data from core surfaces were analyzed using DDS program. Since it was difficult to find an apparently physical parameter as an additional criterion to determine an adequate model from most of core surfaces, only the independence of the residuals (the a_t 's) and the F-test criterion were applied to find an adequate model for each scan.

6.1 Modeling analyses of 1 micron step size scans

Microtexture is a measure of the degree of polishing of a pavement surface or of the aggregate at the surface. It has been realized that microtexture is dependent primarily upon the aggregate petrographic characteristics and the traffic intensity (expressed in terms of commercial-vehicle flows). It is clear that at different stages of a pavement's service life the aggregate surface presents different texture profiles. For example, in the early stage of a pavement service, the influence of an aggregate's polishing characteristics on skid resistance is thought to be minimal because the bituminous matrix has not been worn sufficiently to expose the aggregate. But in the later stage of the pavement service, aggregate surface texture is essential to sustain the skid resistance of the pavement. Therefore, the elevation profiles of aggregate surface may, to some extent, represent the variation of road surface friction and the 1 micron step size scan is adopted to correlate the aggregate surface texture with pavement surface friction. In the this paper, the DDS analysis of 1 micron step size scan is correlated with the British Pendulum Tester numbers (BPN) and Laboratory Friction Tester values (LBF) obtained from the same pavement core, respectively.

The BPN has been wildly used in laboratory and field testing of frictional properties of surfaces. The BPN value measured by the tester represents the amount of kinetic energy lost when a rubber slider attached to the end of a pendulum arm is propelled over a test surface. Many researchers have proved that BPN value has a good correlation with the microtexture measurements. The laboratory friction tester (LBF) was built in a laboratory of Michigan Department of Transportation (MDOT). It is a companion friction tester for the circular

Aggregate Wear Index (AWI) wear track. It can measure an initial peak drag force as a test tire comes into contact with specimen at a special velocity. More details about LBF tests can be found in the paper of Dewey, et al. (2002). The BPN and LBF data were obtained in house by MDOT from all the 212 road surface cores. Figure 6 presents a correlation between BPN and LBF. A very good correlation can be found between BPN and LBF test values (The coefficient of correlation (R^2) is 0.9081).

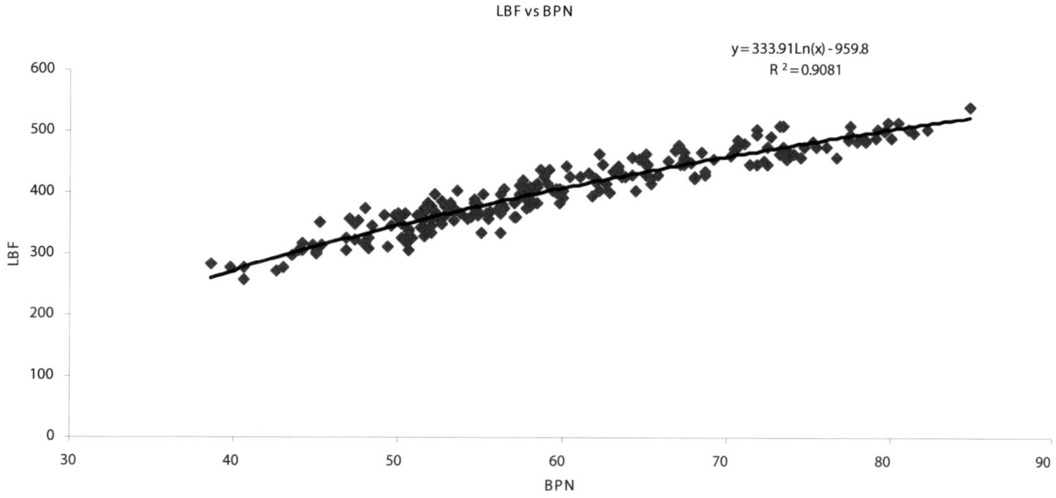

Fig. 6. Correlation between BPN and LBF for all cores

After an adequate model is obtained, the DDS model provides frequency (refers to number of cycles per millimeter), wavelength (is simply the inverse of this spatial frequency), damping ratio (indicates how well a given wavelength component of the profile repeats at that frequency) and contribution (is the averaged squared deviation from a mean height). Generally in an elevation profile, there always exists a dominant peak that represents the largest contribution and the largest wavelength compared to other wave crests or wave troughs and is called dominant contribution. The dominant peak may not repeat in the same elevation profile and will show up as a real root with large wavelength in DDS analysis. In the previous section of this paper, the DDS approach has been utilized to investigate the characteristics of polished and unpolished aggregate surfaces. It was found that for both polished and unpolished aggregate surfaces, the dominant contributions almost all came from real roots and the corresponding wavelengths are attributed to the microtexture. Since the dominant contribution is much larger than other contributions in each scan, it dominates the value of variance and this leads the variances to have the same trend as the dominant contributions. Therefore, the correlation of variances with BPN or LBF is explored for 1 micron step size scan.

Figure 7 and 8 present a correlation between the calculated variances versus BPN values and LBF values for all the 212 pavement surface cores, respectively. The variance data in these figures is obtained by summing the variances from all scans on a given core, and then averaging it. A nearly linear trend can be found from the two figures and LBF values have a slight better correlation with the surface roughness variances than BPN values have. In Figures 7 and 8, since the values of variances are very small (those values are scattered over the range of around 10^{-1} to 10^{-4}) and the variance has units of squared length we have used the square root of variance to further improve the correlations. Moreover the square root of variance has a physical meaning that indicates the averaged roughness of pavement surface with a length unit. Figures 9 and 10 present the correlations with BPN and LBF. The square root of variance is calculated by obtaining the square root of variance from every scan, adding them up for a given core and then averaged for all the scans. Comparing to the Figures 7 and 8, Figures 9 and 10 do show a better correlation. The correlation coefficients (R^2) increase from 0.091 to 0.1962 for BPN and from 0.1227 to 0.2471 for LBF. It can be found again that LBF has a higher R^2 than BPN.

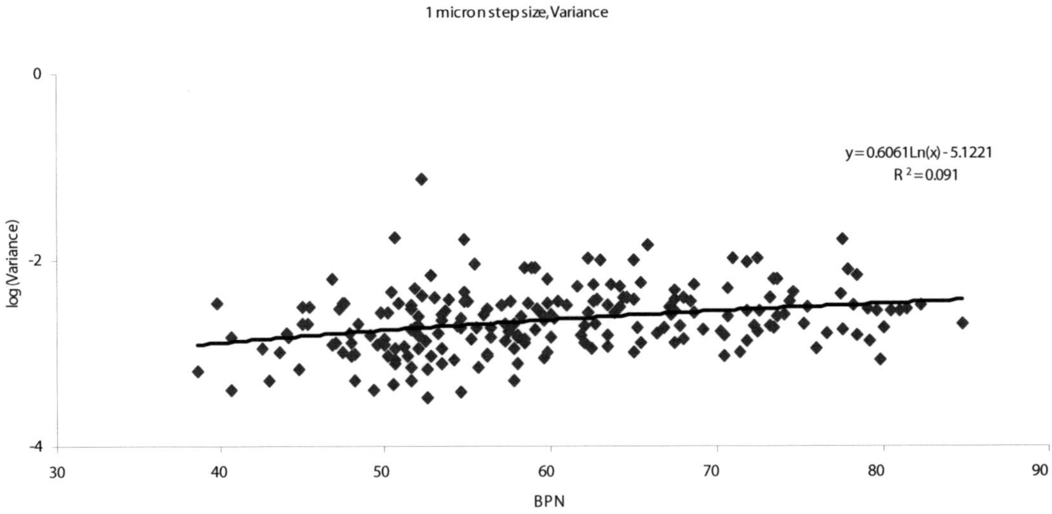

Fig. 7. Correlation between variance and BPN for all cores

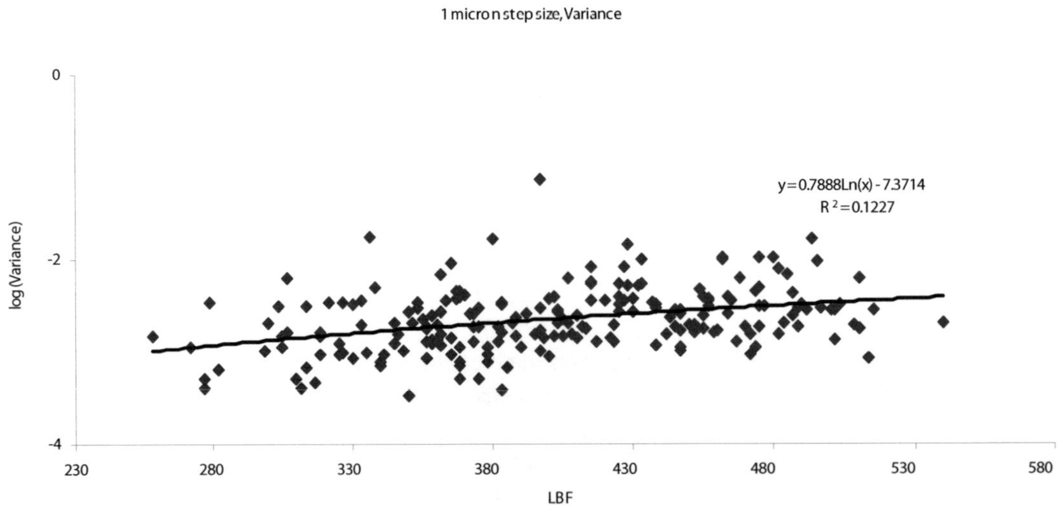

Fig. 8. Correlation between variance and LBF for all cores

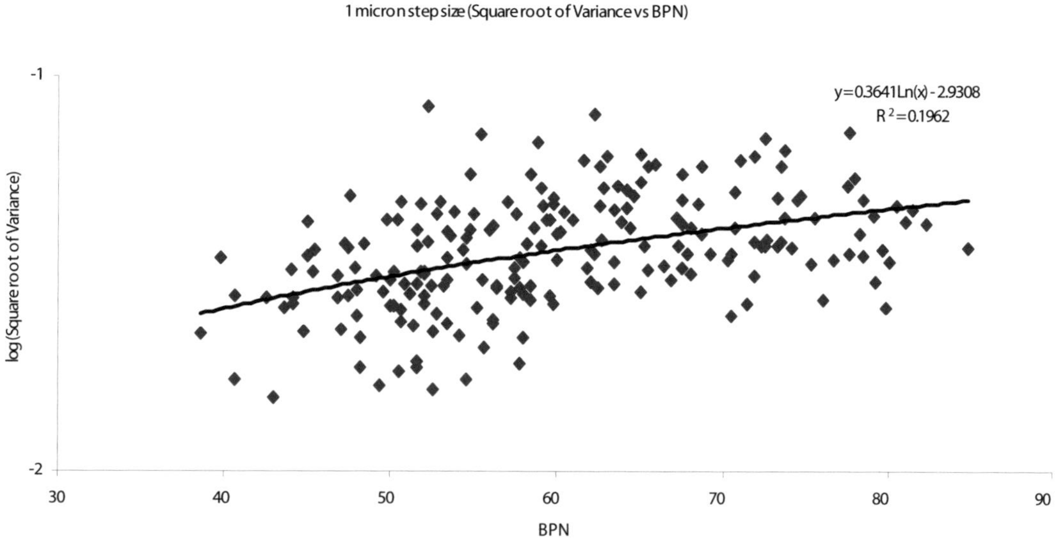

Fig. 9. Correlation of square root of variance with BPN for all cores

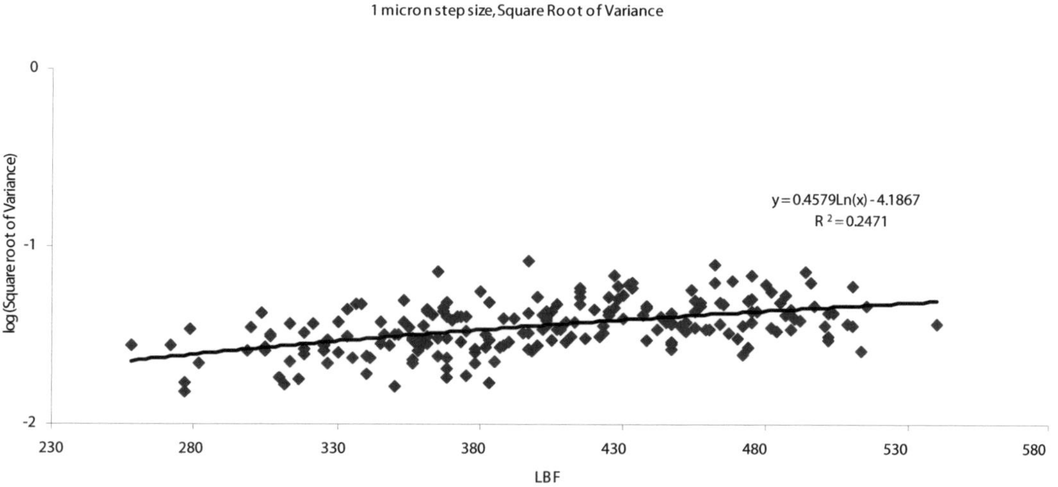

Fig. 10. Correlation of square root of variance with LBF for all cores

In the above plots all the pavement surface samples including those cored from shoulder, both wheel paths and between wheel paths were included. However, those cores from shoulder and between wheel paths are seldom polished by vehicle tire and hence may affect the correlation of the aggregate roughness with BPN and LBF values. In order to explore the effect of tire polishing on road surface friction, Figure 11 and Figure 12 present those variance data collected only from the samples that were cored from both wheel paths. This reduces the total number of cores to 116. Figures 11 and 12 exhibit the correlation of square root of variance from wheel path cores with BPN and LBF. Compared to the correlation with all the cores presented in Figures7 and 8, the correlation with wheel path cores has a better coefficient of correlation, which indicates that the correlation of aggregate roughness with BPN and LBF is really affected by the vehicle tire polishing and, to some extent, the averaged aggregate roughness (square root of variance) does exhibit a relationship with the skid resistance of road surface. On the other hand, we explored some other contributions related to very short wavelengths by damping ratio criterion (see next section for detail of the criterion) and correlated them with BPN and LBF values based on total cores and wheel path cores, respectively. None of these measures gave good correlations.

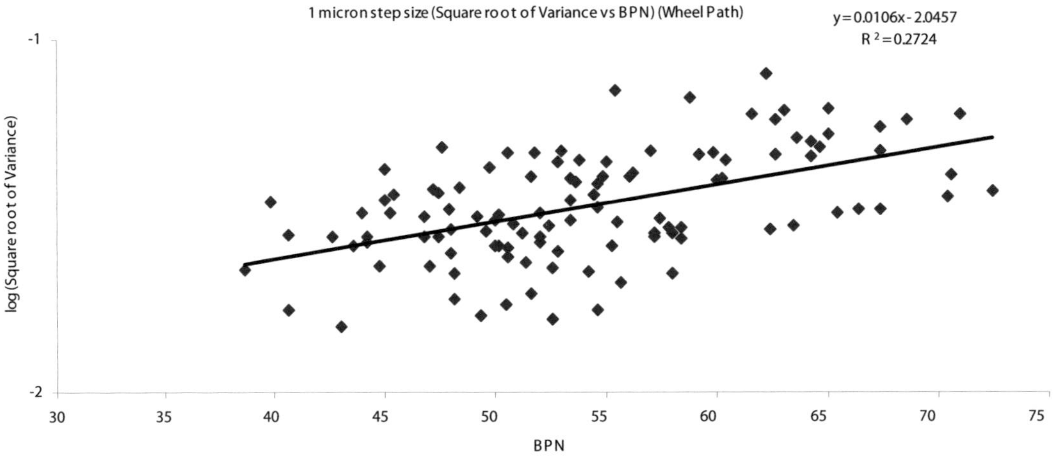

Fig. 11. Correlation of square root of variance with BPN for wheelpath

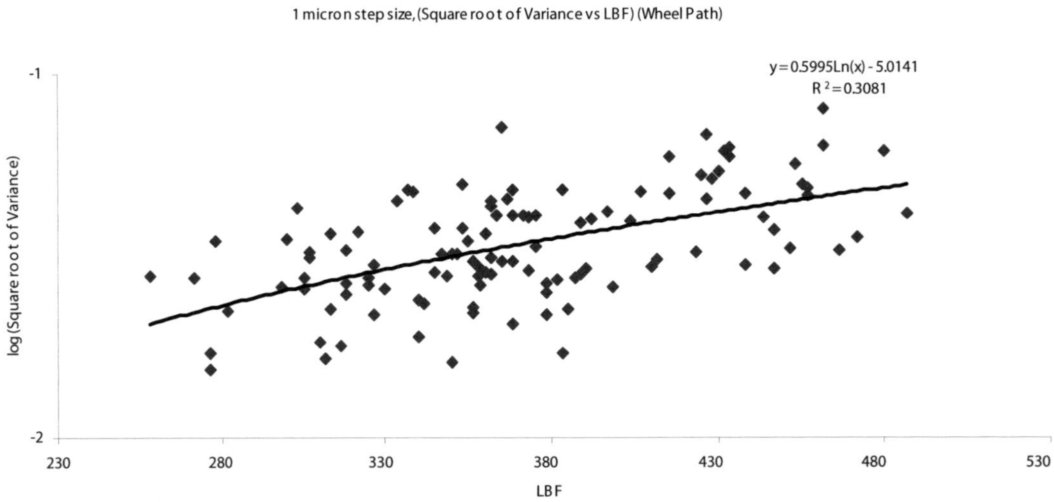

Fig. 12. 7 Correlation of square root of variance with LBF for wheelpath

6.2 Modeling analyses of 45 micron step size scans

Although the 1 micron step size scans collected from aggregate surfaces on road pavement cores present a correlation with pavement skid resistance, it is still necessary to explore a correlation between the pavement surface texture and pavement skid resistance. Because the aggregate surfaces just can provide with microtextural information to correlate with skid resistance, more information about the macrotexture of pavement surface need to be explored.

In the previous section, the 45 micron step size scan, which spans over a length around 46mm, has been found to be a good measurement for the characterization of macro- as well as micro-texture. For most of 45 micron step size scans, the dominant real roots usually present the characteristics of macrotexture and the characteristics of microtexture can be captured by using damping ratio criterion. Similar to the situation in 1 micron step size scans, the dominant contribution is much larger than other contributions in most of scans, so variance should have the same trend as the dominant contribution and represents the characteristics of macrotexture. The damping ratio criterion used here is that when the damping ratio is less than 13%, the corresponding contribution will be selected and then summed over all the contributions that satisfy this damping ratio criterion to obtain a partial contribution in one scan. For any given pavement core, there are twelve elevation scans collected from its surface so a total 12 partial contributions are summed up and then averaged to gain a partial contribution value for the given core. Since most of the wavelengths satisfying the damping criterion are less than 0.5mm, the texture can be depicted as microtexture and the associated partial contributions physically describe the averaged square deviation of microtexture roughness.

Table 10 presents the correlation between DDS analysis and road surface friction BPN and LBF. In Table 10 the numbers represent the coefficients of correlation (R^2), The variance or square root of variance does not have a correlation with either LBF or BPN. It implies that it is almost impossible to correlate macrotexture with road surface friction in 45 micron step size scans. In other words we could not find the correlation between macrotexture measures and road surface friction by the simple variance analysis or the dominant contribution analysis of DDS approach. On the other hand, there dose exist a correlation between microtexture (which is expressed by partial contribution) and the two kinds of road surface friction. One interesting phenomenon is that the correlation with wheel path cores is less than the correlation with all the cores. This implies that the traffic polishing influence on microtexture of road surface matrix is not significant. Here, a question is raised, how can the polishing process of vehicle tire affect road surface matrix? Will it change the macrotexture or something else?

Table 11 presents the correlations of contributions that satisfy a new wavelength criterion with the two kinds of road surface friction. The wavelength criterion is that for every scan a special wavelength is prescribed, if a wavelength is equal to or less than the prescribed wavelength, the corresponding contribution will be picked and then summed up to gain a partial contribution based on the scan, which represents the averaged square roughness corresponding to the prescribed wavelength. For a given core, sum up all the partial contributions coming from all the scans on the given core and then average them based on the scan number. Four scale wavelengths are chosen, they are 0.5, 1.0, 1.5 and 2.0 mm. From the Table 11, when wavelength $\lambda \leq 1.0$ mm, it will have the highest coefficient of correlation with the two kinds of friction and those two values are close to the coefficients of correlation values we found in 1 micron step size scans. Therefore, the polishing process of vehicle tire will mainly affect the 1 mm wavelength texture on road surface matrix. It should be pointed out that bituminous pavement surfaces in our analysis all belong to mixed types of pavement, they are composed of different compositions, such as different kind of aggregates and different size of aggregates. The type of mixed bituminous pavement has exhibited a significant influence on the correlation of pavement surface texture with road surface friction tests. Table 12 and 13 presented those coefficients of correlation between DDS analysis and friction tests for 1 micron and 45 micron step size based on different mixed type surface, respectively.

In Table 12, SQRT(variance) means square root of variance, the square root of variance was obtained from each scan and then was averaged based on each pavement core. mixtype 4 that it was mainly composed of sand and gravel (SG) and had the maximum aggregate size around 12.5mm, has the highest coefficient of correlation with BPN and LBF (the coefficients of correlation (R^2) are 0.5335 and 0.5167, respectively). Therefore, the aggregate profile on this type of pavement surface can be correlated well with friction tests. Mixtype SMA has the second place for the correlation with friction test. SMA was generally composed of SG (43%) and carbonate (57%) and the maximum aggregate size was around 19mm. The poorest coefficients of correlation with friction were mixtype Micro and Slag. It is realized that pavement constructed by slag aggregate usually has no correlation with friction test based on the characteristics of slag. So it is not surprising to obtain a poor correlation for slag pavement in terms of DDS analysis.

Table 13-1 and 13-2 present the correlation of pavement surface texture with friction for 45 micron step size scan analysis. SQRT(partial) indicates the square root of partial contribution and negative values represents an inverse trend. The partial contribution is obtained in terms of the above damping ratio criterion and is calculated based on each scan and is averaged for a given pavement core. In the tables, the column of SQRT(variance) has a weak correlation with both the friction tests for different types of mixtype pavement. It proves again that variance of elevation profile is not an appropriate description for surface texture to correlate with road surface friction. This conclusion is also consistent with the results we obtained in Chapter 2. When the surface texture wavelength is equal to or less than 1mm, mixtype SMA has the highest coefficient of correlation with friction BPN and LBF. For mixtype Micro, different from 1 micron analysis, it presents a good correlation with the frictions when wavelength varies from $\lambda \leq 2$mm, $\lambda \leq 1.5$mm, $\lambda \leq 1$mm, $\lambda \leq 0.5$mm and $\lambda < 0.5$mm (usually, the corresponding wavelengths satisfying the damping ratio criterion are always less than 0.5mm). Comparing Table 13-1 and 13-2, the wavelength $\lambda \leq 0.5$mm has the best correlation with the two friction tests (the R^2 values are 0.5498 with BPN and 0.4375 with LBF). For other mixtype road surfaces, mixtype 4 presents some trend with friction tests and mixtype Slag presents no trend with friction once again.

7. Conclusion

As a research subject, pavement texture has been widely studied for the past several decades. However, most of these research works were limited to characterizing the pavement surface texture qualitatively and are limited to either variance or RMS or MTD. Pavement surface texture has been generally accepted as a combination of microtexture and macrotexture. Friction between tire and road surface is considerably dependent on surface texture. In recent years, more and more researchers revealed that microtexture, in particular, plays a significant role in the road/tire contact friction. In this paper, experimental texture measurements and DDS modeling methodology were introduced to analyze the real road pavement surfaces obtained from Michigan. The elevation profiles collected from real road core surfaces were composed of 1 micron step size scan and 45 micron step size scan by an accurate laser sensor and were modeled by DDS program. Comparison of the modeling results, the following conclusions can be drawn.

1. Aggregate surface on real road pavement, to some extent, presents a correlation with road surface friction. For some mixtype (Such as Mixtype SMA, Mixtype 4 and Mixtype 13) pavements, the aggregate surface has a good correlation with friction tests, but for some mixtype does not.
2. The type of pavement has a significant effect on the correlation between surface texture and friction. Some types of pavement (such as Mixtype micro, Mixtype 4 and Mixtype SMA) present a strong correlation with friction, but some types do not exhibit the correlation with friction. And Mixtype 4 and Mixtype 13 have a consistent correlation from both aggregate surface and pavement surface matrix.
3. Variance of elevation profile is found to be an improper parameter to correlate with surface friction for large step size scan, such as 45 micro step size scan.
4. For some type of pavement, there exists a variation range of wavelengths where the corresponding roughness presents a good correlation with friction and these wavelengths are in the range of around 0.5mm to 2mm.
5. It is worth mentioning that there exist many factors that may affect the road surface roughness, such as the construction date, the amount of daily traffic, etc. After more factors that affect the surface roughness are considered, a better correlation of roughness with friction can be obtained.

Frequency Cycles/mm	Wavelength (mm)	Damping Ratio	Contribution (mm^2)	Variance (mm^2)
434	0.002306	3.09E-02	2.48E-08	2.25E-04
250	0.004006	3.19E-01	4.77E-07	
294	0.003403	4.49E-02	5.18E-08	
168	0.005956	2.40E-02	-1.04E-08	
9.94	0.100563	3.33E-01	2.37E-05	
500	0.002		9.90E-08	
2.31	**0.433839**		**2.01E-04**	

Table 1. DDS model results from a 1mm scan (01a) of a polished aggregate on AWI track

Frequency Cycles/mm	Wavelength (mm)	Damping Ratio	Contribution (mm^2)	Variance (mm^2)
466	0.002146	3.44E-03	1.51E-08	1.04E-03
217	0.004615	5.49E-02	-2.73E-08	
270	0.003711	4.75E-02	3.52E-09	
322	0.003102	5.09E-03	2.49E-10	
415	0.002410	2.37E-02	3.83E-08	
367	0.002727	1.18E-02	1.69E-08	
97.2	0.010288	6.77E-02	9.44E-08	
112	0.008947	2.77E-01	-3.30E-07	
7.02	0.14245	7.22E-01	-3.80E-04	
93.2	0.010725	1.16E-06		
193	0.005195	5.63E-02	-2.63E-08	
2.69	**0.372024**	**1.42E-03**		

Table 2. DDS model results from a 1mm scan (011) of an unpolished aggregate on AWI track

Scan ID		Variance (mm²)		Dominant Wavelength (mm)		Dominant Contribution (mm²)	
Polished	Unpolished	Polished	Unpolished	Polished	Unpolished	Polished	Unpolished
01a	011	2.25E-4	1.04E-3	0.4338395	0.3720238	2.01E-4	1.42E-3
01b	012	7.24E-5	2.75E-3	0.1930502	0.1228199	6.40E-5	2.73E-3
01c	013	1.14E-4	7.86E-4	0.2016536	0.2457606	1.13E-4	7.83E-4
01d	014	2.66E-4	5.42E-4	0.2810568	0.307031	2.65E-4	8.67E-4
01e	015	4.65E-5	1.41E-3	0.2644803	0.4299226	4.47E-5	2.72E-3

Table 3. Comparison of polished and unpolished Surfaces Modeling (1micron)

Frequency Cycles/mm	Wavelength (mm)	Damping Ratio	Contribution (mm²)	Critetion (mm²)	Partial (mm²)	Variance (mm²)
2.10	0.477	2.84E-01	2.60E-04		1.013E-5	8.48E-4
8.70	0.115	3.49E-02	-1.17E-06	-1.17E-06		
12.0	0.084	1.84E-02	3.30E-03	3.30E-07		
16.7	0.060		3.60E-05			
5.54	0.181	6.92E-03	1.10E-05	1.10E-05		
0.597	**1.674**		**5.28E-04**			
10.8	0.093	3.64E-01	1.44E-05			

Table 4. 30a Polished Scan

Frequency Cycles/mm	Wavelength (mm)	Damping Ratio	Contribution (mm²)	Critetion (mm²)	Partial (mm²)	Variance (mm²)
5.09	0.196	6.11E-02	2.89E-05	2.89E-05	1.80E-4	6.85E-2
15.0	0.067	3.81E-03	5.70E-07	5.70E-07		
11.4	0.087	8.60E-02	-4.38E-06	-4.38E-06		
9.44	0.105	9.36E-03	-1.84E-08	-1.84E-08		
1.96	0.510	2.39E-01	3.15E-05			
0.773	1.293		6.56E-05			
7.04	0.142	6.54E-02	3.09E-05	3.09E-05		
4.10	0.244	8.29E-03	1.24E-04	1.24E-04		
12.7	0.079	7.98E-03	-3.47E-07	-3.47E-07		
0.012	**82.44**		**6.83E-02**			
14.1	0.071	2.36E-01	-1.26E-05			

Table 5. 301 Unpolished Scan

Scan ID		Variance (mm²)		Microtexture Wavelength Range (mm)		Partial Contribution (mm²)	
Polished	Unpolished	Polished	Unpolished	Polished	Unpolished	Polished	Unpolished
30a	301	8.48E-04	6.85E-02	0.08~0.18	0.07~0.24	1.01E-05	1.80E-04
30b	302	2.66E-03	9.82E-03	0.07~0.16	0.07~0.76	1.70E-05	1.05E-04
30c	303	1.21E-03	2.24E-03	0.09	0.07~0.66	-3.62E-07	1.60E-04
30d	304	2.17E-03	1.84E-02	0.08	0.07~0.23	-2.80E-06	1.54E-04
30e	305	4.58E-04	1.77E-02	0.15	0.07~0.4	1.21E-05	1.34E-04

Table 6. Comparison of Polished and Unpolished Surfaces Modeling (30micron)

Frequency Cycles/mm	Wavelength (mm)	Damping Ratio	Contribution (mm²)	Critetion (mm²)	Partial (mm²)	Variance (mm²)
0.0961	**10.4**	**4.40E-01**	**1.46**		-5.35E-07	1.46
6.56	0.153	3.25E-02	-5.35E-07	-5.35E-07		
5.31	0.188		-4.96E-03			
11.1	0.09		-1.67E-05			

Table 7. c45 Polished Scan

Frequency Cycles/mm	Wavelength (mm)	Damping Ratio	Contribution (mm²)	Critetion (mm²)	Partial (mm²)	Variance (mm²)
11.1	0.09		-7.55E-04		8.143E-04	0.636
2.12	0.47	2.76E-02	2.05E-04	2.05E-04		
5.89	0.17	1.88E-03	9.08E-06	9.08E-06		
7.57	0.13	2.34E-02	-9.83E-06	-9.83E-06		
9.86	0.10	3.61E-03	2.04E-06	2.04E-06		
3.81	0.26	3.84E-02	1.86E-04	1.86E-04		
11.1	0.09		2.34E-06			
2.98	0.34	4.78E-02	4.21E-04	4.21E-04		
0.15	**6.64**	**4.74E-01**	**6.36E-01**			

Table 8. w45 Unpolished Scan

Scan ID		Variance (mm²)		Dominant Wavelength (mm)		Partial Contribution (mm²)		Microtexture Wave-length Range (mm)	
Polished	Unpolished	Polished	Unpolished	Polished	Unpolished	Polished	Unpolished	Polished	Unpolished
A45	V45	0.336	0.463	4.7	6.14	-9.23E-6	5.44E-4	0.153	0.1~0.37
B45	W45	2.03	0.723	5.26	7.53	1.3E-4	5.37E-4	0.1~0.22	0.1~0.47
C45	X45	1.46	0.636	10.4	6.64	-5.35E-7	8.14E-4	0.1~0.14	0.1~0.47
D45	Y45	0.247	1.55	15.74	5.68	1.26E-5	1.19E-3	0.1~0.23	0.1~0.37
E45	Z45	0.697	0.999	3.9	9.41	-4.65E-5	7.02E-4	0.1~0.13	0.1~0.32

Table 9. Comparison of Polished and Unpolished Surfaces Modeling (45micron)

	BPN		LBF	
	All cores	Wheel Path cores	All cores	Wheel Path cores
Variance (mm^2)	0.0044	No trend	0.0015	No trend
Square root of variance (mm)	0.0104	No trend	0.0059	No trend
Partial contribution (mm^2)	0.1395	0.1156	0.1333	0.101

Table 10. Correlation between DDS analysis and road surface friction

	Contributions satisfy wavelength λ (mm) criterion			
	$\lambda \leq 0.5$	$\lambda \leq 1.0$	$\lambda \leq 1.5$	$\lambda \leq 2.0$
BPN	0.1769	0.2455	0.1187	0.1408
LBF	0.0766	0.2837	0.0362	0.1564

Table 11. Polishing influence on wavelengths of road surface texture

Core Types	Coefficient of correlation (R^2)			
	BPN		LBF	
	variance	SQRT (variance)	variance	SQRT (variance)
All cores	0.0842	0.1835	0.1187	0.2383
Wheel path cores	0.1542	0.2742	0.1763	0.3142
Mixtype Micro	0.0086	0.0377	0.0015	0.0027
Mixtype SMA	0.2589	0.3484	0.1858	0.2263
Mixtype 13	0.1325	0.1695	0.1964	0.221
Mixtype 4	0.3434	0.5335	0.3281	0.5167
Mixtyp SLAG	0.0122	0.0298	0.0355	0.0959

Table 12. Correlations between core types and surface measures for 1 micron step size

Core Types	Coefficient of correlation (R^2)					
	SQRT (variance)	SQRT(Partial)	$\lambda \leq 0.5$	$\lambda \leq 1.0$	$\lambda \leq 1.5$	$\lambda \leq 2.0$
Wheel path cores	0.0613	0.1461	0.1769	0.2455	0.1187	0.1408
Mixtype Micro	0.1264	0.4198	0.5498	0.4672	0.6319	0.5209
Mixtype SMA	0.2198	0.4553	0.5027	0.7122	0.6569	0.585
Mixtype 13	0.0929	0.0291	0.0001	0.0068	0.0016	0.0038
Mixtype 4	0.0015	0.2879	0.1575	0.1057	0.0575	0.048
Mixtyp SLAG	0.1212	0.0021	0.0575	0.0715	0.0905	0.0777

Table 13-1. Correlations between BPN and surface measures for 45 micron step size

Core Types	Coefficient of correlation (R^2)					
	SQRT (variance)	SQRT (Partial)	$\lambda \leq 0.5$	$\lambda \leq 1.0$	$\lambda \leq 1.5$	$\lambda \leq 2.0$
Wheel path cores Mixtype Micro Mixtype SMA Mixtype 13 Mixtype 4 Mixtyp SLAG	0.0603 0.2969 0.0391 0.0792 0.0136 0.0299	-0.002 0.3399 0.1205 0.0595 0.345 0.0526	0.0766 0.4375 - 0.00000 1 -0.1503 0.0073 0.2446	0.2837 0.2544 0.4417 0.0272 0.2407 0.2246	0.0362 0.2032 0.008 -0.2499 0.0752 0.2114	0.1564 0.2023 0.2179 - 0.00009 0.1588 0.2072

Table 13-2. Correlations between LBF and surface measures for 45 micron step size

8. Reference

Chernyak, Yu. B. and A. I. Leonov (1986), On the theory of the adhesive friction of elastomers, Wear 108,105-138

Dewey, G. R., A. C. Robords, B. T. Armour and R. Muethel (2001), Aggregate Wear and Pavement Friction, *Transportation Research Record*, Paper No. 01-3443.

Do, M. T., H. Zahouani and R. Vargiolu (2000), Angular parameter for characterizing road surface microtexture. In Transportation Research Record 1723, TRB, National Research Council, Washington, D. C., 66.

Fülöp, I. A., I. Bogárdi, A. Gulyás and M. Csicsely-Tarpay (2000), Use of friction and texture in pavement performance modeling, *J. of Transportation Engineering*, 126(3), 243-248.

Gunaratne, M., M. Chawla, P. Ulrich and N. Bandara (1996), Experimental investigation of pavement texture characteristics, *SAE 1996 Transactions Journal of Aerospace*, 105(1), 141-146.

Kokkalis, G. (1998), Prediction of skid resistance from texture measurements, Proc. Instn Civ. Engrs Transp., 129, 85-.

Kummer, H. W., Unified theory of rubber friction, Engrg. Res. Bull. B-94, Penn State University, State College, University Park, Pa., (1966)

Pandit, S. M. and S. M. Wu (1983), *Time series and system analysis with applications*, John Wiley.

Pandit, S. M. (1991), *Modal and Spectrum Analysis: Data Dependent Systems in State Space*, Wiley Interscience.

Perera, R. W., S. D. Kohn and S. Bemanian (1999), Comparison of road profilers, *Transportation Research Record*, 1536, 117-124.

Persson, B. N. J. and E. Tosatti (2000), Qualitative theory of rubber friction and wear, *Journal of Chemical Physics*, 112(4), 2021-2029.

Rohde, S. M. (1976), On the effect of pavement microtexture and thin film traction. *Int. J. Mech. Sci.*, 18(1), 95-101.

Taneerananon, P. and W. O. Yandell (1981), Microtexture roughness effect on predicted road-tire friction in wet conditions, *Wear*, 69, 321-337.

Schallamach, A. (1963), Wear 6, 375

Yandell, W. O. and S. Sawyer (1994), Prediction of tire-road friction from texture measurements, Transportation Research Record 1435, Transportation research Board, National Research Council, washinton D. C..

15

Adhesion Theory for Low Friction on Ice

Katsutoshi Tusima
University of Toyama
Japan

1. Introduction

Ice is one of the lowest frictional materials on the earth. Its low friction enables us to utilize for enjoyment of skating, skiing and sledging.

Why friction on ice is so low? It has been known since ancient times that a liquid lubricant such as oil can reduce the friction, and many scientists have analogically guessed that water formed at the interface between ice and a slider may serve as lubricant. Two theories have been proposed to explain the formation of liquid water at the interface: one relates it to pressure melting (Joly, 1887; Reynolds, 1899) and other to friction melting (Bowden & Hughes, 1939). Bowden and Hughes obtained for μ_k between the plates and rotating ice disk a large value of 0.3 at a velocity of 30 mm/s against a small value of 0.04 at a higher velocity of 5 m/s. This experimental result has been essential basis in friction melting theory.

Pressure melting theory has been abandoned because heat must be carried from temperature region higher than real contact area. Friction melting theory has been supported by Bowden (1953, 1955), Shimbo (1961), Barnes et al. (1979), Evans et al. (1976) and other many reseacher to explain their experiments. Also, Huzioka (1962, 1963) observed the refreezed icicles appeared snow grains and Tusima & Yosida (1969) observed the splashed water from interface between a rotating disk of ice and an annular slider at high-speed friction (10~20m/s). Hence, the existence of liquid water has been generally accepted as the cause of the low frictional coefficient of ice. Other speculative theories have been proposed by Weyl (liquid-like layer, 1951), Niven (rotation of ice molecules, 1959), McConica (vapor lubrication, 1959), Huzioka (sintering, 1962), and Tusima (adhesion theory, 1976, 1977).

The frictional melting theory thought that the melted water prevented the direct contact of two surfaces and lubricated between slider and ice as self-lubrication. This speculation introduced by similarity that small coefficient of friction may be inherent to liquid lubrication without examination feasibility of adhesion theory. So that, in many cases, it has been missed the following important property concerning to the friction process of ice: hardness and shear strength of ice, adhesive strength, real contact area, observation of frictional track, qualitative explanation of frictional resistance, etc. Several contradictory report have been presented on μ_k of ice in the absence of liquid water. Tabor & Walker (1970) and Barnes et al. (1971) obtained a low value of 0.05 for μ_k between an ice cone and a stainless steel plane in a velocity from 10^{-5} to 10^2 mm/s. Tusima (1977) obtained 0.005 to 0.1 forμ_k in low velocity range 0.1mm/s. Even if liquid lubrication were exist, we don't know reliable thickness of melt water for lubrication, because one scientist say few μm (Bowden &

Hughes, 1939; Ambach & Mayr, 1981) and other say few nm as thickness of melted water (Evans et al., 1976). However, frictional anisotropy changed unavailable the liquid lubrication. This anisotropy of ice can explain only by adhesion theory.
We can point out logical question for liquid lubrication theory that the water must be melted by frictional heat. Namely, if the friction was too small for production melt-water, the friction should be large in view of the theory. This is clear logical contradiction. Huzioka (1962) indicated high friction coefficient of 0.3 when remarkable icicles were observed around real contact area of snow grains. In speed skating, μ_k is extremely small, nearly 0.005 (Kobayashi, 1973; Koning et al., 1992; Tusima et al., 2000). Under these extremely low friction, skate will slide without lubrication liquid. Therefore (0001) ice rinks could display the properties of crystallographic plane of ice and μ_k became smaller than normal rink. It is clear that low value 0.01 to 0.05 does not mean always the liquid water lubrication. If liquid lubrication appear, μ_k should be the value lower than 0.0001 as pointed out by Evans et al. (1976).
According to classical adhesion theory of friction, frictional coefficient μ_k is given by

$$\mu_k = s/p + \text{(ploughing and other term)}$$

where s is adhesive shear strength of real contact, roughly equal to bulk shear strength of weaker material, p is the pressure of real contact area, nearly equal to the Brinnel hardness of softer material. Ice has extremely small shear strength s (1MPa at -10°C) compared to hardness (100MPa at -10°C). Therefore, μ_k becomes nearly 0.01 under dry friction. This means ice has an inherent low friction materia. In generally, second term is too small and can neglect (however in ice, this term can not always neglect depend on shape of slider.)
The narrow water between ice and material can not apply bulk contact angle and behave abnormal as shown by Hori (1956) and Jellinek (1967). Itagaki & Huber (1989) noticed that the effect of squeeze out will thin water layer in real contact area as shown by Furushima (1972).

2. Physical properties of ice

2.1 Hardness of ice

Fiction occurs at real contact area. When hard steel ball slides on flat plate of ice, real contact area will be formed by the plastic deformation of ice. The pressure of real contact decrease in

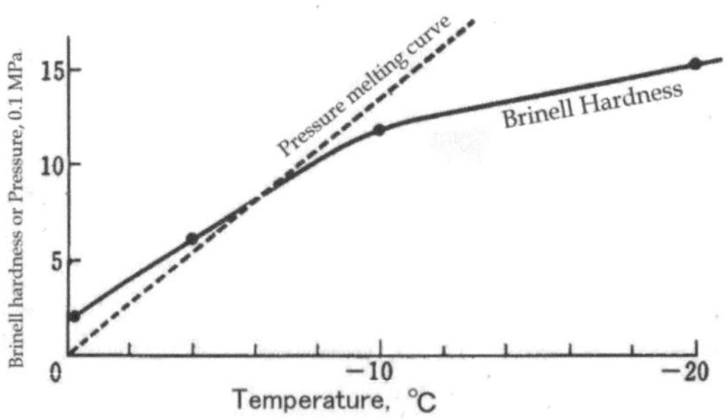

Fig. 1. Brinell hardness of single crystal of ice (Mendenhall Glacier ice) (from Butkovich, 1954), solid line and for polycrystalline of ice (from Barnes & Tabor,1966), dashed line shows pressure melting curve.

the process of plastic deformation, it will attain area depend on the sliding condition. However it is difficult to estimate an exact area in the sliding process on ice. When hard steel ball slides on flat plate of ice, apparent contact area is only one, the area will be equal to real area. The area will be given by the Brinell hardness. Butkovich (1954) measured the Brinell hardness as a function of temperature and loading time, by the use of indenter diameter 3.2mm. The result shows in Fig. 1 and Table 1. Hardness changed by crystallographic plane of ice (parallel and perpendicular to c-axis), temperature and loading time. The value increased with lowering temperature.

Temperature °C	Butkovich[1)]		Barnes & Tabor[2)] polycrystalline	Bowden & Tabor[3)] polycrystalline
	∥ C-axis	⊥ C-axis		
-0.25				18
-2			30	
-4	60	77		
-5			45	34
-10	118		60	
-15	125	111		
-20	145	126	90	60

1) Indenter 1/8"ball, single crystal of Mendenhall glacier, ∥ C-axis 15.4N load, ⊥ C-axis 25.2N load
2),3) Indenter Diameter 50mm, load 1000N 2)Barnes & Tabor (1966) 3)Bowden & Tabor(1964)

Table 1. Brinell hardness of ice, p MPa(=10kgf/cm^2), loading time 1 sec.

Barnes et al. also measured the Brinell hardness of ice under the load of 1000N, diameter of indenter 50mm. The value of hardness becomes lower in larger indenter than smaller one.

2.2 Shear strength of ice

If the bond of real contact area is strong enough, the break will occur in inside of ice in sliding process. In generally, the value will not exceed the shear strength of ice itself. Therefore it is interested in shear strength of ice.

Temperature °C	Jellinek MPa	Raraty & Tabor MPa	Butkovich MPa	Tusima & Fujii(1973)	
				∥ C-axis	⊥ C-axis
-2	0.2			1.8	2.7
-5	0.5	0.8	1.37		
-10	1.2	1.6	1.55	2.2	3.3
-15	1.5				
-20		3.1		2.6	4.3
-30		5.1	2.17	2.9	5.5

Table 2. Shear strength of ice, S MPa

Table 2 shows the measured value in several experiments. The value was very low 0.5~1.4 MPa at -5°C, and 1.2~3.3 MPa at -10°C in comparison to hardness of same temperature. The ratio s/p gives μ_k of ice in adhesion theory. From table 1 and 2, μ_k is estimated about 0.007~0.09 at -5°C and 0.01~0.07 at -10°C.

2.3 Adhesive strength of ice

There are many studies on adhesive strength of ice to other materials. Some results are shown in Table 3. It is noticed that the value of table is 1 order smaller than bulk shear strength of ice (Table 2).

Brunner (1952) MPa	Landy & Freiberger (1967) -12°C, MPa	Jellinek (1970) -4.5°C, MPa
metal 0.95	PTFE 0.32	stainless rough 0.61
polystyrene 0.59	PE 0.26	polish 0.3
paraffin 0.46	PMMA 0.64	mirror 0.06

Table 3. Adhesive shear strength of ice

Jellinek (1970) showed the effect of surface roughness of stainless steel as shown in each surfaces noticeably cleaned. We know that the adhesive strength is smaller than shear strength of ice in experience.

3. Friction of steel ball on single crystal of ice

The sliding of hard spherical surface on flat plate has been used for fundamental study of the mechanism of friction between materials (Bowden & Tabor, 1950). In this sliding, apparent contact area will be equal to real contact area. Therefore it gives to possibility qualitative evaluation for friction.

3.1 Experimental apparatus

The apparatus is shown schematically in Fig. 2. A rectangular-shaped ice sample was onto PMMA (Polymethylmethacrylate) disk A, which was mounted on a metal block M. The block M was driven either forwards or backwards on the upper surface of the thick rigid framework by a motor through reduction worm gears, and the ice sample on it was moved at a constant speed ranging from 1.5×10^{-7} to 7.4×10^{-3} m/s. Apparatus adjusted to 1mm/m by precise level.

A steel ball, 6.4 mm in diameter, contacting the ice surface was mounted and fixed to a brass cylinder, to the top of which a metal lever L was firmly fixed. One end of the lever was free, while the other end was connected to a universal joint. A load which ranged from 0.4 to 31 N, was exerted onto the ice surface by suspending a weight the lever. The weight which corresponds to a given load was immersed in an oil bath that prevented the weight from shaking.

The friction force between the fixed steel ball the moving ice surface was continuously measured by the use of a force-measuring system which consisted of transducer, a bridge box, a strain meter and recorder. The ice sample can be shifted in the transverse direction by moving the mount M so that each friction run may be made on a virgin ice surface. The ice sample can also be rotated into any horizontal orientation by turning the disk A so as to measure the friction force on ice for various crystallographic orientations.

3.2 Ice samples and steel ball

Tyndall figures were artificially produced at a corner of a large single crystal of ice collected from the Mendenhall Glacier, Alaska. By the aid of the Tyndall figures, two rectangular ice pieces were simultaneously cut out from the ice crystal in a way in which the frictional surface of the one was set parallel to the crystallographic basal plane (0001) and that of the

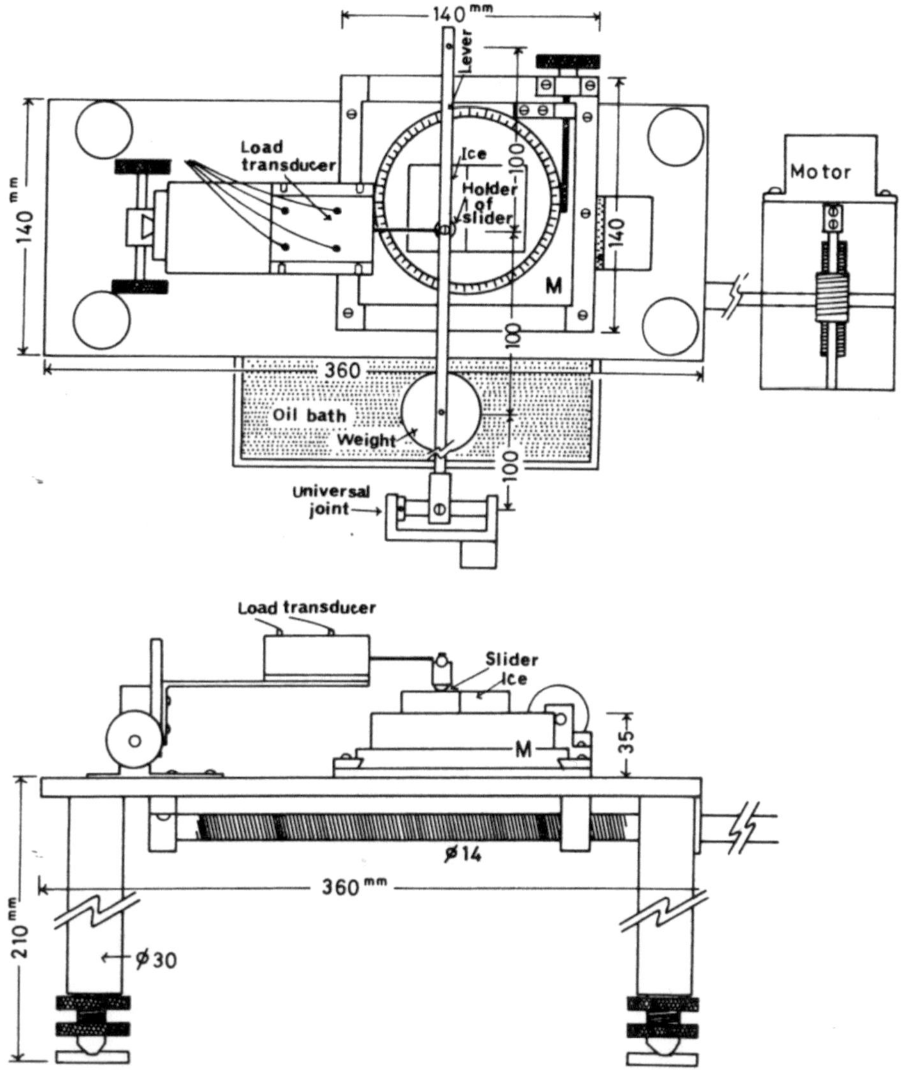

Fig. 2. Schematic diagram of the experimental apparatus.

other parallel to the prismatic plane $(10\bar{1}0)$. These two pieces were placed side by side and frozen to an PMMA disk so as to form a bicrystal sample of ice. This sample was finished by lathe. It was annealed again at -3°C until the turned surface become glossy like a mirror, and then brought into a cold room at an experimental temperature of -0.5 to -30°C. When it was exposed to lower temperature than -10°C, its surface occasionally became cloudy. Such samples were excluded from the experiment, and only glossy surfaces were used experimental studies on friction.

Steel ball with different sizes ranging from 1.6 to 12.7 mm in diameter were used in the experiment. The steel ball was cleaned by immersing it in an ultrasonic cleaning-bath filled with a mixture of alcohol and acetone and then in bath filled with distilled water. The ball was cleaned again by washing it in the bath of distilled water and dried under a heating lamp.

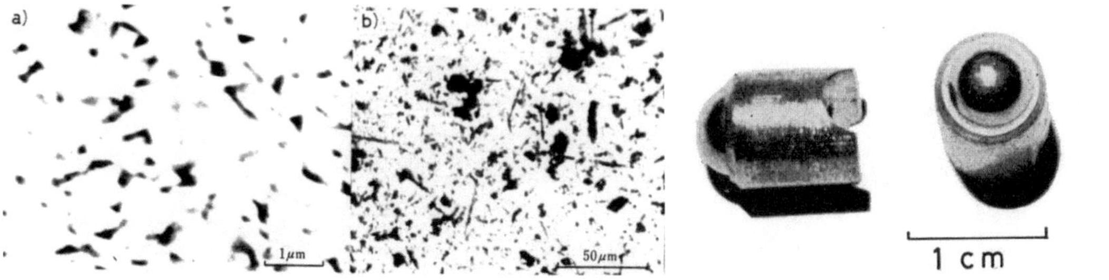

Fig. 3. A steel ball slider mounted on a brass cylinder.Left: Microscopic asperities of a slider 6.4 mm in diameter (a) tungsten carbide ball, (b)steel ball

4. Experimental results

4.1 Anisotropy of friction on crystallographic plane of ice

4.1.1 Friction curve

Steel was slid on flat plate of ice linearly connected 5 single crystal of grains as illustrated in Fig. 4. Velocity was slow as 7.4x10-5m/s, temperature at -10ºC, slider diameter 6.4mm of steel ball. In this condition, melting of ice does not occur. It was observed that the frictional coefficient changed by each grain. However it is noticed that the values were low as from 0.02 to 0.04.

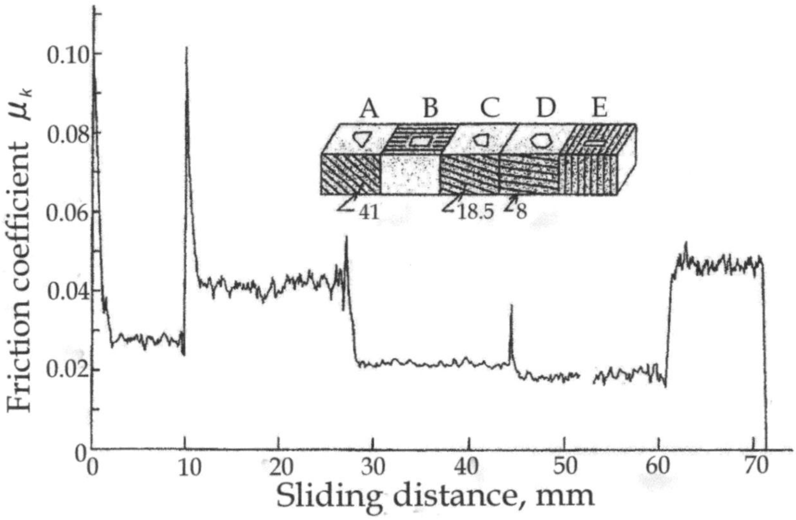

Fig. 4. Anisotropy of friction on crystallographic plane A, B, C, D, and E of ice. Longitudinal axis friction coefficient, horizontal axis sliding distance mm. Inclined lines show (0001) plane of ice. Temperature -10ºC, Velocity 7.4x10-5 m/s, Slider diameter 6.4 mm, Load 4.7N. Inclined line shows (0001) of ice.

Anisotropy in Fig. 4 will not explain by frictional melting theory. This supports adhesion theory because the hardness, shear strength and plowing strength depend on crystallographic plane of ice. Plane (0001) of ice is most hard for vertical load and most weak for shear force because (0001) correspond to crystallographic sliding plane of ice.

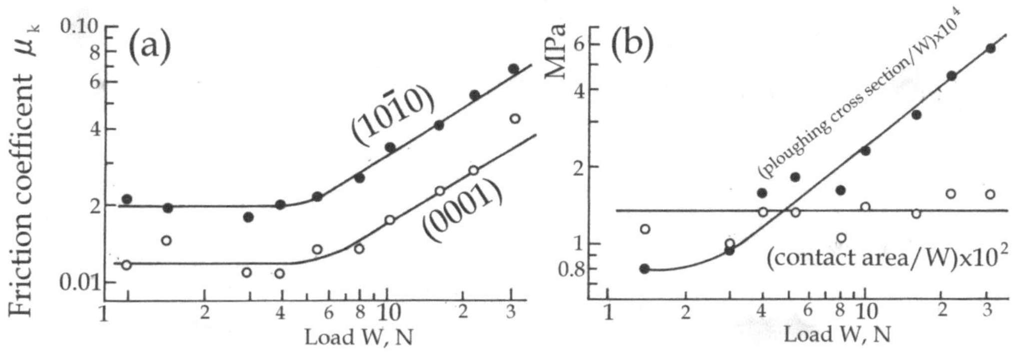

Fig. 5. (a) Dependence of friction on load for a basal and a prismatic plane of ice. (b) Contact area and ploughing cross-section against load. Velocity 7.4x10^{-5} m/s, Temperature -10ºC, slider diameter 6.4 mm. ○ (0001), • $(10\bar{1}0)$ (from Tusima, 1977).

4.1.2 Load effect

As an example, μ_k for both the basal and prismatic planes, at a velocity of 7.4×10^{-5} m/s and at a temperature of -10ºC, was plotted against the lower range of loads, less than 5 N for both cases, while it linearly increased with the increase in load in the higher range of load. A similar tendency to that in Fig. 5 was observed for different sliding velocities as seen in Fig. 9. The friction F in the present experiment is composed of two factors:

$$F = F_s + F_p, \tag{1}$$

where F_s and F_p respectively are concerned with the adhesion of ice and the ploughing of ice.

F_s and F_p are, respectively, proportional to A/W and A*/W, in which W is the load applied, and A and A* are the contact area and the ploughed area, respectively. It was found in the experiment that the ratio A/W is constant for any load, but the ratio A*/W increases with increasing load as shown in Fig. 5(b). Since the ploughing area A* was so small in the lower range of load, the ploughing effect was very small as compared with the sliding effect. It may, therefore, be concluded that the increase of μ_k in the higher range of load may be attributed to the increase of the ploughing effect.

As described before, it is important to measure the width of the sliding track left on the ice for interpreting the experimental results. The track width, the contact area, the average pressure acting on the contact area, and the cross-section ploughed for different loads are summarized in Table 4.

The contact area A can be expressed by using the track width φ as follows;

$$A=\pi(\varphi/2)^2 k \tag{2}$$

where k is a factor which is dependent on the visco-elastic properties of the contact area, the value of k being between 0.5 and 1.0. Fig. 6 shows the real contact area in the process of friction of a glass ball on ice. We know that the value of k is equal to 0.8 from this Fig. 6.

Load W N	Coefficient of friction μ_k	Track width, φ 10^{-3}m	Contact area, A 10^{-6} m^2	Mean pressure MPa	Ploughing area, A* 10^{-8}m^2	Calculated values μ_s	μ_p	$\frac{\mu_s+\mu_p}{\mu_k}$
1.4	0.020	0.16	0.016	9.0	0.011	0.008	0.008	0.8
3.0	0.018	0.22	0.030	97	0.028	0.007	0.009	0.9
4.0	0.021	0.29	0.053	75	0.064	0.009	0.016	1.2
5.5	0.022	0.34	0.073	75	0.10	0.009	0.018	1.2
8.0	0.026	0.37	0.085	92	0.13	0.007	0.016	0.9
10.0	0.036	0.47	0.14	75	0.27	0.009	0.026	1.0
16.0	0.043	0.58	0.21	76	0.51	0.009	0.032	1.0
22.0	0.057	0.75	0.35	62	1.10	0.011	0.058	1.1
31.0	0.070	0.88	0.49	64	1.80	0.011	0.058	1.0

T=-10°C, V=7.4×10^{-5} m/s. glacier ice ($10\bar{1}0$), S=0.7 MPa, *K*=0.8

Table 4. Some experimental results obtained in the experiment on friction of ice and the predicted values of the shear friction μ_s and the ploughing friction μ_p. (after Tusima, 1977)

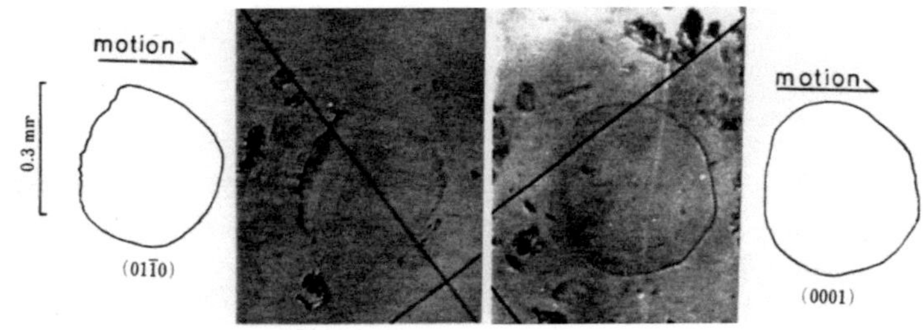

Fig. 6. Real contact area in the process of friction. Sliding of a hemispherical glass slider on a flat plate of ice coated with silicon oil to avoid condensation on the slider. Velocity: 7.4x10^{-2}mm/s, load: 4.75N left ($10\bar{1}0$), at right (0001)

4.1.3 Velocity dependence of friction

In order to clarify the dependence of the friction of ice to velocity, the friction force was measured with velocities for various loads. A typical results obtained is shown in Fig. 7, in which μ_k is plotted against the velocity obtained for both the basal and prismatic planes. As seen in this Figure, μ_k decreases with an increase in the velocity V. The width f of the track of the ball was also measured for each run of the experiment, and a similar tendency was obtained between φ and V to that obtained between μ_k and V. This shows that the larger friction at lower velocities can be attributed to the larger plastic deformation of ice at the contact area.

4.1.4 Temperature dependence of friction

The coefficient μ_k and the width of the sliding track φ are plotted in Fig. 8(a) and (b) against the ice temperature in raising process from -20°C up to -1°C at a rate of 1.5 deg/h. It was found that friction reaches a minimum at a temperature of -7°C when the sliding velocity is 7.4×10^{-5} m/s and the load is 4.8 N. As seen in this figure, the friction at a temperature below minimum friction increases on lowering the temperature, which is due to the increase of shearing strength of ice (Butkovich, 1954; Tusima & Fujii, 1973). The friction at higher temperatures above the temperature of minimum friction markedly increases as the ice

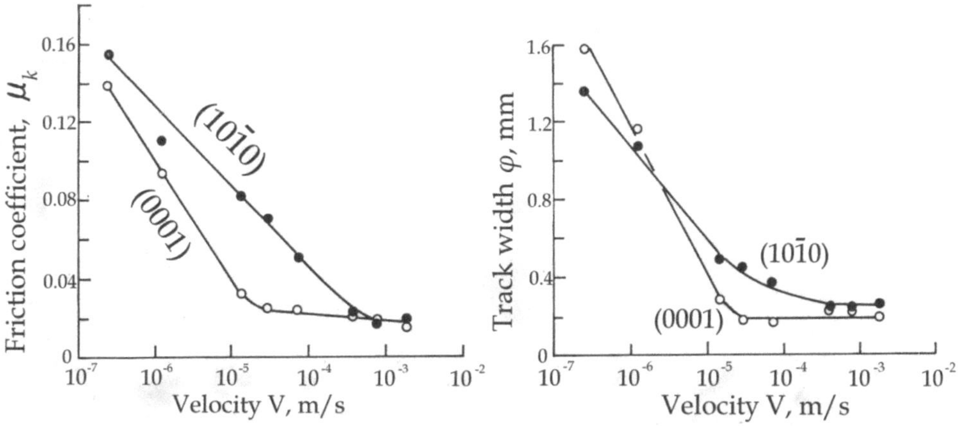

Fig. 7. (a) Dependence of friction an velocity, and (b) width of a sliding track against load. Temperature -10°C, Load 14N, Diameter of slider 6.4mm (after Tusima,1977)

temperature approaches its melting point. This increase may be closely to the increase in the width of the sliding track as shown in Fig. 8(b). We may conclude that the increase in the friction is caused by the ploughing of ice at the contact area. It should be noted that the minimum friction shifted to a higher temperature as the friction velocity was reduced. For example, it was at -4°C and -2°C when the velocity was 1.5×10^{-5} and 1.2×10^{-6} m/s, respectively.

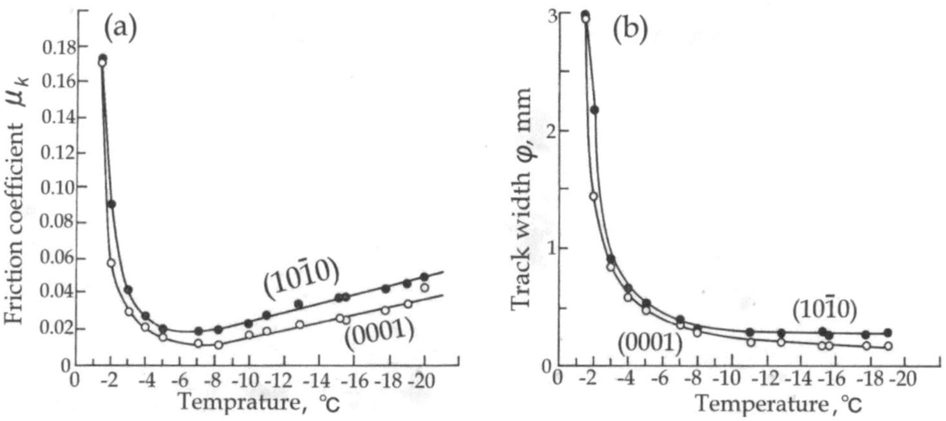

Fig. 8. (a) Dependence of friction on temperature, and (b) dependence of the width of the sliding track on temperature (after Tusima, 1977).

4.1.5 μ_k -V-W diagram

Dependence of the friction coefficient on the sliding velocity and load for a prismatic and a basal plane of ice are respectively summarized in Fig. 9(a) and (b). The coefficient μ_k ranged from 0.005 to 0.16. Though the friction varies with velocity, load and temperature, it is much smaller than those observed for metals. The coefficient μ_k is much smaller for the basal plane than for the prismatic plane for any experimental conditions. This may be due to the fact that the ice is very strong when it is compressed perpendicular to the basal plane, while it is very weak against a shearing force, which will be discussed later again.

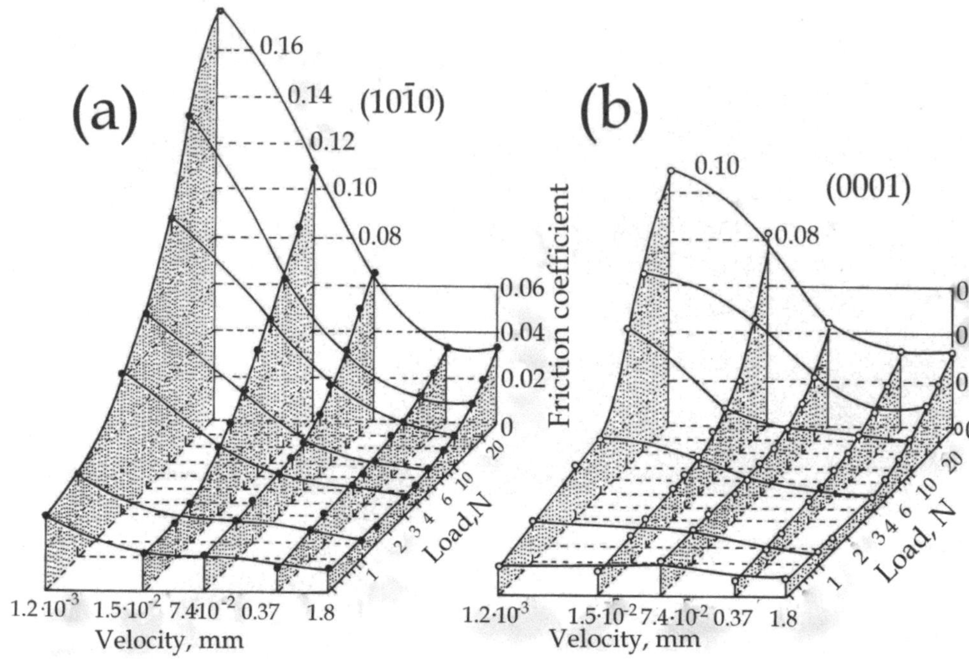

Fig. 9. μ_k-V-W diagram, (a) for a prismatic plane, and (b) for a basal plane. Temperature -10°C, Slider diameter 6.4mm(after Tusima, 1977).

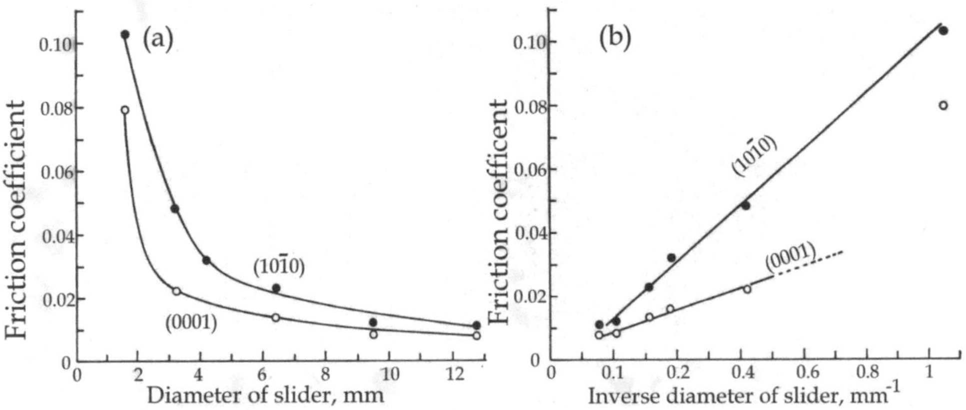

Fig. 10. Size effect of a steel ball on the friction of ice against diameter (a) and inverse diameter (b). Solid circle on the prism plane ($10\bar{1}0$); open circle on the basal plane (0001). Temperature -10°C, velocity 7.4x10^{-5} m/s, load 4.8N. (after Tusima,1977)

4.1.6 Effect of the size of ball

The degree of ploughing of ice by a steel ball may become larger as the ball becomes smaller in size. In order to examine the size effect of ball on friction of ice, steel balls of different diameters ranging from 1.6 to 12.7 mm were used as a slider. The results obtained are shown in Fig. 10. As was expected, μ_k increased with the decrease in size of the ball for a smaller range of diameters than 9.5 mm when the load, the sliding velocity, and the temperature were 4.8 N, 7.4×10^{-5} m/s and -10ºC, respectively. However, μ_k remained unchanged when a steel ball larger than 9.5 mm in diameter were used and this will give pure shear friction.

μ_k-V-1/R diagram is shown in Fig. 11. R is diameter of slider. Fig. 11 shows the same relation to Fig. 10.

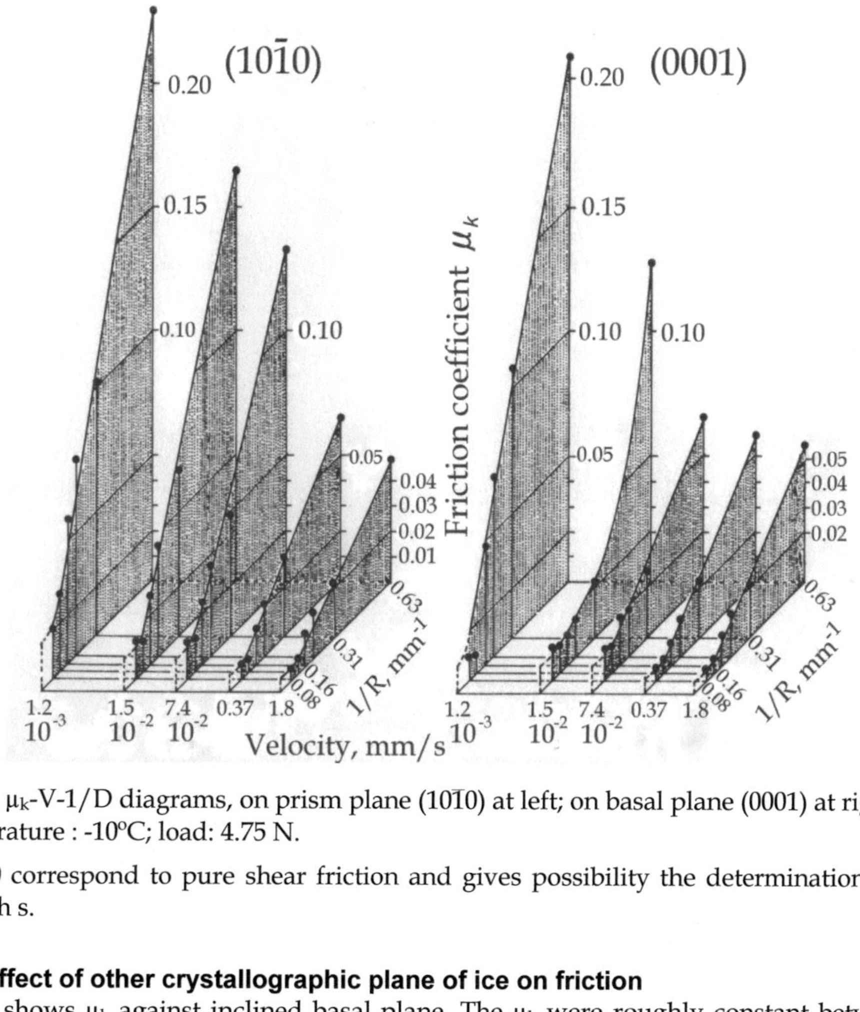

Fig. 11. μ_k-V-1/D diagrams, on prism plane (10$\bar{1}$0) at left; on basal plane (0001) at right. Temperature : -10ºC; load: 4.75 N.

1/R→0 correspond to pure shear friction and gives possibility the determination of shear strength s.

4.1.7 Effect of other crystallographic plane of ice on friction

Fig. 12 shows μ_k against inclined basal plane. The μ_k were roughly constant between 0 to 60°, but μ_k increased to high value in between 70 and 90°. Of course, μ_k changes by sliding orientation even on same plane.

4.1.8 Feature of frictional track of ice

Observation of frictional track of ice as shown Fig. 13 may give information as the solid friction mechanism. Ice has high vapor pressure and the disturbed region was changeable by sublimation, annealing and recrystallization etc. Therefore, the track must be observed quickly after sliding. Fig. 13 shows the groove, recrystallizaion, microcrack, plastic deformation etc.

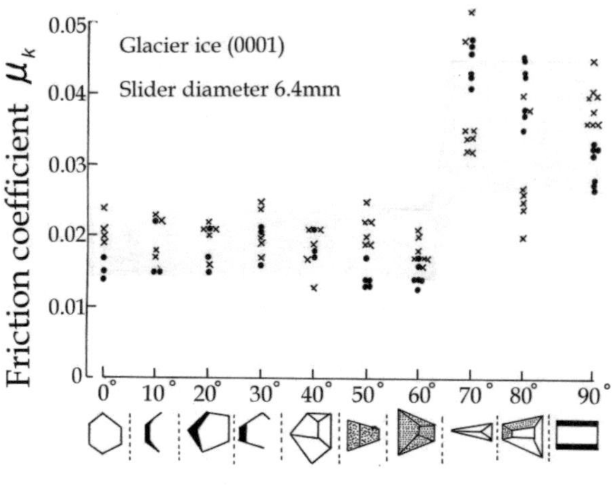

Fig. 12. Coefficient of kinetic friction μ_k against angle of basal plane for ice surface. Temperature -10°C, velocity 7.4x10^{-5}m/s, load 4.7N.

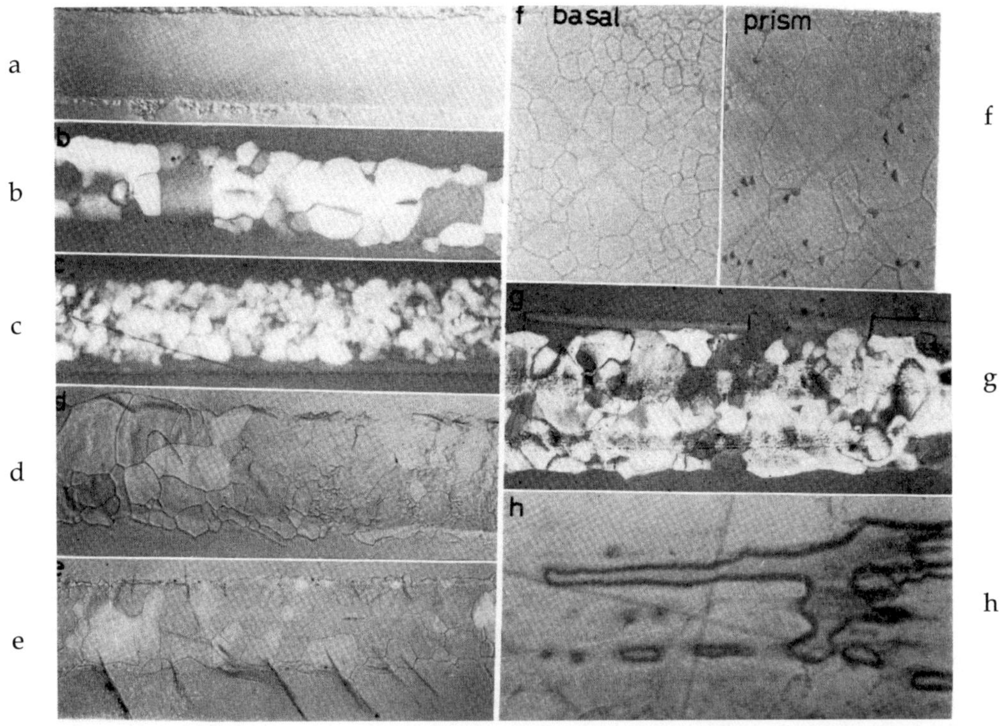

Fig. 13. Frictional track of ice a : low load, b : recrystallization on prism plane of ice, medium load, c : recrystallization and crack, heavy load, d : recrystallization of basal plane of ice, medium load, e : recrystallization and crack of basal plane of ice, heavy load, f :(left) recrystallization of basal plane, (right) recrystallization of prism plane, g : small angle grain boundary, crack and recystallization of prism plane, a~g at the temperature of -10°C, h : rarely pattern like melting at -30°C

It was found that friction on ice is very low even at a very small velocity; namely the μ_k varied from 0.005 to 0.16 for velocities ranging from 1.8×10^{-3} to 1×10^{-7} m/s. Such low friction on ice at extremely small velocities cannot be explained by frictional heating.

The temperature rise of ice, ΔT, due to friction can be expressed as follows (Bowden & Tabor, 1950):

$$\Delta T = \mu_k WV/[4a(k_1+k_2)], \quad (3)$$

Where μ_k is the coefficient of friction, W the load applied, V the friction velocity, a the radius of the contact area and k_1 and k_2 the thermal conductivities of slider and ice. The maximum temperature rise ΔT predicted from this formula is only 0.3 deg even when the maximum values of load W and friction velocity V used in the present experiment and the maximum value of friction coefficient 0.2 were substituted into the formula. It is obvious that the temperature rise due to frictional heating cannot cause melting of the ice.

It was also confirmed that melt water cannot be produced at the contact surface by pressure except at high temperatures.

According to adhesion theory, the frictional force F on ice can be divided into the shear resistance F_s and the ploughing resistance F_p: $F=F_s+F_p$. The coefficient of friction $\mu_k(=F/W)$ can, therefore, be written as the sum of the shear term μ_S and ploughing term μ_p: $\mu_k=\mu_s+\mu_p$. According to Bowden & Tabor (1950), F_s and F_p were respectively given by:

$$F_s=k\pi\varphi^2 s/4, \quad F_p=\varphi^3 p/6R^2 \quad (4)$$

where s and p are respectively the shear and the ploughing strength of ice, R is the diameter of slider, φ is the width of the sliding track and k is a constant. The coefficient of friction μ_k can, therefore, be expressed as

$$\mu_k = k\pi\varphi^2 s/4W + \varphi^3 p/6WR \quad (5)$$

As the first term, $k\pi\varphi^2/4W$, and a part of the second term, $\varphi^3/6WR$, are constant for given load and a temperature, this formula can be simply expressed as

$$\mu_k = A + a/R \quad (6)$$

where A and a are constant.

A linear relation was actually obtained between μ_k and $1/R$ be in the experiment on the effect of slider size (Fig.10(b)). This is evidence that the adhesion theory can be adopted for the friction of ice.

The values of s and p were estimated as follows: As described when considering the size effect, only a shear deformation took place in the contact area when a slider of diameter $R\geqq 9.5$mm was used.

The shear strength s is, therefore, given by $s = 4F/k\pi\varphi^2$, where the value of k is 0.8 as mentioned before. The value of the ploughing strength p was estimated from Equation (5). Since $\mu_p=\varphi^3 p/6WR$, the values of p can be obtained by substituting values of φ, W and R used in the experiment. The values thus obtained for s and p are 0.7 MPa and 75 MPa, respectively. By substituting into the Equation (3) these values of s and p, together with the experimental data obtained. These values are summarized in Table 4, together with some experimental data obtained for various load. As seen in this table, the coefficient of shear friction μ_s does not vary with load, while that of ploughing friction μ_p increases markedly with the increase in the load. The predicted values of $\mu_k(=\mu_s+\mu_p)$ agreed fairly well with those obtained by experiment for any load that ranged from 1.4 to 41 N as seen from the last column of Table 4 in which the ratio of $(\mu_s+\mu_p)$ to μ_k observed in the experiment was given. The fact that the predicted value based on the adhesion theory agreed well with those observed in the experiment. It should be emphasized that ice still exhibits a very low friction even though the ploughing effect is fairly large at very small sliding velocities.

4.2 Anisotropy of friction to sliding direction on same crystallographic plane of ice

4.2.1 Anisotropy in friction and track width on prism planes (10$\bar{1}$0)

Friction was measured every 10° on a prism plane (10$\bar{1}$0). No abrasive fragmentation occurred along the track, thus, friction tracks formed only by plastic deformation of ice.

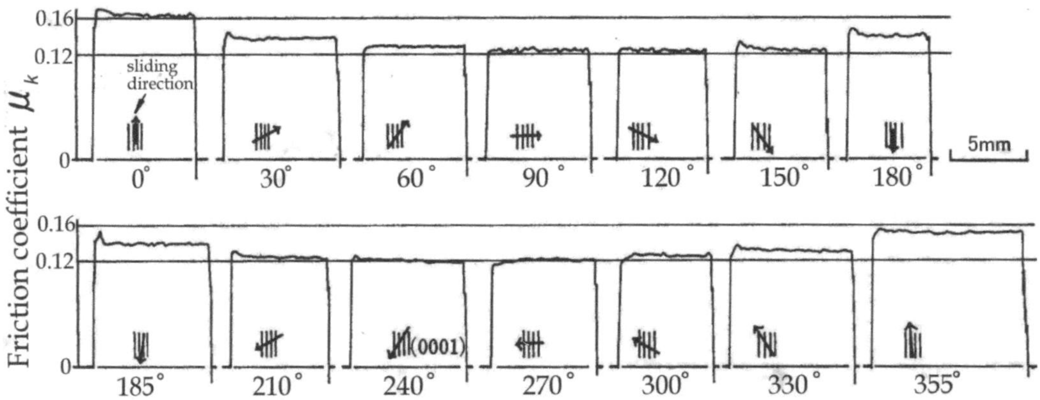

Fig. 14. Friction curve on prism plane (10$\bar{1}$0). Load : 6.9N, velocity : 7.4×10^{-5}m/s, temperature : -25°C, slider diameter 2.34 mm, allow shows direction of sliding.

Fig. 14 shows the record of friction as a function of the angle θ from the [10$\bar{1}$0] direction. Other parameters of the test were: temperature, -25°C; velocity, 7.4×10^{-5} m/s; applied load, 6.9 N; diameter of the slider, 2.34 mm. The coefficient μ_k reached its maximum in the [10$\bar{1}$0] direction and a minimum along [0001]. The value of μ_k ranged from 0.12 to 0.16, the ratio maximum/minimum being 1.3.

Fig. 15 shows photographs of the terminal areas of friction tracks produced by a slider on the prismatic surface at -21°C. The deformed regions are extended beyond the sides of the track revealing mainly horizontal slip lines and microscopic crack produced by the slider (2.34 mm in diameter). Fig. 15(b) shows a deformed bulge that moved in front of the slider parallel to the basal plane. Note that many cracks which are oriented normal to the basal planes propagate ahead of the slider, but that no significant deformation areas were found at the sides of track. The deformed area which formed near the terminus when the slider was moved diagonally to the basal plane (Fig. 15(c)) showed an intermediate pattern between those of Fig. 15(a)and (b). Note that many cracks were created normal to slip lines oriented in the [10$\bar{1}$0] direction.

From inspection of these photographs, we may conclude that when slider is moved parallel to the basal plane (Fig. 15(b)), comparatively higher values of μ_k may be obtained because of bulge formation in front of the slider.

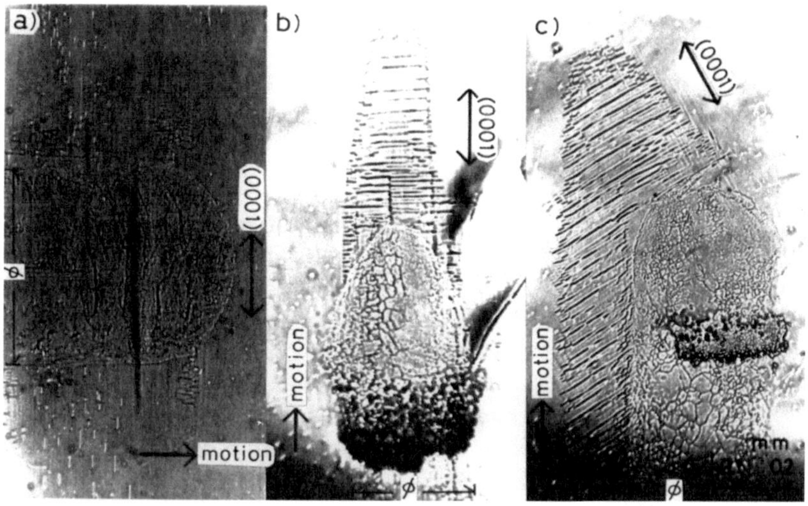

Fig. 15. Traces in track ends of friction on ($10\bar{1}0$). Sliding directions: (a) [0001]; (b) [$10\bar{1}0$]; (c) 30°from [0001]. Note that the basal slip lines extend parallel to [$10\bar{1}0$] and there is a dark line normal to the slip lines. Load 6.7N, velocity 7.4x10^{-5} m/s, Temperature -21°C, →sliding direction. (after Tusima, 1978)

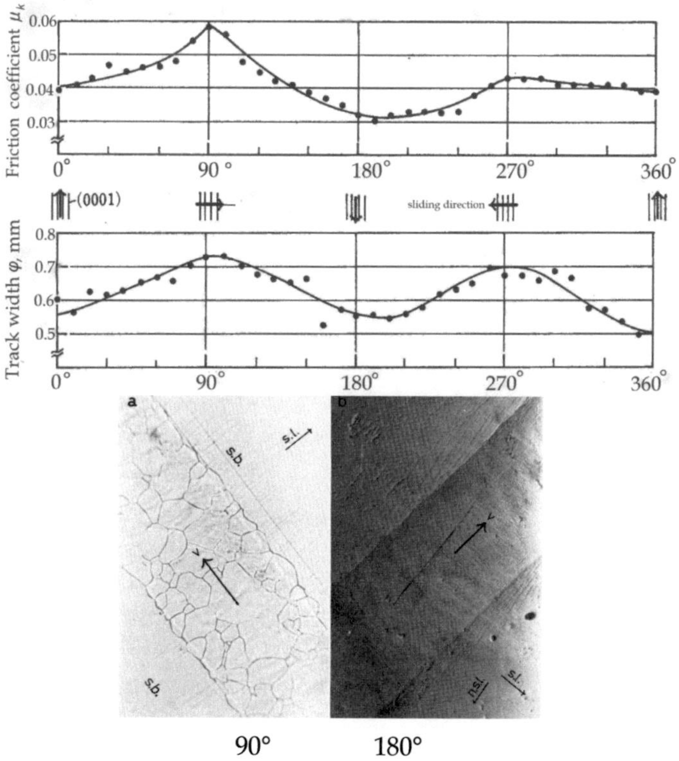

Fig. 16. Anisotropies in friction μ_k and track width f on ($10\bar{1}0$). Temperature: -10°C; load: 14.4N; velocity :6×10^{-5} m/s; slider: 6.4 mm in diameter. (after Tusima, 1978)

The friction coefficient reversed at temperatures of -10°C and above shown in Fig. 16. For this experiment, the temperature was -10°C, the velocity was 6.0×10^{-5} m/s, the load was 14.4 N, and the slider diameter was 6.4 mm. The maxima of the friction coefficient and the track width were observed to be in the same direction, which contrasts with the results obtained at temperatures of -21°C and below.

4.2.2 Anisotropy in friction on inclined surfaces to (0001)

As shown in Figs. 14 and 16, a significant amount of anisotropy in the μ_k value was found on the prismatic surface where the orientation of the basal planes in ice are normal to the test surface, but on the basal surface itself no such anisotropy was observed. This may suggest that the μ_k value measured on a given surface of ice depends on the relationship between the sliding direction and the orientation of the basal planes in ice. Fig. 17 shows anisotropies in the μ_k value and the track width measured on a surface of ice cut diagonally against the basal plane. In this sample the (0001) plane was inclined at approximately 30° to surface. The abscissa is the angle of revolution of the test surface against the slider. When the angle of the test surface was fixed at 0°, the slider moved parallel to the basal plane on the surface.

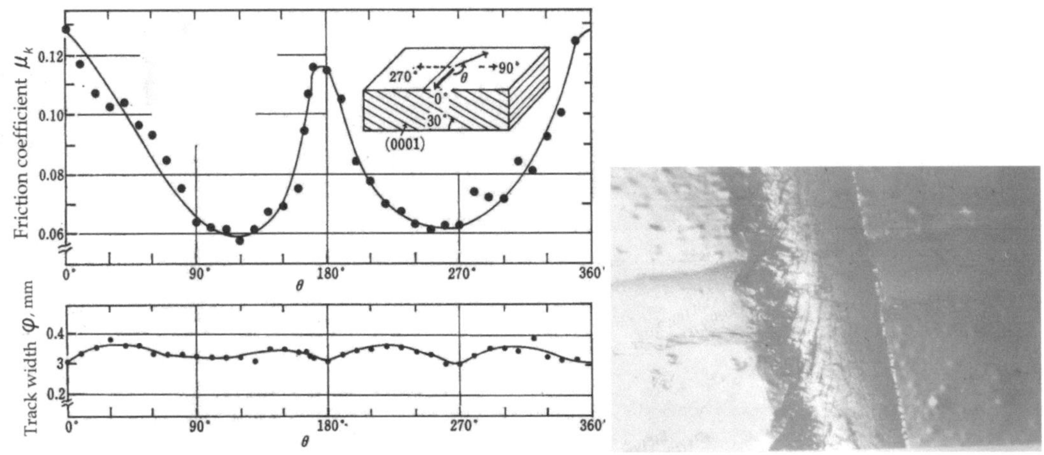

Fig. 17. Anisotropies in friction μ_k and track width φ on (0001) declined at 30°. Temperature: -20°C; load: 6.9N; velocity : 7.4×10^{-5} m/s; slider: 2.34 mm in diameter (right) frictional track (after Tusima,1978)

As shown in Fig. 17, two maxima in μ_k appeared at 0° and 180° (where the slider moved in parallel with the basal plane), but two minima appeared at about 120° and 260° (where the slider moved nearly perpendicular to the basal plane). Although values of μ_k changed significantly with sliding direction, a slight variation in track width was observed.

As seen above, μ_k values on the prismatic plane in diagonally cut surfaces depended on sliding direction. In order to interpret the observed anisotropy, the friction track produced on the specimen surfaces was observed using optical microscopy. According to our observation, recrystallization, cleavage fissures, microcracks, slip lines, and small-angle boundaries were found to have formed along a friction track.

The coefficient of kinetic friction μ_k may be explained solely in terms of prismatic deformation of ice. According to the adhesion theory, μ_k is given by

$$\mu_k = (k\varphi^2 s/4 + k'\varphi^3 p/6R)/W \quad (7)$$

where s is the interfacial shear strength between ice and slider, p the ploughing strength of ice, R the diameter of slider, W the applied load on the slider, φ the width of track produced by friction, k and k' the shape factors (for the sake of convenience, we shall assume that k = 0.8 and that k' =1 (Tusima ,1977).

As seen in Equation (1), μ_k is composed of two terms, the first terms, interfacial shear and the ploughing effect of the slider.

Fig. 10 shows μ_k, measured on the prism surface, as a function of temperature and reciprocal slider diameter. The velocity of the slider and value of applied load are indicated on the figure. μ_k is inversely proportional to the diameter of the slider, suggesting that Equation (1) can be used in the interpretation of our results. According to Equation (5), the effect of slider ploughing disappears if the slider can be considered to have an infinite diameter. The largest slider diameter used in our experiments was 12.5 mm. If we assume that the ploughing term is negligible in value of μ_k measured with this slider, then we can plot the shear strength of ice calculated by the shear term in Equation (4) as a function of temperature (Fig.18).

From Equation (5), the ploughing strength p of ice is expressed as

$$p = 6R(\mu_k W - k\varphi^2 s/4)k'\varphi^3 \quad (8)$$

Since the s is known from Fig. 18, if we substitute numerical values for μ_k and φ measured in the various sliding directions into Equation (8), we can estimate p as a function of direction. Fig. 19 shows the p anisotropy measured on the prism surface $(10\bar{1}0)$, p is one of the indices used to express ice surface hardness.

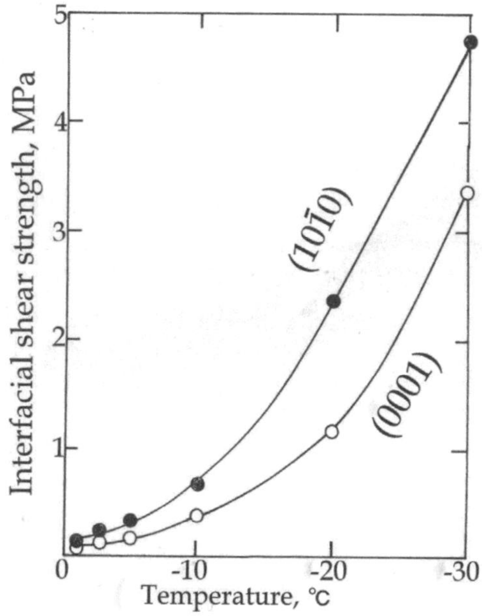

Fig. 18. Interfacial shear strength between ice and steel, plotted against temperatures, for planes $(10\bar{1}0)$ and (0001). (after Tusima, 1978)

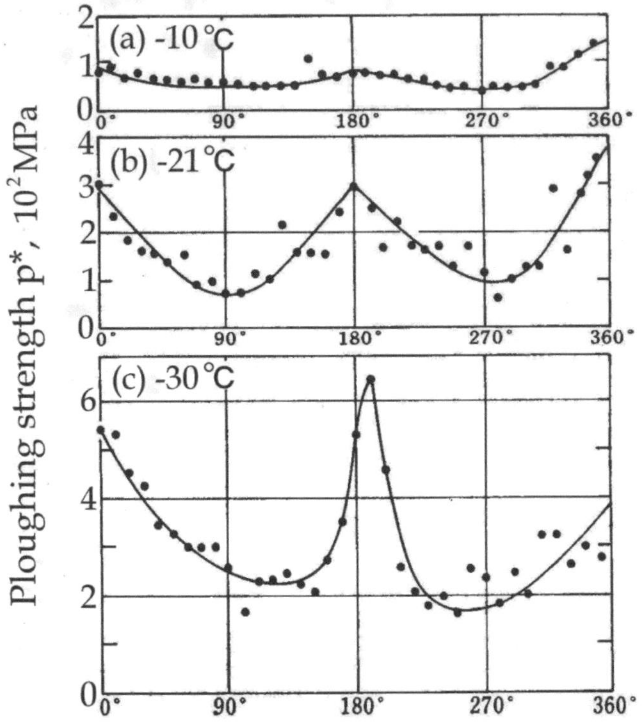

Fig. 19. Ploughing strength on (10$\bar{1}$0) calculated from friction coefficients and track width. (a) Temperature: -10°C; load: 14.4N; velocity: 6.0×10-5 m/s; slider: 6.4 mm in diameter. (b) Temperature: -21°C. (after Tusima, 1978)

Butkovich (1954) and Offenbacher & Roselman (1971) measured the hardness anisotropy of ice single crystals. Offenbacher & Roselman found that, on the prism plane, Knoop hardness measured in the direction normal to the basal plane was smaller than the value obtained parallel to the basal plane. The p anisotropy obtained by the present author seems to agree roughly with the Knoop-hardness values obtained by Offenbacher & Roselman (1971).

The appearance of a maximum in the value of p along [10$\bar{1}$0] may be understood by the bulge formed ahead of slider (Fig.15(b)). The anisotropy in the coefficient of kinetic friction on the prism surface of ice can be explained in terms of the anisotropy in p and φ. Therefore, the anisotropy in μ_k depends on $p\varphi^3$. The value of p and φ showed maxima or minima in the sliding direction [10$\bar{1}$0] and [0001]. Thus, the frictional anisotropy μ_k[10$\bar{1}$0]/μ_k[0001]may be proportional to

$$(p[10\bar{1}0]/p[0001])\times(\varphi[10\bar{1}0]/\varphi[0001])^3.$$

These values are summarized for various temperatures in Table 5. When the ratio of $p\varphi^3$[1120]/$p\varphi^3$[0001] becomes smaller than 1, frictional anisotropy is dominated by the track-width anisotropy. When the ratio is larger than 1, frictional anisotropy is dominated by ploughing-strength anisotropy. Both results agree with the experimental observations. The anisotropy in friction can be well explained by the anisotropies in ploughing strength and track width.

Temperature °C	$(\varphi[10\bar{1}0]/\varphi[0001])^3$	$p[10\bar{1}0]/p[0001]$	$\mu[10\bar{1}0]/\mu[0001]$	
			calculated	observed
-10	0.40	1.8	0.72	0.67
-21	0.39	3.4	1.3	1.3
-30	0.51	2.8	1.4	1.2

Table 5. The ratio of track width, ploughing strength, and friction coefficient between directions of $[10\bar{1}0]$ and [0001] for different temperatures on prism plane $(10\bar{1}0)$. (after Tusima, 1978)

Friction anisotropy on the basal (0001)and prism $(10\bar{1}0)$ planes of ice was measured as a function of track width, the amount of plastic deformation caused by frictional sliding, as so on. It was found that, for the prism planes, friction reaches a maximum in the $[10\bar{1}0]$ sliding direction on the $(10\bar{1}0)$ planes and in the $[10\bar{1}0]$ direction on the $(10\bar{1}0)$ planes. Friction is at a minimum in the [0001] direction for both planes at temperatures below -19°C. At temperatures of -10°C and above, the maximum friction was observed in the [0001] direction and the minimum in the $[10\bar{1}0]$ direction for $(10\bar{1}0)$ planes, and in the $[10\bar{1}0]$ direction for $(10\bar{1}0)$. A remarkable friction anisotropy was $(10\bar{1}0)$ planes, and in the $[10\bar{1}0]$ direction for $(10\bar{1}0)$. A remarkable friction was observed on pyramidal planes, although track-width anisotropy was very small. No marked anisotropy in friction was observed on the basal plane.

The width of the frictional track also varied with the sliding direction on the prism plane; that is, it was at a maximum along [0001] and reached minima along $[10\bar{1}0]$ for the $(10\bar{1}0)$ plane and along $[10\bar{1}0]$ for the $(10\bar{1}0)$ plane, independent of temperature.

4.3 Friction of plastic balls on ice

4.3.1 Experimental method and samples

The apparatus shown in Fig. 2 was used measurements μ_k of polymer balls on a flat plate with crystal orientation (0001) of single-crystal ice.

Polymer balls obtained from Urtraspherics Co.Ltd., USA, were: polymethylmethacrylate (PMMA), polytetrafluoroethylene (PTFE, Teflon), polypropylene (PP), polyethylene (PE), polyamide (PA, Nylon), polyacetal (POM), polycarbonate (PC) and polystyrene (PS). They were also cleaned in the same manner as a steel and tungsten-carbide ball (TC) without acetone.

4.3.2 Friction and frictional track of plastic balls on ice

The relations of μ_k to the load applied are shown in Fig. 20. The values of μ_k which were roughly constant for loads except for PMMA, were high, from 0.04 to 0.15. Constancy in μ_k follows from softness of plastics, compared with ice.

Plastic ball except PMMA showed frictional track wider than steel ball and its μ_k were greater than steel ball. Ice surfaces were disturbed markedly as shown in Fig. 21. Fig. 21 shows many streak, abraded particle, recrystallization, crack. The value of μ_k was the greater, surface damage of ice the heavier.

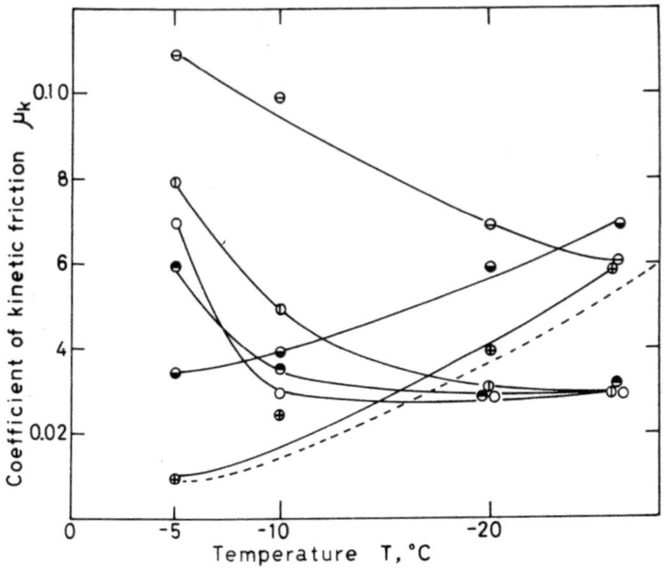

⊖:PA; ⏀:PE; ○:PTFE; ◓:POM; ◒:PP; ⊕:PMMA;

Fig. 20. Dependence of friction on temperature on basal plane (0001) of ice. Load: 4.75 N; velocity: 7.4x10^{-2} mm/s; diameter of slider: 6.4 mm. broken line: steel ball. (after Tusima, 1980)

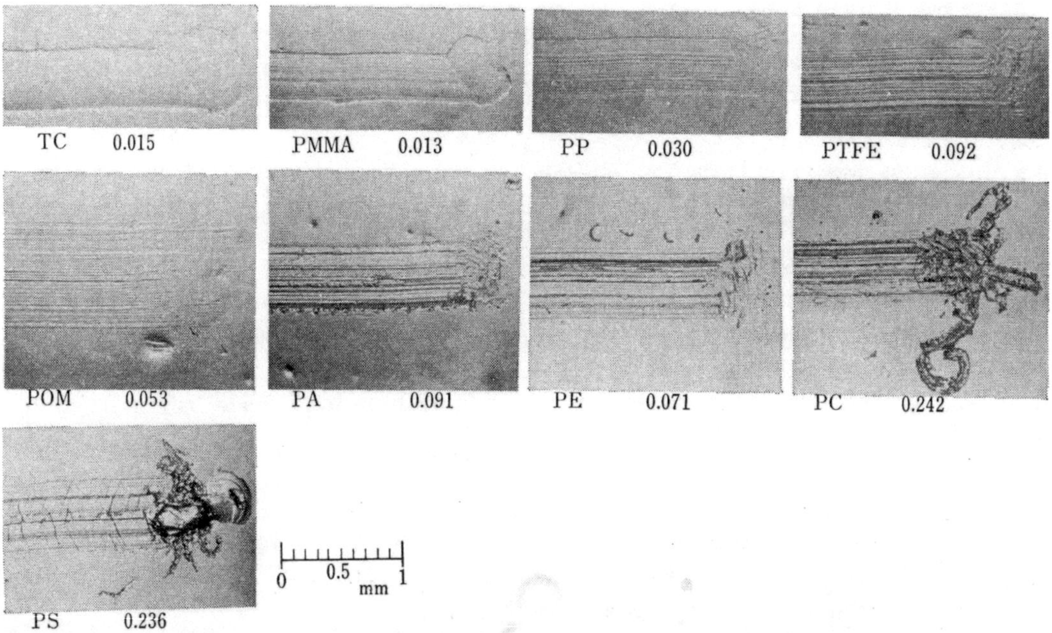

Fig. 21. Trace of a track marked by a slider ball on basal plane (0001) of ice. Figure under each picture shows coefficient of kinetic friction. Temperature:-5°C, load: 47.5N, velocity: 7.4x10^{-2}mm/s, diameter of a slider: 6.4mm (after Tusima,1980)

5. Other frictional properties of ice and snow

5.1 Sliding of speed-skate on speed-skate ice-rink

We knew that friction of ice had a minimum value in (0001) plane of ice at sliding of tungsten carbide ball on single crystal of ice. This relation was confirmed with sliding of speed skate as shown in Fig. 22 and Table 6. At first, ice plates with (0001) was connected in linearly, and test course gained of 1 m long.

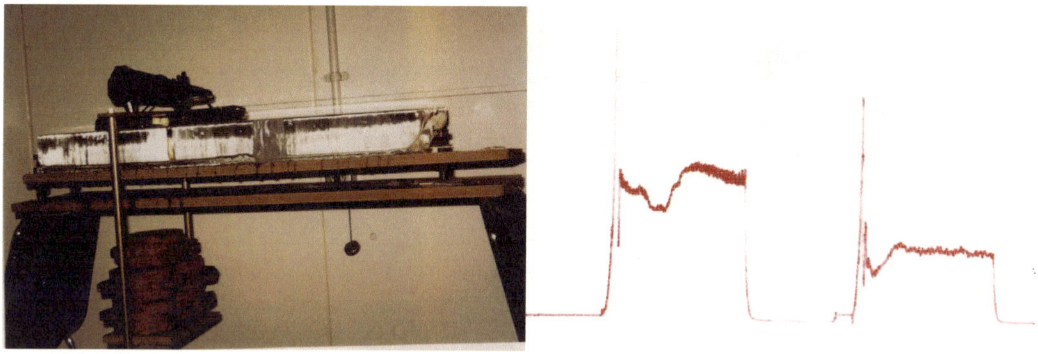

Fig. 22. Sliding test by real skate left: μ0.031 on polycrystalline ice, right: μ 0.021 on (0001) of ice. Temperature: -10°C, velocity :0.05m/s, load: 550N

Ice surface cut carefully by microtome and one blade of speed-skate was slid by pulling constant speed. Friction force detected electrically by load cell and recorded on paper as shown in Fig. 22. Friction coefficient was 30% smaller on (0001) plane of ice than polycrystalline commercial ice.

Next, rink test is planned to control (0001) ice surface for skating rink because low friction of (0001) is confirmed on ice sample. Numerous large single-crystal of ice (diameter 0.1~0.15m, long 0.3~0.4m) was grown in direction parallel to c-axis in temporary prepared cold room of length 190m (Fig. 23). Thin plate of single-crystal (thickness 7mm) was cut by band-saw and the plate was pasted on ice-rink and grown by repeatedly supply a small amount of tepid water (40°C). Vertical thin section of obtained ice was shown in Fig. 25, left figure was view transparent light and right figure the view of cross polaroid. The crystallographic orientation was maintained constant in growing process of oriented over growth. Normal rink is polycrystalline as shown in Fig. 24.

Preliminary test was made in short track of skating rink. Test skate with speed-skating brades was started at the initial speed, v_0, of 1m/s. The distance, ℓ, until stop was measured. Average friction coefficient, μ, was determined as $\mu = v_0^2/2g\ell$.

Obtained results is shown in Table 6 at initial velocity about 1 m/s and ice temperature -3°C, load of 392N. Friction coefficient 0.0064 on (0001) is compared 0.0081 on normal ice rink, the coefficient decreased 21% in (0001) than one of polycrystalline ice rink (normal rink).

Next, main test repeated in full scale speed-skating rink at Olympic Memorial Arena "M-Wave" Nagano, Japan. In this case, all surface round, 400m, changed to (0001) of ice. Therefore, comparison of μ was tried between (0001) ice of speed- skate rink and polycrystalline ice of virgin ice-hockey rink on same floor. Table 6 shows comparison (0001) rink to normal rink at initial starting velocity 1m/s. Friction coefficient 0.0038 on (0001) was 16% smaller than 0.0045 on normal rink. Measured coefficient of friction when initial velocity changed in the range 0.2~3.5m/s was low as 0.003~0.008. At lower than 1m/s, the coefficient increased with decreasing velocity though the value was 0.004~0.006 in higher initial velocity range than 1 m/s.

Fig. 23. Ice making on ice-stalagmite Right: Horizontal section of ice-stalagmite

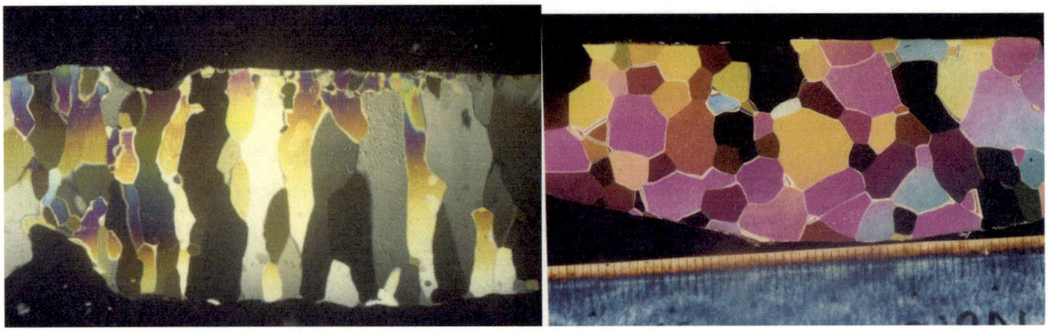

Fig. 24. Normal rink ice left: vertical section, right: horizontal section (scale: minimum division 1mm)

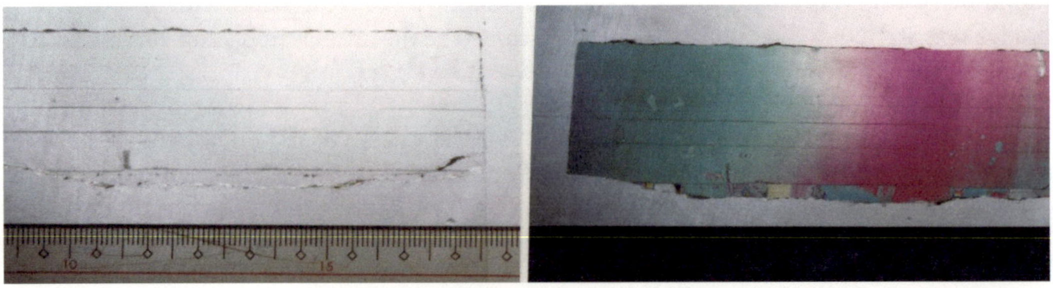

Fig. 25. Ice-stalagmite rink controlled to (0001) surface, Vertical section. Scale: minimum division 1mm

Fig. 26. Test skate by catapult

Frictional melting does not occur in these conditions, any water did not exist. Pure solid friction will attain in these test, therefore it is considered to be reflected in sliding the properties of crystallographic plane of ice. If water concern sliding on ice, the water should be prepared by the frictional melting. The decrease of friction means the decrease of heat, therefore it is impossible to melt in extremely small friction as sliding of speed-skate shown in Fig. 27. It is concluded that anisotropy of friction by crystallographic plane of ice caused in pure solid friction without liquid.

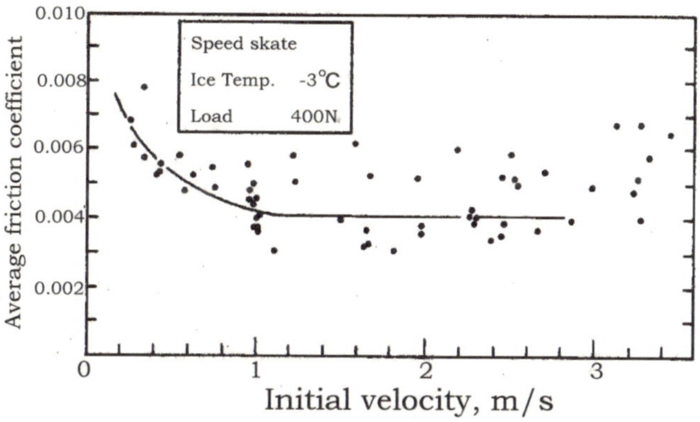

Fig. 27. Relation between initial velocity and average friction coefficient (M-wave, Speed-skating rink, Nagano, Japan) (after Tusima,2009)

Temperature	-3.1°C	-4.1°C
	Speed skate rink (0001)	Ice hockey rink polycrystalline
	0.0040	0.0046
	0.0037	0.0044
	0.0036	0.0046
	0.0037	0.0042
	0.0040	0.0045
mean	0.0038	0.0045

Table 6. Comparison of friction coefficient between test ice (0001) and normal one (M-wave, Nagano) Load: 294N, Initial velocity: 0.9~1.1m/s. (after Tusima et.al.,2000)

It is noticed the increase of μ the decreasing initial velocity. This μ−v relation will give the explanation of curl of curling rock.

5.2 Tribology of curling stone on pebbled ice sheet

5.2.1 Friction of curling stone on curling rink ice

Friction of rock on curling rink ice estimated as following two methods. Average friction coefficient μ was derived from $mV_0^2/2 = \mu mgs$. Fig. 28 shows the average friction. The fiction coefficient increased in low speed region though constant for high speed. In generally, curling rock moves with weak spin. When counter clockwise spin is given, left side will decelerate in velocity therefore friction increases as shown in Fig.29. Anisotropy of friction between left and right will be cause of curl as mentioned by Harrigton (1924) and Denny (1998).

Friction coefficient as a function of velocity will be given as follows; The change of sliding distance by unit time gives velocity; the change of velocity by unit time gives μg (where, g is gravitational acceleration). Therefore if the relation between sliding distance, s, until stop and sliding time, t, could be gained, frictional coefficient μ will be given;

$$F = m\,dv/dt = -\mu mg$$

where F is the frictional resistance, m the mass of rock, v the velocity.

$$\mu = -(1/g)(dv/dt) = -(1/g)(d^2s/dt^2)$$

Fig. 30 shows the log-log relation between sliding distance s and time t until stop of rock.

In generally, stone is given very slow rotation (2~3 rotation until stop), it is wander the extremely low damping of angular velocity. To understand this reason, it is to clear the transverse friction component f_y and angular friction one f_θ.

Rock will almost move without decrease of angular velocity except around stop because friction to rotation becomes extremely small compare to longitudinal one.

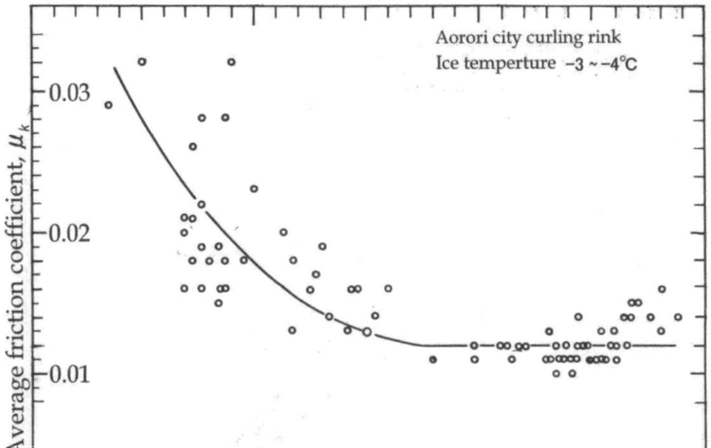

Fig. 28. Average friction coefficient of rock for initial velocity. (after Tusima, 2010)

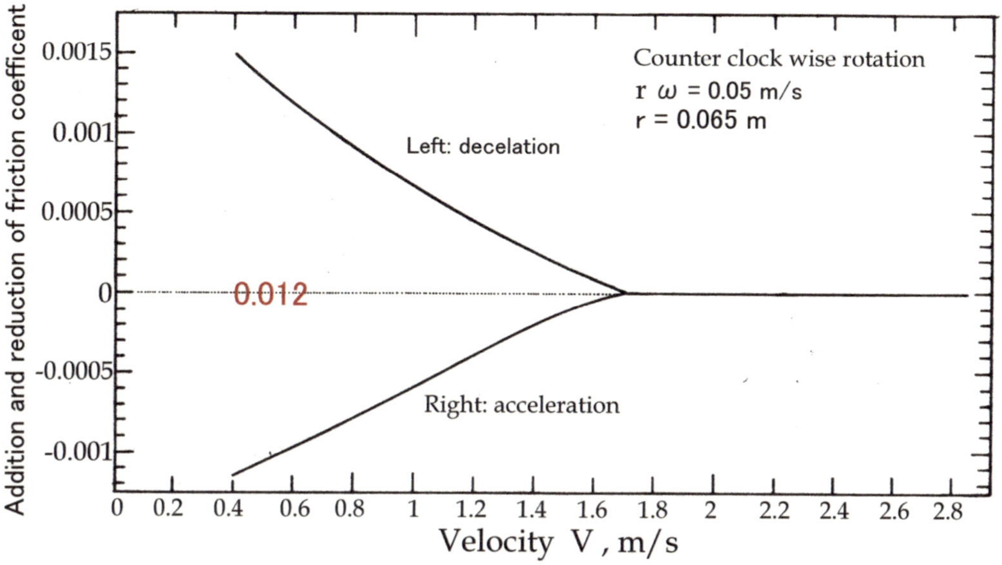

Fig. 29. Addition and reduction of friction by rotation of rock. (after Tusima, 2010)

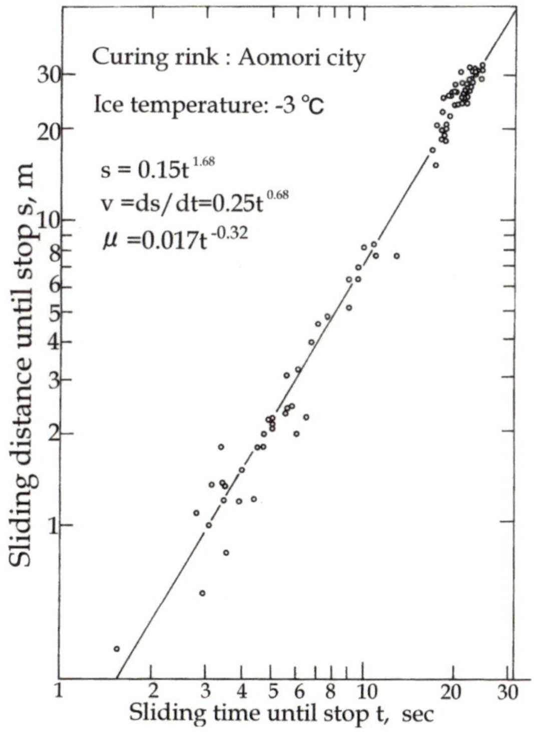

Fig. 30. Relation of sliding distance s to sliding time t until stop of rock. (after Tusima, 2010)

We suppose frictional resistance f_y as moving direction and f_θ as tangential direction of rotation. Total resistance is f. The angle between f and f_y supposed to ß.
θ is angle from x axis.

$$f_\theta = (\sin ß / \sin\theta) f$$

$$\mu_\theta = (f_\theta / f)\mu$$

$$= \{r\omega / [v^2 + 2r\omega v\cos\theta + (r\omega)^2]^{1/2}\}\mu$$

If we suppose μ of 0.01, v of 1 m/s and rω of 0.04 m/s, we will get about 0.0004 for μ_θ. This means angular velocity of rock does not almost decrease at high sliding speed.

6. Conclusion

In this chapter, it was noticed two points connected to friction: the considerably high value of ice hardness and extremely small shear strength of ice compared to hardness. This leads low friction in the view point of adhesion theory.
The friction coefficient of steel ball on flat plate of ice was low even in low speed range. The friction showed remarkable anisotropy on crystallographic plane and sliding orientation. Frictional track on ice also showed remarkable anisotropy by the crystallographic plane and sliding orientation. These frictional features were well explained by adhesion theory. The friction of speed-skate was around 0.005, and this extremely low value gave dry friction, therefore the crystallographic planes of ice reflected to the friction coefficient of skate.
In frictional research history on ice, almost research had overlooked above mechanical characteristics of ice and had thought much of fluid lubrication which based on the analogy of slipperiness and type of friction. Though the thermal analysis have explained well the frictional properties on ice, however it is not astonishing because the most frictional energy converts to heat. Thermal analysis apply to the frictional melting theory, adhesion theory and other.
In frictional melting theory, it must exist liquid water which is prepared by fusion of ice. Therefore its theory needs the heat to melt which is given by frictional heat. If the heat is insufficient (namely, extremely low friction), the theory can not apply. This is logical contradiction of frictional melting theory for slipperiness of ice. Of course, if liquid exist, it will lead to low friction by weakness of contact bond.

7. References

Ambach, W. & Mayr, B. (1981). Ski sliding and water film. *Cold Region Science and Technology*, Vol.5, pp.59-65.

Barnes, P. & Tabor, D. (1966). Plastic flow and pressure melting in deformation of ice I. *Nature*, Vol.210, pp.878-882.

Barnes, P.; Tabor, D., & Walker, J.C.F. (1971). Friction and creep of polycrystalline ice. *Proceedings of the Royal Society London*, Ser. A, Vol.324, No.1557, pp.127-155.

Bowden, F.P. (1953). Friction on snow and ice. *Proceedings of the Royal Society of London*, Ser.A, Vol.217, No.1131, pp.462-78.

Bowden, F.P. (1955). Friction on snow and ice and the development of some fast-running ski. *Nature*, Vol.176, No.4490, pp.946-47.

Bowden, F.P., & Hughes, T.P. (1939). The mechanism of sliding on ice and snow. *Proceedings of the Royal Society of London*, Ser. A, Vol.172, No.949, pp.280-98.
Bowden, F.P. & Tabor, D. (1950). *The friction and lubrication of solids*. [Part 1.] Oxford, Clarendon Press.
Butkovich, T.R. (1954). Hardness of single ice crystals. *U.S.Snow, Ice and Permafrost Research Establishment*. Research Paper 9. 12pp.
Denny, M. (1998). Curling rock dynamics. *Can. J. Phys.* Vol.76, pp.295-304.
Evans, D.C.B.; Nye, J.F., & Cheeseman, K.J. (1976). The kinetic friction of ice. *Proceedings of the Royal Society of London*, Ser.A, Vol.347, No.1651, pp.439-512.
Furushims, T. (1972). Studies on the frictional resistance of skate. *Seppyo*, Vol.34, pp.9-14.
Harrington, L.E. (1924). An experimental study of the motion of curling stones. *Proceedings and Transactions of the Royal Society of Canada*. Vol.18, Sect. 3, pp.247-259.
Hori, T. (1956). On the supercooling and evaporation of thin water film. *Low Temperature Sience*(Hokkaido University), Ser.A, [No.] 15, pp.33-42
Huzioka, T. (1962). Studies on the resistance of snow sledge.V. *Teion-kagaku: Low Temperature Science*, Ser.A, [No.] 20, pp.159-79.
Huzioka, T. (1963). Studies on the resistance of snow sledge.Ⅵ. *Teion-kagaku: Low Temperature Science*, Ser.A, [No.]21, pp.31-44.
Itagaki, K. & Huber, N.P. (1968): Measurement of dynamic friction of ice. *Physics and chemistry of ice*. Eds. by N. Maeno & T. Hondo, pp.212-218.
Jellinek, H.H.G. (1957). Adhesive properties of ice. *SIPRE Research Report*, No.38, 20pp.
Jellinek, H.H.G. (1967). Liquid-like (transition) layer on ice. *Jounal of Colloid and Interface Science*, Vol.25, pp.192-205.
Jellinek, H.H.G. (1970). Ice adhesion and abhesion: A survey. *CRREL Special Report*, No.115, pp.46-77.
Joly, J. (1887). The phenomena of skating and Professor J.Thomson's thermodynamic relation. *Scientific Proceedings of the Royal Dublin Society*, New Series, Vol.5, Pt 6, pp.453-54.
Kobayashi, T. (1973). Studies of the properties of ice in speed skating rink. *ASHRAE Journal*, January, pp.52-56.
Koning, J.J.de; Groot, G.de, & IngenSchnau, G.J.van. (1992). Ice friction during speed skating. *J. Biomechanics*, Vol.25, pp.565-571.
Landy, M. & Freiberger, A. (1967). A studies of ice adhesion. *J. Colloid and Interface Science*, Vol.25, pp.231-244.
McConica, T.H. (1950). Sliding on ice and snow. American Ski Company, Technical Report.
Niven, C.D. (1959). A proposal mechanism for ice friction. *Canadian Journal of Physics*, Vol.37, No.3, pp.247-255.
Offenbacher, E.L. & Roselman, I.C. (1971). Hardness anisotropy of single crystal of ice Ih. *Nature, Physical Science*, Vol.234, No.49, pp.112-113.
Raraty, L.E. & Tabor, D. (1958). The adhesion and strength properties of ice. *Proceedings of the Royal Society of London*, Ser.A, Vol.245, No.1241, p.184-201.
Reynolds, O. (1899). On the slipperiness of ice. *Memoirs and Proceedings of the Manchester Literary and Philosophical Society*, Vol.43, Pt.2, No.5.
Shimbo, M., (1961). The mechanism of sliding on snow. *Union Geodesique et Geophysique Internationale. Association Internationale d'Hydrologie Scientifique.* Assemble generale

de Helsinki, 25-7-6-6 1960. Commission des Neiges et Glaces, pp.101-06. (Publication No.54. de l'Association Internationale d'Hydrologie Scientifique.)

Tabor, D. & Walker, J.C.F. (1970). Creep and friction of ice. *Nature*, Vol.228, pp.137-139.

Tusima, K. (1976). A review of mechanism of friction of ice. *Journal of Japan Society of Lubrication Engineering*, Vol.21, pp.287-294.

Tusima, K. (1977). Friction of a steel ball on a single crystal of ice. *Journal of Glaciology*, Vol.19, No.81, pp.225-235.

Tusima, K. (1978). Anisotropy of the friction on a single crystal of ice. Journal of Glaciology, Vol.21, [No.]85, pp661-668.

Tusima, K. (2009). Tribology in skating, skiing, and curling. *Japanese Journal of Tribology*, Vol.54, [No.]7, pp.470-475.

Tusima, K. (2010). Estimation friction coefficient of stone and mechanism of curl. *Proceedings of Cold Region Technology Conference*. Vol.26, pp.422-427.

Tusima, K. & Fujii, T. (1973). Measurements of shear strength of ice. *Low Temperature Science*, Ser.A, No.31, pp.33-43.

Tusima, K., & Yosida, Z. (1969). Melting of ice by friction. *Low Temperature Science*, Ser.A, [No.]27, pp.17-30

Tusima, K.; Yuki, M., Kiuchi, T., & Shimodaira, S. (2000). Development of a high speed skating rink by the control of crystallographic plane of ice. *Japanese Journal of Tribology*, Vol.45, pp.17-26.

Weyl, W.A. (1951). Surface structure of water and some of its physical and chemical manifestation. *J. Colloid and Interface Science*, Vol.6, pp.389-405.

16

Tribology of 2-Mercaptobenzothiazole in Lithium Complex Grease

B S Nagarkoti
Hindustan Petroleum Corporation Limited, R&D Centre, Plot D - 500,
T T C Industrial Area, Turbhe, Navi Mumbai - 400 705
India

1. Introduction

There has been no much difference in usage pattern of EP/antiwear additive chemistry of additives in greases and other lubricants like cutting fluids and gear oils. These additives perform very well in some greases but not able to perform up to desired level in other applications. Due to high active sulfur content, they need supplementary additive with them for non-ferrous metal corrosion protection. Additive chemistry plays a vital role in performance of greases, none the less base fluid and thickener play an important role. As the greases is worked and churned in bearing, the grease is required to liberate some oil to lubricate race way surfaces. Some of the additives also released with oil but not able to go back in grease's jell structure. As a result depletion of additive takes place gradually and performance level of grease decrease. Thus the life of grease and bearings reduced. In view of this, the usage of solid additives with nano particle size is increasing in high performance greases in recent times. The biggest advantage of solid additives with nano particle size in greases is it provides the structure to grease by working as the thickener/fillers and support the development of lubricating film on friction contact surfaces. Thus reduces the chances of depletion of additives with oil due to break down of grease structure due to high temperature and high load during operation. Analyzing the properties of 2-Mercaptobenzothiazole, we have attempted to study the performance of this additive in lithium complex grease. We have evaluated this additive in wide perspective of tribological behaviour in different combination with antiwear additives. Though the primary use of 2-Mercaptobenzothiazole is vulcanization accelerator for rubber, fungicide, copper corrosion inhibitors in various lubricants and coolants at a very low treat level. 2-Mercaptobenzothiazole has very high sulfur content hence it can also be used as an excellent EP additive. Due to presence of two nitrogen atom in MBT, it is slightly basic in nature which helps to neutralize the acidic components formed in application thus increases the life of grease. Additionally, due to solid nature it can work as a filler in greases and subsequently increases the yield of greases.

2. Experimental details

Lithium complex grease was prepared by conventional method by using 12-hydroxy stearic acid and lithium hydroxide. Dodecanedioic acid was used as complexing agent in Lithium soap grease. An ISO VG 220 mineral oil was selected as a base fluid for better lubricity at high temperature application. The grease consistency was kept in NLGI 2 range. This grease was fortified with commercially available antioxidant, rust inhibitor to meet other requirements i.e. Thermal and oxidation stability, rust & corrosion resistance, water washout resistance property. Series of extreme pressure properties of Lithium complex grease were checked in following combinations of EP and antiwear additives.

1. Lithium complex grease without any EP and antiwear additive
2. Lithium complex grease with only 2.5% 2-Mercaptobenzothiazole (MBT)
3. Lithium complex grease with combination of 2.5% 2-Mercaptobenzothiazole and 1.0% Zinc dialkyl dithiophosphate antiwear additive
4. Lithium complex grease with combination of 2.5% 2-Mercaptobenzothiazole and 1.0% ashless antiwear additive
5. Lithium complex grease with combination of 2.5% conventionally available sulfur-phosphorous chemistry based EP additive and 1.0% Zinc dialkyl dithiophosphate antiwear additive.

The coefficient of friction was tested by universal tribometer in following combinations of EP and antiwear additives.

1. Lithium complex grease with combination of 2-Mercaptobenzothiazole and Zinc dialkyl dithiophosphate antiwear additive
2. Lithium complex grease with combination of 2-Mercaptobenzothiazole and ashless antiwear additive

The testing conditions were as under.

1. Load - 300 N
2. Temperature - 60^0C
3. Frequency - 50 HZ
4. Time - 120 minutes
5. Contact - Pin-on-disk mode

3. Results and discussion

The test results of Lithium complex grease with 2- Mercaptobenzothiazole and ZDDP vis-à-vis Lithium complex grease with conventional EP additive and ZDDP are provided in **Table 1**. The various regular properties of grease with new additive combinations i.e. combination of 2-Mercaptobenzothiazole and Zinc dialkyl dithiophosphate antiwear additive are similar with Lithium complex grease with combination of conventional EP additive and Zinc dialkyl dithiophosphate. This reveals that 2-Mercaptobenzothiazole has compatibility and good additive response in lithium complex grease. The series of extreme pressure properties of greases with different combinations of 2-Mercaptobenzothiazole with Zinc dialkyl dithiophosphate, ashless antiwear additive and grease with commercially available extreme pressure additive in combination with Zinc dialkyl dithiophosphate are provided in **table 2**. The four ball weld load of lithium complex grease 2 (only 2.5% MBT) has 500 kg weld load. The wear scar dia is 0.78 mm. After addition of 1.0% Zinc dialkyl dithiophosphate and 1.0% ashless antiwear additive respectively in this grease the wear scar dia has come down to 0.30 mm in grease 3 and 4 respectively. Additionally, the weld load is increased by 50 kg. Whereas, the grease with conventional EP additive & antiwear additive has shown a weld load of 315 kg and wear scar dia 0.42 mm. The Timken OK load is also slightly higher with MBT compared to conventional EP additive. These higher values

Characteristics	Lithium Complex Grease with 2 - Mercaptobenzothiazole and ZDDP	Lithium Complex grease with conventional EP additive and ZDDP	Test Method
NLGI Grade	NLGI 2	NLGI 2	NLGI
Consistency, @ 25°C Worked, 60 X Worked, 100, 000 X	 281 305	 283 309	ASTM D 217
Drop Point,°C	272	272	ASTM D 566
Copper Corrosion @ 100 °C, 24 hrs	1 a	1a	ASTM D 4048
Heat Stability, @ 100°C, 30 hrs % loss	1.85	1.76	ASTM D 6184
Wheel Bearing test, Leakage by mass, gm Slump test	 2.63 Pass	 2.66 Pass	ASTM D 1263
Water washout @ 80°C, % loss wt.	4.66	4.98	ASTM D 1264
Roll Stability, % change @ ambient, after 16 hrs @ 82°C, after 48 hrs.	 6 16	 7 18	ASTM D 1831
Oxidation Stability, @ 100°C Drop in psi, @ 100 hrs.	6	6	ASTM D 942
Emcor Rust Test, rating	0,0	0,0	IP 220

Table 1. Test Results of Lithium Complex Grease with 2 - Mercaptobenzothiazole and ZDDP antiwear additive vis-à-vis Lithium Complex grease with conventional EP additive and ZDDP antiwear additive

Characteristics	Grease 1	Grease 2	Grease 3	Grease 4	Grease 5	Test Method
Load wear Index, kg	66	132	166	166	112	ASTM D 2596
Four ball weld Point,kg	200	500	550	550	315	ASTM D 2596
Four ball Wear scar dia, mm	0.80	0.78	0.30	0.30	0.42	ASTM D 2266
Timken OK load, lb	15	40	50	50	45	ASTM D 2509

Table 2. Tribological properties of Greases

of four ball weld load, higher Timken load and lower wear scar diameter indicate that 2-Mercaptobenzothiazole has shown better response with lithium complex grease. The results of coefficient of friction of grease 3 (Lithium complex grease in combination of 2-Mercaptobenzothiazole and Zinc dialkyl dithiophosphate antiwear additive) and grease 4 (Lithium complex grease in combination of 2-Mercaptobenzothiazole and ashless antiwear additive) are provided in **Figure 1**. The coefficient of friction of 2-Mercaptobenzothiazole in combination with Zinc dialkyl dithiophosphate and ashless antiwear additive is very low.

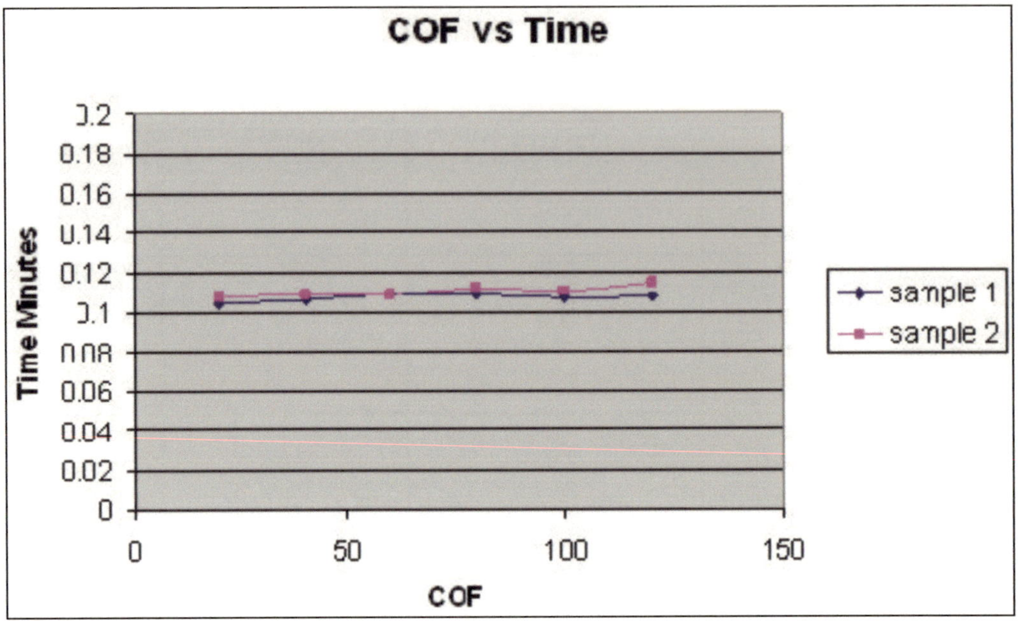

Fig. 1.

4. Conclusions

2-Mercaptobenzothiazole has shown better compatibility with Lithium complex grease and resulting good additive response.

Due to solid nature of 2-Mercaptobenzothiazole the yield of grease has increased marginally.

The series of four ball EP tests and Timken Ok load test reveal that 2- Mercaptobenzothiazole has exceptional load carrying capacity compared to conventional EP additives.

This is supported by low coefficient of Friction of greases compared to conventional EP additives. The Based on these test results, it can be summarized that 2- Mercaptobenzothiazole has exceptional load carrying capacity compared to conventional EP additives.

5. References

[1] Denis Smit & Sam Lane, High Performance Products from Invista C12 Business, Tribology & Lubrication Technology, pages 36-39

[2] Wilfried J. Bartz, The Importance of Synthetic Greases for Future High Performance Applications. ELGI Annual Meeting 2005, Edenburgh

[3] Avery , et al. Organosulfur adducts as multifunctional additives for lubricating oils and fuels and as multifunctional lubricants, United States Patent, April 9, 1991

[4] Kavin J. Chase & Gaston Aguilar, R.T. Vanderbilt Co., Inc. metal Dithiocarmates: A New Approach to Old technology, NLGI Spokesman, Vol. 71 No. 4, July 2007

[5] William c. ward jr., Richard A. denis, Morey Najman & Carlos L. Cerda de Groote: EP Additive Response and Tribochemical film formation in Lithium and lithium complex Grease, NLGI Spokesman, Vol. 71 No. 7, October 2007

[6] Polish, A. F., & farmer, H.H. Dithiocarbamate additives in Lubricating greases, NLGI Spokesman, pg 200-205, 1979

Part 4

Tribology and Lubrication

17

No Conventional Fluid Film Bearings with Waved Surface

Florin Dimofte, Nicoleta M. Ene and Abdollah A. Afjeh
The University of Toledo
USA

1. Introduction

A new type of fluid film bearings called "wave bearing" has been developed since 1990's by Dimofte (Dimofte, 1995 a; Dimofte, 1995 b). The main characteristic of the wave bearings is that they have a continuous wave profile on the stationary part of the bearing.

The wave bearings can be designed as journal bearings to support radial loads or as thrust bearings for axial loads. One of the main advantages of the wave bearings is that they are very simple and easy to manufacture. In most cases they have only two parts. A journal bearing consists of a shaft and a sleeve while a thrust bearing consists of a stationary and a rotating disk. One of the bearing parts is sometimes incorporated into the machine part that is supported by the bearing. For example, the wave bearing can be used to support the gear of a planetary transmission, the bearing sleeve being incorporated into the gear (Dimofte et al., 2000).

Compressible (gases) or incompressible (liquids) fluids can be used as lubricants for both the journal and thrust wave bearings. Tests were conducted with liquid lubricants (synthetic turbine oil, perfluoropolyethers -PFPE-K) and air on dedicated test rigs installed in NASA Glenn Research Centre in Cleveland, OH USA (Dimofte et al., 2000; Dimofte et al., 2005). In this chapter, the wave bearings lubricated with incompressible fluids, commonly known as fluid film wave bearings, are analysed. The performance of both journal and thrust bearings is examined. Because one of the most important properties of the wave journal bearings compared to other types of journal bearings is their improved stability, the first part of the chapter is dedicated to the study of the dynamic behaviour of the journal wave bearings. The wave thrust bearings can be used for axially positioning the rotor or to carry a thrust load. For this reason, the steady-state performance of the thrust wave bearings is analysed in the second part of the chapter.

2. The journal wave bearing concept

For a journal bearing, if the shaft rotates and the sleeve is stationary, then the wave profile is superimposed on the inner diameter of the sleeve. To exemplify the concept, a comparison between a wave bearing having circumscribed a three-wave profile on the inner diameter of the sleeve and a plain journal bearing is presented in Fig. 1. In Fig. 1, the wave amplitude and the clearance between the shaft and the sleeve are greatly exaggerated to better visualize the geometry. Actually, the clearance is around a thousandth of the diameter and the wave amplitude is less than one half of the clearance.

Fig. 1. Comparison between the wave journal bearing and the plain journal bearing

Because the geometry of the wave bearing is very close to the geometry of the plain circular bearing, the load capacity of the wave bearing is close to that of the plain journal bearing and superior to the load capacity of other types of journal fluid bearings. In fact, due to their improved thermal stability, the wave journal bearings can actually carry more load than the plain bearings. The wave bearing concept solves two problems encountered by plain fluid film bearings by stabilizing the shaft (Ene et al., 2008, a) and by giving enhanced stiffness to the bearing (Dimofte, 1995, a). The wave bearings have also important damping properties. They attenuate the vibration of the rotor. Consequently, the additional fluid damping system, usually required when other types of bearings are used to support the shaft, can be eliminated. Due to their damping properties, the wave bearings can be also used to attenuate the noise generated by the gear mesh in a geared transmission (Dimofte & Ene, 2009). The geometrical parameters of a journal wave bearing can be seen in Fig. 2.

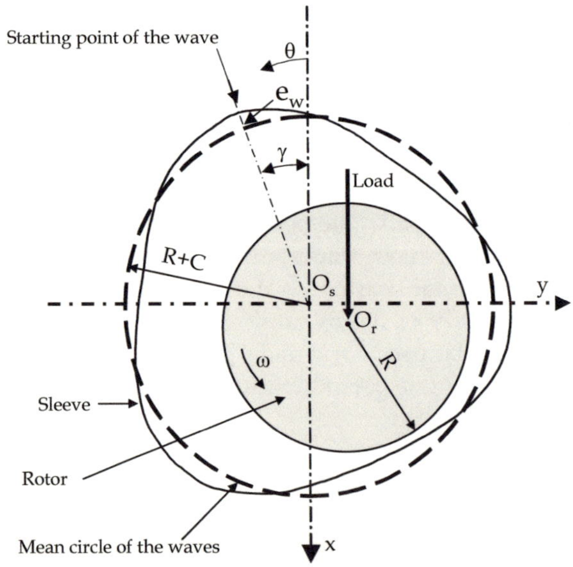

Fig. 2. The geometry of a wave journal bearing

The radial clearance C of the wave bearing is defined as the difference between the radius of the mean circle of the waves, R_{med}, and the radius, R, of the shaft:

$$C = R_{med} - R \tag{1}$$

The radial clearance is usually around one thousandth of the journal radius. For computational purposes, the wave amplitude is usually non-dimensionalised by dividing it by the radial clearance:

$$\varepsilon_w = \frac{e_w}{C} \quad (2)$$

The ratio ε_w is generally called the wave amplitude ratio. The wave amplitude ratio is one of the most important geometrical characteristics of a wave bearing because the performance of the wave bearing is strongly influenced by this ratio (Ene et all., 2008 a). The value of the wave amplitude ratio is usually smaller than 0.5.
The performance of a wave journal bearing also depends on the number of the waves, n_w, and on the wave position angle, γ. The wave position angle is defined as the angle between the starting point of the waves (one of the points where the wave has maximum value) and the load, W (see Fig. 2). Theoretical and experimental studies indicate that the best performance is obtained for a bearing with three waves and a zero wave position angle. (Dimofte, 1995 a; Dimofte, 1995 b).
The load capacity of a wave bearing is due to the rotation of the shaft and to the variation of film thickness along the circumference. In a system of reference O_Sxy fixed with respect to the sleeve (Fig. 2), the film thickness is given by:

$$h = C + x\cos\theta + y\sin\theta + e_w \cos[n_w(\theta+\gamma)] \quad (3)$$

where θ is the angular coordinate starting from the negative Ox axis and (x,y) are the coordinates of the rotor centre.

3. Methods for analysing the dynamic behaviour of wave journal bearings

The analysis of the dynamic behaviour of the journal bearings that support a rotor is of practical importance because under small loads the journal bearings can become unstable. In most of the practical cases, the sleeve is rigid and the rotor rotates freely inside the bearing clearance. When the motion becomes unstable the rotor can touch the sleeve, a phenomenon that can destroy the bearing. There are also other situations when the bearing sleeve is mobile while the shaft is rigid. In this case, when the fluid film becomes unstable, the sleeve can come into contact with the rotor, damaging the bearing. The dynamic behaviour of the wave journal bearing for both types of motions is analysed in the next sections.

3.1 Analyse of the wave bearing dynamic behaviour when the sleeve is rigid and the rotor rotates freely inside the bearing clearance

For this type of motion, the bearing sleeve is considered rigid and the rotor rotates freely inside the bearing clearance. Two different approaches can be used to analyze the dynamic stability of the wave journal bearing in this case:

- the identification of the bearing stability threshold based on the critical mass values (Lund, 1987);
- transient approach based on nonlinear theory (Kirk & Gunter, 1976; Vijayaraghavan & Brewe, 1992; Ene et al., 2008 b).

The critical mass method is very popular because of its simplicity and limited computational time requirements. The main disadvantage of this method is that no bearing information can be obtained after the appearance of the unstable whirl motion. The post-whirl motion can be simulated only with a transient method. The major inconvenience of the transient approach is that it requires large computational time.

Transient analysis

In absence of any external load, the equations of motion of the rotor centre can be written in a fixed system reference O_sxy (Fig. 2) as:

$$\begin{aligned} m\ddot{x} &= F_x + m\rho\omega^2\cos\omega t \\ m\ddot{y} &= F_y + m\rho\omega^2\sin\omega t \end{aligned} \tag{4}$$

where F_x , F_y are the components of the fluid film force, ρ - the shaft run-out, 2m- the rotor mass, ω - the rotational velocity, and (x, y) – the coordinates of the shaft centre.
The components of the fluid force F_x and F_y are obtained by integrating the pressure - p over the entire film:

$$\begin{bmatrix} F_x \\ F_y \end{bmatrix} = R\int_0^L\int_0^{2\pi} p \begin{bmatrix} \cos\theta \\ \sin\theta \end{bmatrix} d\theta dz \tag{5}$$

where R is the shaft radius, L is the bearing length, and θ and z are the angular and axial coordinates, respectively. At a particular moment of time, the pressure distribution is described by the transient Reynolds equation:

$$\frac{1}{R^2}\frac{\partial}{\partial\theta}\left(\frac{h^3}{k_\theta}\frac{\partial p}{\partial\theta}\right) + \frac{\partial}{\partial z}\left(\frac{h^3}{k_z}\frac{\partial p}{\partial z}\right) = 6\mu\omega\frac{\partial h}{\partial\theta} + 12\mu\dot{x}\cos\theta + 12\mu\dot{y}\sin\theta \tag{6}$$

where μ is the oil viscosity, and k_θ and k_z are correction coefficients for turbulent flow. The correction coefficients can be calculated by using Constantinescu's model of turbulence (Constantinescu et. al, 1985; Frêne & Constantinescu, 1975). According to this model, the correction coefficients are function of an effective Reynolds number:

$$\begin{aligned} k_\theta &= 12 + 0.0136\mathrm{Re}_{eff}^{0.9} \\ k_z &= 12 + 0.0044\mathrm{Re}_{eff}^{0.9} \end{aligned} \tag{7}$$

The first signs of turbulence appear when the local mean Reynolds number Re_m is greater than local critical Reynolds number Re_{cr}. The flow becomes dominantly turbulent when the mean Reynolds number Re_m is greater than $2Re_{cr}$. With these assumptions, the effective Reynolds number is:

$$\mathrm{Re}_{eff} = \begin{cases} 0 & \mathrm{Re}_m < \mathrm{Re}_{cr} \\ \left(\frac{\mathrm{Re}_m}{\mathrm{Re}_{cr}} - 1\right)\mathrm{Re}_l & \mathrm{Re}_{cr} £ \mathrm{Re}_m £ 2\mathrm{Re}_{cr} \\ \mathrm{Re}_l & \mathrm{Re}_m > 2\mathrm{Re}_{cr} \end{cases} \tag{8}$$

where:

$$\begin{aligned} Re_{cr} &= \min\left(41.2\sqrt{\frac{R}{h}}, 2000\right) \\ Re_l &= \frac{\rho R\omega h}{\mu} \\ Re_m &= \frac{2\rho q}{\mu} \end{aligned} \tag{9}$$

and q is the total flow.

The numerical and experimental studies show that, due to the pumping effect of the wave profile, the oil flow for the wave bearings is greater than the flow for plain journal bearings. Moreover, the greater the amplitude ratio is, the greater the flow is. Consequently, it can be assumed that the total heat generated in the fluid film is removed exclusively through the fluid transport (convection). The heat removed from the fluid through conduction to the bearing walls can be neglected. Also, the conduction within the fluid itself is neglected. In order to minimize the computation time, a constant mean temperature is assumed to occur over entire film. With these assumptions, the increase of the lubricant temperature (the difference between the temperature of the lubricant entering the film and the constant mean temperature of the film) is given by:

$$\Delta T = \frac{F_f R\omega}{\rho c_v q_{lat}} \tag{10}$$

where c_v is the lubricant specific heat, q_{lat} is the rate of lateral flow and F_f is the friction force.

The bearing trajectory is obtained by integrating the non-linear differential equations of the motion, Eqs. (4). A fourth order Runge-Kutta algorithm is used to integrate the motion equations. At each time step, an initial pressure distribution corresponding to the motion parameters, mean film temperature, and correction coefficients for turbulent flow from the previous moment of time is first obtained by integrating the Reynolds equation, Eq. 6. The Reynolds equation is solved by using a central difference scheme combined with a Gauss - Seidel method. The Reynolds boundary conditions are assumed for the cavitation region.

Next, an energy balance is performed and a new mean film temperature is obtained, Eq. 10. The lubricant properties (viscosity, density and specific heat) are then updated for the new mean film temperature. A new set of correction coefficients corresponding to the new pressure distribution is then calculated, Eqs. 8-9. The Reynolds equation is integrated again for the new values of the correction coefficients and lubricant viscosity. The iterative process is repeated until the relative errors for the correction coefficients are smaller than prescribed values. Furthermore, the fluid film forces are calculated by integrating the final pressure distribution over the entire film, Eqs. 5. Then the equations of motion, Eqs. 4, are integrated to determine the parameters of the motion for the next time step. The algorithm is repeated until the orbit of the journal centre is completed.

The critical mass approach

The bearing stability can be also analysed by evaluating the critical mass. The critical mass represents the upper limit for stability. If the rotor mass is smaller than the critical mass, the system is stable and the rotor centre returns to its equilibrium position. Particularly, in absence of any external load, the rotor centre rotates with a small radius around the bearing centre. The size of the radius depends on the shaft run-out. If the rotor mass is greater than

the critical mass then the rotor centre leaves its static equilibrium position and the system is unstable.
The critical mass is function of the dynamic coefficients of the bearing:

$$m_{cr}=\frac{K_s}{\gamma_s^2} \tag{11}$$

where K_s is the effective bearing stiffness:

$$K_s=\frac{B_{xx}K_{yy}+B_{yy}K_{xx}-B_{xy}K_{yx}-B_{yx}K_{xy}}{B_{xx}+B_{yy}} \tag{12}$$

and γ_s is the instability whirl frequency:

$$\gamma_s=\sqrt{\frac{(K_{xx}-K_s)(K_{yy}-K_s)-K_{xy}K_{yx}}{B_{xx}B_{yy}-B_{xy}B_{yx}}} \tag{13}$$

The dynamic coefficients can be obtained by integrating the pressure gradients:

$$\begin{bmatrix} K_{xx} & K_{xy} \\ K_{yx} & K_{yy} \end{bmatrix}=R\int_{-\frac{L}{2}}^{\frac{L}{2}}\int_{0}^{2\pi}\begin{bmatrix} p_x\cos\theta & p_y\cos\theta \\ p_x\sin\theta & p_y\sin\theta \end{bmatrix}d\theta dz$$
$$\begin{bmatrix} B_{xx} & B_{xy} \\ B_{yx} & B_{yy} \end{bmatrix}=R\int_{-\frac{L}{2}}^{\frac{L}{2}}\int_{0}^{2\pi}\begin{bmatrix} p_{\dot{x}}\cos\theta & p_{\dot{y}}\cos\theta \\ p_{\dot{x}}\sin\theta & p_{\dot{y}}\sin\theta \end{bmatrix}d\theta dz \tag{14}$$

The pressure gradient distributions are obtained by solving the following partial differential equations:

$$\frac{1}{R^2}\frac{\partial}{\partial\theta}\left(\frac{h^3}{k_\theta\mu}\frac{\partial p_x}{\partial\theta}\right)+\frac{\partial}{\partial z}\left(\frac{h^3}{k_z\mu}\frac{\partial p_x}{\partial z}\right)=-\frac{\omega}{2}\left(\sin\theta+3\frac{\cos\theta}{h}\frac{\partial h}{\partial\theta}\right)-\frac{h^3}{4\mu R^2}\frac{\partial p_0}{\partial\theta}\frac{\partial}{\partial\theta}\left(\frac{\cos\theta}{h}\right)$$
$$\frac{1}{R^2}\frac{\partial}{\partial\theta}\left(\frac{h^3}{k_\theta\mu}\frac{\partial p_y}{\partial\theta}\right)+\frac{\partial}{\partial z}\left(\frac{h^3}{k_z\mu}\frac{\partial p_y}{\partial z}\right)=\frac{\omega}{2}\left(\cos\theta-3\frac{\sin\theta}{h}\frac{\partial h}{\partial\theta}\right)-\frac{h^3}{4\mu R^2}\frac{\partial p_0}{\partial\theta}\frac{\partial}{\partial\theta}\left(\frac{\sin\theta}{h}\right)$$
$$\frac{1}{R^2}\frac{\partial}{\partial\theta}\left(\frac{h^3}{k_\theta\mu}\frac{\partial p_{\dot{x}}}{\partial\theta}\right)+\frac{\partial}{\partial z}\left(\frac{h^3}{k_z\mu}\frac{\partial p_{\dot{x}}}{\partial z}\right)=\cos\theta$$
$$\frac{1}{R^2}\frac{\partial}{\partial\theta}\left(\frac{h^3}{k_\theta\mu}\frac{\partial p_{\dot{y}}}{\partial\theta}\right)+\frac{\partial}{\partial z}\left(\frac{h^3}{k_z\mu}\frac{\partial p_{\dot{y}}}{\partial z}\right)=\sin\theta \tag{15}$$

where the steady-state pressure p_0 is given by the steady-state Reynolds equation:

$$\frac{1}{R^2}\frac{\partial}{\partial\theta}\left(\frac{h^3}{k_\theta\mu}\frac{\partial p_0}{\partial\theta}\right)+\frac{\partial}{\partial z}\left(\frac{h^3}{k_z\mu}\frac{\partial p_0}{\partial z}\right)=\frac{\omega}{2}\frac{\partial h}{\partial\theta} \tag{16}$$

The first problem that must be solved when using the critical mass approach is to determine the equilibrium position of the rotor centre. At the equilibrium, in absence of any external force, the static component of the fluid film force must be vertical and equal to the rotor weight. The equilibrium position is determined by integrating the steady-state Reynolds equation, Eq. 16, for different positions of the rotor centre until the resultant reaction load is vertical and equal to the external load. An iterative algorithm based on the bisection method was developed for this purpose. For each position of the shaft, the turbulence correction coefficients are determined by successive iterations using an algorithm similar to that used for the transient approach. The steady-state Reynolds equation, Eq. 16, is discretized with a finite difference scheme. The resultant system of equations is solved with a successive over-relaxation method. The Reynolds boundary conditions are assumed in the cavitation regions. The two ends of the bearing are considered at atmospheric pressure. In the oil supply pockets, the pressure is assumed to be equal to the supply pressure.
Having the equilibrium position of the shaft and the turbulence correction coefficients corresponding to this position, the pressure gradients can now be determined by integrating Eqs. 15 with a finite difference scheme. The pressure gradients are assumed to be zero at the two ends of the bearing, in the pocket regions and in the cavitation regions. The dynamic coefficients are evaluated by integrating the pressure gradients distribution along the fluid film, Eqs. 14, and the critical mass is then determined with Eqs. 11-13.

Numerical simulations

The two methods are used to predict the dynamic behaviour of a three-wave bearing having a length of 27.5 mm, a radius of the mean circle of waves of 15 mm, and a clearance of 35 microns. The rotor mass corresponding to one bearing is 0.825 kg. A 2 micron rotor run-out is considered for the numerical simulations. Synthetic turbine oil Mil-L-23699 is used as a lubricant. The bearing has also three supply pockets situated at 120° one from each other. The theoretical methods are validated by comparing the numerical results obtained with the two methods one to each other and also to experimental data (Dimofte et al., 2004).
The numerical simulations and the experiments show that for wave amplitudes greater than 0.3, the fluid film of the analyzed wave bearing is stable even at speeds of 60000 rpm, supply pressures of 0.152 MPa, and oil inlet temperatures of 190° C. For example the rotor centre trajectory predicted with the transient method for a wave amplitude ratio of 0.305 and a speed of 60000 rpm is presented in Fig. 3. The rotor centre rotates on a closed orbit with a radius almost equal to the run-out. The FFT analysis of the motion is presented in Fig. 4. For comparison, the FFT analysis of the experimental signal is shown in Fig. 5. It can be seen that both FFT diagrams contains only the synchronous frequency. The presence of only the synchronous frequency indicates a stable fluid film. The same conclusion can be drawn from the critical mass approach. The variation of the critical mass with the speed is shown in Fig. 6. Because the critical mass is greater than the rotor mass, it can be concluded that the fluid film is stable for speeds up to 60000 rpm.
For wave amplitudes smaller than 0.3, a stability threshold can be found. The experiments and the numerical simulations show that the threshold of stability depends on the wave amplitude, oil supply pressure and inlet temperature. For instance, for a wave amplitude of 0.075, a supply pressure of 0.276 MPa, and an oil temperature inlet of 126° C, the threshold of stability is around 39000 rpm. The variation of the critical mass with the rotational speed is presented in Fig. 7. The diagram shows that the critical mass is greater than the mass of the shaft related to one bearing for speeds smaller than 39000 rpm. The critical mass is very

close to the rotor mass around 39000 rpm and then it becomes smaller than the rotor mass. Consequently, it may be concluded that the fluid film of the wave bearing is unstable for rotational speeds greater than 39000 rpm.

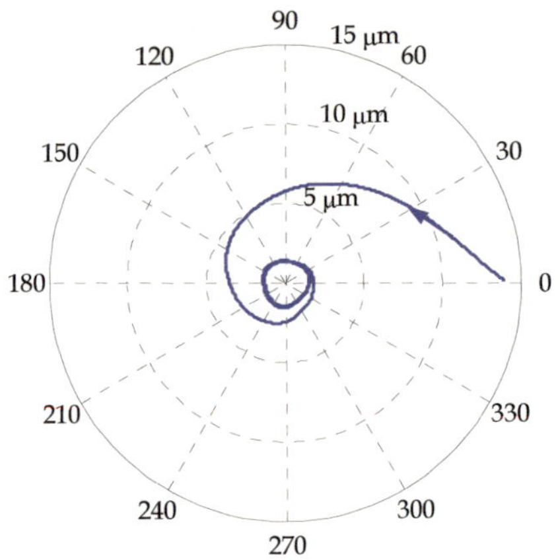

Fig. 3. Trajectory of the rotor centre for a wave amplitude of 0.305, a rotational speed of 60000 rpm, and a supply pressure of 0.152 MPa

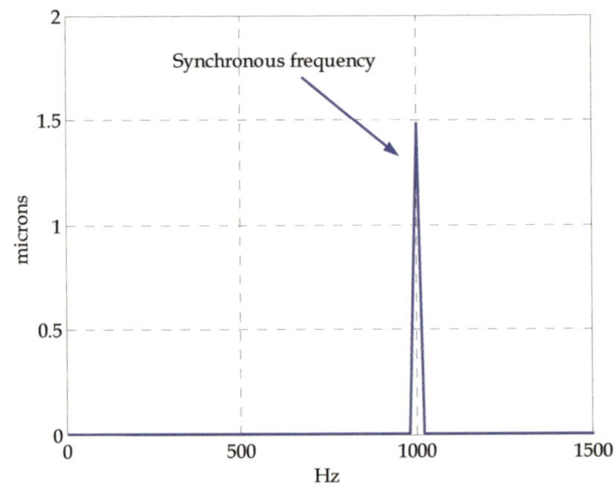

Fig. 4. FFT diagram of the motion for a wave amplitude of 0.305, a rotational speed of 60000 rpm, and a supply pressure of 0.152 MPa

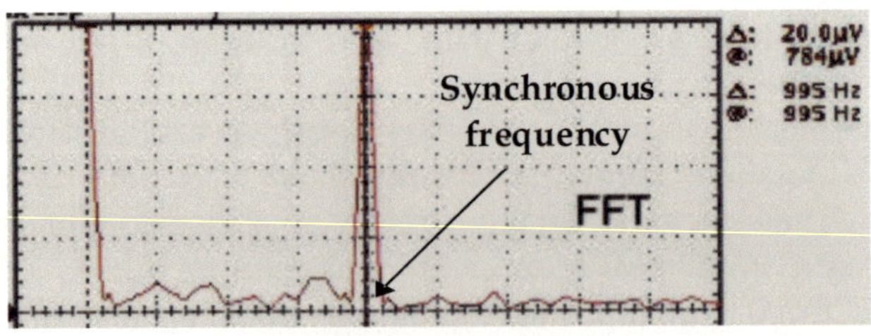

Fig. 5. FFT diagram of the experimental signal for a wave amplitude of 0.305, a rotational speed of 60000 rpm, and a supply pressure of 0.152 MPa

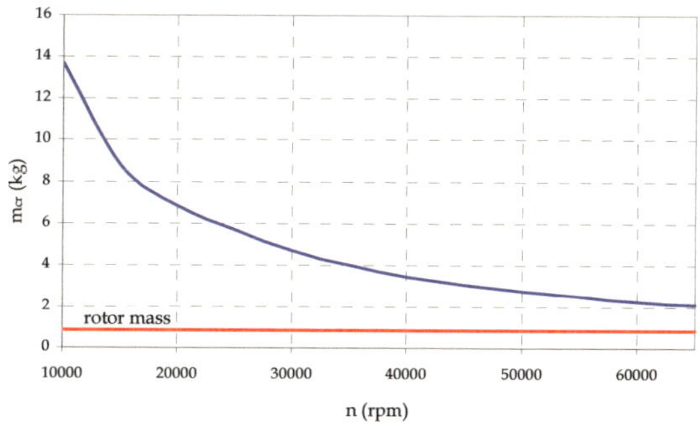

Fig. 6. Critical mass as function of the rotational speed for a wave amplitude of 0.305 and a supply pressure of 0.152 MPa

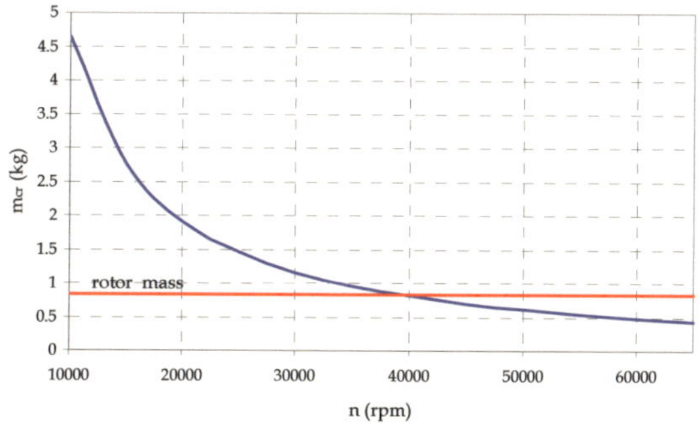

Fig. 7. Critical mass as function of running speed for a wave amplitude of 0.075 and a supply pressure of 0.276 MPa

The trajectories of the journal centre are very different for stable and unstable conditions. For example, the trajectory of the journal centre for a rotational speed of 36000 rpm, which corresponds to a stable condition, is presented in Fig. 8. In this case, the journal centre rotates with one frequency around the sleeve centre on a closed orbit having the radius almost equal to the rotor run-out. The FFT analyses of the numerically simulated motion (Fig. 9) and of the motion recorded from experiments (Fig. 10) reveal that the motion frequency is equal to the synchronous frequency.

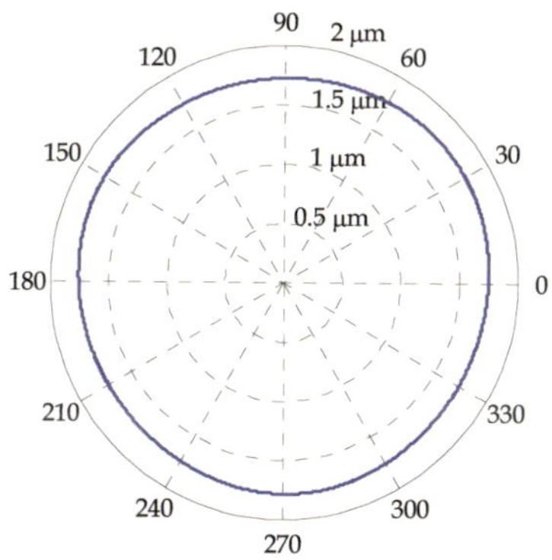

Fig. 8. Trajectory of the rotor centre for a wave amplitude of 0.075, a rotational speed of 36000 rpm and a supply pressure of 0.276 MPa

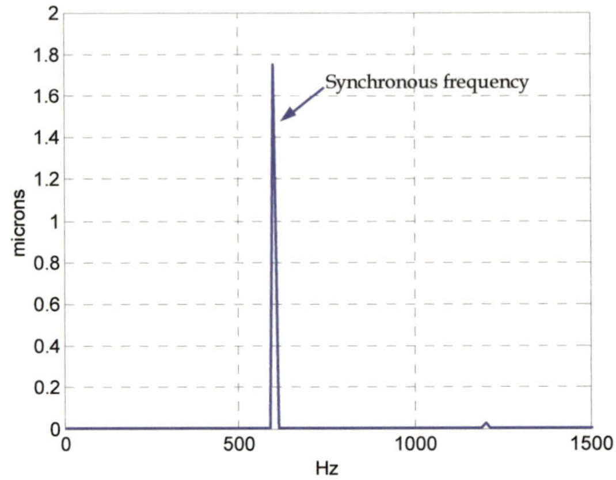

Fig. 9. FFT diagram of the motion for a wave amplitude of 0.075, a rotational speed of 36000 rpm and a supply pressure of 0.276 MPa

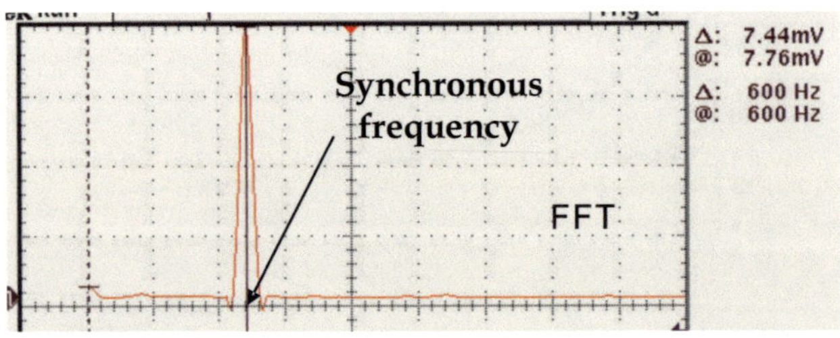

Fig. 10. FFT diagram of the experimental signal for a wave amplitude of 0.075, a rotational speed of 36000 rpm and a supply pressure of 0.276 MPa

If the rotational speed is increased to the stability threshold (39000 rpm), an incipient sub-synchronous frequency can be detected (Figs. 11 and 12). The journal centre rotates in this case on a limit cycle with two frequencies - the synchronous and the sub-synchronous frequency (Fig. 13).

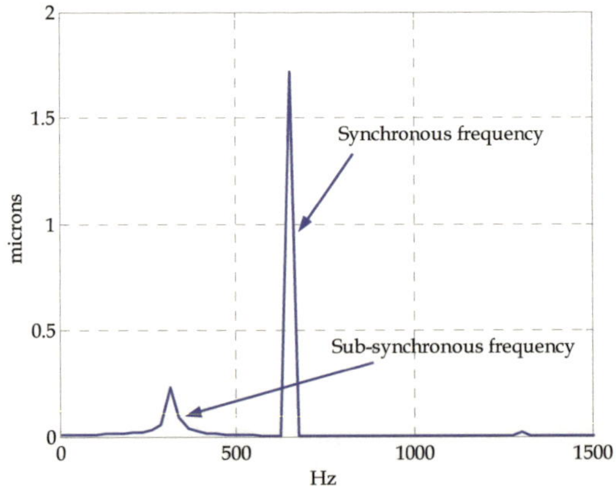

Fig. 11. FFT diagram of the motion for a wave amplitude of 0.075, a rotational speed of 39000 rpm and a supply pressure of 0.276 MPa

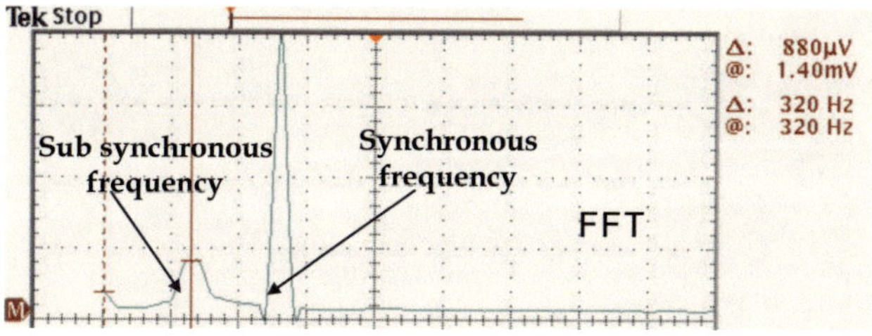

Fig. 12. FFT diagram of the experimental signal for a wave amplitude of 0.075, a rotational speed of 39000 rpm and a supply pressure of 0.276 MPa

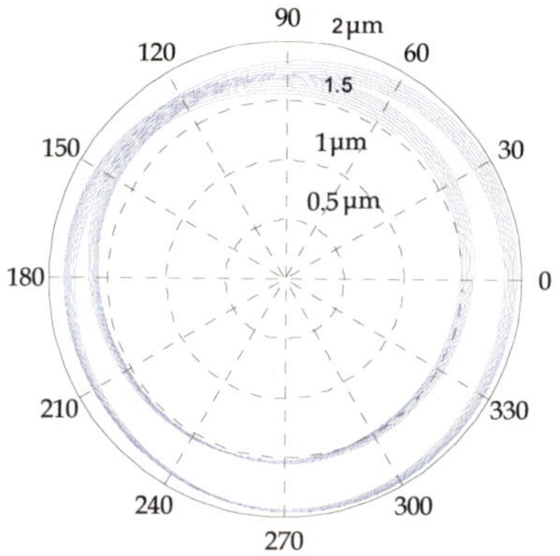

Fig. 13. Trajectory of the rotor centre for a wave amplitude of 0.075, a rotational speed of 39000 rpm and a supply pressure of 0.276 MPa

For speeds greater than 39000 rpm, the sub-synchronous frequency becomes dominant. As an exemplification, the FFT diagrams of the numerically simulated motion and of the experimental signal for a rotating speed of 44000 rpm are shown in Figs. 14 and 15. The trajectory of the rotor centre is presented in Fig. 16. The journal centre rotates on a limit cycle with a radius greater than the rotor run-out. However, due to the particular geometry of the wave bearing, the rotor centre maintains its trajectory inside the bearing clearance.

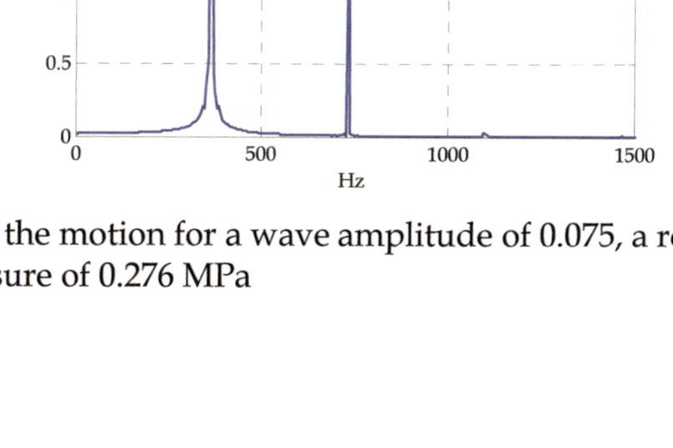

Fig. 14. FFT diagram of the motion for a wave amplitude of 0.075, a rotational speed of 44000 rpm and a supply pressure of 0.276 MPa

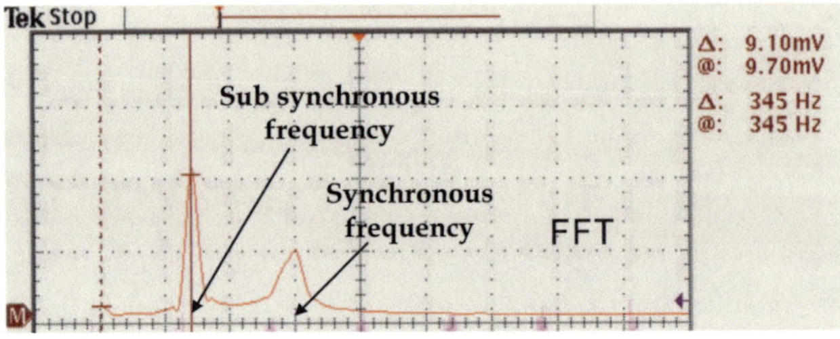

Fig. 15. FFT analysis of the experimental signal for a wave amplitude of 0.075, a rotational speed of 44000 rpm and a supply pressure of 0.276 MPa

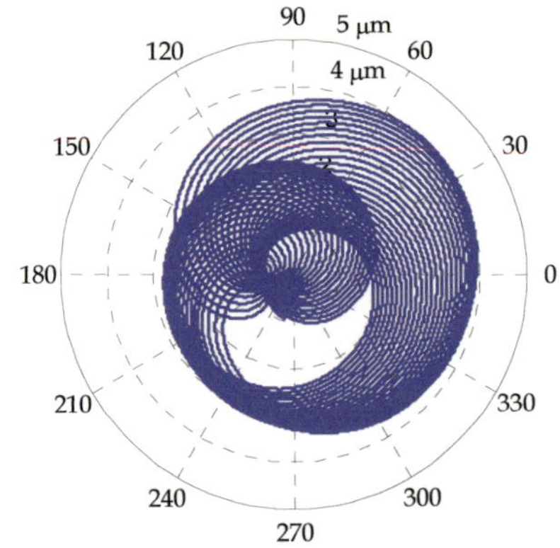

Fig. 16. Trajectory of the rotor centre for a wave amplitude of 0.075, a rotational speed of 44000 rpm and a supply pressure of 0.276 MPa

An increase of the supply pressure to 0.414 MPa makes the fluid film stable for speeds up to 60000 rpm. The critical mass becomes greater than the rotor mass (Fig. 17), the sub-synchronous frequency disappears and the rotor centre rotates with the synchronous frequency on a closed orbit with the radius almost equal to the rotor run-out. As an illustration, the FFT diagrams of the experimental signal, theoretical motion and the trajectory of the rotor centre for a rotating speed of 60000 rpm are shown in Figs. 18, 19 and 20, respectively.

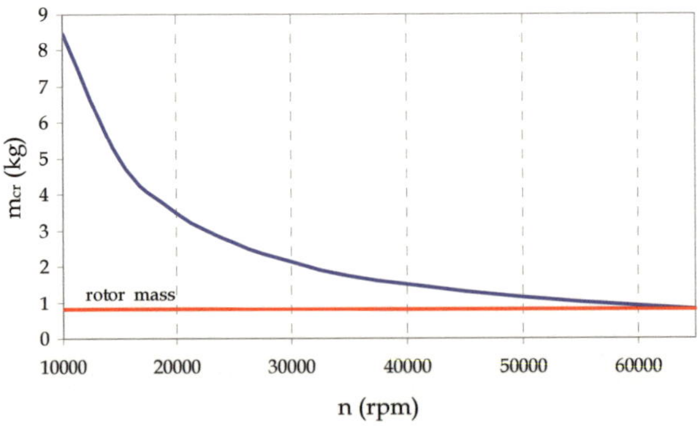

Fig. 17. Critical mass as function of running speed for a wave amplitude of 0.075 and a supply pressure of 0.414 MPa

It can be noticed from the above simulations that both the wave amplitude and the oil supply pressure strongly influence the bearing stability. The bearing stability also depends on the oil inlet temperature (Lambrulescu et al., 2003).

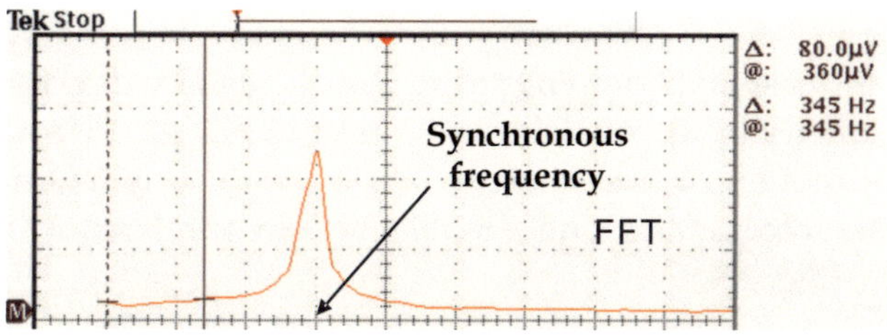

Fig. 18. FFT analysis of the experimental signal for a wave amplitude of 0.075, a rotational speed of 60000 rpm and a supply pressure of 0.414 MPa

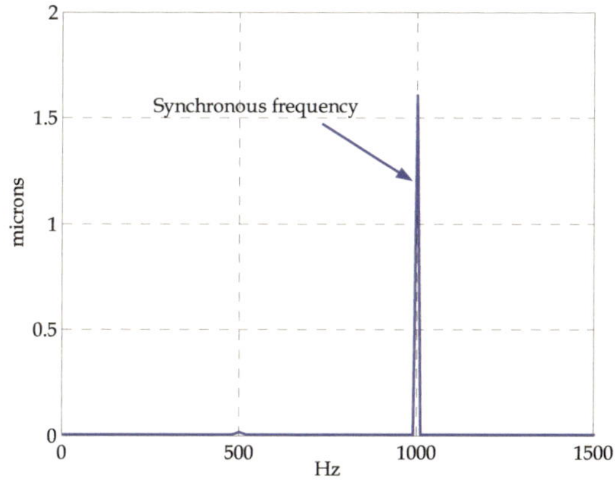

Fig. 19. FFT a diagram of the motion for a wave amplitude of 0.075, a rotational speed of 60000 rpm and a supply pressure of 0.414 MPa

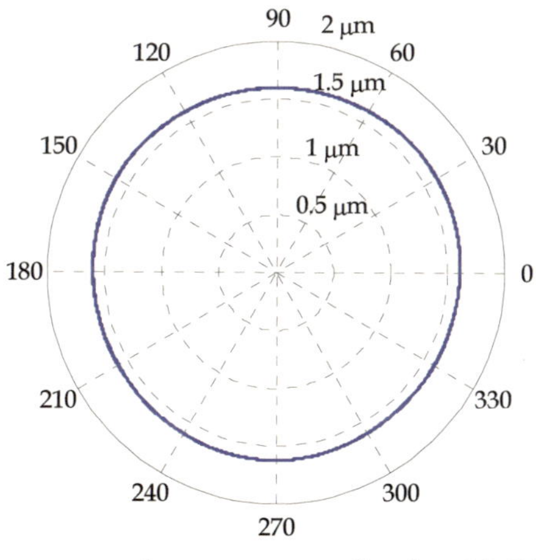

Fig. 20. Trajectory of the rotor centre for a wave amplitude of 0.075, a rotational speed of 60000 rpm and a supply pressure of 0.414 MPa

3.2 Analyse of the wave bearing dynamic behaviour when the bearing sleeve is mobile

For this type of motion, the bearing sleeve is mobile and the rotor rotates around a fixed axis with an angular velocity ω. The rotor has an inherent unbalance characterized by a small run-out ρ. The bearing sleeve is connected with the machine housing by an elastic element (Fig. 21) having the stiffness and damping coefficients in x and y directions k_x, k_y and b_x, b_y, respectively. The geometry of the motion can be seen in Fig. 22. Two systems of reference are used to study the motion: a fixed system Ox′y′ with the origin O situated on the fixed axis around which the rotor rotates, and a mobile system of reference $O_s xy$ with the origin O_s in the sleeve centre (Fig. 22).

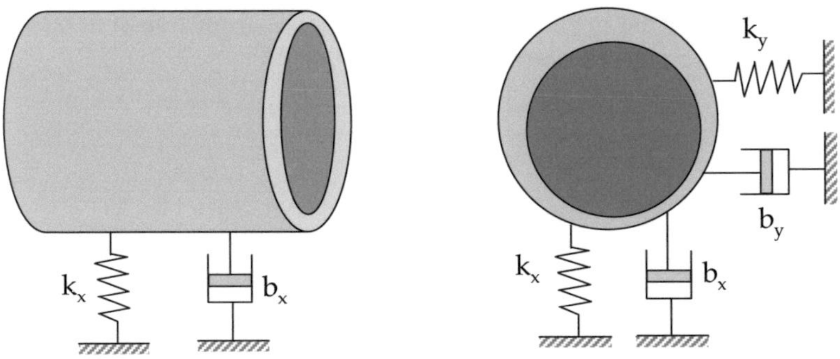

Fig. 21. Wave bearing with the sleeve supported by an elastic element

Because of its run-out ρ, the trajectory of the rotor centre in the Ox′y′ system of reference is a circle with radius ρ. Therefore the coordinates of the rotor centre in the Ox′y′ system are:

$$\begin{aligned} x_R &= x_{R0} + \rho\cos\omega t \\ y_R &= y_{R0} + \rho\cos\omega t \end{aligned} \tag{17}$$

where x_{R0} and y_{R0} are the coordinates of the rotor centre at the initial moment of time.
In the fixed system of reference Ox′y′, the equations of motion for the sleeve centre are:

$$\begin{aligned} m'\ddot{x}_s &= -F_x - k_x x_s - b_x \dot{x}_s \\ m'\ddot{y}_s &= -F_y - k_y y_s - b_y \dot{y}_s \end{aligned} \tag{18}$$

where Fx, F_y are the components of the fluid film force, x_s, y_s– the coordinates of the sleeve centre, and m′ - the mass of the sleeve. The fluid film forces are calculated by integrating the pressure distribution over the entire fluid film. The pressure distribution can be obtained by solving the Reynolds equation which in the mobile system of reference $O_s xy$ has the following form:

$$\frac{1}{R^2}\frac{\partial}{\partial\theta}\left(\frac{h^3}{k_\theta}\frac{\partial p}{\partial\theta}\right)+\frac{\partial}{\partial z}\left(\frac{h^3}{k_z}\frac{\partial p}{\partial z}\right)=6\mu\omega\frac{\partial h}{\partial\theta}+12\mu\left(\dot{x}_r-\dot{x}_s\right)\cos\theta+12\mu\left(\dot{y}_r-\dot{y}_s\right)\sin\theta \tag{19}$$

where the fluid film thickness is given by:

$$h=c+(x_r-x_s)\cos\theta+(y_r-y_s)\sin\theta+e_w\cos[n_w(\theta+\gamma)] \tag{20}$$

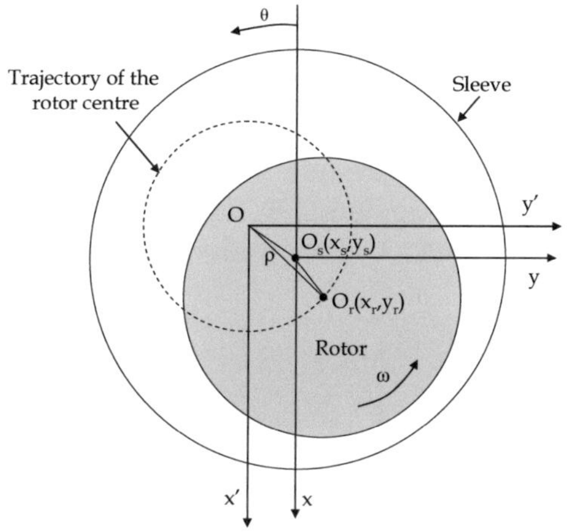

Fig. 22. The geometry of the motion when the bearing sleeve is free to rotate

A numerical algorithm similar to that presented for the previous type of motion was developed to obtain the absolute motion of the sleeve centre and the motion of the sleeve centre relative to the rotor centre. Because the relative motion between the sleeve and the rotor is of practical importance, the numerical simulations will be focused only on this motion.

Numerical simulations

The above model is used to study the influence of the wave amplitude on the dynamic behaviour of the sleeve of a three - wave bearing having the same geometrical characteristics as the bearing analysed in the previous section (length - 27.5 mm, radius of the mean circle of the waves - 15 mm, clearance - 35 microns, rotor run-out – 2 microns). The mass of the sleeve is 0.5 kg. One of the axial ends of the bearing is exposed at a pressure of 0.14 MPa. At the other axial end the pressure is 0.5 MPa. The stiffness and damping coefficients of the sleeve elastic support are 10000 N/m and 10 Ns/m, respectively. To have a better understanding of the influence of the wave profile on the dynamic behaviour of the wave bearing sleeve, a plain

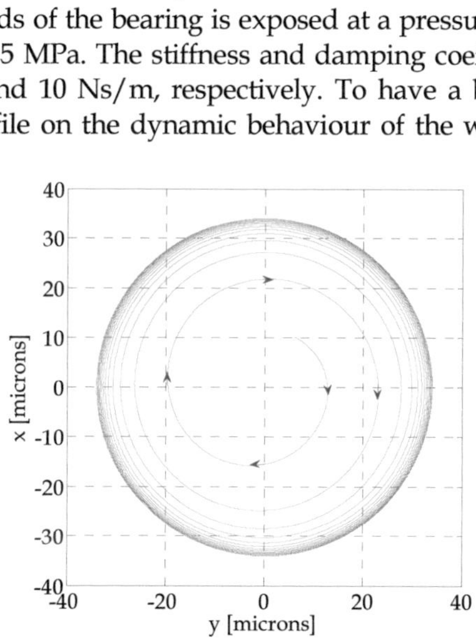

Fig. 23. The trajectory of the sleeve centre of a plain journal bearing with respect to the rotor centre

journal bearing having the same characteristics as the analysed wave bearing was initially considered. The motion of the sleeve centre of the plain journal bearing with respect to the rotor centre for a rotational speed of 60000 rpm is presented in Fig. 23. The sleeve centre rotates around the rotor centre on a large orbit with a radius very close to the bearing clearance. Therefore the sleeve can come into contact with the rotor and the bearing can be destroyed. The FFT analysis of the motion (Fig. 24) indicates the presence of both the half-whirl frequency and the synchronous frequency, the half-whirl frequency being dominant.

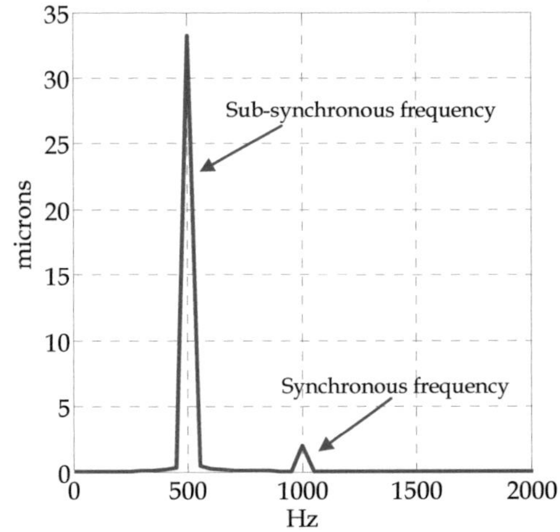

Fig. 24. The FFT diagram of the motion of the plain journal bearing sleeve

For a wave bearing with a wave amplitude ratio of 0.1, the half-whirl frequency is also dominant (Fig. 25) at 60000 rpm. However, the radius of the sleeve centre orbit is much smaller than the bearing clearance (Fig. 26) and the bearing can run safely. It must be also remarked that the trajectory of the sleeve centre with respect to the rotor centre has a shape very similar to shape of the wave bearing profile.

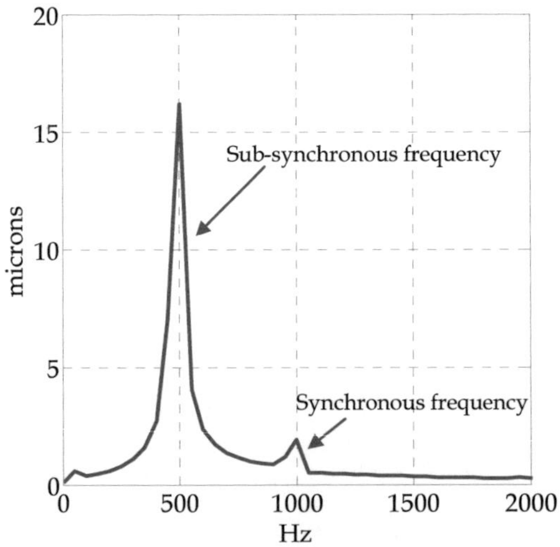

Fig. 25. The FFT diagram of the sleeve centre motion for a wave amplitude ratio of 0.1

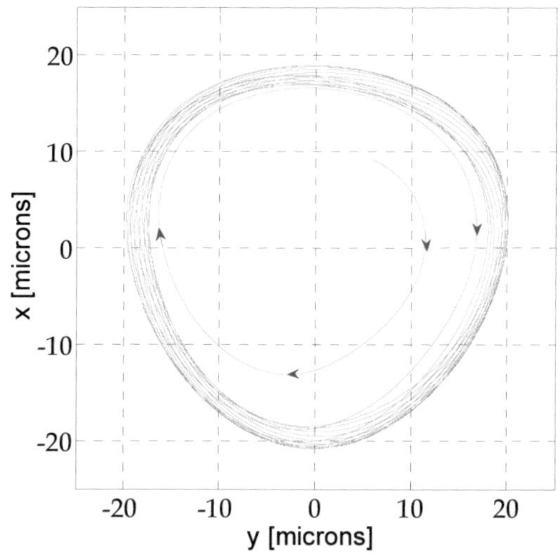

Fig. 26. The trajectory of the sleeve centre with respect to the rotor centre for a wave amplitude ratio of 0.1

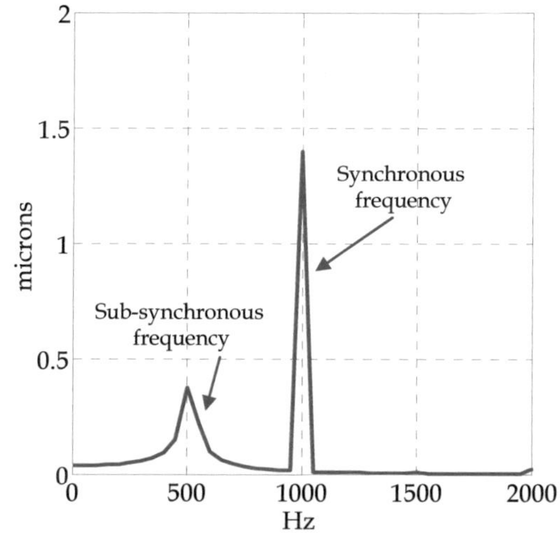

Fig. 27. The FFT diagram of the sleeve centre motion for a wave amplitude ratio of 0.2

For a wave amplitude ratio of 0.2, the motion still contains both the sub-synchronous and synchronous frequencies. In this case, the synchronous frequency is the main frequency (Fig. 27). The sleeve centre approaches the rotor centre and rotates around it with two frequencies on a small orbit (Fig. 28).

The sub-synchronous frequency completely disappears when the wave amplitude ratio is increased to 0.3 (Fig. 29). The sleeve centre rotates around the rotor centre on a closed orbit with a radius closed to the rotor run-out (Fig. 30).

Therefore it can be concluded that the waves have a stabilising effect on the bearing. Even when the fluid film of the wave bearing is unstable, the sleeve centre maintains its trajectory inside the bearing clearance.

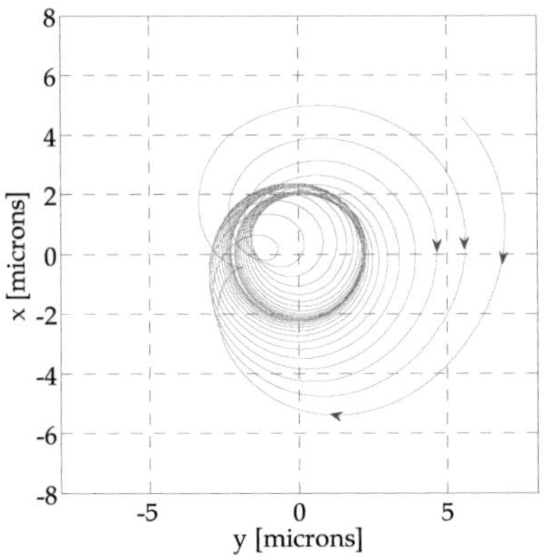

Fig. 28. The trajectory of the sleeve centre with respect to the rotor centre for a wave amplitude ratio of 0.2

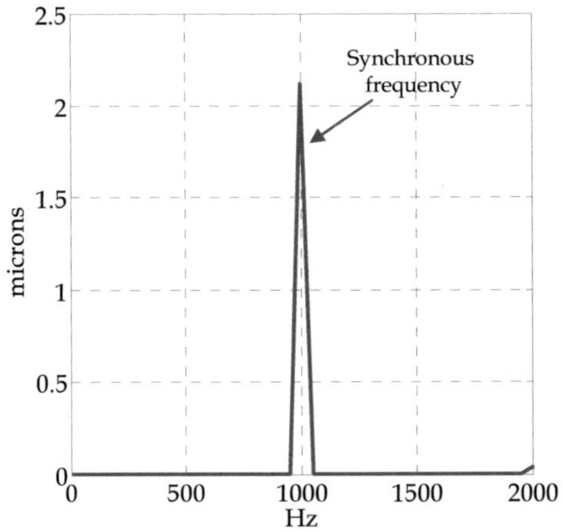

Fig. 29. The FFT diagram of the sleeve centre motion for a wave amplitude ratio of 0.3

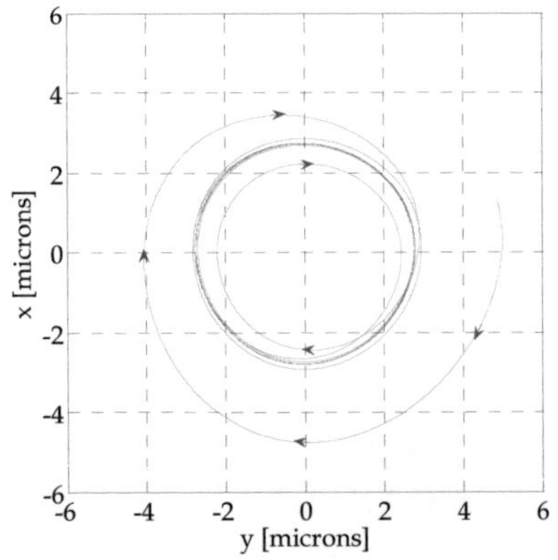

Fig. 30. The trajectory of the sleeve centre with respect to the rotor centre for a wave amplitude ratio of 0.3

4. The thrust wave bearing concept

A typical thrust fluid film bearing used to support an axial load is shown in Fig. 31. As can be seen in Fig. 31, a thrust fluid film bearing consists of a rotating disk (runner) fixed on the shaft and a stationary thrust plate which faces the rotating disk. The gap between the two disks is filled with lubricant. If the active surfaces of both of the disks are flat, no hydrodynamic pressure will be generated in the fluid film and the bearing will not be able to support an axial load. Rayleigh steps or spiral grooves engraved on the active surface of one of the disks can be used, for example, to create hydrodynamic pressure.

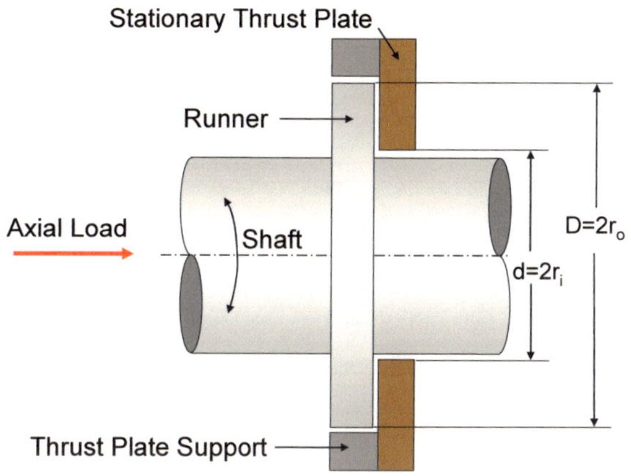

Fig. 31. Typical thrust fluid film bearing

An alternative novel solution is to make a wave profile on the active surface of the stationary disk. As an example, a four-wave thrust plate is shown in Fig. 32. The most important characteristics of a thrust wave bearing are also presented in Fig. 32. The bearing inner radius is r_i and the outer radius is r_o. The wave amplitude e_w is measured from the middle plane of the waves. The lubricant is supplied to the bearing through holes and closed pockets. The pockets are orientated in radial direction.

If the thrust bearing is used for positioning the shaft in the axial direction then two thrust plates located on both sides of the runner disk can be used. In this case, the thrust bearing can be lubricated with the oil that comes out from the nearby journal bearing, and the supply system can be eliminated.

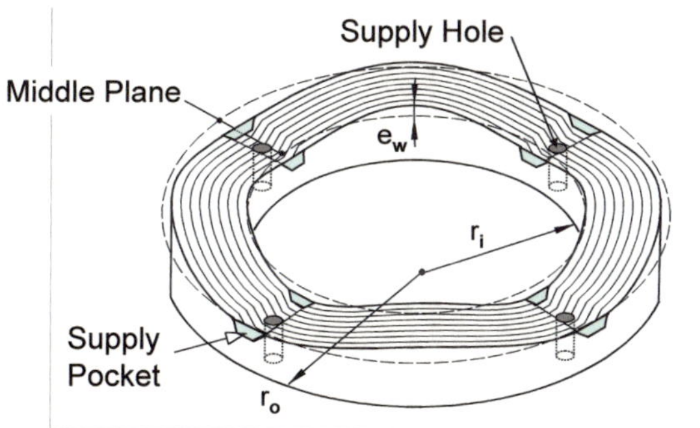

Fig. 32. Wave bearing thrust plate

The hydrodynamic pressure generated in the fluid film can be obtained by solving the Reynolds equation in cylindrical coordinates:

$$\frac{\partial}{\partial\theta}\left(h^3\frac{\partial p}{\partial\theta}\right)+r\frac{\partial}{\partial r}\left(h^3 r\frac{\partial p}{\partial r}\right)=6\mu\omega r^2\frac{\partial h}{\partial\theta} \tag{21}$$

where p is the pressure, h - the film thickness, μ - the oil viscosity, ω - the angular velocity, θ - the angular coordinate, and r- the radial coordinate (see Fig. 33).

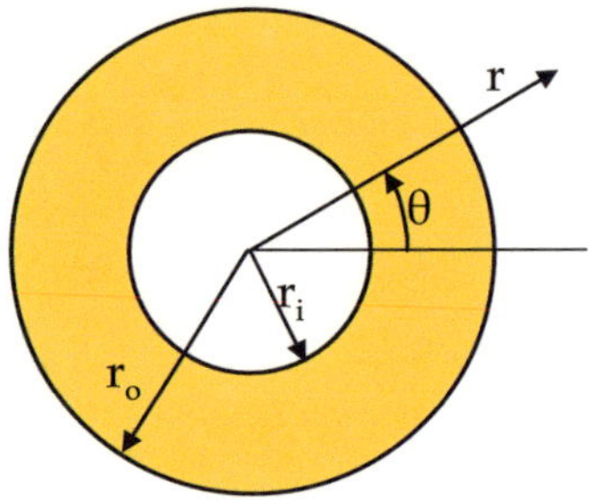

Fig. 33. System of coordinates for the wave thrust bearing

The load W can be then obtained by numerically integrating the pressure over the entire bearing area:

$$W=\int_0^{2\pi}\int_{r_i}^{r_o}(rp)\,dr\,d\theta \tag{22}$$

The increase of the oil temperature when the oil passes through the bearing can be calculated with a method similar to that used for journal bearing.
The influence of the parameters of the wave thrust bearing on its steady-state performance is analysed below. A thrust wave bearing having an inner diameter of 50 mm, an outer diameter of 100 mm, and a wave amplitude of 50 microns was considered. The bearing is supplied with synthetic turbine oil MIL-L-23699 at 65° C. It is well known that at high temperatures, the "oil viscosity collapse phenomenon" can occur. The direct consequence of the oil viscosity collapse is the dramatic lost of the bearing load capacity. For this reason, a maximum limit of 30° C is imposed for the oil temperature increase inside the fluid film. The numerical simulations show that in order to avoid oil temperature increases greater than 30° C, the analysed thrust wave bearing must have the minimum film thickness greater than 50 microns.
The influence of the number of the waves on the average pressure (the ratio between the load and the bearing surface) and oil temperature increase is first analysed. For example, the evolutions of the average pressure and of the oil temperature increase with the rotational speed for a supply pressure of 0.5 MPa, a minimum film thickness of 50 microns and different number of waves are presented in Figs. 34 and 35, respectively.
It can be seen from Fig. 34 that the average pressure has a linear variation with the rotational speed. In addition, the number of the waves has only a small influence on the average pressure (Fig. 34). Nevertheless, the number of the waves can significantly influence the temperature of the oil passing through the bearing. When the number of the waves increases, the number of oil supply holes and pockets increases. An increased number of

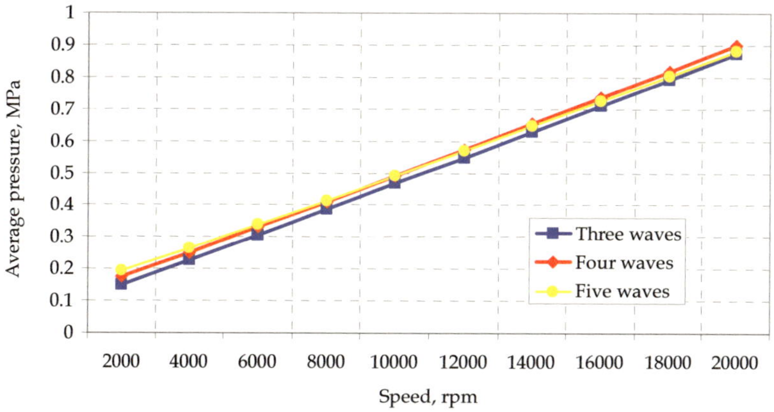

Fig. 34. Average pressure vs. speed for different number of waves (oil supply pressure - 0.5MPa, minimum film thickness - 50 microns)

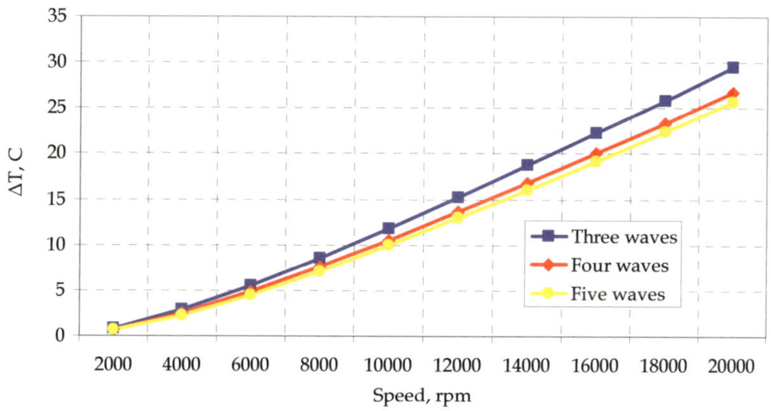

Fig. 35. The increase of the lubricant temperature vs. speed (oil supply pressure - 0.5MPa, minimum film thickness - 50 microns)

supply holes and pockets allows for a better cooling of the bearing. Consequently, the oil temperature increase is smaller for a greater number of waves (Fig. 35). However a significant change of the temperature with the number of the waves can be observed only when the number of waves is increased from three to four.

Next, the influence of the oil supply pressure on the bearing performance was analysed. For example, the evolutions with the rotational speed of the average pressure and oil temperature increase for different supply pressures, a minimum film thickness of 50 microns and a number of five waves are presented in Figs. 36 and 37, respectively.

Fig. 36 shows that the oil supply pressure has only a small influence on the bearing average pressure. The influence of the oil supply pressure on the increase of the oil temperature inside the fluid film can be significant (Fig. 37). From Fig. 37, it can be seen that the oil temperature increase is smaller for higher supply pressures.

Therefore from the numerical simulations it can be concluded that the increase of the oil temperature inside the fluid film can be reduced by increasing the number of the waves or the oil supply pressure. A more significant reduction of the increase of oil temperature

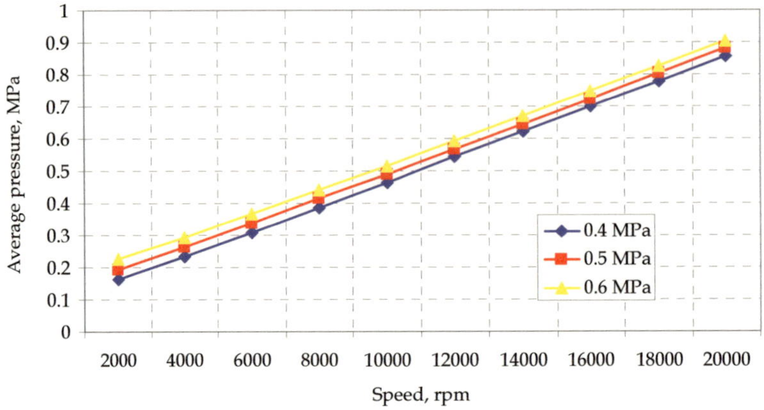

Fig. 36. Average pressure vs. speed for different supply pressures (five waves, minimum film thickness - 50 microns)

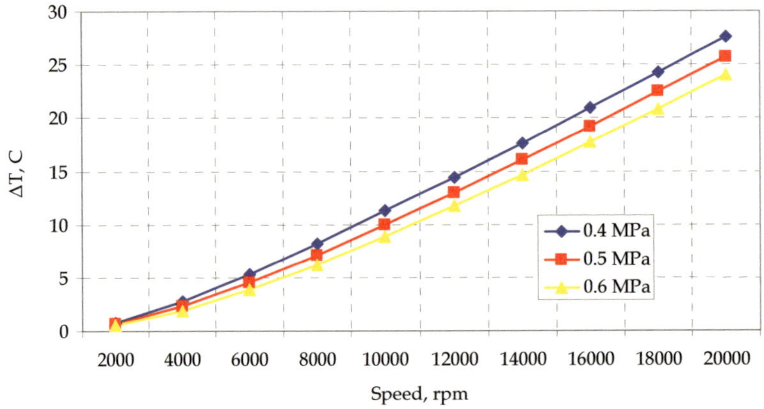

Fig. 37. The increase of the lubricant temperature vs. speed (five waves, minimum film thickness - 50 microns)

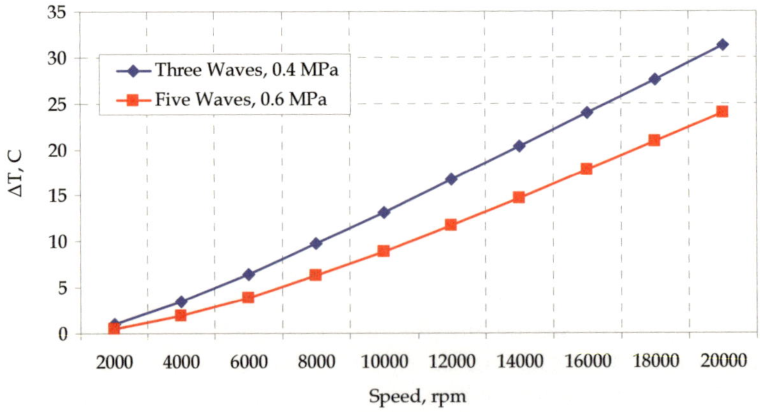

Fig. 38. Oil temperature increase vs. speed for different number of waves and supply pressures (minimum film thickness -50 microns)

through the bearing can be obtained if the two effects are used together, as can be seen in Fig. 38. Fig. 38 shows that the increase of the oil temperature can be with up to 33% smaller for the bearing with five waves supplied with oil at 0.6 MPa than for the bearing with three waves supplied with oil at 0.4 MPa.

5. Concluding remarks

A wave bearing can be a good alternative to any other type of fluid film bearing.

Journal wave bearings offer increased dynamic and thermal stability compared to other types of journal fluid film bearings. Their actual load capacity could be even higher than the load capacity of the plain journal bearings due to the pumping effect of the wave profile that allows for a better cooling of the wave bearing. Besides the bearing clearance, the wave amplitude and the running parameters such as oil inlet pressure and temperature can be used to adjust the bearing performance to the needs of a specific application. The wave bearings are also simple and easy to manufacture. The wave journal bearing steady-state and dynamic performance can be precisely predicted with computer codes validated by experiments on dedicated test rigs.

Thrust wave bearings can be successfully used to carry axial loads or to axially position the shaft. Thrust wave bearings are also very simple compared to any other fluid film thrust bearings. They can have an individual oil supply system or can be lubricated with the oil that leaks laterally from a nearby journal bearing. Their performance can be also precisely predicted. The number of the waves, the wave amplitude, the minimum film thickness, the oil supply pressure and temperature can be used to maximize the bearing performance for a particular application.

6. References

Constantinescu, V. N., Nica, A., Pascovici, M. D., Ceptureanu, G. & Nedelcu S. (1985). *Sliding Bearings*, Allerton Press, ISBN 0-89864-011-3, New York

Dimofte, F. (1995). Wave journal bearing with compressible lubricant - Part I : The wave bearing concept and a comparison to the plain circular bearing, *Tribology Transactions*, Vol. 38, No. 1, pp. 153-160

Dimofte, F. (1995). Wave journal bearing with compressible lubricant - Part II : A comparison of the wave bearing with a groove bearing and a lobe bearing, *Tribology Transactions*, Vol. 38, No. 2, pp. 364-372

Dimofte, F.; Proctor, M. P.; Fleming, D. P. & Keith, T. G. Jr. (2000). Wave fluid film bearing tests for an aviation gearbox, *Proceedings of the 8th International Symposium on Transport Phenomena and Dynamics of Rotating Machinery, ISROMAC- 8*, Honolulu, Hawaii, March 26-30, 2000, NASA/TM-2000-209766, January 2000

Lambrulescu M. I; Dimofte, F. & Sawicki, J., T.(2003). Stability of a rotor supported by wave journal bearings as a function of oil temperature, *Proceedings of 2nd ISCORMA Meeting*, Gdansk, Poland, August 4-8, 2003

Dimofte, F., Proctor, M.P., Fleming, D.P. & Keith, T. G. (2004). Experimental investigations on the influence of oil inlet pressure on the stability of wave journal bearing, *Proceedings of the 10th International Symposium on Transportation Phenomena and Dynamics of Rotating Machinery*, Honolulu, Hawaii, ISROMAC 10-2004-146

Dimofte, F.; Fleming, D. P.; Anderson, W. J. & Klein, R. C. (2005). Test of a fluid film wave bearing at 350°C with liquid lubricants, *STLE Tribology Transactions*, Vol. 48, pp. 515-521

Dimofte, F. & Ene, N.M. (2009). Comparison of the vibration/ noise of a geared transmission with shafts on fluid film bearings and on rolling element bearings, *Proceedings of the 3rd International Conference Power Transmissions '09*, 1-2 October, Chalkidiki, Greece

Ene, N. M; Dimofte, F. & Keith Jr., T. G. (2008). A dynamic analysis of hydrodynamic wave journal bearings, *STLE Tribology Transactions*, Vol. 51, No. 1, pp. 82-91

Ene, N. M; Dimofte, F. & Keith Jr., T. G. (2008). A stability analysis for a hydrodynamic three-wave journal bearing, *Tribology International*, Vol. 41, No. 5, pp. 434-442

Frêne, J. & Constantinescu, V. N. (1975). Operating characteristics of journal bearings in the transition region, *Proceedings of the 2nd Leeds – Lyon Symposium on Tribology*, pp. 121-124

Kirk, R. G. & Gunter, E. J. (1976). Short bearing analysis applied to rotordynamics I: Theory, *ASME Journal of Lubrication Technology*, Vol. 98, pp. 47-56

Lund, J. W. (1987). Review of the concept of dynamic coefficients for fluid film journal bearings, *ASME Journal of Tribology*, Vol. 109, No. 37, pp. 37-41

Vijayaraghavan, D. & Brewe & D. E., (1992). Frequency effects on the stability of a journal bearing for periodic loading, *ASME Journal of Tribology*, Vol. 114, pp. 107-115

18

Identification of Discharge Coefficients of Orifice-Type Restrictors for Aerostatic Bearings and Application Examples

Guido Belforte, Terenziano Raparelli,
Andrea Trivella and Vladimir Viktorov
Department of Mechanics, Politecnico di Torino
Italy

1. Introduction

In this chapter is described an experimental study conducted in order to identify the supply hole discharge coefficients of externally pressurized gas bearings. Tests were carried out over specific hole, feed pocket and air gap size ranges on pneumatic pads with two types of air feeding systems: annular orifices (inherent orifices) and simple orifices with feed pocket. Air consumption and pressure distributions were measured as a function of supply pressure and air gap height. Discharge coefficients were approximated by an experimental formula based on the Reynolds number and the feeding system geometry. The validity of the formulation found in the study was verified by comparing the numerically calculated pressure distribution with the experimental distribution measured on different pad types. The numerical pressure distribution was calculated using equations for air flow through the supply holes and the Reynolds equations for the air gap.

2. Study of externally pressurized gas bearings

There is a high demand for air bearings in all applications where support with almost no friction between parts in relative motion is essential. Unlike oil, air is universally available at no cost, and requires no auxiliary recirculation circuits. In addition, air bearings offer a number of advantages: high precision, speed and positioning repeatability, absence of wear, negligible power losses, high reliability and durability, lower operating costs and zero environmental impact.

As they use clean, filtered air and thus avoid the ecological problems caused by oils and greases, air bearings are often employed in the food processing and textile industries as well as in medical equipment. They are also found in small machine tools and measuring robots, in the microturbines used for distributed electricity generation, and in the gyroscopes employed in inertial navigation systems. Air bearings can withstand high temperatures and radiation, and can thus be installed in nuclear reactors with no need for maintenance over a number of years. Another emerging application is that of high-speed compressors for fuel cell systems.

For a number of years, theoretical and experimental investigations have addressed the static and dynamic behavior of air bearings. Several basic studies have indicated that simulating bearing behavior calls for an understanding of the flow rate and pressure distribution in the air gap: Gross, 1962; Grassam & Powell, 1964; Mori & Miyamatsu, 1969; Poupard & Drouin, 1973; Majumdar, 1980; Kazimierski & Trojnarski, 1980; Belforte et al. ,1999.

Externally pressurized gas bearings can feature different types of supply systems: with calibrated orifices, with orifices and feed pockets, or, in other cases, with porous resistances. In all cases, the behavior of the air flow in the bearing and hence its performance is heavily dependent on supply system type.

Al-Bender &Van Brussel, 1992-a, presents an overview of the basic studies that have been performed and the solution methods that have been used, with particular reference to feeding systems employing annular orifices (also called inherent orifices). In Blondeel et al., 1980, an externally pressurized gas thrust bearing is studied analyzing separately the influence of the clearance and restrictor on stability. For this type of feeding system, Al-Bender & Brussel, 1992-b, investigate tilt motion and calculate tilt stiffness and damping coefficients. Huges et al.,1996, present a gas thrust bearing facility to measure pressure and temperature distributions. Fourka et al., 1996, investigate thrust bearing stability analytically and experimentally.

Belforte, et al., 2006, perform theoretical and experimental studies of air journal bearings with annular orifice feeding systems for high-speed spindles, while Bang & Lee, 2002, investigate the thrust bearing design for a high-speed composite air spindle. For pocket-type feeding systems Li & Ding, 2007, study the performance of an aerostatic thrust bearing with pocketed orifice-type restrictor while Stout et al., 1993 analyze the behavior of flat pad bearings with pocketed orifice restrictors.

For feeding systems with grooves Nakamura & Yoshimoto, 1996; Yoshimoto et al., 1999, prove the benefits of introducing micro-grooves in terms of load capacity and tilt stiffness and investigate the influence of groove position and groove depth on stability.

Grooved gas thrust bearings are simulated with different methods: Chen & Lin, 2002; Chen et al., 2002; Chen et al., 2010, solve the Reynolds equation with the resistance network method; Bonneau et al., 1993, use finite elements for the analysis. Hashimoto & Namba, 2009, present studies of the optimization of spiral groove geometries for dynamic thrust bearings.

Several studies have also addressed air bearings employing porous resistances and their supply systems, e.g., Yoshimoto & Kohno, 2001; Plante et al., 2005; Ng, et al., 2005; Belforte et al., 2007-b; Belforte et al., 2009.

As all studies of bearings and the different types of supply system indicate, choosing the correct identification parameters has a clear influence on the results of the mathematical model describing the system.

For the orifice and feed pocket supply systems considered in this chapter, flow behavior is quite complicated, especially in the immediate vicinity of the air inlet orifices. Figure 1 shows the two most common solutions: annular orifice (1a) and simple orifice with feed pocket (1b).

These systems are generally simulated using lumped parameter models, though this approach has certain limitations.

Each supply hole is considered as an ideal nozzle, through which the mass flow rate G is given by:

$$G = C_d \cdot G_{th} \quad (1)$$

where G_{th} is the theoretical mass flow rate which considers an isentropic expansion of the flow through the nozzle, and C_d is an appropriate discharge coefficient. The major difficulty

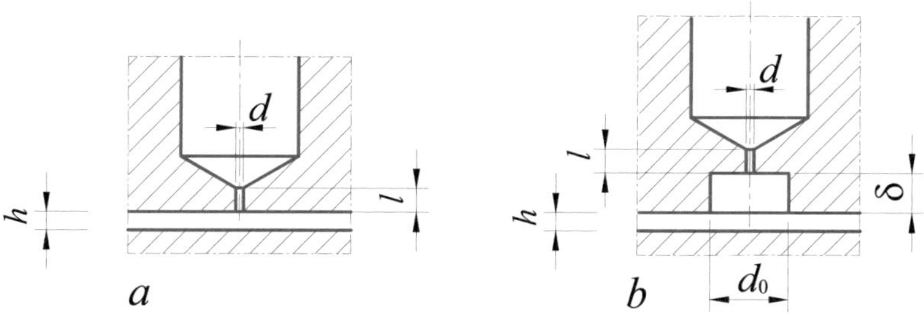

Fig. 1. Example of supply systems: *a*) Annular orifice; *b*) Simple orifice with feed pocket

in this approach lies in evaluating this coefficient, given the significant influence of the system's geometrical parameters (hole and pocket dimension, air gap height) and of the flow parameters (supply pressure, Reynolds number). In certain cases, this approach is simplified by selecting single numbers, which normally vary from 0.6 to 0.8, for this coefficient: Lund, 1964; Bryant et al., 1986; Goodwin, 1989. In other cases, analytical formulas are used: Mori & Miyamatsu 1969; Elrod & Glanfield, 1971; Kazimierski & Trojnarski, 1980. With both methods, however, the validity of the result is limited to a narrow parametric range.

Alternatively, the discharge coefficient can be evaluated experimentally through flow rate and pressure measurements on the actual system Kassab et al., 1997, for example, present an experimental investigation of the effects of varying supply hole dimensions and supply pressure on the performance of aerostatic bearings, where the consumption and pressure distributions thus obtained can be used to study discharge coefficients. On the basis of this idea, Belforte et al., 2007, present an experimental study of annular orifice and simple orifice air bearing feeding systems which was carried out in order to identify the discharge coefficient as a function of the above parameters using analytical formulations. For this type of approach, both the dimensions and configuration of the air supply channels must first be analyzed under the microscope because of the significant influence that feeding system geometry has on the discharge coefficient. The results obtained with this method have been applied in numerical simulation programs for radial and axial bearings used in high-speed spindles: Belforte, et al., 2008. In Belforte et al., 2010-a and 2010-b, the effects of a groove on thrust bearing performance are investigated.

Commercial CFD programs have also been used recently as another method of evaluating discharge coefficients: Renn & Hsiao, 2004. Yoshimoto et al., 2007, present a study of the pressure distribution in circular aerostatic bearings with a single central hole. Using Navier-Stokes equations, experimental and analytic results are compared with CFD tools. In Neves et al., 2010, numerical methods are used to investigate the influence of the discharge coefficient on aerostatic journal bearing performance. In general, the limitation of this type of approach lies in precisely defining the supply system geometry, which must be accomplished on the basis of previous experimental work, as shape defects and imperfections resulting from machining will obviously vary, albeit only slightly, from hole to hole in the actual system. Belforte et al., 2006, demonstrate how small geometrical differences (chamfers, fillets) resulting from the drilling process have a major influence on pressure distribution adjacent to the holes, on thus on the discharge coefficient.

This chapter presents a detailed discussion of the experimental method followed in Belforte, et al., 2007, and the results obtained. Flow rate and pressure distribution in the air gap were measured on a number of flat pad bearings for various supply hole and pocket dimensions and bearing supply pressures. The results made it possible to define two formulations for the discharge coefficient where the air flow is considered as passing both through the circular section of the hole and through the annular section between the hole and the air gap. Finite difference models using these formulations were developed specifically for the cases examined. Comparisons are then presented between the pressure distributions obtained experimentally and numerically, both for the pads used for identification and for other similar pads tested previously: Belforte et al., 2010-c. Experimental tests were then carried out to determine that these formulas can also be applied under certain geometrical conditions to feeding systems with grooves.

3. Test bench and pads under test

3.1 Test bench and instrumentation

The different types of feeding system were tested using bearings with flat thrust surfaces (aerostatic pads), though the results can be extended to cylindrical bearings. For this purpose, a test bench was constructed to measure load capacity, consumption and pressure distribution as a function of air gap height h and supply pressure on the pads under test. A cross sectional view of the bench is shown in Figure 2. The bench frame consists of a base (*1*), three columns (*2*) and a crossmember (*3*).

The air gap h is established between the pad under test (*4*) and the stationary bearing member (*5*). Vertical pad movement is controlled by handwheel (*6*), screw (*7*) and pushrod (*8*). Plate (*9*), which is connected directly to the pad, provides a support surface for three micrometric probes (*10*) used to measure the height of the air gap and equally spaced around the circumference. The plate is connected to the pushrod by means of ball (*11*), which ensures that the bearing axis is vertically aligned with the applied force and that air gap height is constant under the entire pad. It is also possible to vary air gap height by angling the pad slightly by means of two screws (*12*) located above the plate. Thrust on the pad is measured by a load cell (*13*). Dowel (*14*) prevents the load cell from rotating.

The stationary member face is provided with two 0.2 mm diameter holes connected to the associated stationary member outlet ports, which are connected to pressure transducers. The

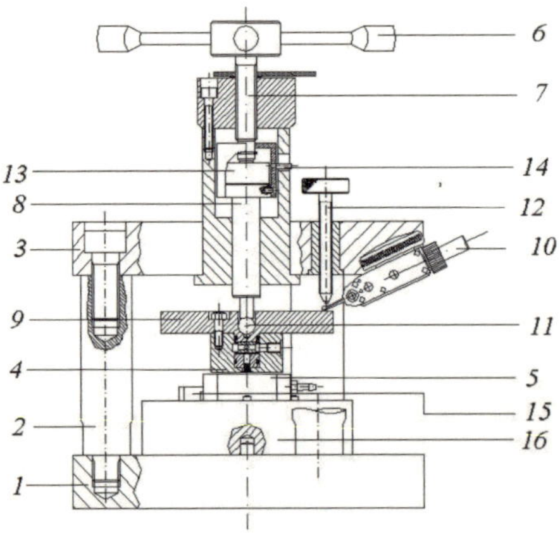

Fig. 2. Test bench schematics

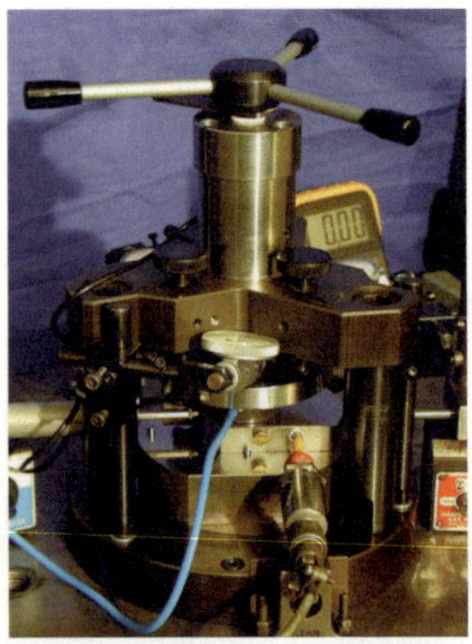

Fig. 3. Photo of test bench

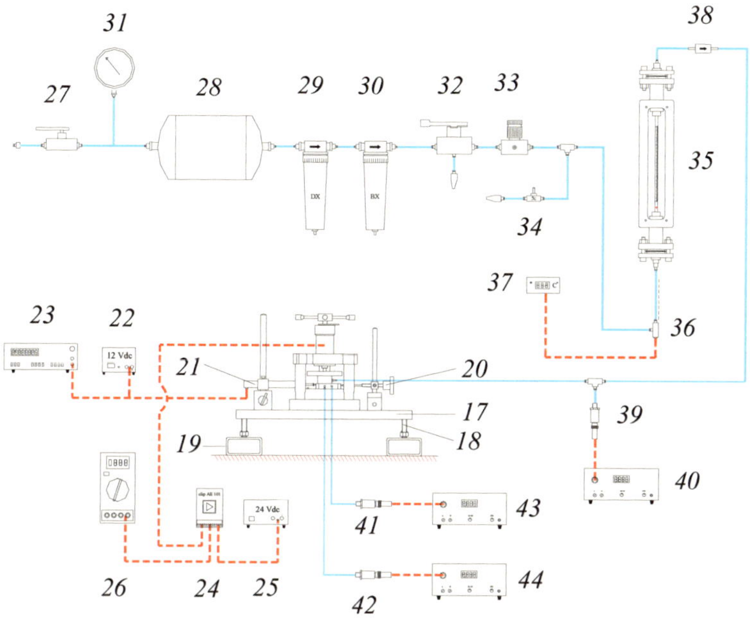

17	Steel plate	30	Air filter
18	Adjuster screws	31	Pressure gauge
19	Steel sections	32	Cutoff valve
20	Pad movement adjusting screw	33	Pressure reducer
21	Linear displacement transducer	34	Variable resistance
22	Displacement transducer supply unit	35	Float type flowmeter
23	Multimeter	36	Thermocouple
24	Load cell amplifier	37	Thermometer
25	Load cell supply unit	38	Air filter
26	Voltmeter	39	Pressure transducer p_S
27	Cutoff valve	43	Amplifier output pressure signal 1
28	Air reservoir	44	Amplifier output pressure signal 2
29	Air filter		

Fig. 4. Bench with measurement instrumentation and compressed air supply circuit

stationary member can be moved radially with respect to the pad under test by means of a guide (*15*) secured to platform (*16*). In this way, pressure distribution along the air gap can be determined. In order to perform measurements in different radial directions, the platform can be rotated around the pad axis. A photograph of the test bench is shown in Figure 3.
Figure 4 shows the measurement instrumentation connected to the test bench and the compressed air supply circuit.

3.2 Pads under test

The discharge coefficient was identified using three types of circular pad designated as types "*a*", "*b*" and "*c*" as shown in Figure 5. All have an outside diameter of 40 mm and a height of 22 mm.

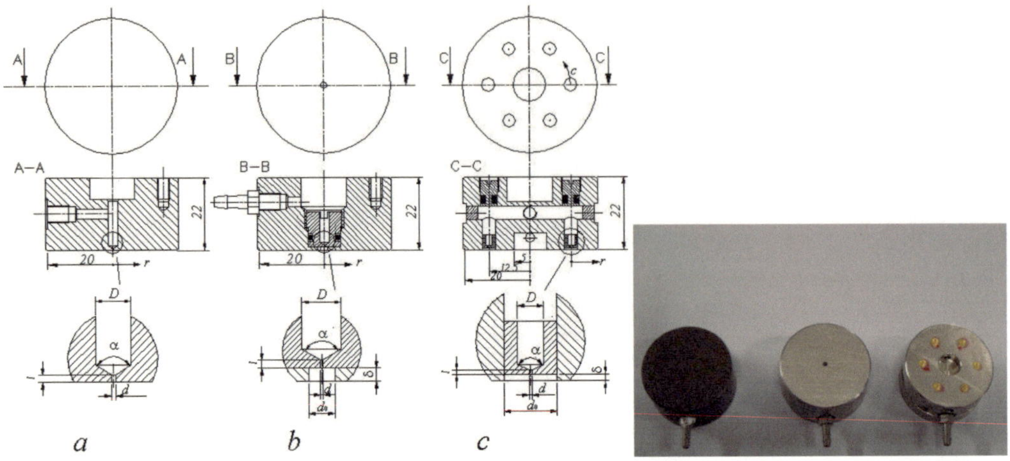

Fig. 5. Schematic view and photo of pad types "*a*", "*b*", "*c*" under test

Type "*a*" pads were constructed for the annular orifice supply system and types "*b*" and "*c*" were constructed for the simple orifice with cylindrical feed pocket. For type "*a*", seven cylindrical anodized aluminum pads with a single central hole were used. For types "*b*" and "*c*", two cylindrical stainless steel pads were constructed, one with a single central hole (type "*b*") and one with six holes equally spaced around the circular space (type "*c*"). For each pad type, Table 1 shows nominal supply system dimension for systems with and without feed pocket. The supply passage diameter and conical angle are represented by D and α. Hole diameter and length are designated as d and l respectively, while for type "*b*" and "*c*" pads, pocket diameter and depth are designated respectively as δ and d_0 .

On pads "*b*" and "*c*", the orifices were machined in interchangeable cylindrical inserts on the pads. For type "*b*", aluminum inserts were screwed into the center of the pad; different sizes of supply hole can be tested while the pocket depth is always equal to 1 mm. For type "*c*",

Pad N.	Pad type	Hole				Pocket	
		α [°]	D[mm]	l [mm]	d [mm]	d_0 [mm]	δ [μm]
1	*a*	118	3	0.3	0.2	-	-
2	*a*	118	3	0.9	0.2	-	-
3	*a*	118	3	0.3	0.3	-	-
4	*a*	118	3	0.9	0.3	-	-
5	*a*	118	3	0.3	0.4	-	-
6	*a*	118	3	0.6	0.4	-	-
7	*a*	118	3	0.9	0.4	-	-
8	*b*	118	3	0.3	0.2	2	1000
9	*b*	118	3	0.3	0.3	2	1000
10	*b*	118	3	0.3	0.4	2	1000
11	*c*	118	2	0.3	0.2	4	0, 10, 20
12	*c*	118	2	0.3	0.3	4	0, 10, 20

Table 1. Nominal supply system dimensions for the three pad types

the inserts consist of brass, and pocket depth can be changed by bonding the insert at different heights.

On all pads, the holes were produced using a twist drill and thus feature a chamfer, with a fillet radius of about 5% of average diameter. Maximum deviation of average hole diameter is around 10% of the nominal dimension. By way of example, Figure 7 shows an enlargement of the hole on pad "*a*", type number "4".

Aluminum and stainless steel pad thrust surfaces have an average roughness of 0.3 μm and 0.1 μm respectively, and a flatness error of less than 1 μm. The stationary bearing member face is steel and has an average surface roughness of 0.1 μm with a flatness error of less than 1 μm.

4. Results and analysis

For all pad types, flow rate and pressure distribution measurements were performed with different air gap heights, maintaining the thrust surface parallel to that of the stationary bearing member. For each air gap height, measurements were carried out for different relative pad supply pressures: p_S= 0.3, 0.4, 0.5, 0.6 and 0.7 MPa. Ambient pressures and pad supply air temperatures were recorded for all tests.

Pad characteristics are heavily influenced by supply pressure and air gap. To ensure that results are repeatable, a suitable procedure was specified for establishing air gap height. The pad was supplied with compressed air and is initially pushed against the stationary bearing member by means of the handwheel with a force sufficient to guarantee contact without producing distortion. In this condition, the flowmeter's sensitivity was such that no flow rate is detected and the displacement transducers are set to zero.

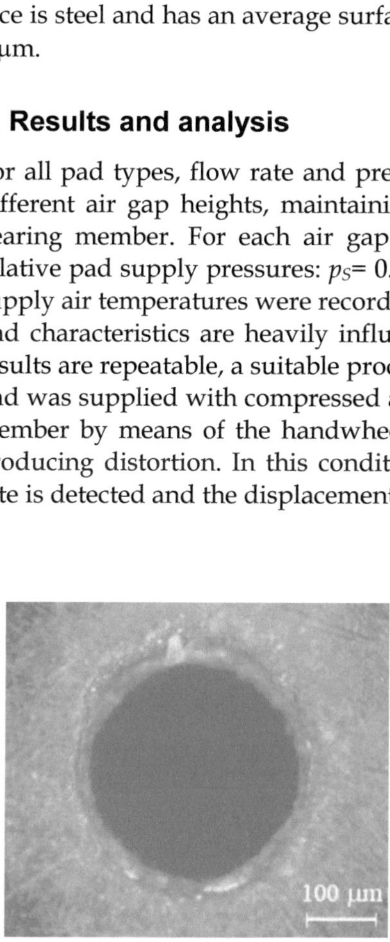

Fig. 6. Enlargement of hole type number "4", pad "*a*"

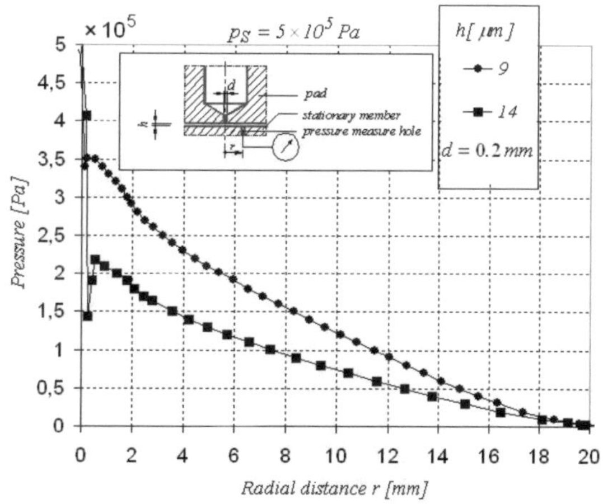

Fig. 7. Radial pressure distribution across pad number "1", type "*a*", supply pressure p_S = 0.5 MPa, orifice diameter d = 0.2 mm, air gap height h = 9 and 14 μm

The handwheel was then turned to decrease the thrust on the pad, and the pressure under the pad, the load capacity and the air flow rate are measured for the set air gap heights. At each height setting, value h was determined as the average of the three transducer readings.

Once the air gap height was established, the supply pressure was corrected using a pressure reducer and a precision gauge in order to keep it constant.
For type "*a*" and "*b*" pads, radial pressure distribution measurements were carried out, while for type "*c*" pads, both radial and circumferential pressure measurements were performed at one selected hole. For pads "*a*" and "*b*", radial distance *r* was measured from an origin point located at the pad center, whereas for pads "*c*", the origin point for radial (*r*) and circumferential (*c*) distances is located at the center of the hole, and pressure distribution was measured relative thereto. Figures 8 through 14 show examples of the relative pressure distribution with different pads. All geometric parameters and supply pressures are shown in the figures and detailed in the captions. As the measurement hole in the stationary bearing member is equal to 0.2 mm, the measured pressure profile does not follow the feed pocket's geometric profile exactly, but is offset by 0.1 mm from the center.
For the annular orifice supply system (type "*a*" pads), behavior can be described by the results shown in Figures 7 and 8. Figure 7 shows the radial pressure distribution obtained with pad "*a*", type number "1", for air gap heights *h* of 9 and 14 μm. Figure 8 shows a detail of this pressure distribution.

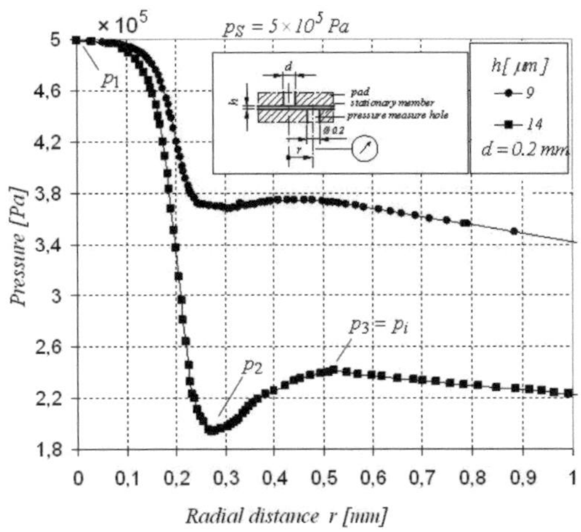

Fig. 8. Detail of Figure 7

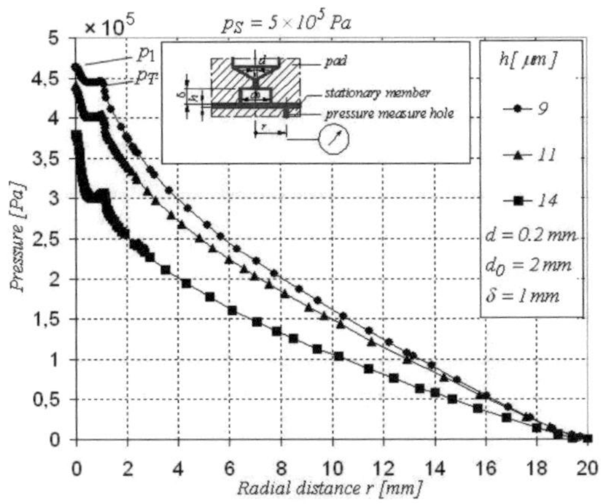

Fig. 9. Radial pressure distribution across the entire pad number "8", type "*b*", supply pressure p_S = 0.5 MPa.

Maximum pressure p_1 occurs immediately below the hole. As the distance from the center of the hole increases, pressure drops and, when the air reaches the inlet to the gap, the flow passage area is drastically reduced, thus causing an inertia phenomenon known as pressure depression (Mori & Miyamatsu, 1969; Yoshimoto et al., 2007) which consists of an initial pressure drop down to a minimum local value p_2, followed by a partial increase to a local maximum p_3. Inside the air gap, viscous friction forces become increasingly predominant as flow moves farther from the inlet hole, finally transforming it into laminar flow. With r_i as the radius at which the distributed viscous resistance zone is completely developed and p_i as the corresponding pressure for these types of supply systems, we have $p_i = p_3$. For all cases considered experimentally for type "*a*" pads, the local maximum values of p_3 are positioned around radius r_i ranging from 0.45 to 0.8 mm, which depends on the orifice diameter *d* and gap height *h*. Results showed that radius r_i can be approximated by the following equation:

$$r_i = \frac{d}{2} + 40\,h \tag{2}$$

For simple orifices with feed pocket (type "*b*" and "*c*" pads), behavior can be described by the results shown in Figures 9 through 13.

For pads with very deep pockets (type "*b*", $\delta = 1$ mm), a good example of radial pressure distribution is shown in Figure 9 for pad "8". Supply pressure p_S is to 0.5 MPa, and the curves refer to air gap heights h of 9, 11 and 14 µm. A detail of the pressure distribution in the area adjacent to the supply orifice is shown in Figure 10.

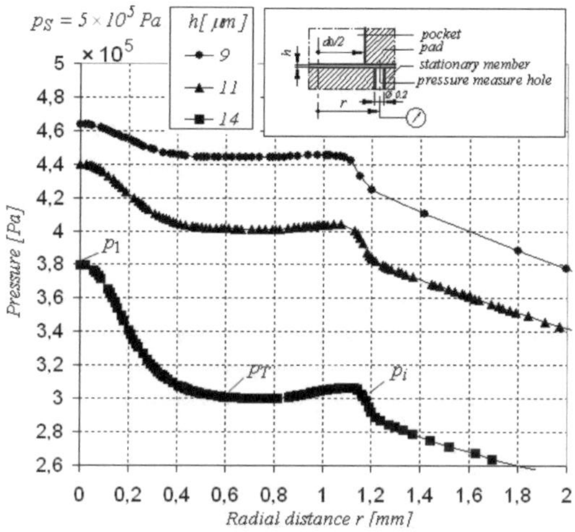

Fig. 10. Detail of Figure 9

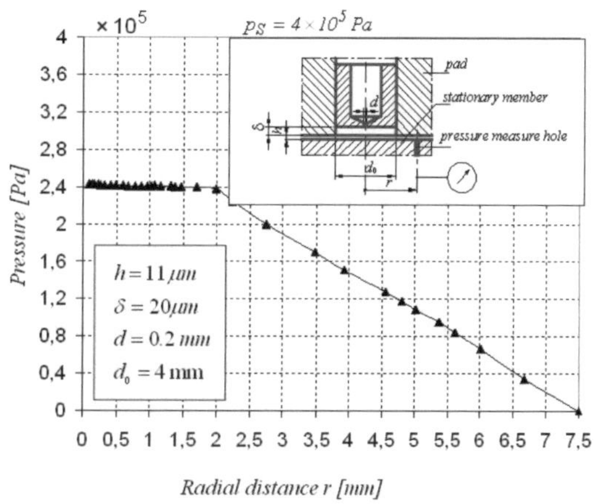

Fig. 11. Radial pressure distribution across pad number "11", type "c", supply pressure p_S = 0.4 MPa.

Maximum pressure p_1 occurs immediately below the air inlet hole. As the distance from the hole inside the pocket increases, the pressure first drops, then rises rapidly. With further increases in distance, the pressure gradually reaches a value which remains practically constant almost to the air gap inlet section. Pressure p_T in the pocket is measured at the point where pressure distribution has zero slope. Pressure starts to rise slightly near the air gap inlet section, an effect which becomes more pronounced as air gap height is increased. At the inlet, pressure drops sharply as a result of the abrupt reduction in the passage section, reaching a value p_i measured at pocket diameter d_0. The pressure then drops again, in this case evenly, as viscous friction forces become predominant.

For shallow pockets (δ = 10 µm and 20 µm, pad type "*c*"), the behavior of the pressure distribution is similar to that of deep pockets. Figure 11 and 12 show examples of pressure distributions in the area adjacent to one of the six orifices of the pad type "11". In particular, Figure 11 shows the radial pressure distributions measured with supply pressure p_S equal to 0.4 MPa, pocket depth δ equal to 20 µm and air gap heights h equal to 11 µm; Figure 12 shows the circumferential pressure distribution with supply pressure p_S equal to 0.5 MPa, pocket depth δ equal to 10 and 20 µm and air gap heights h equal to 11 and 16 or 17 µm. For type "*c*" pads, it is difficult to position the pressure measurement hole exactly below one of the pad supply orifices. As a result, maximum pressure p_1 was not measured in certain tests, including that shown by way of example in Figure 11. However, pressure p_T inside the pocket was measured, as it is useful for identification purposes.

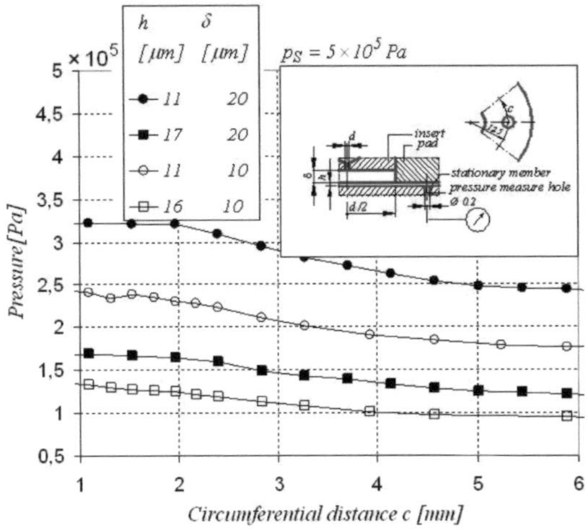

Fig. 12. Circumferential pressure distribution across pad number "11", type "c", supply pressure p_S = 0.5 MPa, orifice diameter d = 0.2 mm, pocket diameter d_0= 4 mm, pocket depth δ = 10 and 20 μm.

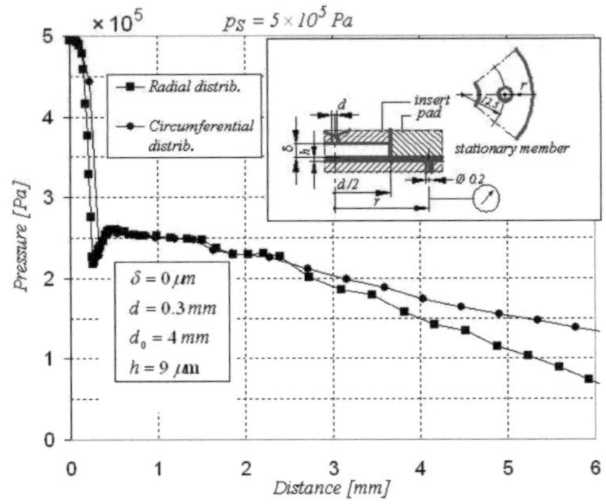

Fig. 13. Radial and circumferential pressure distribution across pad number "12", type "c", supply pressure p_S = 0.5 MPa, orifice diameter d = 0.3 mm, pocket diameter d_0= 4 mm, pocket depth δ=0 μm.

For all the cases experimentally considered for annular orifices (pad types "*b*" and "*c*"), it can be seen that pressure distribution behavior pressure is heavily dependent on δ, as well as on h and d. If $\delta \geq h$ (Figures 10, 11), a uniform pressure is reached in the pocket, whereas if $\delta < h$ (Figure 12), the pocket's resistive effect predominates over the capacitive effect, and there is a slight reduction in the pressure along the pocket. In practice, this reduction is negligible by comparison with the pressure drop in the air gap, as the dimensions of the pocket are so much smaller than those of the pad. If $\delta << h$ (Figure 13), the reduction in the pressure along the pocket is not negligible and the results are similar to those for the annular orifice feeding system. For pad type "*c*", moreover, a comparison of all radial and circumferential pressure distributions indicates that they exhibit a sufficient degree of axial symmetry with respect to the orifice axis; an example is shown in Figure 13.

5. Identification method

Discharge coefficient identification consists of considering the pad as a series of pneumatic resistances sharing the same center: the hole resistance, the gap inlet resistance and the viscous resistance along the gap.

In the case of the annular orifices (type "*a*" pads), the hole resistance and the gap inlet resistance can be grouped in a single resistance which causes pressure drops $p_S - p_i$. It is thus possible to define a discharge coefficient $C_{d,c}$ for this pressure drop, considering the supply orifice's circular section as the air passage section.

For simple orifices with feed pocket (types "*b*" and "*c*" pads), the hole resistance and the gap inlet resistance can be considered independent, as they cause $p_S - p_T$ and $p_T - p_i$ respectively.

Two discharge coefficients can thus be defined, one for each of the two localized pressure drops. As for annular orifice systems, discharge coefficient $C_{d,c}$ for the first pressure drop p_S - p_T is calculated considering the supply orifice's circular section of diameter d as the air passage section. Discharge coefficient $C_{d,a}$ for the second pressure drop p_T - p_i is calculated by taking the annular section of height h and diameter d_0 as the passage section. When the pocket is sufficiently deep ($\delta > 20$ μm), the pressure drop at the air gap inlet is significant by comparison with that across the inlet hole, and in this case both the discharge coefficients must be defined. In all cases with $\delta \leq 20$ μm, $p_T \cong p_i$ and it is possible to define only coefficient $C_{d,c}$.

The theoretical air flow rate through each lumped resistance is given by equation (3):

$$G_t = S \cdot P_u \sqrt{\frac{k}{k-1}\left[\left(\frac{P_d}{P_u}\right)^{\frac{2}{k}} - \left(\frac{P_d}{P_u}\right)^{\frac{k+1}{k}}\right] \cdot \frac{2}{R \cdot T}} \quad \text{if} \quad \frac{P_d}{P_u} \geq 0.528$$

$$G_t = S \cdot P_u \sqrt{\left(\frac{2}{k+1}\right)^{\frac{2}{k-1}} \cdot \left(\frac{k}{k+1}\right) \cdot \frac{2}{R \cdot T}} \quad \text{if} \quad \frac{P_d}{P_u} < 0.528 \qquad (3)$$

where P_u and P_d are the resistances' upstream and downstream absolute pressures, T is the absolute temperature upstream of the nozzle, S is the passage section area, R = 287.1 J/(kg K) is the air constant, and k = 1.4 is the specific heat ratio of air at constant pressure and volume.

As G_t and G are known, the values of $C_{d,c}$ and $C_{d,a}$ were calculated using equation (1).

In order to allow for the effect of geometric parameters and flow conditions on system operation, $C_{d,c}$ and $C_{d,a}$ can be defined as a function of the Reynolds number Re.

Considering the characteristic dimension to be diameter d for the circular passage section, and height h for the annular passage section, the Reynolds numbers for the two sections are respectively:

$$\mathrm{Re}_c = \frac{\rho\, u\, d}{\mu} = \frac{4\,G}{\pi\, d\, \mu} \; ; \; \mathrm{Re}_a = \frac{\rho\, u\, h}{\mu} = \frac{G}{\pi\, d_0\, \mu} \qquad (4)$$

where ρ, u and μ are respectively the density, velocity and dynamic viscosity of air.

Figure 14 shows the curves for $C_{d,c}$ versus Re_c obtained for the pads with annular orifice supply system (type "*a*") plotted for the geometries indicated in Table 1 at a given gap height. Each experimental curve is obtained from the five values established for supply pressure. Results indicate that supply orifice length l in the investigated range (0.3 mm - 1 mm) does not have a significant influence on $C_{d,c}$. By contrast, the effect of varying orifice diameter and gap height is extremely important. In particular, $C_{d,c}$ increases along with gap height, and is reduced as diameter increases, with all other geometric parameters remaining equal. For small air gaps, $C_{d,c}$ generally increases along with Re_c, and tends towards constant values for higher values of Re_c. With the same orifices but larger air gaps, values of Re_c numbers are higher: in this range, the curves for the $C_{d,c}$ coefficients thus obtained have already passed or are passing their ascending section.

Figure 15 shows $C_{d,c}$ versus Re_c obtained for the pads with simple orifices with feed pocket supply system (pads "*b*" and "*c*") for pressure drop p_S - p_T, plotted for the geometries indicated in Table 2 with δ = 10, 20, 1000 μm and at a given gap height. As the effects of

varying orifice length were found to be negligible, tests with feed pocket supply system were carried out only on pads with l = equal to 0.3 mm. The values for $C_{d,c}$ obtained with the simple orifices with feed pocket are greater than the corresponding coefficients obtained with annular orifice system, but the trend with Re_c is similar. In particular, the same results shown in Figure 14 are obtained if δ tends to zero.
If $\delta = 10$, 20 μm $C_{d,c}$ is heavily dependent on h, d, δ: it increases along with gap height h and δ, and is reduced as diameter d increases, with all other geometric parameters remaining equal.
If δ = 1 mm, values for $C_{d,c}$ do not vary appreciably with system geometry, but depend significantly only on Re_c. For $Re_c \to \infty$, the curves tend toward limits that assume average values close to $C_{d,c\ max}$= 0.85.

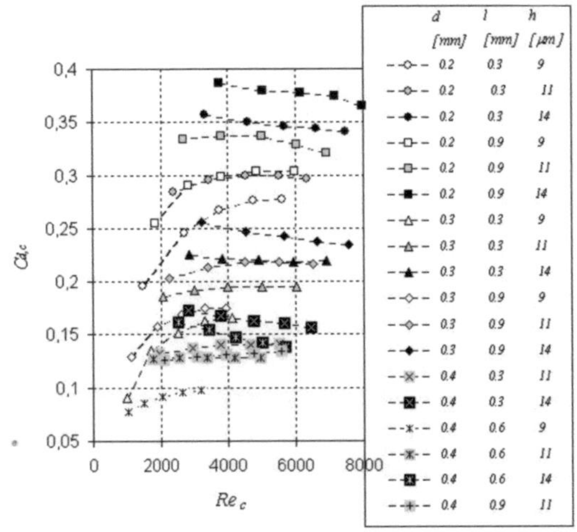

Fig. 14. Experimental values for $C_{d,c}$ versus Re_c obtained for type "*a*" pads

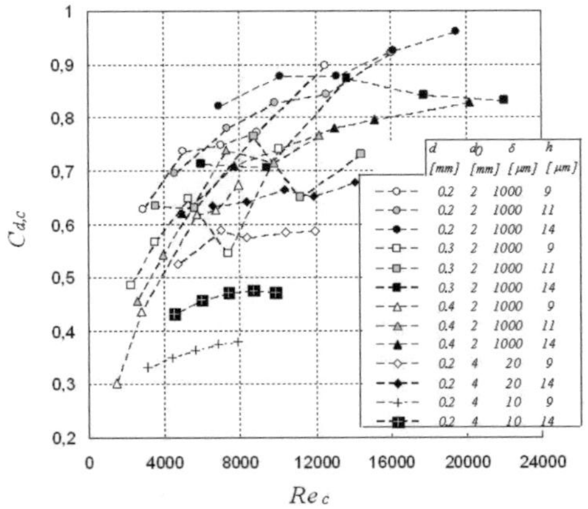

Fig. 15. Experimental values for $C_{d,c}$ versus Re_c obtained for type "*b*" and "*c*" pads, and with l = 0.3 mm

Figure 16 shows $C_{d,a}$ versus Re_a obtained for the pads with simple orifice and feed pocket supply system and with a non-negligible pressure drop p_T – p_i (type "*b*" pads). Here again, values for $C_{d,a}$ depend significantly only on Re_a and tend towards the same limit value slightly above 1. This value is associated with pressure recovery upstream of the inlet resistance.
In order to find formulations capable of approximating the experimental curves of $C_{d,c}$ and $C_{d,a}$ with sufficient accuracy, the maximum values of these coefficients were analyzed as a function of supply system geometrical parameters.
Figures 17 and 18 show the experimental maximum values of $C_{d,c}$ from Figure 14 and Figure 15 respectively, as a function of ratio h/d and $(h+\delta)/d$. Figure 18 also shows the results of Figure 17 ($\delta = 0$) only for l = 0.3 mm. The proposed exponential approximation function is also shown on the graphs:

$$f_1[(h+\delta)/d] = 0.85\left(1 - e^{-8.2((h+\delta)/d))}\right) \tag{5}$$

where $\delta = 0$ for annular orifices. To approximate the experimental values of the discharge coefficients in the ascending sections of the curves shown in Figures 14 and 15, a function f_2 of h, δ, Re_c is introduced:

$$f_2[(h+\delta)/d] = \left(1 - e^{-0.001\,\mathrm{Re}_c \cdot \frac{h+\delta}{h+4\delta}}\right) \tag{6}$$

The complete function proposed to identify $C_{d,c}$ thus assumes the following form:

$$C_{d,c} = f_1 \cdot f_2 = 0.85\left(1 - e^{-8.2((h+\delta)/d)}\right) \cdot \left(1 - e^{-0.001\,\mathrm{Re}_c \cdot \frac{h+\delta}{h+4\delta}}\right) \tag{7}$$

The graphs in Figures 14 and 15 show several curves where all the values of coefficients $C_{d,c}$ are in a range equal to about 5% of the maximum calculated value. For these cases, it is assumed that the $C_{d,c}$ curves have already reached their limit, which is considered to be equal to the average calculated value. In the other curves, the values of $C_{d,c}$ do not reach their limits, to extrapolate these limits the values obtained with the highest Re_c have been divided by the function f_2 and the results are shown in Figure 18.
Experimental data for $C_{d,c\ max}$ can be grouped into three zones: zone I ($(h+\delta)/d$ <0.1) for orifices with no pockets, zone II ($(h+\delta)/d$ = 0.1 to 0.2) for shallow pockets, and zone III ($(h+\delta)/d$ >0.2) for deep pockets. While $C_{d,c\ max}$ depends on h and δ in zones I and II, it reaches a maximum value which remains constant as $(h+\delta)/d$ increases in zone III. In particular, when d is predetermined and δ is sufficiently large, $C_{d,c\ max}$ is independent of h. In this range, the supply system provides the best static bearing performance, as reducing air gap height does not change $C_{d,c\ max}$ and thus does not reduce the hole's conductance. However, excessive values for δ or for pocket volume can cause the bearing to be affected by dynamic instability problems (air hammering), which must be borne in mind at the design stage.
The proposed exponential approximation function for $C_{d,a}$ in simple orifices with feed pocket is an exponential formula which depends only on Re_a:

$$C_{d,a} = 1.05 \cdot \left(1 - e^{-0.005\,\mathrm{Re}_a}\right) \tag{8}$$

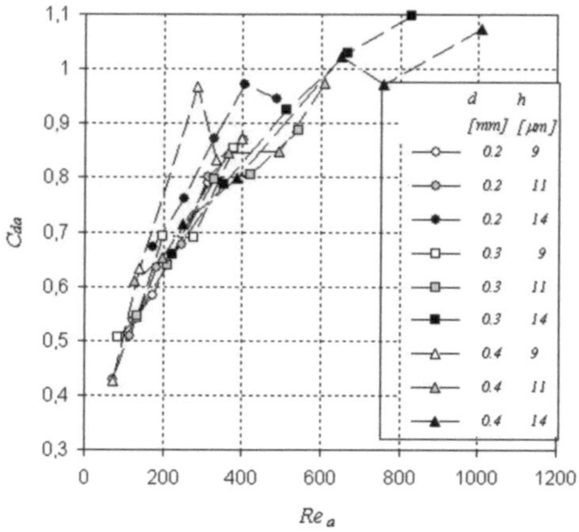

Fig. 16. Experimental values for $C_{d,a}$ versus Re_a, for type "b" pads, with δ = 1 mm and d_0 = 2 mm

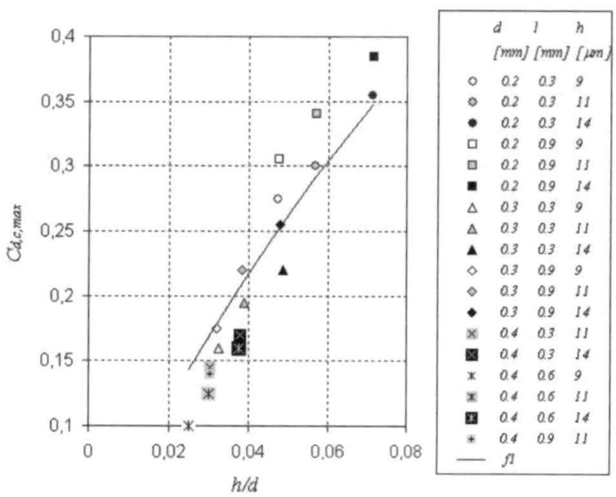

Fig. 17. Maximum experimental values of $C_{d,c}$ and function f_1 (solid line) for pads with annular orifice, versus ratio h/d

The graphs in Figures 19-22 show a comparison of the results obtained with the these approximation functions. Specifically, Figures 19-20 give the values of $C_{d,c}$ for the annular orifices, while Figures 21 and 22 indicate the values of $C_{d,c}$ and $C_{d,a}$ respectively for the simple orifices with pocket.
As can be seen from the comparison, the data obtained with approximation functions (7) and (8) show a fairly good fit with experimental results.

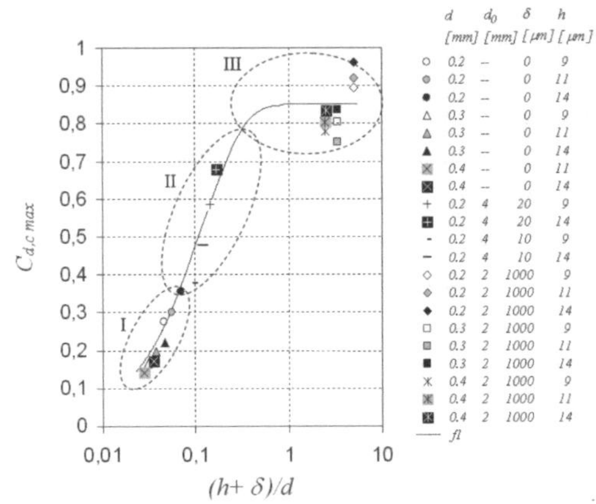

Fig. 18. Maximum experimental values of $C_{d,c}$ and function f_1 (solid line) for the pad with simple orifices and feed pocket, versus ratio $(h+\delta)/d$

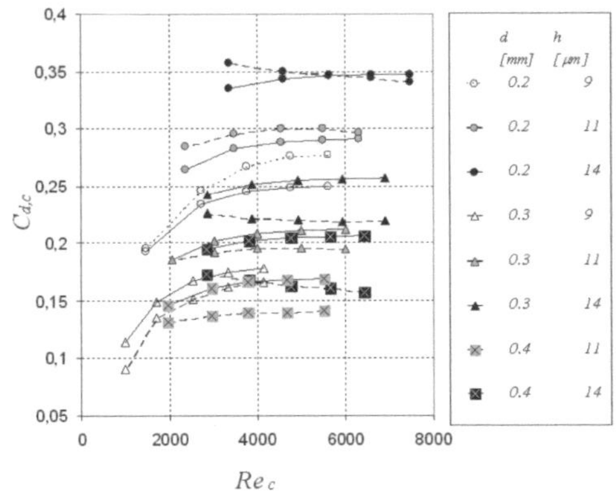

Fig. 19. Experimental values (dotted lines) and approximation curves (solid lines) for $C_{d,c}$ versus Re_c, for type "a" pads with l = 0.3 mm

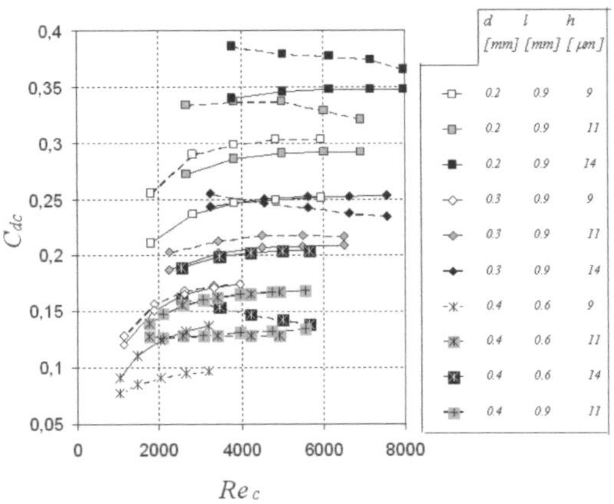

Fig. 20. Experimental values (dotted lines) and approximation curves (solid lines) for $C_{d,c}$ versus Re_c, for type "a" pads with l = 0.6 mm and l = 0.9 mm

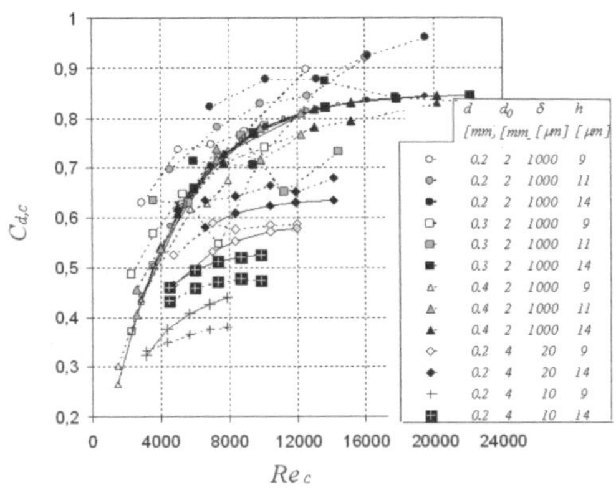

Fig. 21. Experimental values (dotted lines) and approximation curves (solid lines) for $C_{d,c}$ versus Re_c, for type "b" and "c" pads, and with l = 0.3 mm

6. Mathematical model of pads

The mathematical model uses the finite difference technique to calculate the pressure distribution in the air gap. Static operation is examined. As air gap height is constant, the study can be simplified by considering a angular pad sector of appropriate amplitude. For type "*c*" pads, this amplitude is that of one of the supply holes.

Both the equations for flow rate G (3) across the inlet holes and the Reynolds equations for compressible fluids in the air gap (9) were used.

$$\frac{1}{r}\frac{\partial}{\partial r}\left(rh^3\frac{\partial P^2}{\partial r}\right)+\frac{1}{r^2}\frac{\partial}{\partial \vartheta}\left(h^3\frac{\partial P^2}{\partial \vartheta}\right)+24\mu\, R\, T_0\frac{G}{rdrd\vartheta}=0 \tag{9}$$

The Reynolds equations are discretized with the finite difference technique considering a polar grid of "n" nodes in the radial direction and "m" nodes in the angular direction for the pad sector in question. The number of nodes, which was selected on a case by case basis, is appropriate as regards the accuracy of the results. Each node is located at the center of a control volume to which the mass flow rate continuity equation is applied. Because of the axial symmetry of type "*a*" and "*b*" pads, flow rates in the circumferential direction are zero. In these cases, the control volume for the central hole is defined by the hole diameter for type "*a*" pads or by the pocket diameter for type "*b*" pads. The pressure is considered to be uniform inside these diameters. For type "*c*" pads, the center of each supply hole corresponds to a node of the grid. As an example for this type, Figure 23 shows a schematic view of an air gap control volume centered on generic node i,j located at one of the supply holes. Also for this latter type, several meshing nodes are defined in the pockets to better describe pressure trends in these areas.

In types "*b*" and "*c*", the control volumes below the pockets have a height equal to the sum of that of the air gap and pocket depth.

For type "*b*", which features very deep pockets, the model uses both formulations for discharge coefficients $C_{d,c}$ and $C_{d,a}$, whereas for type "*c*" only $C_{d,c}$ is considered.

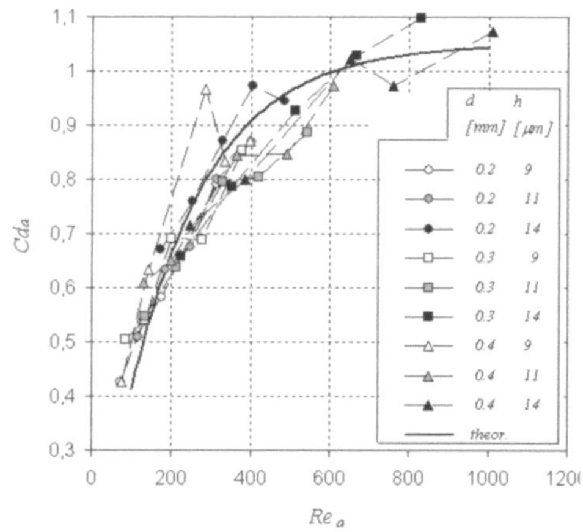

Fig. 22. Experimental values (dotted lines) and approximation curve (solid line) for $C_{d,a}$ versus Re_a, for type "*b*" pads, $\delta = 1$ mm, $d_0 = 2$ mm

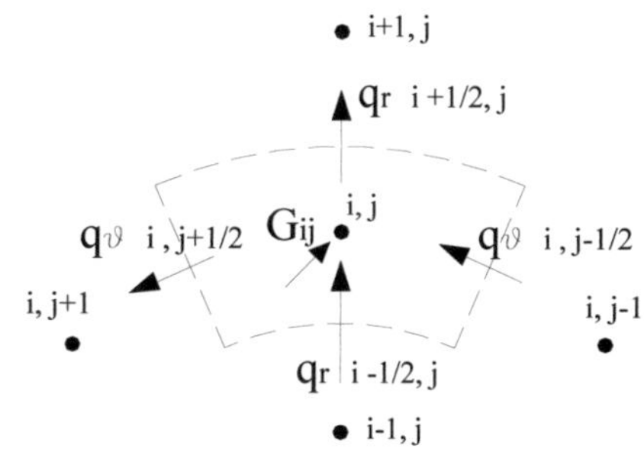

Fig. 23. Control volume below generic node i,j located on a supply hole of type "*c*" pad.

The model solves the flow rate equations at the inlet and outlet of each control volume iteratively until reaching convergence on numerical values for pressure, the Reynolds number, flow rate and discharge coefficients.

7. Examples of application for discharge coefficient formulations: comparison of numerical and experimental results

A comparison of the numerical results obtained with the radial and circumferential pressure distributions indicated in the graphs in Figures 9 – 12 will now be discussed. The selected number of nodes is shown in each case. For all simulations, the actual hole diameters for which pressure distribution was measured were considered. The data obtained with the formulation are in general similar or slightly above the experimental data, indicating that the approximation is sufficiently good. In all cases, the approximation is valid in the points located fairly far from the supply hole, or in other words in the zone where viscous behavior is fully developed, inasmuch as the model does not take pressure and velocity gradients under the supply holes into account. This is clearer for type *"a"* pads (Figure 24) than for type *"b"* and *"c"* pads, where the model considers uniform pressure in the pocket. For cases with deep pockets (type *"b"* pads), the numerical curves in Figure 25 show the pressure rises immediately downstream of the hole and at the inlet to the air gap due to the use of the respective discharge coefficients. For type *"c"* pads with pocket depth $\delta \leq 20$ μm (Figure 26), on the other hand, the pressure drop at the air gap inlet explained in the previous paragraph was not taken into account. In general, the approximation problems were caused by air gap height measurement errors resulting both from the accuracy of the probes and the difficulties involved in zeroing them. Thus, it was demonstrated experimentally that pressure in the air gap and flow rate are extremely sensitive to inaccuracies in measuring h. Significant variations in h entail pressure variations that increase along with average air gap height. Figure 27 shows another comparison of experimental and numerical pressure distributions for type *"c"* pads, this time with different air gap heights and pocket depths.

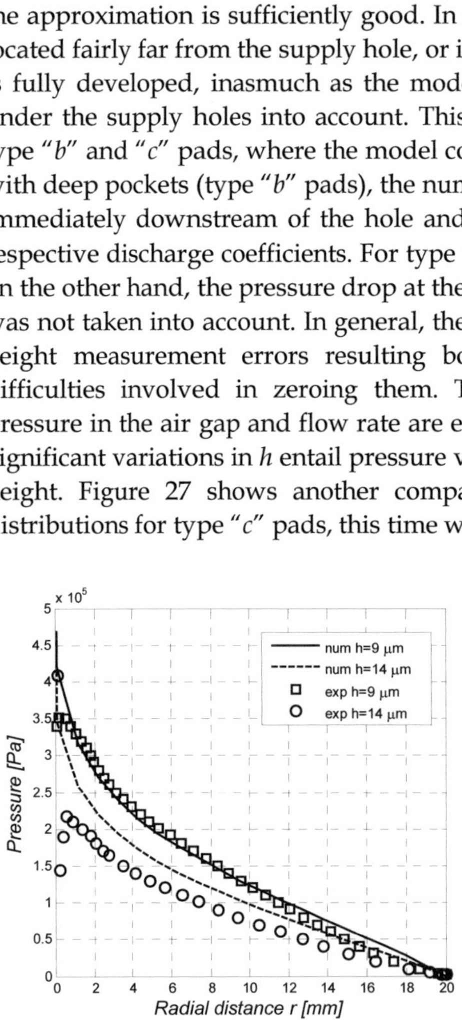

Fig. 24. Numerical and experimental radial pressure distribution across number "1" pad, type "*a*" , supply pressure $p_S = 0.5$ MPa, orifice diameter $d = 0.2$ mm, air gap height $h = 9$ and 14 μm, n×m= 20×20.

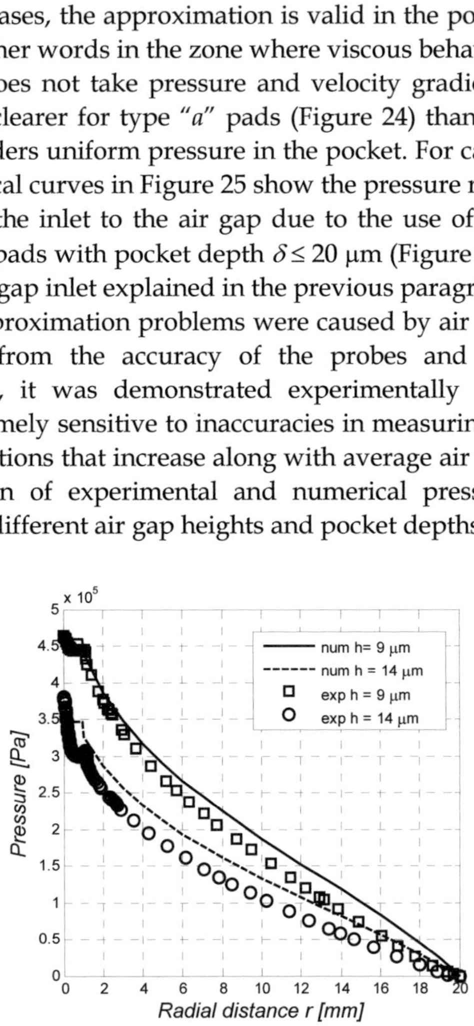

Fig. 25. Numerical and experimental radial pressure distribution across entire pad number "8", type "*b*", supply pressure $p_S = 0.5$ MPa, orifice diameter $d = 0.2$ mm, pocket diameter $d_0 = 2$ mm, pocket depth $\delta = 1$ mm, air gap height $h = 9$ and 14μm, n×m= 20×20.

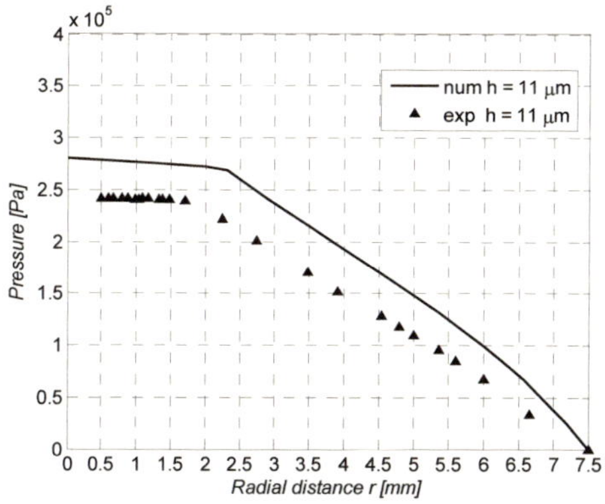

Fig. 26. Radial pressure distribution across pad number "11", type "*c*", supply pressure $p_S = 0.4$ MPa, orifice diameter $d = 0.2$ mm, pocket diameter $d_0 = 4$ mm, pocket depth $\delta = 20$ μm, air gap height $h = 11$ μm , n×m= 20×20.

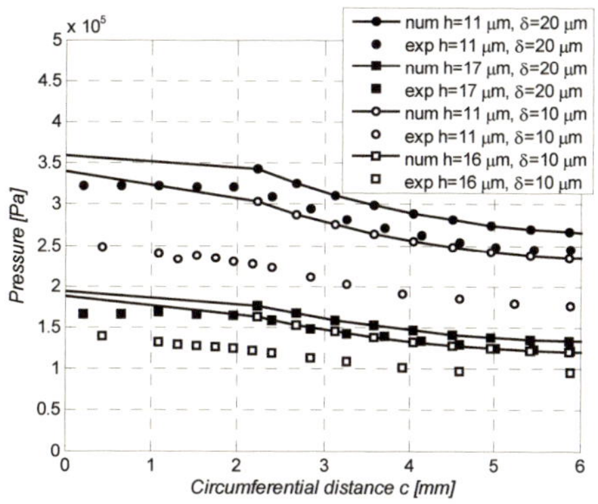

Fig. 27. Numerical and experimental circumferential pressure distribution across pad number "11", type "*c*", supply pressure $p_S = 0.5$ MPa, orifice diameter $d = 0.2$ mm, pocket diameter $d_0 = 4$ mm, pocket depth $\delta = 10$ and 20 μm, n×m= 20×20.

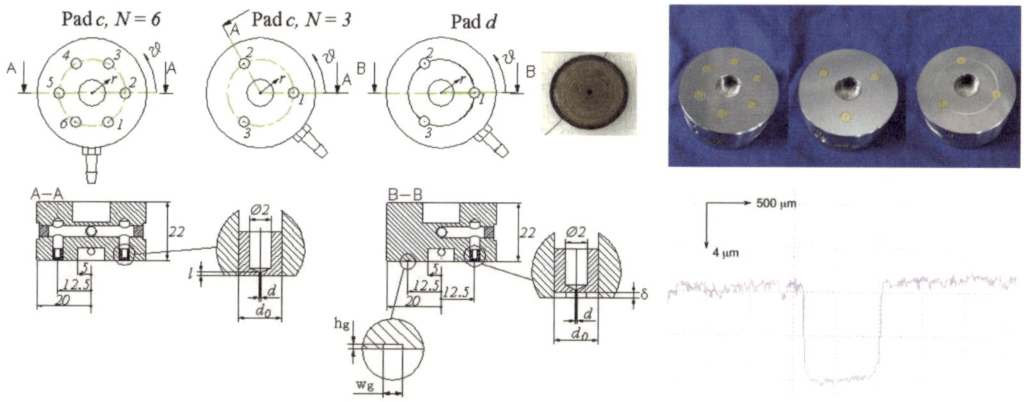

Fig. 28. Further pads tested to verify the discharge coefficient formulation

The discharge coefficient formulation was also verified experimentally on a further three pads as shown in the diagram and photograph in Figure 28, including two type "*c*" pads and one grooved pad (type "*d*"). Table 3 shows the nominal geometric magnitudes for each pad. The first two (13, 14) have a different number of holes and pocket depth is zero. The third (15) features 10 μm deep pockets and a circular groove connecting the supply holes. The groove is 0.8 mm wide and its depth is equal to that of the pockets. The figure also shows an enlargement of the insert and groove for pad 15 and the groove profile as measured radially using a profilometer.

In these three cases, the center of the pads was selected as the origin point for radial coordinate *r* and the center of one of the supply holes was chosen as the origin point of angular coordinate ϑ.

In all cases, the actual average hole dimensions were within a tolerance range of around 10% of nominal values. A mathematic model similar to that prepared for type "*c*" pads was also developed for type "*d*", considering the presence of the groove. Comparisons of the experimental and numerical pressure distributions for the three cases are shown in Figures 29 - 31.

Pad N.	Pad type	n. holes	Insert				Groove	
			Hole		Pocket			
			l [mm]	d [mm]	d_0 [mm]	δ [μm]	w_g[mm]	h_g[μm]
13	c	6	0.4	0.2	4	0	-	-
14	c	3	0.4	0.2	4	0	-	-
15	d	3	0.4	0.3	4	10	0.8	10

Table 2. Nominal dimensions of pads 13, 14, 15.

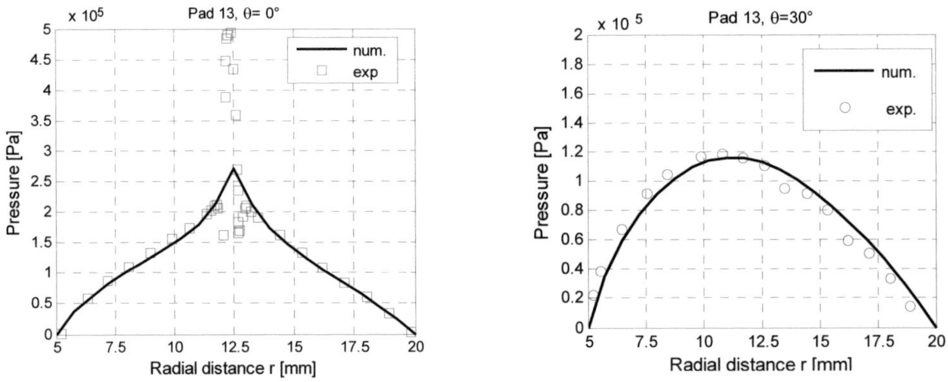

Fig. 29. Numerical and experimental circumferential pressure distribution across pad "13", supply pressure p_S = 0.5 MPa, orifice diameter d = 0.2 mm, pocket diameter d_0= 4 mm, pocket depth δ = 0 μm, air gap height h = 15 μm, θ = 0° and 30°, n×m= 21×72.

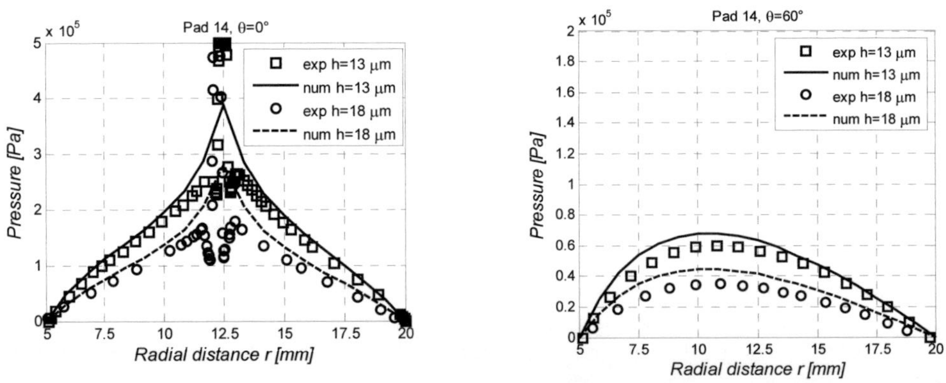

Fig. 30. Comparison of experimental and numerical radial pressure distributions, pad "14", p_S = 0.5 MPa, orifice diameter d = 0.2 mm, pocket diameter d_0 = 4 mm, pocket depth δ = 0 μm, air gap height h = 13 and 18μm, θ = 0° and 60°, n×m= 21×72.

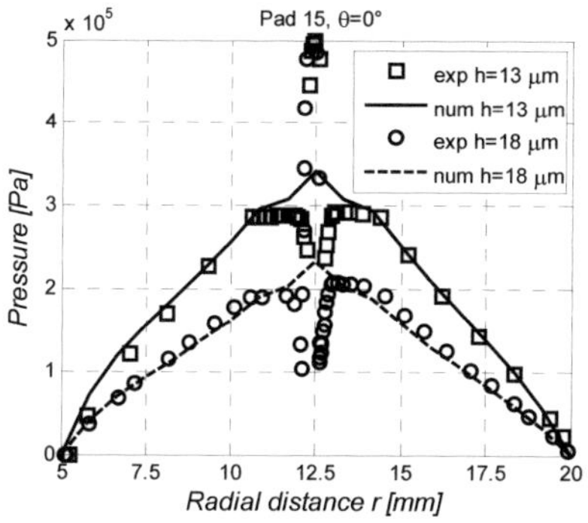

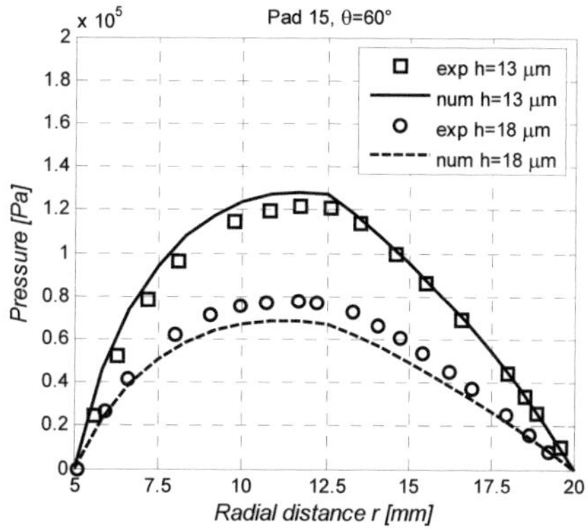

Fig. 31. Comparison of experimental and numerical radial pressure distributions, pad "15", $p_S = 0.5$ MPa, orifice diameter $d = 0.3$ mm, pocket diameter $d_0 = 4$ mm, pocket depth $\delta = 10\mu m$, air gap height $h = 13$ and 18 μm, θ = 0° and 60°, n×m= 21×72.

Here as in the previous cases, the numerical curves correspond with the experimental data or overestimate them slightly.

For the pad with groove and pockets in particular, the width of the groove is slightly greater than the diameter of the supply holes and the pockets are sufficiently large to distance the groove from the holes. In this way, the influence of the groove on the air flow adjacent to the supply holes is negligible, the system's behavior is similar to that of the type "*c*" pad, and the validity of the formulation is also confirmed for this case.

It should be borne in mind, however, that reducing the size of the pockets and groove can have a significant influence on flow behavior around the supply holes. In cases where the formulation is not verified, it will be necessary to proceed with a new identification of the supply system.

8. Conclusions

This chapter presented an experimental method for identifying the discharge coefficients of air bearing supply systems with annular orifices and simple orifices with feed pocket.

For annular orifice systems, it was found that the flow characteristics can be described using the experimental discharge coefficient relative to the circular orifice section, $C_{d,c}$.

For simple orifices with feed pocket, the flow characteristics can be described using two experimental discharge coefficients: $C_{d,c}$ for the circular section of the orifice and $C_{d,a}$ for the annular section of the air gap in correspondence of the pocket diameter. In particular, for deep pockets with $(h+\delta)/d \geq 0.2$, both coefficients apply, while for shallow pockets with $(h+\delta)/d < 0.2$, only coefficient $C_{d,c}$ applies.

Analytical formulas identifying each of the coefficients were developed as a function of supply system geometrical parameters and the Reynolds numbers.

To validate the identification, a finite difference numerical model using these formulations was prepared for each type of pad. Experimental and numerical pressure distributions were in good agreement for all cases examined. The formulation can still be applied to pads with a circular groove if sufficiently large pockets are provided at the supply holes. Future work could address supply systems with grooves and pockets with different geometries and dimensions.

As pad operating characteristics are highly sensitive to air gap height, the identification method used calls for an appropriate procedure for measuring the air gap in order to ensure the necessary accuracy. The method also requires detailed measurement of the pressure distribution adjacent to the supply hole to identify the local maximum p_i. Alternative identification methods are now being investigated in order to overcome the difficulties involved in performing these measurements, and preliminary findings are discussed in Belforte et al., 2010-d.

In general, the proposed formulation is applicable for values of ratio $(h+\delta)/d$ varying from 0.03 to 5. Further developments will address the identification of annular orifice supply systems with ratio h/d under 0.03.

9. Nomenclature

C_d	Discharge coefficient
$C_{d,a}$	Discharge coefficient for annular section
$C_{d,c}$	Discharge coefficient for circular section
D	Supply passage diameter
G	Mass flow rate
G_t	Theoretical mass flow rate
P	Absolute pressure
P_d	Downstream absolute pressure
P_u	Upstream absolute pressure
R	Constant of gas (= 287.1 J/kg K)
Re_a	Reynolds Number for annular section
Re_c	Reynolds Number for circular section
S	Passage section
T	Absolute temperature upstream of the nozzle
T_0	Absolute temperature in normal condition (288 K)
c	Circumferential coordinate
d	Supply orifice diameter
d_0	Pocket diameter
dr	Generic control volume radial length
$d\vartheta$	Generic control volume angular width
k	Specific heat ratio of air (= 1.4)
h	Air gap height
l	Supply orifice length
q_r	Mass flow rate across control volume in the radial direction
q_ϑ	Mass flow rate across control volume in the circumferential direction
r	Radial coordinate
r_i	Radius of completely developed viscous resistance zone
p_i	Relative pressure at radius r_i
p_T	Pocket relative pressure
p_S	Supply relative pressure
u	Air velocity
α	Conicity angle
δ	Pocket depth
μ	Air viscosity in normal condition (= 17.89 10^{-6} Pa s)
ρ	Air density in normal condition (=1.225kg/m^3)
ϑ	Angular coordinate

10. References

Al-Bender, F.; Van Brussel, H. (1992-b). Tilt characteristics of circular centrally fed aerostatic bearings. *Tribology International*, 25 (3), pp. 189-197.

Al-Bender, F.; Van Brussel, H. (1992-a). Symmetric radial laminar channel flow with particular reference to aerostatic bearings. *Journal of Tribology*, 114, (7), pp. 630-636.

Bang, K.G. ; Lee, D.G. (2002). Thrust bearing design for high-speed composite air spindles. *Composite Structures*, 57, (1-4), pp. 149-160.

Belforte, G. ; Raparelli, T., Viktorov, V. (1999). Theoretical Investigation of fluid inertia effects and stability of self-acting gas journal bearings. *Journal of Tribology*, 125 (10), pp. 836-843.

Belforte, G. ; Raparelli, T.; Trivella, A. ; Viktorov, V. ; Visconte, C. (2006). Numerical analysis on the supply hole discharge coefficient in aerostatic bearings, *Proceedings of International Conference on Tribology AITC-AIT 20-22 September 2006*, cd rom ISBN 88-902333-0-3, Parma, Italy.

Belforte, G.; Raparelli, T. ; Viktorov, V. ; Trivella, A. ; Colombo, F. (2006). An Experimental Study of High-Speed Rotor Supported by Air Bearings: Test rig and First experimental results. *Tribology International*, 39, (8), pp. 839-845.

Belforte, G. ; Raparelli, T. ; Viktorov, V. ; Trivella, A. (2007-a). Discharge coefficients of orifice-type restrictor for aerostatic bearings. *Tribology International*, 40 (3), pp. 512-521.

Belforte, G. ; Raparelli, T. ; Viktorov, V.; Trivella, A. (2007-b). Permeability and inertial coefficients of porous media for air bearing feeding systems, *Journal of Tribology*, 129, (10), pp.705-711.

Belforte, G.; Colombo, F.; Raparelli, T.; Trivella, A.; Viktorov V. (2008). High-speed electrospindle running on air bearings:design and experimental verification, *Meccanica*, 43 (6), pp.591-600.

Belforte, G. ; Raparelli, T. ; Trivella, A ; Viktorov, V. (2009). Metal Woven Wire Cloth Feeding System For Gas Bearings. *Tribology International*, 42, (4), pp. 600-608.

Belforte, G.; Colombo, F.; Raparelli, T.; Trivella, A.; Viktorov V. (2010 a). Performance of Externally Pressurized Grooved Thrust Bearings. *Tribology Letters*, 37, (3), pp. 553-562.

Belforte, G.; Colombo, F.; Raparelli, T.; Trivella, A.; Viktorov V. (2010-b). Comparison between grooved and plane aerostatic thrust bearings: static performance. *Meccanica*, DOI 10.1007/s11012-010-9307-y.

Belforte, G.; Colombo, F.; Raparelli, T.; Trivella, A.; Viktorov V. (2010-c). Static behaviour of air plane pads: comparison between different feeding solutions. *Journal of Mechanical Engineering*, 56 (4), pp.261-267, UDC 531.2:621.822.

Belforte, G.; Colombo, F.; Raparelli, T.; Trivella, A.; Viktorov V. (2010-d). A new identification method of the supply hole discharge coefficient of gas bearings, *Tribology and Design, WIT Transaction on Engineering Sciences*, 66, pp.95-105, Online ISSN:1743-3533.

Blondeel, E.; Snoeys, R. ; Devrieze, L. (1980). Dynamic Stability of Externally Pressurized Gas Bearings. *Journal of Lubrication Technology*, 102 (10), pp. 511-519.

Bonneau, D.; Huitric, J.; Tournerie, B. (1993). Finite Element Analysis of Grooved Gas Thrust Bearings and Grooved Gas Face Seals. *Journal of Tribology*, 115, (3), pp. 348-354.

Bryant, M.R. ; Velinsky, S.A. ; Beachley, N.H. ; Froncza, K.F.T. (1986). A design methodology for obtaining infinite stiffness in an aerostatic thrust bearing. *Journal of Mechanisms, Transmissions, and Automation in Design*, 108 (12), pp. 448-56

Chen, M. F.; Chen,Y. P.; Lin, C. D. (2002). Research on the arc type aerostatic bearing for a PCB drilling station. *Tribology International*, 35 (4), pp. 235-243.

Chen, M.F.; Lin, Y.T. (2002). Static Behavior and Dynamic Stability Analysis of Grooved Rectangular Aerostatic Thrust Bearings by Modified Resistance Network Method. *Tribology International*, 35, (5), pp.329-338.

Chen, M. F. ; Huang, W. L. ; Chen, Y. P. (2010). Design of the aerostatic linear guideway with a passive disk-spring compensator for PCB drilling machine. *Tribology International*, 43, (1-2), pp. 395-403.

Elrod H.G. ; Glanfield, G.H. (1971). Computer procedures for the design of flexibly mounted, externally pressurized, gas lubricated journal bearing. *Proceedings of the Gas Bearing Symposium*, pp. 1-37, paper 22, University of Southampton, March 1971.

Fourka, M.; Tian, Y.; Bonis, M. (1996). Prediction of the stability of air thrust bearing by numerical, analytical and experimental methods. *Wear*, 198, (1-2), pp. 1-6.

Goodwin, M.J. (1989). *Dynamics of rotor-bearing systems*, Unwin Hyman, London.

Grassam, N. S.; Powell, J. W. (1964). *Gas Lubricated Bearings*, Butterworths, London, pp. 135-139.

Gross, W.A. (1962). *Gas film lubrication*, Wiley, New York.

Hashimoto, *H.* ; Namba, T. (2009). Optimization of groove geometry for a thrust air bearing according to various objective functions. *Journal of Tribology*, 131, (4), pp. 704-710.

Huges, S. J.; Hogg., S. I.; Jones, T. V. (1996). Analysis of a Gas Lubricated Hydrodynamic Thrust Bearing, *Journal of Tribology*, 118, (7), pp. 449-456.

Kassab, S. Z. ; Noureldeen E.M. ; Shawky A. (1997). Effects of operating conditions and supply hole diameter on the performance of a rectangular aerostatic bearing. *Tribology International*, 30 (7), pp. 533-45.

Kazimierski,Z.; Trojnarski, J. (1980). Investigations of externally pressurized gas bearing with different feeding systems. *Journal of Lubrication Technology*, 102 (1), pp.59-64.

Li, *H.*; Ding, *H.* (2007). Influences of the geometrical parameters of aerostatic thrust bearing with pocketed orifice-type restrictor on its performance. *Tribology International*, 40, (7), pp. 1120-1126.

Lund, J. W. (1964). The hydrostatic gas journal bearing with journal rotation and vibration, *Journal of Basic Engineering*, 86, pp. 328-336.

Majumdar, B.C. (1980); Externally pressurized gas bearings: a review. *Wear* , 62, (2), pp. 299-314.

Mori, *H.*; Miyamatsu, Y. (1969). Theoretical flow-models for externally pressurized gas bearings. *Journal of Lubrication Technology*, 91 (1), pp. 181-193.

Nakamura, T.; Yoshimoto , S. (1996*)*. Static tilt characteristics of aerostatic rectangular double compound restrictors. *Tribology International*, 29, (2), pp. 145-152.

Neves, M. T. ; Schwarz, V. A. ; Menon G. J. (2010). Discharge coefficient influence on the performance of aerostatic journal bearings. *Tribology International*, 43, (4) pp. 746-751.

Ng, S. W. ; Widdowson, G. P., Yao, S. (2005). Characteristics estimation of porous air bearing. Proceedings of the COMSOL Multiphysics User's Conference, Stockholm, cds.comsol.com/access/dl/papers/1043/NG.pdf

Plante, J .; Vogan, J. ; El-Aguizy, T. ; Slocum, A. *H.* (2005). A design model for circular porous air bearings using the 1D generalized flow method. *Precision Engineering*, 29, (7), pp. 336-346.

Poupard, *H.*; Drouin, G. (1973). Theoretical and experimental pressure distribution in supersonic domain for an inherently compensated circular thrust bearing. *Journal of Lubrication Technology*, 95 (3), pp.217-221.

Renn, J.; Hsiao, C. (2004). Experimental and CFD study on the mass flow-rate characteristic of gas through orifice-type restrictor in aerostatic bearings, *Tribology International*, 37 (4), pp. 309-315.

Stout, K. J.; El-Ashkar, S.; Ghasi, V. ; Tawfik, M. (1993). Theoretical analysis of two configurations of aerostatic flat pad bearings using pocketed orifice restrictors. *Tribology International*, 26, (4), pp. 265-273.

Yoshimoto, S.; Tamura, J.; Nakamura, T. (1999). Dynamic Tilt Characteristics of Aerostatic Rectangular Double-Pad Thrust Bearings With Compound Restrictors. *Tribology International*, 32, (12), pp. 731-738.

Yoshimoto, S.; Kohno, K. (2001). Static and dynamic characteristics of aerostatic circular porous thrust bearings. *Journal of Tribology*, 123, (7), pp. 501-508.

Yoshimoto, S., Tamamoto, M.; Toda, K. (2007). Numerical calculations of pressure distribution in the bearing clearance of circular aerostatic bearings with a single air supply inlet, *Transactions of the ASME*, 129 (4), pp. 384-390.

19

Inverse Approach for Calculating Temperature in Thermal Elasto-Hydrodynamic Lubrication of Line Contacts

Li-Ming Chu[1], Hsiang-Chen Hsu[1], Jaw-Ren Lin[2] and Yuh-Ping Chang[3]
[1]*Department of Mechanical and Automation Engineering, I-Shou University,*
[2]*Department of Mechanical Engineering, Nanya Institute of Technology,*
[3]*Department of Mechanical Engineering, Kun Shan University,*
Taiwan

1. Introduction

It is well known by now that pressure, temperature, and film shape definitely play important roles in the failure of heavily loaded non-conformal contacts, such as rolling element bearings, gears, traction drives, or cams and tappets. Furthermore, the effect of heat generated due to the shearing of the high-pressure lubricant is no longer negligible under sliding conditions, as the heat changes the characteristics of the oil flow because of a decrease in viscosity. Therefore, the thermal effect on the film thickness and traction is significant in elastohydrodynamic lubricated contacts. So an accurate estimate of the temperature distribution in the contact zone at various operational parameters is necessary.
Since (Sternlicht et al., 1961) started to consider the thermal effects of line contact in the EHL under rolling/sliding conditions, the inclusion of thermal effects in EHL has been an important subject of research in the field of tribology. Many numerical solutions considering the thermal effects on EHL have been presented, for instance, by (Ghosh & Hamrock, 1985), (Salehizadeh & Saka, 1991), and (Lee & Hsu, 1993); and for thermal point contact problems by (Zhu & Wen, 1984), (Kim & Sadeghi, 1992), and (Lee & Hsu, 1995). With respect to measuring the temperature increase in EHL contacts, (Cheng & Orcutt, 1965), (Safa et al., 1982), and (Kannel et al., 1978) have measured the temperature increase in a sliding surface using a thin film gauge deposited on a disc. (Turchina et al., 1974) and (Ausherman et al., 1976) employed an improved infrared technique to measure the temperature distribution of the oil film and surface. They demonstrated that the temperature was maximum at zones with minimum film thickness in the contact side lobes. Recently, (Yagi et al., 1966) described the mechanism of variations of EHL oil film under high slip ratio conditions. The oil film thickness between a ball surface and a glass disk was measured using optical interferometry, and the temperature of both the surfaces and of the oil film average across it were measured using an infrared emission technique. They demonstrated that the shape of the oil film can be varied by viscosity wedge action which related to pressure and temperature.
During the last decade, optical interferometry has been found to be the most widely used and successful method in measuring oil film. Several studies of an EHL film were carried

out by experiments (Cameron & Gohar, 1966; Foord et al., 1968; Johnston et al., 1991; Gustafsson et al., 1994; Yang et al., 2001). Since the image processing technique requires a calibration, which always introduces errors, the multi-channel interferometry method was proposed by (Marklund et al., 1998) to overcome such problems. (Luo et al., 1996) measured the center lubricated film thickness on point contact EHL by using a relative optical interference intensity technique. Furthermore, (Hartl et al., 1999) presented a colorimetric interferometry to improve over conventional chromatic interferometry in which film thickness is obtained by color matching between interferogram and color-film thickness dependence obtained from Newton rings for static contact.

When the film thickness map is obtained from the optical interferometry, the pressure distribution can be computed by using the elastic deformation and the force balance theories. This pressure can be used in the Reynolds equation to evaluate the viscosity. (Paul & Cameron, 1992) used an impact viscometer to evaluate the pressure distribution and the apparent viscosity. (Wong et al., 1992) measured the apparent viscosity, the shear stress, and shear rate of liquid lubricants using an impact viscometer. Moreover, they developed a new viscosity-pressure relationship which takes the form of a Barus equation at low pressures and reaches a limiting viscosity at high pressure. (Astrom & Venner, 1994) presented a combined experimental/numerical approach to gain insight into such pressure fluctuations. They used the film thickness map obtained by (Gustafsson et al., 1994) to calculate the pressure distribution from the force balance and elastic deformation theories in a grease-lubricated point contact. (Östensen et al., 1996) theoretically investigated the possibility of using optical interferometry for determining pressure and apparent viscosity in an EHL point contact. Results showed that some of the small fluctuations in pressure are due to the discontinuities in film thickness. However, this small pressure fluctuation would result in a large error in calculating the viscosity due to the amplification by the performing pressure differentiation in the Reynolds equation. Hence, (Lee et al., 2002) developed an inverse approach to overcome this problem in EHL line contacts. In this algorithm, only a few measured points of film thickness are sufficient enough to estimate the pressure distribution without any fluctuation.

Recently, an inverse model proposed by (Yang, 1998) has been widely applied in many design and manufacturing problems in which some of the surface conditions cannot be measured. However, this method used in the inverse TEHL (thermal elastohydrodynamic lubrication) problem is still scarce in the literature. Hence, in this paper, the inverse approach is extended to calculate the mean oil film temperature rise and surface temperature rise distributions and to investigate the sensitivity of the temperature rise and the apparent viscosity for the experimental measurement errors. Moreover, the 'exact' solutions, such as pressure, temperature rise and film thickness are obtained from the numerical solution of the TEHL line contacts problem.

3. Theoretical analysis

As shown in Fig. 1, the contact geometry of two rollers can be reduced to the contact geometry as a roller and a flat surface.

For the steady state, thermal EHL line contact problems, the Reynolds equation can be expressed in the following dimensionless form as:

$$\frac{\partial}{\partial X}\left(\frac{\bar{\rho}H^3}{\bar{\eta}}\frac{\partial P}{\partial X}\right)=\lambda\frac{\partial}{\partial X}(\bar{\rho}H) \tag{1}$$

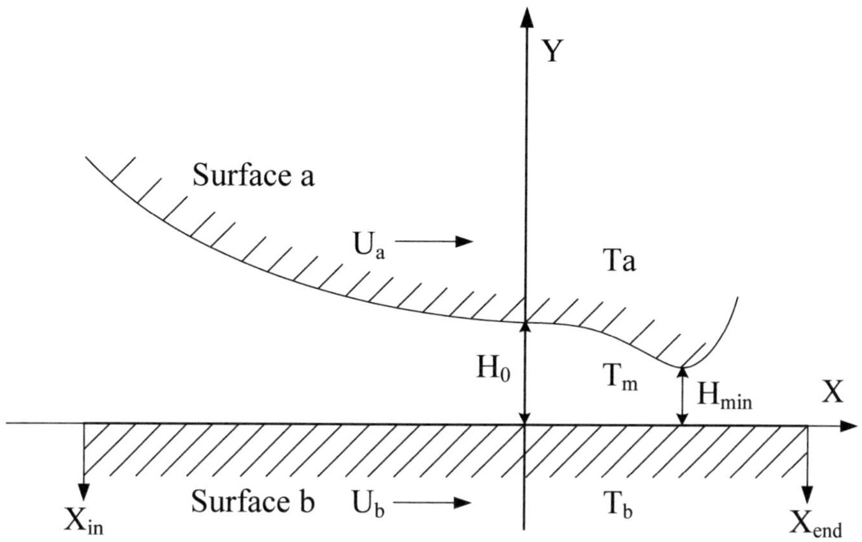

Fig. 1. The geometry of an EHL line contact.

where

$$\lambda = \frac{12\eta_0 \bar{u} R^2}{b^3 p_h}$$

In this equation, the mass density and the viscosity of lubricants related to pressure and temperature can be expressed as:

$$\bar{\rho} = \frac{\rho}{\rho_0} = \left[1 + \frac{0.6\times10^{-9} p}{1+1.7\times10^{-9} p}\right][1-\beta(T_m - T_0)] \tag{2}$$

$$\bar{\eta} = \exp\{(9.67 + \ln\eta_0)[-1 + (1+5.1\times10^{-9} p)^z] - \gamma(T_m - T_0)\} \tag{3}$$

where z is the pressure-viscosity index, β is thermal expansivity of lubricant, γ is temperature-viscosity coefficient of lubricant. If the pressure and mean temperature are given, the apparent viscosity and density can be calculated from equations (2) and (3), respectively.

3.1 Pressure calculation

It has been known that the film thickness in an EHL contact is the sum of the elastic deformation of the surfaces and the gap distance between two rigid surfaces. In the EHL line contact, the film shape in the dimensionless form is given as:

$$H_i = H_0 + \frac{X_i^2}{2} - \frac{1}{\pi}\int_{X_{in}}^{X_{end}} P(X)\ln|X - X|\,dX \tag{4}$$

Discretizing the domain with equi-distant mesh size Δ, Eq. (4) can be approximated by

$$H_i - \frac{X_i^2}{2} - H_0 = -\frac{1}{\pi}\sum_{j=1}^{n} K_{ij} P_j \tag{5}$$

where

$$K_{ij} = \left(i-j+\frac{1}{2}\right)\Delta\left[\left(\ln\left|i-j+\frac{1}{2}\right|\Delta\right)-1\right]-\left(i-j-\frac{1}{2}\right)\Delta\left[\left(\ln\left|i-j-\frac{1}{2}\right|\Delta\right)-1\right] \tag{6}$$

The normal load for the line contact is assumed to be constant, thus the constant H_0 can be obtained from the dimensionless force balance equation as:

$$\int_{X_{in}}^{X_{end}} P(X)\, dX = \pi/2 \tag{7}$$

or in the discretized form

$$\frac{\pi}{2} = \Delta\sum_{j=1}^{n} P_j \tag{8}$$

Once the film shape is measured, the pressure distribution can be calculated from Eqs (5) and (8) or in a matrix form as:

$$\mathbf{H} = \mathbf{DP} \tag{9}$$

or

$$\begin{Bmatrix} H_i - \dfrac{X_i^2}{2} \\ \pi / 2\Delta \end{Bmatrix} = \begin{bmatrix} -\dfrac{1}{\pi}K_{ij} & 1 \\ 1 & 0 \end{bmatrix} \begin{Bmatrix} P_j \\ H_0 \end{Bmatrix} \tag{10}$$

This system equation consists of n+1 equations and unknowns, and the matrix D is a full square matrix. In this paper, if the pressure distribution is calculated from this system equation, then it is called the direct inverse method.

3.2 Inverse approximation for calculating pressure

In order to avoid the small fluctuation in calculating the pressure, the pressure distribution can be assumed to be a polynomial function, the pressure distribution can be represented in the following series form in terms of X as:

$$P_j = \sum_{m=0}^{l} a_m (X_j - c)^m = \sum_{m=0}^{l} a_m f_{jm} \tag{11}$$

where a_m is an undetermined coefficient, l is a positive integer, and c is a constant. In this paper, most of c is set to zero. Substituting this approximation into the system Eq. (10), the governing equation becomes

$$\mathbf{H} = \mathbf{DFA} \tag{12}$$

or

$$\begin{Bmatrix} H_i - \dfrac{X_i^2}{2} \\ \pi / 2\Delta \end{Bmatrix} = \begin{bmatrix} -\dfrac{1}{\pi}K_{ij} & 1 \\ 1 & 0 \end{bmatrix} \begin{bmatrix} f_{jm} & 0 \\ 0 & 1 \end{bmatrix} \begin{Bmatrix} a_m \\ H_0 \end{Bmatrix} \tag{13}$$

In this equation, Hi is the film thickness with n measured point, and D is a $(n+1)\times(k+1)$ matrix, where k is the total node number of the pressure. It is obvious that F is a $(k+1)\times(l+2)$ matrix, and A is a $(l+2)$ vector. For the viscous EHL problem, it has been known that the pressure spike appears in the dimple region of the film shape. Hence, to obtain the pressure distribution using the inverse approach, it is necessary to divide the domain into a few regions due to the singular point at the pressure spike. Therefore, the divided matrix form can be written as:

$$\begin{Bmatrix} [H_i - \dfrac{X_i^2}{2}]_I \\ \bullet \\ \bullet \\ \bullet \\ [H_i - \dfrac{X_i^2}{2}]_N \\ \pi / 2\Delta \end{Bmatrix} = \begin{bmatrix} [-\dfrac{1}{\pi}K_{ij}]_I & \begin{bmatrix} 1 \\ \bullet \\ \bullet \\ \bullet \\ 1 \end{bmatrix} \\ \bullet & \\ \bullet & \\ \bullet & \\ [-\dfrac{1}{\pi}K_{ij}]_N & \\ [1 \ \bullet \ \bullet \ \bullet \ 1] & 0 \end{bmatrix} \begin{bmatrix} [f_{jm}]_I & [0] & [0] & [0] & [0] & [0] \\ [0] & \bullet & [0] & [0] & [0] & [0] \\ [0] & [0] & \bullet & [0] & [0] & [0] \\ [0] & [0] & [0] & \bullet & [0] & [0] \\ [0] & [0] & [0] & [0] & [f_{jm}]_N & [0] \\ [0] & [0] & [0] & [0] & [0] & 1 \end{bmatrix} \begin{Bmatrix} [a_m]_I \\ \bullet \\ \bullet \\ \bullet \\ [a_m]_N \\ H_0 \end{Bmatrix} \tag{14}$$

If the unknown variables a_m and H_0 are given, the estimated film thickness can be calculated from Eq. (14). It is obvious that the estimated film thickness is different from the measured one. Hence, to obtain the smallest error between the measured and estimated film thickness, the linear least squares methods is employed here. The least-squares criterion requires the following function to be minimized.

$$g = (\mathbf{H}^{\text{estimated}} - \mathbf{H}^{\text{measured}})^{\mathrm{T}} (\mathbf{H}^{\text{estimated}} - \mathbf{H}^{\text{measured}}) \tag{15}$$

The best value for A can be determined from

$$\frac{\partial g}{\partial \mathbf{A}} = 0 \tag{16}$$

Then, the governing equation for the best value of A can be obtained as

$$\mathbf{A} = \left[(\mathbf{DF})^{\mathrm{T}} (\mathbf{DF}) \right]^{-1} (\mathbf{DF})^{\mathrm{T}} \mathbf{H} \tag{17}$$

In this equation, $(\mathbf{DF})^T(\mathbf{DF})$ is a $(l+2)\times(l+2)$ matrix. As a result, it can save much computation time in calculating A because l is generally less than 20 in the present study. The variables a_m are given, the pressure distribution can be calculated from Eq. (11). Furthermore, since the pressure has been expressed as the polynomial form, the small fluctuation in the pressure can be eliminated.

3.3 Temperature calculation

The temperature distribution within an oil film can be solved by the energy equation subject to appropriate boundary conditions. Neglecting heat convection across the film and circumferential conduction in the film, the energy equation for line contact problems may be written as:

$$\frac{\partial}{\partial y}\left[\kappa\frac{\partial T}{\partial y}\right]=\rho c_p u\frac{\partial T}{\partial x}-\eta\left(\frac{\partial u}{\partial y}\right)^2-T\beta u\frac{\partial p}{\partial x} \quad (18)$$

Adding the continuity equation to Eq. (18) and integrating the resultant energy equation across the film thickness can be expressed as:

$$\rho c_p\int_0^h(\frac{\partial uT}{\partial x}+\frac{\partial vT}{\partial y})dy=\kappa\frac{\partial T}{\partial y}\bigg|_0^h+\int_0^h\eta\left(\frac{\partial u}{\partial y}\right)^2dy+\int_0^h T\beta u\frac{\partial p}{\partial x}dy \quad (19)$$

The non-linear integral convective terms in Eq. (19) can be linearized and be approximately equivalent to the product of the mean velocity and the temperature.

$$T_m=\frac{\frac{1}{h}\int_0^h uTdy}{\frac{1}{h}\int_0^h udy}=\frac{\frac{1}{h}\int_0^h uTdy}{u_m} \quad (20)$$

Integration of the continuity equation across the film thickness can be expressed as:

$$\frac{\partial u_m h}{\partial x}=0 \quad (21)$$

The substitution of Eqs. (20) and (21) into the energy equation leads to:

$$\rho c_p u_m h\frac{\partial T_m}{\partial x}=\frac{2k(T_a+T_b-2T_m)}{h}+\beta u_m h T_m\frac{\partial p}{\partial x}+\frac{\eta u^{*2}}{h}+\frac{1}{12\eta}(\frac{\partial p}{\partial x})^2h^3 \quad (22)$$

or in dimensionless form as:

$$\bar{\rho}U_m H^2\frac{\partial\bar{T}_m}{\partial X}=2\xi(\bar{T}_a+\bar{T}_b-2\bar{T}_m)+\Omega U_m H^2\bar{T}_m\frac{\partial P}{\partial X}+\Lambda\bar{\eta}U^{*2}+\frac{\Lambda\Phi^2H^4}{12\bar{\eta}}(\frac{\partial P}{\partial X})^2 \quad (23)$$

where

$$\xi=\frac{k\eta_0 R}{\rho_0 c_p E' b^3}$$

$$\Omega=\frac{\beta p_h}{\rho_0 c_p}$$

$$\Lambda=\frac{E' R^3}{\rho_0 c_p b^3 T_0}$$

$$\Phi=\frac{p_h}{E'}(\frac{b}{R})^3$$

The energy equation requires the surface temperatures of the rollers as its boundary conditions. The problem of the surface temperature of a semi-infinite solid subjected to a moving heat source was first solved by Carslaw and Jaeger. Therefore, the boundary conditions for Eq. (23) can be expressed as:

$$\bar{T}_b(X,0) = \bar{T}_{b0} + \frac{\kappa\sqrt{b}}{T_0\sqrt{\pi\rho_b c_b k_b u_b}} \frac{T_0 R}{b^2} \int_{X_{in}}^{X} \frac{2(\bar{T}_m - \bar{T}_b)}{H} \frac{dX'}{\sqrt{X-X'}} \tag{24}$$

$$\bar{T}_a(X,1) = \bar{T}_{a0} + \frac{\kappa\sqrt{b}}{T_0\sqrt{\pi\rho_a c_a k_a u_a}} \frac{T_0 R}{b^2} \int_{X_{in}}^{X} \frac{2(\bar{T}_a - \bar{T}_m)}{H} \frac{dX'}{\sqrt{X-X'}} \tag{25}$$

$$\bar{T}_m = \bar{T}_0 \text{, at } X = -\infty \tag{26}$$

Once the film shape is measured, the variables a_m can be calculated from Eq. (17). Therefore, the pressure distribution can be calculated from Eq. (11), and the estimated film thickness can be calculated from equation (14). As mentioned above, since the mass density and the viscosity of lubricants are the functions of the pressure and mean temperature, so the initial mean temperature must be guessed, and the Gauss-Seidel iteration is employed to calculate the temperature from Eqs. (2), (3) and (23)-(26). If the pressure distribution, film shape, mass density and viscosity of lubricants with are given, then the Reynolds Eq. (1) can be satisfied.

4. Results and discussion

4.1 Direct inverse method

In order to illustrate the validity of the present technique discussed above, several numerical tests have been performed. Typical results for the case of W=1.8182x10^{-5}, U=7.2727x10^{-12}, and G = 3500 with 169 grid points are shown in Figs. 2 and 3. In this study, the 'exact' solution of film thickness, pressure, and temperature rise distribution in the lubricated contact are calculated numerically by using the lubricant and roller properties according to Tables 1 and 2. In solving the thermal EHL line-contact problems, the coupled Reynolds, energy, load balance, film thickness, rheology, and surface temperature equations must be solved simultaneously.

When the measured points are the same as the estimated points, it is not necessary to use the series form to represent the pressure distribution. Hence, the solution obtained from Eqs. (2), (3), (9) and (23)-(26) are called the solution of the direct inverse method. By using the direct inverse method, Fig. 2 shows the effect of the measured points of the film thickness on the pressure distribution, and Fig. 3 shows the effect of the measured points of the film thickness on the temperature rise distribution. In these figures, the input data is the film thickness at different measured points. It is obvious that with increasing the number of measured points, the solution approaches the 'exact' numerical solution. When the 8-point measurements are taken uniformly, results show that the estimated values of the pressure and temperature rise are bad in the whole region. When the 43-point measurements are taken uniformly, results show that the estimated values of the pressure and temperature rise are quite accurate in the inlet and outlet regions, but the pressure and temperature rise spikes are incorrect. When the measured points are increased, results show that the estimated values of the pressure and the temperature rise become quite accurate in the pressure spike region. It is clear that the direct inverse method requires a lot of measured points.

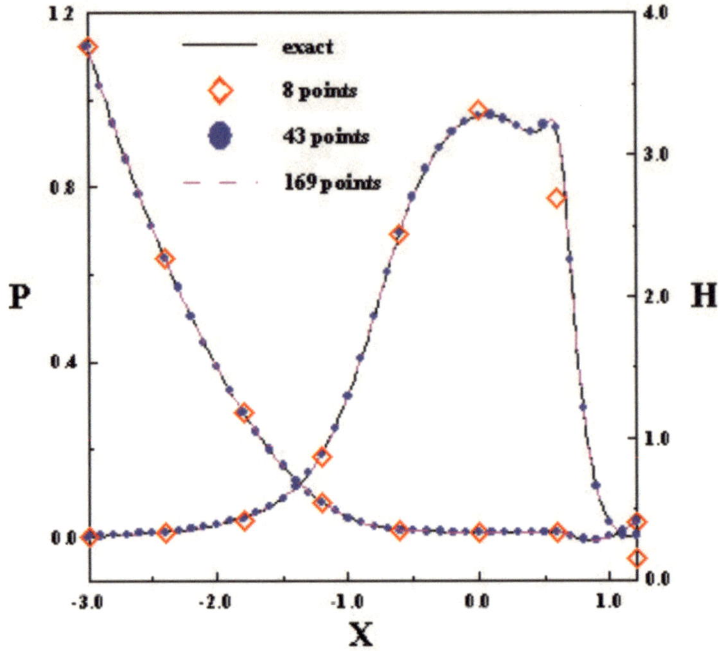

Fig. 2. Pressure distributions for different measured points in the film thickness using direct inverse method.

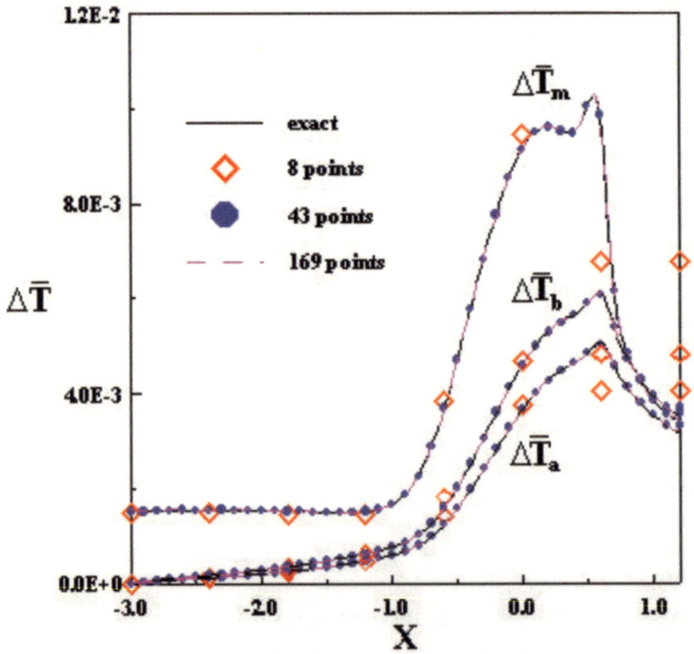

Fig. 3. Temperature rise distributions for different measured points in the film thickness using direct inverse method.

G (Material parameter)	3500.0
Inlet temperature of lubricant, K	313.0
Inlet viscosity of lubricant, Pa-S	0.04
Inlet density of lubricant, kg/m^3	846.0
Pressure viscosity coefficient, 1/GPa	15.91
Temperature-viscosity coefficient, 1/K	6.4x10^{-4}
Thermal conductivity of lubricant, W/m-K	0.14
Specific heat of lubricant, J/Kg-K	2000.0
Pressure-viscosity coefficient (Roelands)	0.4836

Table 1. Properties of lubricant.

Equivalent radius, m	0.02
Thermal conductivity of rollers, W/m-K	47.0
Specific heat of rollers, J/Kg-K	460.0
Density of rollers, kg/m^3	7850.0
Elastic modulus of rollers, GPa	200
Poisson's ratio of rollers	0.3

Table 2. Properties of rollers.

Usually, the measurement error always occurs in the optical interferometry measurement. This minor measurement error makes the estimation away from the 'exact' numerical solution. In this study, the measured film thickness is generated from the pre-selected 'exact' numerical film thickness with the minor measurement error, and it can be expressed as:

$$H_i^{measured} = H_i^{exact} + \sigma H_c \lambda_i \tag{27}$$

where λ_i is a random number, H_c is the dimensionless central film thickness, and σ is the dimensionless standard deviation of the measurement error. Figs. 4 and 5 show the effect of the dimensionless standard deviation of the measurement error on the pressure and temperature rise distribution. It is seen from these figures that the minor measurement error on the oil film significantly influences the pressure and temperature rise distribution. Table 3 also shows that the resolution in film thickness can result in an error in the solution of the pressure and temperature rise distribution. Generally, the film thickness gradient is sensitive to the pressure distribution. As a result, pressure fluctuations can be found everywhere. If this pressure distribution is used to solve the temperature rise distribution, the error in the temperature rise should be also found everywhere.

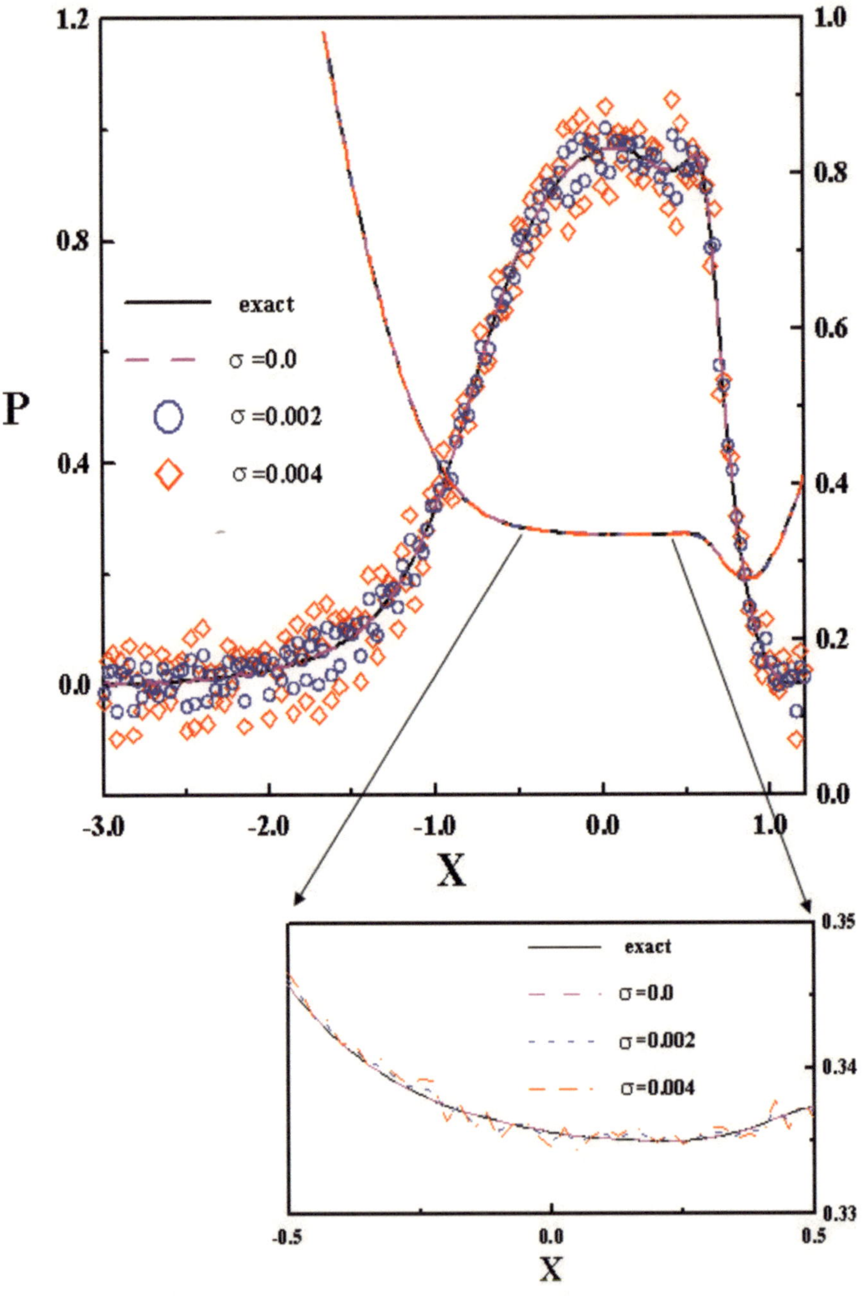

Fig. 4. Pressure distributions for different standard deviations of the measured errors with 169 measured points in the film thickness using direct inverse method.

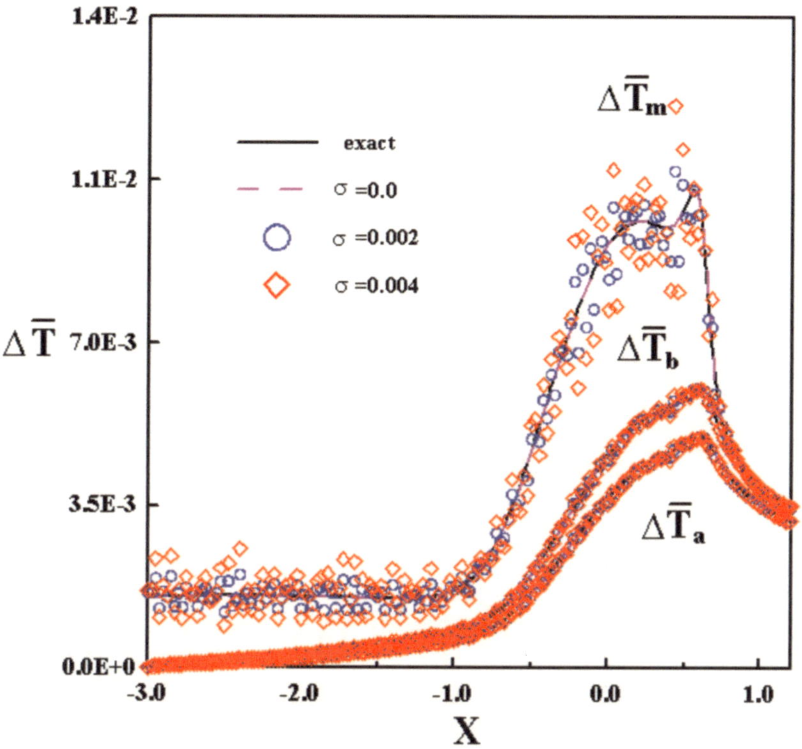

Fig. 5. Temperature rise distributions for different standard deviations of the measured errors with 169 measured points in the film thickness using direct inverse method.

standard deviation of the measured error		$\sigma = 0.002$	$\sigma = 0.004$	$\sigma = 0.006$
$(\frac{H^{measured} - H^{exact}}{H^{exact}})_{max.}(\%)$		-0.230	-0.463	-0.693
$(\frac{P^{present} - P^{exact}}{P^{exact}})_{max.}(\%)$	D.I.	77.467	153.541	229.362
	I.A.	2.355	4.061	5.309
$(\frac{\Delta T_m^{present} - \Delta T_m^{exact}}{\Delta T_m^{exact}})_{max.}(\%)$	D.I.	-12.477	26.587	44.270
	I.A.	3.480	6.601	8.908
$(\frac{\Delta T_a^{present} - \Delta T_a^{exact}}{\Delta T_a^{exact}})_{max.}(\%)$	D.I.	-3.240	7.205	13.283
	I.A.	1.797	3.833	5.571
$(\frac{\Delta T_b^{present} - \Delta T_b^{exact}}{\Delta T_b^{exact}})_{max.}(\%)$	D.I.	-3.527	7.690	13.983
	I.A.	1.858	3.969	5.786

Table 3. Effect of film thickness measurement error on pressure and temperature ($-1 < X < 1$)

4.2 Inverse approach solution for pressure and temperature rise distributions

To overcome the problems of pressure and temperature rise fluctuations, this paper proposes the use of an inverse approach. It is known that a pressure spike appears in the dimple region of the film shape in the viscous EHL problems. Hence, to obtain accurate pressure and temperature rise distributions using the inverse approach, it is necessary to divide the domain into several regions due to the singular point at the pressure spike. In this paper, the domain is divided into three regions including the inlet region, the Hertz contact region, and the outlet region, respectively.

The unknown pressure distribution is approximated by a polynomial form shown in Eq. (11) with $c = -3$ and $a_0 = 0$; and the degrees of these three regions are shown in Table 4. The measured points in the film thickness can be chosen as 12, 21, and 31, within a domain from Xin = -3 to Xend = 1.2, and have been shown in Table 4. By using Eq. (17), the polynomial coefficients can be calculated. Substituting these coefficients into Eq. (11), the estimated pressure distribution can be calculated. It is seen from Figs. 6 and 7 that the number of measured points significantly influences the pressure and temperature rise distributions. With the increasing measured points, the solutions of the pressure and temperature rise distributions approach the 'exact' numerical solution. A twelve-point measurement in the film thickness gives a smooth solution of the pressure and temperature rise distributions, but it still exhibits a few errors on the whole region. For the 21 measured points in the film thickness, the solution of the pressure and temperature rise distributions are almost the same as the 'exact' numerical solution, but it still exhibits a few errors on the pressure and temperature rise in the vicinity of the spike region. Therefore, to obtain accurate pressure and temperature rise estimations, it is necessary to increase the number of measured points in the dimple region and in the pressure spike region, as shown in Fig. 6 and Table 4.

The minor measurement error on the oil film significantly influences the pressure and temperature rise distributions. Therefore, to understand the effect of pressure error on the film shape, the estimated film shape can be calculated from the calculated pressure using Eq. (14). It is seen from Fig. 6 that with a twelve-point measurement, the estimated film shape is a very smooth curve passed through the measured points, but it is different from the 'exact' numerical film shape within the domain from X=-3.0 to X=1.2. For the other two measured cases, the estimated film shape is quite close to the 'exact' numerical film shape. It was discussed above that any small error on the film shape could result in a significant error on the pressure and temperature rise distributions. Hence, with more measured points, the error in the estimated film shape can be reduced, so that the estimated pressure and temperature rise distributions can be closer to the 'exact' numerical solution.

The unknown pressure distribution is approximated by a polynomial form shown in equation (11) with the highest degree of 3, 5 and 7, respectively. It is seen from Figs. 8 and 9 that with the highest degree of 3, the inverse approach can simulate the pressure and temperature rise distributions, but the pressure and temperature rise slight oscillations occur at all regions, the error in the pressure and temperature rise are obvious and the pressure and temperature rise spikes disappear. With increasing polynomial degree, the errors in the pressure and temperature rise decrease.

It was discussed in Figs. 4 and 5 that the minor measurement error in the film thickness can result in pressure and temperature rise fluctuations everywhere. To obtain a smooth curve in the pressure and temperature rise distributions, the inverse approach is employed to solve this case. Figs. 10 and 11 present typical results obtained using the inverse approach when all the operating parameters are the same as those used in Figs. 4 and 5. In these

Item \ Region	I	II	III
X-range	-3.00≦X<-1.00	-1.0≦X<1.0	1.0≦X<1.2
Degrees	7	7	6
Measured points	-3.000 -2.750 -2.500 -2.250 -2.000 -1.750 -1.500 -1.250 -1.025	-1.000 -0.750 -0.500 -0.250 0.000 0.250 0.400 0.425 0.450 0.475 0.500 0.525 0.550 0.625 0.700 0.775 0.850 0.925	1.000 1.075 1.150 1.200

Table 4. The regions and measured points in the film thickness.

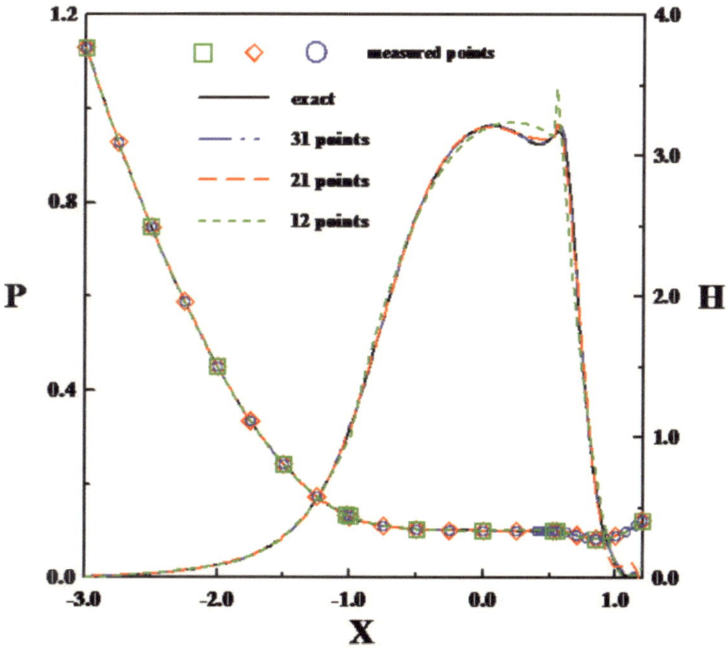

Fig. 6. Pressure distributions and film shapes for different measured points in the film thickness using inverse approach.

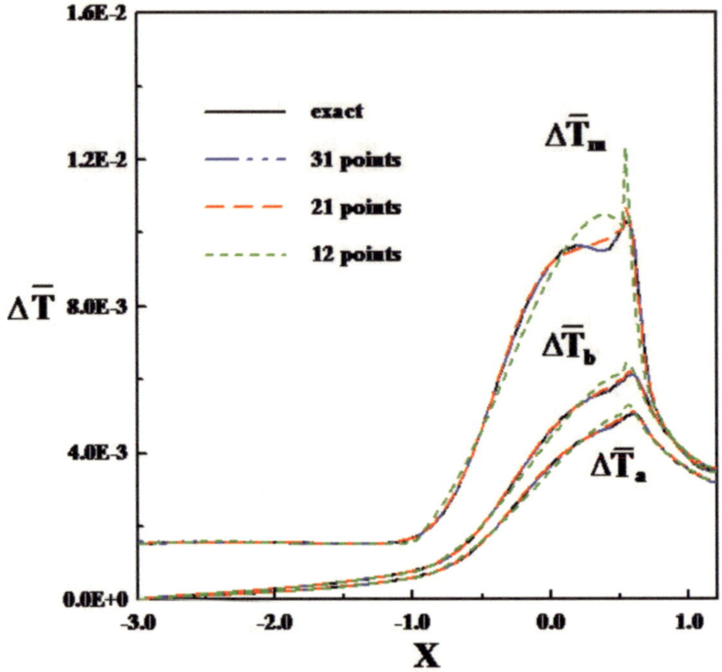

Fig. 7. Temperature rise distributions for different measured points in the film thickness using inverse approach.

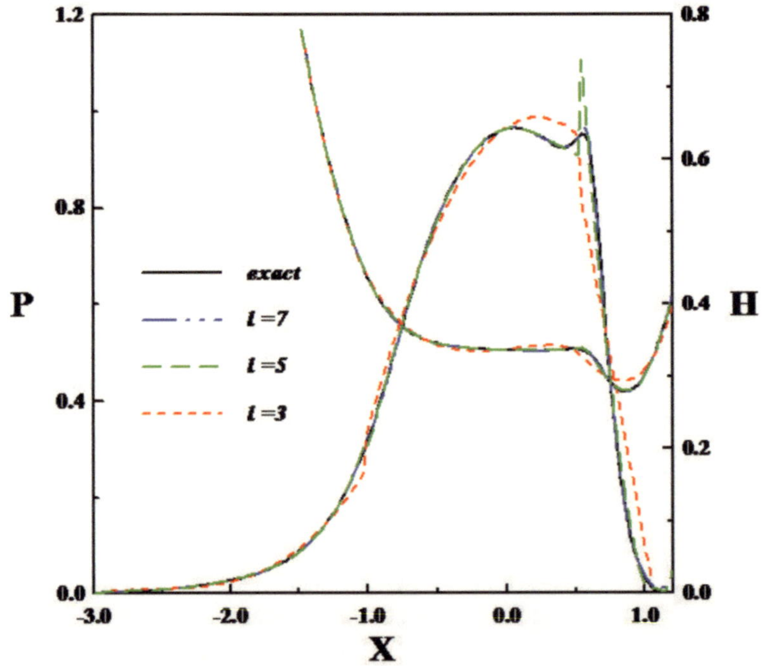

Fig. 8. Pressure distributions and film shapes for different degrees with 31 measured points in the film thickness using inverse approach.

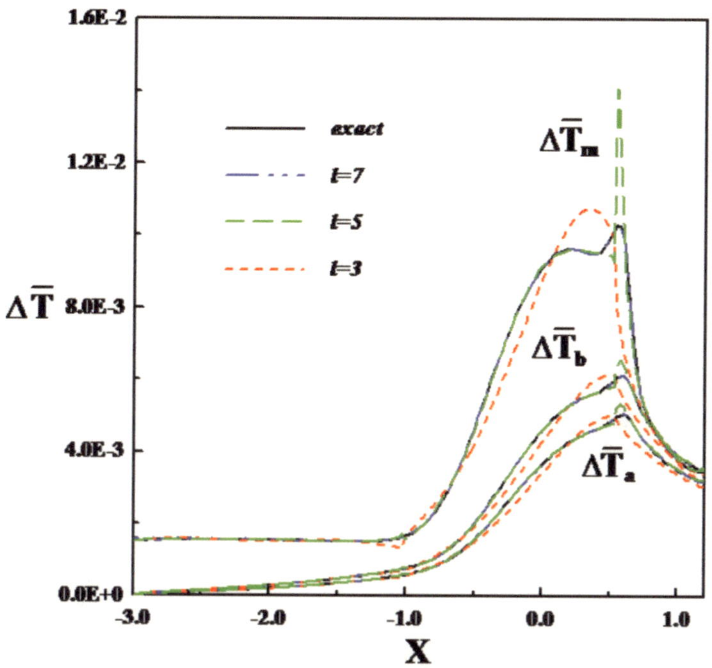

Fig. 9. Temperature rise distributions for different degrees with 31 measured points in the film thickness using inverse approach.

figures, the domain for the estimated pressure and temperature is divided into three regions with a total of 31 measured points in the film thickness. The range, degree of the polynomial form, and distribution of the measured points across these three regions are shown in Table 4. The numerical test tries to overcome the pressure and temperature rise fluctuations resulting from the implemented error in the film thickness. It is seen from Figs. 10 and 11 that with the increasing dimensionless standard deviation of the measurement error, the errors in the pressure and temperature rise distributions also increase, especially at the Hertz contact region. For the case of $\sigma = 0.002$, the results in the pressure and temperature rise distributions are almost equal to the 'exact' numerical solution. These results indicate that the present algorithm can overcome the pressure and temperature rise fluctuations due to the measurement error in the film thickness. Moreover, computing time can be saved due to the small matrix used in Eq. (17).

4.3 Inverse solution for apparent viscosity

Once the film shape is measured, the pressure distribution and estimated film thickness can be calculated from Eqs. (11), (14), and (17). As mentioned in section 3.3, the temperature rise distribution is calculated from Eqs. (2), (3) and (23)-(26) by using the Gauss-Seidel iteration. Furthermore, the mass density and the apparent viscosity of lubricants can also be obtained. Fig. 12 shows the dimensionless apparent viscosity versus the X-coordinate for various simulated measurement errors in the film thickness. The results show that the present algorithm still gives a reasonable solution for the apparent viscosity. By contrast, use of the

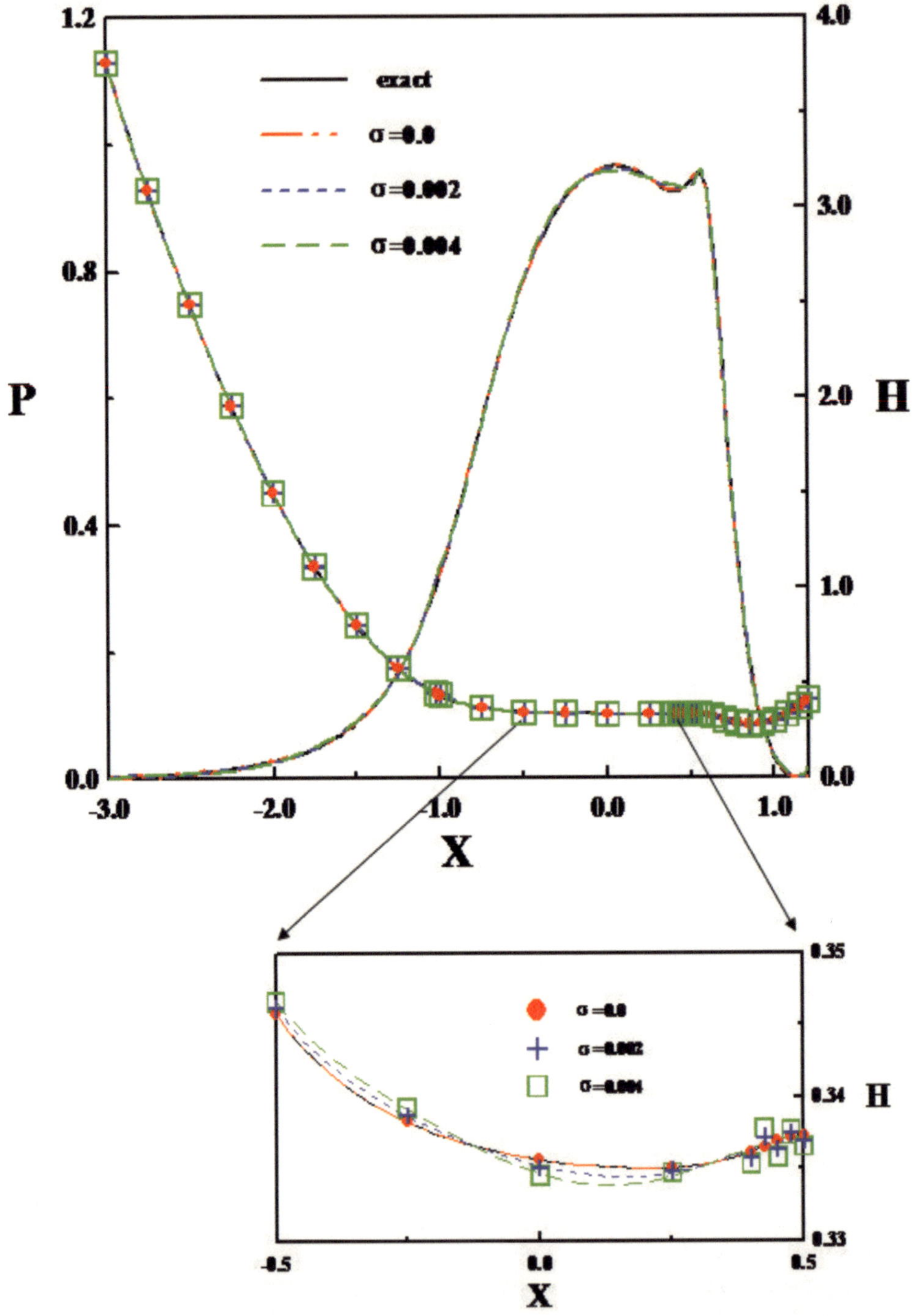

Fig. 10. Pressure distributions and film shapes for different standard deviations of the measured errors with 31 measured points in the film thickness using inverse approach.

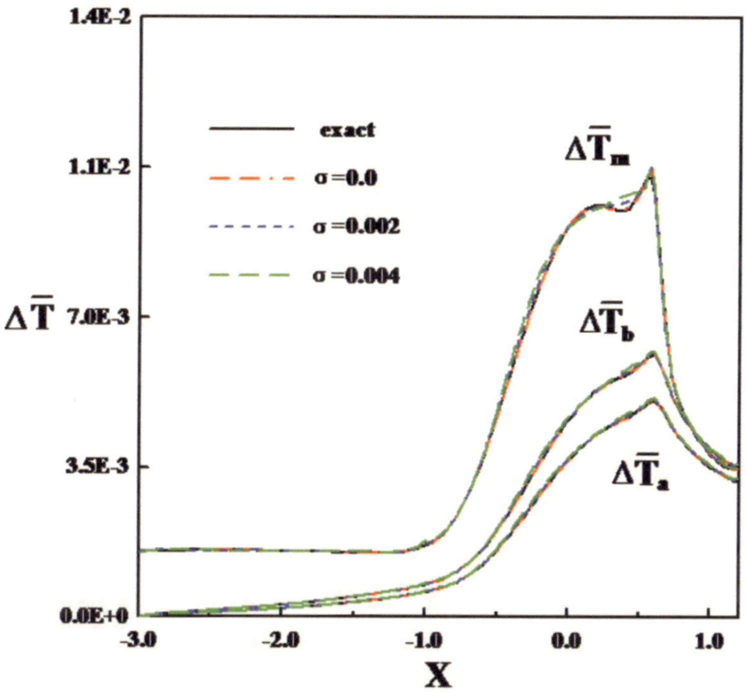

Fig. 11. Temperature rise distributions for different standard deviations of the measured errors with 31 measured points in the film thickness using inverse approach.

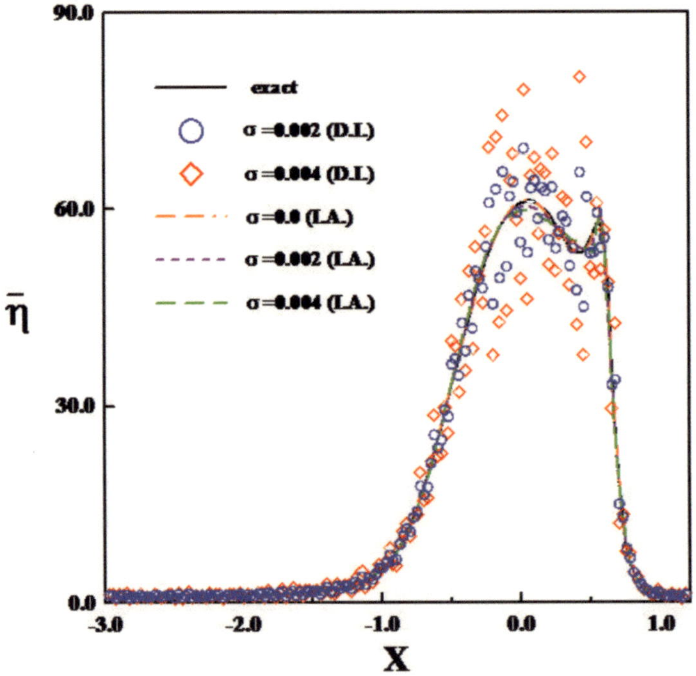

Fig. 12. Comparison of dimensionless apparent viscosity between two algorithms with implemented errors.

direct method results in much greater errors in the apparent viscosity results, especially in the Hertz contact region. This is the consequence of the pressure fluctuation mentioned in Section 4.1. However, in the present algorithm, when an implemented error occurs in the film shape, the estimated film shape can be obtained by using Eq. (14). Hence, the present algorithm provides a smooth curve not only in the pressure distribution, but also in the estimated film shape. Therefore, it produces much more accurate apparent viscosity results. However, the error in the apparent viscosity increases with the increasing standard deviation of the implemented error.

5. Conclusions

This paper describes an inverse approach to estimate the pressure distribution, the temperature rise distribution, and the apparent viscosity distribution in an EHL line contact if the film thickness is given. To obtain an accurate pressure, it is necessary to divide the domain into several regions due to the singularity at the pressure spike. The principle to choose the measured points is proposed. This approach overcomes the problems of pressure and temperature rise fluctuations and generates accurate results of the pressure and temperature rise distribution from a small number of measured points of film thickness, which also saves computing time. The conclusions from the main results can be summarized as follows:

1. The direct inverse method requires a lot of measured points of film thickness to establish the amplitude and location of the pressure and temperature spikes, but the inverse approach can obtain accurate results with only 31 measured points.
2. With the measurement error from the resolution in the film thickness measurement, this approach also presents a smooth curve of the pressure and temperature distributions with a small error in the inlet, pressure spike, and outlet regions.
3. Without the measurement error in the film thickness, this approach provides quite good solution for apparent viscosity. With the minor measurement error in the film thickness, this approach still gives a quite good solution in apparent viscosity, but the direct method provides much larger error in apparent viscosity.

6. Nomenclature

b	semiwidth of hertzian contact (m)
c_p	specific heat of the lubricant (J kg^{-1} K^{-1})
c_a, c_b	specific heat of rollers a and b (J kg^{-1} K^{-1})
E'	equivalent Young's modulus(Pa)
G	dimensionless material parameter, $\alpha E'$
H	dimensionless film thickness, hR/b^2
Hc	dimensionless central film thickness
H_0	dimensionless constant defined in Eq. (4)
K_{ij}	discretized kernel in Eq. (5)
k	thermal conductivity of lubricant (W m^{-1} K^{-1})
k_a, k_b	thermal conductivity of rollers a and b (W m^{-1} K^{-1})

p	pressure (Pa)
p_h	maximum hertzian pressure (Pa)
P	dimensionless pressure, p/p_h
R	equivalent radius of contact (m)
T	temperature (K)
T_m	mean temperature of lubricant film (K)
T_a, T_b	surface temperature of rollers a and b
$\overline{T}$	dimensionless temperature, T/To
u_a, u_b	surface velocity of rollers a and b (m s^{-1})
$\overline{u}$	average rolling velocity (m s^{-1})
U	dimensionless speed parameter, $\eta_0\overline{u} / E'R$
w	load per unit width (N m^{-1})
W	dimensionless load, W= $w / E'R$
x,y	coordinate (m)
X,Y	dimensionless coordinate, X=x/b, Y=y/h,
z	Roelands' pressure-viscosity index
Δx or Δ	distance between two neighboring grid points
η	viscosity of lubricant (Pa-s)
η_0	viscosity of lubricant at ambient pressure and inlet lubricant (Pa-s)
$\overline{\eta}$	dimensionless viscosity, η / η_0
λ	dimensionless viscosity parameter, $\lambda = 12\eta_0\overline{u}R^2 / b^3p_h$
λ_i	random number
ρ	density of lubricant (kg m^{-3})
ρ_0	inlet density of lubricant (kg m^{-3})
ρ_a, ρ_b	density of rollers a and b (kg m^{-3})
$\overline{\rho}$	dimensionless density of lubricant, ρ / ρ_0
β	thermal expansivity of lubricant (K^{-1})
γ	temperature-viscosity coefficient of lubricant (K^{-1})
σ	dimensionless standard deviation of the measured error

7. Acknowledgments

The authors would like to express their appreciation to the National Science Council (NSC-97-2221-E-214-028) in Taiwan, R. O. C. for financial support.

8. References

Sternlicht, B., Lewis, P. & Flynn, P. (1961). Theory of lubrication and failure of rolling con tacts, *ASME J. Basic Eng.*, 83, pp. 213-226.

Ghosh, M. K. & Hamrock, B. J. (1985). Thermal elastohydrodynamic lubrication of line contacts, *Trans. ASLE 28 (2)*, pp. 159-171.

Salehizadeh, H. & Saka, N. (1991). Thermal non-Newtonian elastohydrodynamic lubrication of rolling line contacts, *ASME J Tribol.*, 113, pp. 481-491.

Lee, R. T. & Hsu, C. H. (1993). A fast method for the analysis of thermal-elastohydrodynamic lubrication of rolling/sliding line contacts, *Wear*, 166, pp. 107-117.

Zhu, D. & Wen, S. (1984). A full numerical solution of the thermal elastohydrodynamic lubrication problem of elliptic contacts, *ASME J. Lubrication Technology*, 106, pp. 246-254.

Kim, K. H. & Sadeghi, F. (1992). Three-dimensional temperature distribution in EHD lubrication: Part I -circular contact, *ASME J Tribol.*, 114, 32-41.

Lee, R. T., Hsu, C. H., Kuo, & W. F. (1995). Multilevel solutuion for thermal elastohydrodynamic lubrication of rolling/sliding circular contacts, *Tribol. Int.*, 28(8), pp. 541-552.

Cheng, H. S. & Orcutt, F. K. (1965). A correlation between the theoretical and experimental results on the elastohydrodynamic lubrication of rolling and sliding contacts, *Proc. Instn Mech. Engrs, Part B: J. Engineering Manufacture*, 180, pp. 158–168.

Safa, M. M. A., Anderson, J. C., & Leather, J. A. (1982). Transducers for pressure, temperature and oil film thickness measurement in bearings, *Sens. Actuators*, 3, pp. 119–128.

Kannel, J. W., Zugaro, F. F., & Dow, T. A. (1978). A method for measuring surface temperature between rolling/sliding steel cylinders, *Trans. ASME J. Lubr. Technol.*, 100, pp. 110–114.

Turchina, V., Sanborn, D. M., & Winer, W. O. (1974). Temperature measurements in sliding elastohydrodynamic point contacts, *Trans. ASME J. Lubr. Technol.*, 96, pp. 464–471.

Ausherman, V. K., Nagaraj, H. S., Sanborn, D. M., & Winer, W. O. (1976). Infrared temperature mapping in elastohydrodynamic lubrication, *Trans. ASME J. Lubr. Technol.*, 98, pp. 236–243.

K. Yagi1 r, K. Kyogoku, and Nakahara T. (1996). Experimental investigation of effects of slip ratio on elastohydrodynamic lubrication film related to temperature distribution in oil films, *Proc. IMechE Vol. 220 Part J: J. Engineering Tribology*, pp. 102-112.

Yagi, K., Kyogoku, K. & Nakahara, T. (1997). Relationship Between Temperature Distribution in EHL Film and Dimple Formation, *Transactions of the ASME*, pp. 412-418.

Cameron, A. & Gohar, R. (1966). Theoretical and experimental studies of the oil fim in lubricated point contact, *Proc. R. Soc., Ser. A*, 291, pp. 520-536.

Foord, C. A., Hammann, W. C. & Cameron, A. (1968). Evaluation of lubricants using optical elastohydrodynamics, *ASLE Trans.*, 11(1), pp. 31-43.

Johnston, G. J., Wayte, R. & Spikes, H. A. (1991). The measurement and study of very thin lubricant films in concentrated contacts, *STLE Tribology Trans.*, 34(2), pp. 187-194.

Gustafsson, L., Hoglund, E. & Marklund, O. (1994). Measuring lubricant film thickness with image analysis, *Proc. Instn. Mech. Engrs., Part(J), J. of Eng. Tribology*, 208, pp. 199-205.

Yang, P., Qu, S., Kaneta, M. & Nishikawa, H. (2001). Formation of steady Dimples in point TEHL contacts, *ASME, J. of Tribology*, 123, pp.42-49.

Marklund, O. (1998). A pseudo-phase stepping approach to interferometric measurements, *TULEA:09 Lulea Univ. of Tech.*, S-97187 Lulea, Sweden.

Luo, J., Wen, S. & Huang, P. (1996). Thin Film Lubrication Part I: Study on the Transition between EHL and Thin Film Lubrication Using a Relative Optical interference Intensity Technique, *Wear*, 194, pp. 107-115

Hartl, M., Krupka, I., Poliscuk, R. & Liska, M. (1999). An Automatic System for Real-Time Evaluation of EHD Film Thickness and Shape Based on the Colorimetric Interferometry, *Trib. Trans.*, 42, 2, pp303-309.

Paul, G. R. & Cameron, A. (1972). A high-pressure microviscometer based on refractive index, *Proc. R. Soc., Ser. A*, 331, pp. 171-184.

Wong, P., Lingard, L. S. & Camerson, A. (1992). The high pressure impact microviscometer, *STLE Tribology Trans.*, 35(3), pp. 500-508.

Astrom, H. & Venner, C. H. (1994). Soap thickener induced local pressure fluctuations in a grease lubricated elastohydrodynamic point contact, *Proc. Instn. Mech. Engrs., Part(J), J. of Eng. Tribology*, 208, pp. 191-198.

Östensen, J. O., Larsson, R. & Venner, C. H. (1996) Determination of viscosity and pressure in an elastohydrodynamic rolling contact by using optical interferometry: a theoretical study, *Proc. Instn. Mech. Engrs., Part(J), J. of Eng. Tribology*, 210, pp. 259-268.

Lee, R. T., Chu, H. M. & Chiou, Y. C. (2002). Inverse approach for calculating pressure and viscosity in elastohydrodynamic lubrication of line contacts, *Tribology International*, Vol. 35, pp. 809-817.

Yang, C. Y. (1998). A linear inverse model for the temperature-dependent thermal conductivity determination in one-dimensional problems, *Applied Mathematical Modelling*, 22, pp. 1-9.

Dowson, D. & Higginson, G. R. (1966). Elastohydrodynamic Lubrication, *Pergamon Press*, pp. 88-92.

Roelands, C. J. A., Vlugter, J. C. & Watermann, H. I. (1963). The viscosity temperature pressure relationship of lubricating oils and its correlation with chemical constitution, *ASME Journal of Basic Engineering*, pp. 601-606.

Carslaw, H. S. & Jaeger, J. C. (1959). *Conduction of Heat in Solids*, Oxford University Press.

20

Construction of Various Self-assembled Films and Their Application as Lubricant Coatings

Prof. Dr. Jinqing Wang, Dr. Junfei Ou,
Prof. and Dr. Sili Ren and Prof. Shengrong Yang
State Key Laboratory of Solid Lubrication, Lanzhou Institute of Chemical Physics, CAS
P. R. China

1. Introduction

Recently, self-assembled nanofilms (SANFs), including self-assembled monolayers (SAMs), self-assembled multilayer films (SAMFs), self-assembled inorganic films (SAIFs) and self-assembled organic-inorganic composite films (SAO-ICFs), have generated substantial interest not only for its simple preparation procedure but also for its wide potential applications in many fields, such as surface modification, boundary lubricant coatings, sensors, photoelectronics, and functional bio-membrane modeling, etc (Foisner et al., 2004; Gulino et al., 2004; Hsu, 2004; Love et al., 2005; Ostuni et al., 1999; Song et al., 2008; Ulman, 1996; Wang et al., 2005). As a potential lubricant film for controlling stiction and friction in micro-/nano-electromechanical systems (MEMS/NEMS), SANFs offers distinct advantages over other strategies for lubrication of MEMS/NEMS devices. Especially, its assembly process is rapid, shape independent and needs no complicated equipment (Ulman, 1996). In general, the molecules can be assembled onto the targeted surfaces by a simple solution immersion or vapor phase deposition, even within nanoscale crevices between moving components of MEMS/NEMS.

To date, the adhesion, friction and wear behaviors of SANFs are extensively investigated in nanoscale and macroscale by using various techniques and apparatus. Scanning force microscope (SFM) techniques, mainly including atomic force microscope (AFM) (Butt et al., 2005), interfacial force microscope (IFM) (Houston & Michalske, 1992), and chemical force microscope (CFM) (Noy et al., 1997), are often applied to evaluate the nanotribological performances of thin films. In AFM, the adhesive force between the AFM tip and the surface can be calculated from the "force-distance" curve (Cappella & Dietler, 1999) as follows (Tsukruk & Blivnyuk, 1998; Xiao & Qian, 2000):

$$F_{ad} = k_c \times Z_p \times \frac{k_2}{k_1} \tag{1}$$

where k_c is the force constant of the cantilever, Z_p is the vertical displacement of the piezotube, k1 and k2 are the slopes of of the corresponding lines in Fig. 1.

The friction force in nanoscale can also be measured in a contact scanning mode with a scanning angle of 90° or 270° (Fig. 2a). Such scanning will cause torsion of the cantilever and the resultant lateral force is regarded as the friction force (Fig. 2b). Theoretically speaking, the friction force can be easily calculated by applying the Hooke's law:

$$F_{lateral} = k_{lat} \times \frac{\Delta V}{S_{lat}} \tag{2}$$

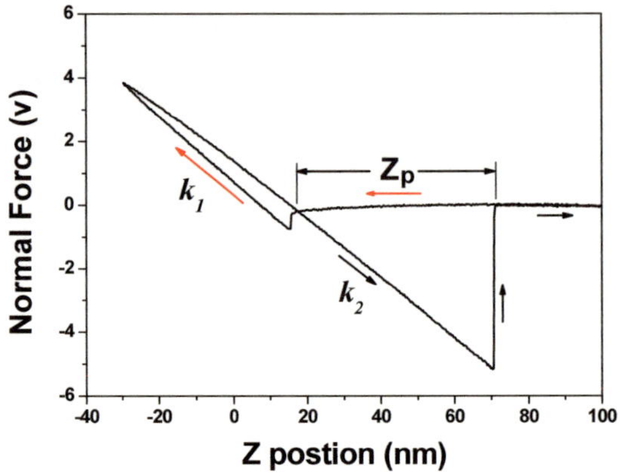

Fig. 1. A typical force-distance curve obtained by AFM.

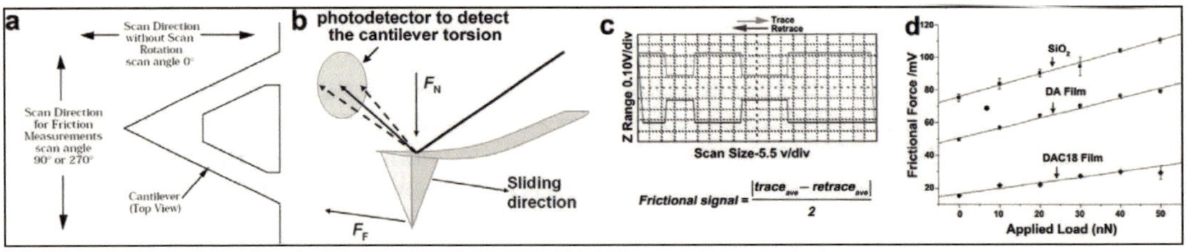

Fig. 2. Scanning angle selection for AFM tip (a); The cantilever torsion at a scanning mode of contact and scanning angle of 90° (b), reproduced from Leggett et al., 2005; A typicaly friction loop (c); Friction versus applied load curves acquired by AFM, reproduced from Song et al., 2008.

where k_{lat} is the lateral spring constant, S_{lat} is the lateral sensitivity of the photodiode, ΔV is the torsion signal. However, the calibration of k_{lat} is still a challenge and therefore the friction force obtained from the friction loops (Fig. 2c) is generally expressed in a raw voltage form in the current studies. By acquiring friction at different applied loads, the friction (F_f)-applied load (F_n) curves can be plotted, which is described by equation (Schwarz et al., 1995):

$$F_f = c_1 F_n^m + c_2 F_n + c_3 \tag{3}$$

where c_1-c_3 are the material-dependent constants and the index m ($0<m<1$) depends on the asperity geometry (Li et al, 1999). However, plenty of studies show that a linear dependence is often observed, equation 3 is therefore simplified to the following form by assuming $c_1 = 0$.,

$$F_f = \mu F_n + F_0 \tag{4}$$

where μ is friction coefficient, and F_0 is assumed to be related with the adhesive force between AFM tip and the surface (Brewer et al., 2001; Foster et al., 2006; Li et al, 1999; Ou et al., 2009; Song et al., 2006; Song et al., 2008; Zhao et al., 2009).

The CFM is distinguished from the usual AFM technique by its probe tip which is chemically immobilized with certain functional molecules (Fig. 3a). This new SFM technique has been used to probe adhesion and friction forces between distinct chemical groups in organic and aqueous solvents. As shown in Fig. 3b, covalent modification of the Si_3N_4 tip with thiols and reactive silanes can be realized by different approaches (Noy et al., 1997).

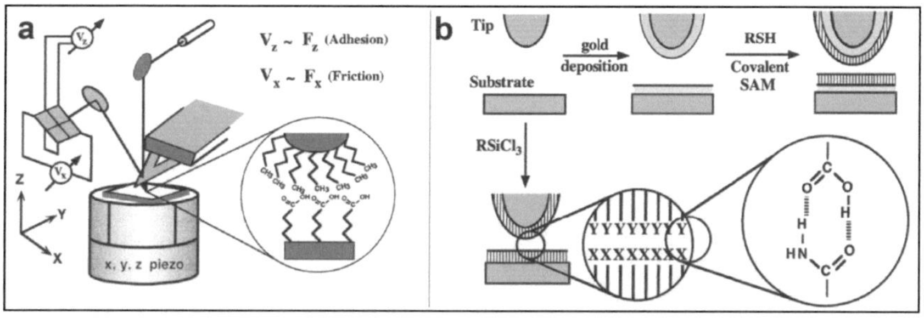

Fig. 3. Schematic drawing of the CFM setup. The inset illustrates the chemically specific interactions between a gold (Au)-coated, CH_3-terminated tip and a COOH-terminated region of a sample (a); Scheme for chemical modification of tips and sample substrates (b). Reproduced from Noy et al., 1997.

Similar to AFM, the IFM is also an ideal tool to investigate the interaction between a scanning tip and a nanoscale surface. The IFM setup is shematically depicted in Fig. 4. As shown in Fig. 4a, a piezo tube acts as a translator to move the mounted sample in *xyz* directions. The probe tip is mounted to a differential capacitor sensor instead of a cantilever in AFM. This special force sensor is mechanically stable and able to determine both the normal and friction forces over the entire range of the interfacial interaction, including the contact and noncontact regions (Fig. 4b, c) (Houston et al., 1992 and 2005).

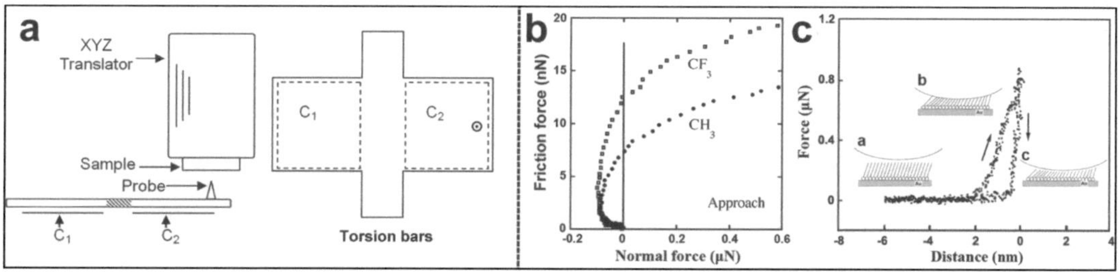

Fig. 4. Schematic of the IFM (a). Averaged IFM data of frictional force vs normal force comparing the behavior of the CH_3- and CF_3-terminated films (b); An interfacial force profile (that is, force versus separation) for a 500 nm radius W probe interacting with an Au sample surface covered by SAMs of n-alkanethiol molecules (c). For b and c, negative values indicate attractive forces while repulsive forces are shown as positive. Reproduced from Houston & Michalske, 1992 (a and c) and Houston et al., 2005 (b).

To investigate the macrotribological behaviors, various ball-on-plate tribometers, such as UMT, are usually applied. The friction coefficient versus sliding time curve of the tested specimen can be recorded automatically as the reciprocating sliding goes on. From this curve, the macroscopic friction coefficient and anti-wear life, which refers to the sliding time at which friction coefficient rises sharply, corresponding to lubrication failure, can be reflected (Fig. 5c).

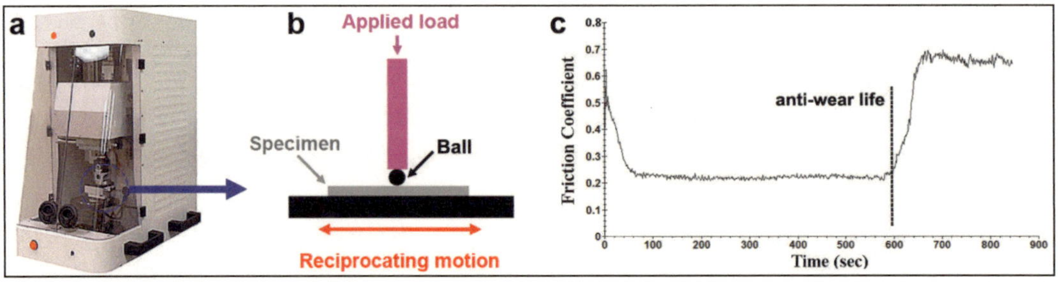

Fig. 5. The photo of a UMT tribometer (a) and the schematic operation principle (b); A friction coefficient versus time curve acquired by a UMT tribometer (c).

As these techniques developed, lots of researches have been done. It is revealed that the tribological properties are structure and composition dependent. Roughly speaking, the alkyl chain length and head/tail group of SAMs have a great influence on its tribological behaviors. For SAMFs, the nature of each layer and the interaction between adjacent layers are key factors. In this chapter, the tribological behaviors of SANFs, including SAMs, SADLs, SAMFs, and SAO-ISFs, are reviewed, aiming at discovering the basic "microstructures-properties" correlation.

2. SAMs

2.1 One component SAMs

SAMs have been widely investigated in the past 20 years because of its potential applications in the field of surface modification, boundary lubricant, sensor, photoelectronics, and functional bio-membrane modeling, etc (Foisner et al., 2004; Gulino et al., 2004; Hsu, 2004; Love et al., 2005; Ostuni et al., 1999; Ulman, 1996; Wang et al., 2005). On the basis of the surface chemical reaction and synthetic approaches, the chemical structures of SAMs can be manipulated easily at molecular level. Generally, two kinds of SAMs, namely, monolayers of alkylsilanes on silicon (Si) wafer surfaces and the monolayers of alkylthiols on Au surfaces (Tsukruk, 2001; Love et al., 2005), have been intensively studied as model lubricants not only for their excellent tribological properties but also for the wide application of Si substrate in MEMS/NEMS and the highly ordered structures of Au wafer. As schematically shown in Fig. 6, the precursor surfactant molecules [X_3Si-$(CH_2)_n$-Y, HS-$(CH_2)_n$-Y, X=-Cl/-OCH_3/-OC_2H_5] of SAMs consist of three parts, viz, head groups (X_3Si- or HS-), alkylchains [-$(CH_2)_n$-], and tail groups (Y). Each part has great effect on the quality and tribological property of SAMs.

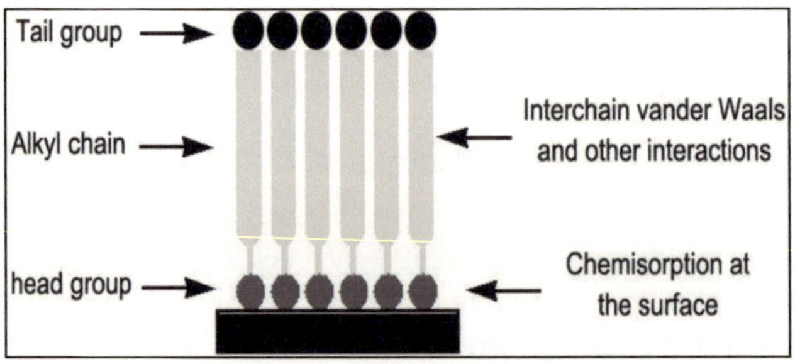

Fig. 6. A schematic view of the formation and forces in a SAMs.

The influence of headgroups

The head groups interact with the active substrate through certain covalent bonding and serve as anchoring points to determine the affinity and stability of SAMs. The superiority of covalent bonding can be reflected by comparing with another popular candidate for MEMS/NEMS lubricant of Langmuir-Blodgett (LB) film, which attaches to the substrate via weak van der Waal force. As expected, SAMs is found to be much more stable against shear stress and possesses better wear resistance as compared with LB film with similar composition and structures (Bliznyuk et al., 1998; Bushan et al., 1995; DePalma & Tillman, 1989; Kim et al., 1999; Overney et al., 1992; Peach et al., 1996; b,Tsukruk et al., 1996; Tsukruk, 2001.). As shown in Fig. 7, C18 SAMs possess much better wear resistance as compared with zinc arachidate LB film (Bushan et al., 1995).

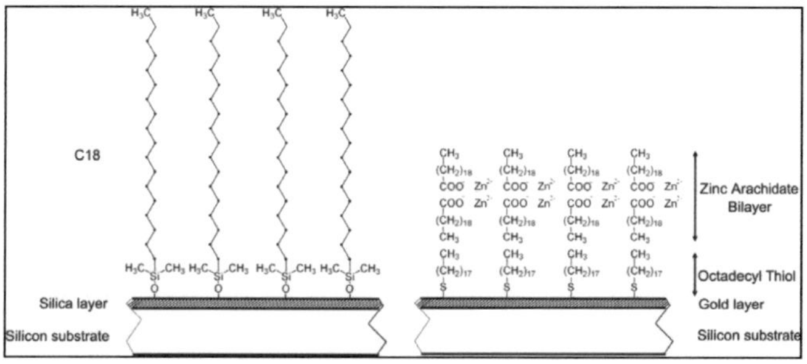

Fig. 7. The structures of C18 SAMs and zinc arachidate LB film. Both of the films possess long alkyl chains of C18. Reproduced from Bushan et al., 1995.

It is generally believed that strong affinity of the molecules to the substrate is a basic requirement for effective boundary lubrication. For SAMs with similar composition and structures, the stronger adhesion is, the better wear-resistance is expected to be achieved. For example, the chemisorption of alkylsilane/alkylthiol on the Si/Au substrate surface is realized by Si-O/Au-S covalent bonding, respectively. The bond energy of Au-S is lower than that of Si-O (Bushan et al., 2005). Thus, alkylsilane SAMs can withstand higher normal loads than the alkylthiol ones with the same alkyl chain and tail group (Bushan et al., 2005; Booth et al., 2009). Moreover, the cross-link of head groups may also play an important role in stabilizing the SAMs. For instance, owing to the intermolecularly cross-link of Si-O-Si between adjacent molecules, n-octadecyltrichlorosilane SAMs (OTS-SAMs, Fig. 8) possess much better wear resistance than n-octadecyldimethylchlorosilane SAMs (ODS-SAMs, Fig. 8) (Booth et al., 2009). However, comparing with alkylthiols, the cross-link of head groups causes chain distortion and the lack of long range order in the silane SAMs, both of which can serve as the energy-dissipating modes to increase the friction (Lio et al., 1997).

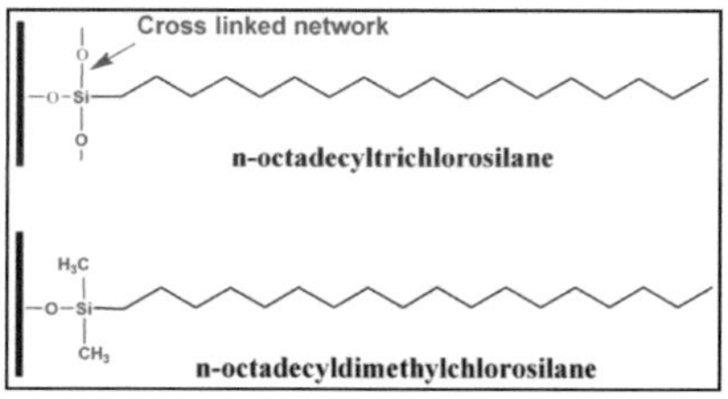

Fig. 8. Schematic structures of OTS-SAMs and ODS-SAMs.

The influence of alkyl chains

It has been proved that the frictional behaviors varied significantly with the alkyl chain length. SAMs with shorter chain length possesses higher friction coefficient (Xiao et al, 1996; McDermott et al., 1997) and lower load affording capability (Xiao et al, 1996; Bushan et al., 2005). To discover the origin of the chain length dependence, lots of works have been done. It is found that the frictional force is proportional to the contact area and shear strength (Tsukruk et al., 2001; b, Wang et al., 2005). Due to the loosely packed and disordered structures of the SAMs with shorter alkyl chain, on one hand, the contact area increases. On the other hand, CH_2-CH_2 backbones in a loosely packed SAMs are exposed to the AFM tip, which increases the van der Waals interaction between the tip and surface and thereby enhances the shear strength. So, it can be concluded that the microstructures, i.e., lower packing density and substantial disorder in SAMs, are key factors for the higher friction. In other words, the shorter chains are apt to form SAMs with more disorders which facilitate the energy-dissipation and give rise to a high friction and friction coefficient. The lower load affording capacity for the shorter chain SAMs is because that the shorter chains are less flexible to tile in response to the applied load (Chandross et al., 2005).

As the chain length increases to a critical value, the inter-chain attraction is large enough to form ordered structures and the chain length dependence is not so distinct. However, as discovered by Liu et al, the ultra-high compact density caused by the very long alkyl chain could result in a higher friction (Liu et al., 1996). For example, monolayers of dieicosyldimethylammonium bromide (Fig. 9, n=20, 22) are believed in a frozen state owing to the strong interchain interaction, while it is in a melted state for the molecules with short chain of C14. It is acordingly found that the monolayers of C14 yield a lower friction due to the compliant of the melted chains.

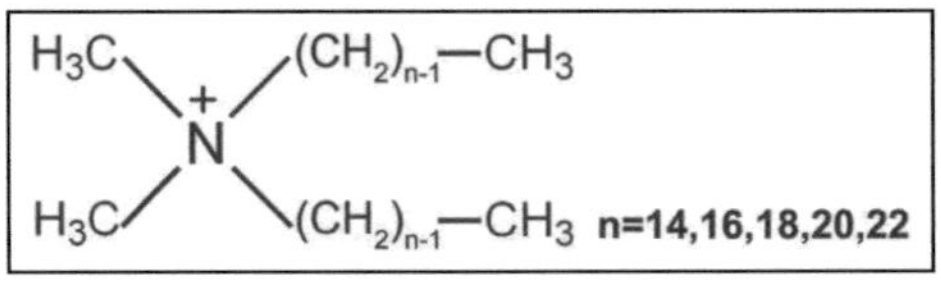

Fig. 9. The molecules investigated in the reference of Liu et al., 1996.

According to above interpretation, it seems that the chain length only has an indirect influence to affect the tribological performances by defects. To understand the direct correlation more clearly, lots of theoretical simulations have been performed (Chandross et al., 2005). It is observed that, for SAMs with no defects, longer chain length can endure heavier applied load. This is because that the longer chains are more flexible to tile in response to the applied load as compared to shorter ones. Moreover, for well-ordered, fully packed SAMs with different chain length of C6, C8 and C12, friction coefficient decreases as the chain carbon number increasing (Chandross et al., 2005). This is because that the effectiveness of stress transmission during sliding is dependent on the chain length–longer chains have more contact with neighboring ones. From the above discussions, it can be summarized that the frictional behaviors are influenced by the chains length due to the formed defects or the intrinsic difference between short and long alkyl chains.

The influence of tail groups

The tail groups exposed to the ambient environment have a significant effect on the surface nature of SAMs, such as wettability and adhesion (Tsukruk et al., 2001). Generally, the adhesion force (F_{ad}) between surfaces includes the capillary force (F_C), van der Waals forces

(F_{vdW}), electrostatic force (F_E), and chemical bonding force (F_B), which is described by equation (5):

$$F_{ad} = F_C + F_{vdW} + F_E + F_B \tag{5}$$

In ambient air conditions, F_C is proportional to the cosine value of water contact angle ($cos\theta$) on the surface and takes main contribution to the adhesion. Generally, lower surface energy corresponds to a hydrophobic surface, which has a high water contact angle (lower $cos\theta$ value) and thereby result in a lower F_C and then a lower F_{ad}. In liquid medium, F_C is eliminated and the adhesive force between different tail groups can be measured by CFM. It is observed that the adhesive forces between -COOH and -CH_3 groups are reduced in the following order: COOH/COOH >CH_3/CH_3 >COOH/CH_3 (Fig. 10a) (Noy et al., 1995). The adhesive force difference between COOH/COOH and CH_3/CH_3 may be result from the item of F_B. Specifically, the COOH polar groups tend to form intermolecolar hydrogen bonding to boost the chemical bonding force. Compared with the asymmetric pair of COOH/CH_3, the adhesive force for CH_3/CH_3 is higher. This can be explained as follows: the adhesive force is the product of tip radius R and adhesive work W_{st}, specifically,

$$F_{ad} = 1.5\pi R W_{st}, (W_{st} = \gamma_s + \gamma_t - \gamma_{st}) \tag{6}$$

where γ_s, γ_t and γ_{st} are the surface energy of sample, CFM tip and interface energy between them, respectively (Noy et al., 1995; Tsukruk et al., 1998). For the symmetric pair of CH_3/CH_3, the null interface energy γ_{st} will result in a higher adhesive work and adhesive force. Correspondingly, friction of different pairs are arranged in the same order, viz, COOH/COOH >CH_3/CH_3 >COOH/CH_3 (Fig. 10b)

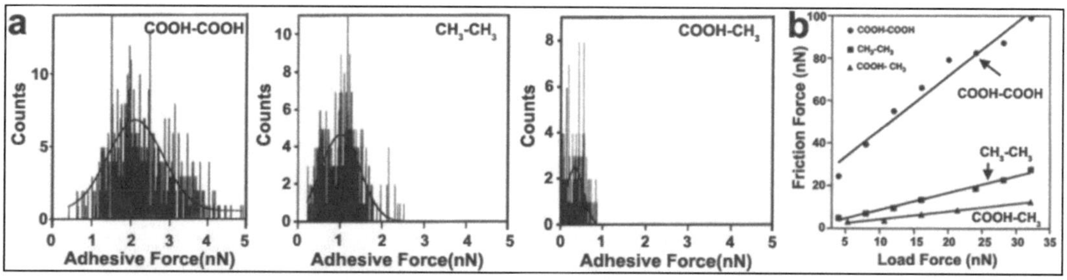

Fig. 10. Adhesion and friction for different pairs. Reproduced from Noy et al., 1995.

F_E is always generated between the charged tip and the charged samples. When tested in liquid medium, F_E is dependent on not only the nature of tail groups but also the pH value of the aqueous solution (Tsukruk & Blivnyuk, 1998). To be specific, in the pH range of pK_1~pK_2 (pK_1 and pK_2 are the isoelectric points of the sample and the tip, respectively, Fig. 11), attraction between the tip and sample is generated, which result in a high friction. In the cases of pH value lower than pK_1 or higher than pK_2, repulsion and lower friction is correspondingly obtained.

The spatial orientation of the tail groups also has a prominent impact on adhesion and friction. For example, an "odd-even" effect is observed for SAMs with same tail groups but different CH_2 number (odd or even) in the alkyl chain (Chang et al., 1994; Lee et al., 2001; Smith & Porter, 1993; Tao, 1993; Wong et al., 1998). As shown in Fig. 12, the spatial orientation of the -COOH tail groups are different for the two SAMs with odd or even number of CH_2 units. As a result, intra-film hydrogen bonds are produced within a film for the pairs of odd-COOH SAMs, while inter-film hydrogen bonds are gegerated between the two surfaces for the pairs of even-COOH SAMs. It is then expectedly found that higher adhesion and friction were obtained for the pairs of even-COOH surfaces due to the inter-film hydrogen bonds. (Kim & Houston, 2000).

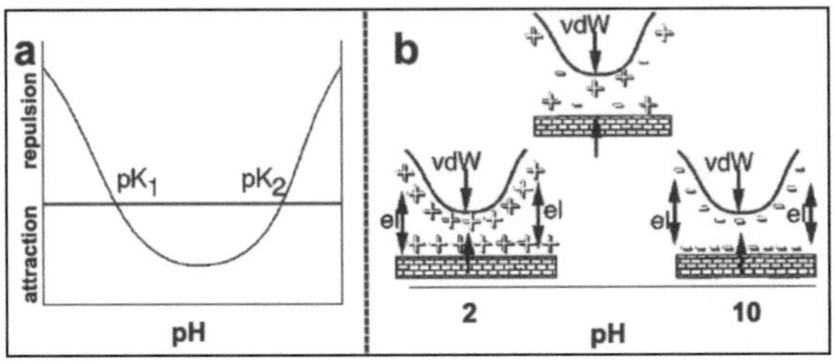

Fig. 11. Expected variation of adhesion-repulsion balance for interacting surfaces with two isoelectric points (a), and a scheme of tip/surface pairs with different surface charge distributions and force balances at different pH values (b). Reproduced from Tsukruk & Blivnyuk, 1998.

Based on the surface energy, tail groups of SAMs can be sorted into two categories of polar terminal groups (such as -OH, $-NH_2$, and -COOH) with high surface energy and apolar terminal groups (such as $-CH_3$ and $-CF_3$) with low surface energy. The SAMs with apolar terminal groups generally possess lower surface energy and relatively weak interaction between two sliding surfaces, which result in a lower adhesion and less energy loss leading to a lower friction force (Liu et al., 1996; Tsukruk et al., 1996; Zhang et al., 2002). However, although the surface energy of $-CF_3$ (12.9 mJ/m^2) is lower than that of $-CH_3$ (~24 mJ/m^2) (Bushan et al., 2005; Luengo et al., 1997), the fluorocarbon SAMs produce higher friction in AFM studies (Bushan et al., 2005; Kim et al., 1997; Peach et al., 1996; Houston et al., 2005). The unexpected higher friction is attributed to the larger size and higher electronegativity of the fluorine atom, which result in two major variations in the surface properties of SAMs (Kim et al., 1997). On one hand, the replacement of $-CH_3$ with larger tail groups of $-CF_3$ into the close-packed $\left(\sqrt{3}\times\sqrt{3}\right)$ R30° lattice gives rise to increased surface steric interactions.

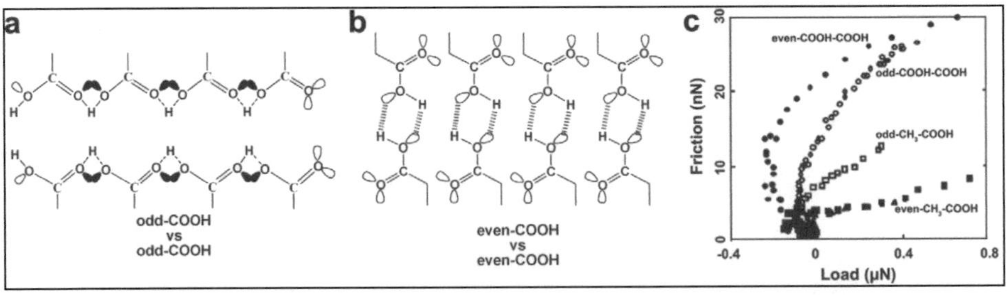

Fig. 12. A schematic representation of the -COOH end group orientations for alkanethiol SAMs having both odd and even numbers of methylene groups (a) and the plots of the lateral friction force vs. interfacial force for various end-group combinations (b). Reproduced from Kim & Houston, 2000.

During sliding, more energy is imparted to the film to overcome the consequent increased steric barriers and then results in higher friction (Kim et al., 1997; Peach et al, 1996). On the other hand, the strong surface dipoles in CF_3-teminated monolayer would cause much higher attractive force between the AFM tip and the surface of SAMs, and eventually cause more energy loss to increase the friction (Houston, 2005). This size effect is also observed between the tail groups of $-CH(CH_3)_2$ and $-CH_3$ (Kim et al., 1999) as well as C60, phenyl and $-CH_3$ (Lee et al., 2001).

2.2 Mixed SAMs

Co-deposition of molecules with different terminal groups or alkyl chain lengths to form mixed SAMs is also extensively studied, which allows an in-depth understanding of the relationship between structure and performance of SAMs. Several reports have revealed the frictional behaviors of the mixed SAMs derived from akanethiols or alkylsilanes. For instance, the mixed monolayers with chemically heterogeneous surface composed of mercaptoundecanoic acid (MUA) and dodecanethiol (DDT) or mercaptoundecanol (MUO) and DDT have been prepared (Brewer & Leggett, 2004; Beake & Leggett, 1999). SFM tips immoblized with COOH or CH_3 groups were applied as the probes to investigate the tribological behaviors. As shown in Table 1, the adhesion for the symmetric pairs (polar-polar or apolar-apolar) is relatively higher, which can be well understood by referring to the equation (6), where γ_{st} is much lower for the symmetric pairs. The surface composition of the mixed SAMs can be reflected by the water contact angle θ. In other words, high fraction of polar group-terminated adsorbate (*e.g.* MUO) will produce a small θ and high $\cos\theta$ value. The relationship between friction coefficient and $\cos\theta$ is depicted in Fig. 13. It can be seen that, when COOH tip is applied, friction coefficient increases with increasing the $\cos\theta$ (i.e., increasing the MUO fraction). Correspondingly, when CH_3 tip is applied, friction coefficient increases with decreasing the $\cos\theta$ (i.e., increasing the DDT fraction). It is therefore concluded that the friction coefficient increases due to the enhanced interaction of the symmetric tip-sample pairs.

Sample	Tip	
	COOH	CH_3
CH_3	0.57±0.17; 0.58±0.26	1.2±0.54
COOH	1.6±0.41; 2.1±0.85	0.78±0.26
OH	1.9±0.34	0.76±0.20

Table 1. Mean adhesion forces (nN) in ethanol between different tip-sample pairs. Obtained from Beake & Leggett, 1999.

When a non-modified Si_3N_4 tip was applied to investigate the tribological behaviors of MUA/DDT mixed SAMs, friction force increased with increasing the relative amounts of MUA in the mixed SAMs (Fig. 13c). This attributes to that higher MUA fraction raises the interaction between the scanning tip and the mixed SAMs, eventually increasing the energy dissipation and friction.

Studies on comparing the tribological properties of one-component SAMs with mixed ones are also performed. As revealed by Whitesides *et al.*, mixed SAMs composed of octadecanethiol (ODT) and dodecanethiol (DDT) on Au substrate exhibit higher friction than one-component SAMs (Fig. 14a) (Bain & Whiteside, 1989). Such difference is attributed to the different structures of the two SAMs. I.e. the one-component SAMs is well-ordered,

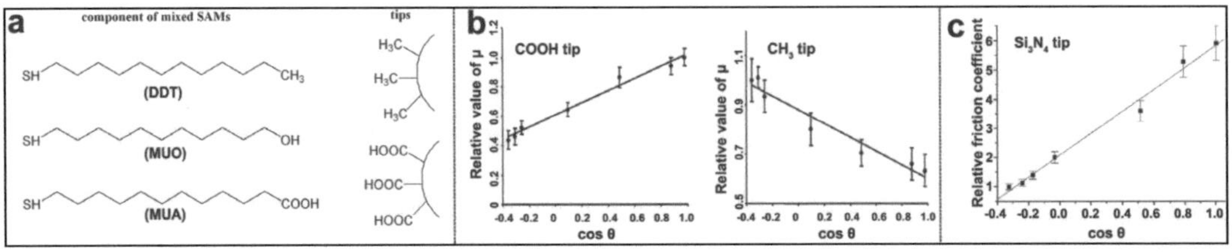

Fig. 13. The molecular structures of ODT, MUO, and MUA and the functionlized tips used to determine the tribological properties of the mixed SAMs (a); Friction coefficient as a function of cosθ for carboxylic acid-terminated tips and methyl-terminated tips (b); Correlation of relative friction coefficient with cosine of the water contact angle for mixed MUA/DDT monolayers (c). Reproduced from Brewer & Leggett, 2004 (b) and Beake & Leggett, 2000 (c).

while the mixed SAMs possess an outer region with disordered structure (Fig. 14b), which would increase the tip-sample interaction greatly and therefore producing a higher friction. However, a different tribological phenomenon has been observed for the mixed SAMs of alkylsilanes with different chain lengths on Si wafer. The friction for the mixed SAMs is lower than that of one-component SAMs. It is explained that the better lubrication performance of the mixed SAMs is attributed to the higher mobility of the tethered molecules in the monolayers, which can be evidenced by the much shorter relaxation time than that of one component SAMs (Zhang & Archer, 2003; Zhang & Archer, 2005).

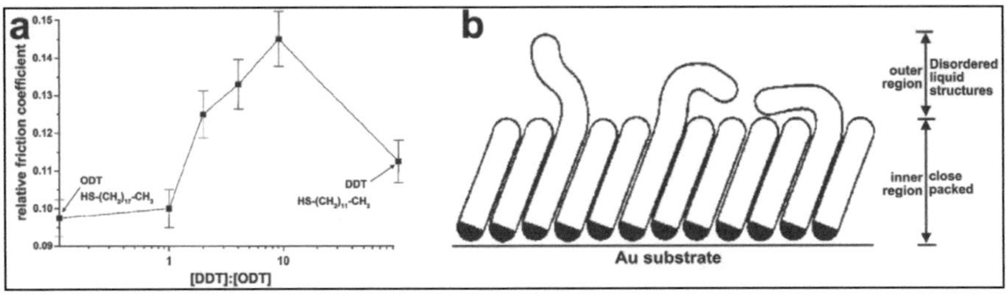

Fig. 14. Variation in relative friction coefficient with composition of mixed DDT/ODT monolayers (a); Structures of the mixed monolayers (b). Reproduced from Beake & Leggett, 2000 (a) and Bain & Whiteside, 1989 (b).

3. SAMFs

3.1 Functional group embedded SAMFs

As a potential lubricant in MEMS/NEMS, SAMs can reduce the adhesion and friction greatly. However, the load-carrying capacity of SAMs is relatively low, which significantly limits its service life. A promising way to further ameliorate the tribological behaviors of SAMs, especially the load-carrying capacity, is to enhance the stability of the films. It is revealed that the SAMFs with synergetic components generally exhibits longer anti-wear life (Ren et al., 2003; Ren et al., 2004; a, Song et al., 2008). The reason for the enhanced wear resistance is ascribed to the special structures of the SAMFs. Generally speaking, there are two approaches to construct SAMFs with unique structures, viz, one-step assembling and multi-step assembling. As to the one-step process, the pre-designed target precursors are assembled onto the substrate (Tam-Chang et al., 1995; Clegg & Hutchison, 1999; Clegg et al.,

1999; Song et al., 2006; Chambers et al., 2005); The multistep method involves common self-assembling and subsequent interface chemical reaction (Ren et al, 2003; Jiao et al., 2006). In most cases, the synthesis and the succedent purification for precursor molecules are too difficult to perform. So, compared to the one-step method, the stepwise strategy is more suitable to construct SAMFs with unique molecular architectures.

For SAMFs, the attachment to the substrate and the tail groups exposed to the ambient environment remain almost the same with SAMs. The most dramatic difference is related to the bulk chain. Specifically, the attraction between adjacent alkyl chains of normal SAMs is the weak van der Waals force. Within SAMFs, the inter-chain interaction is enhanced by functional groups, such as diacetylene (Mowery et al., 1999), peptide (Clegg & Hutchison, 1996; Sabapathy et al., 1998), and sulfone (Evans et al., 1991). It is hypothesized that the functional groups interact laterally taking the form of hydrogen bonding, dipole interaction, π-stacking, or covalent attachment, which are able to influence the integrity, stability and the tribological performances of the film (Ren et al., 2003; a, Song et al, 2008).

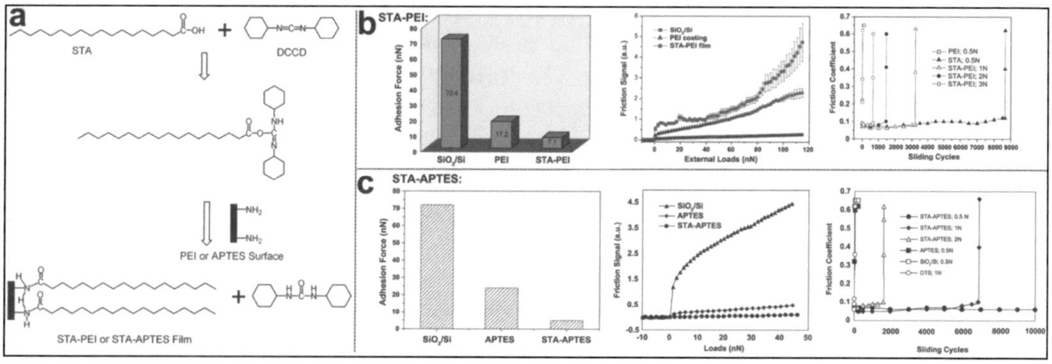

Fig. 15. Generation of an STA monolayer on PEI or APTES coated Si surface by chemical adsorption in the presence of N,N′-dicyclohexylcarbodiimide (DCCD) as a dehydrating agent in the reacting solution (a); Tribological behaviors of STA-PEI (b) and STA-APTES (c). Reproduced from Ren et al., 2004 (a, b), and Ren et al., 2003 (c).

To obtain nano film with promising application in the lubricant system of MEMS/NEMS, much effort in our group has been paid to construct a series of SAMFs with improved tribological properties better than SAMs. For example, taking advantage of amidation reaction between carboxyl (-COOH) and amine ($-NH_2$) groups, SAMFs consisting of 3-aminopropyl triethoxysilane (APTES) (Ren et al., 2003) or polyetherimide (PEI) (Ren et al., 2004) underlayer and stearic acid (STA) outerlayer has been prepared (Fig. 15a). The as-obtained SAMFs of STA-APTES and STA-PEI strongly attach to the substrate and possess a hydrophobic surface and a flexible alkyl chain outerlayers, which make them exhibiting excellent adhesion resistance and low nano-friction (Fig. 15b, c). Moreover, the interaction between adjacent chains intensified by the hydrogen bonding is assumed to be responsible for the improved wear resistance. Comparing with OTS-SAMs, the STA-APTES and STA-PEI SAMFs show much better load carrying and anti-wear capacity, demonstrating that the tribological properties of self-assembled films can be greatly improved by controlling the chemical structure and composition of the SAMFs.

To investigate the influence of underlayer structures on tribological properties, a systematical research has also been done in our group (b, Song et al., 2008). It is found that the structures of underlayer have a great effect on the frictional behaviors. Specifically, SAMFs with different underlayers of APTES, N-[3-(trimethoxylsilyl)propyl]ethylenediamine (DA) or N-[3-(trimethoxylsilyl)propyl]-diethylenetriamine (TA) and indentical outerlayer of lauroyl chloride (coded as C12) have been constructed via amidation reaction (Fig. 16). As attenuated total reflection-Fourier transform infrared spectrometry (ATR-FTIR) analyses

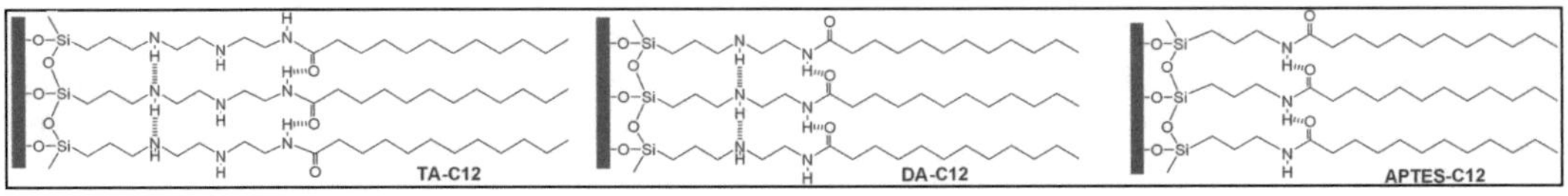

Fig. 16. Schematic structures of TA-C12, DA-C12, and APTES-C12 SAMFs.

indicated, the packing density of the as-prepared films follows the order DA-C12> TA-C12 > APTES-C12. The higher packing density of DA/TA-C12 is due to their longer chains of the underlayers. Even though TA has a longer molecular chain, TA-C12 is slightly less ordered than DA-C12, which is probably due to one more -CH_2CH_2NH- unit contained in the chain of TA (one more -CH_2CH_2NH- means that more random intermolecular hydrogen bonds could be formed between TA molecules). The lower packing density of APTES-C12 results in higher friction coefficient, both in nanoscale and macroscale.

In the above cases, the lateral interaction between adjacent chains is the weak hydrogen bonding. Zhao et al have constructed a triple-layer film (abridged as GAO) with lateral covalent network structures, which is composed of OTS outerlayer, APTES interlayer and 3-glycidoxypropyl-trimethoxysilane (GPTMS) underlayer (Zhao et al, 2009). The structure of the triple-layer film is depicted in Fig. 17a. It is belived that the APTES molecule serves as the linkage to combine the GPTMS with OTS. Specifically, the amine groups of APTES can react with the tail groups of GPTMS-SAMs and the hydroxyl groups formed by the hydroxylation are served as the active points to induce the self assembling of OTS molecules. The as-constructed film shows much better wear resistance as compared with OTS-SAMs, which is ascribed to the lateral Si-O-Si network structures (Fig. 17b).

3.2 Polymer SAMFs

The polymer nano-film with cross-linking network structures can sustain high compression and shear stress (Tsukruk et al., 1999; Luzinov et al., 2000; Luzinov et al., 2001; Maeda et al., 2002). So, polymeric thin film has been used as boundary lubricant coating in many fields including MEMS/NEMS, artificial joints, and computer hard disks, etc. However, physically adsorbed polymer films are easily peeled off during friction. Recently, there are two ways to construct chemically tethered polymer SAMFs, viz, "grafting to" or "grafting from" approaches. In a "grafting to" approach, the presynthesized end-functionalized polymer molecules react with a certain substrate to form polymer brushes. For example, a copolymer of poly[styrene-b-(ethylene-co-butylene)-b-styrene] (coded as SEBS) functionalized with 2% maleic anhydride into the hydrocarbon chains was assembled onto the surface of epoxy-terminated monolayer (Fig. 18a-d) (Luzinov et al., 2001). The as-fabricated films possess low friction coefficient, modest adhesion, low stiction, and good wear stability. To further improve the wear resistance, a SAMF with trilayer sandwiched architecture have been constructed (Fig. 18e, f) (Sidorenko et al., 2002). As expected, the anti wear life is much longer than epoxy composite layer (Fig. 18f).

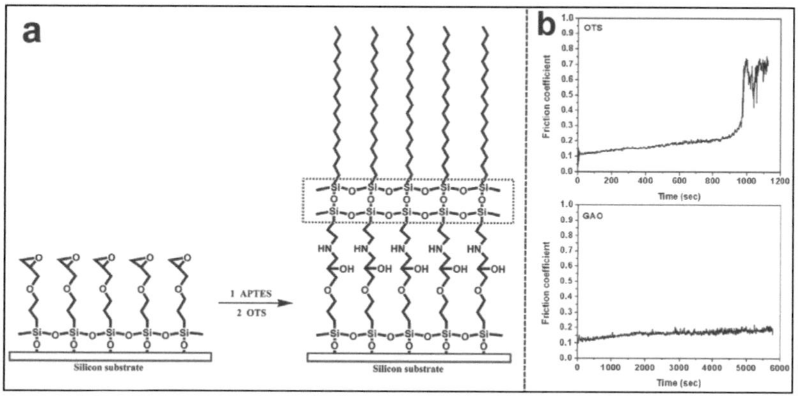

Fig. 17. Schematic structure and the strategy employed to prepare GAO triple-layer film (a); Variation in friction coefficient with time for OTS monolayer and GAO triple-layer film (b). Reproduced from Zhao et al., 2009.

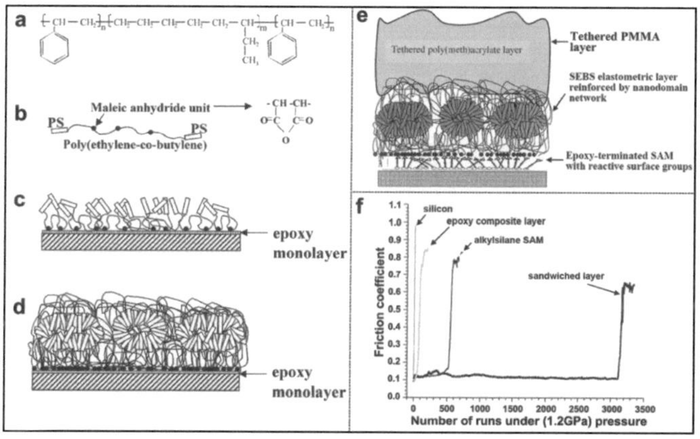

Fig. 18. Chemical (a) and schematic (b) structure of SEBE; SEBS layer with disordered structures and a thickness<2.5 nm (c); SEBS layer with nanodomain morphology and a thickness>2.5 nm (d); Architecture of sandwiched trilayer (e); Friction coefficient versus the number of reciprocal sliding runs for different samples (f). Reproduced from Luzinov et al., 2001 (a-d) and Sidorenko et al., 2002 (e, f).

However, few species of polymers can be immobilized onto the surface by "grafting to" approach due to the following two considerations (Zhao & Brittain, 2000). On one hand, it is difficult to synthesis polymer with functional anchoring groups. On the other hand, the as-synthesized polymer has complicated chain structures, which result in a low grafting density and thick film thickness. To circumvent this problem, "grafting from" approach has been proposed to prepare relatively thicker polymer brush with a higher grafting density. A representative "grafting from" approach, also called surface initiated polymerization (SIP), includes steps of introducing initiators on the substrate surface and succedent in-situ polymerization. Generally speaking, the immobilization of initiators is achieved by forming initiator-containing SAMs on the substrate. For instance, Takahara et al has prepared covalently tethered poly(methyl methacrylate) (PMMA) brushes on the Si wafer immobilized with an initiator SAMs of 2-bromoisobutylate moiety, abridged as DMSB (Fig.

19a) (Sakata et al., 2005). Compared with the spin-coated PMMA film, the self-assembled PMMA brush is found possessing much better wear resistance (Fig. 19b). However, this kind of initiator with complex molecular structure is also difficult to synthesis. Zhou et al. has developed a novel and much easier strategy to prepare PMMA film with comparable thickness based on the surface radical chain-transfer reaction (Zhou et al., 2001). The basic strategy of this novel process is depicted in Fig. 19c. It is clearly shown that, during this simple SIP process, the Si substrate was pre-modified by a SAMs of mercaptopropyltrimethoxylsilane (MPTES) rather than the complex initiators. As revealed by Yan et al, polystyrene (PSt) film can also be prepared by the same procedure (Zhao et al., 2008). The covalently tethered PSt film showed excellent scratch and adhesion resistance (Fig. 19d).

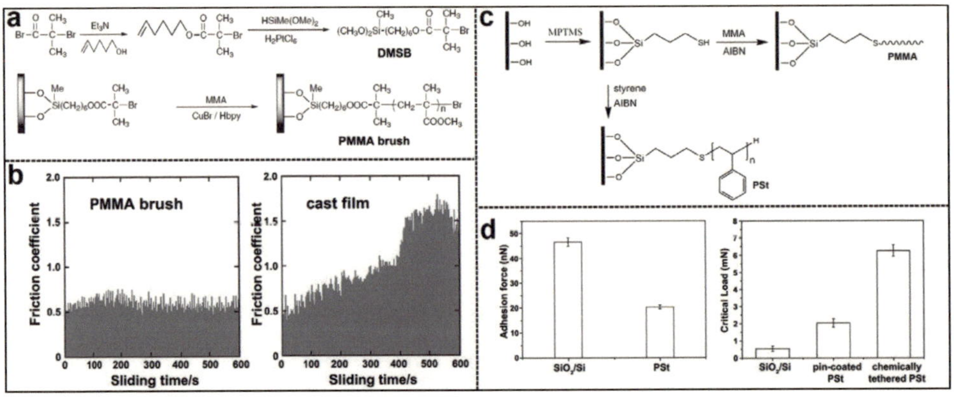

Fig. 19. The synthesis of DMSB and the PMMA brush (a); Friction coefficient versus sliding time for PMMA brush and cast film under a load of 0.49 N at a sliding velocity of 90 mm/min in air (b); A simple strategy to prepare PMMA and PSt brush (c); (d) Adhesion and scratch resistance of the PSt brush. Reproduced from Sakata et al., 2005 (a, b), Zhou et al., 2001 (c) and Zhao et al., 2008 (c, d).

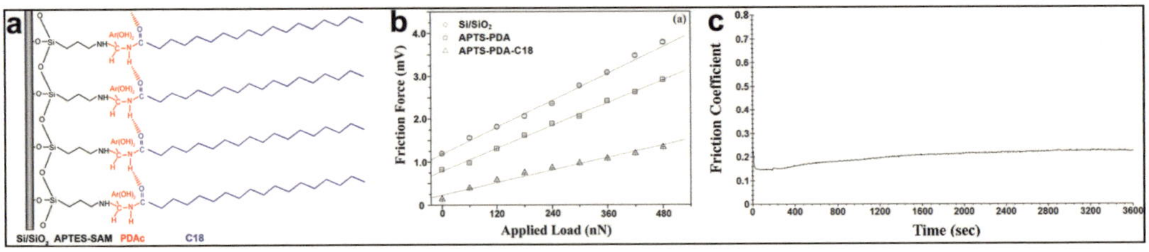

Fig. 20. A schematic view for the formation and combination bonding of the 3-layer film on silicon wafer (a); The nano- and macrotribological behaviors of different samples (b). Reproduced from Ou et al., 2009.

Recently, inspired by the "structure-property" correlation of SAMs, a robust polymer-based film has been constructed in our group by adopting APTES-SAMs as "headgroup", polydopamine (PDA) as "bulk chain", and stearoyl chloride SAMs as "tailgroup" (Ou et al., 2009). The in-situ polymerization of PDA on the APTES-SAMs surface is more like "grafting from" approach. The inherent special chemical structure, viz, the high adhesion to the substrate, the covalent combination between adjacent layers, the cross-linked PDA, as well as the hydrophobic and flexible C18 chain, takes main responsibility for the enhanced load-carrying capacity and lengthened anti-wear life (Fig. 20).

4. SAIFs and SAO-ICFs

Zirconia (ZrO_2) and ZrO_2 based nanocomposite film is a popular candidate for lubricant coating in nano-devices. Several different methods, such as physical vapor deposition and plasma spraying, have been established to prepare ZrO_2 based nanocomposite film (Pakala et al., 1997). High quality films with uniform structure, good compactness and high adherence to the substrate can be obtained by these techniques. However, some strategies need large apparatus and are high energy-consuming. Therefore, lots of efforts have been made to develop novel and feasible technique for deposition of ZrO_2 nanofilm and ZrO_2 based nanocomposite film. Among these researches, aqueous deposition onto SAMs with particular functional tail groups, such as $-SO_3H$ (Wang et al., 2004; Wang et al., 2005; Zlotnikov et al., 2008), $-PO(OH)_2$ (Zhang et al., 2006) and -OH (Ou et al., 2001), are studied intensively. For example, Wang et al. have prepared a crystalline ZrO_2-SAIFs on the Si substrate mediated by a sulfonated MPTES-SAMs (Fig. 21a) (Wang et al., 2004; Wang et al., 2005). As experimental results shown, the as-deposited ZrO_2-SAIFs is characterized by poor mechanical and tribological behaviors (Fig. 21b), which may be ascribed to the loose-packed structures caused by defects. Fortunately, it is found that a simple post-annealing (Fig. 21b) (Wang et al., 2005) or a unique preparation process with high pressure (Fig. 21c, d) (Zhang et al., 2006) can ameliorate the mechanical and tribological properties effectively.

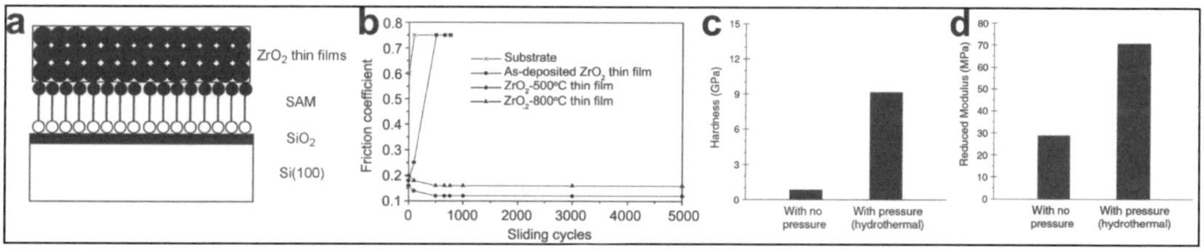

Fig. 21. Schematic of growth of ZrO_2 thin film on SAMs in aqueous medium (a); Friction coefficients as a function of sliding cycles at a load of 0.5 N (b); Mechanical properties of the ZrO_2 films prepared with a hydrothermal process at 135 °C, ~5 atm, for 24 h, as compared to those prepared with no pressure (c, d). Reproduced from Wang et al., 2004 (a), Wang et al., 2005 (b) and Zhang et al., 2006 (c, d).

To prepare ZrO_2 based SAO-ICFs, layer-by-layer (LbL) assembly technique is often applied. Generally speaking, LbL technique is based on sequential adsorption of oppositely charged materials, such as polyelectrolytes and inorganic nanomaterials. For example, Claus *et al* have obtained a superhard ZrO_2/PSS SAO-ICFs by a LbL process which is based on the electrostatic interaction between ZrO_2 nanoparticles (positive charged) and PSS (negative charged) (Fig. 22) (Rosidian et al., 1998).

However, this electrostatic LbL technique is confined to the charged materials. To expand the appliction of the LbL, a novel non-electrostatic layer-by-layer (NELbL) assembly technique has been invented in our group (Ou et al., 2010). The newly-reported PDA is served as the building block for its special nature, viz, high adhesion to almost all surfaces and the active surface with functional groups (such as -OH and $-NH_2$). As schematically illustrated in Fig. 23, PDA can be chemically grafted onto the amine groups of APTES-SAMs

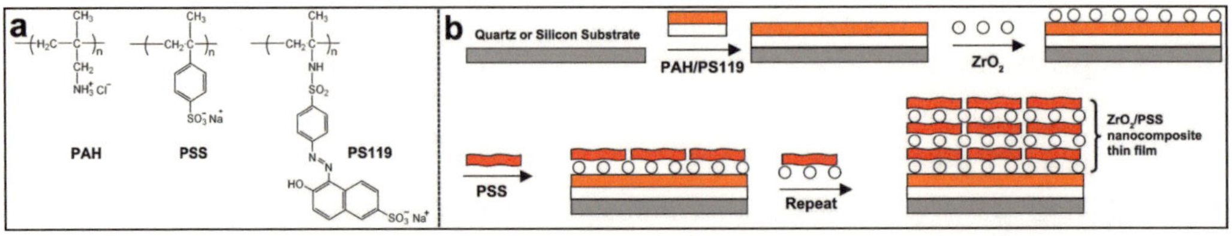

Fig. 22. The molecular structures of PAH, PSS and PS119 (a); Schematic of multilayer fabrication of ZrO_2/PSS SAO-ICF (b). Reproduced from Rosidian et al., 1998.

(Fig. 23, Process II) or hydroxyl groups of ZrO_2 film (Fig. 23, Process IV). Besides, the ZrO_2 clusters formed in the $Zr(SO_4)_2$ solution can deposit onto the PDA surface via chelation (Fig. 23, Process III). Thus, the sequential deposition of ZrO_2 and PDA can present a novel non-electrostatic strategy to construct ZrO_2/PDA SAO-ICFs. The microhardness and elastic modulus of the annealed 15-cycle ZrO_2/PDA film are measured to be as high as 24.10 and 250 GPa, respectively. This microhardness is comparable with that of ZrO_2/PSS SAO-ICF (25.13 GPa) (Rosidian et al., 1998). The outstanding mechanical properties of the ZrO_2/PDA SAO-ICFs can be ascribed to the in-suit deposition and organic-inorganic hybrid microstructures (Ou et al., 2001).

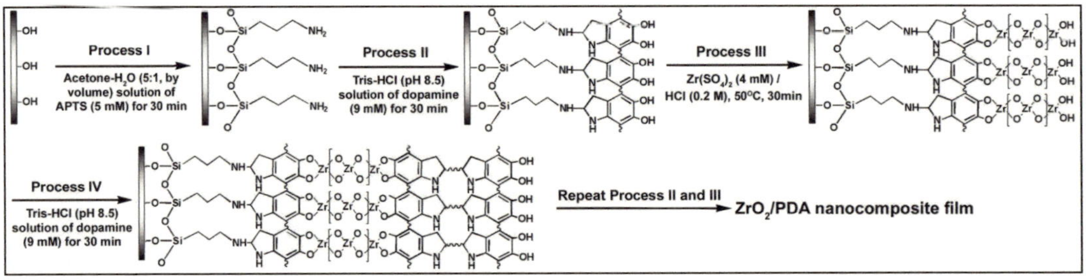

Fig. 23. A schematic view for constructing ZrO_2/PDA SAO-ICFs. Reproduced from Ou et al., 2001.

5. Conclusion

Based on the above discussions, it can be obtained that the tribological behaviors of SANFs are mainly structure dependent. Namely, the interfacial and interfilm interaction is supposed to influence the tribological properties of the prepared SANFs. With the efforts of many researchers, principal dependence between tribological performance and different parts of SAMs, viz, head/tail groups and bulk chains, has been proposed. This can be a basic understanding for us to investigate more complicated systems, such as SAMFs, SAIFs and SAO-ICFs. It is expected that the extracted "structures-properties" correlation can serve as the guidance to direct the further designing of lubricant coatings for MEMS/NEMS and other devices in molecule-level.

6. References

Bain, C. & Whitesides, G. (1989) Formation of monolayers by the coadsorption of thiols on gold-variation in the length of the alkyl chain. *Journal of the American Chemical Society*, Vol. 111, No. 18, (Aug, 1989) 7164-7175, 0002-7863.

Bushan, B.; Kulkarni, A.; Koinkar, V.; Boehm, M.; Odoin, L.; Martelet, C. & Belin, M. (1995) Microtribological characterization of self-assembled and langmuir-blodgett monolayers by atomic and friction force microscopy. *Langmuir,* Vol. 11, No. 8 (Aug, 1995) 3189-3198, 0743-7463.

Bliznyuk, V.; Everson, M. & Tsukruk, V. (1998) Nanotribological properties of organic boundary lubricants: Langmuir films versus self-assembled monolayers. *Journal of Tribology-Transactions of the Asme,* Vol. 120, No. 3 (Jul, 1998) 489-495, 0742-4787.

Beake, B. & Leggett, G. (1999) Friction and adhesion of mixed self-assembled monolayers studied by chemical force microscopy. *Physical Chemistry Chemical Physics,* Vol. 1, No. 14, (Jul, 1999) 3345-3350, 1463-9076.

Beake, B. & Leggett, G. (2000) Variation of frictional forces in air with the compositions of heterogeneous organic surfaces. *Langmuir* Vol. 16, No. 2, (Jan, 2000) 735-739, 0743-7463.

Brewer, N.; Beaker, B.; Leggett, G. (2001) Friction force microscopy of self-assembled monolayers: Influence of adsorbate alkyl chain length, terminal group chemistry, and scan velocity. *Langmuir,* Vol. 17, No. 6, (Mar, 2001) 1970-1974, 0743-7463.

Brewer, N. & Leggett, G. (2004) Chemical force microscopy of mixed self-assembled monolayers of alkanethiols on gold: Evidence for phase separation. *Langmuir,* Vol. 20, No. 10, (May, 2004) 4109-4115, 0743-7463.

Butt, H.; Cappella, B.; & Kappl, M. (2005) Force measurements with the atomic force microscope: technique, interpretation and applications. *Surface Science Reports,* Vol. 59, (Aug, 2005) 1-152, 0167-5729.

Bushan, B.; Kasai, T.; Kulik, G.; Barbieri, L. & Hoffmann, P. (2005) AFM study of perfluoroalkylsilane and alkylsilane self-assembled monolayers for anti-stiction in MEMS/NEMS. *Ultramicroscopy,* Vol. 105, No, 1-4 (Nov, 2005) 176-188, 0304-3991.

Booth, B.; Vilt, S.; McCabe, C. & Jennings, G. (2009) Tribology of Monolayer Films: Comparison between n-Alkanethiols on Gold and n-Alkyl Trichlorosilanes on Silicon. *Langmuir,* Vol. 25, No. 17 (Sep, 2009) 9995-10001, 0743-7463.

Chang, S.; Chao, L. & Tao, Y. (1994) Structures of self-assembled monolayers of aromatic-derivatized thiols on evaporated gold and silver surfaces-implication on packing mechanism. *Journal of the American Chemical Society.* Vol. 116, No. 15, (Jul, 1994) 6792-6805, 0002-7863.

Clegg, R. S. & Hutchison, J. E. (1996) Hydrogen-bonding, self-assembled monolayers: Ordered molecular films for study of through-peptide electron transfer. *Langmuir,* Vol. 12, No. 22, (Oct,1996) 5239-5243, 0743-7463

Clegg, R. & Hutchison, J. (1999) Control of monolayer assembly structure by hydrogen bonding rather than by adsorbate-substrate templating. *Journal of the American Chemical Society,* Vol. 121, No. 22, (Jun, 1999) 5319-5327, 0002-7863.

Clegg, R.; Reed, S.; Smith, R.; Barron, B.; Rear, J. & Hutchison, J. (1999) The interplay of lateral and tiered interactions in stratified self-organized molecular assemblies. *Langmuir,* Vol. 15, No. 26, (Dec, 1999) 8876-8883, 0743-7463.

Cappella, B. & Dietler, G. (1999) Force-distance curves by atomic force microscopy. *Surface Science Reports,* Vol. 34, No. 1, (July, 1999) 1-104, 0167-5729.

Chambers, R.; Inman, C. & Hutchison, J. (2005) Electrochemical detection of nanoscale phase separation in binary self-assembled monolayers. *Langmuir,* Vol. 21, No. 10, (May, 2005) 4615-4621, 0743-7463.

Chandross, M.; Lorenz, C.; Grest, G.; Stevens, M. & Webb III, E. (2005) Nanotribology of anti-friction coatings in MEMS. *Journal of the Minerals, Metals and Materials Society,* Vol. 57, No. 9, (Sep, 2005) 55-61, 1047-4838.

DePalma, V. & Tillman, N. (1989) Friction and wear of self-assembled trichlorosilane monolayer films on silicon. *Langmuir,* Vol. 5, No. 3, (May-Jun, 1989) 868-872, 0743-7463.

Evans, S.; Urankar, E.; Ulman, A. & Ferris, N. (1991) Self-assembled multilayers of omega-mercaptoalkanoic acids-selective ionic interactions. *Journal of the American Chemical Society,* Vol. 113, No. 15, (Jul, 1991) 5866-5868, 0002-7863.

Foisner, J.; Glaser, A.; Leitner, T.; Hoffmann, H. & Friedbacher, G. (2004) Effects of surface hydrophobization on the growth of self-assembled monolayers on silicon. *Langmuir,* Vol. 20, No. 7, (Mar, 2004), 2701-2706, 0743-7463.

Foster, T.; Alexander, M.; Leggett, G.; McAlpine, E. (2006) Friction force microscopy of alkylphosphonic acid and carboxylic acids adsorbed on the native oxide of aluminum. *Langmuir,* Vol. 22, No. 22, (Oct, 2006) 9254-9259, 0743-7463.

Gulino, A.; Mineo, P.; Scamporrino, E.; Vitalini, D. & Fragala, I. (2004) Molecularly engineered silica surfaces with an assembled porphyrin monolayer as optical NO_2 molecular recognizers. *Chemistry of Materials,* Vol. 16, No. 10, (May, 2004), 1838-1840: 0897-4756.

Houston, J. & Michalske, T. (1992) The interfacial-force microscope. *Nature,* Vol. 356, No. 6366, (March, 1992) 266-267, 0028-0836.

Houston, J.; Doelling, C.; Vanderlick, T.; Hu, Y.; Scoles, G.; Wenzl, I.; Lee, T. (2005) Comparative study of adhesion, friction, and mechanical properties of CF_3- and CH_3- terminated alkanethiol monolayers. *Langmuir,* Vol. 21, No. 9 (Apr, 2005) 3926-3932, 0743-7463.

Hsu, S. (2004) Nano-lubrication: concept and design. *Tribology International,* Vol. 37, No, 7, (Jul, 2004), 537-545, 0301-679X.

Jiao, J.; Anariba, F.; Tiznado, H.; Schmidt, I.; Lindsey, J.; Zaera, F. & Bocian, D. (2006) Stepwise formation and characterization of covalently linked multiporphyrin-imide architectures on Si(100). *Journal of the American Chemical Society,* Vol. 128, No. 21, (May, 2006) 6965-6974, 0002-7863.

Kim, H.; Koini, T.; Lee, R. & Perry S. (1997) Systematic studies of the frictional properties of fluorinated monolayers with atomic force microscopy: Comparison of CF_3- and CH_3-terminated films. *Langmuir* Vol. 13, No. 26, (Dec, 1997) 7192-7196, 0743-7463.

Kim, H.; Graupe, M.; Oloba, O.; Koini, T.; Imaduddin, S.; Lee, T. & Perry, S. (1999) Molecularly specific studies of the frictional properties of monolayer films: A systematic comparison of CF_3-, $(CH_3)_2CH$-, and CH_3-terminated films. *Langmuir,* Vol. 15, No. 9 (Apr, 1999) 3179-3185, 0743-7463.

Kim, H. & Houston, J. (2000) Separating mechanical and chemical contributions to molecular-level friction. *Journal of the American Chemical Society.* Vol. 122, No. 48, (Dec, 2000) 12045-12046, 0002-7863.

Liu, Y.; Evans, D.; Song, Q. & Grainger, D. (1996) Structure and frictional properties of self-assembled surfactant monolayers. *Langmuir,* Vol. 12, No. 5, (Mar, 1996), 1235-1244, 0743-7463.

Lio, A.; Charych, D. & Salmeron, M. (1997) Comparative atomic force microscopy study of the chain length dependence of frictional properties of alkanethiols on gold and alkylsilanes on mica. *The Journal of Physical Chemistry B,* Vol. 101, No. 19, (May, 1997) 3800-3805, 1520-6106.

Luengo, G.; Campbell, S.; Srdanov, V.; Wudl, F. & Israelachvili, J. (1997) Measurement of the adhesion and friction of smooth C60 surfaces. *Chemistry of materials,* Vol. 9, No. 5, (May, 1997) 1166-1171, 0897-4756.

Li, J.; Wang, C.; Shang, G.; Xu, Q.; Lin, Z.; Guan, J.; Bai, C. (1999) Friction coefficients derived from apparent height variations in contact mode atomic force microscopy images. *Langmuir,* Vol. 15, No. 22, (Oct, 1999) 7662-7669, 0743-7463.

Lee, S.; Puck, A.; Graupe, M.; Colorada, R.; Shon, Y.; Lee, T. & Perry, S. (2001) Structure, wettability, and frictional properties of phenyl-terminated self-assembled monolayers on gold. *Langmuir,* Vol. 17, No. 23, (Nov, 2001) 7364-7370, 0743-7463.

Luzinov, I.; Julthongpiput, D.; Liebmann-Vinson, A.; Cregger, T.; Foster, M. & Tsukruk, V. (2000) Epoxy-terminated self-assembled monolayers: Molecular glues for polymer layers. *Langmuir* Vol., 16, No. 2, (Jan, 2000) 504-516, 0743-7463.

Luzinov, I.; Julthongpiput, D.; Gorbunov, V. & Tsukruk, V. (2001) Nanotribological behavior of tethered reinforced polymer nanolayer coatings. *Tribology International,* Vol. 34, No. 5, (May, 2001) 327-333, 0301-679X.

Leggett, G.; Brewer, N.; Chong, K. (2005) Friction force microscopy: towards quantitative analysis of molecular organisation with nanometre spatial resolution. *Physical Chemistry Chemical Physics,* Vol. 7, No. 6, (Feb, 2005) 1107-1120, 1463-9076.

Love, J.; Estroff, L.; Kriebel, J.; Nuzzo, R.; Whitesides, G. (2005) Self-assembled monolayers of thiolates on metals as a form of nanotechnology. *Chemical Reviews,* Vol. 105, No. 4 (Apr, 2005) 1103-1169.

McDermott, M.; Green, J. & Porter, M. (1997) Scanning force microscopic exploration of the lubrication capabilities of n-alkanethiolate monolayers chemisorbed at gold structural basis of microscopic friction and wear. *Langmuir,* Vol.13, No. 9, (Apr, 1997) 2504-2510, 0743-7463.

Mowery, M.; Kopta, S.; Ogletree, D.; Salmeron, M. & Evans C. (1999) Structural manipulation of the frictional properties of linear polymers in single molecular layers. *Langmuir,* Vol. 15, No. 15, (Jul, 1999) 5118-5122, 0743-7463.

Maeda, N.; Chen, N.; Tirrell, M. & Israelachvili, J. (2002) Adhesion and friction mechanisms of polymer-on-polymer surfaces. *Science,* Vol. 297, No.5580, (Jul, 2002) 379-382, 0036-8075.

Noy, A.; Frisbie, C.; Rozsnyai, L.; Wrighton, M. & Lieber, C. (1995) Chemical force microscopy-exploiting chemically-modified tips to quantify adhesion, friction, and functional-group distributions in molecular assemblies. *Journal of the American Chemical Society,* Vol. 117, No. 30, (Aug, 1995), 7943-7951, 0002-7863.

Noy, A.; Vezenov, D. & Lieber, C. (1997) Chemical force microscopy. *Annual Review of Materials Science,* Vol. 27, (Aug, 1997) 381-421, 0084-6600.

Ou, J.; Wang, J.; Liu, S.; Zhou, J.; Yang, S. (2009) Self-assembly and tribological property of a novel 3-layer organic film on silicon wafer with polydopamine coating as the interlayer. *The Journal of Physical Chemistry C,* Vol. 113, No, 47, (Nov, 2009) 20429-20434, 1932-7447.

Ou, J.; Wang, J.; Qiu, Y.; Liu, L. & Yang, S. (2010) Mechanical property and corrosion resistance of zirconia/polydopamine nanocomposite multilayer films fabricated via a novel non-electrostatic layer-by-layer assembly technique. *Surface and Interface Analysis,* DOI: 10.1002/sia.3631, 1096-9918.

Ostuni, E.; Yan, L. & Whitesides, G. M. (1999) The interaction of proteins and cells with self-assembled monolayers of alkanethiolates on gold and silver. *Colloids and Surfaces B-Biointerfaces,* Vol 15, No. 1, (Aug, 1999), 3-30, 0927-7765.

Overney, R.; Meyer, E.; Frommer, J.; Brodbeck, D.; Lüthi, R.; Howald, L.; Güntherodt, H.; Fujihira, M.; Takano, H. & Gotoh, Y. (1992) Friction measurements on phase-separated thin-films with a modified atomic force microscope. *Nature,* Vol 359. No. 6391, (Sep, 1992), 133-135, 0028-0836.

Peach, S.; Polak, R. & Franck, C. (1996) Characterization of partial monolayers on glass using friction force microscopy. *Langmuir,* Vol. 12, No. 25 (Dec, 1996) 6053-6058, 0743-7463.

Pakala, M.; Walls, H. & Lin, R. (1997) Microhardness of sputter-deposited zirconia films on silicon wafers. Journal of the American Ceramic Society, Vol. 80, No. 6, (Jun, 1997) 1477-1484, 0002-7820.

Rosidian, A.; Liu, Y. & Claus, R. (1998) Ionic self-assembly of ultrahard ZrO_2/polymer nanocomposite thin films. *Advanced Materials,* Vol. 10, No. 14, (Oct, 1998) 1087-1091, 0935-9648.

Ren, S.; Yang, S. & Zhao, Y. (2003) Micro- and macro-tribological study on a self-assembled dual-layer film. *Langmuir,* Vol. 19, No. 7, (Apr, 2003), 2763-2767, 0743-7463.

Ren, S.; Yang, S.; Wang, J.; Liu, W. & Zhao, Y. (2004) Preparation and tribological studies of stearic acid self-assembled monolayers on polymer-coated silicon surface. *Chemistry of Materials,* Vol. 16, No. 3, (Feb, 2004) 428-433, 0897-4756.

Smith, E. & Porter, M. (1993) Structure of monolayers of short-chain n-alkanoic acids ($CH_3(CH_2)_nCOOH$, n = 0-9) spontaneously adsorbed from the gas-phase at silver as probed by infrared reflection spectroscopy. *The Journal of physical chemistry,* Vol. 97, No. 30, (Jul, 1993), 8032-8038, 0022-3654.

Sabapathy, R.; Bhattacharyya S.; Leavy, M.; Cleland, W. & Hussey, C. (1998) Electrochemical and spectroscopic characterization of self-assembled monolayers of ferrocenylalkyl compounds with amide linkages. *Langmuir* Vol.,14, No. 1, (Jan, 1998) 124-136, 0743-7463.

Sidorenko, A.; Ahn, H.; Kim, D.; Yang, H. & Tsukruk, V. (2002) Wear stability of polymer nanocomposite coatings with trilayer architecture. *Wear,* Vol. 252, No, 11-12, (Jun, 2002) 946-955, 0043-1648.

Sakata, H.; Kobayashi, M.; Otsuka, H. & Takahara, A. (2005) Tribological properties of poly(methyl methacrylate) brushes prepared by surface-initiated atom transfer radical polymerization. *Polymer Journal,* Vol. 37, No. 10, (Oct, 2005) 767-775, 0032-3896.

Song, S.; Ren, S.; Wang, J.; Yang, S.; Zhang, J. (2006) Preparation and tribological study of a peptide-containing alkylsiloxane monolayer on silicon. *Langmuir,* Vol. 22, No. 14, (Jun. 2006) 6010-6015, 0743-7463.

a, Song, S.; Zhou, J.; Qu, M.; Yang, S.; Zhang, J. (2008) Preparation and tribological behaviors of an amide-containing stratified self-assembled monolayers on silicon surface. *Langmuir,* Vol. 24, No. 1, (Jan. 2008) 105-109, 0743-7463.

b, Song, S.; Chu, R.; Zhou, J.; Yang, S.; Zhang, J. Y. (2008) Formation and tribology study of amide-containing stratified self-assembled monolayers: influences on the underlayer structure. *The Journal of Physical Chemistry C,* Vol. 112, No, 10, (Feb, 2008) 3805-3810, 1932-7447.

Tao, Y. (1993) Structural comparison of self-assembled monolayers of n-alkanoic acids on the surfaces of silver, copper, and aluminum. *Journal of the American Chemical Society.* Vol. 115, No. 10, (May, 1993) 4350-4358, 0002-7863.

Tam-Chang, S.; Biebuyck, H.; Whitesides, G.; Jeon, N. & Nuzzo, R. (1995) Self-assembled monolayers on gold generated from alkanethiols with the structure $RNHCOCH_2SH$. *Langmuir,* Vol. 11, No. 11, (Nov, 1995) 4371-4382, 0743-7463.

a, Tsukruk, V.; Everson, M.; Lander, L. & Brittain, W. (1996) Nanotribological properties of composite molecular films: C60 anchored to a self-assembled monolayer. *Langmuir,* Vol. 12, No. 16, (Aug, 1996) 3905-3911, 0743-7463.

b, Tsukruk, V.; Bliznyuk, V.; Hazel, J.; Visser, D. & Everson, M. Organic molecular films under shear forces: Fluid and solid Langmuir monolayers. *Langmuir,* Vol. 12, No. 20 (Oct, 1996) 4840-4849, 0743-7463.

Tsukruk, V. & Blivnyuk, V. (1998) Adhesive and friction forces between chemically modified silicon and silicon nitride surfaces. *Langmuir,* Vol. 14, No. 2, (Jan, 1998) 446-455, 0743-7463.

Tsukruk, V.; Luzinov, I. & Julthongpiput, D. (1999) Sticky molecular surfaces: Epoxysilane self-assembled monolayers. *Langmuir,* Vol. 15, No. 9, (Apr, 1999) 3029-3032, 0743-7463.

Tsukruk, V. (2001) Molecular lubricants and glues for micro- and nanodevices. *Advanced Materials,* Vol. 13, No. 2, (Jan, 2001) 95-108, 0935-9648.

Ulman, A. (1996) Formation and structure of self-assembled monolayers. *Chemical Reviews,* Vol. 96, No. 4, (Jun, 1996), 1533-1554, 0009-2665.

Wong, S.; Takano, H. & Porter, M. (1998) Mapping orientation differences of terminal functional groups by friction force microscopy. *Analytical Chemistry,* Vol. 70, No. 24, (Dec, 1998) 5209-5212, 0003-2700.

Wang, X.; Tsui, O. & Xiao, X. (2002) Dynamic study of polymer films by friction force microscopy with continuously varying load. *Langmuir,* Vol. 18, No. 18, (Sep, 2002) 7066-7072, 0743-7463.

Wang, J.; Yang, S.; Liu, X.; Ren, S.; Guan, F. & Chen, M. (2004) Preparation and characterization of ZrO_2 thin film on sulfonated self-assembled monolayer of 3-mercaptopropyl trimethoxysilane. *Applied Surface Science,* Vol. 221, No. 1-4, (Jan, 2004) 272-280, 0169-4332.

Wang, J.; Liu, X.; Ren, S.; Guan, F. & Yang, S. (2005) Mechanical properties and tribological behavior of ZrO_2 thin films deposited on sulfonated self-assembled monolayer of 3-mercaptopropyl trimethoxysilane. *Tribology Letters,* Vol. 18, No. 4, (Apr, 2005) 429-436, 1023-8883.

Wang, M.; Liechti, K. M.; Wang, Q. & White, J. (2005) Self-assembled silane monolayers: Fabrication with nanoscale uniformity. *Langmuir,* Vol. 21, No. 5, (Mar, 2005) 1848-1857, 0743-7463.

Xiao, X.; Hu, J.; Charych, D. & Salmeron, M. (1996) Chain length dependence of the frictional properties of alkylsilane molecules self-assembled on Mica studied by atomic force microscopy. *Langmuir,* Vol. 12, No. 2, (Jan, 1996) 235-237, 0743-7463.

Xiao, X. & Qian L. (2000) Investigation of humidity-dependent capillary force. *Langmuir,* Vol. 16, No. 21, (Sep. 2000) 8153-8158, 0743-7463.

Zhao, B. & Brittain, W. (2000) Polymer brushes: surface-immobilized macromolecules. *Progress in Polymer Science.* Vol. 25, No. 5, (Jun, 2000) 677-710, 0079-6700.

Zhou, F.; Liu, W.; Chen, M. & Sun, D. (2001) A novel way to prepare ultra-thin polymer films through surface radical chain-transfer reaction. *Chemmical Commununications,* Vol. 23, (Dec, 2001) 2446-2447, 1359-7345.

Zhang, L.; Li, L.; Chen, S. & Jiang, S. (2002) Measurements of friction and adhesion for alkyl monolayers on Si(111) by scanning force microscopy. *Langmuir,* Vol. 18, No. 14, (Jul, 2002) 5448-5456, 0743-7463.

Zhang, Q. & Archer, L. (2003) Boundary lubrication and surface mobility of mixed alkylsilane self-assembled monolayers. *The Journal of Physical Chemistry B,* Vol. 107, No. 47, (Nov, 2003) 13123-13132, 1520-6106.

Zhang, Q. & Archer, L. (2005) Interfacial friction of surfaces grafted with one- and two-component self-assembled monolayers. *Langmuir,* Vol. 21, No.12, (Jun, 2005) 5405-5413, 0743-7463.

Zhang, G.; Howe, J.; Coffey, D. & Blom, D. (2006) A biomimetic approach to the deposition of ZrO_2 films on self-assembled nanoscale templates. *Materials Science and Engineering C: Biomimetic and Supramolecular Systems.* Vol. 26, No. 8, (Sep, 2006), 1344-1350, 0928-4931.

Zhao, J. Chen, M. An, Y. Liu, J. & Yan, F. (2008) Preparation of polystyrene brush film by radical chain-transfer polymerization and micromechanical properties, *Applied Surface Sciences.* Vol. 255, No. 5, (Dec, 2008) 2295-2302, 0169-4332.

Zlotnikov, I.; Gotman, I. & Gutmanas, E. (2008) Characterization and nanoindentation testing of thin ZrO_2 films synthesized using layer-by-layer (LbL) deposited organic templates. *Applied Surface Science,* Vol. 255, No. 5, (Dec, 2008) 3447-3453, 0169-4332.

Zhao, J.; Chen, M.; Liu, J.; Yan, F. (2009) Preparation and tribological studies of self-assembled triple-layer films. *Thin Solid Films,* Vol. 517, No. 13, (May, 2009) 3752-3759, 0040-6090.

21

A Novel Tool for Mechanistic Investigation of Boundary Lubrication: Stable Isotopic Tracers

Ichiro Minami
Iwate University,
Japan

1. Introduction

1.1 Surface analyses of tribological surfaces

It is well understood that lubrication modes can be classified into three categories. Under hydrodynamic conditions, two rubbing surfaces are ideally separated by liquid film derived from lubricating oil. Therefore, the surface chemistry of the rubbing parts is of less importance. When the operational conditions become severe, for example, under conditions of increased load, the rubbing surfaces come in contact. This is defined as the boundary lubrication condition. Under these conditions, the properties of solid surfaces are important. There is an intermediate mode between hydrodynamic lubrication and boundary lubrication; namely mixed lubrication. Direct interaction between surfaces may occasionally take place under these conditions. Tribo-chemistry controls the performance of mixed lubrication and boundary lubrication conditions [1].

Challenges to understanding the surface chemistry of rubbed surfaces by tribologists using instrumental analyses started as early as the 1960s. Tribo-active elements such as phosphorus and sulfur were found by electron-probe microanalysis (EPMA) of rubbed surfaces lubricated with mineral oil containing additives [2]. These reports clearly support the tribo-chemical reaction of anti-wear and extreme pressure additives (AW/EP). In the 1970s, Auger electron spectroscopy (AES) was introduced as a more surface sensitive tool in tribology [3]. EPMA and AES are frequently applied in tribo-chemistry as useful tools. However, these instrumental analyses are elemental analyses of solid surfaces that identify only the elements that exist on the surfaces.

When FeS, FeS_2, and $FeSO_4$ were analyzed, EPMA and AES detected sulfur in the sample but could not identify the chemical states of the sulfur found. X-ray photoelectron spectroscopy (XPS) provides different chemical shifts between iron sulfide (FeS) and iron sulfate ($FeSO_4$) in S 2p spectra [4]. X-ray absorption near edge structure (XANES) identifies iron monosulfide (FeS) and iron disulfide (FeS_2) [5]. These surface analyses mainly detect inorganic compounds. Although carbon is detectable, chemical resolution of carbon by these tools is not always sufficient for identifying the structure of organic compounds in detail.

SIMS has been introduced as another surface sensitive instrumental analysis in tribology [6]. Contrary to AES, EPMA, XANES, and XPS, SIMS does not provide any chemical information regarding the sample directly [7]. It provides the molecular weight of substances that exist on surfaces. On the other hand, it detects all substances that can be ionized by a primary ion. Both organic and inorganic compounds can be analyzed by SIMS.

Therefore, it is a potentially versatile tool for tribo-chemistry in which organic compounds are the major contents in lubricants. One of the most important features of SIMS, and the main subject of this work, is to detect elemental isotopes. The principle of SIMS analysis is described in detail in the following section.

Nomenclature

Abbreviation	Description
AW/EP	anti-wear agents and extreme pressure additives
$C_{18}NH_2$	octadecylamine, $C_{18}H_{37}NH_2$
$C_{18}OH$	octadecyl alcohol, $C_{18}H_{37}OH$
D	deuterium, ^2H
DLC	diamond like carbon
GMO	glycerol monoleate, 2,3-dihydroxypropyl 9(Z)-octadecenoate
m/z	mass to charge ratio
OA-D35	perdeuterio-octadecanoic acid, $C_{17}D_{35}COOH$
OA-H35	octadecanoic acid, $C_{17}H_{35}COOH$
PAO	poly(alpha-olefin)
SIMS	secondary ion mass spectroscopy

1.2 Application of isotopes in tribo-chemical investigation

Isotopes are defined as elements that have the same atomic number but have different mass numbers [8]. For example, hydrogen has three isotopes; they are protium (^1H, usually expressed as H unless otherwise stated) having the mass number of 1, deuterium (^2H, usually abbreviated as D) having the mass number of 2, and tritium (^3H, usually abbreviated as T) having the mass number of 3. The difference in mass number is attributed to the difference in the number of neutrons in the nucleus (Figure 1). Among the three isomers, protium and deuterium are stable isotopes and tritium is a radioactive isotope. Usually, isotopes behave chemically in a similar manner. Therefore, they can be used as tracers in chemical processes if they were effectively detected. Radioactive isotopes are easy to detect using a Geiger counter, even if a small amount exists in the sample. Taking advantage of this high detectability, radioactive isotopes were applied in tribo-chemistry of AW/EP additives before surface analyses were introduced. For example, ^{35}S-labeld organic sulfides were employed as AW/EP additives in mineral oil. Radiation from solid surfaces was detected after rubbing with the labeled lubricants. These results clearly indicate that tribo-chemical reactions of sulfur occurred during rubbing [9]. However, radioactive isotopes are considered to be biological hazards and are difficult to handle. Another limitation of this technique is that it provides merely elemental analysis of surfaces. Therefore, radioactive isotopes are not frequently applied in tribo-chemical investigation.

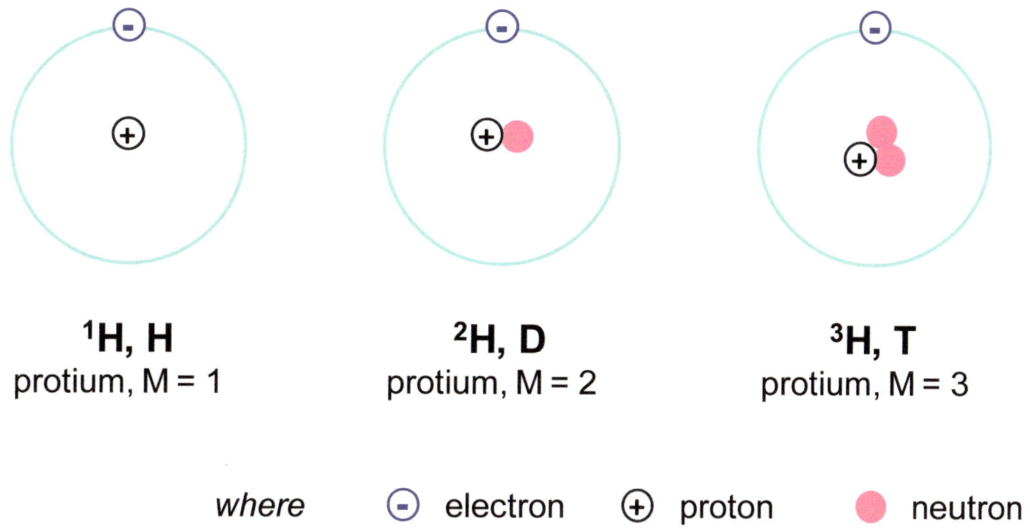

Fig. 1. Schematic model of hydrogen isotopes

Stable isotopes exist everywhere and are not considered to be biological hazards. For example, hydrogen is comprised of ^{1}H and D, and the latter is the minor isotope found in nature. Also carbon is comprised of ^{12}C and ^{13}C, where the later is the minor component (Table 1). High resolution instrumental analyses such as mass spectroscopy can detect and identify these minor isotopes. This led us to expect that additive molecules, which are the minor component in lubricants, could be detected by identifying stable isotopes. In this concern, model lubricants that are enriched with minor isotope(s) (D or ^{13}C) would improve the detectability of a target molecule by instrumental analyses.

Element	Isotope	Mass number	Numbers of		Abundance, atom%
			proton	neutron	
Hydrogen	^{1}H	1	1	0	99.985
	^{2}H	2	1	1	0.015
Carbon	^{12}C	12	6	6	98.9
	^{13}C	13	6	7	1.10
Oxygen	^{16}O	16	8	8	99.76
	^{17}O	17	8	9	0.04
	^{18}O	18	8	10	0.20

Table 1. Natural abundance of isotopes for hydrogen, carbon, and oxygen

1.3 Principle of SIMS analysis

SIMS consists of mass spectroscopy that measures the weight of the molecules in a sample. However, we cannot measure the weight of molecules directly even using an ultra high sensitive balance. Note that the weight of one molecule of water is as light as 3×10^{-23} g. Instead, mass spectroscopy measures the weight of ions. That is, mass spectroscopy ionizes a molecule in a sample followed by separation of generated ions according to m/z. Understanding of the ionization process of a sample is key to successful analysis. In SIMS, as schematically illustrated in Figure 2, irradiation of a primary ion such as Ga^+, Au^+, Bi^+ to the target molecule yields secondary ion(s). If the molecule was simply ionized by exchanging electron(s) with the primary ion (without scission of any chemical bond existing in the molecule) this process yields molecular ion $[M]^+$ or $[M]^-$, where M is the formula weight of the sample molecule. This gives a signal of m/z M in the mass spectrum. Most organic compounds contain C–H bonds. For these compounds, scission of a C–H bond usually occurs during the ionization process. If a proton (H^+) migrated from the original molecule, it yields an $[M\text{-}H]^-$ ion, which causes the loss in the formula weight by one. By contrast, if hydride (H^-) migrated, it results in a $[M\text{-}H]^+$ ion. Since the mass number of hydrogen is 1, the migration process gives m/z M–1 and m/z 1 in the mass spectrum. The migrated proton or hydride are active species and they react with the original molecule that yields $[M+H]^+$ or $[M+H]^-$, respectively. Therefore, the ion of m/z M+1 is frequently found in the mass spectrum of organic compounds. In mass spectroscopy, $[M]^+$ and $[M]^-$ ions are defined as molecular ions. Similarly, $[M\text{-}H]^-$, $[M\text{-}H]^+$, $[M+H]^+$, and $[M+H]^-$ are defined as quasi-molecular ions. These ions are used to identify the sample by its formula weight.

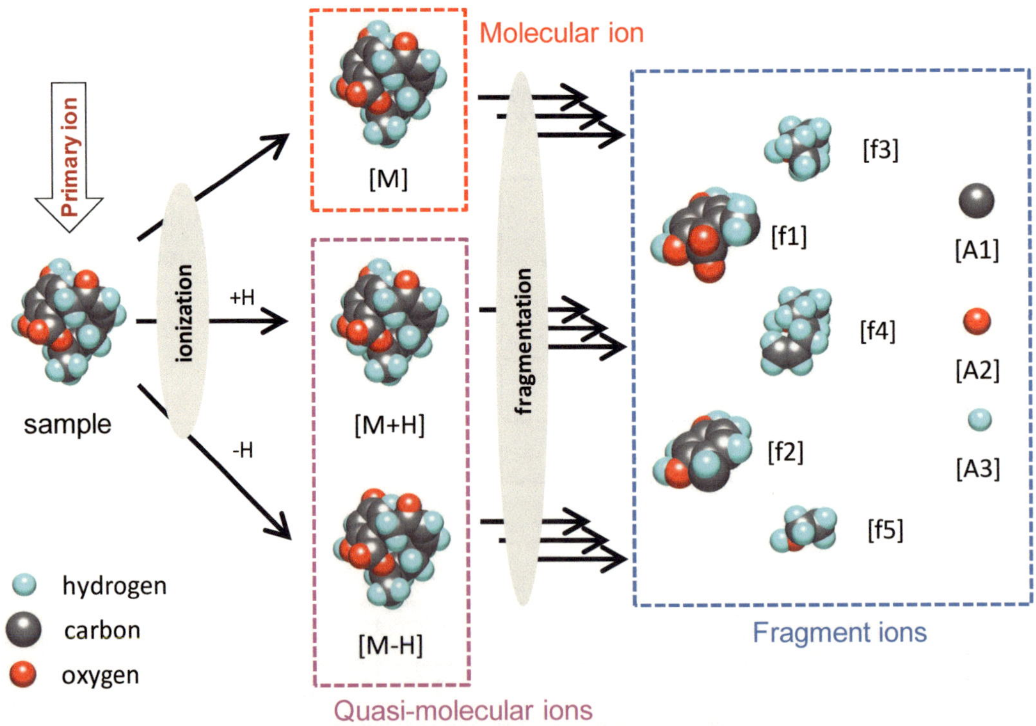

Fig. 2. Generation of secondary ions and the fragmentation in SIMS

Scission of C–C bonds or other bonds (such as C–O) also happens during the ionization process. As a result, the C-C skeleton is destroyed to several moieties. This is defined as fragmentation that yields fragment ions (for example, [f1]–[f5] in Figure 2). The fragmentation is a multistep process; it proceeds further and finally yields atomic ions (for example, [A1]–[A3] in Figure 2). Thanks to the fragmentation of molecule, mass spectroscopy is able to identify atoms that exist in the sample as well as partial structures and functional groups present in the molecule. However, the fragmentation is usually a complicated process. Patterns of fragmentation for typical compounds are available on a database [10]. However, this approach seems insufficient for tribology because tribo-chemistry deals with a complex mixture as lubricants. One of the solutions to this problem is to compare the mass spectrum of sample with that obtained from the authentic sample(s).

A quantitative analysis is possible by SIMS according to the intensities of the ions. For this purpose, the conversion of ionization influences the reliability of the results. Some molecule or moieties therein easily ionize, but others do not. For example, organofluorides provide high intensities of fluorine-containing fragments, according to our experience. Sources of primary ions and their energy are the important factors for quantitative analysis in SIMS.

As already mentioned, SIMS does not provide chemical information of samples directly, but provides only the weight of ions. For example, octadecanoic acid (stearic acid, $C_{17}H_{35}COOH$), which is a well studied model friction modifier, provides m/z values of 283 and 285 as quasi-molecular ions in the mass spectrum. Hydrocarbons that have $C_{21}H_{32}$ moieties are able to provide m/z 283 and 285 as fragment ions. They yield the same quasi-molecule ions in the m/z spectrum. As a result, there are several possible chemical structures that meet the value of m/z. Careful study of the mass spectrum according to the fragmentation of the sample molecule is the usual method in SIMS to identify a molecule. However, tribology deals with complicated mixtures of additives as a solution in base oils. Since the fragmentation of each compound is also complicated, as discussed above, SIMS of tribological surface usually results in a chaotic spectrum. Therefore, novel techniques are highly desired for tribo-chemical research. Taking these fundamental aspects into account, we were inspired to use stable isotopes for tribo-chemistry. Stable isotopes usually behave in a similar manner chemically. Synthesis of model lubricant molecule that labeled with stable isotope(s) is possible. SIMS can trace the molecule during tribological processes by paying attention to the isotopes. This idea provides simple and clear results through the tribo-chemical process.

2. SIMS study of lubricants on surfaces

2.1 Monomolecular layer on nonrubbed surface [11]

Straight-chain fatty acids such as octadecanoic acid are one of the best studied friction modifiers. They reduce the friction between steel/steel contacts when they are added to hydrocarbon oils. The formation of densely packed, well-oriented adsorption films had been widely accepted as their lubrication mechanism [12]. In fact, the monomolecular layer of octadecanoic acid deposited on flat metal surfaces provided low friction [13].

This work began with the simple comparison of the SIMS spectrum of natural octadecanoic acid ($C_{17}H_{35}COOH$, OA-H35) and deuterio-labeled octadecanoic acid ($C_{17}D_{35}COOH$, OA-D35). Thus, the monomolecular films of OA-H35 and OA-D35 were deposited on the Si(100) surface by means of the Langmuir–Blodgett trough [14]. Surface coverage by the adsorbed area was measured by the surface force. As shown in Figure 3, both OA-H35 and OA-D35 provided similar monomolecular films. The SIMS spectrum of OA-H35 (formula weight of 284) deposited on the surface is shown in Figure 4. Quasi-molecule ions at m/z 283 and at

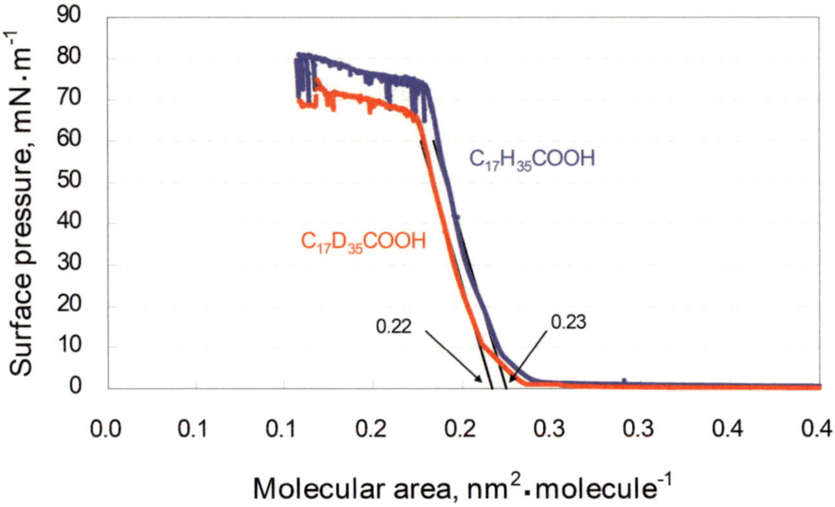

Fig. 3. Diagram of molecular area and surface pressure during the preparation of monomolecula film from octadecanoic acids

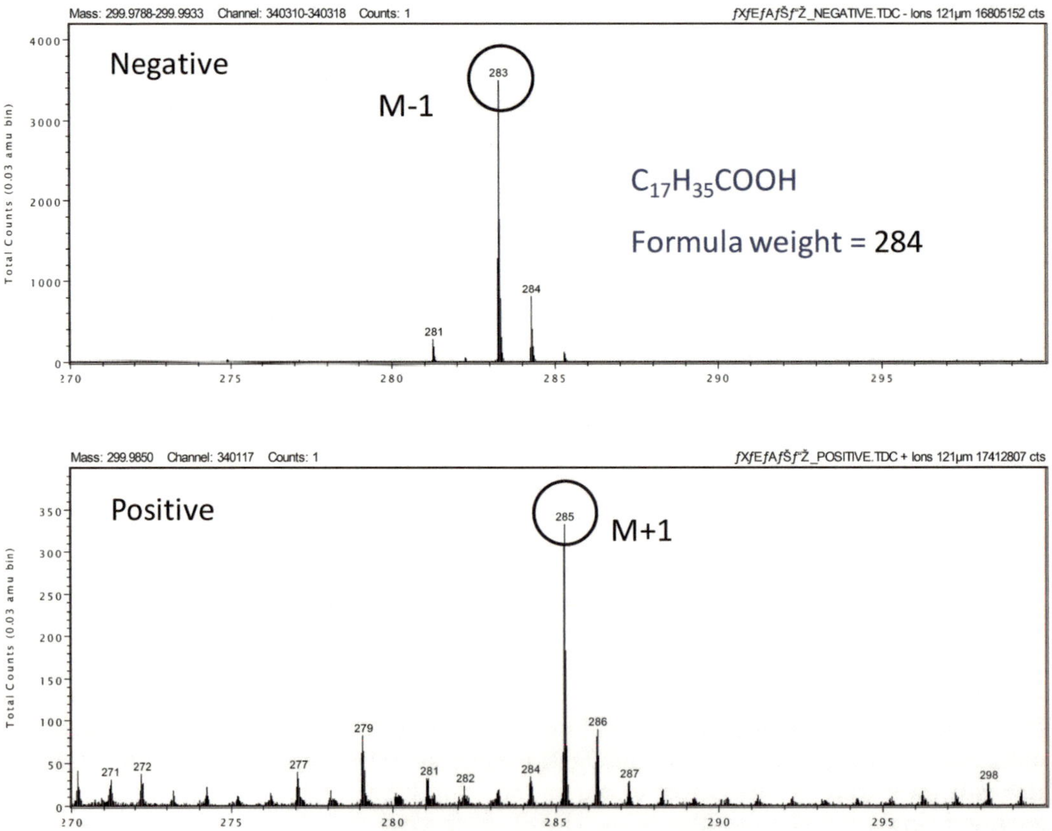

Fig. 4. SIMS spectra of OA-H35 on the Si surface (m/z 270–300)

m/z	Relative intensity, -
284	100
285	20
286	2.4
287	0.17

Table 2. Intensity of molecular ions derived from octadecanoic acid; calculation based on the natural abundance of isotopes

285 that correspond to the $[M\text{-}H]^-$ and $[M+H]^+$ ions respectively were found with high intensities. In addition to these ions, strong signals for negative ions at m/z 284 and positive ions at m/z 286 were also found. These ions are reasonably explained by natural isotopic abundance, as shown in Table 2. Also this table indicates quasi-molecule ions of small signals at m/z 285 in the negative spectra and at m/z 287 in positive spectra.

The SIMS spectrum of OA-D35 (formula weight of 319) provided quasi-molecular ions at m/z 318 in the negative spectra and at m/z 321 in the positive spectra (Figure 5). It is interesting that the m/z of the positive quasi-molecule increased by two from the formula weight of the original molecule, while the m/z of negative quasi-molecular decreased by one from the formula weight of the original molecule. These results can be explained by the mechanism of fragmentation as shown in Figure 6. The migration of the acidic proton (hydrogen in COOH group) from the original carboxylic acid yields a carboxylate anion, which corresponds to a negative quasi-molecular fragment ion $[M\text{-}H]^-$ and a proton H^+ (path A in figure 6). The protonation of the original carboxylic acid (coordination of COOH to the proton) yields an oxonium type cation, which corresponds to a positive quasi-molecular fragment ion $[M+H]^+$ (path B in Figure 6). Not only the acidic hydrogen but also the hydrogen of the alkyl group migrates during the fragmentation process. This process with OA-D35 yields a deuterium cation (D^+) with the mass number of two. Then the coordination of the original acid to the deuterium cation yields an oxonium type cation, which corresponds to a positive quasi-molecular fragment ion $[M+D]^+$ (path C in Figure 6). Therefore, the increase in m/z by two for quasi-molecular ions can be applied as a fingerprint of deuterium labeled compounds.

Further fragmentation yields atomic ions and small size fragment ions. Figure 7 shows positive mass spectra of m/z from 1 to 20. The reference OA-H35 provided fragment ions of m/z 1, 12, 13, 14, and 15, which correspond to H, C, CH, CH_2, and CH_3, respectively. Obviously, OA-D35 provided different mass spectrum from that of OA-H35: m/z values of 1, 2, 12, 14, 16, and 18, which correspond to H, D, C, CD, CD_2, and CD_3, respectively. A considerable intensity of H at m/z 1 was observed in the mass spectra of OA-D35. A plausible explanation is the high conversion of ionization of the acidic proton in OA-D35. This also explains the considerable high intensity of $[M+H]^+$ quasi-molecular ions in the mass spectra of OA-D35.

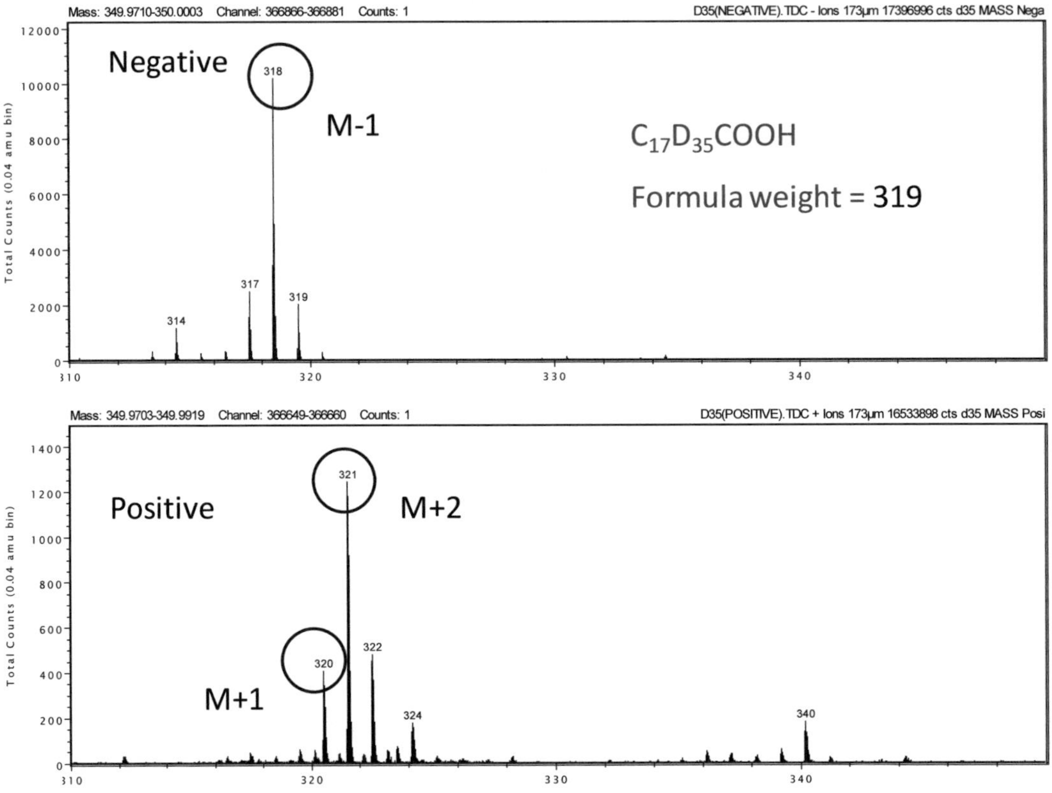

Fig. 5. SIMS spectra of OA-D35 on the Si surface (m/z 310–340)

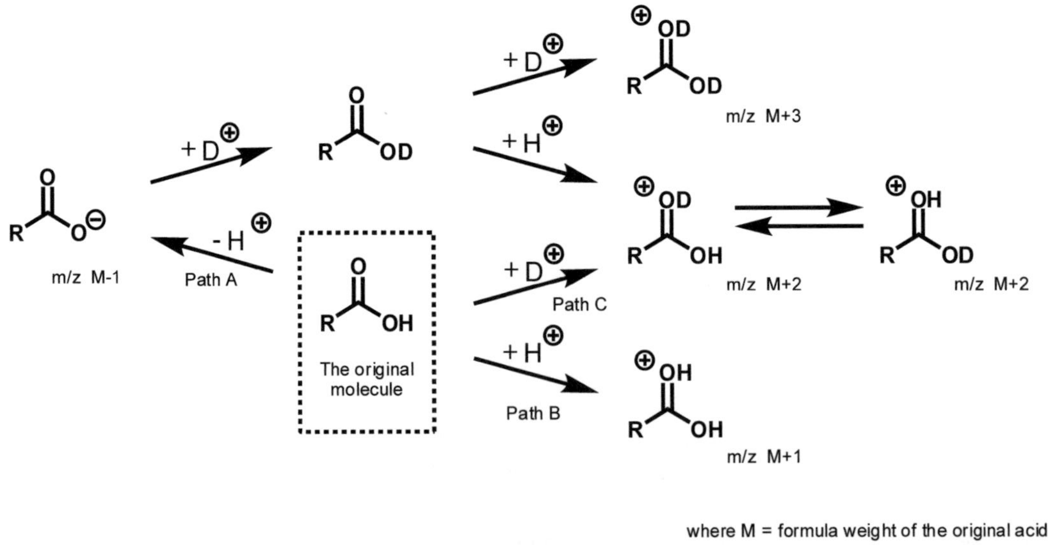

Fig. 6. Generation of quasi-molecular ions from carboxylic acids

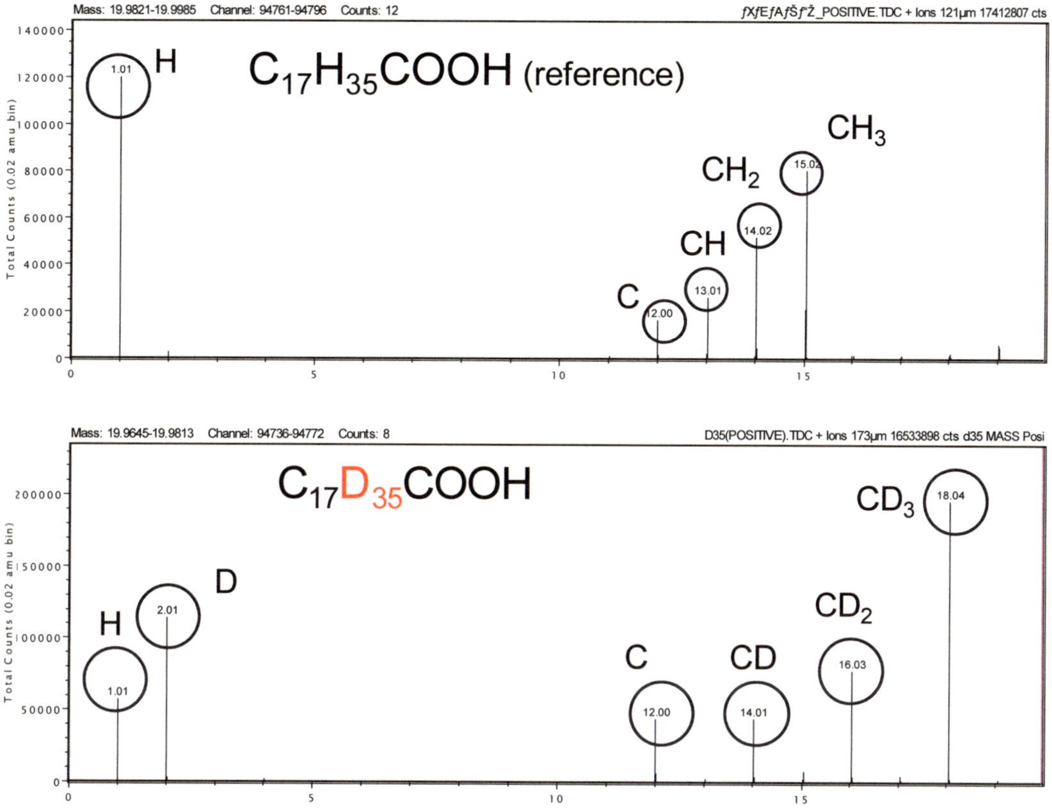

Fig. 7. SIMS spectra of OA-H35 and OA-D35 on the Si surface (m/z 0–20)

2.2 SIMS analysis of rubbed surface covered with monomolecular film [15]

A sapphire ball was slid against a monomolecular film of OA-D35 on a flat Si surface for 50 s at a load of 40 mN. Then, chemical mapping around the rubbed surface was obtained, as shown in Figure 8. The intensity of m/z 28 (Si) increased whereas that of m/z 2 (D) decreased after the tribo-test. Also, the intensity of m/z 27 (Al) increased to some extent. On the ball surface (Figure 9), the intensity of m/z 2 increased and that of m/z 27 decreased. These results can be explained by the removal of the monomolecular film from the flat surface by rubbing. This results in the exposure of Si, which is the base material of the flat surface. The transfer of OA-D35 (removed from the monomolecular film) onto the ball and the transfer of aluminum containing material (worn off from the ball) onto the flat surface took place by rubbing. Changes in surface chemistry by rubbing were also studied by quantitative analysis of the flat surface across the wear track. The intensity of m/z 2 and 28 inside the wear track are clearly different from those outside the wear track. The intensity of m/z 27 inside the wear track slightly increased.

AFM (atomic force microscope) analysis of flat surface around the wear track indicates the differences in the height between the outside and inside of the wear track is approximately 2 nm (Figure 10). Since the molecular size of OA-H35 is 1.8 nm in length, the AFM analysis supports the removal of the monomolecular film from the Si surface. There was a border

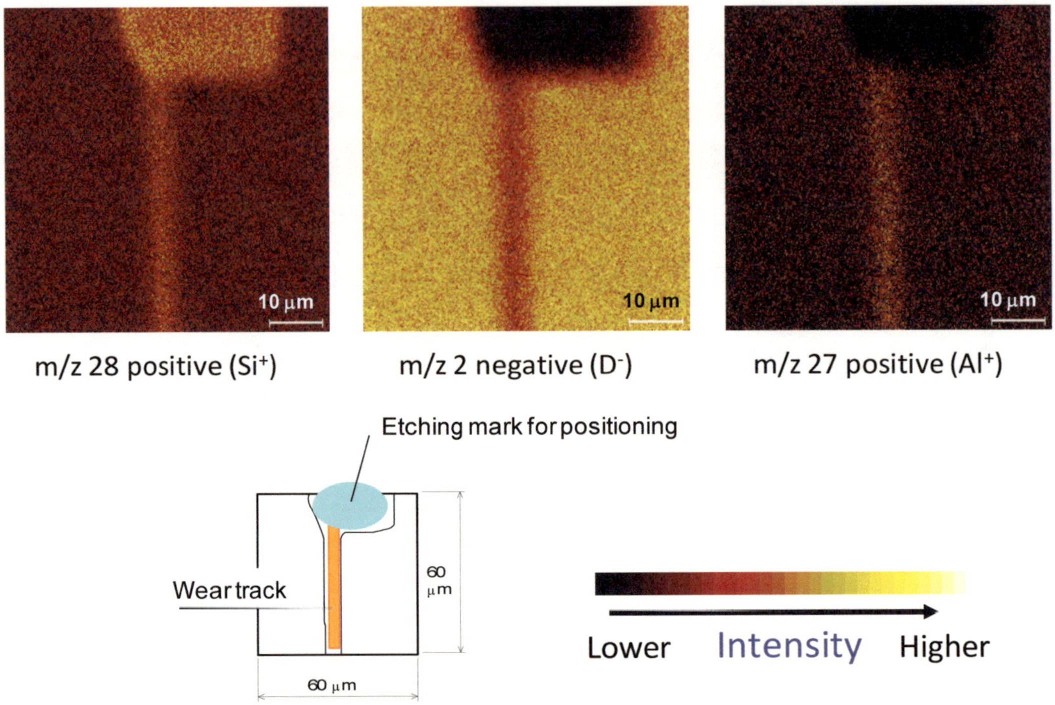

Fig. 8. Chemical mapping of the silicon-flat obtained by SIMS (rubbed against a sapphire-ball for 50 s)

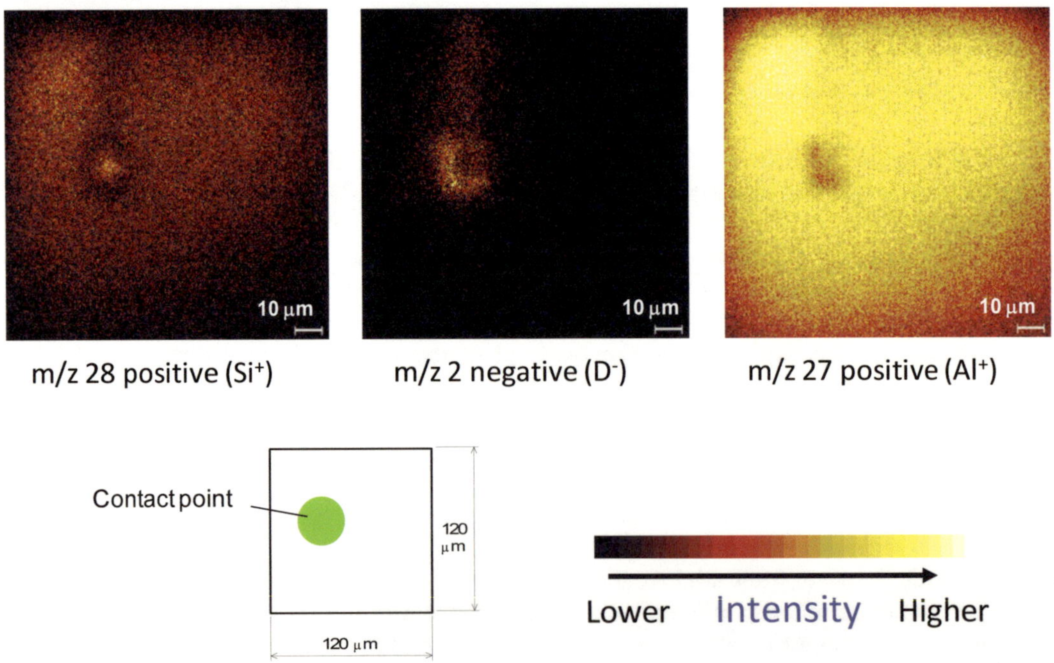

Fig. 9. Chemical mapping of the sapphire-ball obtained by SIMS (rubbed against the silicon surface for 50 s)

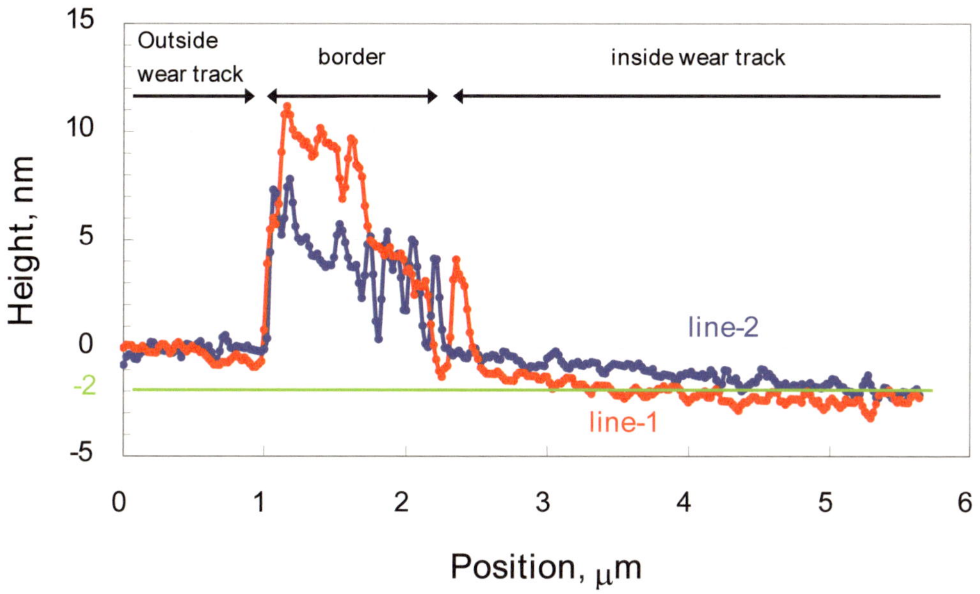

Fig. 10. Profile of the silicon-flat obtained by AFM (rubbed against a sapphire-ball for 50 s)

area between the inside and outside of the wear track. This area is characterized by its considerable height. We are greatly interested in the chemistry of the border area. However, AFM cannot detect any differences in the chemistry of the sample. Unfortunately, the spatial resolution of the SIMS apparatus employed in this work was not sufficient for the small sample area. Wear particles from the tribo-materials and removed OA-D35 might have accumulated during the tribological process. Anyway, we can detect changes in chemistry by rubbing using deuterium labeled lubricant and SIMS analysis of the resultant surface. It should be noted that evidence of wear down to the nanometer level was detected by this technique.

2.3 Influence of tribo-material on the durability of monomolecular films

Steel, sapphire, and glass balls were slid against a silicon surface covered with a monomolecular film of OA-D35. The influences of the ball material on the lifetime of the monomolecular film were observed obviously, as shown in Figure 11. The friction force of the steel ball rose within the test duration of 100 s (2000 cycles). A sapphire ball provides a longer lifetime, approximately 2,100 s. A glass ball provided an excellent lifetime of more than 10,000 s [16]. The glass ball provided a low friction coefficient of approximately 0.06 throughout the tribo-test. The results can be explained by low contact pressure (0.62 GPa for the steel ball, 0.71 GPa for the sapphire ball, and 0.47 GPa for the glass ball) and the good affinity of the carboxylic acid for the ball material. The steel ball also provided a low friction coefficient of 0.06 at the steady state, but the duration was short. The sapphire ball gave a higher friction coefficient in the range of 0.07–0.09. A slight increase in friction during the test with sapphire was observed.

The results of surface analyses suggested the role of the transferred acid onto the ball surface from silicon surface. Pretreatment of the steel ball was conducted to examine the lifetime of the monomolecular film. A fresh ball was rubbed against the monomolecular film

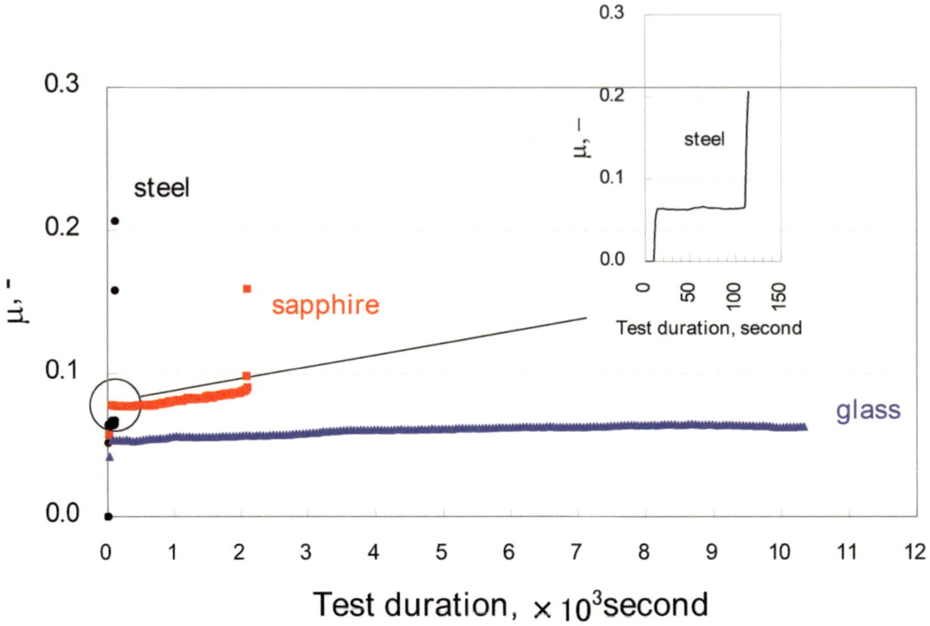

Fig. 11. Effect of the ball material on the durability of the monomolecular film

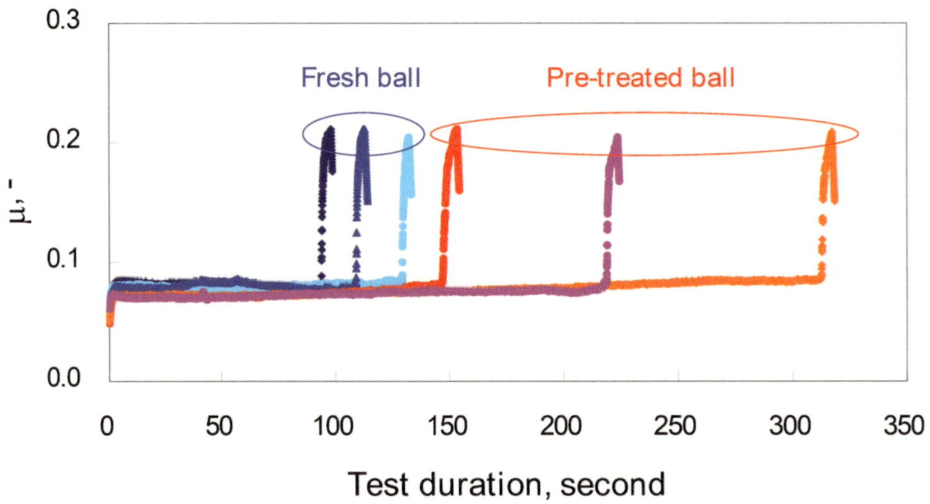

Fig. 12. Effects of pretreatment of the ball surface on the durability of the monomolecular film

deposited on the Si for 10 s (200 cycles). Since the lifetime of the combination of the steel ball and the monomolecular film was 95–130 s (Figure 12), the pretreatment was finished within the lifetime of the monomolecular film. Then, the pretreated ball was slid against a new track on the silicon surface. The pretreatment resulted in 50%–300% improvement in the lifetime [16]. Both surfaces were covered with OA-D35 if a pretreated ball was slid on a new track. This enhances the durability of the monomolecular film for lower friction. However, optimization of the pretreatment has to be investigated.

Taking these results into account, we wish to propose the following mechanism of lubrication by the monomolecular film as expressed in Figure 13. At the initial stage of the tribo-test, OA-D35 on the Si surface was removed by the mechanical stress. A transferred film composed of the acid would be generated if the removed acid adsorbed on the ball surfaces. This leads to a long lifetime of the film with low friction. On the other hand, degradation of the monomolecular film of OA-D35 followed by insufficient generation of the transferred film on the ball surface causes high friction and short lifetime in friction reduction. It was suggested that the affinity of the carboxylic acid in the ball material influences the transfer phenomena.

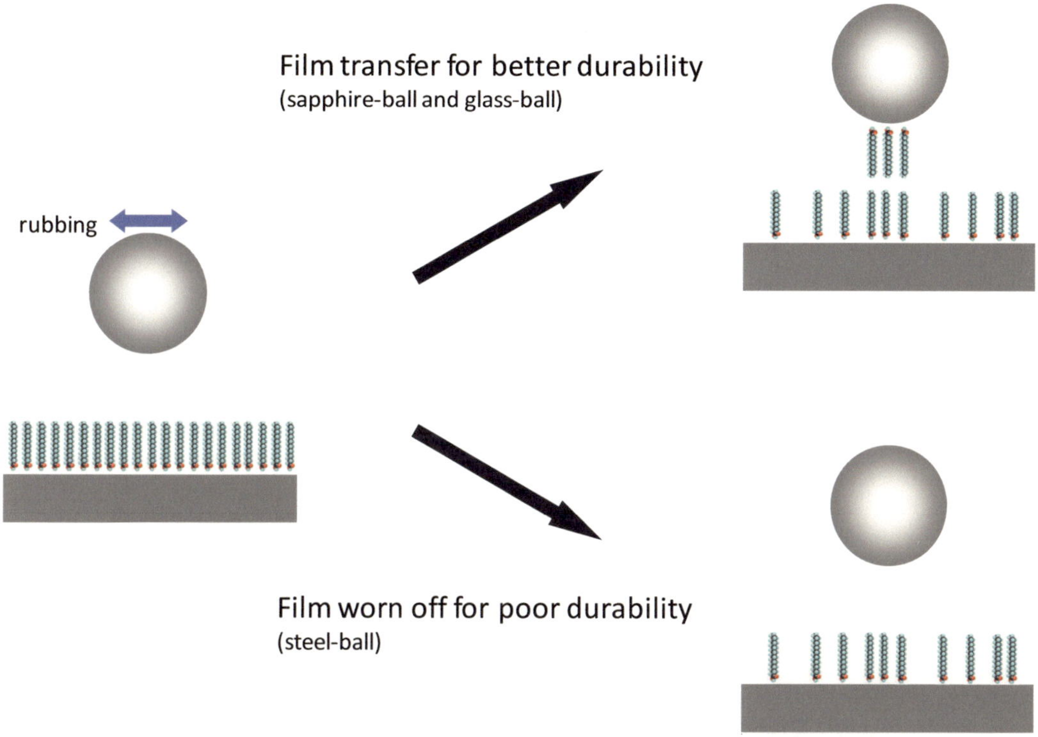

Fig. 13. A proposed mechanism for the durability of the monomolecular film

2.3 The tribological properties of a binary-component monomolecular film [11]

A binary-component monomolecular film was deposited from the equimolecular mixture of octadecylamine ($C_{18}H_{37}NH_2$, $C_{18}NH_2$) and OA-D35 or octadecyl alcohol ($C_{18}H_{37}OH$, $C_{18}OH$) and OA-D35. The tribological properties were obtained by sliding a steel ball against the film. The binary-component film with $C_{18}NH_2$ provided high friction and short durability, while that with $C_{18}OH$ provided low friction and was fairly durable, as shown in Figure 14.
Chemical mapping of the flat surface after the tribo-test for 100 s is shown in Figure 15. For the $C_{18}NH_2$ + OA-D35 binary system, the intensity of the total positive ions on the rubbed area was much greater than those of the nonrubbed area. Careful analysis of the chemical image revealed that the increase in intensity of total ions due to increases in the fragment ion of m/z 28 (Si). At the same time, decreases in the intensities of m/z 2 (D), 270 ($C_{18}NH_2$), and 318 (OA-D35) were observed. The results indicate that the binary-component film was worn off by rubbing.

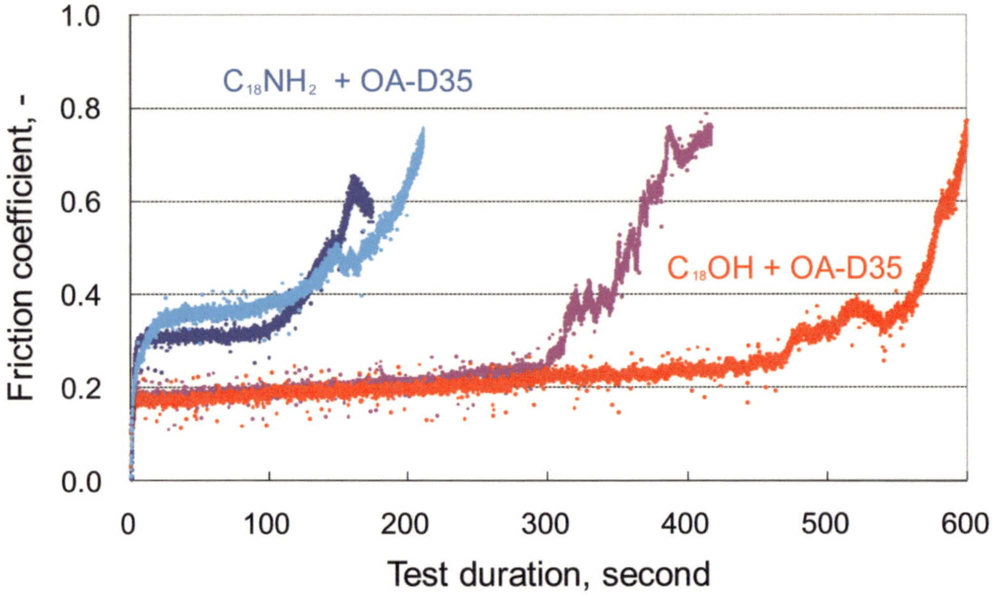

Fig. 14. Friction trace of the binary-component film during the ball-on-rotating-flat test

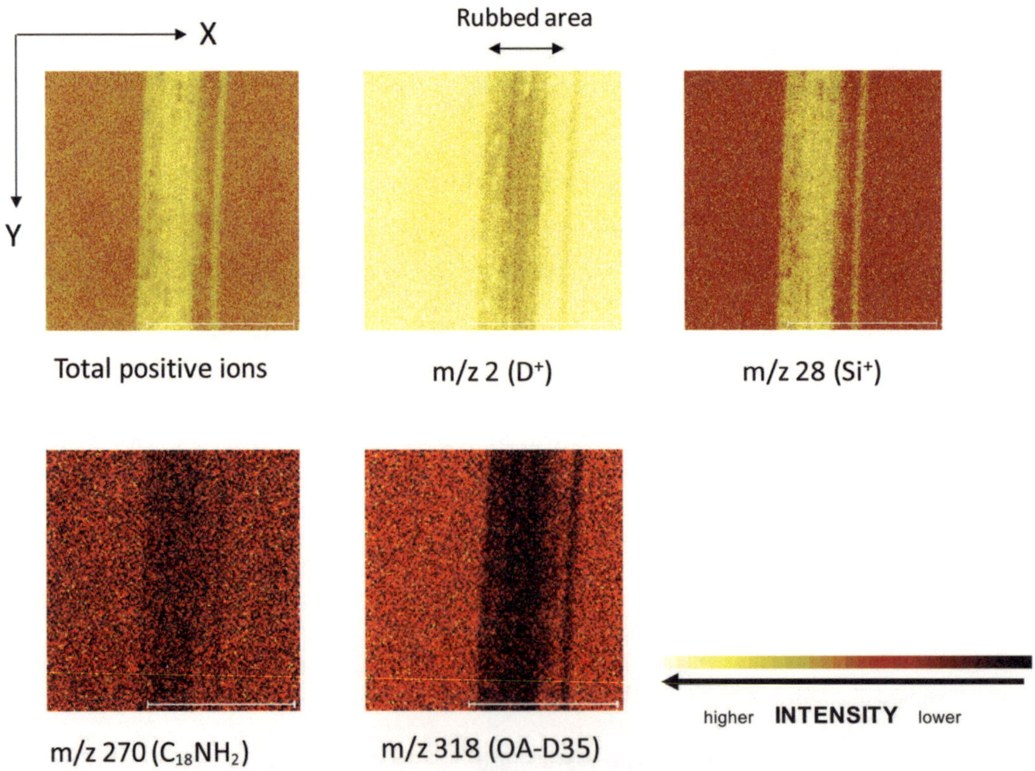

Fig. 15. Chemical mapping of the $C_{18}NH_2$ + OA-D35 binary film after the tribo-test

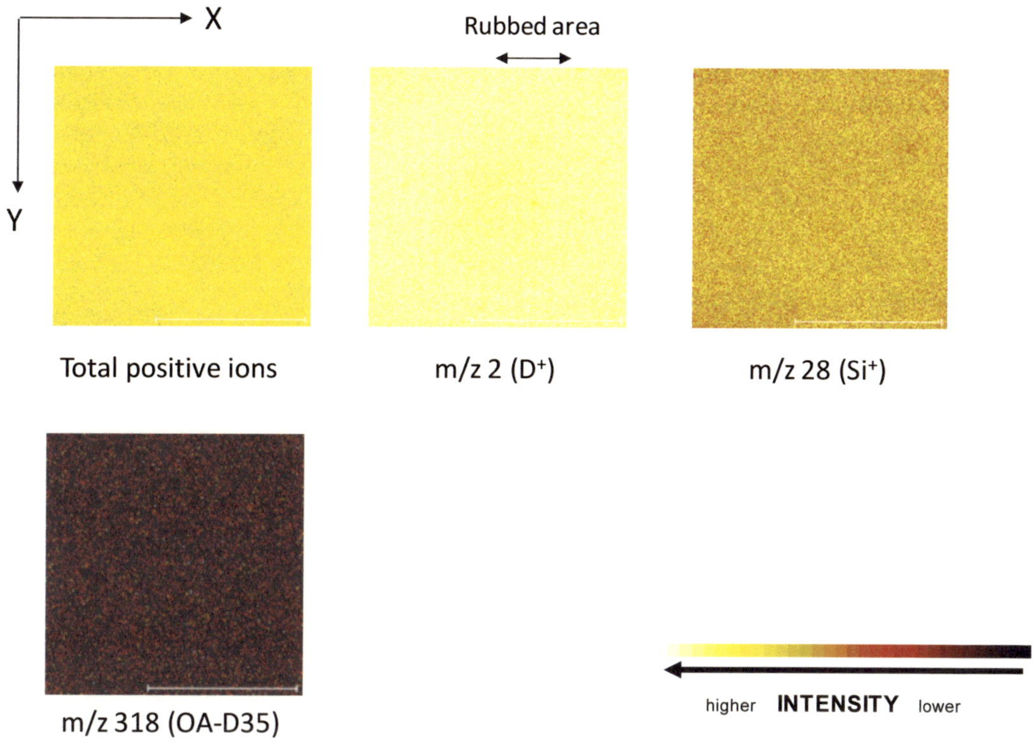

Fig. 16. Chemical mapping of the $C_{18}OH$ + OA-D35 binary film after the tribo-test

By contrast, the binary-component monomolecular film from $C_{18}OH$ + OA-D35 provided a longer lifetime with low friction. Almost no changes in chemical mapping after the tribo-test for 200 s were observed (Figure 16). The results indicate that the $C_{18}OH$ + OA-D35 binary film remained on the surface even after the tribo-test. In consequence, the tribological properties of the binary-component monomolecular film are in good agreement with the results of SIMS analysis.

An interaction between $C_{18}NH_2$ and OA-D35 through the ionic interactions is possible. This makes the adsorption force of OA-D35 on the Si surface, which retards the durability of the film. On the other hand, an interaction of $C_{18}OH$ with OA-D35 is possible through hydrogen bonding between the polar functional groups, alcohol, and carboxyl group. It should be noted that the hydrogen bond is much weaker than the ionic bond. Therefore, the adsorption force of OA-D35 was not weakened by the presence of $C_{18}OH$, which, advantageously, seems to form certain mobile phases on the surface.

2.4 Lubrication mechanism of diamond-like carbon coatings with water [17-18]

Diamond-like carbon (DLC) coatings on metallic materials possess many advantages to tribo-materials such as corrosion resistance, wear resistance, and friction reduction [19]. One of the features of using a DLC coating as a tribo-material is its applicability in humid environments or in water [20]. Furthermore, it has been reported that water improves the tribological properties of DLC. Although the formation of a boundary film by the tribo-chemical reaction of water with DLC has been suggested, a mechanistic investigation based on surface chemistry is difficult. The tribo-chemical reaction of water is supposed to provide hydrogen or oxygen to DLC surfaces. Besides water as lubricating fluid, there are other sources of hydrogen and oxygen atoms under tribological conditions. Examples include oxygen in air or metal oxides, hydrogen in organic contaminants, or DLC itself. Resources of hydrogen or oxygen could not be identified by the usual procedure, even if increments of these elements on surfaces were detected after rubbing. The stable isotopic tracer technique is expected to be powerful tool in studying the tribo-chemistry of DLC. Heavy water, such as D_2O and $H_2{}^{18}O$ were employed in this work. DLC coatings were deposited on the silicon surface by a thermal electron excited chemical vapor deposition procedure using toluene as carbon source. The resultant material was slid against a SUS 440C (JIS) stainless steel ball at a load of 10 N under a reciprocating motion at the frequency of 1 Hz. The three lubricants, H_2O, D_2O, $H_2{}^{18}O$ provided similar tribological properties, as shown in Figure 17.

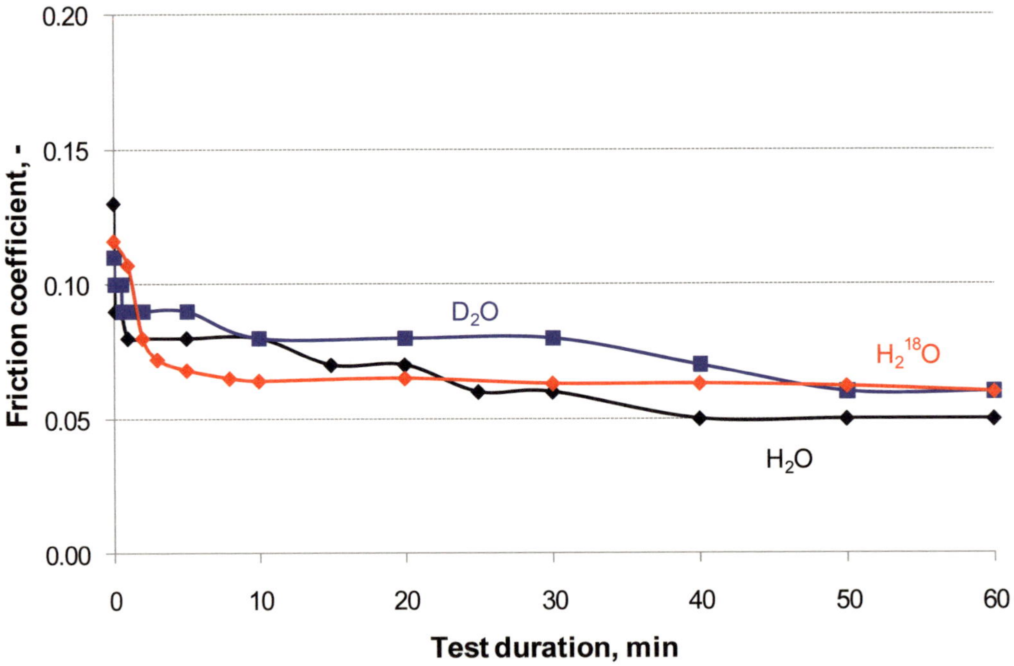

Fig. 17. Friction trace during the tribo-test with water or heavy water

Chemical mapping of the DLC surfaces obtained by SIMS analysis shows a remarkable increase in deuterium content on the rubbed surface with D_2O (Figure 18). Careful analysis of the mapping indicates the deuterium is bonded to oxygen (OD, m/z at 17) and to carbon (CD, m/z at 14). It should be noted that the fragment ion of m/z 14 could be identified as the CD moiety or CH_2 moiety. Presence of the former was supported by another considerable fragment ion of m/z 2, which corresponds to D. An increase in ^{18}O on the surfaces rubbed with $H_2{}^{18}O$ also supports the tribo-chemical reaction of water with DLC. A considerable increase in the fragment ions of m/z 18 and m/z 19, which corresponded to ^{18}O and ^{18}OH, respectively, indicates the formation of a new carbon-oxygen bond. It should be noted that the contents of ^{16}O (regular oxygen) outside the worn surface is higher than that inside worn surface. The results suggest that ^{16}O-containing compounds that existed on nonrubbed surfaces were worn off under the tribological conditions. However, identification of the ^{16}O-containing compound(s) on nonrubbed surface was difficult by the SIMS analysis.

On the basis of SIMS analysis, we wish to propose the mechanism of tribo-chemical reaction of water with DLC as expressed in Figure 19. Both homolysis and heterolysis are possible as the initial step of the reaction when DLC was exposed to mechanical stress. The former yields carbon radicals and the latter yields carbocations and carbanions as active intermediates. Then heavy water reacts with the active intermediates and results in the formation of new C–D and C–OD bonds. In summary, clear evidence of the tribo-chemical reaction of water with DLC was observed using two isotopic tracers, deuterium and ^{18}O.

2.5 Lubrication mechanism of organic friction modifier additives in hydrocarbon oils [21]

GMO (Glycerol monooleate, 2,3-dihydroxypropyl 9(Z)-octadecenoate) is known as one of the organic friction modifiers that improves the tribological properties of hydrocarbon oils. In fact, a solution of GMO in PAO (poly-alpha-olefin) dramatically reduced the friction of steel-DLC [22]. An XPS analysis of the rubbed surface with GMO-PAO indicated the presence of carbonyl compounds on the surface [23]. The results suggest that adsorption of GMO yielded the boundary film on the surface, thereby improving the tribological properties. However, the chemical resolution of XPS analysis for carbon is not sufficient for further investigations. There are two possibilities for the structure of the boundary film. One is an adsorption film of GMO itself and the other is an adsorption film of oleic acid, which is produced by the decomposition of GMO under the tribological conditions. It is difficult to distinguish between GMO and carboxylic acid by their chemical shifts of the carbonyl group.

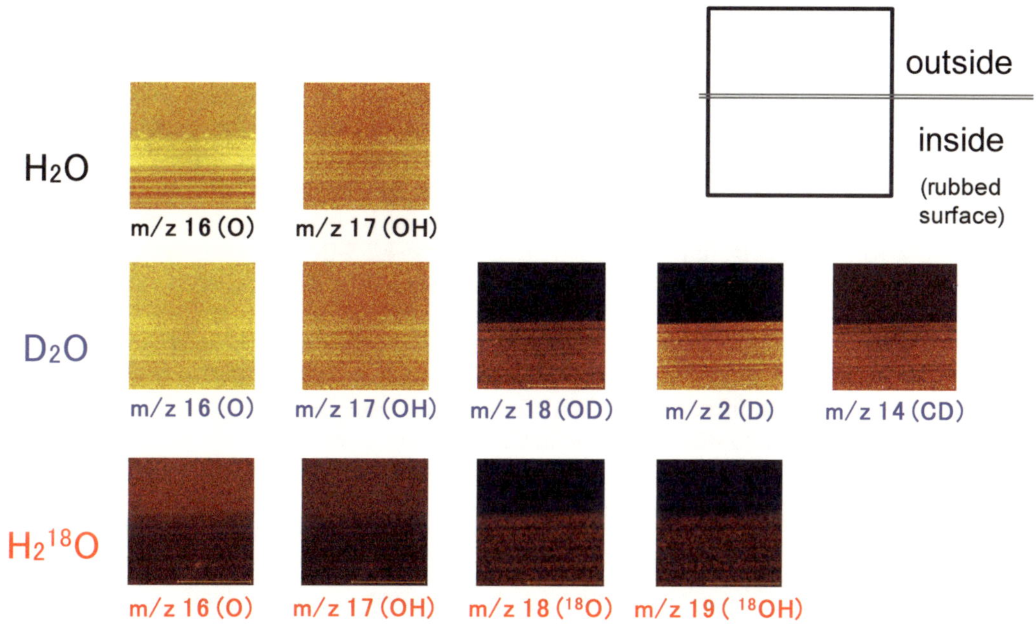

Fig. 18. Chemical mapping of DLC surface after tribo-test with water

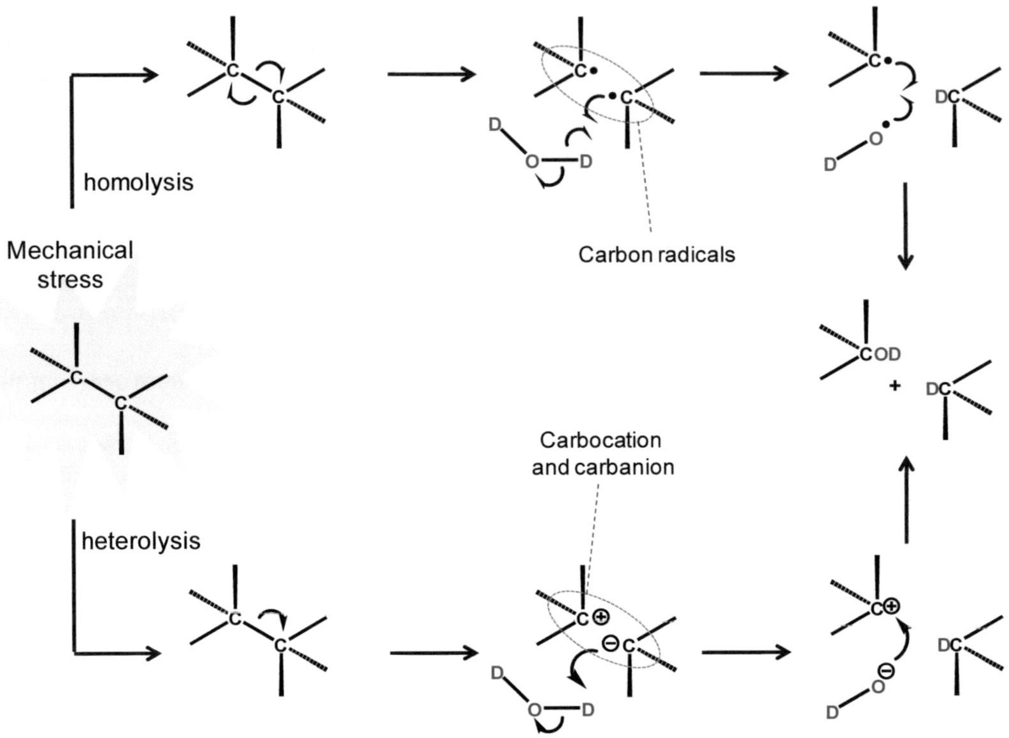

Fig. 19. Reaction mechanism of water with DLC surface under the tribological conditions

There are other ways in which carbonyl compounds are produced on the rubbed surface. An auto-oxidation of hydrocarbons usually occurs during the tribo-test under air. The reaction yields various organic oxides including carbonyl compounds. More troublesome, organic contaminants including carbonyl compounds exist everywhere. Therefore, it is difficult to eliminate carbonyl compounds derived not from GMO, but from sources other than GMO. The present stable isotopic tracer technique would clarify all of these problems. For this purpose, perdeuterio-GMO (in which all hydrogen atoms in GMO are substituted with deuterium) was desired. However, preparation of perdeuterio-GMO seemed difficult because of the availability of its precursors. Generally, GMO is prepared by the esterification of 9(Z)-octadecenoic acid (oleic acid) with propane-1,2,3-triol (glycerol). Commercially available precursors for isotope labeled GMO (where at least one atom is substituted by D or ^{13}C) are listed in Figure 20. Due to limitations in the available precursors, we were required to design isotopic labeled GMO that makes SIMS analysis effective. For this purpose, the fragmentation of GMO in SIMS analysis was considered carefully. GMO has two moieties: an alcoholic moiety with three carbons and the carbonyl moiety with 18 carbons. One of the major fragmentations of GMO is the scission of the ester bond, dividing GMO into a three-carbon moiety and 18-carbon moiety. To identify both fragments as fingerprint fragments in SIMS analysis, we have selected two precursors. One is tri-^{13}C-propane-1,2,3-triol and the other is 1-^{13}C-9(Z)-octadecenoic acid. It should be noted that the former produces the three-carbon fragment with increments of m/z 3 from the natural (unlabeled) propane-1,2,3-triol. The latter produces the 18-carbon fragment with increments of m/z 1 from the natural 9(Z)-octadecenoic acid. In this manner two labeled-GMOs, namely c-GMO and g-GMO, were prepared (Figure 21).

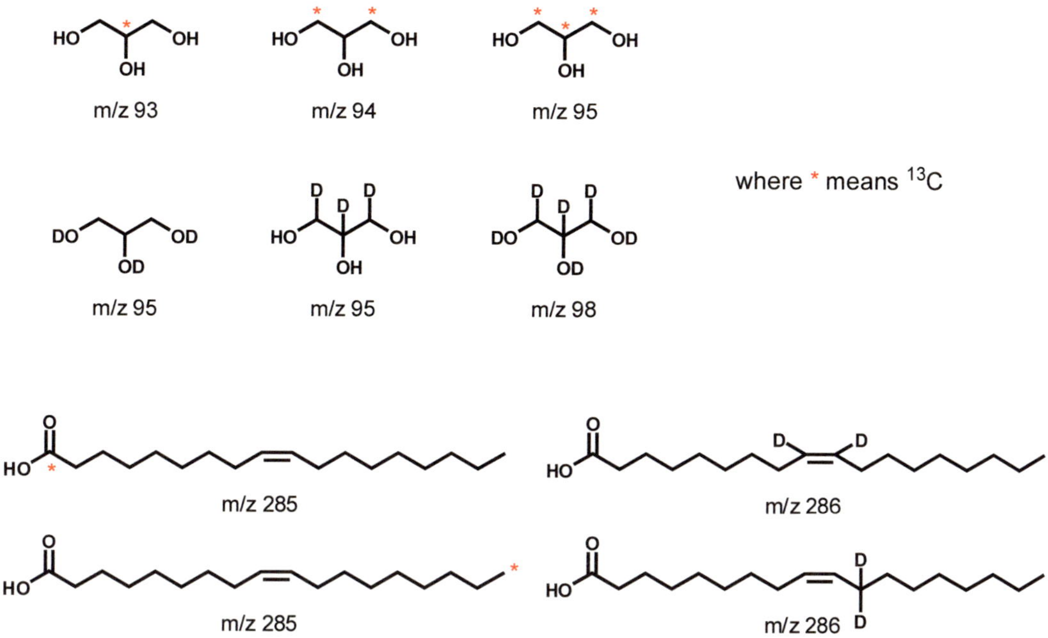

Fig. 20. List of commercially available 9(Z)-octadecenoic acid and propane-1,2,3-triol labeled with stable isotope(s)

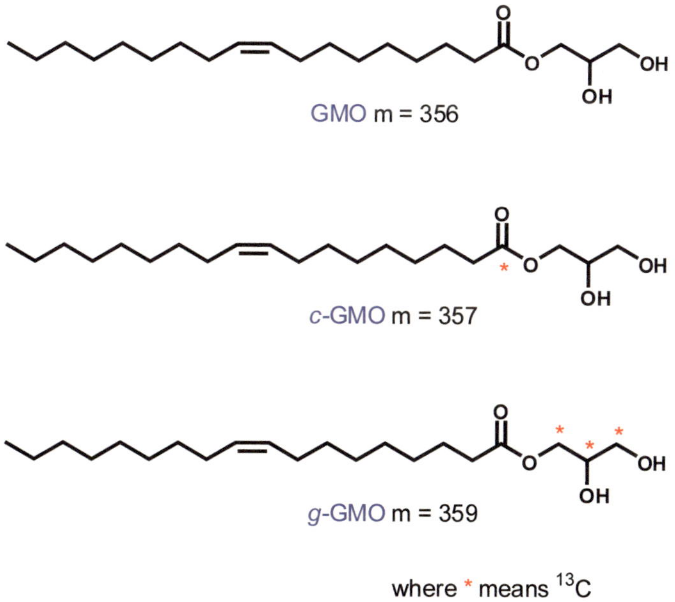

Fig. 21. Structure of labeled GMO as a model additive

We acquired mass spectra of GMOs on nonrubbed surfaces to find out their "fingerprint" fragment ions. GMO yielded fragment ions at m/z 265 and 339 in positive ion spectra and at m/z 280 in negative ion spectra. c-GMO and g-GMO gave fragment ions according to their number of ^{13}C. These fragment ions were generated by scission of the carbon-oxygen bond

in the additive molecule. Here we paid attention to the following fragments of GMO to investigate the boundary film formed on the rubbed surfaces (Figure 22).
"Fragment-A" as the acyl moiety
"Fragment-C" as the carboxyl moiety
"Fragment-E" as an ester dehydroxylated from the original molecule

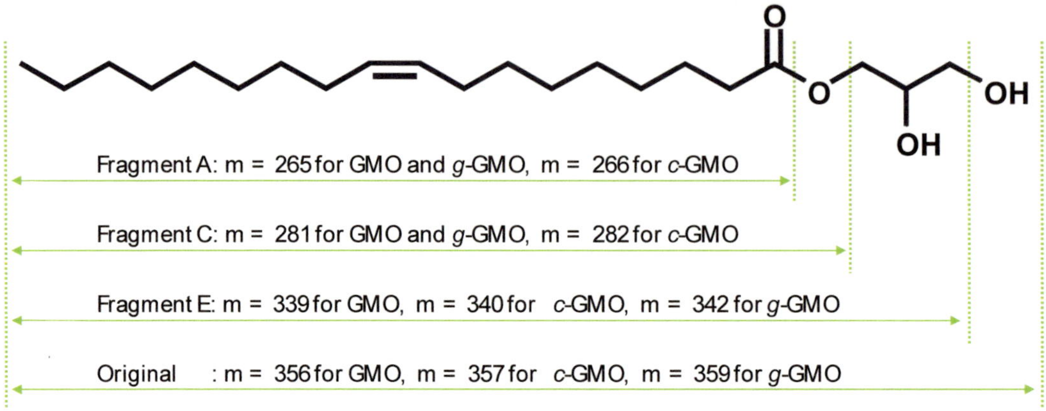

Fig. 22. Formula weight of the fragment ions from GMOs

2.6 SIMS analysis of GMO on DLC

Three GMOs were dissolved in PAO and the solutions were employed for the tribo-test. All GMOs provided similar tribological properties, as shown in Figure 23. The results indicate there are no effects on isotope(s) on the tribological properties, which is essential in the use of an isotope-labeled molecule for chemical analysis.

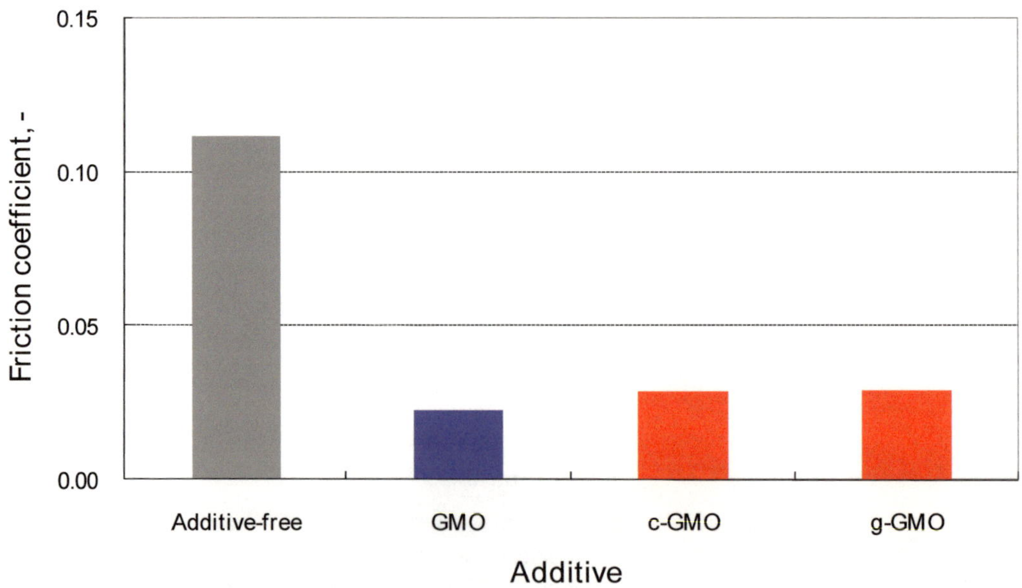

Fig. 23. Results of the tribo-test using labeled GMO in PAO

"Fragment-A" and "Fragment-E" were found in the positive mass spectra, and "Fragment-C" was found in the negative mass spectra of the rubbed surfaces (Figures 24-25). The hydrolysis of GMO yields 9Z-octadecenoic acid (oleic acid), which may adsorb on DLC surfaces. We confirmed analytically whether or not the acid exists on the surfaces. GMO and 9Z-octadecenoic acid afford "Fragment-A" and "Fragment-C." Obviously, "Fragment-E" is attributed to GMO. We compared the relative intensity of (Fragment-A)/(Fragment-E). Our hypothesis is as follows; if the relative intensity of the acid on the wear track is higher than that on the nonrubbed surface, then the acid exists on the rubbed surface. We found the same relative intensities of the fragments. Therefore, the boundary film is mainly composed of GMO as an ester, and if at all, adsorption from the acid constitutes a minor portion. These results indicate that hydroxyl group in GMO is an anchor for interactions with the DLC surfaces.

Finally, we wish to propose the contents of the boundary film that provide low friction upon steel-DLC contact. Adsorption of GMO on the rubbed and on the nonrubbed DLC was detected by SIMS analysis. It has been reported that rubbing can activate DLC surfaces, which result in the adsorption of additives [24, 20]. However, we could not find any clear

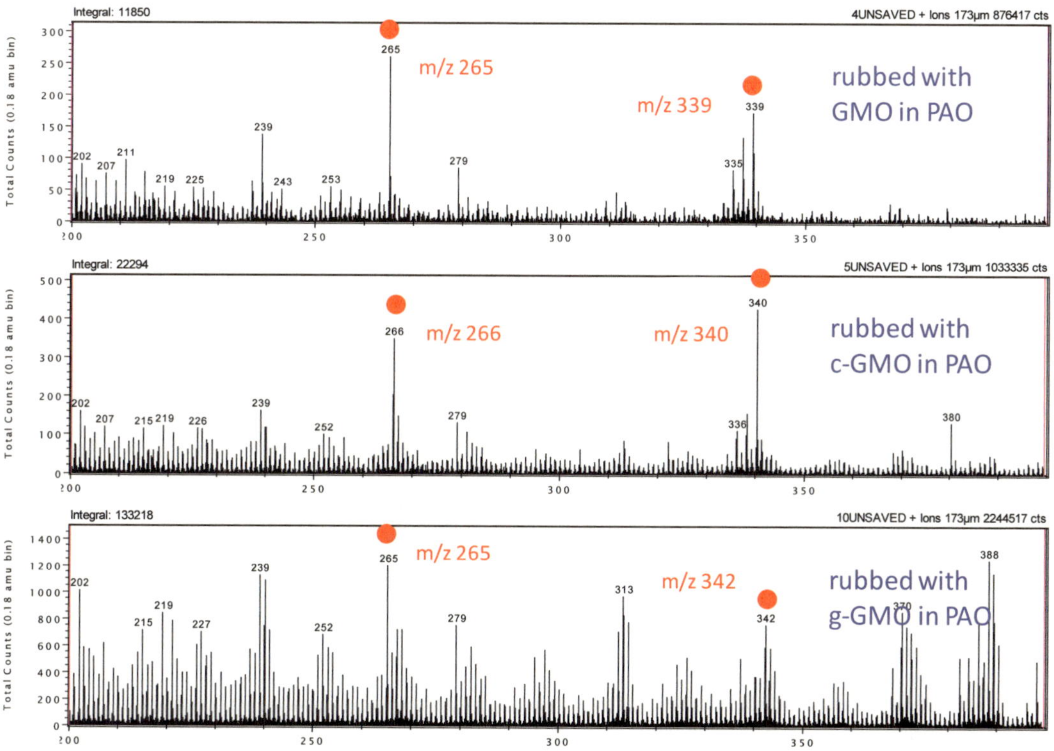

Fig. 24. Mass spectrum of wear track rubbed with GMOs (m/z 200–400, positive)

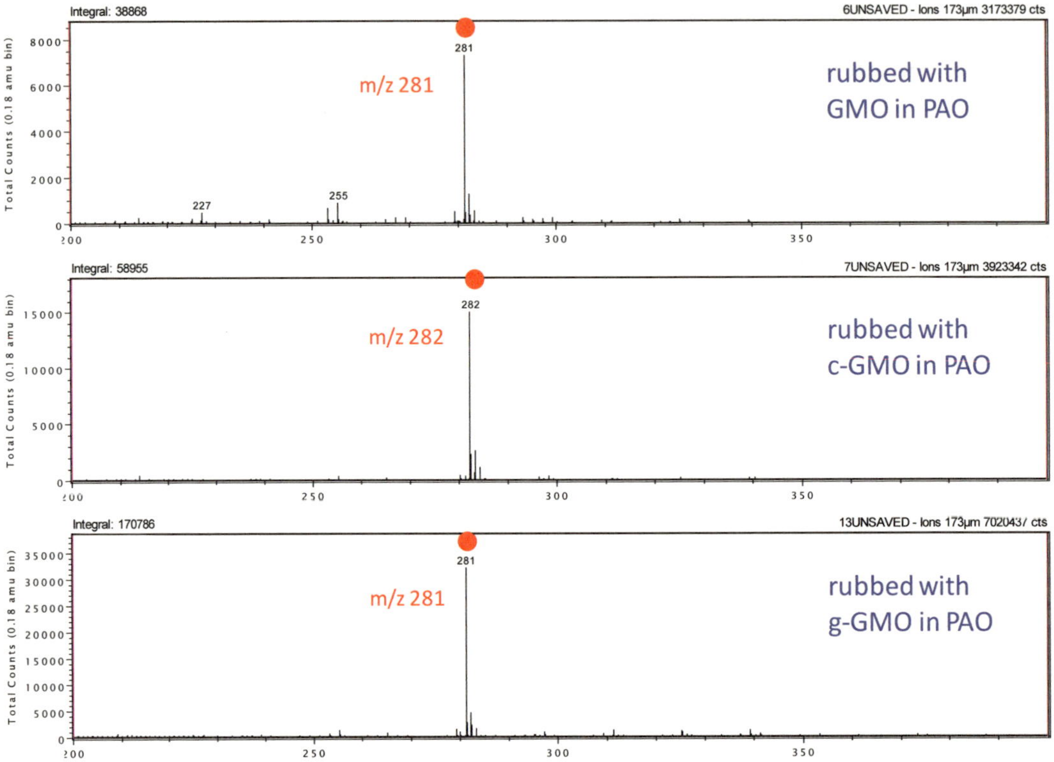

Fig. 25. Mass spectrum of wear track rubbed with GMOs (m/z 200–400, negative)

evidence for the activation of surfaces by rubbing, if the intensities of GMO on the rubbed surfaces were compared with those on the nonrubbed surfaces. Aside from "Fragment-A" and "Fragment E," several broad peaks were found in the mass spectrum. Careful analysis of the spectrum revealed that the cycle of the broad peaks is approximately m/z 14, indicating a methylene (CH_2, m/z 14) unit exists on DLC surfaces. The fragment ions are likely attributed to PAO, which is the major component of the system. A liquid clathrate-type boundary film (which involves an insertion of a branched hydrocarbon moiety in PAO into the adsorption film of GMO) was suggested [21].

3. Scope and limitations

In this article, we introduced a new technique for the investigation of tribo-chemistry that is based on stable isotopic tracers. The potential applicability of this technique would include the following three categories.

1. Degradation of preformed boundary film: Examples include monomolecular film; self-assembled mono-layers (SAMs); and any other surface treatment, including coatings.

For this purpose, isotope-labeled compounds can be employed as precursors of surface treatment.

2. The interaction of additives on the rubbed surface: The results of the binary-component monomolecular film can be applied to study synergism or antagonism for each additive in multicomponent lubricants.
3. Tribo-chemical reaction of base fluids with the surfaces of materials: Simple chemicals in the chemical structure, such as water, are applicable for this purpose at present. The main difficulties lie in availability of isotope-labeled base fluids.
4. The interaction of additive molecules with tribological surfaces: This is still an emerging technique for tracing the target molecule (usually tribo-improving additive) after the tribological process. The most challenging aspect of this technique is to detect small quantities of isotope(s) present in the system. For example, the solution of c-GMO in PAO contains approximately 0.04% of the labeled ^{13}C compared to the total number of carbons in the solution. Note that ^{13}C yields a fragment ion of m/z 13, which is the same m/z as ^{12}CH from hydrocarbons. If a considerable amount of ^{13}C exists in the system, the intensity of the fragment ion at m/z 13 should be obviously increased. The detection of these small quantities of isotopes is difficult. We have solved this problem by tracing major fragment ions derived from two or three precursors. This requires well designed model molecules based on the fragmentation of molecules during SIMS analysis. The introduction of the stable isotope(s) into the appropriate position in the molecule is essential.

It should be pointed out that the stable isotopic technique is highly suitable for organic compounds comprised of only hydrogen and carbon, and possibly also those containing oxygen and/or nitrogen atoms. For these organic compounds, the conventional surface analyses for tribology such as AES, EPMA, and XPS do not provide sufficient chemical resolution. Therefore, it is usually difficult to distinguish between the target molecule and organic contaminants. Usually there are small quantities of the target molecule on the rubbed surface. This makes the surface analysis more difficult. On the other hand, most tribo-active elements in lubricants, such as phosphorus, sulfur, molybdenum, zinc, and chlorine are well identified by the conventional surface analyses in tribology. Although AES, EPMA, and XPS can well detect the heavy elements with high sensitivity, they do not or hardly detect the light elements such as hydrogen and carbon. SIMS detects all elements if they are effectively ionized.

SIMS is a surface sensitive analysis whose analytical depth is as thin as 1–2 nm. Therefore, the sample should contain smooth surfaces to obtain optimal analytical results. The tribological process usually results in surfaces with submicrometer asperities even under a mild wear regime. The present work was achieved by employing wear resistance materials such as DLC. SIMS analysis was performed before wear occurs. This implies the technique is limited to mixed lubrication under low wear conditions. The placement of D, ^{13}C, or ^{18}O enriched atoms at the appropriate position(s) in the molecule is not always available at a reasonable cost. Therefore, isotope-labeled molecule should be designed based on their fragmentation during SIMS analysis.

SIMS approaches in tribo-chemistry using nonlabeled additives have been also achieved by detecting molecular ions or quasi-molecular ions [25-29]. However, signals corresponding to the target molecule are weak. Our isotope labeled approach usually provides clear signals with detailed analysis of the boundary film. In addition to these advantages of SIMS, the combination of multiple analytical tools is highly recommended for studying tribo-chemistry [30-31].

4. References

[1] Pawlak, Z. (2003): Tribochemistry of Lubricating Oils. Elsevier B.V., Amsterdam.
[2] Allum, K.G.; Forbes, E.S. (1968): The Load-carrying Mechanism of Organic Sulphur Compounds. Application of Electron Probe Microanalysis . ASLE Transactions 11 2: 162-175.
[3] Buckley, D.H.; Pepper, St.V. (1972): Elemental Analysis of a Friction and Wear Surface During Sliding Using Auger Spectroscopy . ASLE Transactions 15 4: 252-260.
[4] Moulder, J.F.; Stickle, W.F.; Sobol, P.F.; Bomben, K.D. (1995): Handbook of X-Ray Photoelectron Spectroscopy. Edited by Chastain, J.; King, R.C.Jr. ULVAC-PHI, Inc., Chigasaki, Japan.
[5] Kasrai, M.; Cutler, J.N.; Gore, K.; Canning, G.; Bancroft, G.M.; Tan, K.H. (1998): The Chemistry of Antiwear Films Generated by the Combination of ZDP and MoDTC Examined by X-ray Absorption Spectroscopy. Tribology Transactions 41 1: 69-77.
[6] Benninghoven, A.; Hagenhoff, B.; Niehuis E. (1993): Surface MS: probing real-world samples. Analytical Chemistry 65 14: 630-640.
[7] Vickerman, J.C.; Gilmore, I.S. Surface Analysis - The Principal Techniques, 2nd Edition (2009): John Wiley and Sons, West Sussex.
[8] Criss, R.E. (1999): Principles of Stable Isotope Distribution. Oxford University Press, New York.
[9] Sakurai, T.; Ikeda, S.; Okabe, H. (1962): The Mechanism of Reaction of Sulfur Compounds with Steel Surface During Boundary Lubrication, Using S35 as a Tracer. ASLE Transactions 5 1: 67-74.
[10] Vickerman, J.C.; Briggs, D; Henderson, A. (2007): Static SIMS Library, 4th edition. SurfaceSpectra Ltd. Manchester.
[11] Minami, I.; Kubo, T.; Fujiwara, S.; Ogasawara, Y.; Nanao, H.; Mori, S. (2005): Investigation of tribo-chemistry by means of stable isotopic tracers; TOF-SIMS analysis of Langmuir-Blodgett films and examination of their tribological properties. Tribology Letters 20 3/4: 287-297.
[12] Bowden, F. P.; Tabor, D. (2001): The Friction and Lubrication of Solids. Oxford University Press, New York.
[13] Novotny, V.; Swalen, J.D.; Rabe, J.P. (1989): Tribology of Langmuir-Blodgett Layers. Langmuir 5 2: 485-489.
[14] Nutting, G.C., Harkins, W.D. (1939): Pressure-Area relations of Fatty Acid and Alcohol Monolayers. Journal of American Chemical Society 61 : 1180-1187.
[15] Minami, I.; Furesawa, T.; Kubo, T.; Nanao, H.; Mori, S. (2008): Investigation of tribo-chemistry by means of stable isotopic tracers: Mechanism for durability of monomolecular boundary film. Tribology International 41 11: 1056-1062.

[16] Dominguez, D.D.; Mowery, R.L.; Turner, N.H. (1994): Friction and Durabilities of Well-Orderd, Close-Packed Carboxylic Acid Monolayers Deposited on Glass and Steel Surfaces by the Langmurir-Blodgett Technique. Tribology Transactions 37 1: 59-66.

[17] Wu, X.; Ohana, T.; Tanaka, A.; Kubo, T.; Nanao, H.; Minami, I.; Mori, S. (2007): Tribochemical investigation of DLC coating tested against steel in water using a stable isotopic tracer. Diamond and Related Materials 16 9: 1760-1764.

[18] Wu, X.; Ohana, T.; Tanaka, A.; Kubo, T.; Nanao, H.; Minami, I.; Mori, S. (2008): Tribochemical investigation of DLC coating in water using stable isotopic tracers. Applied Surface Science 254 11: 3397-3402.

[19] Miyoshi, K.; Street, K.W.Jr. (2004): Novel carbons in tribology. Tribology International 37: 865-868.

[20] Li, H.; Xu, T.; Wang, C.; Chen, J.; Zhou, H.; Liu, H. (2005): Tribochemical effects on the friction and wear behaviors of diamond-like carbon film under high relative humidity condition. Tribology Letters 19 3: 231-238.

[21] Minami, I.; Kubo, T.; Nanao, H.; Mori, S.; Okuda, S.; Sagawa, T. (2007): Investigation of Tribo-Chemistry by Means of Stable Isotopic Tracers, Part 2: Lubrication Mechanism of Friction Modifiers on Diamond-Like Carbon. Tribology Transactions 50 4: 477-487.

[22] Kano, M. (2006): Super low friction of DLC applied to engine cam follower lubricated with ester-containing oil. Tribology International 39: 1682-1685.

[23] Ye, J.; Okamoto, Y.; Yasuda, Y. (2008): Direct Insight into Near-frictionless Behavior Displayed by Diamond-like Carbon Coatings in Lubricants. Tribology Letters 29 1: 53-56.

[24] Kano, M.; Yasuda, Y.; Okamoto, Y.; Mabuchi, Y.; Hamada, T.; Ueno, T.; Ye, J.; Konishi, S.; Takeshima, S.; Martin, J.M.; Bouchet, M.I. De Barros; Le Mogne, T. (2005): Ultralow friction of DLC in presence of glycerol mono-oleate (GMO). Tribology Letters 18 2: 245-251.

[25] Dauchot, G.; Castro, E. De; Repoux, M.; Combarieu, R.; Montmitonnet, P.; Delamare, F. (2006): Application of ToF-SIMS surface analysis to tribochemistry in metal forming processes. Wear 260: 296-304.

[26] Wiltord, F.; Martin, J.M.; Le Mogne, T.; Jarnias, F.; Querry, M.; Vergne, P. (2006): Reaction mechanisms of alcohols on aluminum surfaces. Tribologia 37 6: 7-20.

[27] Tohyama, M.; Ohmori, T.; Murase, A. Masuko, M. (2009): Friction reducing effect of multiply adsorptive organic polymer. Tribology International 42 6: 926-933.

[28] Gunsta, U.; Zabelb, W. R.; Vallec, N.; Migeonc, H.-N.; Pollb, G.; Arlinghausa, H.F. (2010): Investigation of laminated fabric cages used in rolling bearings by ToF-SIMS. Tribology International 43 5-6: 1005-1011.

[29] Reichelt, M.; Gunst, U.; Wolf, T.; Mayer, J.; Arlinghaus, H.F.; Gold, P.W. (2010): Nanoindentation, TEM and ToF-SIMS studies of the tribological layer system of cylindrical roller thrust bearings lubricated with different oil additive formulations. Wear 268 11-12: 1205-1213.

[30] Georges, J.M.; Martin, J.M.; Mathia, T.; Kapsa, P.H.; Meille, G.; Montes, H. (1979): Mechanism of boundary lubrication with zinc dithiophosphate. Wear 53 1: 9-34.

[31] Minfray, C.; Martin, J.M.; Esnouf, C.; Le Mogne, T.; Kersting, R.; Hagenhoff, B. (2004): A multi-technique approach of tribofilm characterization. Thin Solid Films 447-448: 272-277.

22

FEM Applied to Hydrodynamic Bearing Design

Fabrizio Stefani
University of Genoa
Department of Mechanics and Machine Design
Italy

1. Introduction

Nowadays hydrodynamic lubrication analysis involves sophisticated models that use a large number of variables. For instance the evaluation of temperatures, which directly determine the viscosity of the lubricant fluid and hence its load carrying capacity, has become a standard procedure and the related analysis type is referred to as *ThermoHydroDynamic* (THD) analysis. *ThermoElastoHydroDynamic* (TEHD) models introduce a further enhancement in lubrication analysis by including the simulation of bearing deformations due to mechanical loads and/or thermal effects.

TEHD lubrication is an interdisciplinary field including structural, thermal, thermo-elastic and hydrodynamic simulations. Hence it requires a multi-physic approach that should be based on a well-rounded discretization technique capable of simplifying the simultaneous management of different interconnected models, which must exchange data between themselves. This task is well-accomplished by the Finite-Element Method (FEM), an overall discretization technique particularly suited for problems with complicated integration domains and non-smooth solutions.

A TEHD model of a kinematic pair simulates the thermal and mechanic interaction among the lubricant film and the lubricated solid members. As far as the fluid film sub-model is concerned, it must rely on the mass and energy conservation principles. The separation of the fluid (cavitation) in the divergent film region and the presence of feed grooves on bearing surfaces encumber the formulation of the conservation principles especially in the finite-element perspective and require special modelling techniques.

The present chapter is aimed to provide the theoretical foundation of FEM mass- and energy-conserving models as well as to report their application to the THD and/or TEHD analysis of different bearing types. Author's original contributions to the simulation methods are explained. They include the FEM groove-mixing theory, the SUPG stabilization of the conservation equations and the "quasi-3D" approach to the thermal problem. The theoretical construct is useful to enable the analysts to manage the models and to understand the responses. The application examples are relevant to both journal and axial bearings with fixed and tilting pads, in order to demonstrate the high flexibility of the method. With few modifications the presented method can be applied to the design of several types of bearings in both steady and dynamic loading conditions. The scope of the paper is anyway limited to the analysis of steadily-loaded bearings working in laminar lubrication regime.

2. State-of-the-art

Modern lubrication analysis methods enable us to assess bearing performances with high accuracy. By taking advantage of detailed THD simulations the maximum deviation between experimental findings and numerical predictions for white metal temperatures may be less than 3-4°C (Banwait & Chandrawat, 1998).
TEHD analysis is compulsory in order to achieve sufficiently reliable results for highly loaded journal bearings (Bouyer & Fillon, 2004), tilting pad journal bearings (Chang et al., 2002), thrust bearings (Brugier & Pasal, 1989) and dynamically loaded supports (i.e. big end bearings of connecting rods for automotive engines) (Piffeteau et al., 2000). TEHD analysis may also be convenient in the case of journal bearings with stiff housing (i.e. for turbomachineries) in order to avoid assumptions about the effective clearance in working conditions.
FEM is more and more often used in lubrication analysis (Booker & Huebner, 1972; Bonneau & Hajjam, 2001). A FEM version of the classic groove-mixing theory (Robinson & Cameron, 1975) is explained in the following. It has been developed by formulating the energy balance for the supply grooves at the element level, in order to deal with all of the lubrication problem details in finite-element terms.
Suitable stabilization techniques are compulsory in lubrication analysis to solve by means of FEM the energy and the cavitation equations, whereas they rule parabolic and hyperbolic differential problems, respectively. The Streamline Upwind Petrov-Galerkin (SUPG) technique (Kelly et al., 1980; Tezduyar & Sunil, 2003), applied by the author to both problems, is fully explained in the following. Although more straightforward upwinding techniques have been initially proposed by other authors (Kumar & Booker, 1991), SUPG is more general, as it does not depend on the element type.
As convection is the main mechanism of heat exchange in the lubricant film, oil temperatures and flows are directly related. Hence a consistent treatment of the thermal problem demands an equally reliable model of film hydrodynamics.
In this perspective the FEM mass-conserving algorithms developed in the last decade by researchers (Kumar & Booker, 1991; Bonneau & Hajjam, 2001) are an essential tool to provide the accurate estimate of the lubricant flow needed by THD and TEHD analysis methods. A mass- and energy-conserving FEM model has been presented by Kumar & Booker (1994). The resulting algorithm is fast as it turns the three-dimensional (3D) thermal problem in a two-dimensional (2D) one by solving the energy equation averaged across the film thickness and by assuming adiabatic walls. Afterwards such a method has been enhanced in a previous work (Stefani & Rebora, 2009), where it is incorporated in a complete 3D TEHD simulation and it is completed with boundary conditions consistent with the continuity of mass and energy throughout the integration domain.
To this purpose the temperature variation across the film thickness is calculated by fitting the temperature profile with a fourth-order polynomial (quasi-3D approach) and the above-mentioned groove-mixing theory is employed. In such an arrangement the algorithm has shown good agreement with experimental results. The computational cost is still reasonable and the algorithm is very flexible and well-suited to different bearing types and geometries. Consequently the model can serve as the basis for codes dedicated to bearing design and verification for industrial purposes.

3. Basic equations

3.1 Thin film mechanics equations

In kinematic pairs working in hydrodynamic and elastohydrodynamic lubrication regime, the lubricant action is exerted through a thin film between two members (with facing surfaces 0 and 1) in motion at velocity $\mathbf{V}_0$ and $\mathbf{V}_1$, respectively (Fig. 1).
The following usual "Reynolds hypotheses" are assumed. The lubricant is Newtonian and it flows in laminar regime in the narrow clearance between the two members, with no-slip conditions at the walls. The film curvature yields negligible effects.

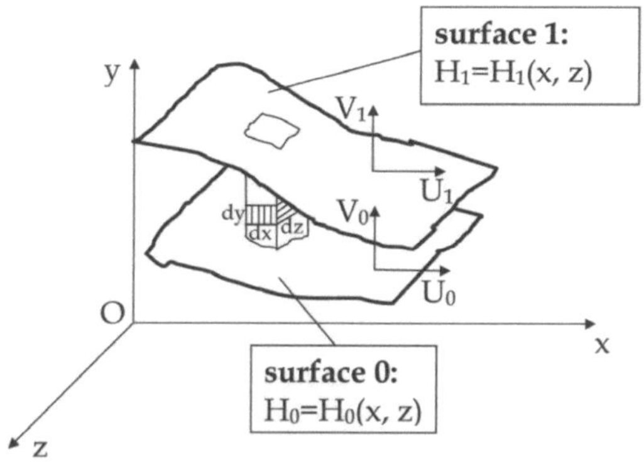

Fig. 1. Reference system O(x, y, z) and flow between the two surfaces 0 and 1 moving with velocity $\mathbf{V}_0$={U_0, V_0, 0} and $\mathbf{V}_1$={U_1, V_1, 0}

The velocity distributions of the fluid into the film thickness can be obtained from the Navier-Stokes equations, simplified by means of the above-mentioned assumptions and integrated with suitable boundary conditions (the fluid velocity at the surfaces 0 and 1 are $\mathbf{V}_0$ and $\mathbf{V}_1$, respectively). After velocity is substituted in the continuity equation, its integration in a columnar element of fluid (Fig. 1) with height $H = H_1 - H_0$ and rectangular basis $dx \cdot dz$ provides the thin film mechanics equation, also referred to as Reynolds generalized equation, (Dowson, 1967). At time t, for a full film of compressible fluid developed in a rectangular domain, where a cartesian coordinate system O(x, y, z) is fixed, it states

$$\frac{\partial}{\partial x}\left(\rho g \frac{\partial p}{\partial x}\right)+\frac{\partial}{\partial z}\left(\rho g \frac{\partial p}{\partial z}\right)=\frac{\partial}{\partial x}\left\{\rho\left[U_1 H_1 - U_0 H_0 - f\left(U_1 - U_0\right)\right]\right\}+\frac{\partial}{\partial t}(\rho H) \qquad (1)$$

where the hydrodynamic pressure p as well as the fluid density ρ are independent of the y coordinate and

$$\left.\begin{aligned} f &= i_1 / i_0 \\ g &= i_2 - i_1^2 / i_0 \end{aligned}\right\} \qquad (2)$$

with

$$i_s = \int_{H_0}^{H_1} \frac{y^s}{\mu}\, dy \qquad (3)$$

The lubricant viscosity μ (eq. (3)) is variable along the film thickness (with the y coordinate) as well as in the bearing surface (with the x and z coordinates).
Although equation (1) is time dependent, it is not referred to as the equation of the thin film dynamics, as inertia and volume forces are negligible in a thin film. Hence the reference frame O(x, y, z) can be chosen regardless of whether it is inertial or not.
In order to express eq. (1) in a form useful for bearing analysis, the plane y=0 is assumed to lie on the surface 0. Therefore equations of surfaces 0 and 1 become $H_0 = 0$ and $H_1 = H$, respectively. In addition, let O(x, y, z) be fixed to surface 0, namely surface 0 is steady in this reference frame. Hence, if U is the relative velocity between surface 1 and surface 0, the kinematic terms in eq. (1) are $U_0=0$ and $U_1=U$.
These assumptions simplify the thin film mechanics equation as follows

$$\frac{\partial}{\partial x}\left(\rho g\frac{\partial p}{\partial x}\right)+\frac{\partial}{\partial z}\left(\rho g\frac{\partial p}{\partial z}\right)=\frac{\partial}{\partial x}\left[\rho U(H-f)\right]+\frac{\partial}{\partial t}(\rho H) \tag{4}$$

Equation (4) may be used for analysis of journal bearings, when their geometry and operating conditions enable Reynolds hypotheses to be fulfilled. As curvature effects are neglected the x axis is set in circumferential direction (x = R ϑ, where R is the shaft radius).
In order to avoid the simulation of moving grooves, the surface where feed holes and/or lubricant supply grooves are machined is chosen as surface 0. Usually surface 0 and surface 1 lie on the bush and the journal respectively, as in the case of rotor journal bearings, submitted to steady loads and fed through suitable grooves in the bushing. In big end bearings of connecting rods for internal combustion engines, working under dynamic loads, the feed holes are machined in the crankshaft. Hence the sleeve wall may be chosen as surface 0 only if it houses a circumferential groove. Otherwise surface 0 is chosen on the journal, and U becomes the velocity of the bearing with respect to the journal. Nevertheless, as this chapter focus more specifically on rotor bearings, the reference surface 0 for radial bearings will be always on the sleeve in the following. Consequently, if ω is the shaft rotation speed (with respect to the bearing), U = ω R.
The relative surface velocity U (or ω) may either depend or not depend on the x (or ϑ) coordinate as for journal bearings submitted to either dynamic or steady loads, respectively. In the former case such a dependency yields higher order infinitesimal in eq. (4) and it can be neglected in the simulation of the thin film mechanics.
The resulting form of the thin film mechanics equation for journal bearings is

$$\frac{1}{R^2}\frac{\partial}{\partial \vartheta}\left(\rho g\frac{\partial p}{\partial \vartheta}\right)+\frac{\partial}{\partial z}\left(\rho g\frac{\partial p}{\partial z}\right)=\omega\frac{\partial}{\partial \vartheta}\left[\rho(H-f)\right]+\frac{\partial}{\partial t}(\rho H) \tag{5}$$

In the annular domain, where the lubricant film develops in the case of thrust bearings, the coordinate frame used to locate a generic point Q (Fig. 3) is the cylindrical coordinate system O(r, y, ϑ). An analogous integration of the Navier-Stokes and continuity equations in the reference frame O(r, y, ϑ), leads to

$$\frac{1}{r}\frac{\partial}{\partial \vartheta}\left(\frac{\rho g}{r}\frac{\partial p}{\partial \vartheta}\right)+\frac{1}{r}\frac{\partial}{\partial r}\left(\rho g r\frac{\partial p}{\partial r}\right)=\omega\frac{\partial}{\partial \vartheta}\left[\rho(H-f)\right]+\frac{\partial}{\partial t}(\rho H) \tag{6}$$

By taking advantage of a conformal mapping technique (Wang et al., 2003), in agreement with the coordinate transformation z = R ln(r/R) (R is the inner pad radius, shown in Fig. 3), the end face of the thrust bearing can be transformed from its annular (physical) domain to a rectangular (computational) domain. Accordingly, by substituting r=R exp(z/R) in equation (6), it is turned into the following form

$$\frac{1}{\gamma^2}\left[\nabla \cdot (\rho g \nabla p)\right] = \omega \frac{\partial}{\partial \vartheta}\left[\rho (H - f)\right] + \frac{\partial}{\partial t}(\rho H) \tag{7}$$

where $\nabla=\{\partial/(R\ \partial\vartheta), \partial/\partial z\}=\{\partial/\partial x, \partial/\partial z\}$ is the gradient operator and $\gamma=\gamma\ (z)=\exp(z/R)$ is the conformal mapping operator. In the computational domain (ϑ, z), for $\gamma=1$ equation (7) is the same as the thin film mechanics equation for journal bearings (eq. (5)). Hence equation (7) is the universal thin film mechanics equation for journal and thrust bearings, provided that $\gamma=1$ for journal bearings and $\gamma=\exp(z/R)$ for thrust bearings. If the fluid viscosity is considered constant across the film thickness and equal to the local mean viscosity (the one calculated at the cross-film averaged temperature), equations (2) can be easily integrated

$$\left.\begin{aligned} f &= H/2 \\ g &= H^3/(12\mu) \end{aligned}\right\} \tag{8}$$

By means of substitution of eq. (8) into eq. (7), the universal Reynolds equation for journal and thrust bearings is obtained

$$\frac{1}{\gamma^2}\left[\nabla \cdot \left(\rho \frac{H^3}{12\mu} \nabla p\right)\right] = \frac{\omega}{2}\frac{\partial}{\partial \vartheta}(\rho H) + \frac{\partial}{\partial t}(\rho H) \tag{9}$$

3.2 Film thickness equations

As TEHD models take into account the deformations (due to both mechanical and thermal actions) of the two members of the pair, in order to calculate the film thickness H, the relative displacement of the two facing surfaces must be known.

Let d_i be the displacement of a point on the surface i (i = 0, 1) in the normal direction (roughly the y direction for both the walls). Such a direction becomes radial in the case of a journal bearing, due to the curvature of the x axis. For very compliant bearings (i.e. for connecting rod applications) d_i is the radial component of the displacement deprived of the rod rigid body motion, i.e., the mean displacement among points located on the sleeve surface. As explained in the previous paragraph, the reference frame is put on surface 0 and, precisely, in its real (deformed) configuration. This clarification implies that the equations of surface 0 and 1 can be expressed respectively by $H_0 = 0$ and $H_1 = H = h + d_1 - d_0$, where h is the ideal film thickness measured between the surfaces in undeformed state. The present paragraph deals with the assessment of the ideal film thickness h, while thermoelastic displacements d_i are focused in paragraph 5.5.

Starting from a cylindrical (complete) journal bearing (Fig. 2) with no misalignment, the classical expression for the ideal film thickness, obtained by neglecting higher order infinitesimal terms, is $h = c_b + e \cdot \cos\theta$, where c_b is the small radial clearance and e is the journal center eccentricity O_bO_j (the norm of the vector **e**). As it evaluates h in the reference system O′(x′, y′) fixed to the center-line (the dash-dotted line in Fig. 2) that moves together with the journal, the above-mentioned classical expression is not suitable to deal with TEHD analysis by means of FEM. Indeed, structural models can only evaluate the displacements of discrete points (nodes) on the bearing surface localized in a reference frame fixed to the bush. Hence the ideal film thickness equation must be also referred to a coordinate system fixed to the bearing, i.e. the reference frame O(x, y) shown in Fig. 2, by means of the following equation

$$h = c_b + e_X \cos\vartheta + e_Y \sin\vartheta \tag{10}$$

where the journal center location is given by the Cartesian coordinates (e_X, e_Y) in the reference frame O_b(X, Y) fixed at the geometrical center of the shell.

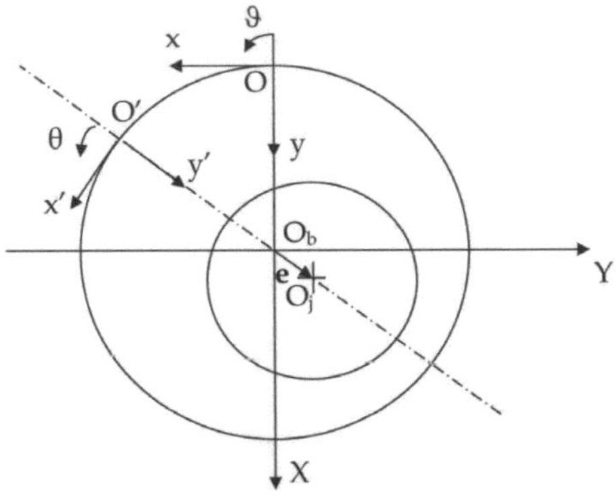

Fig. 2. Cylindrical pair (complete journal bearing) with ideal (rigid) members and reference systems

In a tilting pad thrust bearing assembly (Fig. 3), the geometrical center overlaps the origin of the reference system O(r, y, ϑ) used for the thin film mechanics equation (6). Hence O(X, Z) and its polar counterpart O(r, ϑ) are the references employed to measure the coordinates that rule the relative position of the assembly members. By moving their origin in the pivot

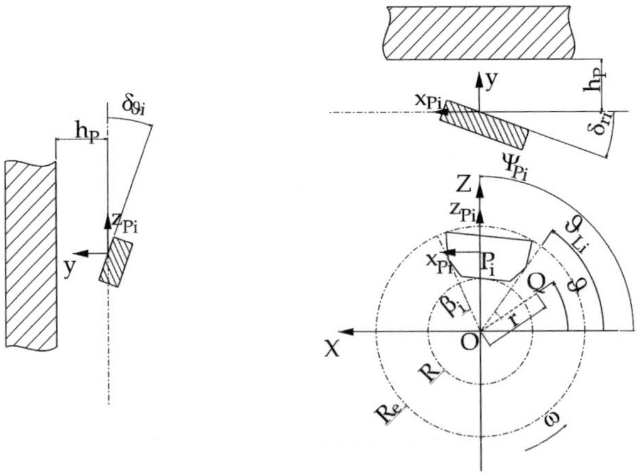

Fig. 3. Tilting pad-collar pair (thrust bearing) with ideal (rigid) members and reference systems

P_i of the ith pad, the reference frame $P_i(x_{Pi}, z_{Pi})$ shown in Fig. 3 is obtained (the z_{Pi} axis is oriented in radial direction), in order to easily express the ideal film thickness at pad i as

$$h_i = h_p - \delta_{ri} r \sin(\vartheta - \psi_{Pi}) + \delta_{\vartheta i}\left[r\cos(\vartheta - \psi_{Pi}) - r_{Pi}\right] \tag{11}$$

where the coordinate pair (r_{Pi}, ψ_{Pi}) gives the polar location of the pivot that supports the ith pad, δ_{ri} and $\delta_{\vartheta i}$ are the tilt angles of pad i around the radial axis z_{Pi} and the tangential axis x_{Pi} respectively (for line-contact pivots $\delta_{\vartheta i} = 0$), h_P is the film thickness at pivot.
By substituting $r = \gamma(z)$ R and/or $\vartheta = x/R$ in eqs. (10) and (11), the relevant ideal film thickness expressions in the mapped (computational) thin film domain O(x, z) are obtained.

3.3 Kinematic pairs motion equations

When the time history of the external load acting on the bearing is given instead of the relative position of the pair members, the resulting problem is referred to as *indirect problem* instead of *direct problem*. The indirect problem requires more relations than the sole thin film mechanics equation in order to determine the pressure field evolution, once the viscosity distribution is known by solving the energy equation. The additional relations are the equations of motion for the (moving) members of the kinematic pair.
One of the most effective iterative methods for solving the indirect problem (the coupled thin film mechanics and motion equations with the relevant initial and boundary conditions) is based on the Newton-Raphson procedure and it is explained, for different type of bearings, in many papers (i.e.: Chang et al., 2002).
Steadily loaded bearings are analyzed by means of the same method as dynamically loaded ones, whereas the mass-conserving approach (paragraph 4.1) retains the transient terms of eq. (7). In such a case, the simulated transitory evolving from an arbitrary initial condition is not meaningful, and only the steady conditions, reached after a sufficient number of time steps, are considered simulation results.
Unfortunately the resort to rotational equilibrium equations, which are simpler than momentum of momentum ones and might be sufficient to produce the fictitious transitory needed to reach the steady state, may cause the iterative procedure not to converge. Hence, rotational equilibrium equations are disregarded in the following and angular inertia is treated as a stabilization parameter for steady-state analyses.
Let $\mathbf{F} = \{F_X, F_Y, 0\}$ be the external load acting on the moving member of the pair.
In the case of a journal bearing (Fig. 2), the equilibrium equations of the journal are

$$\left.\begin{aligned} F_X + \int_\Omega p\cos\vartheta\,\gamma^2 d\Omega = 0 \\ F_Y + \int_\Omega p\sin\vartheta\,\gamma^2 d\Omega = 0 \end{aligned}\right\} \tag{12}$$

where Ω is the mapped domain (the union of the pad domains in a thrust bearing assembly) and $\gamma^2\, d\Omega = \gamma^2\, dx\, dz$ is an infinitesimal element of the physical domain. Equation (12) holds for steadily loaded and also dynamically loaded journal bearings, i.e. in a connecting rod big end bearing the inertia force acting on the journal is carried by the crankshaft bearing. In the case of a tilting pad thrust bearing (Fig. 3, F_X=0), the collar equilibrium implies

$$F_Y - \int_\Omega p\gamma^2 d\Omega = 0 \tag{13}$$

while the momentum of momentum equations for the (frictionless) i[th] pad motion are

$$\left.\begin{aligned}
&-\int_{\Omega} p\gamma R\sin(\vartheta-\psi_{Pi})\gamma^2 d\Omega - I_{zPi}\frac{\delta_{ri}-2\delta_{ri}^{\,t-\Delta t}+\delta_{ri}^{\,t-2\Delta t}}{\Delta t^2}=0\\
&\int_{\Omega} p\left[\gamma R\cos(\vartheta-\psi_{Pi})-r_{Pi}\right]\gamma^2 d\Omega - I_{xPi}\frac{\delta_{\vartheta i}-2\delta_{\vartheta i}^{\,t-\Delta t}+\delta_{\vartheta i}^{\,t-2\Delta t}}{\Delta t^2}=0
\end{aligned}\right\} \tag{14}$$

where I_{xPi} and I_{zPi} are the (mass) moment of inertia around the x_{Pi} and z_{Pi} axes, respectively, or the stabilization parameters of pad i.

4. The mass-conserving lubrication model

4.1 Integration domain and basic assumptions

Mass conserving cavitation models are based on the so-called JFO theory for moderately and highly loaded bearings, which assumes an infinite number of streamers in the cavitated region. Fig. 4 shows the thin film (mapped) computational domain Ω which is divided into an active (or pressurized) region Ω_a and an inactive (or cavitated) region Ω_c in such a way that $\Omega = \Omega_a \cup \Omega_c$. Let Γ_e be the external boundary of Ω, Γ_c the boundary between active and inactive film regions, Γ_{e1} the eventual portion of Γ_e that bounds the active film region and Γ_{e2} the remaining part so that $\Gamma_e = \Gamma_{e1} \cup \Gamma_{e2}$. The unit vectors $\mathbf{n_a}$ and $\mathbf{n_c}$ denote the outwards normals respectively to Ω_a and Ω_c.

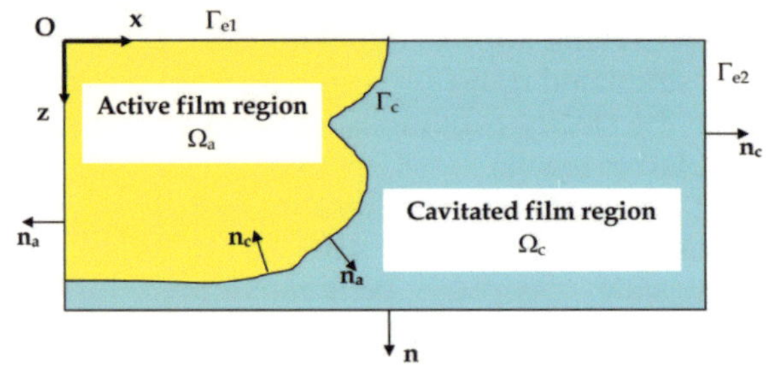

Fig. 4. Integration domain

It is assumed that the lubricant behaves like an incompressible fluid when hydrodynamic pressure build is allowed, and like a fictitious gas-liquid mixture with variable density and constant kinematic viscosity when cavitation occurs and pressure can be considered constant ($p = p_c$), so that mixture density ρ and viscosity μ are related to liquid density ρ_L and viscosity μ_L by

$$\frac{\mu}{\mu_L}=\frac{\rho}{\rho_L} \tag{15}$$

The complete film region ($\rho = \rho_L$) includes Ω_a and in some cases, the part of Ω_c where pressure cannot rise and density is going to decrease, due to the divergence of the film, while the incomplete film region ($\rho < \rho_L$) is a portion of Ω_c.

For the applications at the hand, the lubricant density ρ_L is considered constant and the lubricant viscosity μ_L is assumed to depend solely on film temperature. In order to approximate the temperature-viscosity dependence of the lubricant in a range of temperature that is quite narrow but still reasonable for steadily-loaded bearings, the following simple equation is often used

$$\mu_L = \mu_{L0} \exp\left[-\beta\left(T - T_0\right)\right] \tag{16}$$

where μ_{L0} is the lubricant viscosity at the reference temperature T_0 and β a viscosity-temperature coefficient. By taking into account the assumptions about the lubricant behavior, the universal thin film mechanics equation (7) becomes

$$\Delta m = \frac{\rho_L}{\gamma^2}\left[\nabla\cdot\left(g_L \nabla p\right)\right] - \mathbf{U}\cdot\nabla\left[\rho\left(H - f_L\right)\right] - \frac{\partial}{\partial t}\left(\rho H\right) = 0 \tag{17}$$

where $\mathbf{U} = \{\omega R, 0\}$ is the relative velocity, Δm the residual mass flow (per unit area) and

$$\left.\begin{aligned} f_L &= i_{L1} / i_{L0} = f \\ g_L &= i_{L2} - i_{L1}^2 / i_{L0} = g\,\rho/\rho_L \end{aligned}\right\} \tag{18}$$

with

$$i_{Ls} = \int_0^H \frac{y^s}{\mu_L}\,dy \tag{19}$$

Equation (17) ensures the continuity of the mass, by imposing that the difference between the lubricant flow into and out of the columnar element shown in Fig. 1 balances the variation of the mass per unit time in the same volume. The mass flow through the walls of the columnar element (per unit length) is

$$\mathbf{m} = -\frac{\rho_L\, g_L \nabla p}{\gamma} + \gamma\, \rho\left(H - f_L\right)\mathbf{U} \tag{20}$$

4.2 Classic Kumar and Booker type differential formulation

Assuming $\rho = \rho_L$ on region Ω_a and $p = p_c$ on region Ω_c, the simulation of both the film regions may be performed by means of eq. (17), which becomes an elliptic equation in the unknown pressure p on Ω_a and a hyperbolic equation in the unknown density ρ on Ω_c. The method for determining the partitioning of region Ω takes advantage of a complementarity principle (Murty, 1974; LaBouff & Booker, 1985) that allows dividing the complete active from the complete inactive region, where pressure p and density derivative $\partial\rho/\partial t$ are calculated, respectively. Afterwards a time integration technique is used to compute the density ρ of the film in such a way that the incomplete inactive region extent is immediately determined at each time step (Kumar & Booker, 1991).

4.3 Bonneau and Hajjam type differential formulation

An alternative formulation (Bonneau & Hajjam, 2001) turns out to be more accurate than the classic one in the case of dynamic loading conditions. In this approach the gas film content is defined as

$$\nu = -H\left(\rho_L - \rho\right) \tag{21}$$

Expressing the thin film mechanics equation (17) in terms of such variable yields

$$\Delta m = \frac{\rho_L}{\gamma^2}\nabla\cdot\left(g_L \nabla p\right) - \rho_L \mathbf{U}\cdot\nabla\left[H\left(1-\frac{f_L}{H}\right)\right] - \rho_L\frac{\partial H}{\partial t} - \mathbf{U}\cdot\nabla\left[\nu\left(1-\frac{f_L}{H}\right)\right] - \frac{\partial \nu}{\partial t} = 0 \qquad (22)$$

Equation (22) must be integrated with the following constraints

$$\left.\begin{array}{l} \nu = 0 \quad and \quad p > p_c \quad on\, \Omega_a \\ p = p_c \quad and \quad \nu \le 0 \quad on\, \Omega_c \end{array}\right\} \qquad (23)$$

that match the requirements of both regions Ω_a and Ω_c mentioned in the previous paragraph.

4.4 JFO cavitation conditions

The classic Jakobbson, Floberg and Olsson (JFO) conditions (Floberg & Jakobsson, 1957; Olsson, 1965) impose the continuity of the flow through the cavitation boundary Γ_c. They can be obtained by means of the flow balance suggested by Fig. 5, where $\mathbf{V}_\Gamma$ denotes the

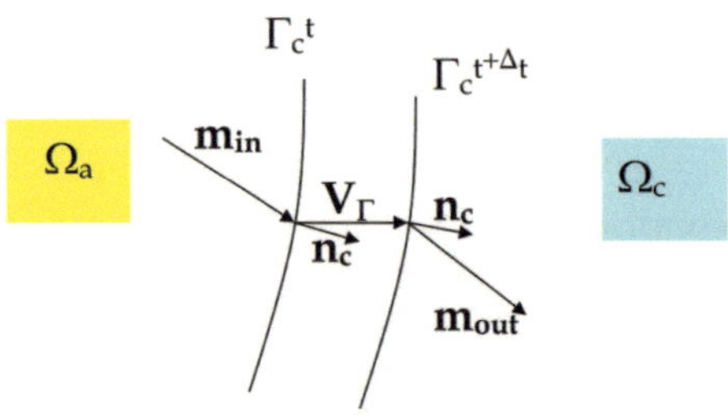

Fig. 5. Mass flows through the moving cavitation boundary

velocity of the moving boundary of Γ_c, crossed by the hydrodynamic mass flows $\mathbf{m}_{in}$ and $\mathbf{m}_{out}$, respectively leaving and entering the active film. Both such flows can be computed on the basis of eq. (20), particularized for Ω_a and Ω_c. Therefore mass continuity through Γ_c is ensured by the equation

$$\left.\begin{array}{l} \mathbf{m}_{in}\bullet\mathbf{n}_c - \mathbf{m}_{out}\bullet\mathbf{n}_c - H(\rho_L-\rho)\gamma\mathbf{V}_\Gamma\bullet\mathbf{n}_c = \\ = \left[\gamma(H-f_L)(\rho_L-\rho)\mathbf{U} - \frac{\rho_L\, g_L}{\gamma}\nabla p - \gamma\mathbf{V}_\Gamma H(\rho_L-\rho)\right]\bullet\mathbf{n}_c = 0 \end{array}\right\} \qquad (24)$$

where all of the variables must be assessed on the boundary Γ_c. In this form, the JFO boundary conditions can be coupled with eq. (17) that is the Kumar and Booker's formulation.

In another way, in terms of the gas film content variable, by taking advantage of eq. (21), the same condition can be written

$$\left\{\gamma\nu\left[\left(1-\frac{f_L}{H}\right)\mathbf{U} - \mathbf{V}_\Gamma\right] + \frac{\rho_L g_L}{\gamma}\nabla p\bullet\mathbf{n}_c\right\} = 0 \qquad (25)$$

which is the boundary condition on Γ_c for the Bonneau and Hajjam's formulation of the mass conserving lubrication problem.

4.5 Strong differential hydrodynamic problem

The strong form of the differential problem is solved by finding the unknown pressure and gas content fields (respectively p and ν) that fulfill eq. (22) on Ω together with the relevant constraints eq. (23), the corresponding cavitation boundary conditions on Γ_c (eq. (25)) and the essential boundary conditions $\nu = \nu_e$, $p = p_e$ on Γ_e.

4.6 Weak integral hydrodynamic problem

The film domain Ω is suitably discretized by a finite element mesh with n nodes. The differential equations must be turned into discrete systems of integral relationships employing the weighted residual method. Integration must be performed in the physical domain, which infinitesimal area is $\gamma^2\, d\Omega = \gamma^2\, dx\, dz$. Let W_i be a weighting function associated with node i. After taking into account the different integration regions in Ω, the integral strong form of the problem can be expressed by the weighted equation

$$\left.\begin{aligned}
&\int_{\Omega_a} W_i \left\{ \frac{\rho_L}{\gamma^2} \nabla\cdot(g_L \nabla p) - \rho_L \mathbf{U}\cdot\nabla\left[H\left(1-\frac{f_L}{H}\right)\right] - \rho_L \frac{\partial H}{\partial t} \right\} \gamma^2 d\Omega + \\
&+\int_{\Omega_c} W_i \left\{ -\rho_L \mathbf{U}\cdot\nabla\left[H\left(1-\frac{f_L}{H}\right)\right] - \rho_L \frac{\partial H}{\partial t} - \mathbf{U}\cdot\nabla\left[\nu\left(1-\frac{f_L}{H}\right)\right] - \frac{\partial \nu}{\partial t} \right\} \gamma^2 d\Omega + \\
&+\int_{\Gamma_c} W_i \left\{ \gamma\nu\left[\left(1-\frac{f_L}{H}\right)\mathbf{U} - \mathbf{V}_\Gamma\right] + \frac{\rho_L g_L}{\gamma} \nabla p \right\} \cdot \mathbf{n}_c \gamma\, d\Gamma
\end{aligned}\right\} \tag{26}$$

for i = 1 to n, together with the constraint eq. (23) and the essential boundary conditions on Γ_e. Applying the divergence theorem (eq. (A1), see appendix) and the Reynolds transport theorem (eq. (A2)) to the strong form (eq. (26)) yields

$$\left.\begin{aligned}
&-\rho_L \int_{\Omega_a} \nabla W_i \cdot g_L \nabla p\, d\Omega + \rho_L \int_{\Gamma_c \cup \Gamma_{e1}} W_i\, g_L \nabla p \cdot \mathbf{n}_a\, d\Gamma + \\
&+\rho_L \int_{\Omega_a} \nabla W_i \cdot H\left(1-\frac{f_L}{H}\right)\gamma^2 \mathbf{U}\, d\Omega - \rho_L \int_{\Gamma_c \cup \Gamma_{e1}} W_i H\left(1-\frac{f_L}{H}\right)\gamma^2 \mathbf{U}\cdot\mathbf{n}_a\, d\Gamma - \rho_L \int_{\Omega_a} W_i \frac{\partial H}{\partial t}\gamma^2 d\Omega + \\
&+\rho_L \int_{\Omega_c} \nabla W_i \cdot H\left(1-\frac{f_L}{H}\right)\gamma^2 \mathbf{U}\, d\Omega - \rho_L \int_{\Gamma_c \cup \Gamma_{e2}} W_i H\left(1-\frac{f_L}{H}\right)\gamma^2 \mathbf{U}\cdot\mathbf{n}_c\, d\Gamma - \rho_L \int_{\Omega_c} W_i \frac{\partial H}{\partial t}\gamma^2 d\Omega + \\
&+\int_{\Omega_c} \nabla W_i \cdot \nu\left(1-\frac{f_L}{H}\right)\gamma^2 \mathbf{U}\, d\Omega - \int_{\Gamma_c \cup \Gamma_{e2}} W_i \nu\left(1-\frac{f_L}{H}\right)\gamma^2 \mathbf{U}\cdot\mathbf{n}_c\, d\Gamma + \\
&-\frac{\partial}{\partial t}\int_{\Omega_c} W_i \gamma^2 \nu\, d\Omega + \int_{\Gamma_c \cup \Gamma_{e2}} W_i \gamma^2 \nu \mathbf{V}_\Gamma \cdot \mathbf{n}_c\, d\Gamma + \\
&+\rho_L \int_{\Gamma_c} W_i\, g_L \nabla p \bullet \mathbf{n}_c d\Gamma + \int_{\Gamma_c} W_i \gamma^2 \nu\left[\left(1-\frac{f_L}{H}\right)\mathbf{U} - \mathbf{V}_\Gamma\right] \bullet \mathbf{n}_c d\Gamma = 0
\end{aligned}\right\} \tag{27}$$

that considers the generic case shown in Fig. 4, where Ω_a is bounded by Γ_c and a portion Γ_{e1} of the external boundary. In such case the essential boundary conditions must be consistent, namely the gas film content has to vanish on Γ_{e1}. The relationship

$$\left.\begin{aligned}
\mathbf{F}\cdot\mathbf{n_a} &= -\mathbf{F}\cdot\mathbf{n_c} \quad on\, \Gamma_c \\
\mathbf{n} = \mathbf{n_a} &= \mathbf{n_c} \quad on\, \Gamma_e
\end{aligned}\right\} \tag{28}$$

holds for whatever field vector **F** is chosen, as evidenced by Fig. 4. Therefore, by taking into account eq. (28), equation (27) for $\Gamma_e = \Gamma_{e1} \cup \Gamma_{e2}$, $\Omega = \Omega_a \cup \Omega_c$ and $\nu = 0$ on Γ_{e1}, is turned into the relation

$$\left.\begin{aligned}&-\rho_L \int_\Omega \nabla W_i \bullet g_L \nabla p \, d\Omega + \rho_L \int_{\Gamma_e} W_i g_L \nabla p \bullet \mathbf{n} \, d\Gamma + \\ &+\rho_L \int_\Omega \nabla W_i \bullet H \left(1 - \frac{f_L}{H}\right) \gamma^2 \mathbf{U} \, d\Omega - \rho_L \int_{\Gamma_e} W_i H \left(1 - \frac{f_L}{H}\right) \gamma^2 \mathbf{U} \bullet \mathbf{n} \, d\Gamma - \rho_L \int_\Omega W_i \frac{\partial H}{\partial t} \gamma^2 \, d\Omega + \\ &+\int_\Omega \nabla W_i \bullet \nu \left(1 - \frac{f_L}{H}\right) \gamma^2 \mathbf{U} \, d\Omega - \int_{\Gamma_e} W_i \gamma^2 \nu \left[\left(1 - \frac{f_L}{H}\right) \mathbf{U} - \mathbf{V}_\Gamma\right] \bullet \mathbf{n} \, d\Gamma - \frac{\partial}{\partial t} \int_\Omega W_i \gamma^2 \nu \, d\Omega = 0\end{aligned}\right\} \quad (29)$$

The lubricant flow, given by eq. (20), can be expressed in terms of the variable ν instead of ρ by means of eq. (21) as follows

$$\mathbf{m} = \gamma \, \rho_L H \left(1 - \frac{f_L}{H}\right) \mathbf{U} + \gamma \, \nu \left(1 - \frac{f_L}{H}\right) \mathbf{U} - \frac{\rho_L g_L}{\gamma} \nabla p \quad (30)$$

Then, by evidencing the expression of the flow (eq. (30)) and assuming that the external boundary is fixed with reference to surface 0 ($\mathbf{V}_{\Gamma e} = \mathbf{0}$), equation (29) becomes

$$\left.\begin{aligned}&\int_{\Gamma_e} W_i \mathbf{m} \bullet \mathbf{n} \, \gamma \, d\Gamma = -\rho_L \int_\Omega \nabla W_i \bullet \frac{g_L}{\gamma^2} \nabla p \, \gamma^2 d\Omega + \\ &+\rho_L \int_\Omega \nabla W_i \bullet H \left(1 - \frac{f_L}{H}\right) \mathbf{U} \gamma^2 d\Omega - \rho_L \int_\Omega W_i \frac{\partial H}{\partial t} \gamma^2 d\Omega + \\ &+\int_\Omega \nabla W_i \bullet \nu \left(1 - \frac{f_L}{H}\right) \mathbf{U} \gamma^2 d\Omega - \frac{\partial}{\partial t} \int_\Omega W_i \nu \gamma^2 d\Omega\end{aligned}\right\} \quad (31)$$

Equation (31) together with the constraint eq. (27) and the essential boundary conditions on Γ_e completely defines the hydrodynamic problem in weak formulation. It allows to implicitly fulfill the continuity boundary conditions on Γ_c (eq. (25)), which are embedded in Eq. (31) as just proved. The corresponding strong form of the problem is the Bonneau and Hajjam type differential formulation presented at paragraph 4.3.
By a numerical point of view the transient term in cavitated region can be evaluated as follows

$$\frac{\partial}{\partial t} \int_\Omega W_i \nu \gamma^2 d\Omega \cong \int_\Omega W_i \frac{\nu - \nu^{t-\Delta t}}{\Delta t} \gamma^2 d\Omega \cong \int_\Omega W_i \frac{\partial \nu}{\partial t} \gamma^2 d\Omega \quad (32)$$

where $\nu^{t-\Delta t}$ is the gas film content calculated at the previous time step.
Substitution of eqs. (32) and (21) in eq. (31) yields

$$\left.\begin{aligned}&\int_{\Gamma_e} W_i \mathbf{m} \bullet \mathbf{n} \, \gamma \, d\Gamma = -\rho_L \int_\Omega \nabla W_i \bullet \frac{g_L}{\gamma^2} \nabla p \, \gamma^2 d\Omega + \\ &+\int_\Omega \nabla W_i \bullet \rho (H - f_L) \mathbf{U} \gamma^2 d\Omega - \int_\Omega W_i \rho \frac{\partial H}{\partial t} \gamma^2 d\Omega - \int_\Omega W_i H \frac{\partial \rho}{\partial t} \gamma^2 d\Omega\end{aligned}\right\} \quad (33)$$

By means of eq. (33), which is roughly equivalent to eq. (31), the hydrodynamic problem is solved in the unknowns p, $\partial\rho/\partial t$ (instead of p, ν), as explained in paragraph 4.2. The corresponding strong form of the problem is the Kumar and Booker type differential formulation presented at paragraph 4.2.
Due to the different integration schemes, the two types of weak formulations are not numerically equivalent.

4.7 FEM formulation and stabilization

The weighting functions are usually chosen according to Galerkin method when elliptic problems are involved, i.e. $W_i = N_i$, where N_i is the shape function of node i. The same expedient produces wiggles when the hyperbolic part of eq. (31) is solved in Ω_c. Stabilization can be performed by means of the streamline upwind / Petrov-Galerkin (SUPG) method, which is element-independent and, in addition, sharply simplifies numerical implementation in comparison with the classical upwinding methods (Bathe, 1996).
In the present case, SUPG stabilization is achieved by adding numerical diffusion only along x direction. Hence, by introducing the usual finite element approximations for the unknown fields variables ($p = p_j N_j$, $\nu = \nu_j N_j$), equation (31) can be written in synthetic form as

$$\Delta M_i = K^p{}_{ij}\, p_j + K^\nu{}_{ij}\, \nu_j + M^e{}_i + M^{t-\Delta t}{}_i = 0 \tag{34}$$

where

$$\left.\begin{aligned}
K^p{}_{ij} &= -\rho_L \int_\Omega \frac{g_L}{\gamma^2} \nabla N_i \bullet \nabla N_j\, \gamma^2 d\Omega \\
K^\nu{}_{ij} &= \int_\Omega \left[\nabla N_i \bullet \left(1 - \frac{f_L}{H}\right) \mathbf{U} N_j - \frac{N_i N_j}{\Delta t} - \frac{\partial N_i}{\partial x} \tau^{SUPG} U^2 \left(1 - \frac{f_L}{H}\right)^2 \frac{\partial N_j}{\partial x} \right] \gamma^2 d\Omega \\
M^e{}_i &= \rho_L \int_\Omega \left[\nabla N_i \bullet H \left(1 - \frac{f_L}{H}\right) \mathbf{U} - N_i \frac{H - H^{t-\Delta t}}{\Delta t} \right] \gamma^2 d\Omega \\
M^{t-\Delta t}{}_i &= \int_\Omega \frac{N_i}{\Delta t} \nu^{t-\Delta t} \gamma^2 d\Omega \\
\Delta M_i &= \int_{\Gamma_e} N_i \mathbf{m} \bullet \mathbf{n}\, \gamma\, d\Gamma
\end{aligned}\right\} \tag{35}$$

In eq. (35), Δt is the time step, $H^{t-\Delta t}$ and $\nu^{t-\Delta t}$ are respectively the film thickness and the gas film content calculated at the previous time, and τ^{SUPG} is the stabilization parameter

$$\tau^{SUPG} = \alpha^{up} \beta^{up} \frac{\Delta s}{|\mathbf{u}_s|} \tag{36}$$

where α^{up}, β^{up} are dimensionless coefficients used to control the added diffusion, Δs is the length of the finite element along the streamwise direction, and $\mathbf{u}_s$ the convective flow velocity. In eq. (36) the term on the right hand side has the effect of weighting the convection operators towards the upstream direction.

In order to solve eq. (34), the following set of values is used in eq. (36): $\alpha^{up} = 1$, $\beta^{up} = 0.5$, $\Delta s = \Delta x$ (element length along the x direction) and $u_s \approx U / 2$, whereas the convective flow in the cavitated region moves in x direction with the mean surface velocity.
Some authors (Fatu et al., 2006) assume that viscosity is independent of cross film direction y in the cavitated region so that $f_L/H = 0.5$ in Ω_c.

5. Fast energy-conserving thermal models

5.1 2D model

A simplified energy conservation equation for thin fluid films was derived by averaging in the cross-film direction the classic 3D energy equation (Kumar & Booker, 1994). Although the resulting 2D model is very fast by a computational point of view, it is unable to assess the temperature gradient $\partial T/\partial y$ at the walls, which must be considered adiabatic. By applying the conformal mapping defined in paragraph 3.1, the 2D energy equation is turned into the following relation

$$\Delta q = \frac{1}{\gamma^2}\nabla\bullet\left(K H \nabla T_m\right) - \rho\frac{c}{\gamma} H\,\mathbf{u_m}\bullet\nabla T_m - \rho c H\frac{\partial T_m}{\partial t} + H\Phi_m = 0 \tag{37}$$

where the subscript m denotes a variable averaged in the transverse direction of the film y, Δq the residual heat flow (per unit area), $\mathbf{u} = \{u, v, w\}$ the lubricant velocity vector, K the thermal conductivity of the lubricant film mixture, c the specific heat and Φ the power dissipation density function.
Equation (37) ensures the continuity of the energy, by imposing that the net heat exchanged by the columnar element of fluid shown in Fig. 1 balances the variation of internal energy and the heat dissipation in the same volume. The heat flow through the walls of the columnar element is

$$\mathbf{q} = -\frac{H}{\gamma} K \nabla T_m \tag{38}$$

The conduction in the oil film plane (x, z) is not taken into account in the original equation devised by Kumar & Booker, as the problem solution is dominated by convection. The conductive term has been introduced in eq. (37) only for numerical convenience, in order to take advantage of SUPG stabilization (paragraph 5.3).

5.2 Quasi-3D model

Equation (37) can be modified in order to take into account the heat flux from the lubricant film into the two members of the pair. In this case, the heat absorbed by the film is equal to the difference between the viscous dissipation and the total heat exchange.
Therefore, the following equation is used in the thin film region Ω

$$\Delta q = \frac{1}{\gamma^2}\nabla\bullet\left(K H \nabla T_m\right) - \rho\frac{c}{\gamma} H\,\mathbf{u_m}\bullet\nabla T_m - \rho c H\frac{\partial T_m}{\partial t} + H\Phi_m - q_0 - q_1 = 0 \tag{39}$$

where q_0 and q_1 are the heat transfers (per unit area) to surface 0 and 1, respectively.
They are calculated by obtaining the temperature gradients at the walls, in the hypothesis that the temperature profile across the film thickness is a fourth-order polynomial.

According to the procedure explained in detail in the previous reference work (Stefani & Rebora, 2009), the five coefficients of this polynomial can be calculated by:
a) equating the film temperature at the walls with the temperatures T_0 and T_1 calculated by the heat conduction equations on surface 0 and 1 respectively; b) imposing that the average temperature in the y direction is equal to T_m, namely the unknown of eq. (39). The resulting heat exchange at the wall surfaces is

$$\left.\begin{aligned} q_0 &= K\left.\frac{\partial T}{\partial y}\right|_{y=0} = \frac{K}{H}\left(5T_m - \frac{7}{2}T_0 - \frac{3}{2}T_1 - \frac{1}{8}T_0''H^2 + \frac{1}{24}T_1''H^2\right) \\ q_1 &= K\left.\frac{\partial T}{\partial y}\right|_{y=H} = \frac{K}{H}\left(5T_m - \frac{3}{2}T_0 - \frac{7}{2}T_1 + \frac{1}{24}T_0''H^2 - \frac{1}{8}T_1''H^2\right) \end{aligned}\right\} \quad (40)$$

where T''_0 and T''_1 are the derivatives $\partial^2 T/\partial y^2$ at the surfaces 0 and 1, respectively.
They can be obtained by projecting the 3D energy equation on both the kinematic pair surfaces. Particularly in steady-state conditions ($\partial T/\partial t=0$, $\partial H/\partial t=0$), under the further assumptions that the temperature on surface 1 does not depend on the x coordinate ($\partial T_1/\partial x=0$), evaluating the 3D energy equation on surfaces 0 and 1 yields respectively

$$\left.\begin{aligned} T_0'' &= \left.\frac{\partial^2 T}{\partial y^2}\right|_{y=0} = -\frac{1}{K}\Phi(x,0,z) = -\frac{\Phi_0}{K} \\ T_1'' &= \left.\frac{\partial^2 T}{\partial y^2}\right|_{y=H} = -\frac{1}{K}\Phi(x,H,z) = -\frac{\Phi_1}{K} \end{aligned}\right\} \quad (41)$$

The hypothesis $\partial T/\partial x=0$ on surface 1 is extensively adopted in the analysis of journal bearings, where the shaft rotation flattens the circumferential variations of temperature (Kucinski et al., 2000). An analogous phenomenon may occur in thrust bearings due to the collar rotation.
The hypothesis $\partial T/\partial t=0$ complies with the analysis of steadily loaded bearings. It is also sufficiently realistic in the simulation of dynamically loaded bearings, as the variations of temperature distribution in the bearing structure over one load cycle are not very significant in comparison with film temperature changes (Kim & Kim, 2001). Nevertheless, equation (41) does not take into account the heat convection due to squeeze and it must be modified to deal with dynamic loading conditions. In addition a full transient thermal analysis ($\partial T/\partial t\neq 0$) requires special techniques (Fatu et al., 2006).
Equation (41) implicitly includes the velocity boundary conditions $\mathbf{u} = 0$ for $y = 0$ as well as $v = \gamma\, U\, dH/dx$, $w=0$ for $y = H$, so that the derivatives T''_0, T''_1 become independent of heat convection. Such a deduction, proved in the previous reference work (Stefani & Rebora, 2009) dealing with journal bearings, has been extended to thrust bearings by taking advantage of the reference system choice (paragraph 3.1).

5.3 FEM formulation and stabilization
The following SUPG-stabilized integral form of eq. (39) (and eq. (37) as particular case) is here proposed

$$\Delta Q_i = K^T{}_{ij}\,T_{mj} + Q^{\Phi}{}_i + Q^{t-\Delta t}{}_i - Q^0{}_i - Q^1{}_i = 0 \tag{42}$$

where

$$\left.\begin{aligned}
K^T{}_{ij} &= -\int_\Omega \left(\frac{1}{\gamma^2} \nabla N_i\,\mathbf{K}\nabla N_j + N_i \frac{\rho c}{\gamma} H\,\mathbf{u_m}\bullet\nabla N_j + \rho c H \frac{N_i N_j}{\Delta t} \right)\gamma^2 d\Omega \\
Q^{\Phi}{}_i &= \int_\Omega N_i H\,\Phi_m\,\gamma^2 d\Omega \\
Q^{t-\Delta t}{}_i &= \int_\Omega \rho c H \frac{N_i}{\Delta t} T_m{}^{t-\Delta t}\,\gamma^2 d\Omega \\
\Delta Q_i &= -\int_{\Gamma e} N_i H\,K\,\nabla T_m \bullet \mathbf{n}\,d\Gamma = \int_{\Gamma e} N_i \mathbf{q}\bullet\mathbf{n}\,\gamma\,d\Gamma \\
Q^0{}_i &= \int_\Omega N_i q_0\,\gamma^2 d\Omega \\
Q^1{}_i &= \int_\Omega N_i q_1\,\gamma^2 d\Omega
\end{aligned}\right\} \tag{43}$$

In eq. (43) $T_m{}^{t-\Delta t}$ is the (averaged) film temperature calculated at the previous time and **K** is the following diffusivity matrix

$$\mathbf{K} = \begin{bmatrix} KH + cC\,\tau^{SUPG} u_m{}^2 & cC\,\tau^{SUPG} u_m\,w_m \\ cC\,\tau^{SUPG} w_m\,u_m & KH + cC\,\tau^{SUPG} w_m{}^2 \end{bmatrix} \tag{44}$$

where C is a reference clearance value. Specifically $C=c_b$ for complete cylindrical journal bearings (eq. (10)) and $C = h_p^{ref}$ (reference film thickness at pivots) for thrust bearing assemblies (eq. (11)). As in the last case the film thickness at pivot h_p is a variable and a fixed parameter for all the time steps is required in the stabilization procedure, a suitable reference thickness at pivot h_p^{ref} is chosen (i.e. for most thrust bearings a good value of h_p^{ref} is 50 μm).
In order to avoid wiggles in the temperature distributions retrieved from eq. (42), the τ^{SUPG} stabilizing parameter is calculated by means of eq. (36), with $\alpha^{up}=1$, $\beta^{up}=0.5$, and $u_s=|\mathbf{u_m}|$.

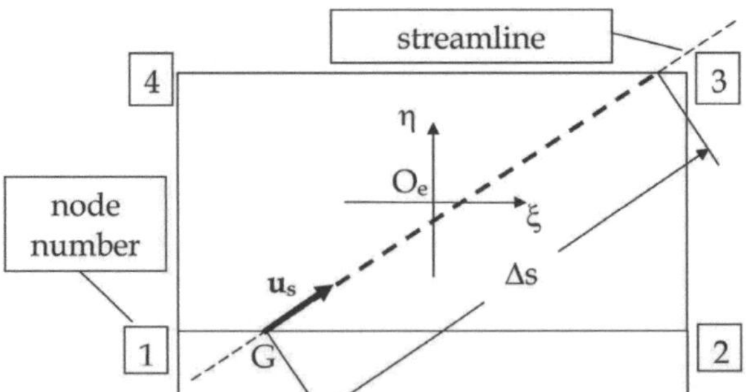

Fig. 6. Streamline across a 4-node isoparametric finite element, where $O_e(\xi, \eta)$ is the element reference system and G is a Gauss point

As the integrals in eq. (43) are evaluated by means of the 2-points Gauss quadrature, the length Δs in eq. (36) is calculated along the streamline across every Gauss point G, as shown in Fig. 6. Since convection typically dominates diffusion along the flow direction, the added streamwise diffusion does not compromise the accuracy of the solution in this direction. In addition no deleterious cross-stream diffusion is added.

5.4 FEM groove mixing theory

If the region Ω_g is the projection in the (x, z) plane of the gth groove fed with supply oil at temperature T_s, the average film temperatures $T_{m\Omega g}$ and $T_{m\Gamma g}$ are assumed to be uniform throughout Ω_g and its boundary Γ_g, respectively. The proposed algorithm is based on the simple idea that the residuals of the integral energy equations calculated at nodes belonging to the grooves must be involved in the calculation. This enables us to assess the entry temperature, which experimental evidence suggests is higher than the oil-feed temperature. An approximated energy balance provides the additional equation needed to evaluate the unknown entry temperature $T_{m\Gamma g}$.

N_i being the ith shape function, n the number of the nodes in Ω, n_g the number of the nodes in Ω_g, and n_{bg} the number of nodes lying on the groove boundary Γ_g, the energy-balance equation for the oil flows that mix in Γ_g is given by

$$\begin{aligned}&\sum_{i=1}^{ng}\int_{\Omega g} N_i \Delta q_g \gamma^2 d\Omega + \sum_{i=1}^{n}\int_{\Gamma g} N_i \mathbf{q}\bullet\mathbf{n}\gamma d\Gamma + c\int_{\Gamma g} N_i \mathbf{m}\bullet\mathbf{n}\gamma d\Gamma\left(T_{m\Gamma g} - T_s\right) = \\ &= \sum_{i=1}^{nbg}\Delta Q_i^{\,g} + \Delta Q_i + c\,\Delta M_i\left(T_{m\Gamma g} - T_s\right) = 0\end{aligned} \tag{45}$$

where Δq_g and ΔQ_i^g are the local and the integral residual at node i, respectively, of the energy balance in the groove domain Ω_g, described below. In eq. (45) ΔQ_i and ΔM_i are the weighted residual (calculated at node i) of the energy and thin film mechanics equations integrated in Ω according to eqs. (42) and (34), respectively. Ω_g is not a thin film region and therefore it is *not* a part of the integration domain Ω. Hence in Ω_g the specific energy balance equation at node i is required to conserve energy and to determine $T_{m\Omega g}$

$$\Delta Q_i^{\,g} = \int_{\Omega g} N_i \Delta q_g \gamma^2 d\Omega = 0 \tag{46}$$

In eq. (46) Δq_g evaluates the net heat flow in the region Ω_g, calculated by

$$\Delta q_g = \Phi_{gm} H_g - 2\alpha_0 q_{0g} - q_{1g} \tag{47}$$

where H_g is a groove equivalent thickness, Φ_g is the groove power dissipation density function, α_0 is the ratio between the effective conductive area of the grooves (lying on surface 0) and their total area projected on surface 0 and 1, and q_{0g}, q_{1g} are the heat flows exchanged with the surface 0 and 1, respectively, in the region Ω_g; q_{0g} and q_{1g} are given by

$$\left.\begin{aligned} q_{0g} &= K\left.\frac{\partial T}{\partial y}\right|_{y=0} = \frac{K}{H_g}\left(5T_{m\Omega g} - \frac{7}{2}T_0 - \frac{3}{2}T_1 + \frac{1}{8}\frac{\Phi_{g0}}{K}H_g^{\,2} - \frac{1}{24}\frac{\Phi_{g1}}{K}H_g^{\,2}\right) \\ q_{1g} &= K\left.\frac{\partial T}{\partial y}\right|_{y=Hg} = \frac{K}{H_g}\left(5T_{m\Omega g} - \frac{3}{2}T_0 - \frac{7}{2}T_1 - \frac{1}{24}\frac{\Phi_{g0}}{K}H_g^{\,2} + \frac{1}{8}\frac{\Phi_{g1}}{K}H_g^{\,2}\right)\end{aligned}\right\} \tag{48}$$

Equation (48) has been obtained by combining eqs. (40) and (41). Hence, eq. (48) does not hold for a thick film in a general case, but remains valid if the groove hydrodynamic behavior is simulated as an equivalent Couette flow (in laminar regime with negligible body forces) into a gap of uniform thickness H_g with each wall at uniform temperature. In this hypothesis, if the relevant shear stress is denoted by τ_g, the groove power dissipation density function Φ_g at the walls and its mean value Φ_{gm} are given by

$$\Delta q_g = \Phi_{g0} = \Phi_{g1} = \Phi_{gm} = \tau_g \frac{U}{H_g} = \frac{1}{2} \frac{\lambda \rho U^3}{H_g} \tag{49}$$

where the friction factor λ characteristic of the groove type, is defined by

$$\lambda = \frac{\tau_g}{\frac{1}{2} \rho U^2} \tag{50}$$

Application of the FEM groove mixing approach to the study of different types of grooves is made possible by means of a reasonable choice of the characteristic parameters λ and H_g. The Wendt (Wendt, 1933) and the surface drag losses (Khonsari & Booser, 2008) empirical formulas have been used to assess the friction factor λ for journal and thrust bearings, respectively.

5.5 Structural and thermal simulation of the pairs

Steady-state elasticity and heat conduction equations must be solved in the structures of the kinematic pair members to assess the displacements due to mechanical as well as thermal actions and the temperature of the bearing surfaces due to heat dissipation in the lubricant film. By taking advantage of the related problem linearity as well as of FEM condensation procedures, the thermoelastic displacements d_i and the thermal fields T_i of the two surfaces (i=0, 1) are obtained by means of linear operators, which numerical counterparts are represented by suitable vectors and matrices. They may be calculated in a preprocessing phase, by means of a separate (commercial) FEM code, i.e. Ansys, which is very flexible and suitable to perform complex operations thanks to the APDL language.
The displacements due to mechanical actions $d^m{}_i$ and due to thermal dilatation $d^t{}_i$ are evaluated separately, so that $d_i=d^m{}_i+d^t{}_i$.
In order to compute mechanical displacements $d^m{}_i$ (in the y direction) of the two surfaces (i=0, 1), the following linear expression is used

$$d^m{}_i = d^{me}{}_i + C^m{}_i A p \tag{51}$$

where $d^{me}{}_i$ is the additional displacement due to external mechanical actions, $C^m{}_i$ is the compliance operator of the structure and A is a suitable area operator.
A typical external action is the bolt preload in connecting rod journal bearings, which leads to the $d^{me}{}_i$ contribution. For journal bearings the displacement direction is radial and the mean displacement of the bearing surface must be subtracted (paragraph 3.2). The assumption $d^m{}_1=0$ is very usual for both journal and thrust bearings, as shaft and collar are much stiffer than bush and pads. Details about the calculation procedure can be found in many papers (i.e.: Bonneau & Hajjam, 2001).
In order to compute the thermal field T_i on the two surfaces (i=0, 1), the linear equation is

$$T_i - T_e = T^e{}_i + S_i A q_i \tag{52}$$

where T_e is the external temperature, $T^e{}_i$ is the temperature field due to essential boundary conditions on a part of the thermal model, and S_i is referred to as thermal sensitivity operator. T_e is the bulk temperature of the air surrounding the structure, whereas heat convection boundary conditions are used to model the heat exchange with the environment. Essential boundary conditions on the groove surface ($T_0=T_s$ on Ω_g) may be useful in order to simulate the heat exchange between cold lubricant and hot metal, which is influential when T_s value is not much higher than T_e. In such a case $T^e{}_i$ must be computed, unless the temperature T_e and T_s are assumed to be equal as in the plain bearing studied in the reference work (Stefani & Rebora, 2009). The same paper deals with the procedure required to calculate the tensors for the discretization of eq. (52), except $T^e{}_i$.
In order to compute thermal displacements $d^t{}_i$ (in the y direction) of the two surfaces (i=0, 1), the linear expression is

$$d^t{}_i = d^{te}{}_i + C^t{}_i A q_i \tag{53}$$

where $d^{te}{}_i$ is the additional displacement due to essential temperature boundary conditions, $C^t{}_i$ is the thermal compliance operator. In eq. (53) $d^{te}{}_i$ is the displacement due to the temperature field $T^e{}_i$ discussed above. The assessment of $C^t{}_i$ is discussed in the above-cited reference work, where it is expressed as the product of two tensors, evaluated in separate thermal and structural analyses. Indeed $C^t{}_i$ may be also estimated directly by resorting to coupled-field solid elements.

6. Application examples

6.1 Journal bearing

In order to show the degree of reliability of the above-explained methods, a comparison among experimental and numerical results is presented for the two-axially grooved journal bearing studied by Lund (Lund & Tonnesen, 1984). Extensive bibliography, simulation details, assumed bearing data and boundary conditions are reported in the reference work (Stefani & Rebora, 2009). Fig. 7 shows the trends of the temperature T_0 (on the bush surface) along the bearing centerline for a shaft rotation speed of 5000 rpm and an external load of 5600 N. The responses of the 2D THD model and the quasi-3D TEHD model (with α_0=5 in eq. (47)) are compared.
Particularly, the 2D model (dotted line) strongly underestimates the experimental temperature, as explained in the reference work. The dash-dotted line is obtained by assuming, in agreement with the reference and the Lund's work, that the temperature of the effective area 2 α_0 Ω_g exchanging heat with the oil film is constant and equal to the supply oil temperature T_s (hypothesis A). Such an assumption means that $T_0=T_s$ is imposed in eq. (48), but the groove surface temperature in the thermal model is free ($T^e{}_0=0$ in eq. (52)), so that the dash-dotted line exceeds Ts=50°C in the groove regions ($85° \le \vartheta \le 95°$ and $265° \le \vartheta \le 275°$) in Fig. 7. Although such response seems to be realistic by the physical point of view, as a limited variation of the white metal temperature through the groove region is expected, the hypothesis A requires different assumptions in different sub-models. Differently the solid line has been obtained by imposing $T_0=T_s$ in Ω_g coherently in all of the sub-models (hypothesis B). In the present bearing study, the supply temperature is greater than the

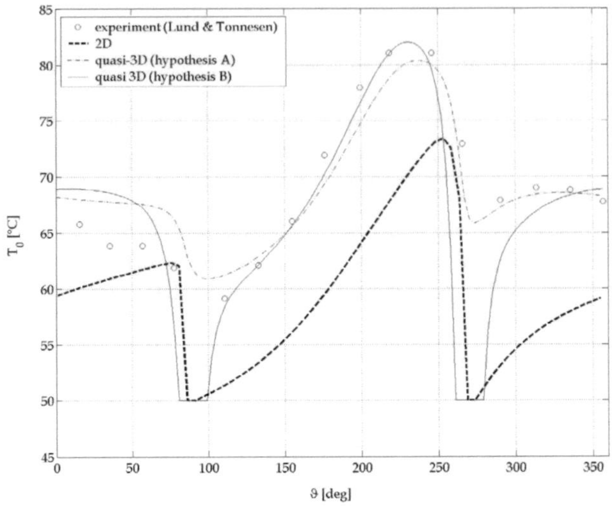

Fig. 7. Centerline temperatures on the bearing surface calculated by means of the 2D model and the quasi-3D model (two-axial groove journal bearing, ω=5000 rpm, F_X=5600 N)

external temperature (T_s=50°C, T_e=20°C) and, therefore, the thermal simulation requires the calculation of the term T^e_0 in eq. (52) in order to fulfil hypothesis B. Accordingly T_0=T_e has been imposed on Ω_g while calculating the corresponding sensitivity matrix (term S_0 in eq. (52)).
The adoption of either hypothesis A or B enables the heat exchange between the structure of the bush and the lubricant in the grooves to be simulated. The lack of such modelling detail causes, in the present case, a bush peak temperature overestimation of roughly 10°C. Although hypothesis A yields a better temperature trend than hypothesis B, it leads to an underestimation of the journal temperature that is evident in the cited Lund's paper too. A comparison among journal temperature results is reported in Table 1, where axially averaged values are calculated for the one-dimensional journal sub-models adopted in the quasi-3d analyses (see the reference paper for details).

	Experiment (Lund & Tonnesen)	Quasi-3D (hypothesis B)	Quasi-3D (hypothesis A)	Lund calculation
T_1 [°C]	77.8	78.7	74.5	73.1

Table 1. Journal temperatures: experimental vs. numerical results (average values)

6.2 Thrust bearing

Another comparison among experimental and numerical results is presented for the tilting-pad thrust bearing studied by Glavatskikh, who published accurate experimental results (Glavatskikh, 2001). This publication also furnishes the bearing data needed by the analyses. The thermal and the thermoelastic behavior of the thrust bearing is taken into account (in the quasi-3D analysis) by means of a 3D model of the pad and 2D axisymmetric model of the shaft-collar assembly (Fig. 8). The choice of an axisymmetric collar model complies with the assumption $\partial T_1/\partial x=0$ (paragraph 5.2).

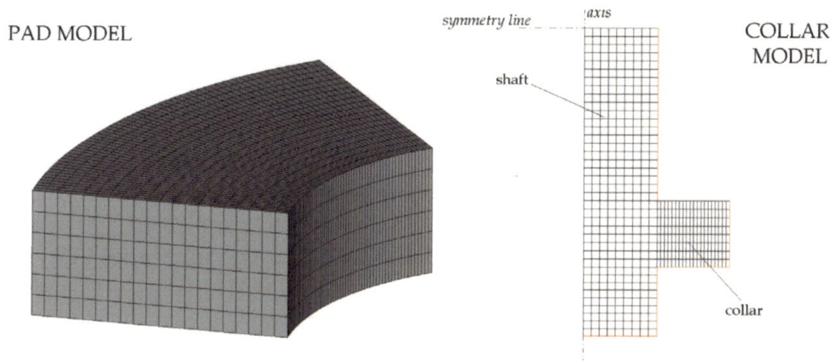

Fig. 8. Finite element models of the thrust pads and the shaft-collar assembly

In the pad thermal and thermoelastic models, used to calculate the thermal sensitivity S_0 (eq. (52)) and the thermal compliance C^t_0 (eq. (53)) respectively, convective heat transfer boundary conditions are assumed for all of the thrust pad surfaces except the babbitt face (surface 0) where the nodal heat flows are applied.

Analogous temperature boundary conditions are adopted for the collar (in Fig. 8 the red lines locate the convective surfaces) and, in addition, symmetry (natural) boundary conditions are reasonably assumed in order to model only a quarter of the shaft, as the test rig is made by two opposite thrust bearings. The heat transfer coefficient and the bulk temperature are set to 50 W m^{-2} $°C^{-1}$ and 20°C, respectively, for all of the pad and shaft-collar convective surfaces.

Different displacement boundary conditions are imposed in the assessment of the thermal compliance C^t_i and the structural compliance C^m_i (eqs. (53) and (51), respectively).

In the former case, the outer surfaces of both the thrust pad and the shaft-collar assembly are left free to expand and only the in-plane displacements of a suitable number[1] of nodes lying on these surfaces have been constrained to avoid ill-conditioning. In the latter case, a small portion of the pad bottom surface around the pivot has been constrained from moving and, in addition, the radial displacements of all of the pad nodes located on the outer and inner diameters are constrained. The collar is assumed to be much stiffer than the pad (C^m_1=0) and for both the pads and the collar no external mechanical action is considered (d^{me}_0=d^{me}_1=0 in eq. (51)). The shaft-collar symmetry constraint is enforced in the thermoelastic model by locking the axial displacement of all of the nodes lying on the symmetry line (Fig. 8).

As the metal on the bottom of the groove cavity (the carrier ring) is not simulated, the lubricant in the inner groove region is assumed to remain, on the average, at the supply oil temperature ($T_{m\Omega g}$=T_s), while the groove boundary mean temperature $T_{m\Gamma g}$ is still an unknown of the problem. Consequently, no essential temperature boundary conditions on the groove surface Ω_g are required on both thermal and thermoelastic models (T^e_i=d^{te}_i=0 in eqs. (52) and (53)).

The universal Reynolds equation (eq. (9)) has been used to obtain the results reported below for 1 MPa specific load (F_Z=26130 N) and T_s=40°C supply temperature. Radial tilt of pads is neglected ($\delta_{\vartheta i}$=0). For the quasi-3D model, the coefficient α_0 (eq. (47)) is increased from 0 to 5 when the rotation speed decreases from 3000 rpm to 1500 rpm, as the heat transfer in the

[1] 4 nodes for the pad 3D model, 3 nodes for the shaft-collar 2D model

grooves is more efficient at low speed, when the lubricant flow rate entering the fully flooded pad inlet is low in comparison with the (constant) flow rate supplied to the test rig.
Fig. 9 shows the variation with the shaft rotation speed of the pad (surface 0) temperature $T_{75/75}$, assessed at a location r_{75}=R+0.75 (R_e-R), $\vartheta_{75}=\vartheta_{Li}+0.75\beta_i$ (Fig. 3), of the collar (surface 1) temperature T_{75} at the location r=r_{75}, and of the shaft temperature T_s at the location r=0 (on the pivot plane). For the 2D model, which does not evaluate wall temperatures, only the mean film temperature at the point (r_{75}, ϑ_{75}) is reported in Fig. 9. Again, in comparison with the reference experimental results, the 2D model strongly underestimates the temperatures, while the quasi-3D model ensures a good level of reliability. In the same Fig. 9 the calculated oil flow rate M is also given. The oil flow rate supplied to the test rig is 15 l/min. Excess flow, beyond what is needed hydrodynamically, acts to provide overall cooling to the bearing. Therefore the assumed variation of α_0 with the rotation speed is well-justified.

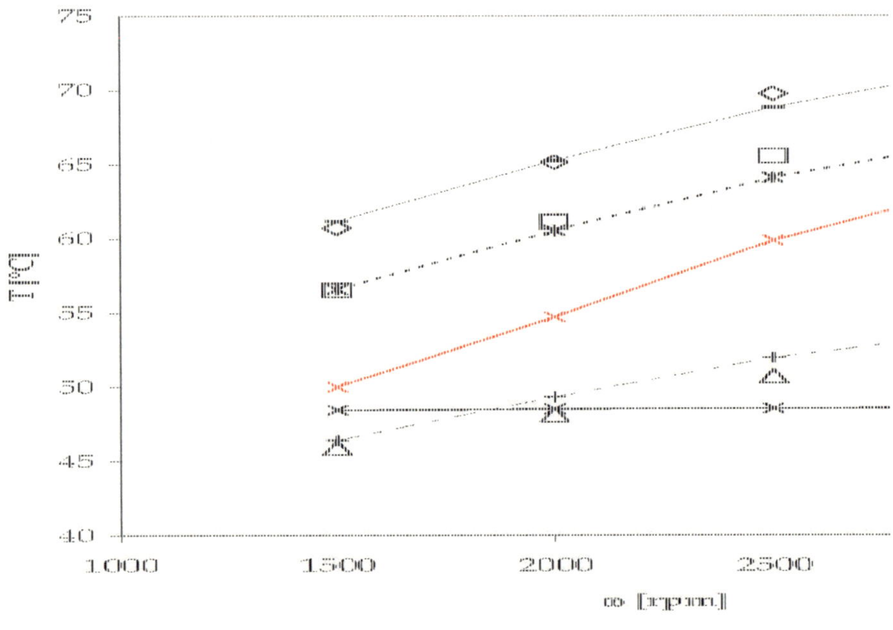

Fig. 9. Pad ($T_{75/75}$), collar (T_{75}), shaft (T_s) temperature and calculated flow rate (M)

The corresponding variations of the power loss P and of the film thickness at the pad inlet and outlet (h_1 and h_3, respectively) are given in Fig. 10. The temperature underestimation of the 2D model leads to unreliable predictions of the film thickness.
Fig. 11 a) compares the experimental and theoretical (quasi-3D model) circumferential variations of pressure p_{25} and p_{75} evaluated at r_{25} and r_{75}, respectively, for ω=3000 rpm. By removing the constraint $\delta_{\vartheta i}$=0, the differences between theoretical and experimental peak pressures increase. For the same rotation speed, Fig. 11 b) shows the pressure distribution calculated by means of the quasi-3D model on the 6 pads of the bearing. The maximum value of this distribution is 2.4 MPa, which is very close to the peak of the experimental pressure p_{75} (2.3 MPa).
The overall agreement between numerical and experimental results might be further improved by taking into account the variation of the viscosity across the film thickness.

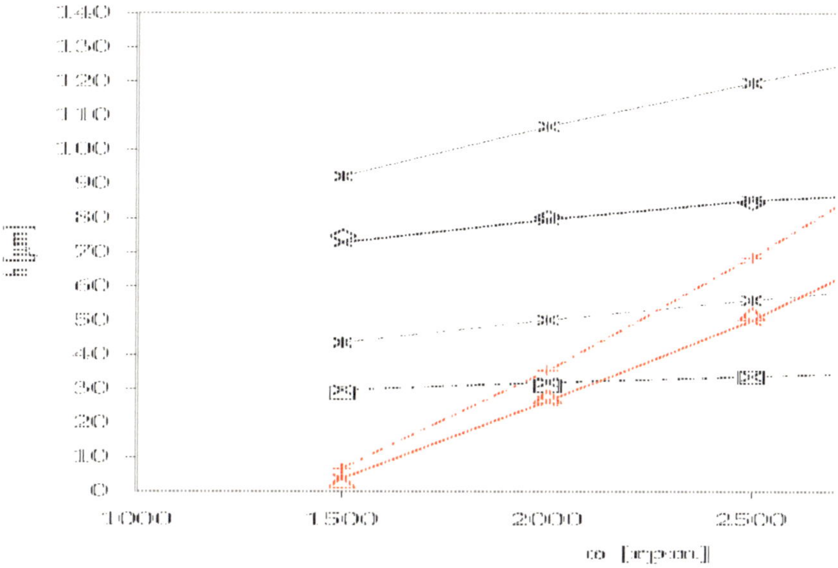

Fig. 10. Power loss (P) and film thickness at pad inlet (h_1) and outlet (h_3)

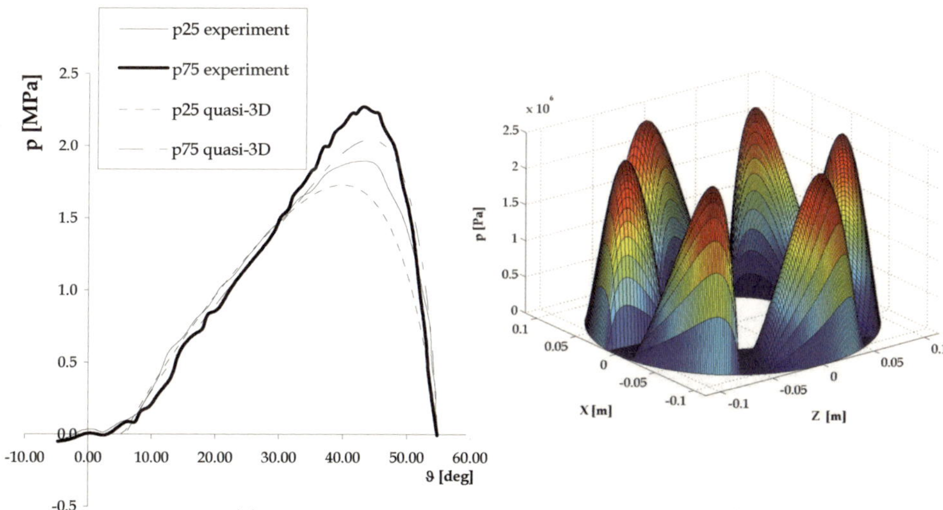

(a) (b)

Fig. 11. Hydrodynamic pressure for ω=3000 rpm: a) circumferential variations (p_{25}, p_{75}); b) pressure field computed by means of the quasi-3D model

7. Conclusion

A general-purpose FEM approach to the TEHD analysis of hydrodynamic bearings has been described, focusing on the theoretical aspects, whereas the relevant numerical procedures are reported in a large number of papers. Particularly, the most used FEM formulations of

the mass-conserving lubrication problem have been proved, while an original approach to the thermal problem has been explained.
The numerical examples show how the quasi-3D approach has enhanced the reliability of the mass- and energy-conserving lubrication analysis proposed by Kumar and Booker.
Indeed, TEHD models are very sensitive to boundary conditions, which choice is particularly difficult in all of the multi-physics simulations.
Future work will adapt the devised method to detailed transient analyses and it will further extend the model flexibility by including advanced turbulent lubrication theory.

8. Appendix

Let f and **F** be scalar and vector-valued functions respectively. A variant of the divergence theorem states

$$\int_{\Omega}\left(\mathbf{F}\bullet\nabla f + f\,\nabla\bullet\mathbf{F}\right)d\Omega = \int_{\Gamma} f\,\mathbf{F}\bullet\mathbf{n}\;d\Gamma \qquad \text{(A1)}$$

where Γ is the boundary of Ω oriented by the outward-pointing unit normal **n**.
If $\mathbf{V}_{\Gamma}$ is the Eulerian velocity at the boundary Γ, the Reynolds transport theorem generalizes the Leibniz's rule to multidimensional integrals as follows

$$\frac{\partial}{\partial t}\left(\int_{\Omega} f\,d\Omega\right) = \int_{\Omega}\frac{\partial f}{\partial t}d\Omega + \int_{\Gamma} f\,\mathbf{V}_{\Gamma}\bullet\mathbf{n}\;d\Gamma \qquad \text{(A2)}$$

9. References

Banwait, SS. & Chandrawat, HN. (1998). Study of thermal boundary conditions for a plain journal bearing. *Tribol. Int.*, Vol. 31, No. 6, pp. 289–296, ISSN: 0301-679X

Bathe, K.-J. (1996). *Finite Element Procedures*, Prentice-Hall, ISBN: 0-13-301458-4 1, Upper Saddle River, New Jersey

Booker, J. F. & Huebner, K. H. (1972). Application of Finite Element Methods to Lubrication: An Engineering Approach. *ASME J. Lubr. Technol.*, Vol. 94, pp. 313–323 , ISSN: 0022-2305

Bonneau, D. & Hajjam, M. (2001). Modélisation de la rupture et de la formation des films lubrifiants dans les contacts élastohydrodynamiques. *Revue Européenne des Eléments Finis,* Vol. 10, No. 6-7, pp. 679-704, ISSN : 1250-6559

Bouyer, J. & Fillon, M. (2004). On the Significance of Thermal and Deformation Effects on a Plain Journal Bearing Subjected to Severe Operating Conditions. *ASME J. Tribol.*, Vol. 126, No. 4, pp. 819-822, ISSN: 0742-4787

Brugier, D. & Pasal, M.T. (1989). Influence of elastic deformations of turbo-generator tilting pad bearings on the static behavior and on the dynamic coefficients in different designs. *ASME J. Tribol.*, Vol. 111, No. 2 , pp. 364–371, ISSN: 0742-4787

Chang, Q.; Yang, P.; Meng, Y. & Wen, S. (2002). Thermoelastohydrodynamic analysis of the static performance of tilting-pad journal bearings with the Newton–Raphson method. *Tribol. Int.*, Vol. 35, No. 4, pp. 225-234, ISSN: 0301-679X

Dowson, D. (1967). A Generalized Reynolds Equation for Fluid-Film Lubrication. *Int. J. Mech. Sci.*, Pergamon Press Ltd., Vol. 4, pp. 159-170

Floberg, L. & Jakobsson, B. (1957). The finite journal bearing considering vaporization. *Transactions of Chalmers University of Technology*, Vol. 190, Gutenberg, Sweden

Fatu, A.; Hajjam, M. & Bonneau, D., (2006). A new model of thermoelastohydrodynamic lubrication in dynamically loaded journal bearings. *ASME J. Tribol.*, Vol. 128, pp. 85–95, ISSN: 0742-4787

Glavatskikh, S. (2001). Steady State Performance Characteristics of a Tilting Pad Thrust Bearing, *ASME J. Tribol.*, Vol. 123, No. 3, pp. 608-616, ISSN: 0742-4787

Kelly, D.W.; Nakazawa, S. & Zienkiewicz, O.C. (1980). A Note on Upwinding and Anisotropic Balancing Dissipation in Finite Element Approximations to Convective Diffusion Problems. *Int. J. Numer. Meth. Eng.*, Vol. 15, pp. 1705-1711, ISSN: 0029-5981

Kim, B.J. & Kim, K.W. (2001). Thermo-elastohydrodynamic analysis of connecting rod bearing in internal combustion engine, ASME J. Tribol., Vol. 123, pp. 444–454, ISSN: 0742-4787

Khonsari, M.M. & Booser, E.R. (2008). *Applied tribology: bearing design and lubrication*, Second Edition, Wiley & Sons, ISBN: 9780470057117, Chichester, UK

Kumar, A. & Booker, J.F. (1991). A finite element cavitation algorithm: Application/validation. *ASME J. Tribol.*, Vol. 107, pp. 253-260, ISSN: 0742-4787

Kumar, A. & Booker, J.F. (1994). A Mass and Energy Conserving Finite Element Lubrication Algorithm. *ASME J. Tribol.*, Vol. 116 , No. 4, pp. 667-671, ISSN: 0742-4787

LaBouff, G.A. & Booker, J.F. (1985). Dynamically Loaded Journal Bearings: A Finite Element Treatment for Rigid and Elastic Surfaces. *ASME J. Tribol.*, Vol. 107, pp. 505-515, ISSN: 0742-4787

Lund, J.W. & Tonnesen J. (1984). An approximate analysis of the temperature conditions in a journal bearing. Part II: Application. *ASME J. Tribol.*, Vol. 106, pp. 237–245, ISSN: 0742-4787

Kucinski, B.R.; Fillon, M.; Frêne, J. & Pascovici, M. D., (2000). A transient Thermoelastohydrodynamic study of steadily loaded plain journal bearings using finite element method analysis, *ASME J. Tribol.*, Vol. 122, pp. 219-226, ISSN: 0742-4787

Murty, K.G. (1974). Note on a Bard-type Scheme for Solving the Complementarity Problem. *Opsearch*, Vol. 11, pp. 123-130

Olsson, K. O. (1965). Cavitation in dynamically loaded bearing. *Transactions of Chalmers University of Technology*, Vol. 308, Guthenberg, Sweden

Piffeteau, S.; Souchet, D. & Bonneau, D. (2000). Influence of Thermal and Elastic Deformations on Connecting-Rod End Bearing Lubrication Under Dynamic Loading. *ASME J. Tribol.*, Vol. 122, No. 1, pp. 181-191, ISSN: 0742-4787

Robinson, C.L. & Cameron, A. (1975). Studies in hydrodynamic thrust bearings. *Philos. Trans.*, Vol. 278, No. 1283, pp. 351–395, ISSN: 1364-503X

Stefani, F. & Rebora, A. (2009). Steadily loaded journal bearings: Quasi-3D mass–energy-conserving analysis. *Tribol. Int.*, Vol. 42, No. 3, pp. 448-460, ISSN: 0301-679X

Stieber, W. (1933). Das Schwimmlager. *VDI*, Berlin

Swift, H. W. (1932). The stability of lubricating films in journal bearings. *Proc. Inst. Civil Eng.*, Vol. 233, pp. 267–288

Tezduyar, T. & Sunil, S. (2003). Stabilization Parameters in SUPG and PSPG formulations. *Journal of Computational and Applied Mechanics*, Vol. 4, No. 1, 7 pp. 1-88, ISSN: 15862070

Wang, Y.; Wang, Q.J. & Lin, C. (2003). Mixed Lubrication of Coupled Journal-Thrust-Bearing Systems Including Mass Conserving Cavitation. ASME *J. Tribol.*, Vol. 125, pp. 747-756, ISSN 0742-4787

Wendt, F. (1933). Turbulente stromungen zwischen zwei rotierenden konaxialen zylindern. *Ingenieur-Archiv*, Vol. 4, No. 3, pp. 577–595

23

Comparison between Different Supply Port Configurations in Gas Journal Bearings

Federico Colombo, Terenziano Raparelli and Vladimir Viktorov
Politecnico di Torino, Department of Mechanics
Italy

1. Introduction

Because of their precision, gas bearings are widely used for very high speed spindle applications. Compared to conventional oil bearings, gas bearings generate less heat and do not pollute the environment. Air viscosity is three orders of magnitude lower than oil, so the power dissipated in gas bearings is very low. The major disadvantage of these bearings is rotor whirl instability, which restricts the possible range of applications.

Researchers have studied this problem using different methods since the '60s. Gross first applied a perturbation method to evaluate the stability of an infinitely long journal bearing (Gross & Zachmanaglou, 1961). Galerkin's method was used by others to calculate rotor speed and mass at the stability threshold (Cheng & Pan, 1965). Lund investigated the stiffness and damping coefficients of hydrostatic gas bearing, and used these coefficients to investigate whirl instability (Lund, 1968). Wadhwa et al. adapted the perturbation method to calculate the dynamic coefficients and to study the stability of a rotor supported by orifice compensated gas bearings (Wadhwa et al., 1983). Results show that aerostatic bearings have a larger load capacity and higher stability than plain journal bearings. Han et al. proved that more circumferential supply ports result in increased stiffness coefficient but reduced damping (Han et al., 1994). Others found that orifice-compensated and shallow-pocket type hybrid gas journal bearings offer better stability than eight-orifice type bearings (Zhang & Chang, 1995).

Also porous journal bearings were studied (Sun, 1975) and compared against hybrid gas bearings with multi-array entries (Su & Lie, 2006), (Heller et al., 1971). Despite the fact that damping is generally higher in porous bearings than in aerostatic bearings, the results of (Su & Lie, 2006) suggest that at high operating speeds, multi-array entry bearings are more stable than porous bearings.

Other studies (Andres, 1990), (Sawcki et al., 1997), (Yoshikawa et al., 1999) considered various pressurized air compensated configurations, but very few papers analysed the influence of the number and location of entry ports.

In (Su & Lie, 2003) hybrid air journal bearings with multi-array supply orifices were compared to porous bearings. One to five rows of orifices were considered. It was found that five rows of supply orifices perform as well as porous bearings, whilst supply orifice feeding has the advantage of consuming less power than porous feeding. Paper (Yang et al., 2009) compared bearing systems with double-array orifice restrictions to three and six entry

systems. Results show that the stability threshold is better with six-ports than with three ports.
In (Colombo et al., 2009) the authors analysed two externally pressurized gas bearings, one with a central row of supply orifices, the other with a double row. The supply port downstream pressure was found to be proportional to the critical mass. At this pressure reading, the second bearing type was 30% stiffer and 50% more stable.
The aim of this work is to compare three externally pressurized gas journal bearings at given air consumption rates. The idea was to investigate which offers the best spatial distribution of supply orifices under the same pneumatic power. The study compared radial stiffness and pressure distribution for the three bearing types, also evaluating the damping factor and the whirl ratio of the shaft. The stability threshold was calculated for different restriction parameters so that the proposed bearing types could be compared.

2. Description of the problem

The object of the study was a rigid rotor supported by two identical gas journal bearings situated symmetrically with respect to the journal centre. The rotor, with diameter D=50 mm, was considered to be perfectly balanced. The radial air clearance was h_0=20 μm and the bearings had L/D ratio equal to unity.
Three bearing types were considered, as illustrated in figure 1. Bearing type 1 featured four supply ports situated in the centre plane of the bearing; bearing type 2 featured two sets of supply ports, situated at $z=L/4$ and $z=3L/4$; bearing type 3 also featured a central vented circumferential chamber.
The three bearing types were comparable in terms of stiffness and damping coefficients, air consumption and stability. In (Colombo et al., 2009) the authors compared bearing types 1 and 2 (see figure 1) considering the same supply port diameter d_s. The bearing with double array entries (bearing type 2) was found to be 30% stiffer than the one with a single central array (bearing type 1) but the air consumption was two times as much. Moreover, bearing 2 was more stable: the rotor mass at incipient whirl instability was about 50% greater.
Another point of interest was which bearing type was to be preferred for the same level of air consumption. In this paper the bearings illustrated in figure 1 were compared considering different supply port diameters in order to have the same air consumption.

3. Lubrication analysis

3.1 Mathematical model

The two-degree-of-freedom rotor equations of motion are shown in (1). The rotor mass is m. As the shaft was assumed to have cylindrical motion, gyroscopic effects and tilting inertia moments are non-existent. The second member of the equations is zero because the rotor was assumed to be perfectly balanced and there were no external forces applied to it. This was the most unstable condition, as shown in (Belforte et al., 1999).

$$\begin{cases} m\ddot{x} + 2\int_0^L \int_0^{2\pi} p(z,\theta)\cos\theta r d\theta dz = 0 \\ m\ddot{y} + 2\int_0^L \int_0^{2\pi} p(z,\theta)\sin\theta r d\theta dz = 0 \end{cases} \tag{1}$$

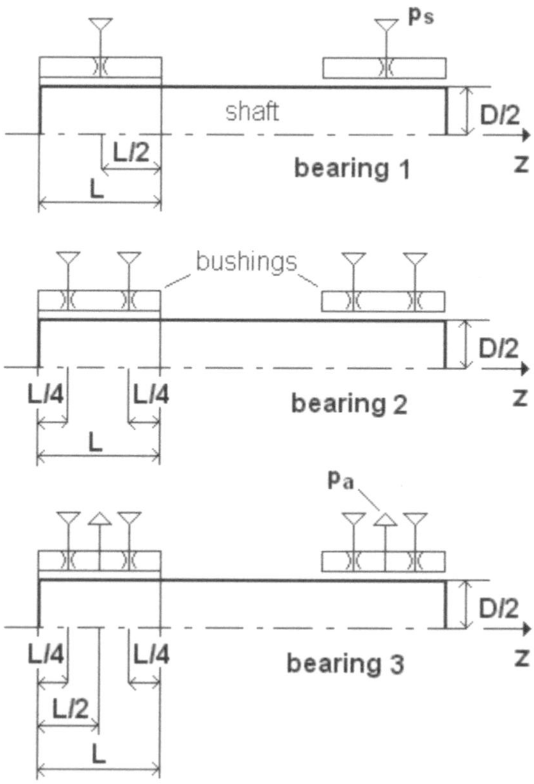

Fig. 1. Bearing types under study

The pressure distribution in clearance h was calculated solving the distributed parameters problem described by the Reynolds equation for a compressible-fluid-film journal bearing (2), assuming isothermal gas expansion.

$$\frac{\partial}{\partial z}\left(ph^3\frac{\partial p}{\partial z}\right)+\frac{\partial}{r\partial\theta}\left(ph^3\frac{\partial p}{r\partial\theta}\right)+12\mu R^0T^0\frac{G}{rdrd\theta}=6\mu\omega\frac{\partial(ph)}{\partial\theta}+12\mu\frac{\partial(ph)}{\partial t} \tag{2}$$

Mass flow rate G at supply orifice was calculated in accordance with the isentropic expansion formula (3), corrected by experimentally identified discharge coefficient c_d, expressed by eq. (4). Reynolds number at the supply hole was calculated as per equation (5). Formula (4) is the result of an extensive set of experimental tests carried out on air pads with different inherence parameters (Belforte et al., 2008).

$$G=c_d\pi\frac{d_s^2}{4}p_s\sqrt{\frac{k}{k-1}\left[\left(\frac{p_c}{p_s}\right)^{\frac{2}{k}}-\left(\frac{p_c}{p_s}\right)^{\frac{k+1}{k}}\right]\frac{2}{RT}}\quad if\ \frac{p_c}{p_s}\geq b$$

$$G=c_d\pi\frac{d_s^2}{4}p_s\sqrt{\frac{k}{k+1}\left(\frac{2}{k+1}\right)^{\frac{2}{k-1}}\frac{2}{R^0T^0}}\quad \text{if } \frac{p_c}{p_s}<b \tag{3}$$

$$c_d = 0.85\left(1 - e^{-8.2\frac{h}{d_s}}\right)\left(1 - e^{-0.001Re}\right) \tag{4}$$

$$Re = \frac{4G}{\pi d_s \mu} \tag{5}$$

Assuming a cylindrical shaft motion, the clearance may be expressed by the following:

$$h(z) = h_0\left(1 - \varepsilon_x \cos\theta - \varepsilon_y \sin\theta\right) \tag{6}$$

3.2 Solution method

The Reynolds equation was discretized using a finite difference method along directions z and θ for integration over the fluid film. A rectangular grid with equi-spaced nodes in both directions was considered. The number of nodes in the axial (index i) and circumferential (index j) directions were n and m respectively. Equation (2) may be written for each node as follows:

$$\begin{aligned} &p_{i+1,j}^2\left(a_{i,j} + b_{i,j}\right) + p_{i-1,j}^2\left(a_{i,j} - b_{i,j}\right) + p_{i,j+1}^2\left(c_{i,j} + d_{i,j}\right) + p_{i,j-1}^2\left(c_{i,j} - d_{i,j}\right) + \\ &-2p_{i,j}^2\left(a_{i,j} + c_{i,j}\right) + \left(p_{i,j+1} - p_{i,j-1}\right)e_{i,j} - p_{i,j}\left(g_{i,j} + 24\mu\frac{h_{i,j}^t - h_{i,j}^{t-1}}{\Delta t}\right) + \\ &+G_{i,j}\frac{24\mu R^0 T^0}{r\Delta\theta\Delta z} = 24\mu h_{i,j}^t \frac{p_{i,j}^{t+1} - p_{i,j}^t}{\Delta t} \end{aligned} \tag{7}$$

where,

$$a_{i,j} = \frac{h_{i,j}^3}{\Delta z^2} \qquad b_{i,j} = \frac{3h_{i,j}^2}{2\Delta z}\left(\frac{\partial h}{\partial z}\right)_{i,j}$$

$$c_{i,j} = \frac{h_{i,j}^3}{r^2\Delta\theta^2} \qquad d_{i,j} = \frac{3h_{i,j}^2}{2r^2\Delta\theta}\left(\frac{\partial h}{\partial \theta}\right)_{i,j}$$

$$e_{i,j} = 6\mu\omega\frac{h_{i,j}}{\Delta\theta} \qquad g_{i,j} = 12\mu\omega\left(\frac{\partial h}{\partial \theta}\right)_{i,j}$$

At the supply port $G_{i,j}$ was calculated using equation (3), whereas elsewhere it was zero. The boundary conditions imposed were:

- $p=p_a$ at $z=0$ and $z=L$; for bearing type 3 $p=p_a$ also at $z=L/2$
- periodic condition at $\theta=0$ and $\theta=2\pi$.

The Euler explicit method was used, so equation (7) becomes:

$$p_{i,j}^{t+1} = p_{i,j}^t + \Delta t \cdot f\left[p_{i,j}^t, p_{i,j+1}^t, p_{i,j-1}^t, p_{i+1,j}^t, p_{i-1,j}^t, h_{i,j}^t, h_{i,j}^{t-1}, \left(\frac{\partial h}{\partial \theta}\right)_{i,j}^t, \left(\frac{\partial h}{\partial z}\right)_{i,j}^t\right] \tag{8}$$

The system of nxm equations (8) was solved together with equations (3) to (6) and rotor equations of motion (1).

The solution procedure started with a set of input data (shaft diameter, radial clearance, bearing axial length, position and diameter of supply orifices, shaft speed).
To calculate the static pressure distribution, h was maintained constant in time and the system was solved with initial condition $p_{i,j}=p_a$ for each node.
Pressure distribution was evaluated at each time step and the bearing forces acting on the shaft were updated in equation (1). Thus, the rotor trajectory was determined starting with the initial static pressure distribution and using the following set of initial conditions:

$$x(0) = h_0\varepsilon_x(0)\,;\; y(0) = h_0\varepsilon_y(0)$$

$$\dot{x}(0) = h_0\dot{\varepsilon}_x(0)\,;\; \dot{y}(0) = h_0\dot{\varepsilon}_y(0)$$

3.3 Mesh size and time step definition

Calculations were made with different mesh sizes and the results were compared for optimum trade-off between computational time and accuracy of the solution.
The grids are detailed in table 1.

nxm	*Δz (mm)*	*rΔθ (mm)*
13x24	4.17	6.54
17x32	3.12	4.91
25x48	2.08	3.27
49x96	1.04	1.64

Table 1. Mesh sizes used in calculations; r=25 mm, L/D=1

Figure 2 shows the axial and circumferential pressure distributions obtained for bearing type 1 with different numbers of grid points. If the number of grid points is increased, the pressure distribution becomes more clearly defined, but the difference is almost negligible. Only at the supply ports, where pressure gradients are high, the difference is more marked. The grid selected for calculation was n=49, m=96.

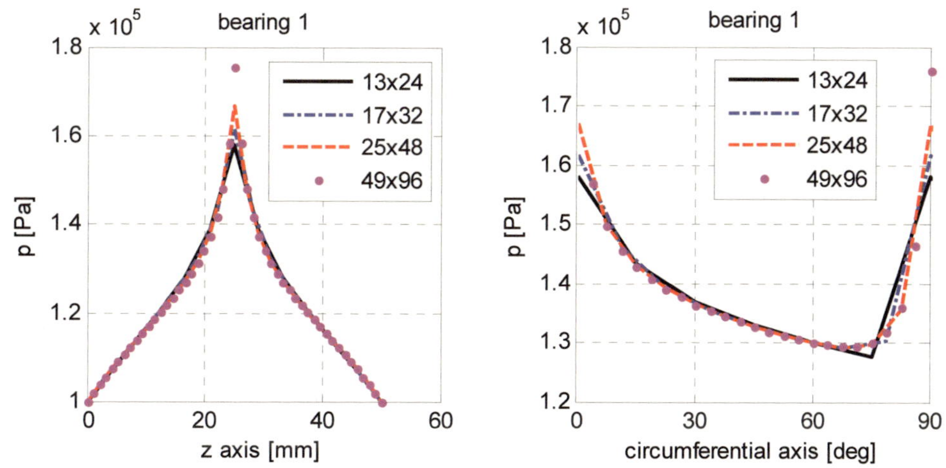

Fig. 2. Axial and circumferential pressure distributions for bearing type 1 obtained with different mesh grids; h_0=20 μm, p_s=5·10^5 Pa rel., d_s=0.1 mm, ω=60 krpm, ε=0

Euler explicit method was used to solve the time progression of the system. The rotor trajectories obtained with different time steps Δt are compared in figure 3.
The rotor initial conditions were:

$$\varepsilon_x(0)=0;\varepsilon_y(0)=0$$

$$\dot{\varepsilon}_x(0)=0;\dot{\varepsilon}_y(0)=0$$

The trajectories are increasingly adjacent with decreasing Δt. The time step used in the paper was Δt=10^{-7} s.

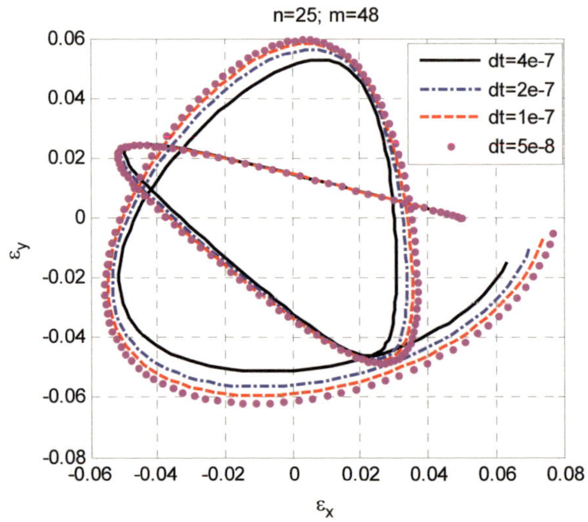

Fig. 3. Rotor trajectories with bearing type 1 obtained with different time steps and grid 25x48; initial conditions specified by $\varepsilon_x(0)$=0.05, $\varepsilon_y(0)$=0, $\dot{\varepsilon}_x(0)=0,\dot{\varepsilon}_y(0)=0$, h_0=20 μm, p_s=5·10^5 Pa rel., d_s=0.1 mm, ω=60 krpm

4. Discussion and results

4.1 Resistance analysis

The air supply system may be described with an equivalent lumped parameters system, illustrated in figure 4.
Orifice restriction resistance R_s is related to the supply ports and decreases with increasing diameter d_s. It may be calculated using linearizing expression (3) with respect to downstream pressure p_c. Clearance resistance R_h depends on clearance h_0, on bearing dimensions size and on the arrangement of the supply ports. It is obtained by solving the distributed parameters problem and calculating pressure distribution in the clearance.
Imposing mass continuity in the lumped parameters system of figure 4, supply port downstream pressure p_c can be obtained by

$$p_c = p_s - \frac{R_s}{R_s + R_h}(p_s - p_a) \tag{9}$$

This pressure depends both on the supply system and on clearance: at reduced d_s, supply port downstream pressure p_c approximates ambient pressure p_a, whereas with increased d_s it approaches supply pressure p_s.
Analysis of resistances at different supply pressures with the shaft rotating in central position was performed for bearings 1 and 2 in (Colombo et al., 2009) which shows the relationship between supply port diameter d_s and downstream pressure p_c, confirming that the influence of bearing number Λ on p_c with rotor in centred position is almost negligible, and air consumption is almost independent of speed.

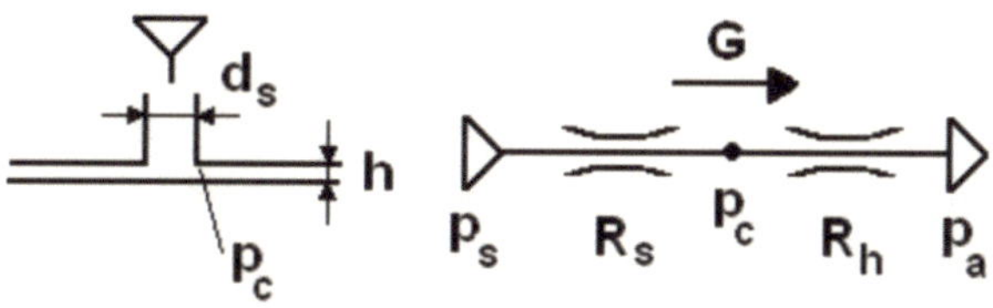

Fig. 4. Lumped parameters model of the restriction and clearance resistances

4.2 Air consumption

The three bearings of figure 1 were compared in terms of air consumption, as shown in figure 5. The air mass flow was calculated as a function of the clearance for different supply port diameters. At reduced d_s, the air consumption for bearing types 2 and 3 was quite identical. Only for d_s=0.2 mm a difference was noted at reduced clearance. The air flow in different bearings (for different resistance R_h) was found to be the same for supply orifices in critical conditions, when air flow is only a function of p_s.
As air consumption is a function of d_s and h_0, the supply ports diameter is determined at specific rates of air consumption G, as shown in table 2.
Bearing type 1 was not considered for the last two values of G because the volume of air passing through its orifices when p_c=p_s (in this condition R_s=0) was lower than these values.

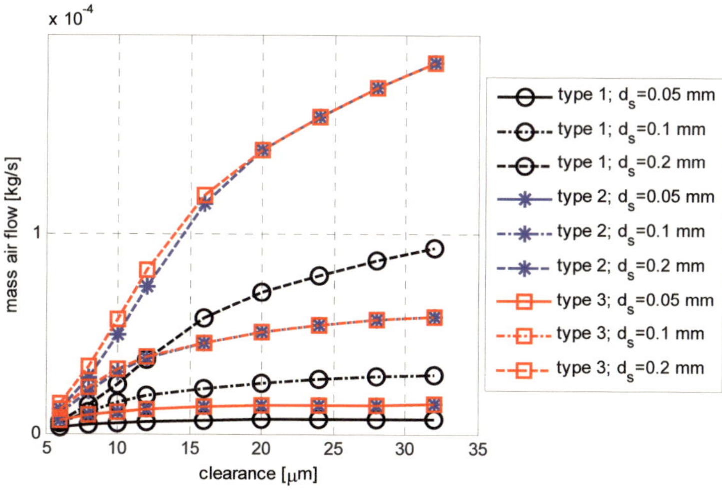

Fig. 5. Air consumption of the three bearings vs. air clearance for different supply port diameters; calculations are for Λ=0 and with rotor in central position; p_s=$5 \cdot 10^5$ Pa rel.

bearing type	diameter d_s [mm]	air flow $G \cdot 10^4$ [kg/s]
1	0.155	
2	0.1	0.5
3	0.1	
1	0.383	
2	0.2	1.42
3	0.2	
1	0.8	
2	0.282	2.14
3	0.275	
2	0.4	2.94
3	0.372	
2	0.6	4.28
3	0.8	

Table 2. Supply port diameter d_s considered in calculations for the three bearings at different air consumption G; p_s=5·10[5] Pa rel.

4.3 Pressure distribution

Figures 6 and 7 compare the axial and circumferential pressure distributions in the three bearings with rotor in central position and restriction parameters specified in table 2. Bearing type 1 shows a lower ratio R_s/R_h than the other bearings because its maximum pressure is the highest. At G=0.5·10^{-4} kg/s all bearings have orifices in sonic conditions, being $p_c/p_s<b$. At G=2.14·10^{-4} kg/s bearing type 1 is near saturation condition ($p_c \simeq p_s$). Speed stretches the circumferential pressure profile toward the direction of rotation, as visible in figure 7.

4.4 Bearing stiffness

Bearing stiffness was calculated by imposing a shaft displacement of 1 μm along direction x and evaluating the bearing reaction force.
Bearing stiffness k was

$$k = \sqrt{k_{xx}^2 + k_{xy}^2} \tag{10}$$

where the stiffness coefficients calculated in steady-state conditions were

$$k_{xx} = \frac{F_x}{h_0 \varepsilon_x} = \frac{\int_0^L \int_0^{2\pi} p(z,\theta)\cos\theta r d\theta dz}{h_0 \varepsilon_x}$$

$$k_{xy} = \frac{F_y}{h_0 \varepsilon_x} = \frac{\int_0^L \int_0^{2\pi} p(z,\theta)\sin\theta r d\theta dz}{h_0 \varepsilon_x}$$

Non-dimensional stiffness k^*, given by

$$k^* = k \frac{h_0}{p_a L D} \tag{11}$$

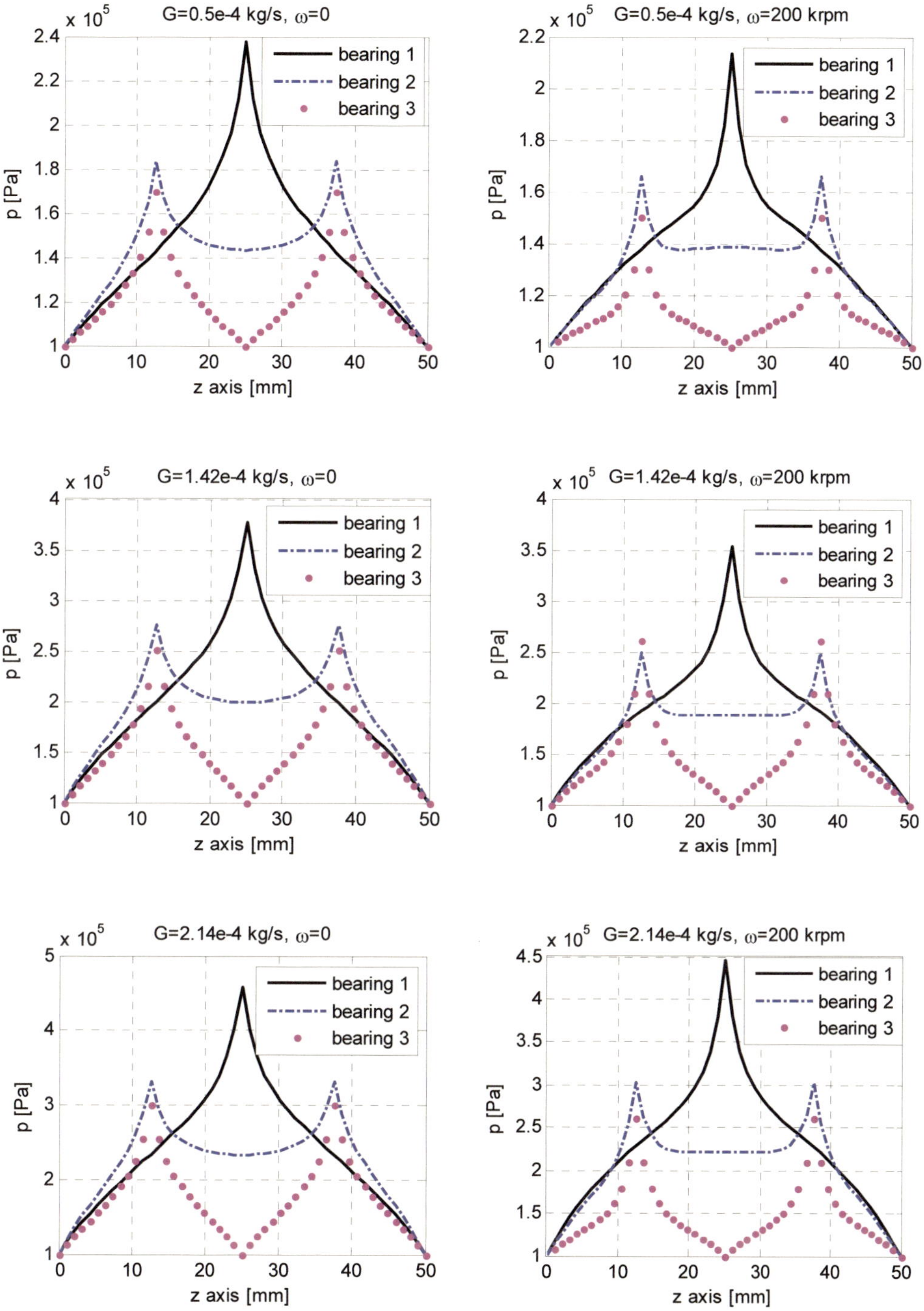
G=0.5e-4 kg/s, ω=0
G=0.5e-4 kg/s, ω=200 krpm
G=1.42e-4 kg/s, ω=0
G=1.42e-4 kg/s, ω=200 krpm
G=2.14e-4 kg/s, ω=0
G=2.14e-4 kg/s, ω=200 krpm
x 10^5
p [Pa]
z axis [mm]
bearing 1
bearing 2
bearing 3

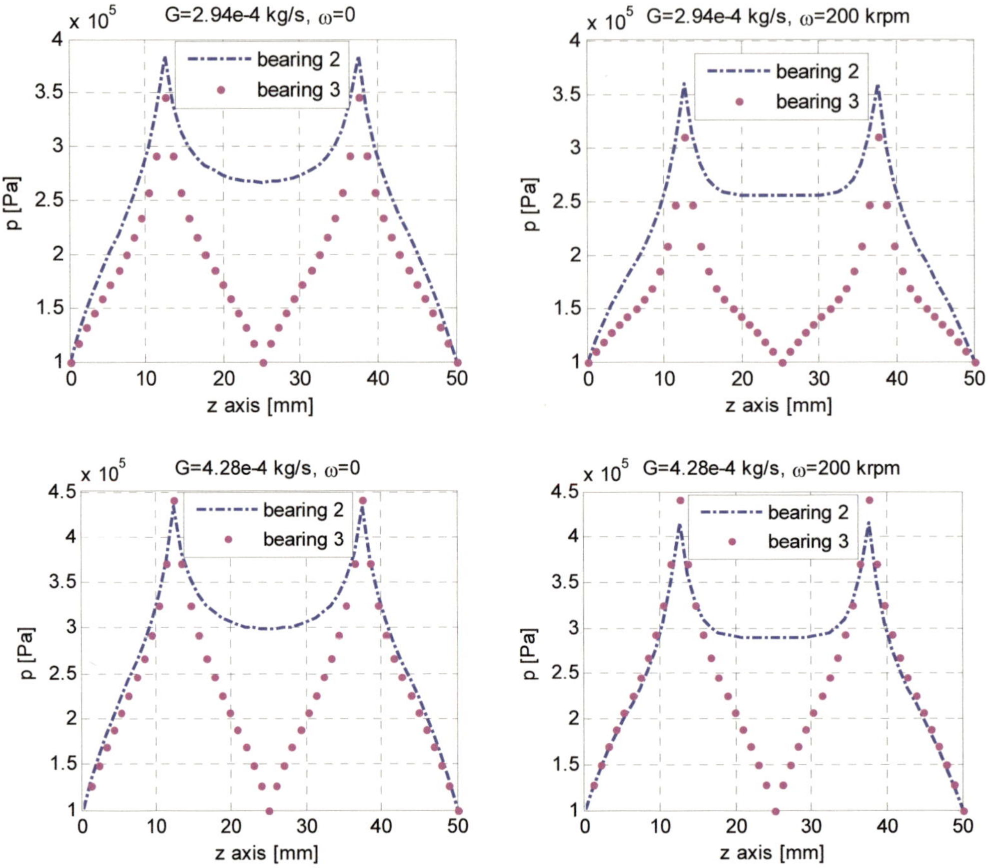

Fig. 6. Axial pressure distribution in the three bearings with ω=0 and ω=200 krpm for five different air consumption rates; restriction parameters are specified in table 2, h_0=20 μm, p_s=5·10^5 Pa rel., ε=0

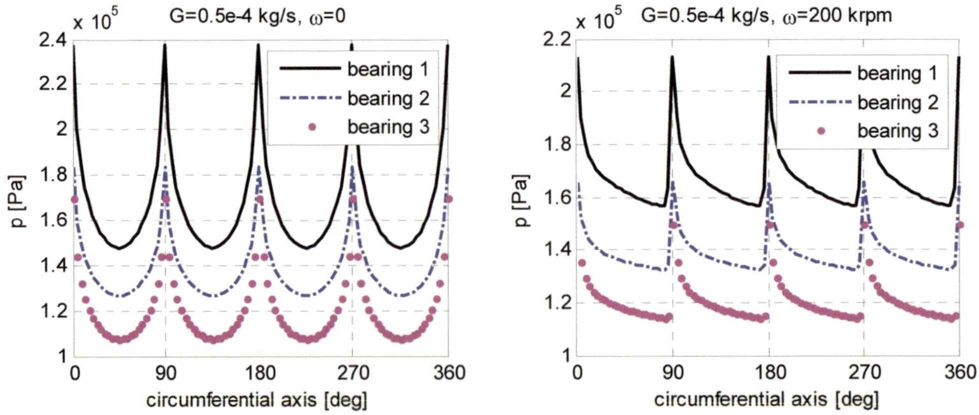

Fig. 7. Axial pressure distribution in bearing type 1 with ω=0 and ω=200 krpm for G=0.5·10^{-4} kg/s; h_0=20 μm, p_s=5·10^5 Pa rel., ε=0

is shown in figure 9 vs. Λ for the three bearings, considering different restriction parameters. Figure 9 also shows steady-state attitude angle β, calculated as follows:

$$\beta = \tan^{-1}\frac{k_{xy}}{k_{xx}} \tag{12}$$

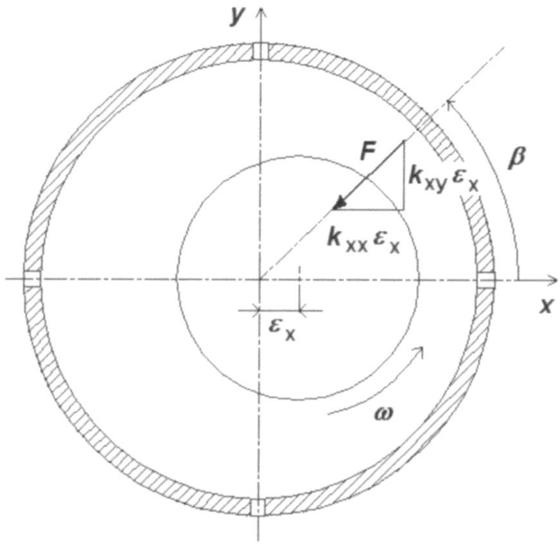

Fig. 8. Bearing reaction force on the journal in steady-state conditions due to shaft displacement along direction x

Stiffness increased with Λ up to saturation (Λ>100). At G=0.5·10^{-4} kg/s bearing type 1 was found to be stiffer than the other two, regardless of Λ, but at higher air consumption bearing type 2 exhibited greater stiffness at low speeds (Λ<9).
With the three bearings in sonic conditions (G=0.5·10^{-4} kg/s) stiffness trends do not intersect and their difference was almost constant. When bearing type 1 approached saturation ($p_c \simeq p_s$), its stiffness at low speed dropped (see case with G=1.42·10^{-4} kg/s). This happened also for bearing type 2, but at greater air consumptions. Stiffness at high speeds (Λ>100) always increased with G. At G=4.28·10^{-4} kg/s, stiffness at low speeds for bearing types 2 and 3 coincided at very low values, due to saturation of bearings.
The attitude angle, with increasing Λ, also increased from zero to a maximum and then returned to zero. The extent of maximum depended on the difference between bearing stiffness at low and high speeds: where this difference was high, also maximum β was high. Table 3 shows ratio $k^*(\Lambda>100)/k^*(\Lambda=0)$ for the three bearings to highlight this relationship.

4.5 Rotor trajectories

The whirl motion of the perfectly balanced rotor during rotation is represented in figure 10. The motion can be stable or unstable. In the former case the rotor is attracted toward the centre of the bushing after initial disturbance; in the latter case the bearing forces move the rotor away from central position.

bearing type	$k^*(\Lambda>100)/k^*(\Lambda=0)$	air flow $G\cdot10^4$ [kg/s]
1	2.62	
2	2.71	0.5
3	3.75	
1	2.54	
2	1.82	1.42
3	2.4	
1	5.8	
2	2	2.14
3	2.2	
2	2.5	2.94
3	2.26	
2	5.33	4.28
3	3.08	

Table 3. Ratio $k^*(\Lambda>100)/k^*(\Lambda=0)$ for the three bearings given different air consumptions G

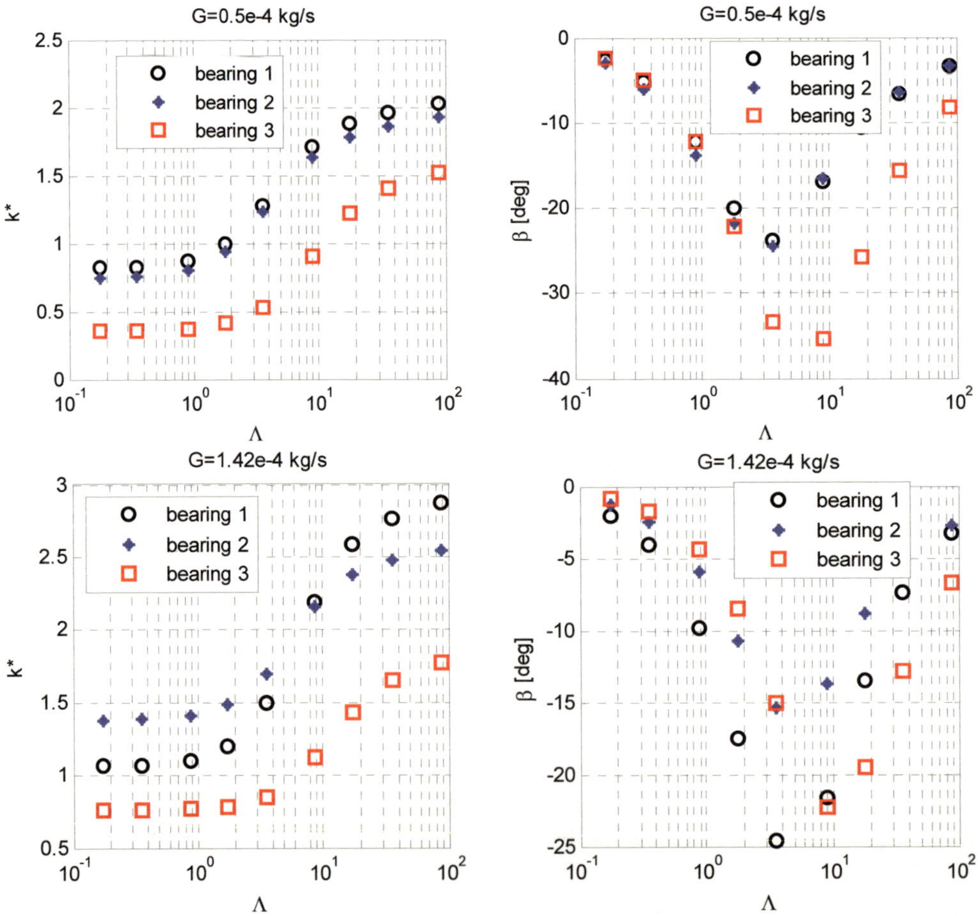

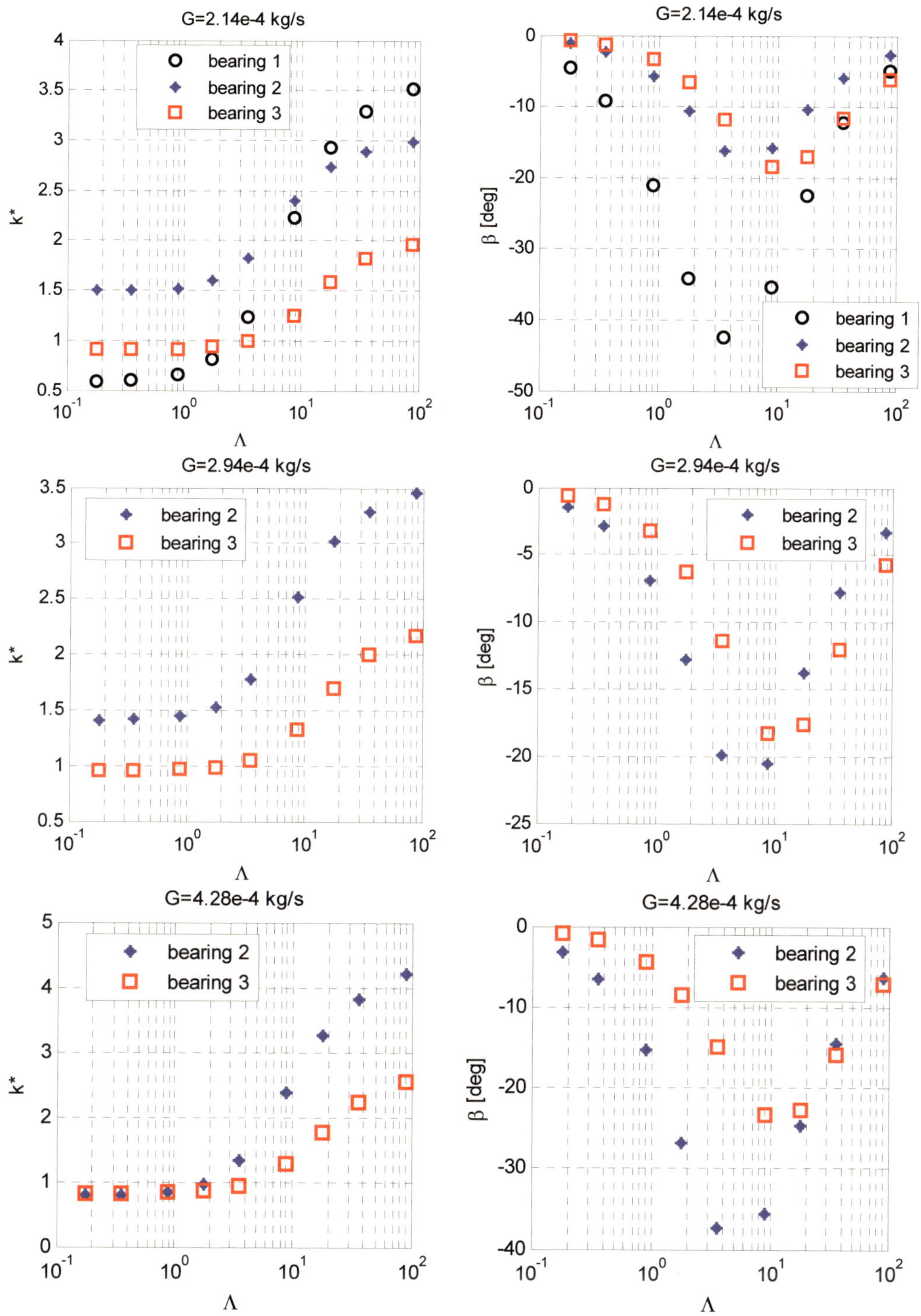

Fig. 9. Non-dimensional bearing stiffness k^* and attitude angle β vs. bearing number Λ for the three bearings

The initial condition used in the following curves are specified by

$$\varepsilon_x(0) = 0.05; \varepsilon_y(0) = 0$$

$$\dot{x}(0) = 0; \dot{y}(0) = x(0)\sqrt{\frac{k_{xx}}{m}}$$

Initial tangential speed was imposed on the rotor to produce a centrifugal force equal to the static radial force. This particular condition was adopted to decrease the simulation time required to distinguish stability from instability. In fact, with a different initial condition on $\dot{y}$, the trajectory would have been less circular, necessitating simulation of a longer transient. Stability decreased with increasing rotor mass m: figure 10 shows a comparison of rotor trajectories obtained for the same initial condition but at different values of m. The rotor-bearing system became unstable when the dynamic attitude angle turned negative, as shown in figure 11. In the stable condition the rotor angular moment, calculated relative to the centre of the bushing, decreased with time. In unstable conditions, the mechanical work done by bearing forces was found to be positive and the rotor angular moment increased (see figure 11b). The curves in figure 11 help distinguish stable versus unstable conditions, as resulting when compared to figure 10.

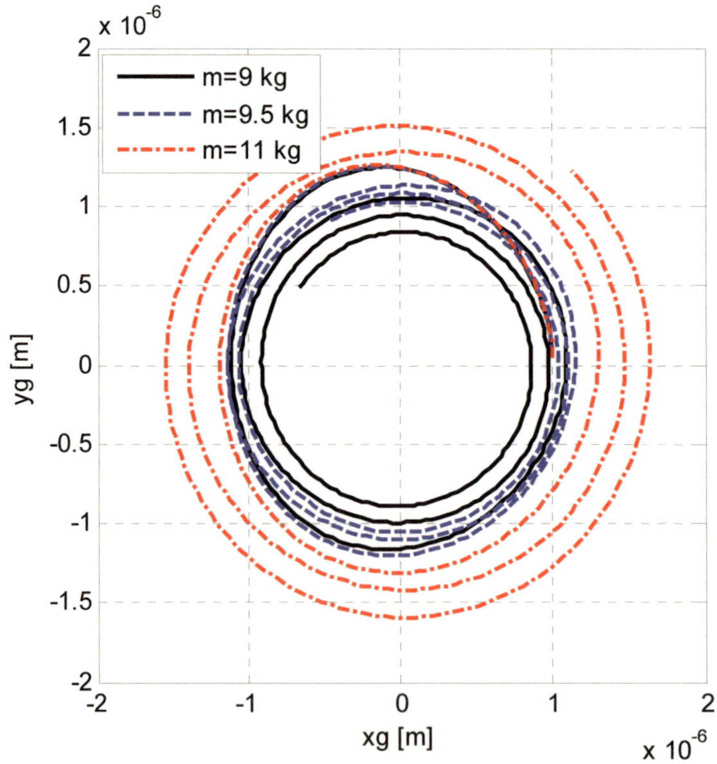

Fig. 10. Rotor trajectories with different rotor masses and initial condition $x(0)$=1 μm, $dy/dt(0)=x(0)\cdot(k_{xx}/m)$^(0.5); ω=20 krpm, bearing type 1

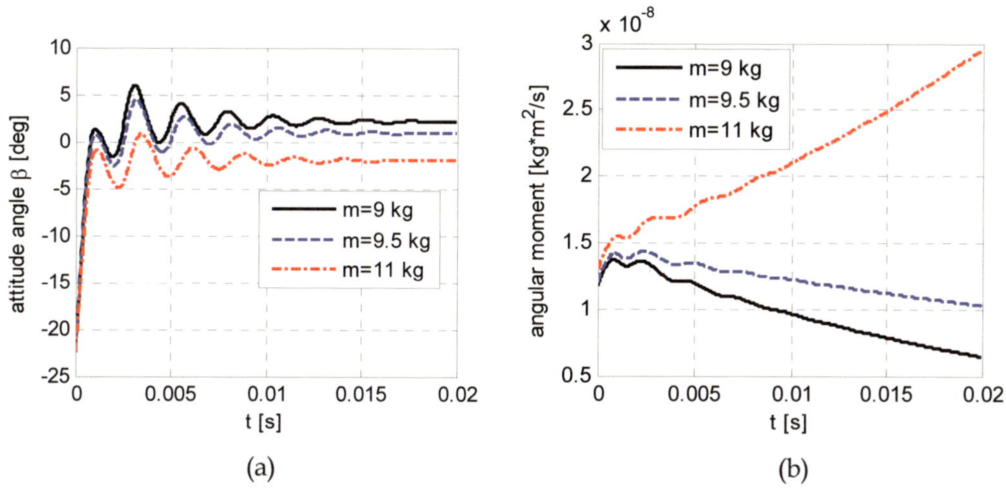

Fig. 11. Attitude angle vs. time (a) and rotor angular moment vs. time (b) with different rotor masses and initial condition $x(0)$=1 μm, $dy/dt(0)=x(0)\cdot(k_{xx}/m)$^(0.5); ω=20 krpm, bearing type 1

The three bearings are compared in figures 12 and 13, showing the rotor trajectories for identical initial condition, the attitude angle vs. time and the rotor angular moment vs. time. In this case bearing types 1 and 2 are very similar, while bearing type 3 is unstable.

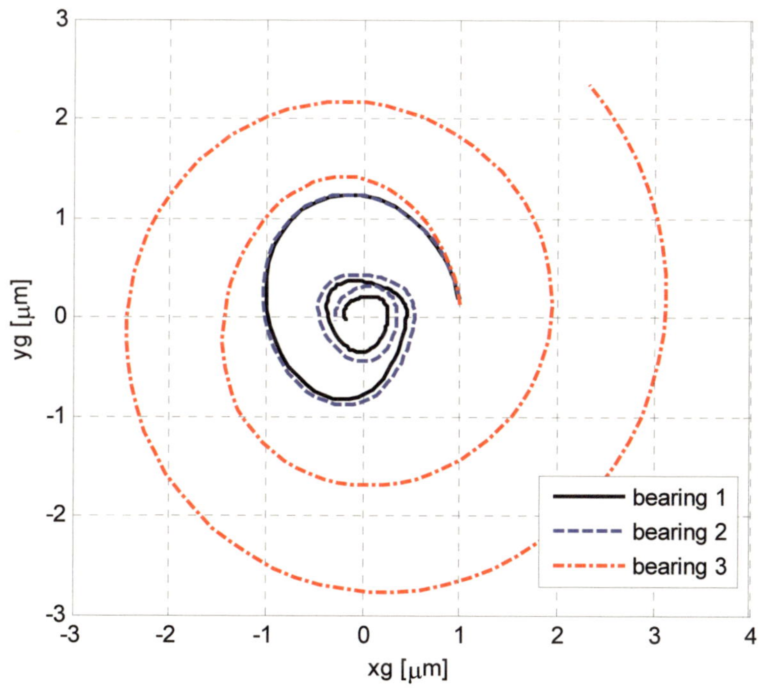

Fig. 12. Rotor trajectories with the three bearing types; m=1 kg, ω=50 krpm; initial conditions $x(0)$=1 μm and $dy/dt(0)=x(0)\cdot(k_{xx}/m)$^(0.5).

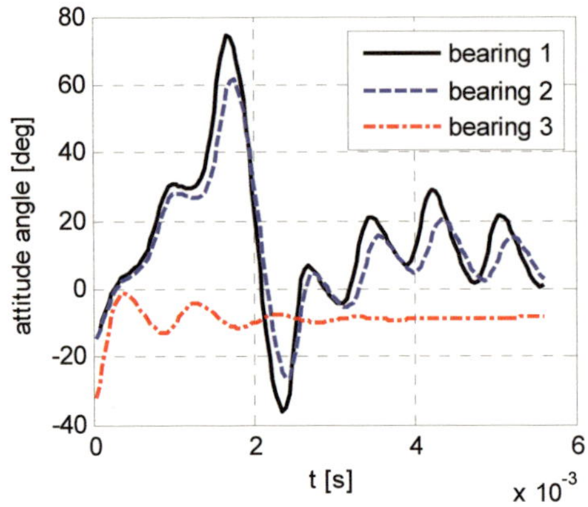

(a)

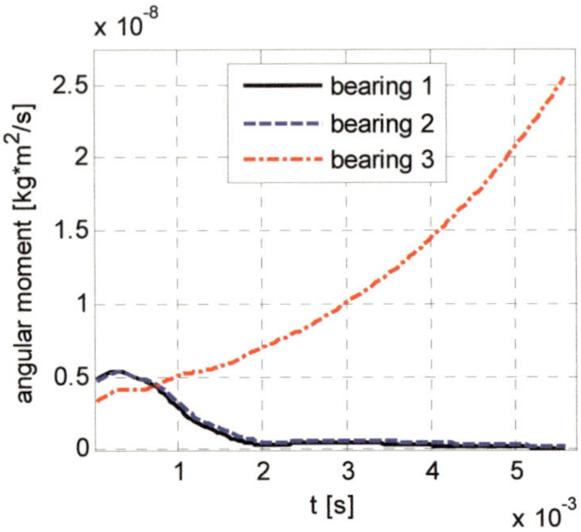

(b)

Fig. 13. Attitude angle vs. time a) and rotor angular moment vs. time b) for the three bearing types; m=1 kg, ω=50 krpm; initial conditions $x(0)$=1 μm and $dy/dt(0)=x(0)\cdot(k_{xx}/m)$^(0.5).

4.6 Bearing damping factor

Stiffness and damping coefficients of gas bearings are known to depend on bearing number Λ and also on whirl frequency ν. Stability may also be evaluated through the equivalent damping factor calculated by identifying the system with a second-order differential equation having constant coefficients:

$$m\dot{x} + c\dot{x} + kx = 0 \tag{13}$$

The damping factor is expressed by

$$\zeta = \frac{c}{2\sqrt{km}} \tag{14}$$

and the radial coordinate of the journal centre is

$$r = r(0)e^{-\zeta\omega_n t} \tag{15}$$

where the natural frequency is

$$\omega_n = \sqrt{\frac{k}{m}} \tag{16}$$

The journal motion is stable when described by a spiral which decreases with time. In this case ζ is positive. When the damping factor is negative the spiral increases with time.
Figure 14 shows damping factor ζ vs. m for G=0.5·10^{-4} kg/s. In this case bearing type 3 exhibited lower damping capacity than the other bearings.

4.7 Whirl ratio

The shaft whirl frequency vs. m is shown in figure 15 for G=0.5·10^{-4} kg/s. The whirl frequency decreases with m and increases with ω. The rotor mass at stability threshold is

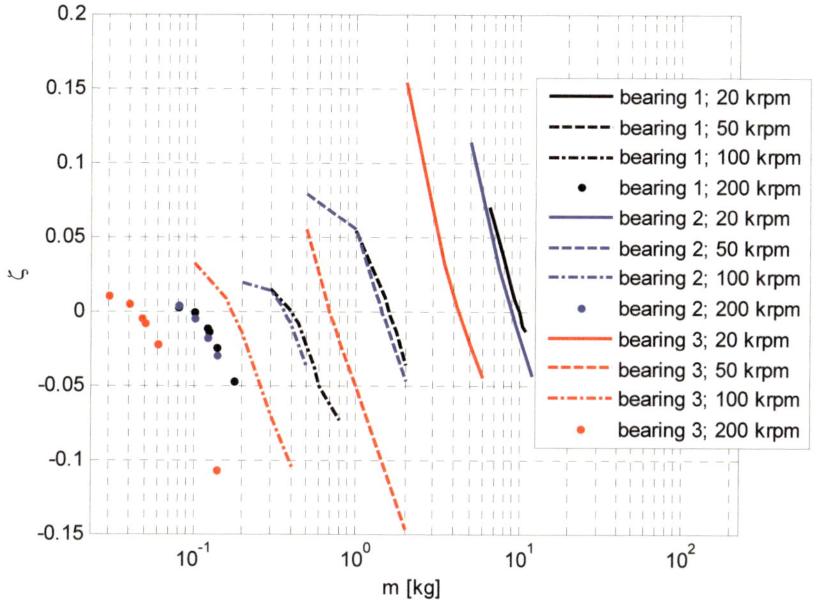

Fig. 14. Damping factor vs. rotor mass at different rotating speeds; G=0.5·10^-4^ kg/s

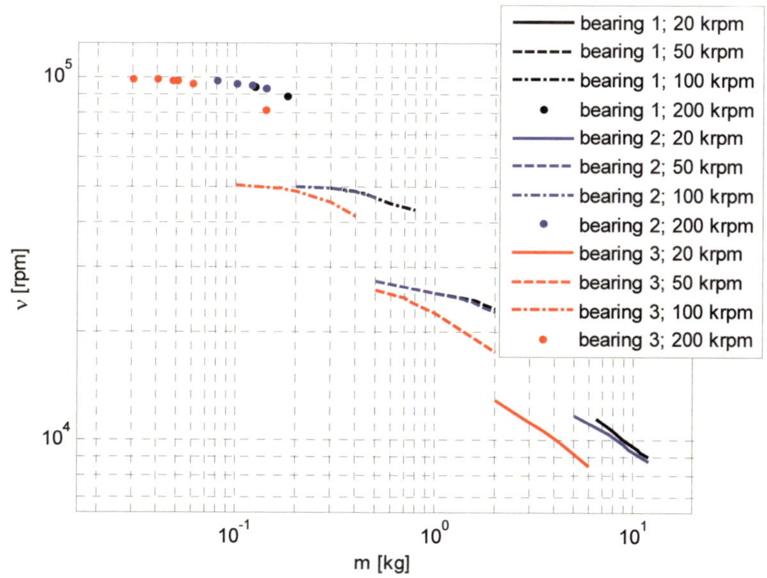

Fig. 15. Whirl frequency ν vs. m at different rotating speeds; $G=0.5\cdot10^{-4}$ kg/s

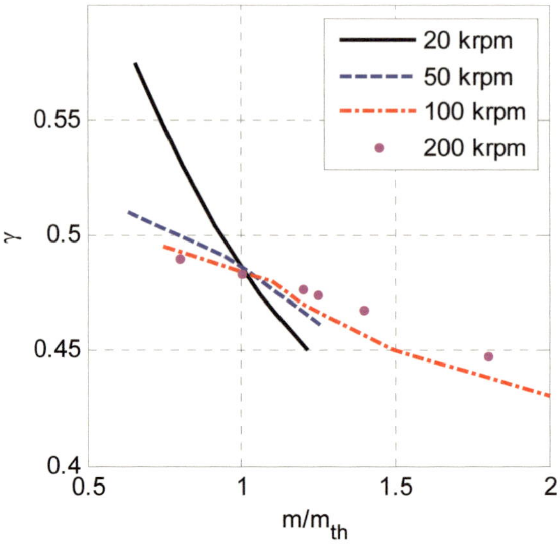

Fig. 16. Whirl ratio γ vs. m/m_{th} at different speeds; bearing type 1, G=$0.5\cdot10^{-4}$ kg/s

indicated as m_{th}. Figure 16 shows whirl ratio γ vs. ratio m/m_{th}. At the stability threshold it is slightly lower than 0.5 and decreases with shaft mass m.

4.8 Stability threshold

Figure 17 shows rotor mass m vs bearing number Λ at the stability threshold for the three bearings. On logarithmic axes the curves are linear and may be expressed by

$$\log_{10} m - \log_{10} m_0 = \alpha\left(\log_{10}\Lambda - \log_{10}\Lambda_0\right) \quad (17)$$

where m_0 and Λ_0 refer to a reference condition. Angular coefficient α is -2 approx. From this equation we obtain the following relation:

$$\frac{m}{m_0} = \left(\frac{\Lambda}{\Lambda_0}\right)^{\alpha} \quad (18)$$

The stability thresholds with different inherence parameters were found to be similar, but translated to different mass values.

4.9 Comparison of bearing types at different restriction parameters

Figure 18 shows the trends of bearing stiffness vs. G for ω=0 rpm and ω=200 krpm, and figure 19 shows critical journal mass m_{th} vs. G. The order of preference of the bearings changes when different air consumption rates are considered.

If stiffness at low bearing numbers is the most important parameter, bearing type 1 is the best option only for G≤$0.5\cdot10^{4}$ kg/s, in other cases bearing type 2 is to be preferred. If it is important to maximize the bearing stiffness at high bearing numbers bearing type is to be chosen.

Considering the stability threshold, bearing type 2 is the best one for G>$0.5\cdot10^{4}$ kg/s, while for G≤$0.5\cdot10^{4}$ kg/s bearing 1 is to be preferred.

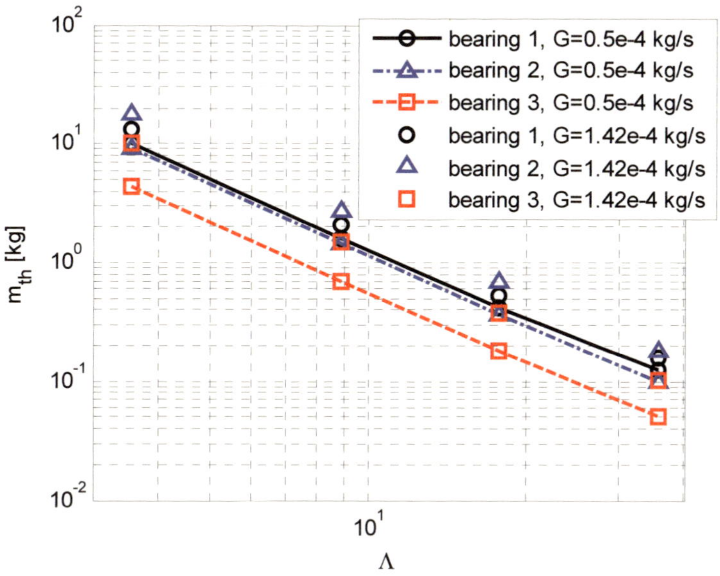

Fig. 17. Rotor mass m at stability threshold vs. bearing number Λ for the three bearings

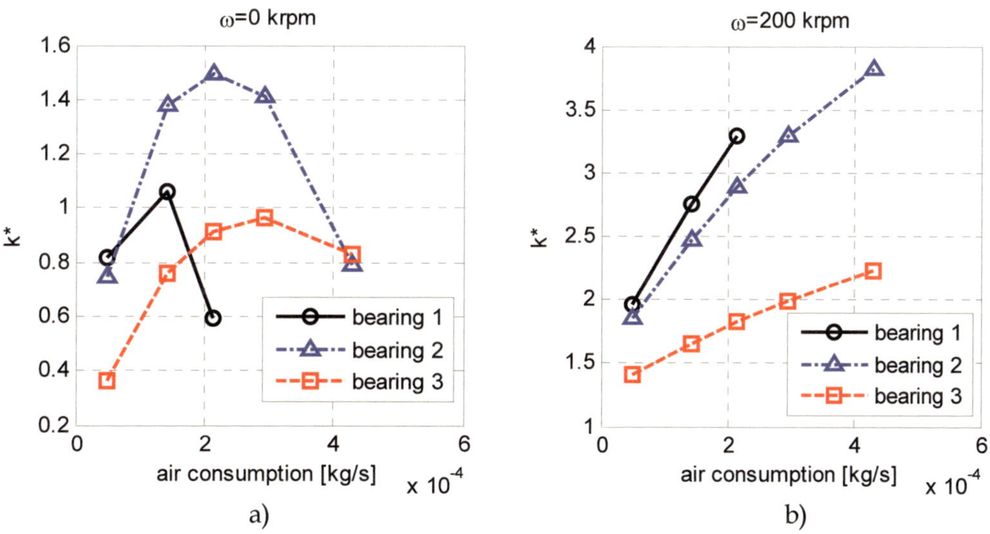

Fig. 18. Bearing stiffness k^* vs. air consumption for the three bearings; a) ω=0 rpm, b) ω=200 krpm

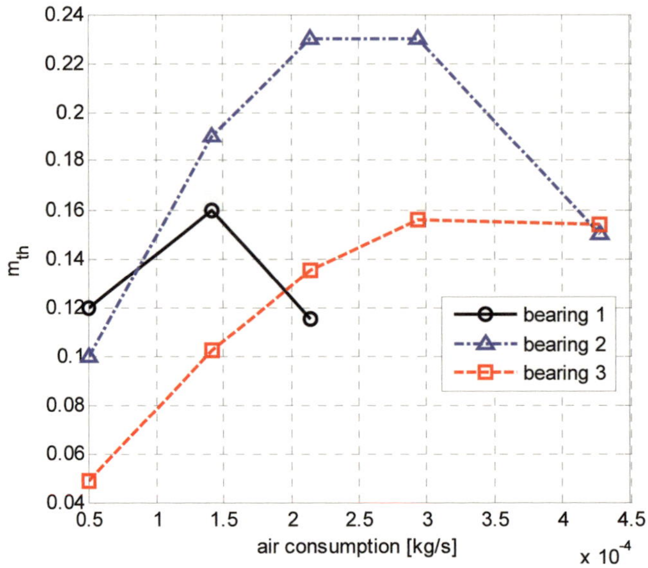

Fig. 19. Rotor mass at stability threshold vs. air consumption for the three bearings

5. Conclusion

Three bearing types were compared for different restriction parameters.
Bearing type 1 featured four supply ports situated in the bearing centre plane. Bearing type 2 featured two sets of supply ports, situated at $z=L/4$ and $z=3L/4$. Bearing type 3 also featured a central vented circumferential chamber.
The following conclusions were drawn:

- bearing type 2 in general is to be preferred to the other bearing types because of the higher stiffness and stability threshold at equal air consumption;
- with increasing Λ, the attitude angle went from zero to max. subsequently returning to zero; max. value was proportional to the difference between bearing stiffness at low and at high speeds;
- at the stability threshold the whirl ratio was slightly lower than 0.5;
- the curve of m_{th} vs. Λ on the logarithmic axes was linear and with changing restriction parameters the shaft critical mass changed by a factor regardless of speed.

6. List of symbols

D	bearing diameter
F	bearing force on journal
G	air mass flow rate
L	bearing axial length
R_s	pneumatic resistance of the supply hole
R_h	pneumatic resistance of clearance
R^0	gas constant, in calculations R^0=287.6 m^2/s^{2}K
Re	Reynolds number
T^0	absolute temperature, in calculations T^0=288 K

b	ratio of critical pressure to admission pressure, b=0.528
c	damping coefficient
c_d	supply hole discharge coefficient
h	local air clearance
h_0	clearance with rotor in centred position
k	bearing radial stiffness
k^*	non-dimensional bearing radial stiffness
m	rotor mass
m_{th}	rotor mass at stability threshold
n,m	number of nodes along axial and circumferential directions
x,y,z	cartesian coordinates
p_a	ambient pressure
p_c	supply hole downstream pressure
p_s	bearing supply pressure
r,θ,z	cylindrical coordinates
t	time
Λ	bearing number, $\Lambda=6m\omega/p_a\cdot(D/2h_0)^2$
β	steady attitude angle
γ	whirl ratio, $\gamma=\nu/\omega$
ε	eccentricity ratio
μ	dynamic viscosity, in calculations μ=17.89·10^{-6} Pa·s
ν	whirl frequency
ζ	bearing damping factor
ω	rotor angular speed

7. References

Andres, L.S. (1990). Approximate analysis of turbulent hybrid bearings, static and dynamic performance for centered operation. *ASME Journal of Tribology*, Vol. 112, 692-698.

Belforte, G.; Raparelli, T.; Viktorov, V. (1999). Theoretical investigation of fluid inertia effects and stability of self-acting gas journal bearings. *ASME Journal of Tribology*, Vol. 121, 836-843.

Belforte, G.; Raparelli, T.; Viktorov, V.; Trivella, A. (2008). Discharge coefficients of orifice-type restrictor for aerostatic bearings, *Tribology International*, Vol. 40, 512-521.

Cheng, H.S.; Pan, C.H.T. (1965). Stability analysis of gas-lubricated, self-acting, plain, cylindrical, journal bearings of finite length, using Galerkin's method. *ASME Journal of Basic Engineering*, 185-192.

Colombo, F.; Raparelli, T.; Viktorov, V. (2009). Externally pressurized gas bearings: a comparison between two supply holes configurations. *Tribology International*, Vol. 42, 303-310.

Gross, W.A.; Zachmanaglou, E.C. (1961). Perturbation solutions for Gas lubricating films. *Trans ASME, Journal of Basic Engineering*, Vol. 83, 139-144.

Han, D.C.; Park, S.S.; Kim, W.J.; KimJ.W. (1994). A study on the characteristics of externally pressurized gas bearings. *Precision Engineering*, Vol. 16, No. 3, 164-173.

Heller, S.; Shapiro, W.; Decker, O. (1971). A porous hydrostatic gas bearing for use in miniature turbomachinery. *ASLE Transactions*, 144-155.

Lund, J.W. (1968). Calculation of stiffness and damping properties of gas bearings. *ASME Journal of Lubrication Technology,* 793-803.

Sawicki, J.T.; Capaldi, R.J.; Adams, M.L. (1997). Experimental and theoretical rotordynamic characteristics of a hybrid journal bearing. *ASME Journal of Tribology,* Vol. 119, 132-141.

Su, J.C.T.; Lie, K.N. (2003). Rotation effects on hybrid air journal bearings. *Tribology International,* Vol. 36, 717-726.

Su, J.C.T.; Lie, K.N. (2006). Rotor dynamic instability analysis on hybrid air journal bearings. *Tribology International,* Vol. 39, No. 1, 238-248.

Sun, D.C. (1975). Stability of gas-lubricated, externally pressurized porous journal bearings. *ASME Journal of Lubrication Technology,* 494-505.

Wadhwa, S.S.; Sinhasant, R.; Singh, D.V. (1983). Analysis of orifice compensated externally pressurized gas bearing. *Tribology International,* Vol. 16, No. 4, 203-211.

Yang, D.W.; Chen, C.H.; Kang Y.; Hwang, R.M.; Shyr, S.S. (2009). Influence of orifices on stability of rotor-aerostatic bearing system. *Tribology International,* Vol. 42, 1206-1219.

Yoshikawa, H.; Ota, T.; Higashino, K.; Nakai, S. (1999). Numerical analysis on dynamic characteristics of cryogenic hydrostatic journal bearing. *ASME Journal of Tribology,* Vol. 121, 879-885.

Zhang, R.Q.; Chang, H.S. (1995). A new type of hydrostatic/hydrodynamic gas journal bearing and its optimization for maximum stability. *Tribology Transactions,* Vol. 38, No. 3, 589-594.